Sensory Abilities
of Cetaceans
Laboratory and
Field Evidence

NATO ASI Series

Advanced Science Institutes Series

A series presenting the results of activities sponsored by the NATO Science Committee, which aims at the dissemination of advanced scientific and technological knowledge, with a view to strengthening links between scientific communities.

The series is published by an international board of publishers in conjunction with the NATO Scientific Affairs Division

A	Life Sciences	Plenum Publishing Corporation
B	Physics	New York and London
C	Mathematical and Physical Sciences	Kluwer Academic Publishers Dordrecht, Boston, and London
D	Behavioral and Social Sciences	
E	Applied Sciences	
F	Computer and Systems Sciences	Springer-Verlag
G	Ecological Sciences	Berlin, Heidelberg, New York, London,
H	Cell Biology	Paris, and Tokyo

Recent Volumes in this Series

Volume 190—Control of Metabolic Processes
edited by Athel Cornish-Bowden and María Luz Cárdenas

Volume 191—Serine Proteases and Their Serpin Inhibitors
in the Nervous System:
Regulation in Development and in Degenerative and
Malignant Disease
edited by Barry W. Festoff

Volume 192—Systems Approaches to Developmental Neurobiology
edited by Pamela A. Raymond, Stephen S. Easter, Jr.,
and Giorgio M. Innocenti

Volume 193—Biomechanical Transport Processes
edited by Florentina Mosora, Colin G. Caro, Egon Krause,
Holger Schmid-Schönbein, Charles Baquey, and Robert Pelissier

Volume 194—Sensory Transduction
edited by Antonio Borsellino, Luigi Cervetto, and Vincent Torre

Volume 195—Experimental Embryology in Aquatic Plants and Animals
edited by Hans-Jürg Marthy

Volume 196—Sensory Abilities of Cetaceans: Laboratory and Field Evidence
edited by Jeanette A. Thomas and Ronald A. Kastelein

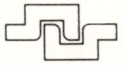

Series A: Life Sciences

Sensory Abilities of Cetaceans

Laboratory and Field Evidence

Edited by
Jeanette A. Thomas
Western Illinois University
Macomb, Illinois

and
Ronald A. Kastelein
Harderwijk Marine Mammal Park
Harderwijk, The Netherlands

Plenum Press
New York and London
Published in cooperation with NATO Scientific Affairs Division

Proceedings of a NATO Advanced Research Workshop and
Symposium of the Fifth International Theriological Congress on
Sensory Abilities of Cetaceans,
held August 22-29, 1989,
in Rome, Italy

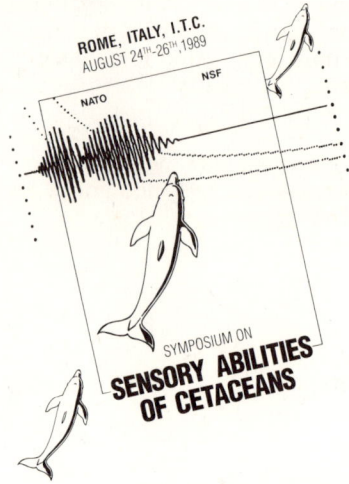

Library of Congress Cataloging-in-Publication Data

```
Sensory abilities of cetaceans : laboratory and field evidence /
  edited by Jeanette A. Thomas and Ronald A. Kastelein.
       p.    cm. -- (NATO ASI series. Series A, Life sciences ; vol.
  196)
    "Proceedings of a NATO advanced research workshop and symposium of
  the Fifth International Theriological Congress on Sensory Abilities
  of Cetaceans, held August 22-29, 1989, in Rome, Italy"--T.p. verso.
    "Published in cooperation with NATO Scientific Affairs Division."
    Includes bibliographical references and index.
    ISBN 0-306-43695-7
    1. Cetacea--Sense organs--Congresses.  2. Cetacea--Behavior-
  -Congresses.  3. Senses and sensation--Congresses.   I. Thomas,
  Jeanette A.  II. Kastelein, Ronald A.  III. North Atlantic Treaty
  Organization. Scientific Affairs Division.  IV. International
  Theriological Congress on Sensory Abilities of Cetaceans (5th : 1989
  : Rome, Italy)  V. Series: NATO ASI series. Series A, Life sciences
  ; v. 196.
  QL737.C4S445  1991
  599.5'04182--dc20                                         90-47395
                                                                CIP
```

© 1990 Plenum Press, New York
A Division of Plenum Publishing Corporation
233 Spring Street, New York, N.Y. 10013

All rights reserved

No part of this book may be reproduced, stored in a retrieval system, or transmitted in any form or by any means, electronic, mechanical, photocopying, microfilming, recording, or otherwise, without written permission from the Publisher

Printed in the United States of America

To my husband, Victor Ramos, and my children, Julienne
and Galen Thomas-Ramos, who provide endless support and
encouragement because they realize that my passion for them
is only rivaled by my scientific curiosity.
—J.A.T.

To Brigitte Slingerland, for her support during the last
two years. I apologize for the many evenings and weekends
that I confined myself to my study room.
—R.A.K.

PREFACE

This book evolved through the efforts of several organizations and the dedication of many individuals. In 1987, we received a request to propose a workshop topic for the Fifth International Theriological Congress (ITC) to be held in August 1989 in Rome, Italy. After looking up the meaning of the word "theriological" in the dictionary and discovering that it pertains to mammalian behavior, we decided a symposium on sensory abilities of whales and dolphins would be an interesting topic. The ITC convenes only every five years and has the distinction of being very well attended by scientists from around the world. We thought that hosting a workshop in conjunction with the ITC would attract a variety of international scientists that rarely have the opportunity to interact.

Fortunately for all involved, our prediction was correct. The first two days of the workshop, 23-24 August 1989, were held in conjunction with ITC and the nearly 1,000 attending scientists were able to view our posters and listen to lectures. The third day was limited to only about 65 invited scientists who were divided into topical working groups chaired by a rapporteur.

This book is organized by the working group topics of Sensory Anatomy and Physiology, Acoustic Senses, Vision/Tactition/Chemoreception, Communication, and Other Senses. The working groups took on very different characters, but each generated a great deal of enthusiasm, promising research ideas, and established a network of colleagues for future collaborations. The success of these working groups is attributed to the rapporteurs: Peter Morgane, Christopher Clark, Frank Awbrey, Karen Pryor, and Herbert Roitblat. The Concluding Comments by these rapporteurs in each section of the book gives you a "flavor" of the dynamic exchanges in these groups and the skill of the rapporteurs in stimulating ideas.

Drs. Luigi Boitani and Ernesto Campana, organizers for the 5th ITC, from the University of Rome provided endless advice and logistic support in coordinating our workshop at the ITC. The success of our workshop and this resulting book is due largely to their hard work.

Our workshop was selected as an Advanced Research Workshop (ARW) funded by the Scientific Affairs Division of NATO in Belgium. Dr. Craig Sinclair, the Director for the

Advanced Study Institutes Programme, encouraged our workshop and gave us a great deal of support. Additional NATO funds were obtained to support the participation of Portuguese scientists through the efforts of Mr. A. Trigo de Abreu.

A large number of US and USSR scientists were able to attend the meeting in Rome because Dr. Herbert Roitblat from the University of Hawaii generously obtained travel funds through a grant from the National Science Foundation.

Our colleagues at the Naval Ocean System Center deserve special recognition for the many ways that they contributed to the development of the workshop. We benefited immensely from the advice provided by Paul Nachtigall, Whitlow Au, Patrick Moore, Bill Friedl, and Sam Ridgway based on the experiences they gained when they organized a similar conference. We acknowledge Winifred Chrismer for her dedicated and careful work in coordinating the large amount of correspondence related to this workshop and book.

The preparation of this book was funded generously by a NATO publication grant from the Scientific Affairs Division in Belgium.

We thank Ruud de Clercq, director of the Harderwijk Marine Mammal Park, for providing the generous logistic support, which largely contributed to the success of the symposium. We also thank him for allowing one of us (RAK) to allocate a large part of his time to the organization of the symposium.

Lastly, we thank Western Illinois University and Harderwijk Marine Mammal Park for their full support in the task of editing the manuscripts in this book. Both organizations provided release time from other duties so that we could edit these papers in a timely manner.

DOLFINARIUM
HARDERWIJK

Jeanette A. Thomas
Western Illinois University
Macomb, Illinois   USA

Ronald A. Kastelein
Marine Mammal Park
Harderwijk, The Netherlands

CONTENTS

SENSORY ANATOMY AND PHYSIOLOGY

Forehead Anatomy of Phocoena phocoena and Cephalorhynchus commersonii: 3-Dimensional Computer Reconstructions with Emphasis on the Nasal Diverticula 1
    Mats Amundin and Ted Cranford

Structure and Thalamocortical Relations of the Cetacean Sensory Cortex: Histological, Tracer and Immunocytochemical Studies 19
    Laurence J. Garey and Alexander V. Revishchin

A Potential Neural Substrate for Geomagnetic Sensibility in Cetaceans 31
    Nicholaas M. Gerrits and Ronald A. Kastelein

Immunocytochemistry of Neurotransmitters in Visual Neocortex of Several Toothed Whales: Light and Electron Microscopy Study 39
    Ilya I. Glezer, Peter J. Morgane, and Csaba Leranth

Evolution of the Nasal Anatomy of Cetaceans 67
    John E. Heyning and James G. Mead

Three-dimensional Reconstructions of the Dolphin Ear 81
    Darlene R. Ketten and Douglas Wartzok

Sensory Neocortex in Dolphin Brain 107
    Peter J. Morgane and Ilya I. Glezer

Evolutionary Morphology and Acoustics in the Dolphin Skull 137
    Helmut A. Oelschläger

Tactile Sensitivity, Somatosensory Responses, Skin Vibrations, and the Skin Surface Ridges of the Bottlenose Dolphin, Tursiops truncatus 163
    Sam H. Ridgway and Donald A. Carder

A Potential System of Delay-lines in the Dolphin Auditory Brainstem 181
    John M. Zook and Ralph A. DiCaprio

Concluding Comments on Sensory Anatomy and Physiology . . 195
    Peter J. Morgane

ACOUSTIC SENSES
A. Echolocation/Sound Production

Target Detection in Noise by Echolocating Dolphins . . . 203
   Whitlow W. L. Au

Preliminary Notes on Behaviour of a Blindfolded Free-swimming Dolphin Performing a Target Echolocation Task in a Pool 217
   Massimo Azzali and Gabriele Buracchi

On the Two Auditory Subsystems in Dolphins 233
   Nikolai A. Dubrovskiy

A Proposed Echolocation Receptor for the Bottlenose Dolphin (Tursiops truncatus): Modelling the Receive Directivity from Tooth and Lower Jaw Geometry 255
   A. David Goodson and M. Klinowska

Studies on Echolocation of Porpoises Taken in Salmon Gillnet Fisheries 269
   Yoshimi Hatakeyama and Hideo Soeda

Very High-frequency Acoustic Emissions from the White-beaked Dolphin (Lagenorhynchus albirostris) . 283
   Ronald B. Mitson

High Intensity Narwhal Clicks 295
   Bertel Møhl, Annemarie Surlykke, and Lee A. Miller

Investigations on the Control of Echolocation Pulses in the Dolphin (Tursiops truncatus) 305
   Patrick W. B. Moore and Deborah A. Pawloski

Purposeful Changes in the Structure of Echolocation Pulses in Tursiops truncatus 317
   Evgeniy V. Romanenko

Echolocation Characteristics and Range Detection Threshold of a False Killer Whale (Pseudorca crassidens) 321
   Jeanette A. Thomas and Charles W. Turl

B. Hearing

Preliminary Hearing Study on Gray Whales (Eschrictus robustus) in the Field 335
   Marilyn E. Dahlheim and Donald K. Ljungblad

Inferences about Perception in Large Cetaceans, Especially Humpback Whales, from Incidental Catches in Fixed Fishing Gear, Enhancement of Nets by "Alarm" Devices, and the Acoustics of Fishing Gear 347
   Jon Lien, Sean Todd, and Jacques Guigne

Formation of an Adaptive Structure of the Peripheral
    Part of the Auditory Analyzer in Aquatic,
    Echo-locating Mammals during Ontogenesis 363
  Galina N. Solntseva

Frequency-selectivity of the Auditory System in the
    Bottlenose Dolphin, *Tursiops truncatus* 385
  Alexander Supin and Vladimir Popov

Masked Hearing Abilities in a False Killer Whale
    (*Pseudorca crassidens*) 395
  Jeanette A. Thomas, Jeffrey L. Pawloski,
  and Whitlow W. L. Au

Electrophysiological Studies on Hearing in Some
    Cetaceans and a Manatee 405
  Vladimir Popov and Alexander Supin

Localization of the Acoustic Window at the
    Dolphin's Head 417
  Vladimir Popov and Alexander Supin

Concluding Comments on Cetacean Hearing and
    Echolocation 427
  Frank T. Awbrey

CHEMICAL/TACTILE/VISUAL SENSES

Preliminary Results from Psychophysical Studies on
    the Tactile Sensitivity in Marine Mammals 435
  Guido Dehnhardt

Taste Reception in the Pacific Bottlenose Dolphin
    (*Tursiops truncatus gilli*) and the California
    Sea Lion (*Zalophus californianus*) 447
  William A. Friedl, Paul E. Nachtigall, Patrick W. B.
  Moore, Norman K. W. Chun, Jeffrey E. Haun, Richard
  W. Hall, and James L. Richards

Cognitive Performance of Dolphins in Visually-guided
    Tasks 455
  Louis M. Herman

Anatomical and Histological Characteristics of the
    Eyes of a Month-old and an Adult Harbor
    Porpoise (*Phocoena*) 463
  Ronald A. Kastelein, Ruud C. V. J. Zweypfenning,
  and Henk Spekreijse

Chemical Sense of Dolphins: Quasi-olfaction 481
  Vitaly B. Kuznetzov

Best Vision Zones in the Retinae of Some Cetaceans . . . 505
  Alla Mass and Alexander Supin

Visual Ecology and Cognition in Cetaceans 519
  Joseph R. Mobley, Jr. and David A. Helweg

Non-acoustic Communication in Small Cetaceans: Glance,
    Touch, Position, Gesture, and Bubbles 537
  Karen W. Pryor

Visual Displays for Communication in Cetaceans 545
  Bernd Würsig, Thomas Kieckhefer,
  and Thomas A. Jefferson.

Concluding Comments on Vision, Tactition, and
    Chemoreception 561
  Karen Pryor

COMMUNICATION

Acoustic Behavior of Mysticete Whales 571
  Christopher W. Clark

Acoustic Behavior in a Local Population of Bottlenose
    Dolphins 585
  Manuel E. dos Santos, Giorgio Caporin,
  H. Onofre Moreira, António J. Ferreira,
  and J. L. Bento Coelho

Organization of Communication System in
    Tursiops truncatus Montagu 599
  Vladimir I. Markov and Vera M. Ostrovskaya

Signalization of the Bottlenose Dolphin during the
    Adaptation to Different Stressors 623
  Irina E. Sidorova, Vladimir I. Markov,
  and Vera M. Ostrovskaya

Concluding Comments on Acoustic Communication 635
  Christopher W. Clark

OTHER SENSES

Geomagnetic Sensitivity in Cetaceans: An Update
    with Live Stranding Records in the
    United States 639
  Joseph L. Kirschvink

Geomagnetic Orientation in Cetaceans: Behavioural
    Evidence 651
  Margaret Klinowska

Attention and Decision-making in Echolocation
    Matching-to-Sample by a Bottlenose Dolphin
    (*Tursiops truncatus*): The Microstructure
    of Decision-making 665
  Herbert L. Roitblat, Ralph A. Penner,
  and Paul E. Nachtigall

Stimulus Equivalence and Cross-modal Perception: A
    Testable Model for Demonstrating Symbolic
    Representations in Bottlenose Dolphins 667
  Ronald J. Schusterman

The Ability of Bottlenose Dolphins, <u>Tursiops</u> <u>truncatus</u>,
    to Report Arbitrary Information 685
    Alexander Zanin, Vladimir I. Markov
    and Irina E. Sidorova

Concluding Comments on Other Sensory Abilities 699
    Herbert L. Roitblat

INDEX 703

FOREHEAD ANATOMY OF PHOCOENA PHOCOENA AND CEPHALORHYNCHUS
COMMERSONII: 3-DIMENSIONAL COMPUTER RECONSTRUCTIONS WITH
EMPHASIS ON THE NASAL DIVERTICULA

Mats Amundin and Ted Cranford[1]

Zoological Institute, Department of Functional
Morphology, University of Stockholm and Kolmården
Zoo, Kolmården, Sweden
[1]Institute of Marine Science, University of
California, Santa Cruz, USA

INTRODUCTION

Sound generation in odontocetes takes place in the upper nasal passage, powered by air pressure created within the bony nares (Norris et al., 1971; Diercks et al., 1971; Hollien, 1976; Dormer, 1979; Ridgway et al., 1980; Amundin and Andersen, 1983; Cranford, 1988). Cranford (1988) describes two pairs of fatty bodies, labelled dorsal bursae, situated in the lateral corners of the spiracular cavity in the spinner dolphin, which may function as "sound transducers and/or conductors of sound" produced at the bursae or in their vicinity. Two posterior branches of the melon connect to the anterior bursae. It also has been suggested that air sacs around the spiracular cavity may guide the emitted sonar pulses rostrally (Norris, 1964; Dormer, 1974; Giro and Dubrovskii, 1975; Alcuri, 1980). Additional sonar beam formation is thought to take place in the melon, as a result of the sound velocity topography of special "acoustic" fats (Norris and Harvey, 1974; Litchfield et al., 1979; Varanasi et al., 1982).

The most powerful sonar clicks in the Harbour porpoise (Phocoena phocoena) and the Commerson's dolphin (Cephalorhynchus commersonii) are, relative to those of most other Delphinids, narrow banded, multicyclic pulses, with all sound energy above 100 kHz, and with the peak at 120-140 kHz (Dubrovskii et al., 1971; Möhl and Andersen, 1973; Kamminga and Wiersma, 1981; Kamminga and Wiersma, 1982; Evans and Awbrey, 1988). A much weaker, low-frequency component at around 2 kHz has been observed in Phocoena (Busnel et al., 1965; Schevill et al., 1969; Amundin in prep). A similar low-frequency, low-intensity component, although with major emphasis at around 800 Hz, were observed in Cephalorhynchus heavisidii, C. eutropia and C. commersonii (Watkins et al., 1977). In addition, Kamminga and Wiersma (1981) reported a 20 kHz component, only 9 dB weaker than the 120 kHz pulse, from a stranded Phocoena. This component has not been reported by other studies. Neither species have been observed to whistle (Busnel and Dziedzic, 1966; Watkins et al., 1977; Evans and Awbrey, 1988).

The first, strong amplitude peak of the high frequency sonar click component in both Phocoena and Cephalorhynchus is followed by at least one peak of lower amplitude, separated from the first by a phase shift with very low amplitude (Möhl and Andersen, 1973; Kamminga and Wiersma, 1981). There is almost no variation in this complex waveform over a large number of subsequent clicks, if recorded in the core of the sonar beam (Kamminga and Wiersma, 1981). Dudok van Heel (1981), Kamminga and Wiersma (1981), and Wiersma (1982) suggest that the low amplitude components are reverberations, due to reflections on the skull bones and/or the air sacs, resulting in multiple propagation paths within the porpoise head. The stable click waveform found in the sonar beam core then indicate that, if the "acoustic mirror(s)" are air spaces, they are fixed in relation to the sound generator during sound production.

Watkins et al. (1977) and Dudok van Heel (1981) point to similarity in sound output in Cephalorhynchus and Phocoena, and suggest that it may be the result of convergent evolution, due to a similar ecological conditions, i.e. life in a coastal or estuarial, shallow water habitat. This suggestion, together with evidence that sounds are generated with forehead, soft-tissue structures (possibly involving the fatty bursa), are compelling reasons to undertake a detailed comparative study of their forehead anatomy.

There are numerous anatomical descriptions of the upper nasal pathways for a number of odontocete species, which present drawings, rubber casts, sections, or schematics (Lawrence and Schevill, 1956; Fraser and Purves, 1960; Hosokawa and Kamiya, 1965; Purves, 1966; Schenkkan, 1973; Evans and Maderson, 1973; Mead, 1975; Green et al., 1980; Gurevich, 1980; Alcuri, 1980; Heyning, 1989). Schenkkan (1973) carried out a comparative study of nasal anatomy in 20 odontocete species, including Phocoena phocoena, and Cephalorhynchus hectori, a relative of C. commersonii. To our knowledge, the only published report on the anatomy of the upper nasal tract of C. commersonii is that of Amundin et al (1988).

Although some of these studies provided detailed anatomical descriptions, which hint at the complexity of the system, most have gleaned their results from tissue geometries, which are commonly distorted with traditional techniques. The complex geometry of the odontocete nasal anatomy is likely an important aspect of its function.

A few recent studies have attempted to retain tissue geometries by combining millimeter scale serial sectioning with computer reconstruction and graphics capabilities. Amundin et al. (1988) presented preliminary results with 3-dimensional computer aided reconstructions, based on serial cryomicrotome sections of the upper nasal anatomy in Phocoena phocoena. Cranford (1988) used a similar technique, based on serial CT-scans, to outline the shape of the melon and its connection to the dorsal bursae. He also prepared a color movie (1985), based on the computer reconstructions, showing the fatty structures in the context of the gross geometry of a spinner dolphin's head.

In this presentation, we approach the problem of describing complex 3-D anatomy by using high resolution, computer generated stereograms based on serial, 2-D sections which depict structures that we believe are important to sonar signal generation in

Phocoena phocoena and *Cephalorhynchus commersonii*. Stereograms allow the viewer's brain to reconstruct the morphology in three dimensions.

MATERIAL AND METHODS

A frozen head of a subadult female harbour porpoise, *Phocoena phocoena*, was embedded in Carboxy Methyl Cellulose (CMC) and sectioned parasagittally by means of a cryo-microtome. At 1 mm intervals, the cut surface was wiped with methanol to enhance the color of the various tissues, and photographed. For more details on this method, refer to Ullberg (1977) and Rauschning (1979). Two holes drilled through the entire head, perpendicular to the cutting plane, provided registration points for the computer reconstructions (cf below).

Another *Phocoena* head (MVZ #172408, male, 161.8 cm total body length) was scanned in a General Electric 9800 CT scanner, at the University of California Medical School in San Francisco. Seventy-nine serial cross-sectional images were generated, 39 over a thickness of 3 mm and collected at 3 mm intervals from 30 to 145.5 mm from the tip of the rostrum (including most of the melon), and 40 averaged over 1.5 mm and collected at 1.5 mm intervals from 145.5 to 204 mm from the tip of the rostrum (covering the area with the bony nares approximate to the caudal margin of the nasal bones). The scale on the CT-images provided registration points for the computer reconstructions.

A *Cephalorhynchus commersonii* head ("Gauchu", male, 145 cm total body length) was obtained from Duisburg Zoo on the condition that the skull was to be left intact. Therefore, the soft tissues above the skull, i.e. the melon and the upper nasal pathways with adjoining tissues, were removed carefully, placed back on top of the skull into a shape as close as possible to the original, and then placed into a freezer. This frozen soft tissue specimen was then embedded in CMC, sectioned parasagittally in the cryomicrotome (each slice 50 µm) and photographed at 1 mm intervals. Two corners of the CMC block were used as registration points during the computer digitization (cf below).

The serial photographs and serial tomograms were then projected on a Hipad EDT-11H, 300 by 300 mm digitizing tablet (Houston Instrument). Selected structures were traced manually with a cross-hair cursor and fed into a computer database (see below).

The 3-D reconstructions were generated with a computer program, based upon an IBM Professional Graphics Display (PGD), with a resolution of 640 pixel by 480 lines, attached to an IBM PC/AT. The system yields high resolution color images, where selected structures are identified by different colors. Depth cues are given by the color shades. The software can compute 3-D rotations, polygon areas, structure volumes, distances (in any dimension), or scaling. The volume of a structure was calculated by adding the area of selected polygons from a number of cross sections according to the following formula:

$$\sum_{A=1}^{n} [\text{Area of polygon A} \times \text{slice thickness} \times 0.5]$$

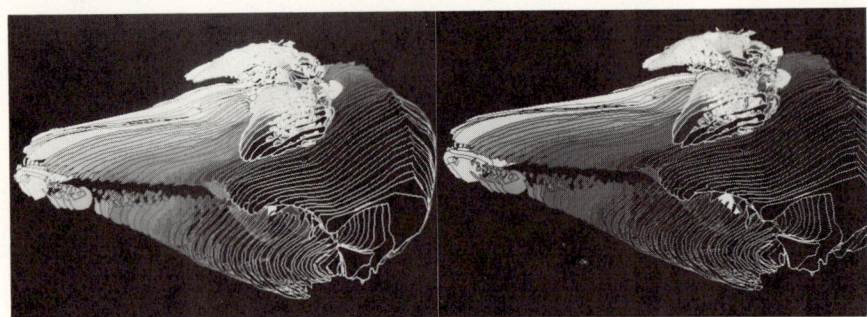

Fig 1. <u>Phocoena</u>. Computer reconstructed stereogram of a skull with the spiracular cavity and all diverticula. This figure and all stereograms below should be viewed through a lens stereoscope, which will produce a stereo image of the shown structure.

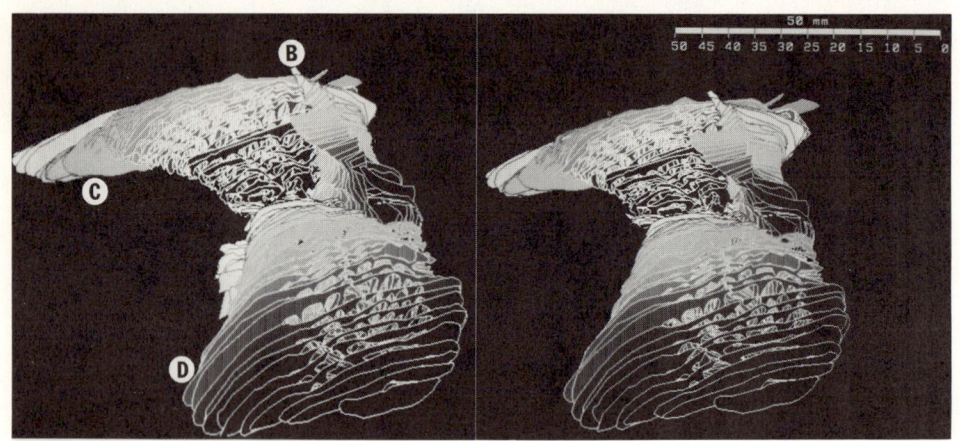

Fig 2. <u>Phocoena</u>. Magnification of the air diverticula in Figure 1. B=blowhole; C=right vestibular sac; D=left vestibular sac

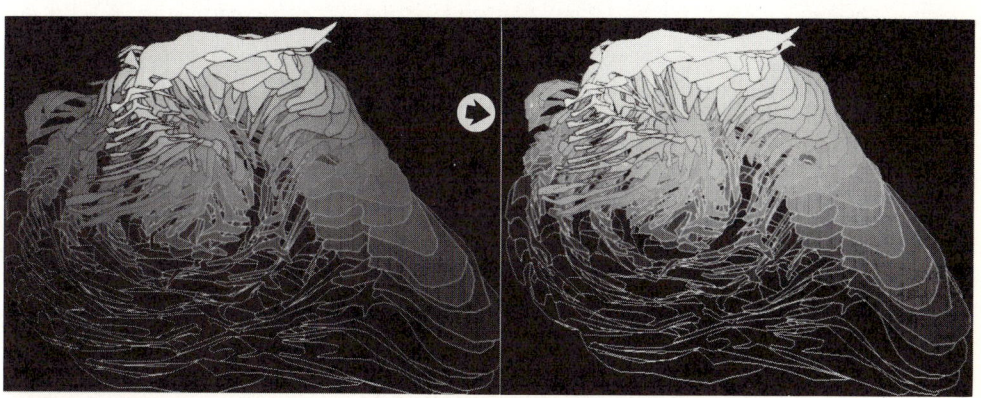

Fig 3. <u>Phoncoena</u>. A view from below of the left vestibular sac showing the complex folded floor. Arrow shows direction to rostrum.

When viewed through a stereoscope, stereograms produce
3-D geometry. Each stereogram consists of two photos of the
screen, one with the object rotated 15 degrees relative to the
other around the computer screen Y-axes. For more information
on stereogrammetry, see American Society of Photogrammetry
(1981), referred to in Nämnden för Skoglig Flygbildsteknik
(1980).

RESULTS

Phocoena

The general outline of the cryosectioned Phocoena skull is
shown (Fig 1), with the nasal diverticula on top. The vestibular
sacs are situated on either side of, and anterior to the
spiracular cavity. This is shown more clearly in Figure 2, which
is a magnification of these sacs. Their maximum rostrocaudal
dimensions are 50 mm for the right sac and 48 mm for the left
one. The maximum transverse (left-right) dimensions are 43 mm and
35 mm, respectively.

The lining of the lateral margins and the roof of both
vestibular sacs are smooth and unfolded, whereas their floor is
complex and folded, forming transversely oriented, concentric,
branched ridges of connective tissue (Fig 3).

A fold between two of the central ridges in each vestibular
sac is extended into a slitlike passage that leads to, and opens
into the anterior wall of the spiracular cavity. The lateral
width of the passages is 19 and 16 mm, respectively, thus
covering almost the full 40 mm lateral width of the spiracular
cavity.

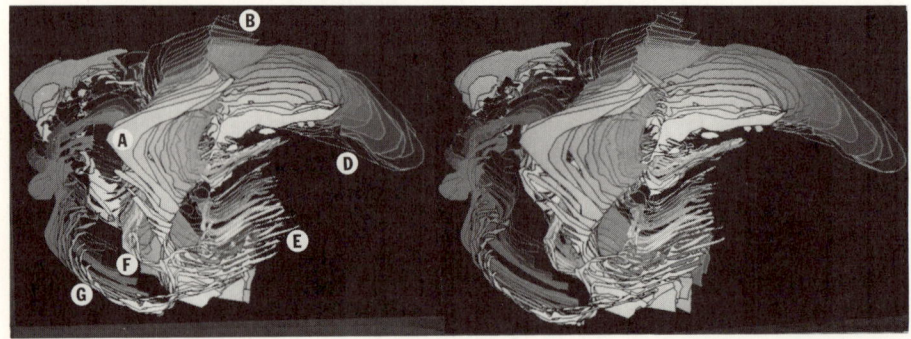

Fig 4. A close-up of the Phocoena air diverticula, with the right
vestibular sac removed to show the nasofrontal sacs (E and F),
that surround the spiracular cavity (A). B=blowhole; D=left
vestibular sac; G=caudal nasal sac.

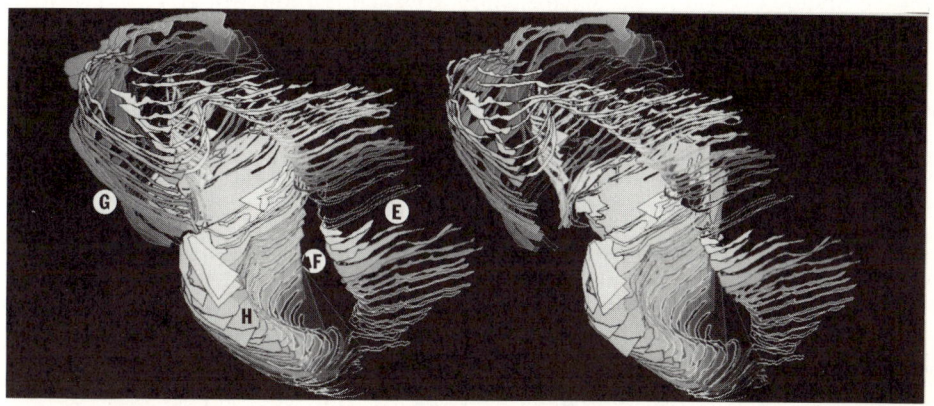

Fig 5. <u>Phocoena</u>. The left "hintere Klappe" (H) surrounded by the caudal nasal sac (G) together with the posterior nasofrontal sac (F). In this image the right "hintere Klappe" has been removed, to show more clearly the sacs on the right side. The spiracular cavity, passing between the anterior and posterior nasofrontal sacs, has also been removed for clarity; E=ant. nasofrontal sacs.

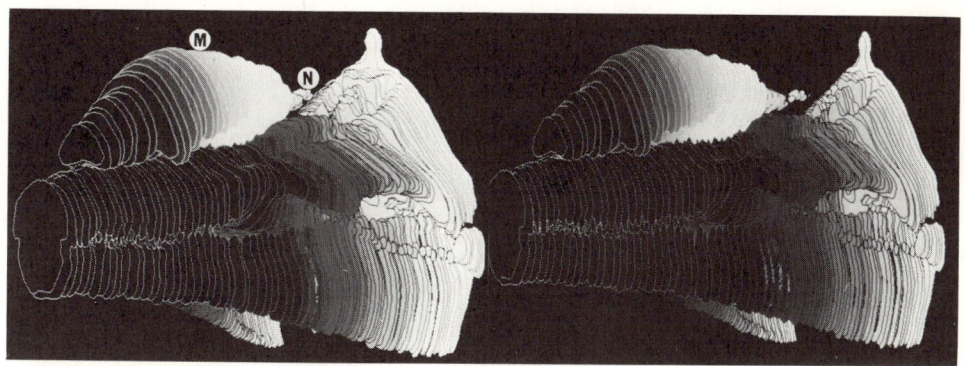

Fig 6. <u>Phocoena</u>. THe CT scanned skull with melon (M) on top. The melon core terminates in the middle of the nasal plug area. Caudolaterally of the melon end are seen two pairs of bursae. N=right bursae.

In a horizontal plane, at the level of the anterior nasofrontal sacs, the spiracular cavity is curved, with the convex side towards the rostrum. In the vertical plane, the lateral margins of spiracular cavity, above the nasal plugs, make a prominent caudal turn. The vestibular sacs are situated so that their caudal roof, when the sacs are inflated, press the anteroventral wall of the upper spiracular cavity against the posterodorsal wall, thereby sealing the blowhole. Lawrence and Schevill (1956) describe a similar pneumatic seal in Tursiops, although here involving the anterior nasofrontal sacs.

In Figure 4, with the right vestibular sac removed, it is clearly seen how the nasofrontal sacs surround the spiracular cavity. The anterior nasofrontal sacs are flattened dorsoventrally, and extended rostrocaudally. The vertical distance from the anterior nasofrontal sacs to the passage between the vestibular sacs and the spiracular cavity is 1.5-6.0 mm.

The posterior nasofrontal sacs and the inferior vestibule are divided by a transverse outgrowth from the roof, termed "hintere Klappe" (posterior flap) by Gruhl (1911, in Schenkkan, 1973, see also Amundin et al., 1988; and Heyning, 1989). Figure 5 shows the outline of the resulting paired, thin sacs, completely surrounding the left and right "hintere Klappe" (the right "hintere Klappe" has been removed for clarity). In the present paper, the terminology of Moris (1969) is used, terming the more caudal of these sacs the caudal nasal sacs, and the more rostral the posterior nasofrontal sacs. The "hintere Klappe" extend anteroventrally to the posterior edge of the bony nares. The joint entrance to the posterior nasofrontal sac and caudal nasal sac from the bony nares/nasal passage, is situated at the level of the nasal plug's posteroventral and lateral margins. Together the right and left side entrances cover the full width of the nares, except for a septum in the middle.

Reconstructions from tomograms of the Phocoena skull are shown in Figure 6, with the outline of the low density melon core (cf Cranford, 1988) above the skull.

The melon core narrows towards its caudal end, terminating blindly in the middle of the nasal plug area. In this region, the cross section of the melon core has a slightly curved, oval shape, approximately 13 mm transversely and 2.4 mm dorso-

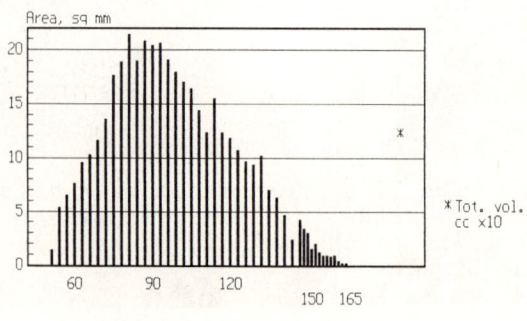

Fig 7. Phocoena. A graph showing the cross section area of each melon section, throughout the whole melon. The first 32 sections were collected at 3 mm, and the last 13 at 1.5 mm intervals.

ventrally, with an area of 18-20 mm². Figure 7 shows graphically the cross-section area in mm² of each melon section throughout the whole melon. The total volume of the melon is approximately 122 cc.

In the same horizontal plane, but 12 mm caudally of the posterior terminus of the melon core, two pairs of transversely elongated, low density structures are found (Fig 6). These are the dorsal bursae of Cranford (1988). Figure 8 shows a horizontal plane through these structures, computed from the original tomogram cross sections. The bursae on each side consist of one anterior and one posterior part. The right bursae pair is approx 20% larger than the left one. Table I shows the dimensions, measured from cross-sections.

As indicated above, there are no branches from the melon continous with the bursae. Figure 9 shows the melon dorsal to the skull, as viewed from above and behind, indicating the relationship between the bursae and the bony nares. The cross-section slices immediately caudal to the dorsal bursae have been removed for clarity. By combining the geometric information from reconstructions based on the cryo-microtome sections and the tomograms respectively (although using two different specimens), a rough estimate can be obtained of the bursal position in relation to the air cavities. The bursae are found at the lateral corners of the spiracular cavity, immediately ventral to the entrances into the passages that lead to the vestibular sac. The spiracular cavity passes between the anterior and posterior bursae on both the right and the left side.

Cephalorhynchus

In Figure 10, we combined a stereo photograph of the intact Cephalorhynchus skull with a 3-D computer stereogram of the spiracular cavity and the air sacs. A magnification of the

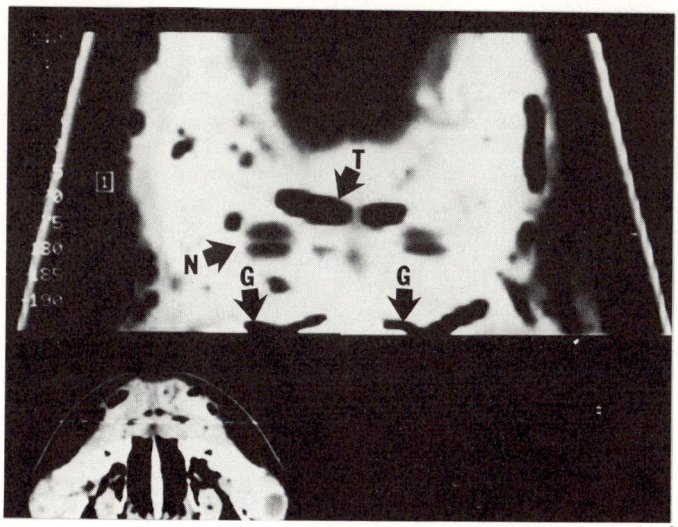

Fig 8. Phocoena. Horizontal tomogram through the two bursae pairs. The spiracular cavity passes between the anterior and posterior part of each bursae pair. G=caudal nasal sac; T=bony nares; N=right bursae.

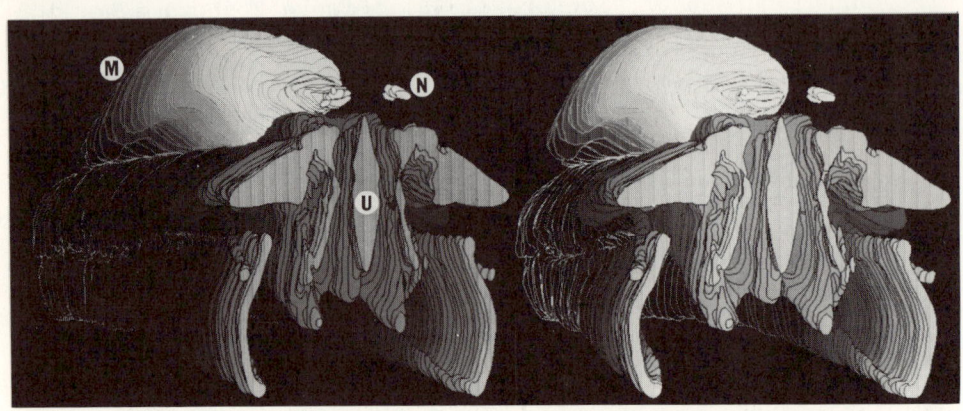

Fig 9. <u>Phocoena</u>. View from behind of the skull with melon (M) on top. Sections caudal to the bursae have been removed, to reveal the bursal position in relation to the bony nares. G=right bursae; U=nasal septum.

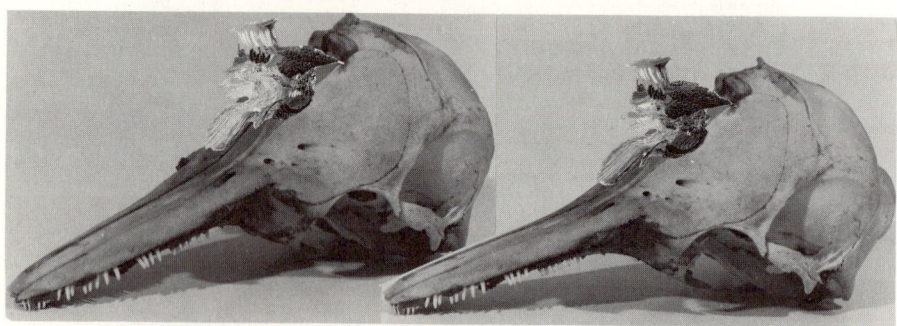

Fig 10. <u>Cephalorhynchus</u>. The skull combined with computer graphics reconstructions of the spiracular cavity and the air diverticula. No exact fitness has been attempted.

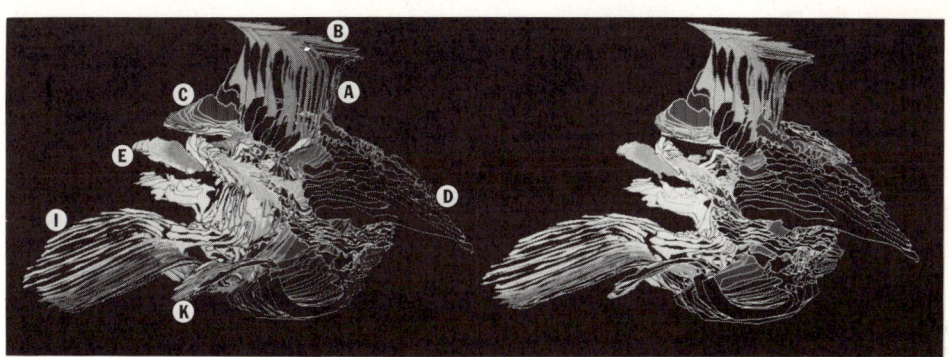

Fig 11. <u>Cephalorhynchus</u>. A magnification of the air diverticula shown in Figure 10. A=spiracular cavity; B=blowhole; C=right vestibular sac; D=left vest. sac; E=ant. nasofrontal sac; I=right premax. sac; K=left premax. sac.

Table I. _Phocoena_. Bursae Dimensions from Cross-section Tomograms. Measures in parentheses are less certain, due to blurry contours.

Slice No	Right side		Left side	
	Vertical "height"	Lateral width	Vertical "height"	Lateral width
1	2.73	6.97	2.42	5.44
2	3.88	9.6	2.34	6.81
3	2.78	7.22	2.28	6.3
4	3.29	9.23	2.83	7.55
5	2.83	7.6	(2.39)	(2.58)
MEAN	3.1	8.12	2.47	6.53

reconstruction (Fig 11) reveals large vestibular sacs that open mainly in the lateral and caudal walls of the spiracular cavity. The anterior portion of the left vestibular sac, partly cut in Figure 11, is dorso-ventrally branched. Figure 11 also demonstrates that the nasofrontal sacs completely surround the spiracular cavity. As in _Phocoena_, the anterior nasofrontal sacs are dorso-ventrally flattened, and rostrocaudally extended. The inferior vestibules are narrow, rostrocaudally flattened tubes, passing from folds in the nasal passage, beneath the blowhole ligament. Prominent nasal plug lateral lips fit into the folds, blocking the entrances to the inferior vestibules (Fig 12).

The right premaxillary sac is considerably larger than the left (Fig 10 and 11), covering 705 mm^2 and 243 mm^2, respectively of the skull rostral and lateral to the bony nares. It corresponds to a similar, but less prominent asymmetry of the underlying premaxillary bones. The area, delineated caudally by the nares, mesially by the mesethmoid groove, laterally by the raised edge of the premaxillary bones, and rostrally by a transverse line drawn through the small trigeminal nerve orifice inside these edges, are approximately 1125 and 830 mm^2 on the right and left side, respectively.

DISCUSSION

Both species have relatively large vestibular sacs compared to other delphinids (cf Schenkkan 1973). In _Phocoena_ the vestibular sacs lie entirely antero-laterally to the spiracular cavity, with separate, slit-like entrances to the spiracular cavity. Our study shows that in _Cephalorhynchus_, the vestibular sacs are situated mainly caudal and lateral to the spiracular cavity, to which they open directly and mainly caudally and laterally. In _Phocoena_ the floor of the vestibular sacs have prominent connective tissue ridges, which are lacking entirely in _Cephalorhynchus_.

Schenkkan (1973) investigated _Cephalorhynchus hectori_ and found vestibular sacs that seem to be similar to those of our _C. commersonii_. He did not, however, report any vertical bifurcation of the rostral edge of the left vestibular sac, as we found in our _C. commersonii_.

Norris (1964) suggests that the air sacs might function as sound reflectors, involved in guiding sound rostrally through the melon and into the water. The size of the vestibular sacs of these two species certainly allow them to be reflectors, being more than 4 times larger than the wavelength of the peak frequency in the high frequency sonar pulse. In Phocoena the vestibular sacs would effectively shield radiation of sound in a vertical and lateral direction.

In a pool situation, Phocoena often rest floating at the surface, with the vestibular sacs visibly inflated (Andersen, pers. comm., cited in Purves and Pilleri, 1983, Amundin, pers. obs.). The sacs are large and certainly could serve as effective floatation devices, keeping the blowhole area above the water surface. This function for the vestibular sacs has been suggested by Sibson (1948, quoted in Mead 1974), and by Purves (1967). The slit-like, partly horizontal entrances from the spiracular cavity into the vestibular sacs may be sealed by inflation of the anterior nasofrontal sacs. This could be an effective method of keeping the vestibular sacs inflated.

This possible pneumatic lock of the vestibular sac entrances in Phocoena gives rise to a question about the role of the nasofrontal sacs during sound production. If the necessary sealing of the spiracular cavity is achieved by inflating the nasofrontal sacs, as suggested by Lawrence and Schevill (1956), would also seal the entrance to the vestibular sacs, it would prevent the vestibular sacs from collecting air used for sound production, and in effect prevent sound production (cf Norris et al., 1971; Dormer, 1974; Ridgway et al., 1980; Amundin and Andersen, 1983). This would indicate, that in Phocoena, the nasofrontal sacs do not function as a pneumatic lock during sound production.

The inferior vestibules in Phocoena are expanded laterally, and have become an inseparable part of the posterior nasofrontal sacs. In the Cephalorhynchus commersonii, as well as in most

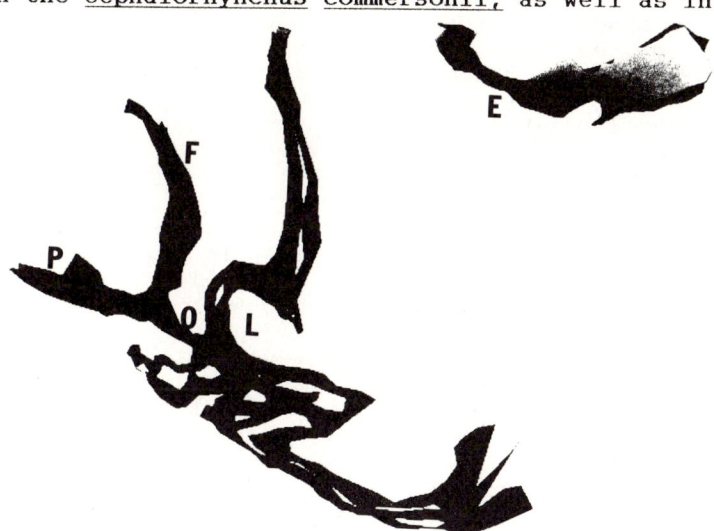

Fig 12. Cephalorhynchus. A detail showing the lateral lips (L) of the nasal plug, the inferior vestibule (O), accessory sac (P), anterior nasofrontal sac (E) and posterior nasofrontal sac (F).

delphinids, the inferior vestibules are rostrocaudally flattened and narrow (6 and 9 mm wide on the left and right side, respectively; Amundin et al., 1988; cf Mead, 1972). They are blocked by the well-developed lateral lips of the nasal plug. It is uncertain if the small, ventro-lateral margins of the nasal plugs in Phocoena are able to block the entrance from the spiracular cavity, and thus, to regulate the amount of inflation of the nasofrontal sacs and/or the caudal nasal sacs (Schenkkan 1973, cf above). A possible blocking action has to work against a considerable air pressure. Amundin and Andersen (1983) measured up to 54 KPa in the bony nares of a phonating Phocoena.

The tomograms show the low density tissue in the melon as dark (cf Cranford, 1988). The 3-D reconstruction of the Phocoena melon is based on manual tracing of this dark area, and thus only is an approximation of the melon. The melon core narrows symmetrically and terminates in the middle of the nasal plug area in our Phocoena. This apparently is the case in Phocoena dioptrica too (Purgue, 1987, pers. comm.), as well as in Cephalorhynchus hectori (Mead, 1975) and C. commersonii (Heyning, 1989). It is a peculiar trait in common with the porpoises, since all of the delphinids we have investigated so far have a laterally branched posterior melon which leads to each set of bursae (Cranford, 1988). This common trait may be a clue to the similarities in the sounds produced by these two different groups of delphinoids.

Although the exact position of the Phocoena bursae is based only on a rough estimate, they seem to be situated as in the spinner dolphin (Cranford, 1988), in the lateral corners of the spiracular cavity, near the region where sounds are supposed to be generated (Ridgway et al., 1980; Amundin and Andersen, 1983; Cranford, 1988). If these bursae are involved in sound production as transducers, as suggested by Cranford (1988) in Stenella, the significance of their missing low density tissue connection with the melon remains to be explained. Also in Stenella the left bursae was found to be rotated as compared to the right bursae, which was not the case in our Phocoena. The significance of this difference for sound production is not clear.

The caudal nasal sacs have a smooth, cup-like shape, lining two concavities of the skull just ventral to the nasal bones. Also the posterior nasofrontal sacs are cup-like in shape, with the concave side facing the rostrum. At sound production, when the air sac system is pressurized, these sacs most likely are inflated, and then provide large sound reflecting surfaces, one behind the other. How this may affect the sound propagation is not clear, because the possible dynamics of these sacs are unknown.

The premaxillary sacs have been suggested to be sound reflectors (Norris, 1964), and in Cephalorhynchus commersonii and C. hectori (Schenkkan, 1973), where these sacs are comparably large, they probably play an important role in sound guiding. In Phocoena, however, the premaxillary sacs are much smaller (Schenkkan, 1973), indicating a less significant role in reflection, if they play any role in sound guiding at all. Another major function of these sacs may be to facilitate the nasal plug to retract during breathing (cf Lawrence and Schevill, 1956).

CONCLUSION

This study has shown that, although the two investigated species produce similar, narrow band, high frequency echolocation clicks, their nasal anatomy, believed to be involved in sound generation, differs in many respects. The Phocoenids possess two unique connective tissue structures, the so called "hintere Klappe", dividing the left and right posterior nasofrontal sacs and inferior vestibules into two pairs of thin, cup shaped diverticula. These are not seen in Cephalorhynchus, in which the posterior nasofrontal sacs connect with the spiracular cavity through narrow inferior vestibules, like in other delphinids. According to Schenkkan (1979) and Heyning (1989), a configuration similar to Phocoena's, although with a varying degree of developement, is found in Delphinapterus leucas, Inia geoffrensis, Platanista sp, Pontoporia blainvillei, Berardius bairdii and Hyperoodon sp. The nasal plug lateral lips are prominent in Cephalorhynchus. In Phocoena they are small, and probably not able to control air flow into the nasofrontal sacs or caudal nasal sacs. The vestibular sacs in both species are large, but differ in shape, structure and position relative to the spiracular cavity. The most prominent difference is the deep, transversely oriented folds found in the floor of the vestibular sacs in Phocoena but not in Cephalorhynchus.

There are, however, also some similarities between the two species. In both Phocoena and Cephalorhynchus the anterior nasofrontal sacs are rostrocaudally extended and dorsoventrally compressed, in contrast to the tubelike shape found in other delphinids (Schenkkan 1973). The posterior melon in Phocoena lacks direct low density tissue connections with the two sets of bursae, situated in the lateral corners of the spiracular cavity, and suggested to be involved in sound production. This trait separates Phocoena from the delphinids (Cranford, 1988). The shape of the melon core was not investigated in our Cephalorhynchus commersonii, but according to Heyning (1989) the melon of this species, as well as of C. hectori (Mead, 1975), terminates in the middle of the nasal plug area, without lateral branches. This would be a major similarity with Phocoena, but will best be confirmed by means of CT scanning.

ACKNOWLEDGMENTS

This study was sponsored by IBM Sweden and Kolmårdens Djurpark, Sweden. Warm thanks to Sten Kallin, IBM Sweden, for improving the computer software, and to Anders Ingvast and Ulf Andersson, Ulleråkers Sjukhus, Uppsala, Sweden, for letting us use their Cryo-Microtome. Thanks also to Dr Wolfgang Gewalt, Duisburg Zoo, West Germany, for providing the Commerson's dolphin tissue material. Many thanks to Dr Anders Magnusson, Akademiska Sjukhuset, Uppsala, Sweden, for his enthusiasm and great help in connection to interpreting the CT-scans, and to Drs. Chris Cann and Henry Goldberg of University of California, San Francisco Medical School for their help and generosity. Thanks also to Alejandro Purgue, University of Utah, USA, for many helpful comments to earlier drafts of this paper.

REFERENCES

Alcuri, G., 1980, The role of cranial structures in Odontocete sonar signal emission, in: Animal Sonar Systems, Fish, J F. (Ed.), Plenum Press, New York.

Altes, R. A., Evans, W. E., and Johnson, C. S., 1975, Cetacean echolocation signals and a new model for the human glottal pulse, J. Acoust. Soc. Am., 57: 1221.

Amundin, M., and Andersen, S. H., 1983, Bony nares air pressure and nasal plug muscle activity during sound production in the harbour porpoise, Phocoena phocoena and the bottlenosed dolphin, Tursiops truncatus, J. Exp. Biol., 105: 275.

Amundin, M. (in prep), The effect of substituting air with a Helium/Oxygen mixture on the click spectrum of the harbour porpoise, Phocoena phocoena.

Amundin, M., Kallin, E., and Kallin, S., 1988, The study of the sound production apparatus in the harbour porpoise, Phocoena phocoena, and the jacobita, Cephalorynchus commersonii, by means of serial cryo-microtome sectioning and 3-D computer graphics. in: Animal Sonar. Processes and Performance, P. E. Nachtigall, P.W.B. Moore, (eds), Plenum Press, New York, pp 61-66.

Busnel, R-G., and Dziedzic, A., 1966, Acoustic signals of the Pilot whale, Globicephala melaena and of the porpoises Delphinus delphis and Phocoena phocoena, in: Whales, dolphins and porpoises, K. S. Norris, (ed), Univ. Calif. Press, Berkeley and Los Angeles, pp 607-646.

Busnel, R-G., Dziedzic, A., and Escudie, B., 1969, Autocorrélation et analyse spectrale des signaux "sonar" de deux espèces de Cétacéa Odontocétes utilisant les basses fréquences. C. R. Acad. Sc. Paris, 269(3):365-367.

Cranford, T. W., 1985, Cephalic symbols II, Movie film shown at the Conference of Marine Mammalogy in Miami, Florida.

Cranford, T. W., 1988, The anatomy of acoustic structures in the spinner dolphin forehead as shown by X-ray computed tomography and computer graphics. in: Animal Sonar: Processes and Performance, P. E. Nachtigall, P. W. Moore, (eds), Plenum Press, New York, pp 67-77.

Diercks, K. J., Trochta, R. T., and Greenlaw, C. F., 1971, Recording and analysis of dolphin echolocation signals, J. Acoust. Soc. Am. 49: 1730.

Dormer, K. J., 1974, The mechanism of sound production and measurement of sound processing, Doctoral dissertation, University of California, Los Angeles.

Dormer, K. J., 1979, Mechanisms of sound production and air recycling in delphinids: Cineradiographic evidence, J. Acoust. Soc. Am. 65: 229.

Dubrovskii, N. A., Krasnov, P. S., and Titov, A. A., 1971, On the emission of echolocation signals by the Azov Sea harbor porpoise, Sov. Phys. Acoust., 16: 444-447 (Akust. Zh. 16:521-525, 1970).

Dudok van Heel, W. H., 1981, Investigations on Cetacean sonar. III. A proposal for an ecological classification of Cetacetes in relation to sonar, Aquatic Mammals 8: 65-68.

Evans, W. E., and Maderson, P. F. A., 1973, Mechanisms of sound production in Delphinid Cetaceans: A review and some anatomical considerations. Am. Zool. 13: 1205-1213.

Evans, W. E., and Awbrey, F. T., 1988, Natural history of marine mammal echolocation: feeding strategies and habitat, in: Animal Sonar. Processes and Performance, P. E. Nachtigall and P. W. B. Moore (eds), Plenum Press, New York, pp 521-534.

Fraser, F. C., and Purves, P. E., 1960, Hearing in Cetaceans, Bull. Br. Mus. Nat. Hist. 7: 1-140.

Giro, L. P., and Dubrovskii, N. A., 1975, Possible role of the pericranial diverticula in the production of dolphin echolocation signals, Sov. Phys. Acoust., 20(5): 428-430.

Green, R. F., Ridgway, S. H., and Evans, W. E., 1980, Functional and descriptive anatomy of the bottlenosed dolphin nasolaryngeal system with special reference to the musculature associated with sound production, in: Animal Sonar Systems, R-G. Busnel and J. F. Fish (eds), Plenum Publ. Corp. New York, pp 199-238.

Gruhl, K., 1911, Beiträge zur Anatomie und Physiologie der Cetaceen nase, Jena Z. Naturw., (N.F.), 47: 367-414.

Gurevich, V. S., 1980, A reconstructing technique for the nasal air sacs system in toothed whales, in: Animal Sonar Systems, J. F. Fish (ed), Plenum Press, New York.

Heyning, J. H., 1989, Comparative facial anatomy of beaked whales (Ziphiidae) and a systematic revision among the families of extant Odontoceti, Contributions in Science, Nat. Hist. Mus. of Los Angeles County, 405: 1-64.

Hollien, H., Hollien, P., Caldwell, D K., and Caldwell, M C., 1976, Sound production by the Atlantic bottlenosed dolphin, Tursiops truncatus, Cetology 26: 1-8.

Hosokawa, H., and Kamiya, T., 1965, Sections of the dolphin's head, (Stenella coeruleoalba), Sci. Repts. Whales Res. Inst., 19: 105.

Kamminga, C., and Wiersma, H., 1981, Investigations on Cetacean sonar II. Acoustical similarities and differences in Odontocete sonar signals, Aquatic Mammals, 8: 41.

Kamminga, C., and Wiersma, H., 1982, Investigations on Cetacean sonar V. The true nature of the sonar sounds of Cephalorhyncus commersonii, Aquatic Mammals, 9: 95.

Lawrence, B., and Schevill, W. E., 1956, The functional anatomy of the delphinid nose, Mus. Comp. Zool. Bull., 114: 103.

Litchfield, C., Karol, R., Mullen, M E., Dilger, J P., and Luthi, B., 1979, Physical factors influencing refraction of the echolocative sound beam in delphinid Cetaceans, Marine Biology, 52: 285.

Mead, J. G., 1972, Anatomy of the external nasal passage and facial complex in the Delphinidae (Mammalia: Cetacea), Doctoral dissertation, University of Chicago.

Mead, J. G., 1975, Anatomy of the external nasal passage and facial complex in the Delphinidae (Mammalia: Cetacea), Smithsonian Contributions to Zoology, 207: 1.

Moris, F., 1969, Etude anatomique de la region cephalique du marsouin, Phocaena phocaena L. (Cetace Odontocete), Mammalia, 33: 666-726.

Möhl, B., and Andersen, S. H., 1973, Echolocation: high frequency component in the click of the harbour porpoise (Phocoena phocoena L.), J. Acoust. Soc. Am. 54: 1368.

Norris, K. S., 1964, Some problems in echolocation in Cetaceans. in: Marine Bio-Acoustics, W. N. Tavolga (ed), Pergamon Press, New York, pp 317-336.

Norris, K. S., Dormer, K. J., Pegg, J., and Liese, G. J., 1971, The mechanism of sound production and air recycling in porpoises: a preliminary report, in: "Proc. 8th Annual Conf. Biol. Sonar and Diving Mammals", Menlo Park, California., pp 113.

Norris, K. S., and Harvey, G. W., 1974, Sound transmission in the porpoise head, J. Acoust. Soc. Am., 56: 659-664.

Nämnden för skoglig flygbildsteknik, 1980, Flygbildsteknik och fjärranalys, Solna, Sweden.

Purves, P. E., 1966, Anatomical and experimental observations on the Cetecean sonar system, in: Proc. Sym. Bionic Models Animal Sonar System, R-G Busnel (ed), Frascati, Italy.

Purves, P. E., 1967, Anatomical and experimental observations on the Cetecean sonar system, in: "Animal Sonar Systems, Biology and Bionics", R-G Busnel (ed), Imprimerie Louis Jean, GAP (Hautes Alpes), France.

Purves, P. E., and Pilleri, G. E., 1983, "Echolocation in whales and dolphins", Academic Press, London.

Rauschning, W., 1979, Serial cryosectioning of human knee-joint specimen for a study of functional anatomy, Science Tools, The LKB Instrument Journal, Special Issue, 26: 47.

Ridgway, S. H., Carder, D. A., Green, R. F., Gaunt, A. S., Gaunt, S. L. L., and Evans, W. E., 1980, Electromyography and pressure events in the nasolaryngeal system of dolphins during sound production, in: "Animal Sonar Systems", R-G. Busnel and J. F. Fish (eds), Plenum Publ. Corp., New York.

Schenkkan, E. J., 1973, On the comparative anatomy and function of the nasal tract in Odontocetes (Mammalia, Cetacea), Bijdr. Dierk., 43: 127.

Schevill, W. E., Watkins, W. A., and Ray, C., 1969, Click structure in the porpoise, Phocoena phocoena, J. Mammal., 50: 721-728.

Sibson, F., 1848, On the blow-hole of the porpoise, Roy. Soc. Lond., Phil. Trans. pp 117-123.

Ullberg, S., 1977, The technique of whole body autoradiography cryosectioning of large specimen, Science Tools, The LKB Instrument Journal, Special Issue.

Varanasi, U., Markey, D., and Malins, D. C., 1982, Role of isovaleroyl lipids in channeling of sound in the porpoise melon, Chemistry and Physics of lipids, 31: 237.

Watkins, W. A., Schevill, W. E., and Best, P. B., 1977, Underwater sounds of Cephalorhynchus heavisidii (Mammalia: Cetacea), J. Mammal. 58(3):316-320.

Wiersma, H., 1982, Investigations on Cetacean Sonar IV. A comparison of wave shapes of Odontocete sonar signals, Aquatic Mammals 9(2): 57-66.

STRUCTURE AND THALAMOCORTICAL RELATIONS OF
THE CETACEAN SENSORY CORTEX: HISTOLOGICAL,
TRACER AND IMMUNOCYTOCHEMICAL STUDIES

Laurence J. Garey and Alexander V. Revishchin[1]

Department of Anatomy, Charing Cross and
Westminster Medical School, London W6 8RF

[1]Laboratory of Comparative Neurobiology,
Severtsov Institute of Evolutionary Morphology
and Animal Ecology, Academy of Sciences,
33 Leninsky Prospect, Moscow 117071, USSR

INTRODUCTION

   The cerebral cortex of cetaceans has interested neuroanatomists for many years because of its remarkable appearance and volume. The ratio of cortical area to brain volume is higher in dolphins than in humans (Elias and Schwartz, 1969; Hofman, 1985; Ridgway, 1986), while the total cortical volume and ratio of brain to body weight almost reach human levels (Ridgway, 1986). Early descriptions of cetacean cortex were of its gross morphology (for review see Flanigan, 1972), but later workers investigated its microscopical structure (Riese, 1925; Rose, 1926; Langworthy, 1932; Pilleri et al., 1968; Morgane and Jacobs, 1972; Entin, 1973; Kesarev et al., 1977; Morgane et al., 1980, 1985, 1986; Jacobs et al., 1984; Garey et al., 1985; Garey and Leuba, 1986; Ferrer and Perera, 1988) while some studied functional localization (Lende and Akdikmen, 1968; Lende and Welker, 1972; Sokolov et al., 1972; Ladygina et al., 1978). One of the striking features of cetacean cortex compared with, for example, that of the primate, is the relative lack of structural differentiation between functional areas (Kesarev et al., 1977).

   Our studies have aimed to describe the cellular architecture of certain neocortical areas in the bottlenose dolphin (<u>Tursiops</u> <u>truncatus</u>) and the Black Sea porpoise (<u>Phocoena</u> <u>phocoena</u>) in terms of the distribution of neurons as seen in Nissl preparations, their shape as determined from Golgi studies, the distribution of neurons containing certain putative neurotransmitters, and the thalamocortical relations of the principal sensory areas.

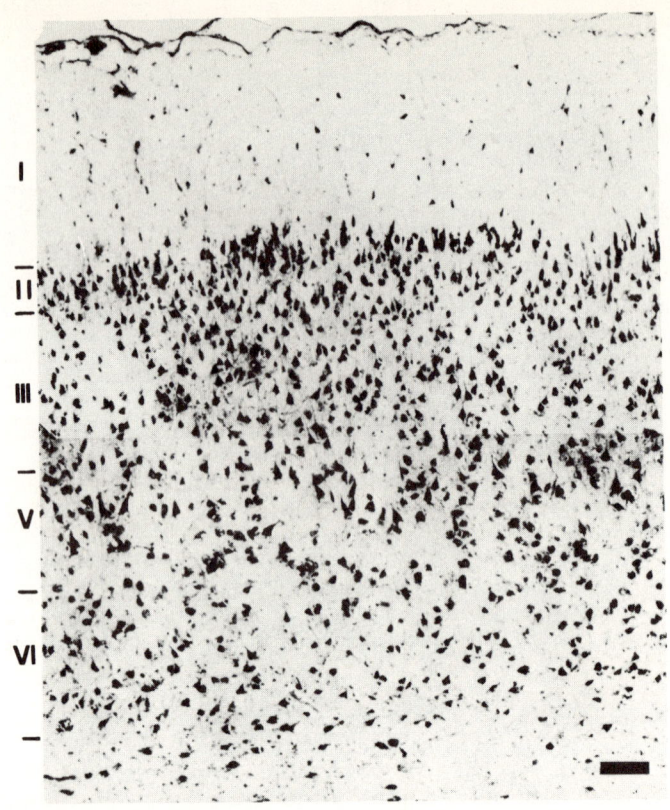

Fig. 1. Nissl section from the lateral gyrus of an adult dolphin to show layers I to VI. Scale bar = 100um. From Garey et al., 1985

NISSL AND GOLGI STUDIES

Materials and Methods

The brains of seven bottlenose dolphins that had died of natural causes were fixed in 10% formalin. Two of them were young (aged 18 days and 3 years), and the others were mature (between 12 and 33 years old). Blocks were removed and prepared for Nissl staining by embedding in paraffin, cutting sections at 20um in an approximately coronal plane, and staining them with cresyl violet. Other blocks were prepared by a modified Rapid Golgi method (Valverde, 1970) and cut at 100um.

Results

The visual cortex (Sokolov et al., 1972; Ladygina et al., 1978) of the lateral gyrus of the adult bottlenose dolphin is thin compared with that of large land mammals (about 1300um) (Fig. 1). Layers I, III and VI are wide, occupying about three quarters of the total cortical

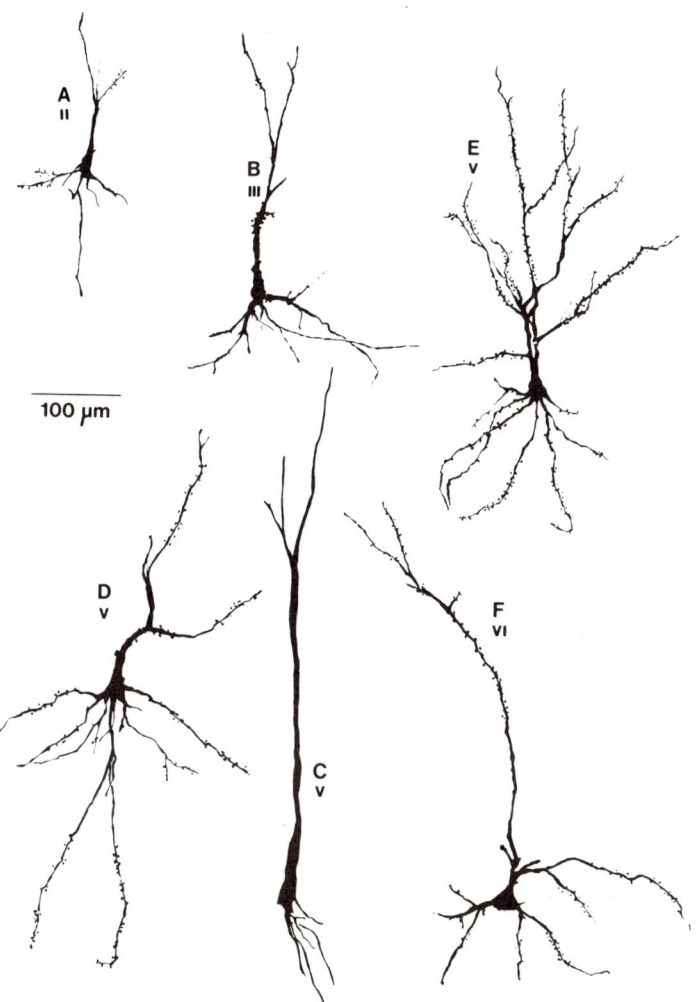

Fig. 2. Golgi-impregnated pyramidal neurons from the dolphin lateral gyrus. The Roman numerals indicate the layer in which the soma is situated. From Garey et al., 1985.

thickness. Layer I contains few neuronal somata, while III and VI have a variety of pyramidal and non-pyramidal cells. Layers II and V are narrow with pallisades of darkly staining pyramidal cells that are largest in V. No clear layer IV is present in the adult dolphin visual cortex. Many of the neurons are typical of pyramids in other mammals, and usually bear highly spiny dendrites (Fig. 2). Others are atypical in having bifurcated or oblique apical dendrites. Large and small, spiny and non-spiny stellate neurons are found mainly in layers III and VI (Fig. 3). In addition, various forms of spindle-shaped, bipolar and multipolar neurons are found throughout the cortex. In the first postnatal months the cortex is thinner, and a granular layer IV is visible (Garey et al., 1985).

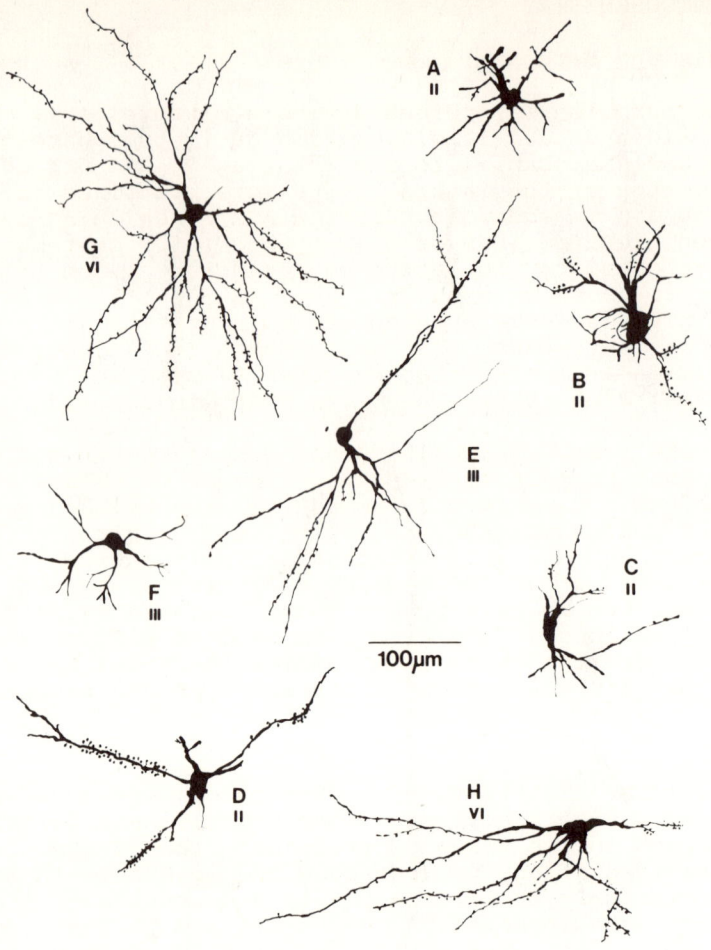

Fig. 3. Non-pyramidal neurons, as in Fig. 2.

Neuronal numerical density measurements in the lateral gyrus of the bottlenose dolphin allow two distinct parts to be identified in the adult, one anterior in which neuronal density is 23,000/mm^3, the other more posterior with almost double this density (Garey and Leuba, 1986). In the neonate, the neuronal density is about twice that of the adult. In the porpoise, although the brain is much smaller (some 350gm in the adult compared with over 1300gm in the bottlenose dolphin), the thickness of the visual cortex of the lateral gyrus is rather similar, as is the relative thickness of the different layers, and the neuronal density of the anterior portion of this area was found to be close to that of the dolphin (Garey et al., 1989).

IMMUNOCYTOCHEMISTRY

Materials and Methods

One porpoise was perfused with a mixture of 1% paraformaldehyde and 1% glutaraldehyde in phosphate-buffered saline. Small blocks, 1 to 2mm^3, through the depth of the visual cortex of the lateral gyrus were postfixed in osmium tetroxide and embedded in Durcopan ACM. "Semithin" sections were then prepared (0.5 or 0.75um thick) for postembedding immunocytochemistry to determine the distribution of neurons positive to gamma-aminobutyric acid (GABA). The resin was removed with sodium ethylate before incubation in goat serum, followed by anti-GABA, goat anti-rabbit IgG and peroxidase-antiperoxidase complex that was then reacted in diaminobenzidine (DAB) with 0.01% hydrogen peroxide.

Blocks from both auditory and visual cortex and thalamus of two other porpoises and one bottlenose dolphin were also prepared, this time for pre-embedding immunocytochemistry using antibodies to GABA, neuropeptide Y, somatostatin, substance P and vasoactive intestinal polypeptide (VIP). Sections were cut at 100um using a Vibratome, before incubating them in primary antibody, followed by biotinylated secondary antibody. They were then reacted with an avidin-biotin horseradish peroxidase complex (ABC) and the reaction visualized with a solution of DAB and hydrogen peroxide.

Results

A similar distribution of GABAergic neurons was found in visual and auditory cortex of both porpoise and dolphin (Fig. 4A). Neurons in the visual cortex that reacted positively to GABA were identified, photographed and measured. In layers I and II relatively small GABA-positive neurons were present, but larger ones appeared in layers III and V. Their somata were round or oval, and often were bipolar, with dendrites pointing towards the pial surface and the white matter. Others had more than two stem dendrites. Near the surface some of the bipolar neurons were oriented horizontally. Overall, these GABAergic neurons fitted into the classification of non-pyramidal, as estimated by comparison with material from the Golgi study mentioned above. The mean density of GABAergic neurons in this study was 2728/mm^3, with the highest values in layers II and III. Overall, some 20% of all neurons were GABA-positive.

Neurons reacting positively to neuropeptide Y were found in all layers (Fig. 4C,D). They were commonly small and bipolar. The dendrites were often filled for some distance from the soma, and a dense plexus of beaded axons was stained in the neuropil. Relatively small numbers of VIP-positive cells were found in layers II, III and V of visual and auditory cortices, mainly small and either round or bipolar in the superficial layers, with larger multipolar cells in layer V. More neurons reacted positively for somatostatin, and they appeared in all layers. Again, round, bipolar and multipolar forms were detected, but some layer V pyramids were immunopositive (Fig. 4B). Most substance P activity involved medium-sized multipolar cells in layers V and VI.

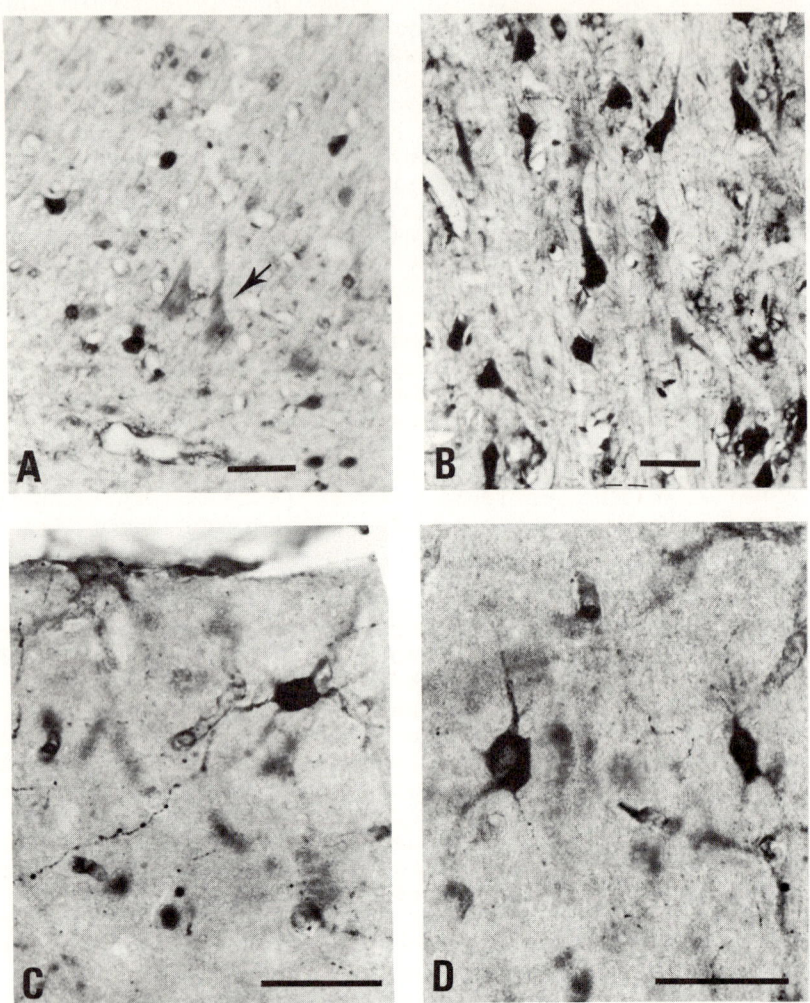

Fig. 4. A. GABAergic neurons in layer V of the lateral gyrus of the porpoise. Note the small size compared with the lightly stained pyramidal cells (arrow, not immunopositive). B. Pyramidal and non-pyramidal cells immunopositive to somatostatin in layer V of porpoise lateral gyrus. C, D. NPY-positive cells in layers I/II and V respectively of dolphin lateral gyrus. Scale bars = 50um.

In the lateral and medial geniculate nuclei a small percentage of neurons proved GABAergic. They were, in general, small or medium in size and round, or multipolar in form. Similar neurons also reacted positively for somatostatin, substance P and VIP.

THALAMOCORTICAL CONNEXIONS

Until recently almost no information existed about connectivity within the cetacean brain. There is axonal degeneration in the lateral geniculate nucleus (LGN) of the thalamus after removal of an eye (Jacobs et al., 1975), and in the cortex after lesions of the medial geniculate nucleus (MGN) (Krasnoshchekova and Figurina, 1980), but the only studies of thalamocortical relations with modern tracer techniques were those of Revishchin (1983) and Voronov et al. (1985). We have now investigated some of the thalamic connexions of the neocortex of the Black Sea porpoise (Revishchin and Garey, 1990). In view of physiological studies that had localized those parts of the cortex that gave visual, auditory and somatosensory evoked potentials (Sokolov et al., 1972; Ladygina et al., 1978), the we concentrated on the suspected connectivity of the lateral, suprasylvian and ectosylvian gyri and the perisylvian cortex.

Materials and Methods

We employed retrograde axonal transport of tracer substances to identify the neurons of origin of the thalamocortical projection. Eight adult porpoises were used. As the experiments were conducted at a Black Sea field station, gaseous anaesthesia was not possible. The animals were restrained in a hammock-like stretcher and tranquillized with intramuscular diazepam, so that they did not react to potentially painful stimuli. They were carefully monitored for signs of distress throughout all further manipulations. The skin and subcutaneous tissues were infiltrated with Novacain and then a minute craniotomy made, which caused no sign of distress, to allow the injection of tracer. After return to the water, all animals showed no signs of behavioural disturbance. Both cerebral hemispheres were used in most cases, and different tracers were employed in each hemisphere, including horseradish peroxidase (HRP) and various fluorescent dyes, to obviate confusion as to which filled neurons projected to which injection sites. The injection sites were spaced widely over the cerebral hemispheres, and the volume of traced used was from 2 to 4ul.

After a survival time of a few days the animal was deeply anaesthetized with Nembutal, that caused cardiorespiratory collapse, and perfused with 10% buffered formalin or a mixture of 1% paraformaldehyde and 1% glutaraldehyde in phosphate-buffered saline. Frozen coronal sections were cut at 30 or 60um over the whole cerebral hemisphere except the frontal and occipital poles. Series of sections were mounted and stained with cresyl violet for examination of injection sites and identification of thalamic nuclei. Outlines of thalamic nuclei were traced, using the nomenclature of Kruger (1959), on a digital graphics tablet from which their coordinates were stored in a computer. The positions of individual labelled neurons were also drawn or entered in the computer as points. This enabled a three-dimensional reconstruction of each thalamus to be made and rotated so that a graphic view of the distribution of labelled neurons resulting from each injection was obtained.

Results

Labelled neurons were visible in the thalamus in all experiments. Their morphology was basically similar in all nuclei studied. The soma size was variable, the largest, approximately 60 x 30um, being found in the LGN, MGN and ventrobasal nucleus, nuclei that also contained the largest unlabelled neurons in Nissl sections, but smaller neurons (20 x 12um) were also present in these nuclei. The somata were usually round or triangular and the commonest dendritic arbors were multipolar, with three to six stem dendrites radiating from the soma. Some dendritic fields were roughly circular, but others were elongated and bitufted. Especially in the LGN these elongated dendritic fields tended to be aligned along the axon bundles crossing the nucleus, that is perpendicular to the long axis of the nucleus.

Three injections were in the visually excitable cortex of the lateral gyrus. All three are near the middle of the visual area and are associated with neuronal labelling in the anteroposterior middle of the LGN. There is a suggestion that the most lateral part of the LGN projects relatively medially in the lateral gyrus, while medial parts project laterally. This is supported by the observation that the lateral spread of tracer from these injections caused islands of filled cells in the adjacent lateral parts of the pulvinar and that the lateral zone of the LGN was spared or only lightly marked after an anterior injection that was laterally placed. No injections were made in the posterior third of the lateral gyrus, except for one very posteriorly after which no LGN neurons were filled. The anterior injections resulted in labelling of the ventral half of the lateral pulvinar nucleus, probably due to lateral spread into the suprasylvian gyrus. The dorsal half was labelled after the very posterior injection, suggesting that dorsal lateral pulvinar projects to the posterior part of the border between the lateral and suprasylvian gyri and ventral lateral pulvinar to the anterior border zone.

Seven injections were in the auditory area of the suprasylvian gyrus. They confirm that ventral pulvinar, especially the inferior pulvinar, projects to the suprasylvian gyrus with little topographical organization. The most lateral margin of the lateral pulvinar, close to the LGN, projects medially in the suprasylvian gyrus, but the medial pulvinar seems only to project to the ectosylvian gyrus and perisylvian cortex. The most anterior suprasylvian injections caused filling at relatively anterior levels of the pulvinar, whereas more posterior injections filled cells in posterior pulvinar. Thus lateral displacement across the suprasylvian gyrus, from the lateral towards the ectosylvian gyrus, tends to give filling of pulvinar cells progressively more medially, while anteroposterior displacement along the suprasylvian gyrus involves cells in an anteroventral to posterodorsal direction. In addition, injections in the suprasylvian gyrus caused labelling in the MGN, and particularly, but not exclusively, its ventrolateral part, including the magnocellular portion and the ventral part of the parvocellular division.

Injections in the ectosylvian gyrus tend to cause filling of cells in the dorsal part of the MGN, restricted to the parvocellular division, with the extreme caudal pole projecting to posterior ectosylvian gyrus. Posterior parvocellular MGN also projects to the temporal operculum. Thus there appear to be three cortical regions that receive projections from the MGN: the suprasylvian and ectosylvian gyri, and the temporal operculum, the first from magnocellular MGN and ventral parvocellular MGN, the last two from dorsal and caudal parvocellular MGN, although there is overlap in dorsal magnocellular MGN. Thus the projection from the presumed auditory thalamus is to a relatively wider area of cortex than to visual or somatosensory cortex from their respective thalamic nuclei, perhaps reflecting the importance of audition to cetaceans.

Two injections involve extreme rostral cortex, one in the anterior part of the suprasylvian gyrus and another in the orbital operculum. They both confirm that the anterior parts of the thalamic domain studied here project to this rostral cortex. Ventrobasal and ventroposterior nuclei project to the anterior portion of the suprasylvian gyrus where somatosensory evoked potentials have been elicited (Lende and Welker, 1972; Sokolov et al., 1972; Ladygina et al., 1978). When the orbital cortex was injected, more dorsal thalamus, but still at the anteroposterior level of the ventrobasal and ventroposterior nuclei, is involved, namely the lateral complex and the posterior nucleus.

As no information about the precise levels of termination of thalamic axons in the cerebral cortex can be deduced from the experiments involving retrograde tracers injected in cortex, we have also studied the distribution of cytochrome oxidase activity in the cortex as an indirect indicator of possible thalamocortical axon localization. Sections of the cortex of one porpoise and one bottlenose dolphin were incubated in a solution of diaminobenzidine and cytochrome C to demonstrate cytochrome oxidase. In the lateral and suprasylvian gyri distinct peaks of activity occurred in layers I and III. In ectosylvian and perisylvian cortex the same peaks were present, but less strikingly. It is known that cytochrome oxidase-rich areas in primate brain are those that receive the thalamocortical afferents (Hendrickson, 1985) and this finding may therefore indicate that layers I and III are such recipient zones.

DISCUSSION

We conclude that the cetacean sensory cortex contains many of the features found in other mammals, including primates, as far as neuronal architecture and immunocytochemical distribution of putative neurotransmitters are concerned. However, it differs in its thickness, laminar organization and in the structure of many of its neurons. The somewhat "primitive" features of cetacean cortex have led

authors to conclude that it is more related to the lower
mammalian species, such as insectivores, or to the allocortex
of higher mammals (Kesarev et al., 1977; Morgane et al.,
1985; Ferrer and Perera, 1988). As there is no clear layer IV
in the adult, the problem arises as to where thalamocortical
afferents terminate. Based on our results concerning the
peaks of cytochrome oxidase activity (Garey and Revishchin,
1988) and GABA-positive neurons (Garey et al., 1989) in layer
III we suggest that this layer is a likely candidate for the
thalamorecipient zone, as is supported by evidence from
anterograde degeneration experiments (Krasnoshchekova and
Figurina, 1980).

The basic topography of the porpoise thalamocortical
radiation is that cell groups from lateral to medial in
thalamus project progressively from medial to lateral in
cortex, with the more anterior thalamic levels projecting
anteriorly in cortex. In general terms, a rather
similar organization can be discerned in various other
animals, including carnivores and primates. For instance, in
them the LGN is posterolateral in thalamus and projects to
the posteromedially situated visual cortex. The MGN is more
medial, and its cortical target is more lateral.
Somatosensory cortex is relatively anterior and receives
afferents from ventral thalamic nuclei, also fairly anterior.

ACKNOWLEDGEMENTS

We are grateful to Professor V.E. Sokolov and Dr. L.
Mukhametov for help and encouragement. Some of the early
studies were made possible thanks to Professor T.H. Bullock
and Dr. S.H. Ridgway. Drs. K. Brauer, J. Hámori, G. Leuba,
W.Y. Ong, J. Takács and E. Winkelmann collaborated at various
stages of the work described.

REFERENCES

Elias, H., and Schwartz, D., 1969, Surface areas of the
    cerebral cortex of mammals determined by stereological
    methods, Science. 166:111-113.
Entin, T. I., 1973, Histological study of the occipital
    cortex in the dolphin brain, Arkh. Anat. Gistol.
    Embriol. 65:92-100.
Ferrer, I., and Perera, M., 1988, Structure and nerve cell
    organisation in the cerebral cortex of the dolphin
    Stenella coeruleoalba a Golgi study, J. Comp. Neurol.
    178:161-173.
Flanigan, N. J., 1972, The Central Nervous System. Cetacea,
    in: "Mammals of the Sea", S.H. Ridgway ed., Charles C.
    Thomas, Springfield, pp. 215-246.
Garey, L. J., and Leuba, G., 1986, A quantitative study of
    neuronal and glial numerical density in the visual
    cortex of the bottlenose dolphin: evidence for a
    specialized subarea and changes with age, J. Comp.
    Neurol. 247:491-496.
Garey, L. J., and Revishchin, A. V., 1988, Laminar
    distribution of cytochrome oxidase activity in porpoise
    neocortex, Dokl. Akad. Nauk SSSR. 302:1486-1489.

Garey, L. J., Takács, J., Revishchin, A. V., and Hámori, J., 1989, Quantitative distribution of GABA-immunoreactive neurons in cetacean visual cortex is similar to that in land mammals, Brain. Res. 485:278-284.

Garey, L. J., Winkelmann, E., and Brauer, K., 1985, Golgi and Nissl studies of the visual cortex of the bottlenose dolphin, J. Comp. Neurol. 240:305-321.

Hendrickson, A. E., 1985, Dots, stripes and columns in monkey visual cortex, Trends in Neurosci. 8:406-410.

Hofman, M. A., 1985, Size and shape of the cerebral cortex in mammals. I. The cortical surface, Brain Behav. Evol. 27:28-40.

Jacobs, M. S., Galaburda, A. M., McFarland, W. L., and Morgane, P. J., 1984, The insular formations of the dolphin brain: Quantitative cytoarchitectonic studies of the insular component of the limbic lobe, J. Comp. Neurol. 225:396-432.

Jacobs, M. S., Morgane, P. J., and McFarland, W. L., 1975, Degeneration of visual pathways in the bottlenose dolphin, Brain Res. 88:346-352.

Kesarev, V. S., Malofeyeva, L. I., and Trykova, O. V., 1977, Ecological specificity of cetacean neocortex, J. Hirnforsch. 18:447-460.

Krasnoshchekova, E. I., and Figurina, I. I., 1980, Cortical projections of the dolphin cerebral geniculate body, Arkh. Anat. Gistol. Embriol. 78:19-24.

Kruger, L., 1959, The thalamus of the dolphin (Tursiops truncatus) and comparison with other mammals, J. Comp. Neurol. 111:133-194.

Ladygina, T. F., Mass, A. M., and Supin, A. Y., 1978, Multiple sensory projections in the dolphin cerebral cortex, Zh. Vyssh. Nerv. Deyat. 28:1047-1053.

Langworthy, O. R., 1932, A description of the central nervous system of the porpoise (Tursiops truncatus), J. Comp. Neurol. 54:437-499.

Lende, R. A., and Akdikmen, S., 1968, Motor field in cerebral cortex of the bottlenose dolphin, J. Neurosurg. 29: 495-499.

Lende, R. A., and Welker, W. I., 1972, An unusual sensory area in the cerebral neocortex of the bottlenose dolphin, Tursiops truncatus, Brain Res. 45:555-560.

Morgane, P. J., and Jacobs, M. S., 1972, Comparative anatomy of the cetacean nervous system, in: "Functional Anatomy of Marine Mammals", Vol. 1, R. J. Harrison, ed., Vol. 1., Academic Press, London, pp. 117-244.

Morgane, P. J., Jacobs, M. S., and Galaburda, A., 1985, Conservative features of neocortical evolution in dolphin brain, Brain Behav. Evol. 26:176-184.

Morgane, P. J., Jacobs, M. S., and Galaburda, A., 1986, Evolutionary aspects of cortical organization in the dolphin brain, in: "Research on Dolphins", M. M. Bryden and R. Harrison, eds., Clarendon Press, Oxford, pp. 71-98.

Morgane, P. J., Jacobs, M. S., and McFarland, W. L., 1980, The anatomy of the brain of the bottlenose dolphin (Tursiops truncatus). Surface configurations of the telencephalon of the bottlenose dolphin with comparative anatomical observations in four other cetacean species, Brain Res. Bull. 5 (Supp. 3):1-107.

Pilleri, G., Kruas, C., and Gihr, M., 1968, The structure of the cerebral cortex of the Ganges dolphin Susu (Platanista) gangetica Lebeck 1801, Z. Mik. Anat. Forsch. 79:373-388.

Revishchin, A. V., 1983, Study of thalamocortical projections in dolphins Phocaena phocaena by fluorescent-dye retrograde transport, Dokl. Akad. Nauk SSSR. 271:973-976.

Revishchin, A. V., and Garey, L. J., 1990, The thalamic projection to the sensory neocortex of the porpoise, Phocoena phocoena, J.Anat. (In Press).

Ridgway, S. H., 1986, Dolphin Brain Size, in: "Research on Dolphins", M. M. Bryden and R. Harrison, eds., Clarendon Press, Oxford, pp. 59-70.

Riese, W., 1925, Formprobleme des Gehirns. Zweite Mitteilung: Uber die Hirnrinde der Wale, J. Psychol. Neurol. 31:275-280.

Rose, M., 1926, Der Grundplan der Cortextektonik beim Delphin, J. Psychol. Neurol. 32:161-169.

Sokolov, V. E., Ladygina, T. F., and Supin, A. T., 1972, Localization of sensory zones in dolphin brain cortex, Dokl. Akad. Nauk SSSR. 202:490-493.

Voronov, V. A., Krasnoshchekova, E. I., Stosman, I. M., and Figurina. I. I., 1985, Morpho-functional organization and cortical projections of the medial geniculate body in the dolphin Phocoena phocoena. Zh. Evol. Biok. Fiziol. 21:55-60.

A POTENTIAL NEURAL SUBSTRATE FOR GEOMAGNETIC SENSIBILITY IN CETACEANS

Nicolaas M. Gerrits, and Ronald A. Kastelein*

Department of Anatomy, Erasmus University
P.O. box 1738, NL-3000 DR Rotterdam, The Netherlands

*Harderwijk Marine Mammal Park
P.O. box 9114, NL-3841 AB Harderwijk, The Netherlands

INTRODUCTION

Recently, data have been presented which might demonstrate a behavioral response of cetaceans towards the earth's magnetic field. Different researchers have noted a correlation between live strandings of dolphins and characteristics of the magnetic field, e.g. direction of the field lines and the fieldstrength (Kirschvink et al., 1986; Klinowska, 1988). In the search for a magneto-sensitive organ, considerable effort has been directed towards the distribution of biogenic magnetite particles (Kirschvink et al., 1985). Magnetite particles have been discovered in many taxa including birds and mammals. Relatively high levels of magnetic material have been found in some parts of the brain (cerebellum, midbrain, corpus callosum) and the dura mater of cetaceans and the rhesus monkey (Bauer et al., 1985; Kirschvink, 1981). However, a sensory organ in which such crystals might be expected in a concentrated form has not been demonstrated anatomically. Kirschvink and Gould (1981) proposed a theoretical model in which the rotation of magnetite crystals changes the membrane resistance of a receptor cell to modulate the neuronal discharge frequency, but such a mechanism has not been demonstrated experimentally yet. An alternative mechanism for the detection of magnetic force has been proposed by Jungerman and Rosenblum (1980) who speculated that a magnetically induced electromotive force (EMF) in the labyrinthine semicircular canals could be detected by the sensory haircells in the ampullae. Both theories mentioned above have deficiencies; the distribution of magnetite particles in functionally diverse brain structures lacks the neuronal specificity which is found in all receptor systems, while the EMF would only be induced if the cupulae close the semicircular canals, e.g. a flow of endolymph, and thus rotational movement, is not allowed.

The Magnetic Induction Sensitive Neuronal System

As a possible explanation for the alleged capabilities of Cetaceans to detect magnetic force, we would like to introduce the concept of the Magnetic Induction Sensitive Neural System (MISNS). The concept departs from two important assumptions. The first of which is that the earths magnetic force induces changes in a specific configuration of "primary" receptor neurons inside the central nervous system; changes that are subsequently detected by a set of relay neurons. The second assumption is that the primary receptor, the relay neurons, and the concomitant fiber connections are organized according to the general principles of most sensory systems investigated sofar. The first assumption requires that:
- the primary receptor consists of a population of neurons with their axons organized in a (large) loop, in which an EMF can be induced, when this loop moves through the earths magnetic field.
- it consists of neurons with a spontanuous discharge activity, which can be modulated by the EMF. It seems likely that the magnetic field is strong enough to modulate neural activity rather than that it will generate action potentials in basically inactive neurons.

Secondary requirements are those which apply to the principles according to which other sensory systems are organized, especially in terms of connections to various parts of the brain where different types of sensory information are integrated. The MISNS should therefore meet the following requirements:

- a bilateral presence of the primary receptor,
- sensory relay neurons to process the primary input,
- commissural connections at the level of the (primary) sensory relay,
- connections to the cerebellum, the thalamus, and the cerebral cortex

In the search for the primary receptor, the sensory relay, and the subsequent connections, it seems reasonable to look for structures that are strongly or even extremely well-developed in cetaceans. Somehow, a similar development may be present in Pinnipedes as well, since these are faced with comparable orientation difficulties in the same opaque medium. The organization of the MISNS and its subsequent connections is tentatively illustrated in a diagram below.

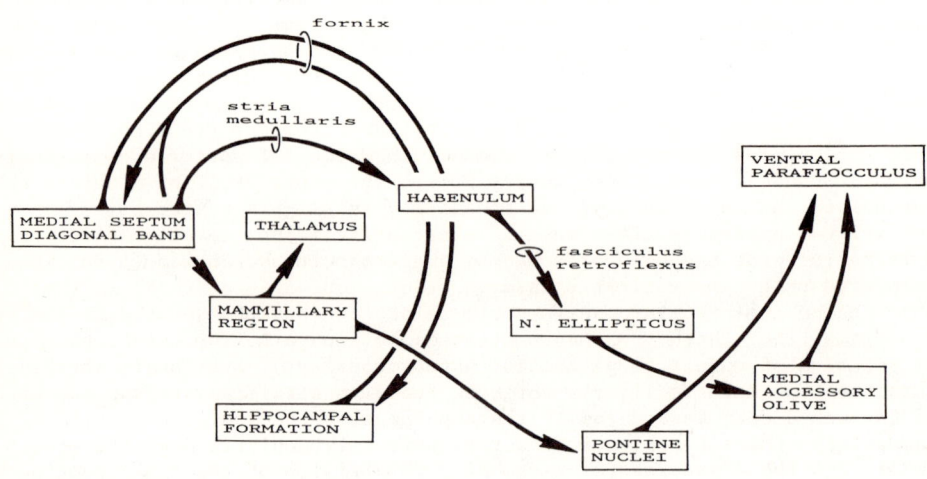

The Primary Receptor and the Sensory Relay

A potential place for the primary receptor is in the basal forebrain. There are a number of fiber bundles present with a circular configuration, that seem suited to be influenced by flux changes in the magnetic field. In cetaceans and also in pinnipedes, especially the fornix, and to a lesser degree the stria medullaris, are well-developed (Breathnach, 1960; Verhaart, 1970). Each of these bundles could contain the axons of spontaneously active neurons. It has been demonstrated that neurons in the medial septal nucleus and in the diagonal band send their axons through the fornix (Amaral and Cowan, 1980; Wyss et al., 1979), and, in addition, have pacemaker properties that are responsible for the theta rhythm of the hippocampus (Apostol and Creutzfeldt, 1974; Stumpf, 1965; Wilson et al., 1976). In this situation the sensory relay neurons would be located in the hippocampus, but an alternative setup could be that the primary neurons are located in the hippocampus and the relay neurons are to be found in the septo-hypothalamic nuclei which receive hippocampal fibers.

The hippocampal formation (hippocampus, subiculum, presubiculum) is a host to an overwhelming variety of afferent and efferent connections (Nauta, 1956). Moreover, its activity has been correlated with different behavioral responses. If magnetic information is processed in relay neurons in the hippocampal formation, the exact configuration of the processing network will be an enigma for the time being, as it is the case for most of the functions attributed to this part of the brain. One of the arguments to prefer a localization of the relay neurons somewhere in the hippocampal formation, is the fact that the left and right hippocampus are interconnected intimately by commissural fibers (as is the case for all other sensory relay nuclei), which are the main constituents of the dorsal and ventral hippocampal commissures (Amaral et al., 1984; Demeter et al., 1985; Swanson et al., 1978).

The Cerebellar Connection

All exteroceptive and proprioceptive information monitored through the spinal cord, as well as the sensory information that enters the nervous system through the cranial nerves is relayed through the cerebellum, which acts as an integrator for these different sensory modalities. For example, different aspects of vestibular, visual and neck muscle signals converge onto different parts of the cerebellum, which, by means of efferent pathways, influences ocular and postural reflexes. Afferent information reaches the cerebellum through two different systems: by climbing fibers, relayed in the inferior olivary nucleus (IO), and by mossy fibers, relayed in a number of nuclei in the brainstem. Electrophysiological studies have demonstrated a connection from the hippocampal formation to the cerebellum (Fanardjan and Donhoffer, 1963; Fox et al, 1967; Newman and Reza, 1979; Sager et al, 1970; Saint-Cyr and Woodward, 1980). The recorded potentials have a bimodal latency: one group 8-10 ms, the other 20-30 ms after hippocampal stimulation. The short latency response was attributed to mossy fiber activation, the longer latencies represent climbing fiber activation. Consequently, the climbing fiber pathway has more synaptic relays than the mossy fiber pathway.

Mossy Fiber Pathway. The shortest brainstem path from the hippocampal formation to the cerebellum has two synaptic relays. Subicular fibers terminate in the mammillary bodies and the supramammillary region (Nauta, 1956; Swanson and Cowan, 1977; Irle and Markowitsch, 1982). From the mammillary bodies a projection to the nucleus reticularis tegmenti pontis (NRTP) and the dorso-rostral paramedian subdivision of the basal pontine nuclei (pPN) was demonstrated (Cruce, 1977; Aas, 1989; Aas and Brodal, 1989). This part of the pPN has an exclusive projection to the uvula and the ventral paraflocculus (Gerrits et al., 1984; Aas and Brodal, 1989). The NRTP sends a heavy mossy fiber projection to the same lobules, but, in addition, innervates almost all other lobules of the cerebellum. The ventral, and also the dorsal paraflocculus are extremely large in cetaceans (Larsell, 1970).

Climbing Fiber Pathway. The two most strikingly developed structures in the cetacean brainstem, as compared to terrestrial mammals, are the elliptic nucleus (EN) and the medial accessory olive (MAO) (Fig. 1A). The EN, which measures up to 8 mm in diameter, is present in the central grey matter at the transition of the rostral mesencephalon into the diencephalon (Ogawa, 1935). The MAO is one of the subdivisions of the IO (Kooy, 1920). In cetaceans, the two nuclei are connected by a very prominent medial tegmental tract (de Graaf, 1967). In mammals, the MAO receives input from the nucleus of Darkschewitsch (DW) through this tract, as has been demonstrated experimentally (Walberg, 1974). Unfortunately, it is not known whether the EN and DW are homologous nuclei, but cetaceans certainly lack a distinct DW (de Graaf, 1967).

The MAO is the origin of the climbing fibers of the C2 module, one of the parasagittal compartments of the cerebellum (Voogd, 1964; Groenewegen et al., 1979). This module includes one of the central cerebellar nuclei (Voogd, 1964; Groenewegen et al., 1979), the posterior interposed nucleus, who's size in cetaceans matches that of the MAO (Korneliussen and Jansen, 1965). In the ventral and dorsal paraflocculus of mammals, only two modules are present: the C2 module with its MAO origin, and the D module with an origin in the principal subdivision of the IO. Since the latter subdivision of the IO is poorly developed in cetaceans (Verhaart, 1970) as compared to other, highly cephalized mammals such as primates (Fig. 1B), it is to be expected that the D module will be small, and the ventral paraflocculus will be dominated by the C2 module.

From the latency of climbing fiber responses in the cerebellum it may be inferred that there are at least three to four synaptic relays in the pathway from the hippocampal formation. Therefore, one or two relays must be present between the hippocampal formation and the EN. Descending input may reach the EN from either the mammillary region directly by way of the medial forebrain bundle, or from the nucleus of the diagonal band and the substantia innominata region (which both receive a strong fornix input) via the stria medullaris and the fasciculus retroflexus, with a possible synaptic relay in the lateral habenular nucleus. The structures mentioned as the second possibility are well-developed in cetaceans (de Graaf, 1967; Verhaart, 1970). Detailed knowledge of the hippocampo-cerebellar climbing fiber pathway is not available, since its components have not been traced with experimental anatomical tracer techniques.

Thalamic and Cortical Connections

Although direct connections from the hippocampal formation to the anterior nuclear complex have been described (Siegel and Tassoni, 1971; Swanson and Cowan, 1977), the major efferent pathway is the projection from the subiculum to the mammillary bodies (Swanson and Cowan, 1977; Meibach and Siegel, 1977), and subsequently to the anterior nucleus of the thalamus. The mammillary bodies in Cetaceans are surprisingly small compared to the thickness of the fornix (Breathnach, 1960; Verhaart, 1970). The Cetacean cerebral cortex seems to be organized quite different compared to the general mammalian plan (Morgane and Jacobs, 1972), and an analysis of the thalamo-cortical relations is lacking.

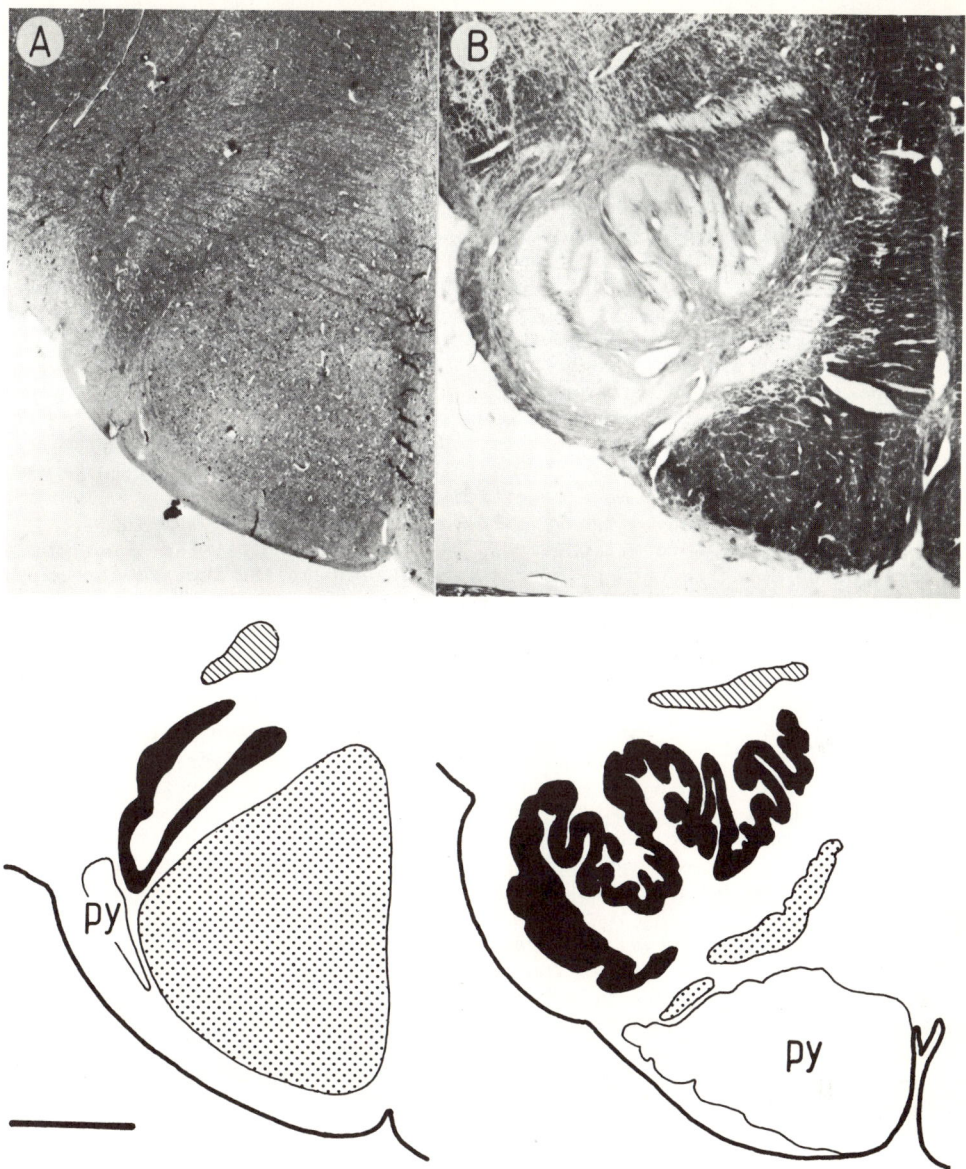

Fig. 1 Photographs and drawings of transverse sections through the left inferior olivary nucleus of Balaenopterus acutorostrata (A, Haggqvist stain) and Homo sapiens (B, Kluver-Barrera stain). Note the extreme difference in development of the principal olive (black) and the medial accessory olive (stippled). The midline is indicated with a broken line. Dorsal accessory olive (hatching); py: pyramidal tract. Bar represents 2 mm

DISCUSSION

Before going into a more detailed discussion of different aspects of the present hypothesis two, more or less related questions arise. The first is: "Does the cetacean nervous system posses unique features that sets them apart from the other mammals ?" To our present state of knowledge, all fiber tracts and nuclei in the cetacean brainstem and basal forebrain described sofar also are present in other mammals (Breathnach, 1960; de Graaf, 1967; Jelgersma, 1934; Morgane and Jacobs, 1972; Ogawa, 1935). The only exception seems to be the elliptic nucleus, although in Phoca, Zalophus, and Elephas, a nucleus of considerable size is present in a similar position (Verhaart, 1970), which could eventually turn-out to be the homologue of the nucleus of Darkschewitsch. Therefore, the answer to this question is a tentative "no", leading to the second question: "Could magneto-sensitivity represent a basic property of a specific part of the mammalian nervous system, which in cetaceans has undergone a strong (functional) development ?" Direct animal responses to magnetic force have not been reported, but some properties of the septal nuclei, the hippocampal formation, and the mamillary region, collected in different species, are of special interest concerning this question and with respect to the hypothesis presented in this paper. Electrophysiological studies of the hippocampus have demonstrated a relation with motoractivity in general (Vanderwolf, 1969), and more specifically during the transition of one movement into the other (Ranck, 1973). Lesions of the medial septal nucleus and the hippocampal formation impair behavior that requires spatial information for orientation (Mitchell et al, 1982); the impairment being attributed to memory deficits. In the rhesus monkey, the hippocampus and the mammillary bodies clearly seem involved in spatial memory and orientation (Rosenstock et al., 1977; Rupniak and Gaffan, 1987). Of course, a plausible explanation could be that the hippocampus contains a cognitive map (O'Keefe and Nadel, 1978) based on a variety of afferent inputs. Unfortunately, connections from areas of the brain, known to process visual, vestibular, auditory and somatosensory information seemingly vital for the construction of such a map or reference frame, are not detected easily among the hippocampal afferents. If somewhere in the septo-hippocampal system magnetic force can be detected directly, as suggested in this paper, this could explain at least part of the experimental data mentioned before. Moreover, it also would be a tentative "yes" to the second question.

Detection of magnetic force as such would be meaningless for an organism unless such information could be compared and integrated with other sensory modalities. The speed, and orietation towards gravity of the animals (i.c. the induction loop) determines the value of the magnetic information. Therefore, the present hypothesis stresses the importance of climbing and mossy fiber pathways to those parts of the cerebellum that also receive vestibular information. In the assumption that the cerebellar connections follow the general mammalian plan, the ventral paraflocculus of cetaceans receives input almost exclusively from the dorso-rostral paramedian subdivision of the basal pontine nuclei (Gerrits et al., 1984) and from the vestibular nuclei (Gerrits and Voogd, 1989).

Thus, the present hypothesis could provide an attractive explanation for the strong development of the fornix, the elliptic nucleus, the medial accessory subdivision of the inferior olivary nucleus, and the ventral paraflocculus of the cerebellum. Since there are no principal differences between the brainstem of cetaceans and other mammals, it also offers an alternative view on the functional relation between the hippocampal formation and spatial orientation.

ACKNOWLEDGEMENT
    The authors wish to thank Dr. E. Marani of the University of Leiden, for his permission to use the Comparative Anatomy Brain-slice Collection of the Department of Physiology.

REFERENCES

Aas, J.-E., 1989, Subcortical projections to the pontine nuclei in the cat, J. Comp. Neurol., 282: 331-354

Aas, J.-E., and Brodal, P., 1989, Demonstration of a mamillo-ponto-cerebellar pathway. A multi-tracer study in the cat, Eur. J. Neurosci. 1: 61-74

Amaral, D. G., Insausti, R., and Cowan, W. M., 1984, The commissural connections of the monkey hippocampal formation. J. Comp. Neurol., 224: 307-336

Apostol, G., and Creutzfeldt, O. D., 1974, Cross-correlation between the activity of septal units and hippocampal EEG during arousal. Brain Res., 67: 65-75

Bauer, G. B., Fuller, M., Perry, A., Dunn, J. R., and Zoeger, J., 1985, Magnetoreception and biomineralization of magnetite in Cetaceans, in: "Magnetite Biomineralization and Magnetoreception in Animals: A new Biomagnetism," J. L. Kirschvink, D. S. Jones, and B. J. McFadden, eds., Plenum Press, New York, pp. 489-507

Breathnach, A. S., 1960, The cetacean central nervous system, Biol. Rev. Cambridge Phil. Soc., 35: 87-230

Cruce, J. A. F., 1977, An autoradiographic study of the descending connections of the mammillary nuclei in the rat, J. Comp. Neurol., 176: 631-644

Demeter, S., Rosene, D. L., and Van Hoesen, G. W., 1985, Interhemispheric pathways of the hippocampal formation, presubiculum, and entorhinal and posterior parahippocampal cortices in the rhesus monkey: The structure and organization of the hippocampal commissures, J. Comp. Neurol., 223: 30-47

Fanardjian, V. V., and Donhoffer, H., 1970, An electrophysiological study of cerebello-hippocampal relationships in the unrestrained cat, Acta Physiol. Acad. Sci. Hung., 24: 321-333

Gerrits, N. M., Epema, A. H., and Voogd, J., 1984, The mossy fiber projection of the nucleus reticularis tegmenti pontis to the flocculus and adjacent ventral paraflocculus in the cat Neuroscience 11: 627-644

Gerrits, N. M., and Voogd, J., 1989, The topographical organization of climbing and mossy fibers afferents in the flocculus and the ventral paraflocculus in rabbit, cat and monkey, Exp. Brain Res. Series, 17: 26-29

Graaf, A. S. de, 1967, Anatomical aspects of the Cetacean brain stem, Van Gorcum, Assen

Groenewegen, H. J., Voogd, J., and Freedman, S. L., 1979, The parasagittal zonation within the olivocerebellar projection. II. Climbing fiber distribution in the intermediate and hemispheric parts of the cat cerebellum, J. Comp. Neurol., 183: 551-602

Irle, E., and Markowitsch, H. J., 1982, Connections of the hippocampal formation, mamillary bodies, anterior thalamus and cingulate cortex. A retrograde study using horseradish peroxidase in the cat, Exp. Brain Res., 47: 79-94

Jelgersma, G., 1934, Das Gehirn der Wassersaugetiere, Verlag J.A. Barth, Leipzig

Jungerman, R. L., and Rosenblum, B., 1980, Magnetic induction for sensing of magnetic fields by animals - an analysis, J. Theor. Biol., 87: 25-32

Kirschvink, J. L., 1981, Ferromagnetic crystals (magnetite?) in human tissue, J. Exp. Biol., 92: 333-335

Kirschvink, J. L., Dizon, A. E., and Westphal, J.A. 1986, Evidence from strandings for geomagnetic sensitivity in Cetaceans, J. Exp. Biol., 120: 1-24

Kirschvink, J. L., and Gould, J. L., 1981, Biogenic magnetite as the basis of magnetic field sensitivity in animals, Biosystems, 13: 181-201

Klinowska, M., 1988, Cetacean "navigation" and geomagnetic fields, J. Navigation, 41: 52-71

Kooy, F. H., 1920, The inferior olive in Cetacea, Folia Neurobiol. (Leipzig), 11: 647-664

Korneliussen, H. K., and Jansen, J., 1965, On the early development and homology of the central cerebellar nuclei in Cetacea. J. Hirnforsch., 8: 47-56

Larsell, O., 1970, The comparative anatomy and histology of the cerebellum from monotremes through apes, Univ. of Minnesota Press, Minneapolis

Meibach, R. C., and Siegel, A., 1977, Efferent connections of the hippocampal formation in the rat. Brain Res., 124: 197-224

Mitchell, S. J., Rawlins, J. N. P., Steward, O., and Olton, D. S., 1982, Medial septal area lesions disrupt theta rhythm and cholinergic staining in medial entorhinal cortex and produce impaired radial arm maze behavior in rats J. Neurosci., 2: 292-302

Morgane, P. J., and Jacobs, M. S., 1972, Comparative anatomy of the cetacean nervous system, in: "Functional Anatomy of Marine Mammals," Vol. 1, R. J. Harrison, ed., Academic Press, New York, pp. 117-224

Nauta, W. J. H., 1956, An experimental study of the fornix system in the rat, J. Comp. Neurol., 104: 247-271

Newman, P. P., and Reza, H., 1979, Functional relationships between the hippocampus and the cerebellum: an electrophysiological study of the cat, J. Physiol., 287: 405-426

Ogawa, T., 1935, Ueber den Nucleus ellipticus und den Nucleus ruber beim Delphin, Arb. Anat. Inst. Sendai, 17: 55-61

O'Keefe, J., and Nadel, L., 1978, The hippocampus as a cognitive map, Clarendon Press, Oxford

Ranck, J. B. Jr., 1973, Studies on single neurons in dorsal hippocampal formation and septum in unrestrained rats, Exp. Neurol., 41: 462-531

Rosenstock, J., Field, T. D., and Greene, E., 1977, The role of mammillary bodies in spatial memory, Exp. Neurol., 55: 340-352

Rupniak, N. M. J., and Gaffan, D., 1987, Monkey hippocampus and learning about spatially directed movements, J. Neurosci., 7: 2331-2337

Sager, O., Florea-Ciocoiu, V., and Rogozea, R., 1970, Study of auditory projections to cerebellum and some cerebello-hippocampo-neocortical circuits, Int. J. Neurol., 7: 218-231

Saint-Cyr, J. A., and Woodward, D. J., 1980, Activation of mossy and climbing fiber pathways to the cerebellar cortex by stimulation of the fornix in the rat, Exp. Brain Res., 40: 1-12

Siegel, A., and Tassoni, J. P., 1971, Differential efferent projections from the ventral and dorsal hippocampus of the cat, Brain Behav. Evol., 4: 18-200

Stumpf, C., 1965, Drug action on electrical activity of the hippocampus, Int. Rev. Neurobiol., 8: 77-138

Swanson, L. W., and Cowan, W. M., 1977, An autoradiographic study of the organization of the efferent connections of the hippocampal formation in the rat, J. Comp. Neurol., 172: 49-84

Swanson, L. W., Wyss, J. M., and Cowan, W. M., 1978, An autoradiographic study of the organization of intrahippocampal association pathways in the rat, J. Comp. Neurol., 181: 681-716

Verhaart, W. J. C., 1970, Comparative anatomical aspects of the mammalian brain stem and spinal cord, Van Gorcum, Assen

Vanderwolf, C. H., 1969, Hippocampal electrical activity and voluntary movement in the rat, Electroenceph. Clin. Neurophysiol., 26: 407-418

Voogd, J., 1964, The cerebellum of the cat. Structure and fibre connections, Van Gorcum, Assen

Walberg, F., 1974, Descending connections from the mesencephalon to the inferior olive: an experimental study in the cat, Exp. Brain Res. 20: 145-156

Wilson, C. L., Motter, B. C., and Lindsley, D. B., 1976, Influences of hypothalamic stimulation upon septal and hippocampal electrical activity in the cat, Brain Res., 197: 55-68

Wyss, J. M., Swanson, L. W., and Cowan, W. M., 1979, A study of subcortical afferents to the hippocampal formation in the rat, Neuroscience, 4: 463-476

IMMUNOCYTOCHEMISTRY OF NEUROTRANSMITTERS IN VISUAL NEOCORTEX

OF SEVERAL TOOTHED WHALES: LIGHT AND ELECTRON MICROSCOPIC STUDY

Ilya I. Glezer, Peter J. Morgane[1] and Csaba Leranth[2]

CUNY Medical School/CCNY of Biomedical Education, NY, NY 10031, USA; [1]Worcester Foundation for Expt. Biology Shrewsbury, MA, 01545, USA; [2]Yale University School of Medicine, New Haven, CT, 06510, USA

INTRODUCTION

A peculiar combination of evolutionary conservative and advanced morphological features in cetacean neocortex has been investigated by us using traditional (Nissl, Golgi, electron microscopy) and computerized image analysis techniques (Morgane et al., 1985, 1986a,b, 1988, 1990; Glezer et al., 1988; Glezer and Morgane, 1990). Although, there are significant logistical and technical problems in obtaining adequately fixed cetacean brains, we have succeeded in acquiring well-preserved brains of several toothed whales (Stenella coeruleoalba, Phocoena phocoena, Globicephala melaena and Tursiops truncatus). This has permitted us to examine the cetacean neocortex in considerable detail, including its intrinsic microcircuitry.

Qualitative and computerized quantitative analysis of synaptology in the visual neocortices of the dolphin (Stenella coeruleoalba) and pilot whale (Globicephala melaena) revealed that, in terms of synaptoarchitectonic and ultrastructure of synapses, there exist some peculiar features differentiating cetacean neocortex from the neocortices of terrestrial mammals (Glezer and Morgane, 1990; Morgane and Glezer, 1990). Thus, we showed that some of the important qualitative and quantitative synaptic features in the visual neocortex of the dolphin are present in special combinations of prototypical and advanced synaptic features not seen in advanced terrestrial mammals. Among the prototypical synaptic features are an abundance of synapses en passage, as well as a significant prevalence of synapses, in terms of their percentages and total areas of synaptic boutons, in superficial cortical layers I and II over that seen in layers V and VI and, particularly, in incipient layer IV (Glezer and Morgane, 1990). The upper two layers of the dolphin visual cortex contain approximately 70% of all synapses counted through the cortex from layers I through VI. On the other hand, the percentage of synapses in incipient layer IV is diminished dramatically and comprises only 1 % of the total number of synapses in a radial cortical slab (Glezer and Morgane, 1990). This marked difference of cetacean neocortex from those of advanced mammalian species, e.g., Primates, in quantitative distribution of synapses, along with other findings, has led to our view that the upper two layers of cetacean neocortex possibly comprise the main input layers from subcortical areas as well as from other cortical areas. Additionally, another interesting feature is that most synapses in cetacean visual cortex show a great variability of synaptic vesicles in terms of their shape and size within the same terminal type boutons or <u>en passage</u> type boutons in axodendritic and, to a lesser extent, in axosomatic synapses.

In the present study we focused on the distribution of some important neurotransmitters within the visual neocortex as revealed by immunocytochemical methods. We concentrated our investigations on the neuropeptide family of putative neurotransmitters with some studies on one catecholamine neurotransmitter (tyrosine hydroxylase). The recent explosion of information on neuropeptides has revealed not only a great chemical heterogeneity of neuron populations in the cerebral cortex of various species, but also an early evolutionary appearance of neuropeptides and their consistent presence in evolution starting from the most non-derived vertebrates (Emson and Marley, 1983; Acher, 1985). In view of our light and electron microscopic findings showing a strongly conservative character of organization of the cetacean neocortex (Morgane et al., 1985, 1986a, b, 1988, 1990; Glezer et al. 1988; Glezer and Morgane, 1990; Morgane and Glezer, this volume), it is of special interest to analyze with immunocytochemical methods the laminar distribution and typology of neurons containing certain neuropeptides, in particular neuropeptide Y and cholecystokinin in cetacean visual neocortex. We have also concentrated on correlating our light-microscopic cytoarchitectonic, Golgi and immunocytochemical findings with electron microscopic data on localization of products of immunostaining in the neuronal perikarya, axons, dendrites and, especially, synapses. This type of data is of importance for understanding the chemical microcircuitry of cetacean neocortex. Because of our interests in comparative neuromorphology of the neocortex we also have used immunostaining procedures on neocortices of other conservatively organized terrestrial eutherian mammals, such as the bat and hedgehog, in order to provide comparative data on the immunocytochemistry of the neocortex in a series of archetypal mammals.

MATERIALS AND METHODS

In this study the visual neocortices of the following species have been investigated: (1) Cetacea: <u>Globicephala melaena</u>, <u>Phocoena phocoena</u>, <u>Tursiops truncatus</u>, <u>Stenella coeruleoalba</u>; (2) Chiroptera: <u>Eptesicus fuscus</u>; (3) Insectivora: <u>Erinaceus europaeus</u>.

<u>Fixation</u>

In the case of <u>Globicephala melaena</u>, a stranded, live animal was euthanized by the veterinary consultant at the New York Aquarium using intravenous injection of agent T-61 (American Veterinary Society approved mixture of non-barbiturate anesthetics) and the brain was removed and immersed in situ into fixing solution of 2.5% glutaraldehyde and 4% paraformaldehyde in 0.1M phosphate buffer five minutes after death. In the case of <u>Stenella coeruleoalba</u>, the stranded, live animal was euthanized by the veterinary consultant of the New York Aquarium in the same manner as in the case of <u>Globicephala melaena</u>. However, in the case of <u>Stenella coeruleoalba</u> the brain was flushed with isotonic saline and then gravity perfused <u>in situ</u> via the descending aorta with a 2.5% glutaraldehyde and 4% paraformaldehyde mixture. The brain then was removed from the skull and immersed in a mixture of 2.5% of glutaraldehyde and 4% of paraformaldehyde in cacodylate buffer. The cortical samples from <u>Tursiops truncatus</u> and <u>Phocoena phocoena</u> were obtained from Dr. A. Revischchin of the Severtzov Institute for Evolutionary Morphology and Ecology, USSR Academy of Sciences.

The animals were gravity perfused after they were anesthetized with Nembutal. The vascular system of both animals was flushed with isotonic saline and then perfused with 2.5% glutaraldehyde and 4% paraformaldehyde solution in 0.1 M phosphate buffer. In case of the hedgehog (<u>Erinaceus europaeus</u>) two animals were anesthetized with ether and Nembutal, perfused intracardially with isotonic 0.1 M phosphate buffer and then with 2.5% glutaraldehyde and 4% paraformaldehyde in 0.1 phosphate buffer. The cortical samples from the bat (<u>Eptesicus fuscus</u>) was received from Dr. J. Zook of the Ohio University. Four animals were anesthetized with ether and Nembutal and then perfused intracardially with isotonic 0.1 M phosphate buffer and then with 2.5% glutaraldehyde and 4% paraformaldehyde in 0.1 M phosphate buffer.

Immunocytochemical procedures

Both pre-embedding and post-embedding immunostaining protocols were used in this study. Most of the data were obtained by the pre-embedding procedure using a modification of the method of Leranth and Feher (1983) and Leranth and Frotcher(1986). Forty μm sections of visual neocortex from all specimens were obtained using a vibratome. In cetacean material, sections were made from the mid-posterior part of the lateral gyrus in the region of the entolateral sulcus. This area was found to have the shortest latency visual evoked potentials (Sokolov et al., 1972; Supin et al., 1978) and has been defined by us as heterolaminar cortex (Morgane et al., 1988). In the bat and hedgehog, we obtained frontal vibratome sections from the occipital pole of the brain, i.e., from the region which is identified morphologically and physiologically as the primary visual cortical area. After washing of sections (6 times each for 10 min) in cold 0.1 M phosphate buffer they were transferred in 10% sucrose and frozen in liquid nitrogen for 5-10 sec. After slow thawing, the sections were washed with 0.1 M phosphate buffer (3 times each for 10 min) and transferred to 0.1% borohydrate to suppress non-specific endogenous H_2O_2. The sections then were washed in 0.1 M phosphate buffer and incubated in primary serum for 48 hours in a cold room. The dilution of all primary antibodies was 1:2000. The following primary antibodies were used: anti-neuropeptide Y (anti-NPY), anti-cholecystokinin (anti-CCK) and anti-tyrosine hydroxylase (anti-TH). RAS7172N anti-NPY serum (lot # 015575-4) was purchased from Peninsula Laboratories (Belmont, CA 94002, USA). Anti-CCK serum was obtained from Dr. M.C. Beinfeld of University of Wisconsin Veterinary Medical School (for immunological characteristics of the serum see reference: Beinfeld et al., 1981). Anti-TH serum was obtained from Dr. T. Joh of Burke Rehabilitation Center (for immunological characteristics of the serum see reference: Pickel et al., 1975). At this stage, half of the sections were treated with 0.3% Triton. The sections were then used only for light microscopy. The other half of the sections were not treated with Triton and were used for electron microscopy. After incubation in primary serum the sections were washed thoroughly (6 times each for 10 min) and incubated for 1.5 hours at room temperature in secondary serum (diluted 1:250 in normal serum in each case). After washing in 0.1 M phosphate buffer (3 times each for 10 min), the sections were incubated with avidin-biotin for the next 1.5 hour and then stained with 0.05% di-amino-benzidine (DAB)-0.1% hydrogen-peroxide (H_2O_2) for 5-7 minutes. Specificity of the immuno-staining was assessed by blocking the primary antibodies with corresponding antigens and by omitting incubation in primary antibodies. Both methods of control reveal that antibodies used in this study were extremely specific.

Sections used for light microscopy were dehydrated in ascending series of alcohols and then mounted on glass slides. Sections used for electron microscopy were osmicated in 0.5% osmium tetroxide and embedded in araldite/epon (flat embedding). These sections then were mounted on epoxy-resin blocks, photographed and their specific regions containing immuno-stained neurons were cut with the ultratome LKB-3. Ultrathin sections were then mounted on copper grids, double stained with lead-citrate and uranyl-acetate and studied using the Philips-301 transmission electron microscope.

Golgi impregnation and Nissl staining

The sandwich method of Golgi impregnation of Freund and Somogyi (1983) modified by Frotscher and Leranth (1986) was used in our study. Sections for Golgi impregnations were cut with the vibratome from the same blocks as sections for immuno-staining, with their thickness being 100-150 μm. Sections were osmicated in 0.5% osmium tetroxide for 15 min. Then they were incubated in 2.5% of potassium dichromate for one hour. Sections were then arranged in blocks in which they were separated by parafilm. Each block of 10-15 sections was sealed with hot agar. Blocks were then incubated in 1% silver nitrate for 24 hours. Sections were separated in glycerol, dehydrated in alcohols and mounted on glass slides. Sequential thin sections (30-40 μm) were stained with toluidine-blue and used as cytoarchitectonic controls.

Computerized image-analysis

After immunostaining light microscope slides were investigated with the Microcomp image analysis system using a Planar Morphometry program. Density of immunostained neurons per cubic mm was measured semi-automatically as described previously (Morgane et al., 1988).

RESULTS

Laminar Distribution and Cell Typology of Neuropeptide Y (NPY) Neurons in Visual Neocortex of Cetaceans

In the visual cortex of all four investigated species of Cetacea (Stenella coeruleoalba, Tursiops truncatus, Phocoena phocoena and Globicephala melaena) NPY neurons are found sparsely dispersed throughout the cortical plate (Fig. 1B). However, these neurons are located mostly in the lower part of layer III as well as in layers V and VI. In layer II, we also found NPY perikarya but their occurrence is much scarcer than in lower cortical layers (Fig. 1B). All NPY neurons were found to be non-pyramidal cells with extremely variable shapes of the perikaryon and diverse distribution of their dendritic trees (Fig. 2).

At least two major types of NPY perikarya were found: one more or less round or slightly ovoid in shape (Fig. 2A, C) and the other elongated or fusiform in shape (Fig. 2B, D). Neurons with round or ovoid perikarya are much more numerous than cells of irregular (Fig. 3A) or elongated fusiform shape (Fig. 2D). The latter are found mainly in layer VI (Fig. 2D). Most of the NPY cells are large (25 x 20 μm) or giant (40 x 30 μm) stellate cells with rectilinear type dendrites (Fig. 2A). These neurons appear to be isodendritic type cells. In cases where the axon of the cell is immuno-stained (Fig. 3A-B) it collateralizes widely, and projects usually in an ascending direction (Fig. 3B). Immuno-staining of the NPY neurons in cetacean neocortex is extremely intensive in the perikaryon as well as in the cell processes (axons and dendrites). Often this staining appears like a Golgi impregnation, though more refined in terms of details of axonal and dendritic branching. The NPY immunostaining reveals multiple varicosities along the course of dendrites and axons (Figs. 2A-D, 3A-B). The fine mesh of NPY axons and dendrites actually extends throughout most cortical layers (Fig. 3A, C). Although we did not find NPY perikarya in layer I, there is a well-developed net of NPY-containing dendrites and axons in this layer

(Fig. 3C). In most cases NPY-containing dendrites and axons in layer I reveal a different distribution in comparison with lower layers. In the lower layers the mesh consists of processes ascending, descending or making three-dimensional loops, whereas in layer I most of the NPY-containing dendrites and axons are tangential to the pia.

Electron microscopy of NPY neurons revealed extremely convoluted nuclear membranes and, as a result, a "twisted" and deeply-folded nuclear shape as well as the presence of numerous, large secretory granules, an extremely developed Golgi apparatus and a rough endoplasmic reticulum (Fig. 4). In cytoplasm of these cells there are numerous multivesicular bodies and primary lysosomes.

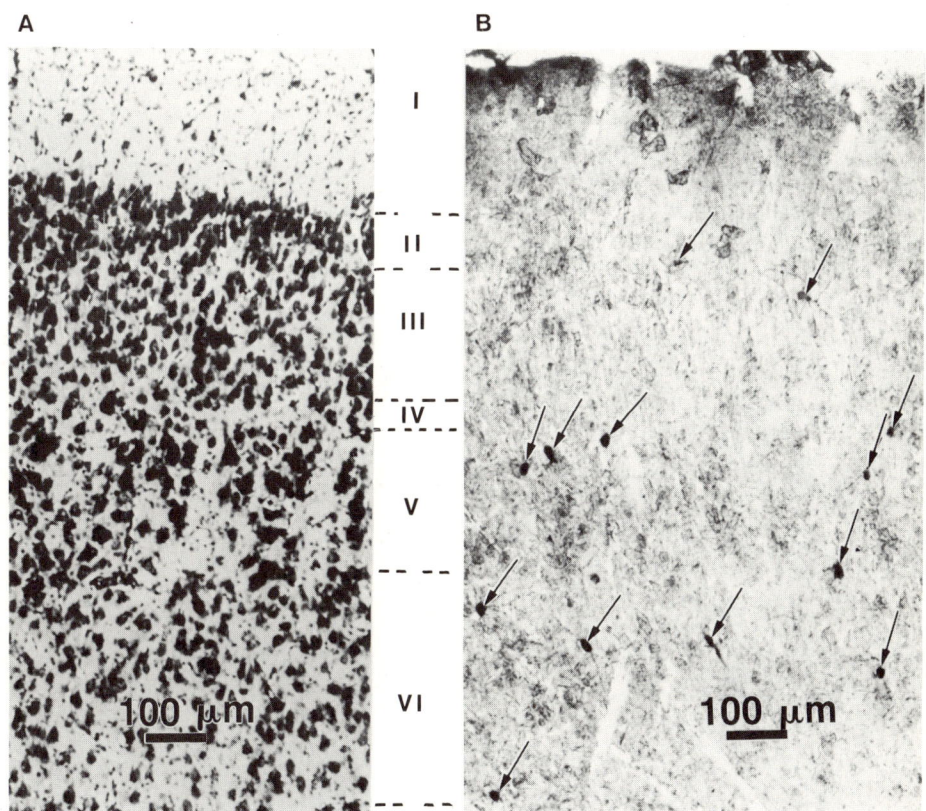

Fig.1. Photomicrographs showing the general laminar distribution of neuronal perikarya and the distribution of NPY-positive perikarya of neurons in visual cortex of pilot whale (Globicephala melaena). A. The general distribution of perikarya in heterolaminar visual cortex in a Nissl stained section. Note wide layer I, accentuation of layer II, relative homogeneity of laminar structure of the cortex, large pyramidal neurons in layer V and incipient layer IV. B. Section from the same block as A processed for immunostaining of NPY neurons. Note that major concentration of the NPY perikarya (arrows) occurs in lower layers (V and VI) of the cetacean neocortex.

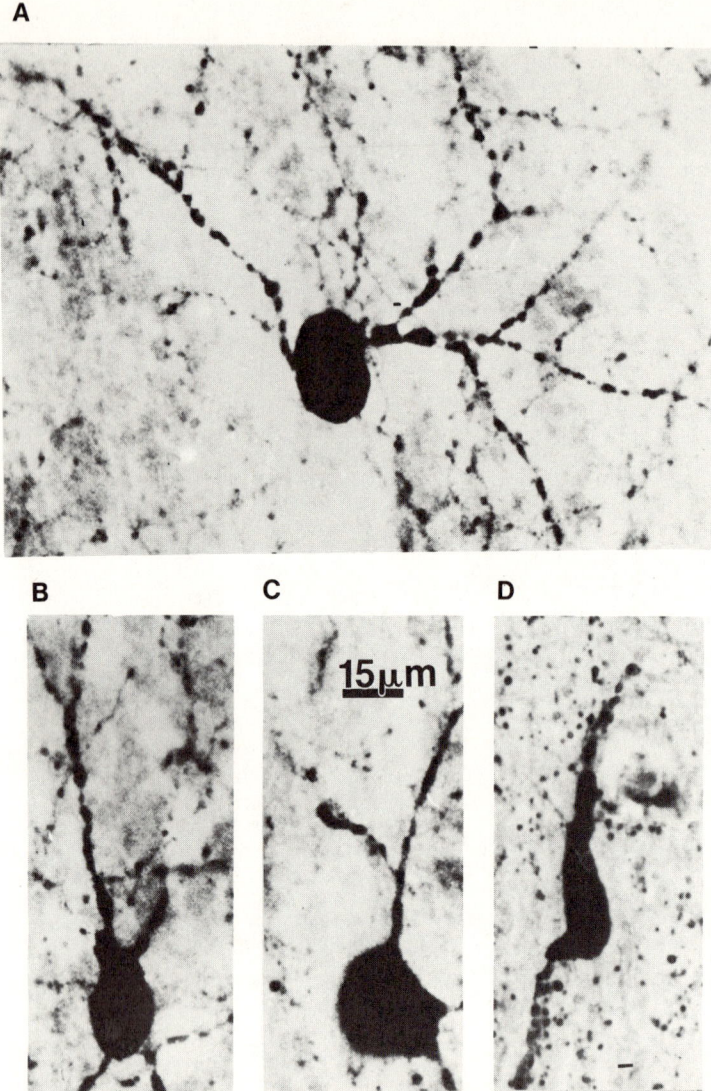

Fig. 2. Photomicrographs showing variably shaped perikarya and dendrites in typical NPY neurons in the visual cortex of Cetacea. Note that all perikarya belong to the large (25 x 30 μm) or giant (30 x 40 μm) non-pyramidal neurons. Immunostaining is localized in perikarya of neurons as well as in dendritic varicosities. A. Ovoid stellate cell with long rectilinear dendrites in layer IIIb (Globicephala melaena). B. Vase-like stellate cell in layer Va (Tursiops truncatus). C. Stellate cell, of nearly spherical type in layer Va (Phocoena phocoena). D. Fusiform stellate cell in layer VI (Stenella coeruleoalba).

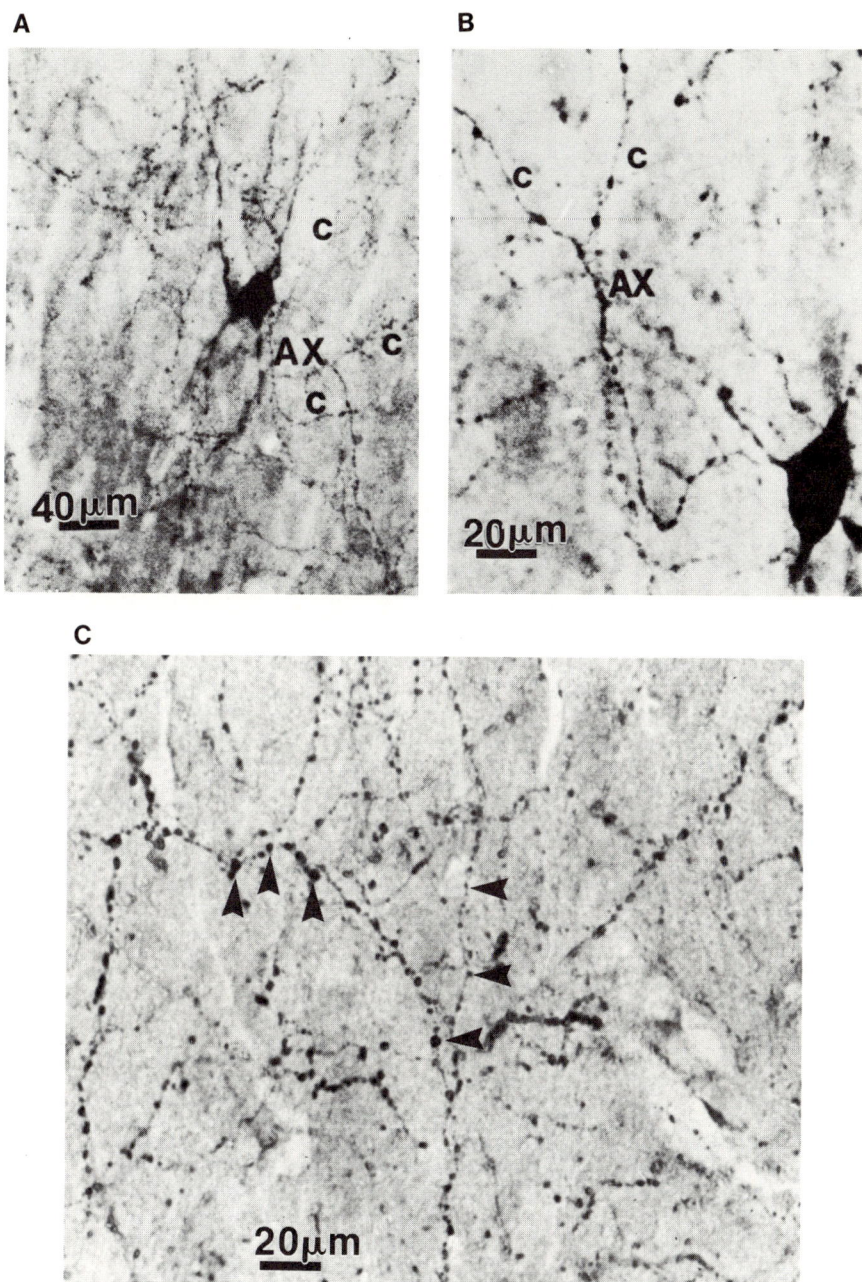

Fig. 3. Photomicrographs showing axonal distribution and dendritic varicosities of NPY neurons in the visual cortex of cetaceans. A. NPY neuron in layer II with the descending axon (AX) forming multiple widely branched collaterals (c) in layers I, II and IIIa (Stenella coeruleoalba). B. NPY neuron in layer IIIa with ascending axon (AX) projecting to layer I (Tursiops truncatus). C. Labelled dendritic and axonal varicosities in layer I are indicated by arrowheads (Globicephala melaena).

45

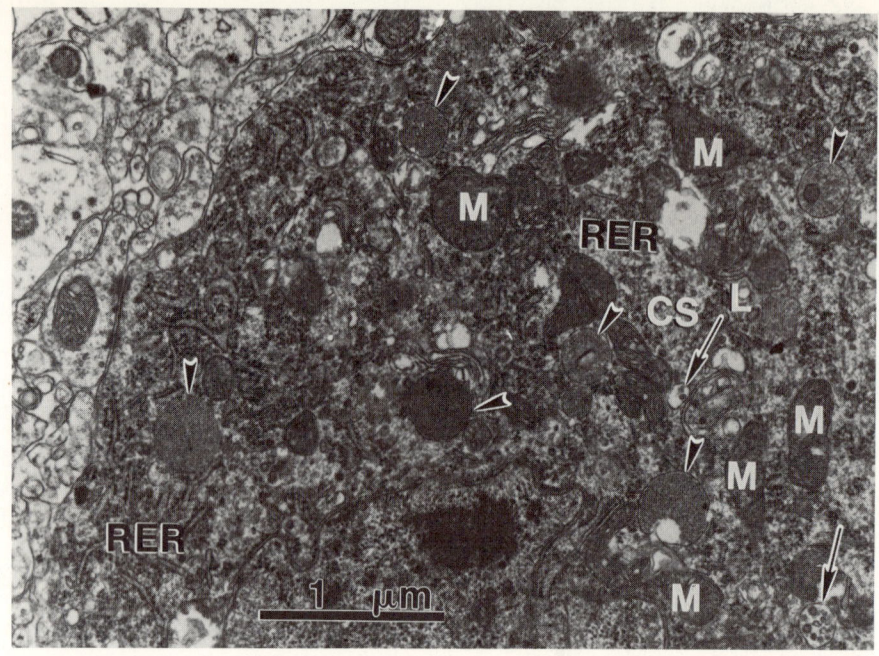

Fig. 4. Photomicrograph showing the ultrastructure of the perikaryon of NPY neuron in layer IIIa of the visual cortex in Tursiops truncatus. The light microscopic photo of this neuron is shown in Fig. 3B. Note the presence of multiple secretory granules (arrowheads), mitochondria with immunostained membranes (M), densely packed ribosomes and membranes of rough endoplasmic reticulum (RER), primary lysosomes(L) and multivesicular bodies (arrows). Also the cytoplasmic fibrillar skeleton is very prominent (CS). Immunostaining is localized mainly in constellations of polysomes and in cristae of some mitochondria.

2. Laminar Distribution and Cell Typology of Cholecystokinin (CCK) Neurons in Visual Neocortex of Cetaceans

Similar to NPY reactive neurons, we found CCK reactive neurons in visual cortex of all four investigated cetacean species. However, the distribution and typology of the CCK cells is quite different compared with NPY neurons. Thus, the major concentration of the CCK perikarya is found in layers I and II, whereas in lower cortical layers CCK perikarya are sparse (Fig. 5B). These cells form cytoarchitectonic radial columns perpendicular to the pia (Fig. 9A). They also were found in Golgi preparations of layer I where they are characterized by long dendrites extending from opposite poles of the perikaryon (Fig. 7B) or dendrites with rich tufts of branches on opposing perikaryal poles. Overall, immuno-staining of these cells and their processes was much weaker compared with NPY neurons though some axons, dendrites and their varicosities (presumably large synapses) can be visualized at the light microscopic level. In most cases, only the perikaryon and bipolar-oriented initial segments of dendrites show a sufficient reaction for both light and electron microscopic visualization (Figs. 7A-B, 8A-B). Most of CCK neurons are spindle-like, bipolar or bitufted cells of medium (20 x 10 μm) and large (30 x 12 μm) size (Figs. 6, 7A,C). Electron microscopy of these cells revealed ultrastructures which are almost identical to those found in NPY neuronal perikarya, i.e., twisted, irregular shape of the nucleus, numerous secretory granules, multiple multivesicular bodies and densely packed polysomes (Figs. 7B, 8A-B).

An exceptionally large concentration of CCK perikarya is found in layer I (Fig. 6). The orientation of these cells in layer I was found to be peculiar, i.e., their long axis is perpendicular to the pial surface (Figs.7A,C, 9A), although there are also some horizontal and irregularly oriented perikarya. The finding of a large number of vertically oriented CCK

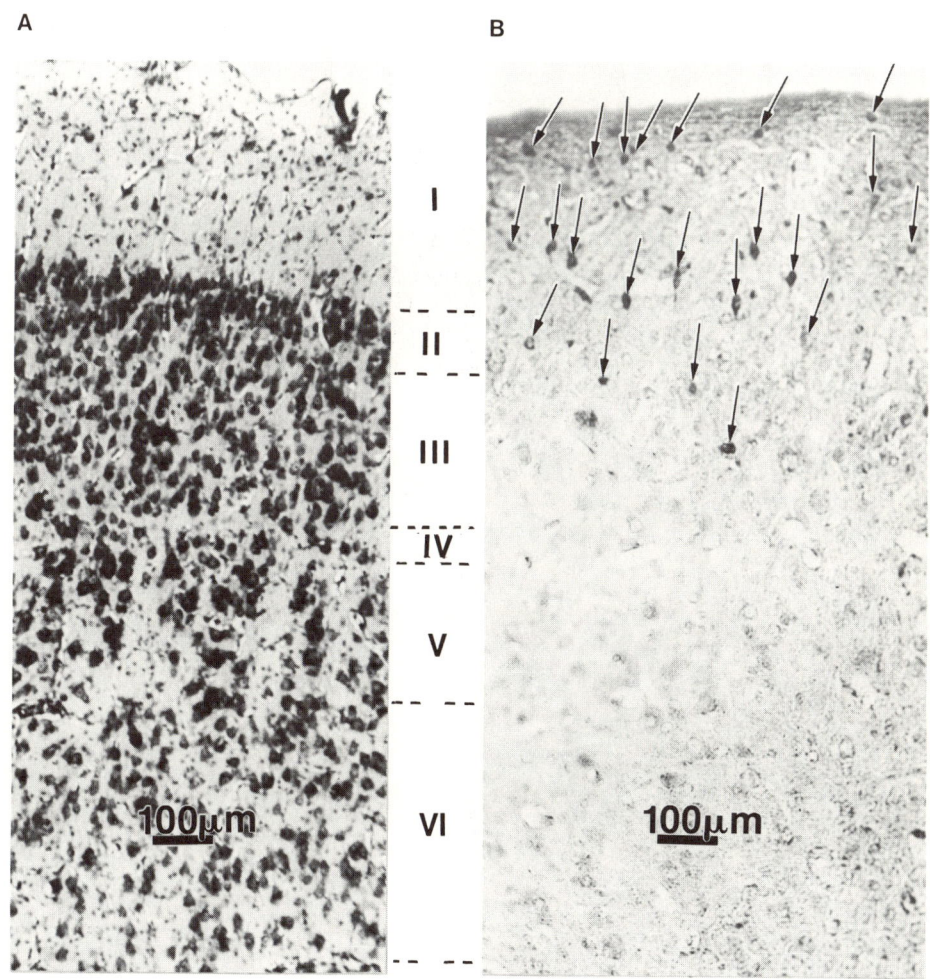

Fig.5. Photomicrographs showing the general distribution of neuronal perikarya and the distribution of the CCK perikarya of neurons in the visual cortex of pilot whale (Globicephala melaena). A. The general distribution of perikarya in heterolaminar visual cortex in the section stained by Nissl method. Note a wide layer I, accentuation of layer II, relative homogeneity of laminar structure of the cortex, large pyramidal neurons in layer V and incipient layer IV. B. Section from the same block as in A, processed for immunostaining of CCK neurons. Note that major distribution of the CCK perikarya (arrows) is distributed mainly in layers I and II of the visual cortex.

perikarya in layer I of the cetacean neocortex was unexpected, because in most other mammals vertically oriented cells are extremely rare or absent (Eccles,1983). The cells present in layer I in most species are horizontally oriented (Morrison and Magistrelli, 1983). We examined the cytoarchitectonic Nissl stained series of cetacean brains in our collection, as well as control cytoarchitectonic sections made from blocks which were used for immuno-staining. These clearly showed the presence of vertically oriented perikarya in layer I (Fig. 9). The presence of the unusual, vertically oriented, CCK neurons in layer I in the cetacean visual neocortex led to our study of the neuronal composition of this same layer in bat (Eptesicus fuscus) and hedgehog (Erinaceus europaeus). We have found that in both of these archetypal mammalian species there are numerous CCK neurons in the upper two layers of the visual cortex, though their number is less than in the cetacean neocortex (Figs. 9B-C, 10A-C).

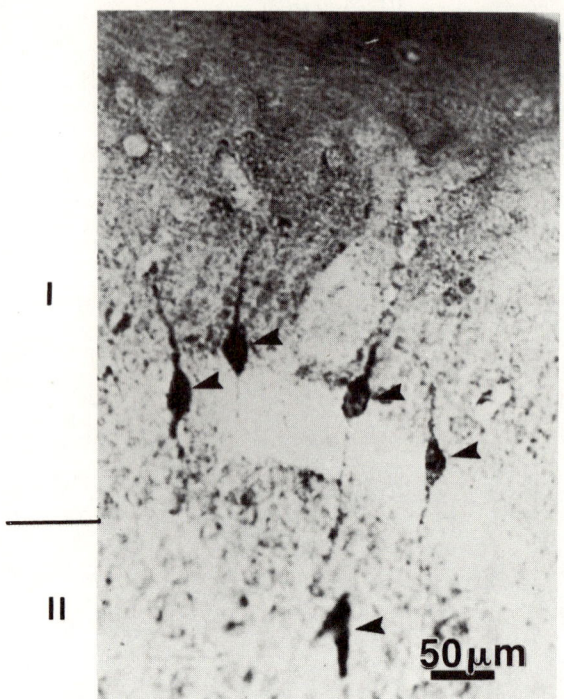

Fig.6. Photomicrograph showing a group of CCK neurons in layers I and II in the visual cortex of pilot whale (Globicephala melaena). Note that all cells are of the fusiform/bipolar type with two main dendrites originating from the opposite poles of the perikaryon.

In hedgehog visual cortex, we found bipolar and bitufted vertically-oriented CCK neurons in layer I (Figs. 9B, 10B). In Nissl preparations of bat visual cortex, most of the neurons in layer I are characterized by an ovoid shape of the perikaryon, although CCK immuno-staining and Golgi material shows that these cells are bipolar or bitufted cells (Figs. 9C, 10C). Thus, vertically-oriented CCK neurons identical to those in layer I of cetacean neocortex also were found to be typical for layer I in neocortices of archetypal terrestrial species of mammals (bats and hedgehogs).

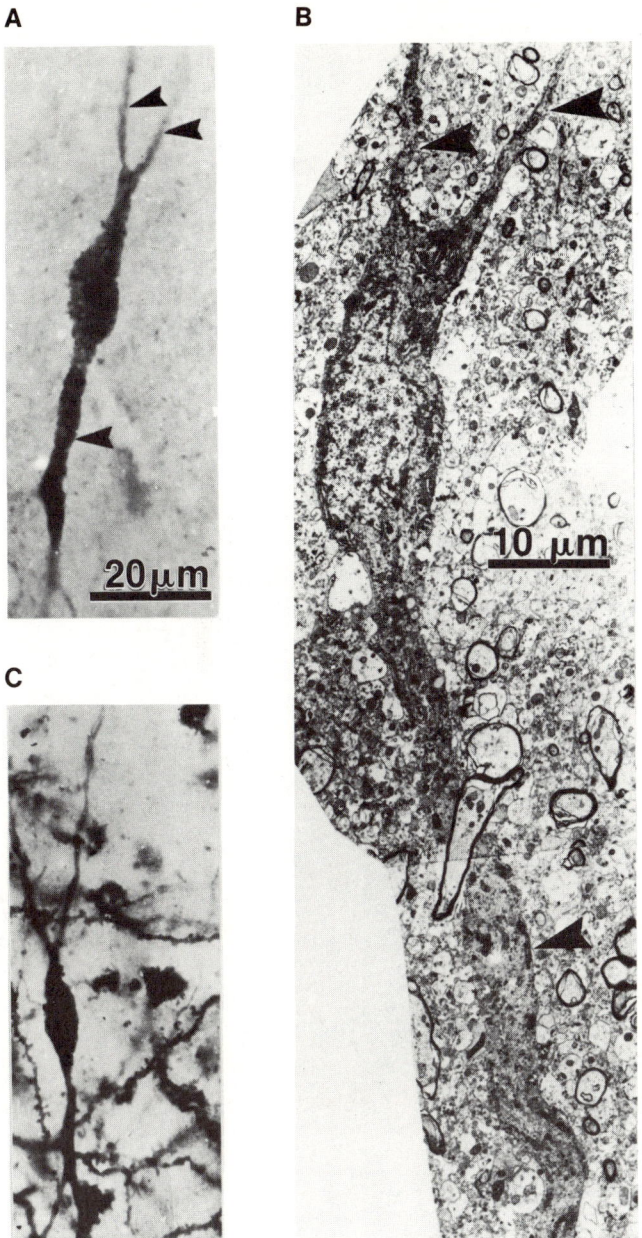

Fig.7. Photomicrographs showing light microscopic and ultrastructural features of the CCK-positive neuron in layer I the in visual cortex of Tursiops truncatus. A. Immunostained fusiform/bipolar CCK neuron. Note two main groups of dendrites (arrowheads) originating from opposite poles of the perikaryon. The neuron is oriented radially, i.e., along the vertical axis of the cortical plate and perpendicular to the pia. B. Montage of electron microscopic photos of the same neuron as in A. Note the concentration of the immuno-staining in the cell perikaryon and dendrites (arrowheads) and the relatively light cell nucleus. C. The same aspinous neuron as seen in the Golgi preparation. Note the absence of dendritic spines.

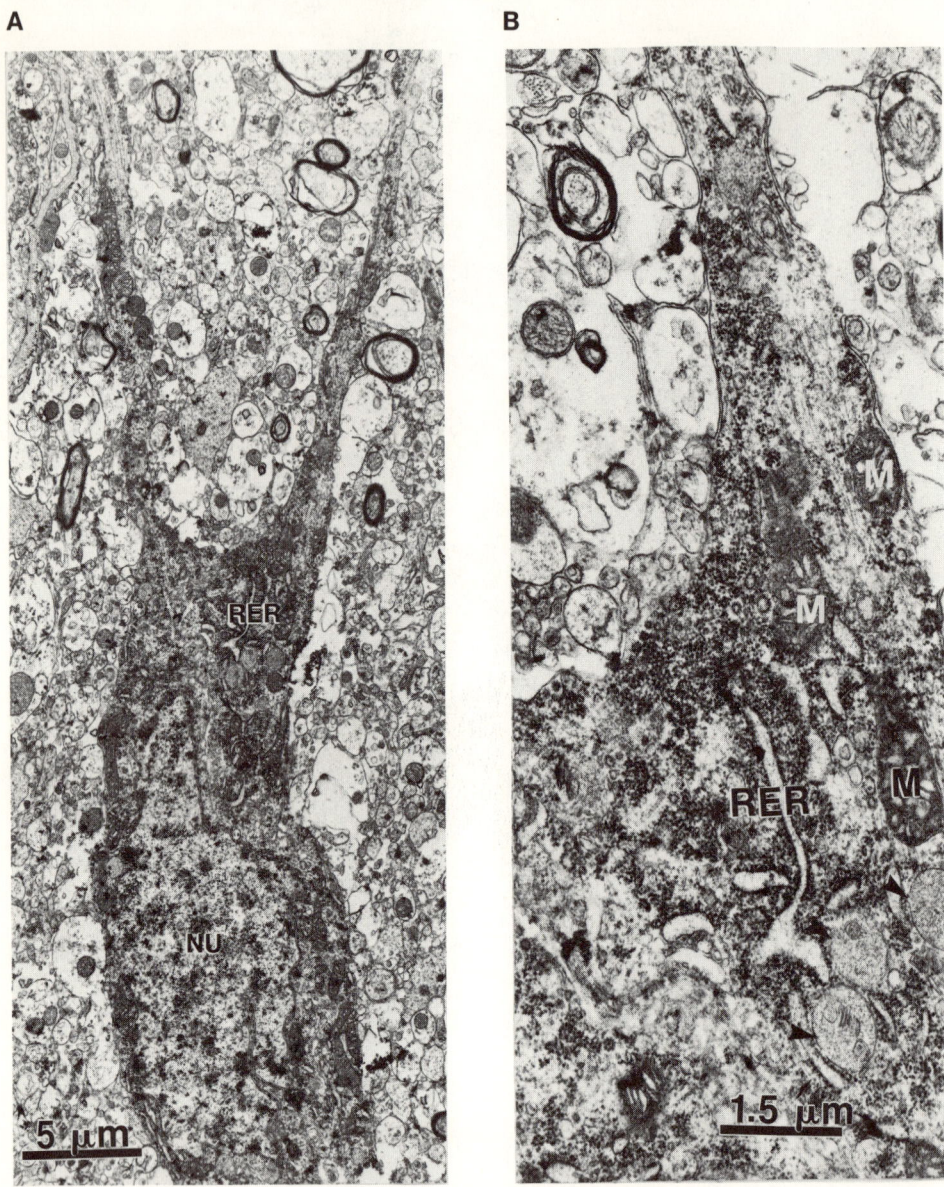

Fig. 8. Photomicrographs showing at higher magnifications ultrastructural organization of the perikaryon and dendrites of a CCK neuron which is shown in on Figs. 7A-B. A. Perikaryon and two dendrites ascending into the upper part of layer I. Note the irregular contour of the nucleus (NU) and densely packed ribosomes and membranes of rough endoplasmic reticulum (RER). B. Ultrastructures in the upper part of the perikaryon and ascending dendrite shown in Fig. 8A. Note the immunoreactive mitochondria (M), rough endoplasmic reticulum (RER) and secretory granules (arrowheads).

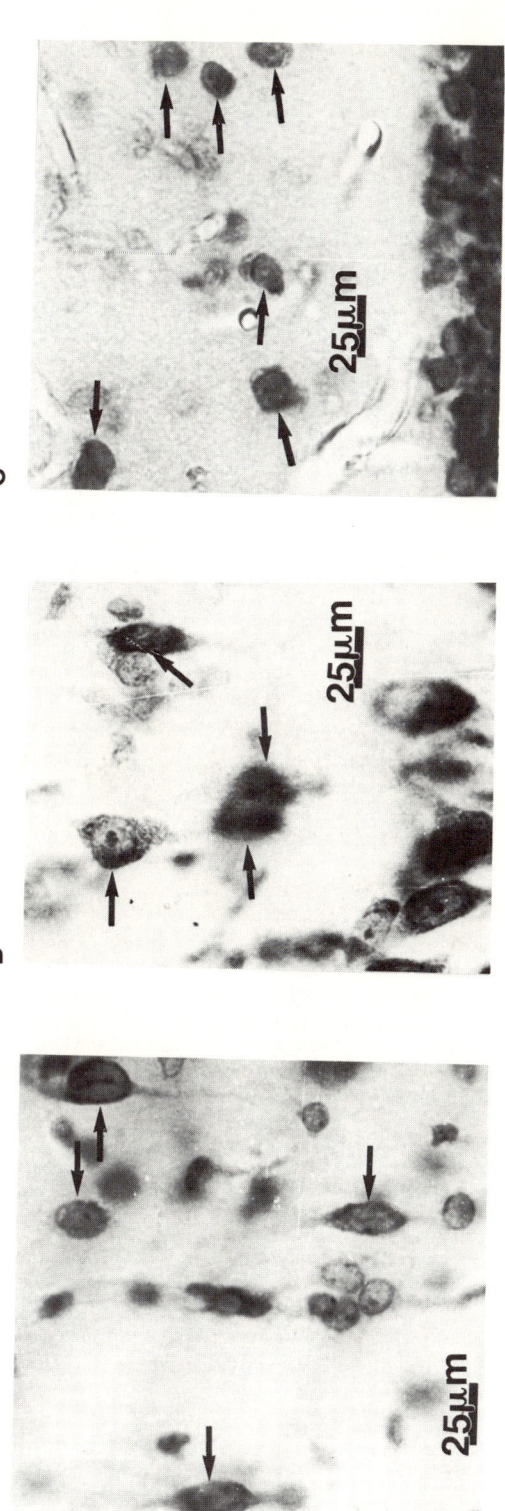

Fig.9. Photomicrographs showing numerous fusiform and ovoid perikarya (arrows) of neurons in layer I of the visual cortex in Cetacea, Insectivora and Chiroptera in Nissl material. A.Pilot whale (<u>Globicephala melaena</u>). All fusiform and ovoid perikarya are oriented vertically, i.e., perpendicular to the pia. Note that many of these cells are arranged into vertical cytoarchitectonic columns. B. Hedgehog (<u>Erinaceus europaeus</u>). Note a general similarity in vertical orientation and shape of the fusiform neurons with those in the pilot whale. C. Bat (<u>Eptesicus fuscus</u>). Note the rounded and ovoid shape of the neuronal perikarya. The latter in Nissl cytoarchitectonic sections are different from typical fusiform cells found in the pilot whale and hedgehog. In immunoreactive preparations, however, these cells show a clear vertical orientation (Fig. 10C).

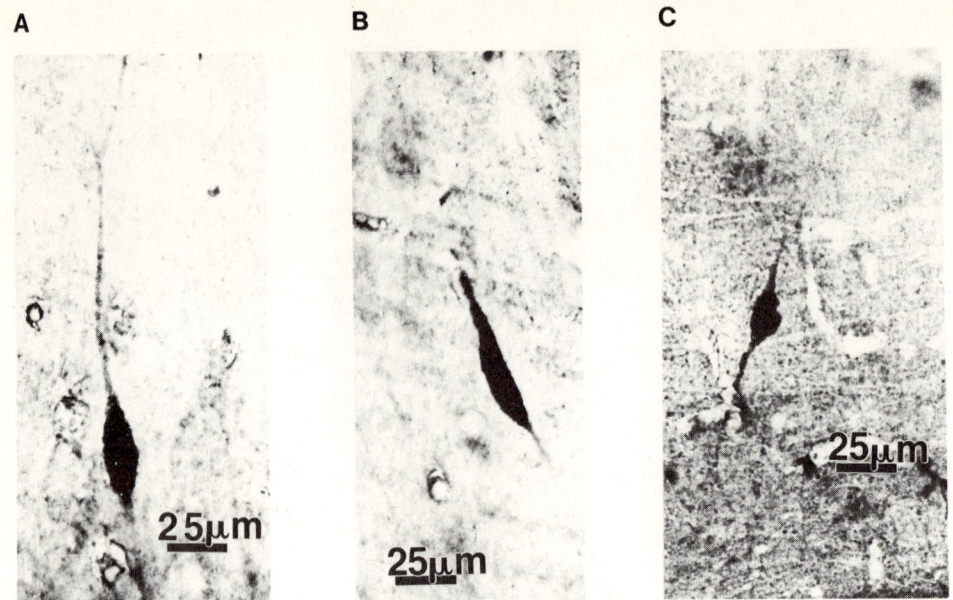

Fig. 10. Photomicrographs showing immunoreactive CCK neurons in layer I in visual cortex of Cetacea: Globicephala melaena (A), Insectivora: Erinaceus europaeus (B) and Chiroptera: Eptesicus fuscus (C). Note close resemblance in shape of the vertically-oriented perikarya in all three species.

Laminar Distribution and Cell Typology of Tyrosine Hydroxylase (TH) Neurons in Visual Neocortex of Cetaceans.

At the light microscopic level, the most prominent TH activity is found in long tangential processes in the lower part of layer I of the cetacean visual cortex (Fig. 11A). These processes were initially were thought to be long monoamine-containing afferents. They are present roughly to the same extent in all four species of cetaceans that we investigated. However, electron microscopic investigations revealed that TH-positive processes in layer I are not axons, but are well developed dendrites running long distances parallel to the pial surface (Fig. 11A). These dendrites have sparsely dispersed spines with long necks which form typical asymmetric synapses with terminal boutons, containing round clear vesicles (Fig. 11B). A closer examination showed the presence of small fusiform (5 x 9 μm) or multipolar (5 x 5 μm) perikarya in layer I which also display a positive reaction to TH (Fig. 12A-B). Fusiform cells are oriented mostly horizontally (Fig.12A). The ultrastructure of the perikarya of these neurons is similar to that described for NPY and CCK perikarya (Fig. 12C). Thus, although very different in size, Golgi structure and laminar distribution, the neurons containing three positively identified neurotransmitters have basically the same ultrastructural features of their perikarya (Figs. 4, 8, 12C).

Synaptology of Neuropeptide and Tyrosine Hydroxylase Neurons

Both axosomatic and axodendritic synapses were found to be positive to CCK and NPY immuno-staining. NPY-positive axosomatic synapses are found on pyramidal cells in cortical layers III and V. Synapses of this type contain several large mitochondria occupying almost all the synaptic bouton volume (Fig. 13A). Synaptic vesicles are small, clear and round in shape in

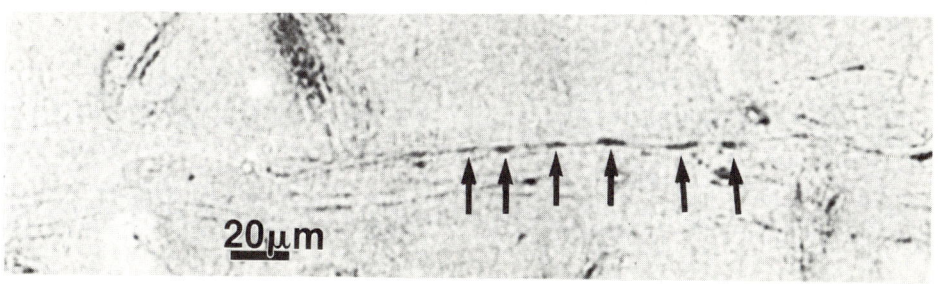
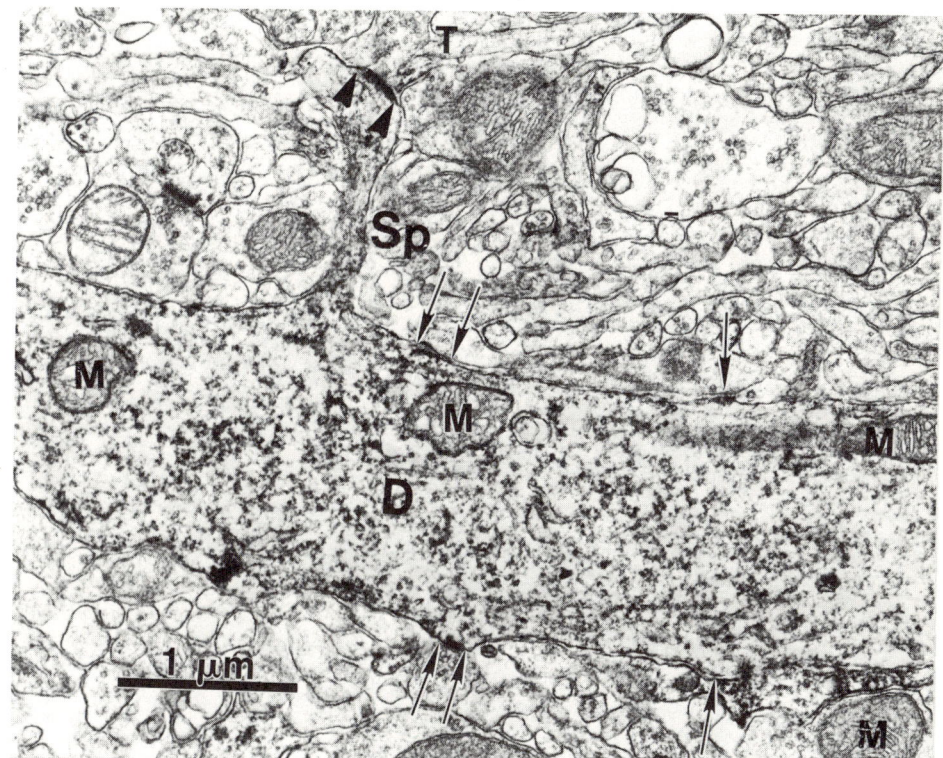

Fig.11. Photomicrographs showing tyrosine hydroxylase-positive dendrites in the visual cortex of the pilot whale (<u>Globicephala</u> <u>melaena</u>). A. Light microscope photo of long TH-positive dendrite in lower part of layer I. Note the presence of varicosities (arrows) along the dendritic shaft, and orientation of the dendrite tangential to pial surface. B. THe ultrastructure of TH-dendrite (D). Note the presence of mitochondria (M). A spine with a long neck is shown (SP) and forms an asymmetric synapse <u>en passage</u> (arrowheads) with a passing elongated bouton (T). There are also symmetric synapses <u>en passage</u> (arrows) with a dendritic shaft. Note that dendritic shaft (D) and spine (SP) are immuno-stained.

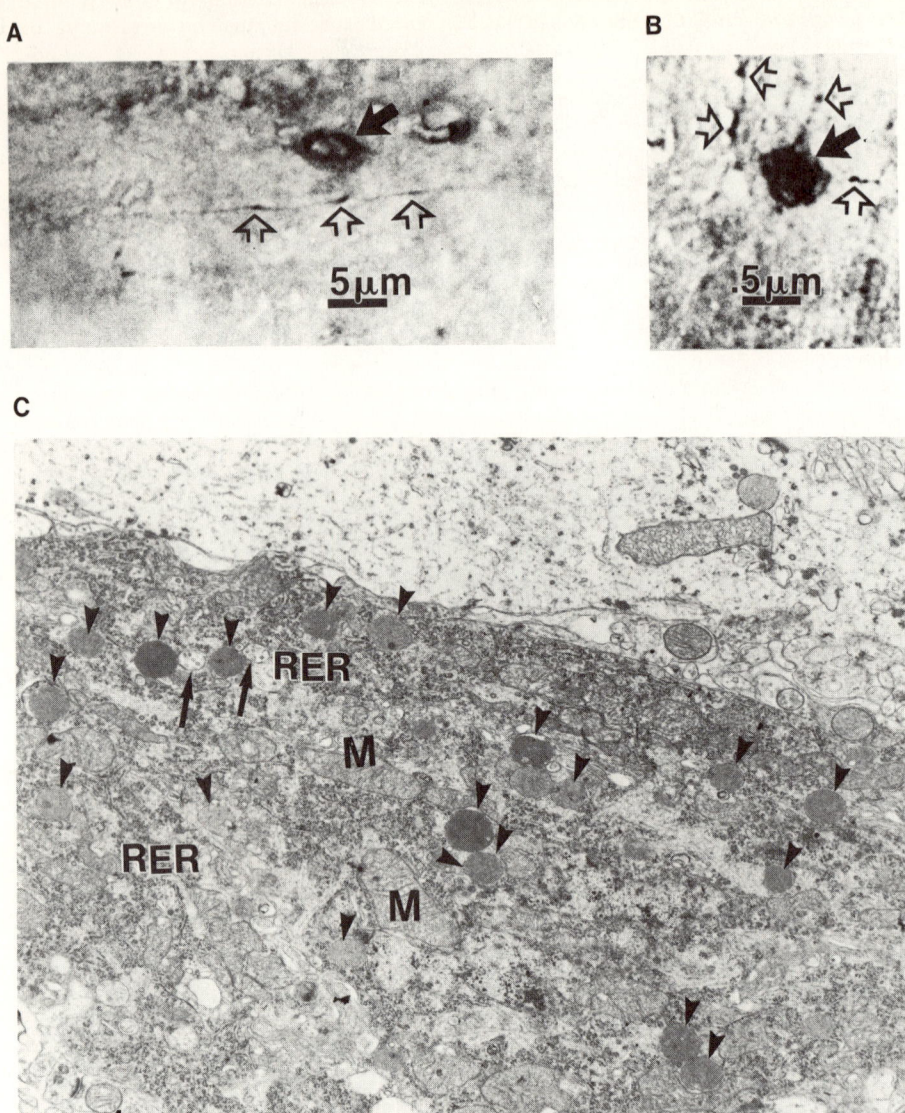

Fig. 12. Photomicrograph showing tyrosine hydroxylase perikarya in layer I of the visual cortex in the pilot whale (Globicephala melaena). A. A fusiform TH perikaryon (large black arrow) tangentially oriented to the pial surface. Note a TH dendrite with varicosities along its shaft (white arrows). B. A multipolar TH perikaryon (large black arrow) with several dendrites (white arrows) originating from perikaryon. C. Ultrastructure of TH perikaryon shown in Fig. 12A. Numerous secretory granules (arrowheads) are shown. Also, mitochondria (M), multivesicular bodies (arrows) and densely packed rough endoplasmic reticulum (RER) are shown. The ultrastructure of the TH-positive perikarya is similar to those of peptide neurons (see Figs. 4 and 8).

unstained material but shown in Fig. 13A as dark immunoreactive vesicles. NPY-positive synaptic boutons also are found contacting the perikarya of the NPY-positive neurons, possibly representing some type of feed-back connections within the NPY population of neurons. All axosomatic NPY-positive synapses belong to the symmetric type as do axodendritic NPY-positive synapses on the shafts of the dendrites of pyramidal neurons (Fig. 13B). Synaptic vesicles of the axodendritic synapses are numerous, variable in size and mostly of the round, clear type in unstained material. The shape of vesicles are often obscured by immuno-reactive deposits in synaptic boutons (Fig. 13B).

CCK-positive synaptic boutons have the same ultrastructural characteristics as NPY-boutons (Fig. 14A-B). However, we found largely CCK-positive axo-spinous boutons, whereas NPY boutons were found to only synapse dendritic shafts. These CCK-positive axo-spinous boutons belong to the asymmetric type of synapses and contain clear, round vesicles in unstained material (Fig. 14B). Although CCK and NPY-positive synapses are located in

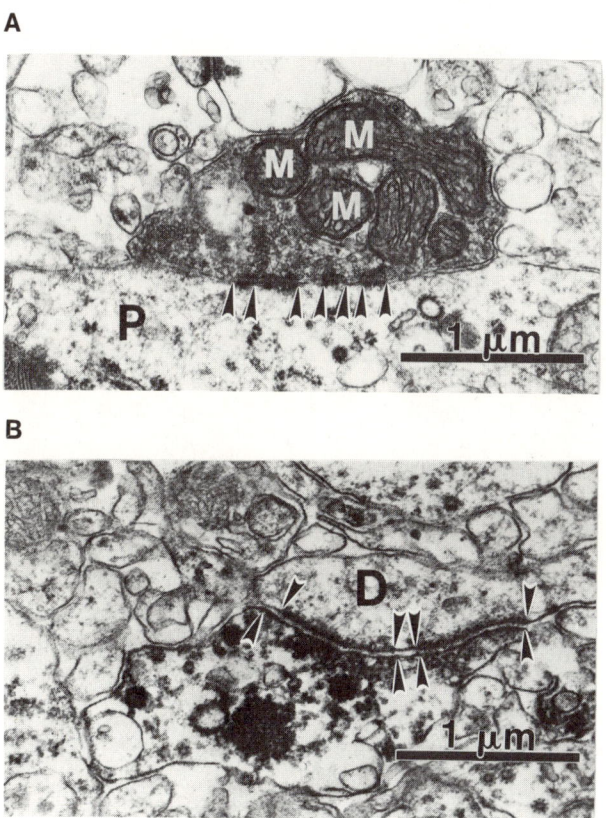

Fig.13. Photomicrographs showing the ultrastructure of NPY-positive synapses in the visual cortex of dolphin (<u>Tursiops truncatus</u>). A. NPY-positive bouton of axosomatic synapse of symmetric type (arrowheads) contacting a pyramidal cell (P) in layer III. Immuno-staining is localized in membranes of synaptic vesicles and in the outer membrane of mitochondria (M). B. A NPY-positive bouton of an axodendritic synapse contacting a dendritic shaft (D) in layer II. Note the symmetric type of synapse (arrowheads).

all cortical layers, CCK-positive synapses, both axosomatic and two types of axodendritic (synapses on shafts and on spines), tend to concentrate in the upper cortical layers, particularly in layers I and II. On the other hand NPY synapses are spread more widely and can be seen rather evenly distributed in all cortical layers. Both cell bodies and dendrites of NPY and CCK-positive cells receive many synapses that are negative to our two peptide antibodies and an example of this for a NPY cell is seen in Fig.15. These peptide-negative synapses are symmetric in the case of axosomatic synapses and asymmetric in case of axodendritic synapses.

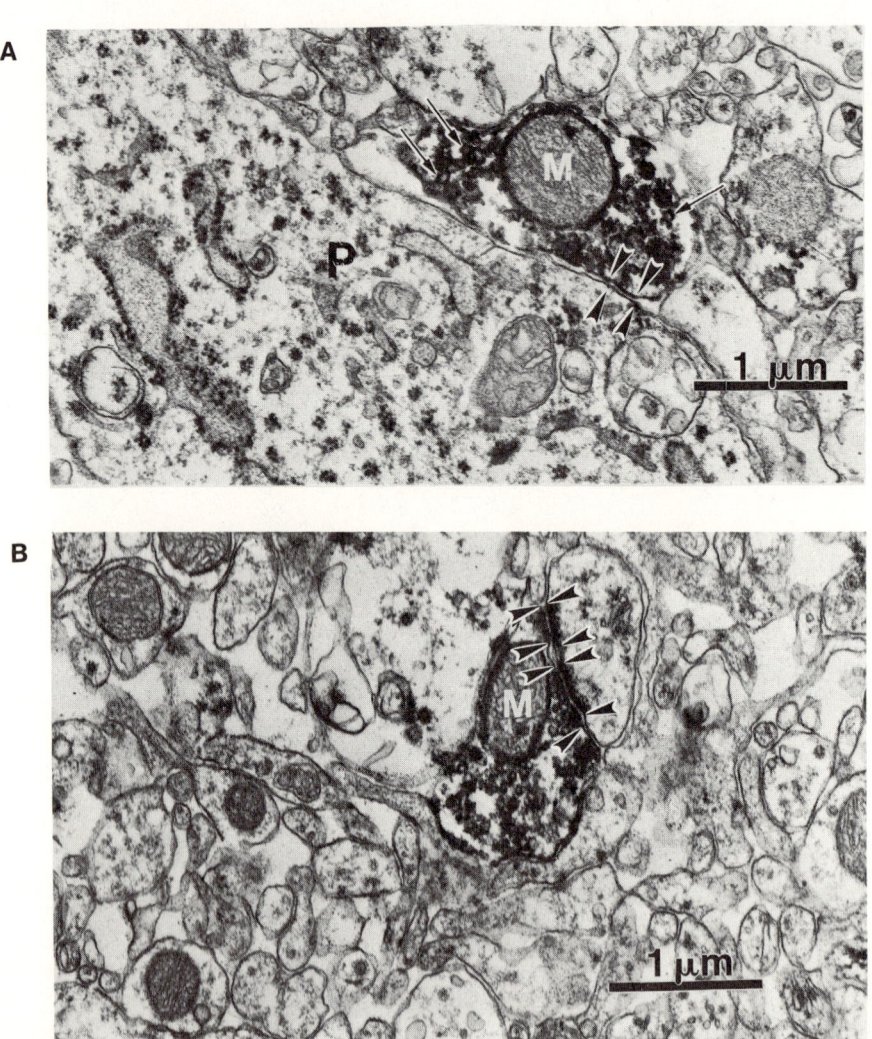

Fig. 14. Photomicrographs showing the ultrastructure of CCK-positive synapses in visual cortex of dolphin (Tursiops truncatus). A. A CCK-positive bouton of an axosomatic synapse of symmetric type (arrowheads) contacting a pyramidal cell (P) in layer II. Immunostaining is localized in membranes of synaptic vesicles (arrows) and in the outer membrane of mitochondria (M). B. CCK-positive bouton of axodendritic synapse contacting two spines in layer I. Note the asymmetric type of synapse (arrowheads).

In tyrosine hydroxylase-positive neurons, only postsynaptic structures (perikarya and dendrites) are revealed by the immuno-staining, whereas, in NPY and CCK neurons both pre and postsynaptic structures are immuno-stained. Long dendrites, tangential to the pial surface with sparsely distributed spines show intensive TH immuno-reactivity (Fig. 11A). Axosomatic synapses on the perikarya of TH-positive neurons are symmetrical and contain round clear vesicles. Synapses formed on the TH-positive spines are asymmetric and all their synaptic boutons are TH-negative (Fig. 11B).

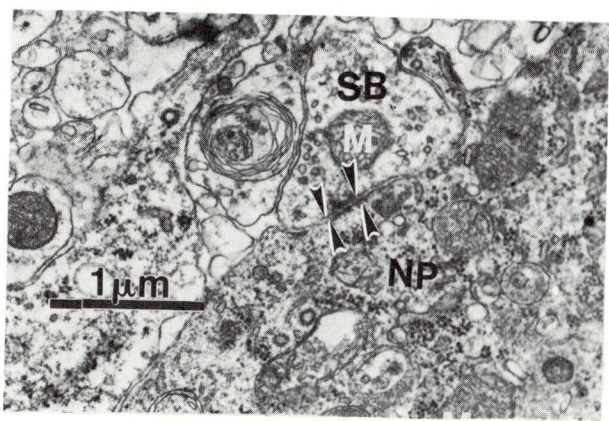

Fig. 15. Photomicrograph showing a NPY-negative synaptic bouton (SB) forming axosomatic synapse (arrowheads) contacting the plasma membrane of a NPY-positive non-pyramidal (NP) perikaryon. Note the symmetric type of synapse.

Quantitative Image Analysis Data on Numerical Density of NPY and CCK-positive Neurons in Visual Neocortex of Cetaceans

The major finding from our image analysis studies was a laminar reciprocal topographical distribution of the CCK and NPY neurons (Table 1). The density of the CCK neurons per cubic mm diminishes from layer I to the layer VI, while the density of the NPY neurons increases in the same direction (Table 1 and Figs. 16 and 17). The overall percentage of these neuropeptide neurons in different laminae varies from 0.23 to 69 % for CCK neurons and from 0 to 1.86 % for NPY neurons. Thus, in most cortical layers only a small fraction of neurons are positive for CCK and NPY. An interesting exception to this is in layer I where approximately 70 % of the cells are CCK-positive (Table 2).

Table 1. Density of CCK and NPY-positive Perikarya in Cetacean visual Cortex per mm^3.

Cortical Layers	Globicephala melaena		Phocoena phocoena		Tursiops truncatus	
	NPY	CCK	NPY	CCK	NPY	CCK
	M±SD	M±SD	M±SD	M±SD	M±SD	M±SD
I	0	480±207	0	1223±722	0	1040±493
II	445±136	741±413	315±132	1345±428	420±211	1763±438
III	528±247	325±210	409±206	305±134	540±234	467±270
IV	0		25±19	40±14	18±13	35±23
V	598±183	0	558±221	32±25	598±278	318±144
VI	705±495	10±7	418±168	58±41	540±248	160±130
Mean	461±216	260±140	418±168	501±227	353±164	545±127

Table 2. Percentage of NPY and CCK-positive Perikarya in Cetacean Visual Cortex.

Layers	NPY %	CCK %
I	0	69
II	0.35	1.36
III	0.64	0.73
IV	0.17	0.23
V	1.86	1.10
VI	0.98	0.76
Mean (excluding Layer I)	0.81	0.84

Quantitative analysis of NPY-positive varicosities, which are likely to be a mixture of dendritic and axonal varicosities, and thus, may indicate an overall spreading of the NPY neuronal processes, shows a distinctive disparity between different lamina. Thus, although the major concentration of NPY neurons is found in lower layers (Fig. 16), the highest density of the NPY axons and dendrites is found in layers I and II with significantly lower densities in layers III, V and VI ($p < 0.001$) (Table 3, Fig. 18). On the other hand, dendritic and axonal varicosities of the CCK neurons are much less conspicuous than NPY varicosities and are localized mainly in layers I and II, i.e., in the same location as the perikarya of the CCK neurons.

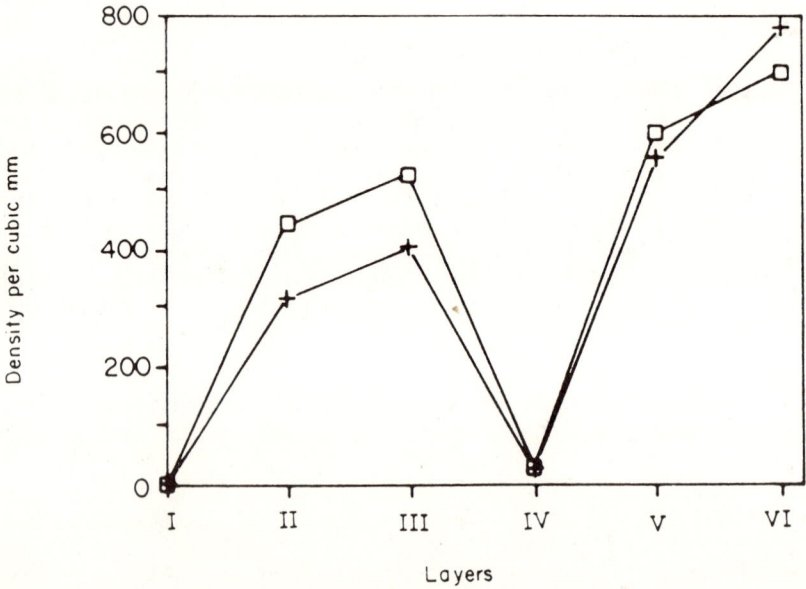

Fig. 16. Graph showing the laminar density of NPY perikarya in visual cortices of the pilot whale (<u>Globicephala melaena</u>) indicated by a white square and harbour porpoise (<u>Phocoena phocoena</u>) indicated by a plus sign. Since layer IV is lacking NPY perikarya, we note the increase of NPY neuronal density from layer II to layer III and from layer V through layer VI.

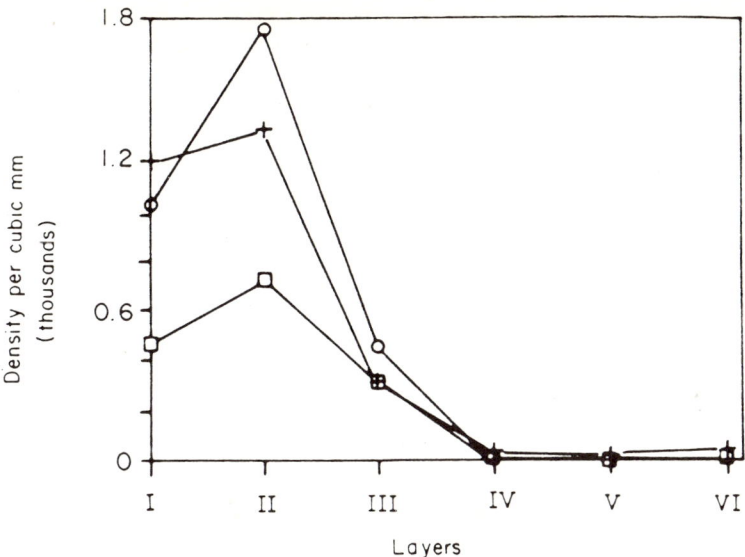

Fig. 17. Graph showing the laminar distribution of CCK perikarya in the visual cortices of the pilot whale (Globicephala melaena) indicated by a white square, the harbour porpoise (Phocoena phocoena) indicated by a plus sign and the bottlenose dolphin (Tursiops truncatus) indicated by a white circle. The highest concentration of CCK perikarya is located in layers I and, especially, II and there is a sharp decrease in density of CCK perikarya in the lower cortical layers (IV, V and VI).

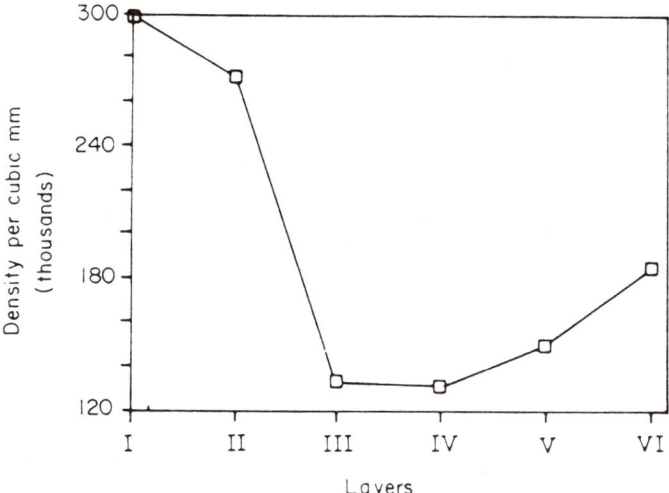

Fig. 18. Graph showing the laminar distribution of NPY varicosities in the visual cortex of bottlenose dolphin (Tursiops truncatus). The highest density of NPY varicosities occurs in layers I and II and there is a sharp decrease in density of NPY varicosities in layers III and IV. In layers V and VI the density of NPY varicosities increases in comparison to layers III and IV. This density is significantly less ($p<0.001$) in comparison with layers II and I.

Table 3. Density of NPY Varicosities in Cetacean Visual Cortex (thousands per mm^3).

Layers	M+SD
I	300±88.8
II	273±41.2
III	133±45.0
IV	132±42.4
V	150±43.7
VI	185+39.1
Mean	196+50.0

DISCUSSION

The major objective of this study was to characterize, using light and electron microscopic immunocytochemistry, the distribution of selected neuropeptides (neuropeptide Y and cholecystokinin) and tyrosine hydroxylase in visual cortex of toothed whales and to correlate the arrangement of the neuropeptides and TH immunoreactivity with our previous findings using Nissl, Golgi and electron analyses of cetacean neocortex (Morgane et al., 1988, 1990; Glezer and Morgane, 1990; Morgane and Glezer, this volume). We have found that the laminar distribution of the investigated neuropeptides and TH in cetacean neocortex have both general mammalian features, as well as some features which appear rather selective to these aquatic mammals and to some prototypical terrestrial mammals (hedgehogs and bats). In our cortical material of four toothed whales the population of CCK neurons is concentrated in the upper cortical layers, whereas the population of the NPY-positive perikarya is concentrated in the lower cortical layers. On the other hand, NPY perikarya in terrestrial mammals such as macaque monkey, rats, hamsters and cats are found in all cortical layers of different areas of the neocortex (Emson and Hunt, 1981; Jones, 1986; Demeulmeester et al., 1988; Kuljis and Rakic, 1989). According to Kuljis and Rakic (1989) in macaque monkey the major concentration of NPY neurons is found in layers II and III of the visual cortex. In our material of toothed whale visual cortex, NPY neurons are not present in layer I, only scarcely present in layers II and IIIA and B, while their major concentration is found in layers IIIc, V and VI, Also, in primate visual cortex NPY perikarya are present in layers IVA and B, whereas in cetaceans, as expected, they are virtually absent in incipient layer IV. In this feature the cetacean primary visual cortex is similar to the so-called association cortices of terrestrial mammals which also have no NPY cells in layer IV (Kuljis and Rakic, 1989). Another important difference of cetacean cortex from primate cortex is in typology of NPY cells. In macaque visual cortex there is a wide spectrum of Golgi types of NPY cells, which include small granule cells forming local circuits and several types of aspinous and semi-spinous non-pyramidal cells, including multipolar and horizontal bipolar cells (Kuljis and Rakic, 1989). In cetacean neocortex almost all of NPY cells belong to aspinous or semi-spinous large and giant stellate cells of the isodendritic type, similar to cells of the reticular core of the brainstem. These have been discussed in our Golgi material (Morgane et al., 1988; Glezer and Morgane 1990; Morgane et al., 1990; Morgane and Glezer, this volume). We have found that prototypical isodendritic neurons with NPY activity, although relatively small in number (approximately 1 % of total neuronal population in the cetacean visual neocortex) provide a vast mesh-like net of axons and dendrites which spread in loops throughout all cortical layers. The density of NPY axonal and dendritic varicosities is especially high in layers I and II and may provide a strong input to large groups of cortical pyramidal cells through axodendritic and axosomatic synapses. We have found that both of these types of NPY-positive synapses occur in layers I and II of the cetacean visual neocortex.

The second neuropeptide we investigated, cholecystokinin, also shows some peculiar features in cetacean neocortex when compared with its distribution in terrestrial mammals. In the latter, CCK cells are concentrated mainly in layers II and III and are rarely seen in layer I while there is a second peak in layer VI (Emson and Hunt; 1981, Morrison and Magistrelli, 1983; Dockray, 1985; Jones, 1986). In cetaceans the main concentration of CCK perikarya is in layers I and II, with the number of CCK cells being two to four times less in lower cortical layers. It is of interest that there are some differences in numerical density of CCK cells in different species of toothed whales. In Globicephala melaena, CCK neurons are virtually absent below layer IIIA. In Phocoena phocoena, Stenella coeruleoalba and Tursiops truncatus there is a significantly higher numerical density per mm^3 of CCK cells than in the visual cortex of the pilot whale (Globicephala melaena). As noted almost all CCK neurons are bipolar, vertically oriented cells, and this feature is especially prominent in layer I. The presence of vertically oriented cytoarchitectonic columns of bipolar or bitufted non-pyramidal cells in layer I of the cetacean neocortex was confirmed by the Nissl and Golgi preparations in sections adjacent to the sections used for immuno-staining. It is well known that in advanced terrestrial mammals, including man, cells in layer I are mostly horizontal and extremely rare in occurrence (Ramón y Cajal, 1911; Eccles, 1983). The recent immunohistochemical studies supported this early data and showed that in more advanced terrestrial mammals, as compared to hedgehogs and bats, neurons in layer I are mostly horizontal cells and contain CCK (Morrison and Magistrelli, 1983; Demeulmeester et al., 1988).

Our finding of a large population of vertically-oriented CCK neurons in layer I of cetacean neocortex might be interpreted in the general framework of evolutionary and functional features of this layer in evolutionary conservative and advanced mammals. Layer I is a phylogenetically old formation in all cortices, including neocortex. It is known that layer I is the main input layer in archi- and paleocortices and that it is highly developed in bat and hedgehog neocortices (Sanides and Sanides, 1974). Investigations on dolphins have shown that it constitutes about one-third of the whole thickness of the dolphin visual cortex, (Morgane and Jacobs, 1972; Kesarev et al., 1977; Morgane et al., 1988). The prevalence of layer I over other layers in cetacean neocortex correlates with our findings in Golgi material and in quantitative electron microscopic synaptology showing that this layer is the location of an extremely developed dendritic plexus of so-called extraverted neurons of layer II. It contains about 40 % of all axodendritic synapses of the dolphin visual cortex (Glezer and Morgane, 1990). For comparison, in primate visual cortex this layer contains only 7 % of axodendritic synapses (O'Kusky and Colonnier, 1981). Our data on structural and functional significance of layer I in cetacean neocortex are in agreement with data of Krasnoschekova and Figurina (1980) and Garey and Revischin (1989) who showed that layer I is an important input layer for thalamocortical afferents. It should be emphasized that extremely well-developed thalamo-cortical inputs from specific nuclei of the thalamus to layer I was found also in the neocortex of the hedgehog (Ebner, 1969; Valverde et al., 1986). The similarity of the cetacean neocortex to the neocortices of prototypical mammals in development of layer I is indicated also by the present immunocytochemical findings. We have shown that in archetypal mammals such as hedgehog and bats there are also numerous vertical bipolar CCK neurons in layer I thus indicating a possible relation between the organization of the neocortex in the dolphin compared to that of archetypal mammals. In this regard, Antonopoulos et al. (1987) reported that in hedgehog neocortex CCK neurons are more numerous in layers I-III than in lower layers IV-VI.

Beside our comparative immunocytochemical evidence, there are data showing that in ontogenesis of some terrestrial mammals initially numerous neuroblasts and neurons in layer I steadily decline and are rare in the adult stage (Rakic, 1974; Eccles, 1983). The presence of a large population of specialized vertically oriented CCK neurons in layer I in the adult stage of cetaceans as well as their presence in other prototypal mammals may be one more sign of an evolutionary conservative characteristic of their neocortices. In this regard, Poliakov (1958) emphasized that the presence of large numbers of vertically oriented fusiform neurons in limbic neocortex above layers VI and VII is one of the conservative features of this area in mammals. It is difficult to interpret the functional significance of this population of the CCK cells in layer I in prototypical mammals because of the limited physiological data on this neuropeptide. Most of the known data shows that CCK neurons exert excitatory action on postsynaptic structures and receptors for CCK are found in layers I,II, III and VI (Dockray, 1985; Shaw et al., 1986; Jones, 1987). Thus, we may tentatively propose that the CCK population of neurons in layers I and II of the cetacean neocortex is one of the groups of interneurons processing information from thalamic afferents and then exerting influence, mostly excitatory, on the perikarya in lower cortical layers and apical dendrites of the pyramidal cells extending into layer I.

The presence of horizontally-oriented tyrosine hydroxylase-positive neurons with long sparsely-spined dendrites deserves a special discussion. Strong monoamine innervation of the cerebral cortex from the subcortical centers is well known (Morrison et al., 1979; Olshowka et al., 1981; Parnavelas and Popadopoulos, 1989). Recent investigations show that monoamine innervation is extremely specific in terms of regional, laminar and intracortical localization (Parnavelas and Popadopoulos, 1989). In most mammals catecholamine afferents are present in all layers of the neocortex with an specially dense concentration in layer I. These latter consist of long horizontal axons running parallel to the pia (Morrison et al.,1979; Parnavelas and Popadopoulos, 1989). Based on this data, we expected to find the specific immuno-staining to tyrosine hydroxylase in the form of long horizontal axons in layer I of the dolphin neocortex. However, electron microscopic investigation revealed that long horizontally running dendrites rather than axons were labeled. We also found small TH-positive perikarya in layer I. These data is supported by findings obtained on mammalian neocortices where tyrosine hydroxylase-positive perikarya were found in layers I and II, as well as in other cortical layers (Kosaka et al., 1987; Gaspar et al., 1987; Hornung et al., 1989). The cell bodies of the TH neurons show close ultrastructural similarities to the CCK and NPY neurons and exhibit many features of neuroendocrine secretory cells, such as the presence of large numbers of secretory granules, a hypertrophied Golgi apparatus, multiple primary lysosomes and multivesicular bodies.

In comparing NPY and CCK populations of neurons in the cetacean visual cortex, we found that the NPY neuronal system is diffuse, and comprises polymorphic non-pyramidal neurons. Many of these neurons are giant stellate cells of the isodendritic type. The CCK perikarya are more vertically oriented in the columnar-like arrangements and are much more localized, being mainly in upper two cortical layers. The projections of these two neuropeptide groups are also different: the NPY system forms synapses on shafts of dendrites, whereas the CCK system forms synapses mostly on dendritic spines. However, both peptide neuronal groups have their axonal and dendritic varicosities extending into layers I and II in the dolphin. It appears that CCK and NPY peptides are localized in phylogenetically more conservative neuronal types in Cetacea. This may represent additional evidence of the overall conservative character of the cetacean neocortex.

ACKNOWLEDGEMENTS

This research was supported by National Science Foundation grants BNS-84-14532, 87-47032 and BNS-8903717 and by NIH grants HD 06364, HD 23830, NS 26068, MBRS/CRS grant of CCNY and the New York Aquarium of New York Zoological Society. We thank Dr. L. Garibaldi, Director of the Osborn Laboratories of Marine Sciences of New York Aquarium for providing various facilities for this project.

REFERENCES

Acher, R., 1985, Principles of evolution: the neural hierarchy model, in: "Brain Peptides", D.T. Krieger, M.J. Brownstein and J.B. Martin eds., John Wiley and Sons, New York, 136-163.
Antonopoulos, J., Karamanlides, A. N., Papadopoulos, G. C, Michaloudi, H., Dinopoulos, A., Parnavelas, J.G., 1989, Neuropeptide-like immunoreactive neurons in the hedgehog (Erinaceus europaeus) and sheep (Ovis aries) brain. J. Hirnforsch., 30:349-360.
Beinfeld, M. C. and Palkovits, M., 1982, Distribution of cholecystokinin (CCK) in the rat lower brain stem nuclei, Brain Res., 238:260-265.
Beinfeld, M. C., Meyer, D. K., Eskay, R. L., Jensen, R. T. and Brownstein, M. J., 1981, The distribution of cholecystokinin immunoreactivity in the central nervous system of the rat as determined by radioimmuno-assay, Brain Res., 212:51-57.
Colonnier, M. L., 1966, The structural design of the neocortex, in: "Brain and Conscious Experience", J. C. Eccles, ed., Springer Verlag, New York, 1-20.
Demeulmeester, H., Vandesande, F., Orban G. A., Brandon, C. and Vanderhaeghen, J. J., 1988, Heterogeneity of GABAergic cells in cat visual cortex, J. of Neuroscience, 8:988-1000.
Dockray, G. J., 1985, The Cholecystokinin, in: "Brain Peptides", D. T.,Kriger, M. J. Brownstein and J. B. Martin, eds., John Wiley and Sons, New York, 852-867.
Eccles, P. C., 1983, The horizontal (tangential) fibers system of lamina I of the cerebral cortex, Acta Morphologica Hungarica, 31:261-284.
Ebner, F. F., 1969, A comparison of primitive forebrain organization in metatherian and eutherian mammals, Ann. N.Y Acad. Sci., 167:241-257.
Emson, P. C. and Marley, P. D., 1983, Cholecystokinin and vasoactive intestinal polypeptide, in: "Neuropeptides", L.L. Iversen, S.D. Iversen and S. H. Snyder, eds. Plenum Press, New York, 255-306.
Emson, P. C. and Hunt, S. P., 1981, Anatomical chemistry of the cerebral cortex, in: "The Organization of the Cerebral Cortex", F. Schmitt, F. Worden, G. Adelman and S. Dennis, eds. Cambridge, 325-345.
Ferrer, I., 1987, The basic structure of the neocortex in insectivorous

Ferrer, I., 1987, The basic structure of the neocortex in insectivorous bats (Miniopterus sthreibersi and Pipistrellus pipistrellus). A Golgi study., J.Hirnforsch., 2: 237-243.

Freund, T. and Somogyi, P., 1983, The section-Golgi impregnation procedure. I. Description of the method and its combination with histochemistry after intracellular iontophoresis or retrograde transport of horseradish peroxidase. Neuroscience., 9:463-474.

Frotscher, M., and Leranth, C., 1986, The cholinergic innervation of the rat fascia dentata: identification of target structures on granule cells by combining choline acetyltransferase immunocytochemistry and Golgi impregnation., J. Comp. Neurol., 243:58-70.

Garey, L. J. and Revishchin, A. V., 1990, Structure and thalamo-cortical relations of the cetacean sensory cortex: histological, tracer and immunocytochemical studies., in: "Sensory Abilities of Cetaceans", J. A. Thomas and R. A. Kastelein, eds., Plenum Press, New York.

Gaspar, P., Berger, B., Febvret A., and Vigny A., 1987, Tyrosine-hydroxylase-immunoreactive neurons in the human cerebral cortex: a novel catecholaminergic group?, Neurosci. Letters, 5:257-262.

Glezer, I. I., Jacobs, M. S. and Morgane, P. J., 1988, The "initial" brain concept and its implications for brain evolution in Cetacea., Behav. Brain Sciences, 11:75-116.

Glezer, I. I. and Morgane, P.J., 1990, Ultrastructure of synapses and Golgi analysis of neurons in the neocortex of the lateral gyrus (visual cortex) of the dolphin (Stenella coeruleoalba) and the pilot whale (Globicephala melaena), Brain Res. Bull., 24:401-427.

Hornung, J. P., Tork, I., and De Tribolet, N., 1989, Morphology of tyrosine hydroxylase-immunoreactive neurons in the human cerebral cortex., Exp. Brain Res., 76:12-20.

Jones, E. G., 1986, Neurotransmitters in the cerebral cortex, J. Neurosurg., 65: 135-153.

Kesarev, V. S., Malofeyeva, L. I. and Trykova, O. V., 1977, Structural organization of the cerebral cortex in cetaceans, Arkhiv Anat. Gistol. Embriol., 73:23-30.

Kosaka, K., Hama K., Nagatsu I., 1987, Tyrosine hydroxylase-immunoreactive intrinsic neurons in the rat cerebral cortex, Exp Brain Res., 68:393-405.

Krasnoshchekova, E. I. and Figurina, I. I., 1980, The cortical projection of the medial geniculate body of the dolphin brain. Arkhiv Anat. Gistol. Embryol., 78:19-24.

Kuljis R. O. and Rakic P., 1989, Neuropeptide Y-containing neurons are situated predominantly outside cytochrome oxidase puffs in macaque visual cortex, Visual Neuroscience, 2:57-62.

Leranth, C. and Frotcher M., 1986, Synaptic connections of cholecystokinin-immunoreactive neurons and terminals in the rat fascia dentata: A combined light and electron microscopic study, J. Comp. Neurol., 254:51-64.

Leranth, C. and Feher E., 1983, Synaptology and sources of vasoactive intestinal polypeptide (VIP) and substance P(SP) containing axons of the cat coeliac ganglion. An experimental electron microscopic immunohistochemical study. Neuroscience, 10:947-958.

Leranth, C., Frotcher M. and Rakic P., 1988, CCK-immunoreactive terminals form different types of synapses in the rat and monkey hippocampus, Histochemistry, 88: 343-352.

Morgane, P. J. and Jacobs, M. S., 1972, Comparative anatomy of the cetacean nervous system, in:"Functional Anatomy of Marine Mammals", R. J. Harrison, ed.,Academic Press, London, 117-244.

Morgane, P. J., Jacobs, M. S., and Galaburda, A.M., 1985, Conservative features of neocortical evolution in dolphin brain, Brain, Behavior and Evolution, 21:176-184.

Morgane, P. J., Jacobs, M. S. and Galaburda, A. M., 1986a, Evolutionary morphology of the dolphin brain, in: "Dolphin Cognition and Behavior, A Comparative Approach", Schusterman R, Woods F., Thomas J., eds. L. Erlbaum Associates, Hillsdale, New Jersey, 5-29.

Morgane, P. J., Jacobs, M. S., Galaburda, A. M., 1986b, Evolutionary aspects of cortical organization in the dolphin brain, in: "Research on Dolphins", R.J. Harrison and M Bryden, eds., Oxford University Press, Oxford, pp. 71-98.

Morgane, P. J., Glezer, I. I., Jacobs, M. S., 1988, Visual cortex of the dolphin: an image analysis study. J. Comp. Neurol., 273:3-25.

Morgane, P. J., Glezer, I. I., Jacobs, M. S., 1990, in press, Comparative and evolutionary anatomy of visual cortex of dolphin, in: "Cerebral Cortex, Vol. 8. Evolution and Comparative Anatomy of Cerebral Cortex." E.G. Jones and A. Peters, eds. Plenum Press, New York.

Morgane, P. J., Glezer, I. I., 1990, Sensory neocortex in dolphin brain, in: "Sensory Abilities of Cetaceans". J. A. Thomas and R. A. Kastelein, eds. Plenum Press, New York.

Morrison, J. and Magistrelli, P. J., 1983, Monoamines and peptides in cerebral cortex. Contrasting principles of cortical organization. Trends in Neurosci. 4:146-151.

Morrison, J. H., Molliver, M .E. and Grzanna, R. R., 1979, Noradrenergic innervation of the cerebral cortex: widespread effects of local cortical lesions, Science, 205:313-316.

O'Kusky, J., Colonnier, M., 1982, A laminar analysis of the number of neurons, glia and synapses in the visual cortex (area 17) of adult macaque monkeys, J. Comp. Neurol., 210:278-290.

Olschowka, J. A., Molliver, M. E. and Grzanna, R. R., 1981, Ultrastructural demonstration of noradrenergic synapses in the rat central nervous system by dopamine-hydroxylase immunocytochemistry, J. Histochem. Cytochem., 29:271-280.

Parnavelas, J. G. and Papadopoulos, G. C., 1989, The monoaminergic innervation of the cerebral cortex is not diffuse and nonspecific, Trends in Neurosci., 12:315-319.

Peters, A. and Kimerer, L. M., 1981, Bipolar neurons in rat visual cortex: a combined Golgi-electron microscopic study. J.Neurocyt., 16:23-38.

Pickel, V. M., Joh T. H. and Reis D.J., 1975, Ultrastructural localization of tyrosine hydroxilase in noradrenergic neurons of brain, Proc. Natl. Acad. Sci. USA, 72:659-663.

Poliakov, G. I., 1958, Some characteristics of neuronal structure complexity in the cerebral cortex of man, monkey and other mammals. Soviet Anthropology, 2:69-85.

Rakic, P., 1974, Neurons in rhesus monkey visual cortex: Systematic relation between time of origin and eventual disposition, Science, 183:425-427.

Ramón y Cajal, S., 1911, Histologie du Système Nerveux de l'Homme et des Vertébrès, vol II. N. Maloine, Paris.

Sanides, F. and Sanides, D., 1974, The "extraverted neurons" of the mammalian cerebral cortex. Z. Anat. Entw. Gesch., 136:272-293.

Shaw, C., Wilkinson, M., Cynader, M., Needler, M. C., Aoki C. and Hall, S.E., 1986, The Laminar distributions and postnatal development of neurotransmitter and neuromodulator receptors in cat visual cortex, Brain Res. Bull., 16:661-671.

Sokolov, V.E, Ladygina, T.F. and Supin, A. Ya., 1972, Localization of sensory zones in dolphin brain cortex, Dokl. Akad. Nauk SSSR, 202:490-493.

Supin, A. Ya., Mukhametov, L. M., Ladygina, T. F., Popov, V. V., Mass, A. M., and Poliakova, I. G., 1978, Electrophysiological Study of the Dolphin Brain, Nauka, Moscow, 29-85.

Valverde, F., De Carlos J. A., López-Mascaraque L. and Donate-Oliver F., 1986, Neocortical layers I and II of the hedgehog (Erinaceus europaeus). II. Thalamo-cortical connections, Anat. Embryol., 175:167-179.

EVOLUTION OF THE NASAL ANATOMY OF CETACEANS

John E. Heyning and James G. Mead [1]

Natural History Museum of Los Angeles County,
900 Exposition Blvd., Los Angeles, CA. 90007, USA

[1] National Museum of Natural History, Washington, D.C. 20560, USA

INTRODUCTION

The nasal morphology of extant cetaceans has evolved significantly from the typical architecture of terrestrial mammals. The need to occlude the nasal passages from water as these animals dived provided strong selection for such changes in morphology. The position of the external nares has shifted posteriorly to a more dorsal position on the head to facilitate respiration with a minimum of head movement as the animal surfaces. The telescoping of the cranial bones as described by Miller (1923) resulted primarily by the repositioning of the nares (Raven and Gregory, 1933). In addition to these modifications of a purely respiratory nature, odontocetes have further evolved a complex series of nasal diverticula superficial to the cranium that are involved with sound production. Based on our dissections of numerous cetaceans (Mead, 1975; Heyning, 1989), we attempt to provide an overview of the salient points in the evolution of the nasal morphology and speculate about some of the selection pressures that might have created the present diversity of nasal morphologies.

In order to establish the evolutionary history of nasal morphology, we use here a previously published cladistic analysis of the relationships among the families of extant odontocetes (Fig. 1; Heyning, 1989). Based on this analysis, Platanista gangetica is placed in the family Platanistidae, and the remaining extant genera of "river" dolphins, Inia, Lipotes, and Pontoporia, are placed within the family Iniidae. We also work on the strongly founded assumption that the two extant cetacean suborders, the Odontoceti and Mysticeti, are monophyletic (Van Valen, 1968; Árnason, 1969, 1974; Árnason et al., 1984; Duffield-Kulu, 1972; Barnes and Mitchell, 1978; Fordyce,1980; De Jong, 1982; Goodman et al., 1982), and that both suborders evolved from the extinct paraphyletic suborder Archaeoceti (Barnes and Mitchell, 1978). Based on this phylogeny, we assume that structures shared by two groups evolved in the common ancestor to both groups, the archaeocetes. Thus, by examining which features are shared by both mysticetes and odontocetes, we can speculate that these structures evolved in at least the advanced archaeocetes that gave rise to the two modern suborders.

GENERAL CETACEAN NASAL MORPHOLOGY

The nasal anatomy of mysticetes is not well documented in the literature. A few short descriptions are available for Balaena mysticetus (Eschricht and Reinhardt, 1866), Balaenoptera acutorostrata (Carte and MacAlister, 1869), B. musculus (Kükenthal, 1893), and B. borealis (Kernan, 1916). Lawrence and Schevill (1956) described briefly the nasal plugs and nasal plug muscles in a specimen of Balaenoptera acutorostrata. Based on our dissections of specimens representing

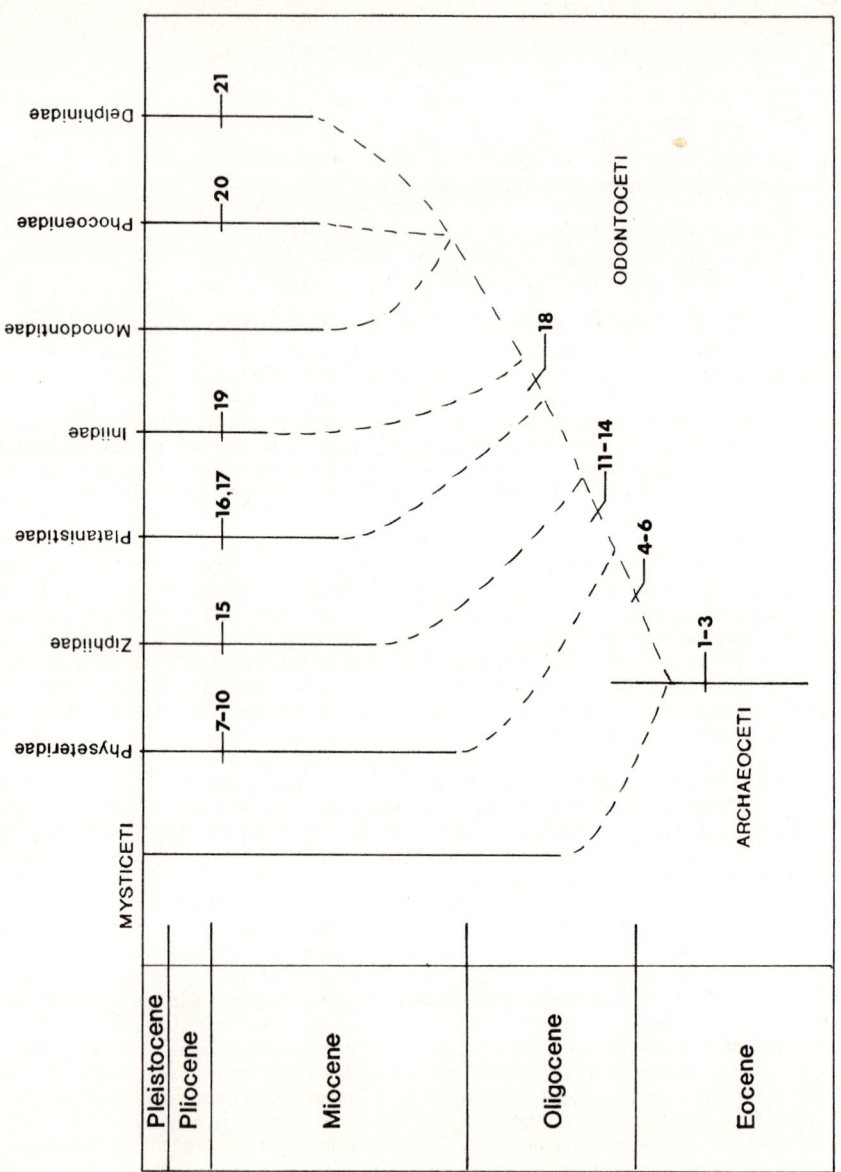

Fig. 1. Phylogeny of extant cetaceans based on Heyning (1989) and Barnes and Mitchell (1978). Solid lines indicate fossil record and dashed lines indicate suggested relationships. The evolution of facial structures within lineages are as follows: (1) nasal plugs, (2) nasal plug muscles, (3) melon, (4) facial asymmetry, (5) single blowhole, (6) nasal bones do not roof over bony nares, (7) spermaceti organ, (8) frontal sac, (9) distal sac, (10) museau de singe, (11) nasal passages confluent, (12) premaxillary sacs, (13) blowhole ligament, (14) inferior vestibule/nasofrontal sac/posterior nasal sacs, (15) loss of posterior aspect of nasofrontal sac in most species, (16) nasofrontal sacs curve dorsally, (17) pneumatisized maxillary crests, (18) vestibular sac, (19) hypertrophy of vestibular sac, (20) rigid and folded floor of vestibular sac, and (21) loss of posterior nasal sac.

three of the four mysticete families (Balaenidae, Eschrichtiidae, and Balaenopteridae) we found they posses well-developed nasal plugs that occlude the nasal passages distal to the bony nares. These fleshy nasal plugs consist of connective tissue, adipose tissue, and a well-developed nasal plug muscle. Relative to odontocetes, the nasal plugs of mysticetes contain less fatty tissue, but are otherwise similar in general morphology.

In the relaxed condition, the nasal plugs occlude the narial passage in all cetaceans. During respiration, the nasal plugs are retracted by the bilaterally paired nasal plug muscles. These muscles originate primarily from the premaxillae and have fibers that are directed posteromedially into the nasal plugs. Contraction of the nasal plug muscles pulls the nasal plugs anterolaterally. In mysticetes, there are cup-shaped cartilages that originate from the skull along the posteromedial wall of the narial passage. These cartilages form a rigid posterior and medial wall to the narial passage so that when the nasal plugs are retracted, the narial passage remains fully open during exhalation and inhalation. We think these cartilages are homologous to the nasal septal and alar cartilages of terrestrial mammals.

The construction of the cranium in the narial region of advanced archaeocetes (see Kellogg, 1936), mysticetes, and primitive odontocetes is comparable. In all these groups the primitive condition exists of the nasal bones roofing over the bony nares. Dorsally, the premaxillae form the medial components of the rostrum and their dorsal surface is slightly to extremely convex. In the narial region, the premaxillae splay laterally and the dorsal surface shifts to a horizontal plane, which forms part of the lateral wall of the bony nares. This change in bony architecture is to accommodate the nasal passages. It is along the anterolateral wall of the premaxilla in the bony nares that the nasal plug muscles originate in all mysticetes. Although the bony nares of archaeocetes are situated more anteriorly on the rostrum, the general morphology of the premaxillae are similar, which indicates that archaeocetes may have developed nasal plugs and nasal plug muscles similar to the general pattern seen in living mysticetes (Fig. 2). Thus, by at least the end of the Eocene, it is probable that the basic pattern of cetacean nasal morphology had evolved. This stage included the shifting of the nares posteriorly and the occluding of the nasal passages with fleshy nasal plugs that are retracted by specialized nasal plug muscles. All these developments are related entirely to respiration in these fully aquatic mammals.

Contrary to statements previously published (Heyning, 1989), mysticetes posses a fatty structure just anterior to the nasal passages that appear to be homologous to the melon of odontocetes. We have found this structure to be well-developed in minke whales (Balaenoptera acutorostrata), right whales (Eubaena glacialis, Fig. 3), and gray whales (Eshrichtius robustus). The melon in these specimens is rather small compared to that of odontocetes, but is of the same consistency of adipose tissue and located in exactly the same position as the melon of toothed whales. Because this unique structure is found in both mysticetes and odontocetes, it provides additional evidence for the monophyly of cetaceans and suggests that the melon first evolved in archaeocetes.

ODONTOCETE NASAL MORPHOLOGY

All modern odontocetes have a complex series of nasal diverticula and

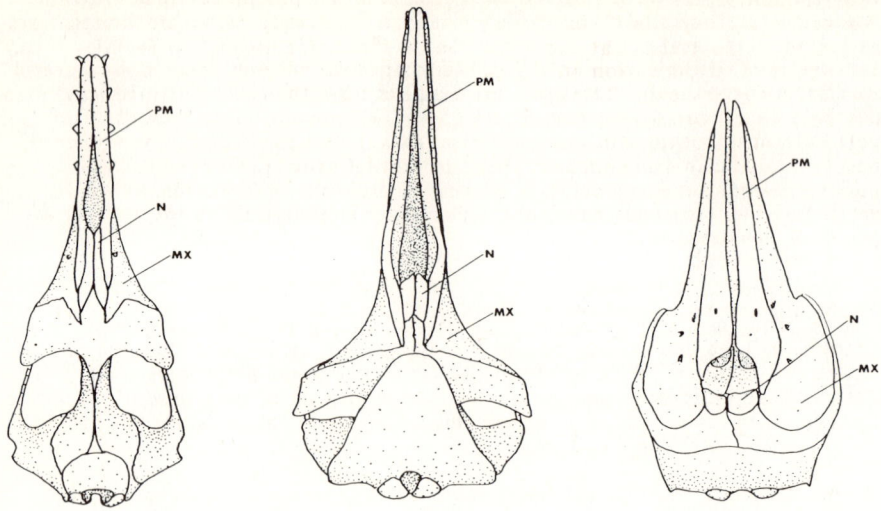

Fig. 2. Skulls of cetaceans from left to right: an archaeocete (<u>Zygorhiza kochii</u>), a mysticete (<u>Balaena mysticetus</u>), and an odontocete (<u>Tursiops truncatus</u>). Note positions of the nares and nasal bones in all three suborders (not to scale). N= nasal bones, PM= premaxilla, MX= maxilla.

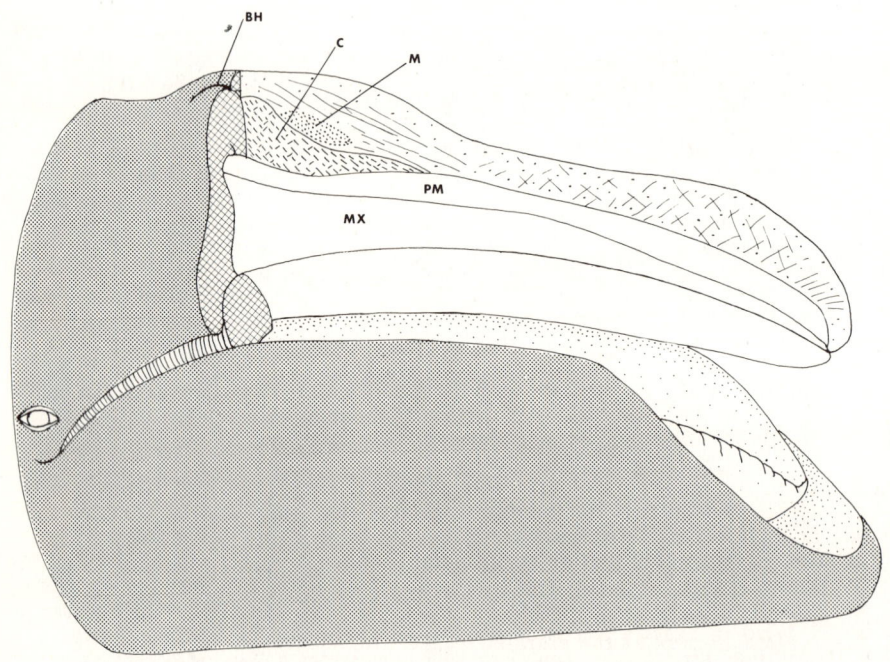

Fig. 3. Transverse section through the rostral soft anatomy of a neonate right whale, <u>Eubalaena glacialis</u>. Note small melon just anterior to the nares. BH= blowhole, C= cartilage, M= melon, MX= maxilla, PM= premaxilla.

associated musculature (Lawrence and Schevill, 1956; Mead, 1975, Heyning, 1989). Because only odontocetes have evolved these complex morphological structures and only odontocetes are capable of producing the high-frequency clicks associated with echolocation, it is theorized that the evolution of these structures is correlated with the development of echolocation in the common ancestor of the modern odontocetes (see Mead, 1975 for review). The epicranial region near the bony nares has been proven to be a sound-producing source in at least one species of delphinid (Ridgway et al., 1980; Amundin and Andersen, 1983), one phocoenid (Amundin and Andersen, 1983), and one monodontid (Ridgway and Carder, 1988). Because the general pattern of nasal morphology is similar in ziphiids, platanistids, and iniids, it would be logical to assume that sound production is generated in the narial region in these groups as well.

The nasal pattern of physeterids is quite complex but not easily comparable to the morphology of other odontocetes. Most researchers have suggested that the epicranial nasal morphology of physeterids also is related to sound production (Norris and Harvey, 1972). An alternative hypotheses is that the spermaceti organ and the nasal passages function in buoyancy control (Clarke, 1970, 1978, 1979). Schenkkan and Purves (1973) argued that the small size of the spermaceti organ in Kogia spp. would not function well as a buoyancy control organ. Ridgway (1971) pointed out that at depth, the lungs of sperm whales should collapse and thus make the animal negatively buoyant, negating Clarke's argument that the spermaceti oil can increase the density of the positively buoyant whale and allow it to be neutrally buoyant. In conclusion, the most evidence marshalled suggests that the nasal morphology of physeterids is a sound generating structure similar to all other modern odontocetes.

In the early evolutionary radiation of odontocetes, species of the Oligocene genera Agorophius, Xenorophus, and Archaeodelphis exhibit, at least to some degree, the primitive condition of having the narial passage partially roofed over by the nasal bones. In all more advanced odontcetes, both fossil and modern, the nasal bones become extremely reduced, knob-like, and do not roof over the narial passage.

The two most distinctive morphological changes in the evolution of odontocetes were the coalescing of the two external nares into a single blowhole and the evolution of asymmetrical soft anatomy surrounding the narial region. It is unclear what advantage there is to having a single external naris and, thus, we cannot speculate as to its evolution.

Asymmetry

The hypertrophy of facial structures and nasal diverticula on the right side of all modern odontocetes has been the subject of some speculation. Earlier in this century, researchers suggested that the asymmetry was due to slightly different functions between the right and left nares (Raven and Gregory, 1933). Asymmetry is correlated with the morphological region associated with at least some sound production in odontocetes. It has been suggested that this asymmetry functions somehow with echolocation sound production, perhaps by the right side more specialized for producing the clicks of echolocation (Mead, 1975). Heyning (1989) hypothesized that there may have been selection for a single sound generating source in the narial region in order to avoid interference generated from two sound sources. Thus, the right side enlarged as it became the primary sound generating location. In any event, because the asymmetry always involves the enlargement of structures on the right side, it is most parsimonious to suggest that it evolved only once in the common ancestor to all modern odontocetes.

The degree of asymmetry in the cranium appears to be correlated with the degree of elevation present at the cranium's vertex. That is, those species with high cranial vertices, such as sperm whales and beaked whales, tend to have the most asymmetrical craniums. We suggest that this is because the functional component of asymmetry pertains to the facial soft anatomy. Therefore, a high cranial vertex has been thrust, though evolution, into a region of soft anatomy that is asymmetrical, thus modifying the hard structures (Fig. 4). For example, within the living Ziphiidae, the genus Berardius has species with the lowest vertices and exhibits the least degree of cranial asymmetry. Members of the genera Mesoplodon

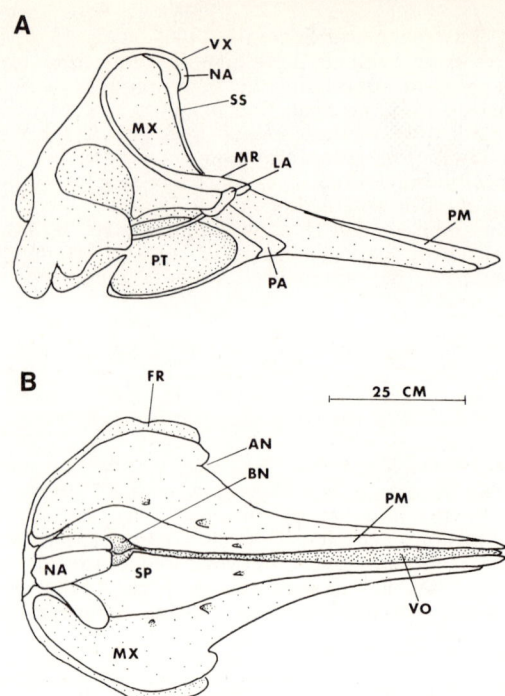

Fig. 4. Lateral view (A) and dorsal view (B) of the skull of an adult female beaked whale, Ziphius cavirostris. Note the highly elevated vertex and extreme asymmetry in this species. AN= antorbital notch, BN= bony nares, LA= lacrimal, MR= maxillary ridge, MX= maxilla, NA= nasal, PA= palatine, PM= premaxilla, PT= pterygoid, SP= spiracular plate of premaxilla, SS= spiracular surface of vertex, VO= vomer, VX= vertex. (From Heyning, 1989)

and Hyperoodon have moderately high vertices and moderate cranial asymmetry. Ziphius cavirostris exhibits the extremes within the family of both vertex height and cranial asymmetry. This phenomenon would explain the condition found in the river dolphin Pontoporia blainvillei in which the cranium is almost symmetrical, yet the facial soft anatomy is extremely asymmetrical (Mead, 1975; Heyning, 1989). Pontoporia blainvillei has a cranium that is the least elevated among modern odontocetes. Thus, a secondarily derived condition for the modern species P. blainvillei of a reduced vertex would result in a cranium that has secondarily lost most of its asymmetry. We believe this explanation is more parsimonious than other explanations of cranial asymmetry arising independently numerous times within the Odontoceti (Barnes, 1978, 1984, 1985).

Physeteridae

The Physeteridae consist of two modern genera, Kogia and Physeter. Physeterids are noted for several unique features of there nasal anatomy. One is the presence of the spermaceti organ, an structure consisting of an oily core surrounded by a tough connective tissue sac. The spermaceti organ is situated posterodorsal to the nasal passages and thus is not homologous to the melon in other odontocetes (Schenkkan and Purves, 1973). The spermaceti "junk" of P. catodon is homologous with the melon of Kogia spp. and all other odontocetes.

Physeterids are also unique among extant odontocetes in that the nasal passages remain as separate tubes to just deep of the blowhole (Fig. 5). As separate nasal passages are the condition found in mysticetes and all other mammals, this condition found in physeterids has been interpreted as primitive among living odontocetes (Heyning, 1989). Although primitive in that aspect, the physeterids are

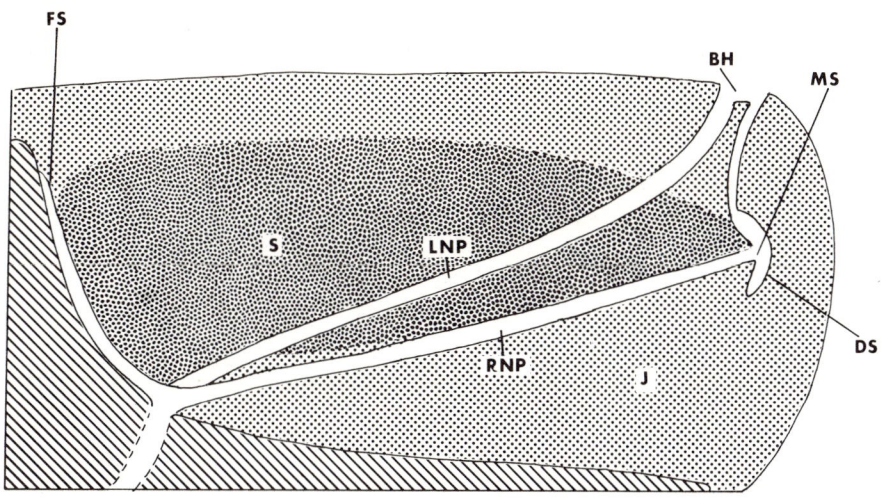

Fig. 5. Parasagittal diagram of the nasal morphology of the sperm whale, Physeter catodon. Note that the nasal passages remain separate until the blowhole. BH= blowhole, DS= distal sac, FS= frontal sac, J= junk, LNP= left nasal passage, MS= museau de singe, RNP= right nasal passage, S= spermaceti organ. (from Heyning, 1989)

among the most highly specialized of cetaceans.

The right nasal passage of sperm whales has two well developed diverticula. The deepest one originates off the nasal passage just distal to the bony nares and forms a saucer-shaped sac at the posterior surface of the spermaceti organ. It has been described by numerous terms, but is most widely called the frontal sac. The superficial diverticula of physeterids is termed the distal sac. The left nasal passage either is devoid of diverticula (Physeter) or only has a small distal diverticulum (Kogia spp.).

Ziphiidae

The remaining odontocete groups, the Ziphiidae, Platanistidae, Iniidae, and the Delphinoidea, are similar in their nasal morphology (Fig. 6). In all these taxa, the nasal passages become confluent just distal to the bony nares, a derived feature. The premaxillary sacs are well-developed and rather conservative in form within this grouping. A blowhole ligament is also present; often with a cartilaginous section, and the development of the nasal diverticula complex (inferior vestibule/nasofrontal sacs/posterior nasal sacs) found posterior to the blowhole ligament. The primitive condition of these diverticula is paired, bilateral openings along the posterior wall of the nasal passages just dorsal to the bony nares. The lower aspect of these diverticula are termed the inferior vestibules. Dorsally, these transversely oriented sacs are divided into an anterior and posterior compartment by a transverse, fleshy fold of tissue termed the hintere klappe (Kukenthal, 1893). The sac anterior to this tissue-fold is the posterior aspect of the nasofrontal sac and the posterior compartment is the posterior nasal sac. In most species, there is an anterior aspect of the nasofrontal sac which gives these sacs a U-shape with medially directed apices when viewed from above (Fig. 6).

The most notable feature of the nasal morphology of ziphiids is the wide nasal vestibule leading to the blowhole and the lack the most superficial diverticula of delphinoids, the vestibular sac (Fig. 6). In all other features, the morphology of the nasal passages and associated soft structures is quite similar to delphinoids. Berardius spp. and Hyperoodon spp. retain the primitive condition of having both a posterior nasal sac and a posterior aspect of the nasofrontal sac separated by the hintere klappe. In Ziphius cavirostris and all species of Mesoplodon we have dissected, the posterior aspect of the nasofrontal sac is lost and the posterior nasal sac is enlarged.

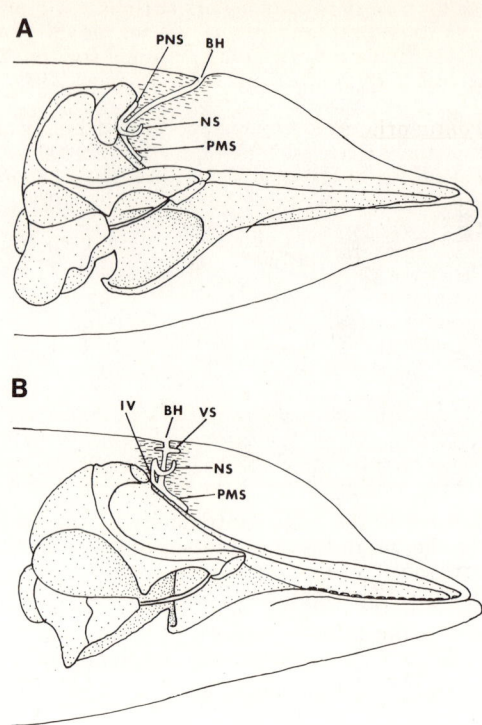

Fig. 6. Diagrammatic view of the nasal passages and diverticula of A, Mesoplodon carlhubbsi, and B, Tursiops truncatus (not to scale). BH= blowhole, IV= inferior vestibule, NS= nasofrontal sac, PMS= premaxillary sac, PNS= posterior nasal sac. (from Heyning, 1989)

The secondary loss of anterior aspect of the nasofrontal sac is distributed widely within the Ziphiidae. In M. carlhubbsi, M. bidens, and M. stejnegeri, the anterior components of the nasofrontal sacs are present, but the right sac is significantly larger than the left sac, especially in M. stejnegeri. In M. densirostris, Berardius bairdii, Ziphius cavirostris, and Hyperoodon ampullatus, the entire anterior portions of the nasofrontal sacs have been secondary lost on both sides.

Platanistidae

River dolphins in the genus Platanista are the only living representative of the family Platanistidae. The most remarkable aspect of the facial anatomy of Platanista spp. are the greatly expanded maxillary crests that are directed anteromedially. These crests are lined on their inner surface with a thin air sac derived from the middle ear air sinus system (Fraser and Purves, 1960). Thus, these pneumatised structures are not derived from any nasal passage diverticula, yet may have a similar acoustic function as the vestibular sacs of iniids and delphinoids.

Platanista spp. also lacks a true vestibular sac and has a longitudinally oriented blowhole and nasal vestibule. The anterior aspects of the nasofrontal sacs are reported to curve dorsally (Purves and Pilleri, 1973).

Iniidae

The remaining odontocete groups, the Iniidae and Delphinoidea, have evolved the vestibular sac. This most distal of the nasal diverticula is a broad, flat sac just deep to the blowhole. Experimental evidence indicates that at least one function of this sac is as a terminal air reservoir during phonations. Air passing by the nasal plugs is stored in the vestibular sac and recycled back down at the end of phonating (Norris et al., 1971; Dormer, 1979). Another suggested function of this sac is as an acoustic reflector (Norris et al., 1971; Dormer, 1979).

The genera Inia, Pontoporia, and Lipotes are united by a couple of anatomical features, one of which is the extreme hypertrophy of the vestibular sac (Heyning, 1989). This enlargement of the vestibular sac is manifested more extensively on the right side.

Delphinoidea

The Delphinoidea contains the extant families Delphinidae, Phocoenidae, and Monodontidae. This group contains the majority of living species and virtually all the animals upon which experimental work has been done on sound production and reception.

Within this group, the nasal anatomy is rather conservative. In Phocoenids, the vestibular sac is derived from the primitive state in that the floor of the sac is rather rigid and thrown into transverse folds. The lumen of the vestibular sac is also unique. Instead of the typical delphinoid pattern of a broad, unpaired opening from the nasal vestibule, small, bilaterally-paired apertures are present in phocoenids.

In the Delphinidae, the posterior nasal sac is lacking (Heyning, 1989) and, thus, the inferior vestibule terminates into the well formed, U-shaped nasofrontal sac. The general pattern within the Delphinidae varies regarding the amount of melon tissue, muscle masses, and to a lesser degree the development of the nasal diverticula (Mead, 1975). The only exception is that in Grampus griseus the anterior aspect of the nasofrontal sac is lacking on the left side.

FUNCTION

The repositioning of the external nares posteriorly on the head is an obvious advantage for respiration during swimming. This change of position of the nostrils occurred primarily during the Eocene, early in cetacean evolution. Most advanced archaeocetes have bony nares that are situated about two-thirds of the snout posterior to the tip. Primitive mysticetes such the Late Oligocene Aetiocetus cotylalveus Emlong, 1966 and primitive odontocetes such as Archaeodelphis patrius Allen, 1921 have bony nares that are situated somewhat more posterior on the cranium to that of their presumed archaeocete ancestors.

The melon of odontocetes has been suggested to function as a type of acoustical lens by Norris (1964) and numerous anatomical and experimental studies have been undertaken to test this hypothesis (see Mead, 1975 for review; Cranford, 1988; and Amundin and Cranford, this volume). Mysticetes do not appear to make high frequency sounds associated with echolocation nor posses the nasal diverticula that odontocetes use to produce such high frequency sounds. Yet, mysticetes posses a small melon. We believe that the original function of the melon in cetaceans was to allow the free movement of the nasal plugs as they are retracted by the nasal plug muscles during respiration. Such a fatty body allows the muscles to contract with a minimal amount of resistance and provides for a smooth exterior contour to the head. Fatty deposits are common around joints for exactly this function. This would explain the presence of this fat body within mysticetes. Early in Odontocete evolution, this fatty body could have hypertrophied as it obtained a secondary function related to sound production.

With the exception of the Physeteridae, all modern odontocetes have well-formed premaxillary sacs that are extremely conservative in morphology. In Physeterids and mysticetes, the tissue separating the nasal plugs from the premaxillae consists of loose epithelium and connective tissue which allows for rather free movement of the nasal plugs as they are drawn posteriorly over the premaxillae during the contraction of the nasal plug muscles. We suggest that the original or primary function of the premaxillary sacs is to provide for an even freer movement of the nasal plugs over the underlying premaxillae. It also is

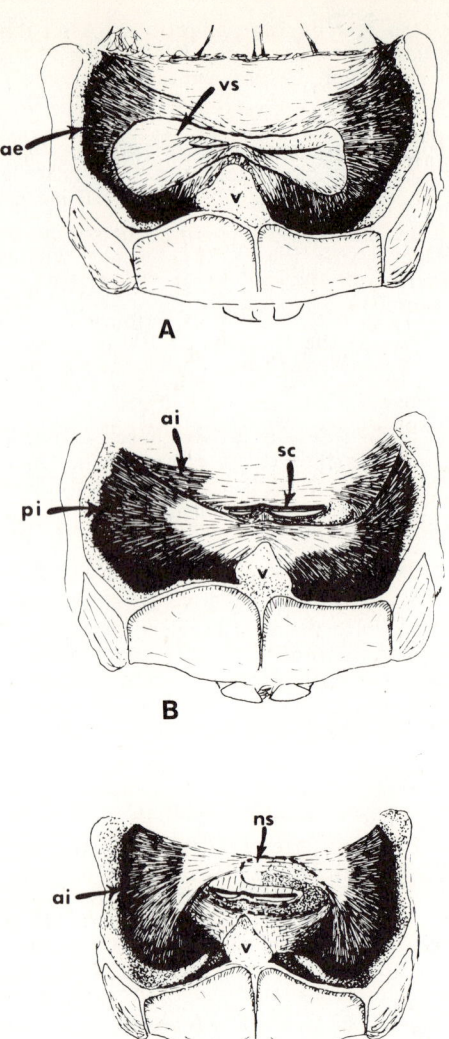

Fig. 7. Facial musculature associated with the narial region of the delphinid, <u>Grampus griseus</u>. ae= pars anteroexternus, ai= pars anterointernus,, ns= nasofrontal sac, pi pars posterointernus, sc=spiracular cavity, v= vertex, vs= vestibular sac. (from Mead, 1975)

possible or probable that the premaxillary sacs function as acoustical reflectors. Experimental evidence indicates that the premaxillary sac at least partially inflates during the first phase of sound production (Dormer, 1979).

The nasofrontal sac is perhaps the most complicated of all the nasal diverticula. These bilaterally-paired U-shaped diverticula encircle the nasal vestibule in most delphinoids. This morphology prompted Lawrence and Schevill (1956) to suggest that inflating these sacs would provide additional pressure on the vestibule in order to form a tighter seal against water. Although this seems reasonable in most delphinoids, it does not explain the lack of the anterior aspect of this sac on the left side in the delphinid <u>Grampus griseus</u>, or the bilateral loss of the anterior aspect of this sac in several ziphiids and iniids. One would assume that if it had a significant function in providing a better seal that all species would have it, especially deep diving species such as ziphiids. It is clear that more experimental evidence is needed to fully resolve the function of the nasal anatomy of odontocetes.

During sound production in toothed whales air is passed by the nasal plugs, or in the case of physeterids, passed by the nasal plugs and through the museau de singe. To avoid losing this air during a dive, air must be stored in the respiratory system distal to the site of sound production. Due to the various morphologies, different families of toothed whales probably use different nasal diverticula for this function. In Iniids and delphinoids, air is stored in the vestibular sac (Norris et al., 1971; Dormer, 1979; Mackay and Liaw, 1981). Sperm whales could store air in their large distal sacs (Norris and Harvey, 1972). Beaked whales probably store air in their posterior nasal sacs (Heyning, 1989).

The musculature associated with the narial region is large and complex (Lawrence and Schevill, 1956; Mead, 1975; Heyning, 1989). Several layers of muscles originate from the broad facial fossa and insert onto the nasal passages and diverticula (Fig. 7). The complex nature of the numerous layers and intricate insertions along the narial airways suggests that odontocetes have extremely good muscular control of the passages and diverticula. The relatively large muscle masses situated more anteriorly on the cranium must be involved with controlling the shape of the melon, perhaps assisting with sound beam formation.

Lawrence and Schevill (1956) found a small intrinsic muscle surrounding the nasofrontal sacs in delphinids. Phocoenids have a unique intrinsic muscle to their highly evolved vestibular sacs (Mead, 1975). It is probable that these intrinsic muscles may provide additional fine control to airflow within these sacs.

CONCLUSIONS

Mysticetes have structures that are homologous to the nasal plug, nasal plug muscle, and melon of odontocetes. These structures function in the opening and closing of the nasal passage during respiration. Odontocetes have evolved further specializations in the form of nasal diverticula and the hypertrophy of the melon as structures related to sound propagation. The condition seen in mysticetes is primitive and was probably evolved in archaeocetes during the Eocene. Cranial and soft anatomy asymmetry is present only in odontocetes and its function is probably related to sound production.

The physeterids are specialized in the development of a spermaceti organ and are unique among extant odontocetes in that the nasal passages remain separate to just deep to the blowhole. The remaining odontocete groups share a similar pattern of complex nasal diverticula with some components lacking in certain lineages. The homologies of these sacs to those found in physeterids is hypothetical.

ACKNOWLEDGEMENTS

We thank L. G. Barnes for his comments on an early version of this manuscript.

REFERENCES

Amundin, M., and Andersen, S. H. 1983, Bony nares air pressure and nasal plug muscle activity during click production in the harbour porpoise, Phocoena phocoena, and the bottlenose dolphin, Tursiops truncatus. J. Exper. Biol. 105:275-282.

Amundin, M., and Cranford, T. (this volume) Forehead anatomy of Phocoena phocoena and Cephalorhynchus commersonii; 3-dimensional computer reconstruction with emphysis on the nasal diverticula. in: "Sensory Abilities of Cetaceans", J. Thomas and R. Kastelein, eds, Plenum Press, New York.

Árnason, U., 1969, The karyotype of the fin whale, Hereditas 62:273-284.

Árnason, U., 1974, Comparative chromosome studies in Cetacea, Hereditas 77:1-36.

Árnason, U., Hoglund, M. and Widegren, B. 1984, Conservation of highly repetitive DNA in cetaceans, Chromosoma 89:238-242.

Barnes, L. G., 1978, A review of Lophocetus and Liolithax and their relationship to the delphinoid family Kentriodontidae (Cetacea: Odontoceti), Nat. Hist. Mus. Los Angeles Co., Sci. Bull. 28:1-35.

Barnes, L. G., 1984, Fossil odontocetes (Mammalia:Cetacea) from the Almejas Formation, Isla Cedros, Mexico, PaleoBios 42: 42 pp.

Barnes, L. G., 1985, Fossil pontoporiid dolphins (Mammalia:Cetacea) from the Pacific coast of North America. Contr. in Sci. 363: 34 pp. Nat. Hist. Mus., Los Angeles Co.

Barnes, L. G., and Mitchell, E. D. 1978, Cetacea, pp. 582-602, in: "Evolution of African Mammals", V.J. Maglio and H.B.S Cooke, eds., Harvard Univ. Press, Cambridge.

Carte, A., and MacAlister, A. 1869, On the anatomy of Balaenoptera rostrata. Phil. Trans. Roy. Soc. Lond. 158:201-261 + pls IV-VII.

Clarke, M. C., 1970, Function of the spermaceti organ of the sperm whale, Nature 228:873-874.

Clarke, M. C., 1978, Buoyancy control as a function of the spermaceti organ in the sperm whale, J. Mar. Biol. Ass. U. K. 68:27-71.

Clarke, M. C., 1979, The head of the sperm whale, Sci. Am. 240(1):128-132, 134, 136-141.

Cranford, T. W., 1988, The anatomy of acoustic structures in the spinner dolphin forehead as shown by X-ray computed tomography and computer graphics. pp. 67-77 in: "Animal Sonar: Processes and Performance", P.E. Nachtigall and P. W. B. Moore, eds., Plenum Press, New York.

De Jong, W. W., 1982, Eye lens proteins and vertebrate phylogeny. pp. 75-114 in: "Macromolecular Sequences in Systematic and Evolutionary Biology", M. Goodman, ed., Plenum Press, New York.

Dormer, K.J., 1979, Mechanism of sound production and air recycling in delphinids: cineraradiographic evidence, J. Acoust. Soc. Am. 65:229-239.

Duffield-Kulu, D., 1972, Evolution and cytogenetics, pp. 503-527, in: "Mammals of the Sea, Biology and Medicine", S. H. Ridgway, ed., Charles C. Thomas, Springfield, Illinois.

Eschricht, D. F., and Reinhardt, J. 1866, On the Greenland right-whale (Balaena Mysticetus), pp. 3-150 in: "Recent Memoirs on the Cetacea", W. H. Flower, ed. Ray Society, London.

Fordyce, R. E., 1980, Whale evolution and the Oligocene Southern Ocean environment. Palaeogeography, Palaeoclimatology, Palaeoecology, 31:319-336.

Fraser, F. C., and Purves, P. E. 1960, Hearing in cetaceans, Bull., Brit. Mus (Nat. Hist.) Zool. 7, 140 pp. + 53 pls.

Heyning, J. E., 1989, Comparative facial anatomy of beaked whales (Ziphiidae) and a systematic revision among the families of extant Odontoceti, Contr. in Sci. 405: 64 pp. Nat. Hist. Mus. Los Angeles Co.

Kellogg, R., 1936, A review of the Archaeoceti, Carnegie Inst., Washington, Publ. 482, xv + 366pp. + 37 pls.

Kernan, J. D., 1916, The respiratory passages, pp. 435-438, in: "Monographs of the Pacific Cetacea, II.-Anatomy of a foetus of Balaenoptera borealis", H. von W. Schulte, ed.

Kükenthal, W., 1893, Vergleichend-anatomische und entwickelungsgeschichtliche Untersuchungen an Walthieren, Denkschriften der Medicinisch-Naturwissenschaftlichen Gesellschaft zu Jena 3(2):VIII + 223-448 + pls. 14-25.

Lawrence, B. and Schevill, W. E. 1956, The functional anatomy of the delphinid nose, Bull. Mus. Comp. Zool. 114(4): 103-151 + 30 figs.

Mackay, R. S., and Liaw, H. M. 1981, Dolphin vocalization mechanisms, Science 212:676-678.

Mead, J. G., 1975, Anatomy of the external nasal passages and facial complex in the Delphinidae (Mammalia:Cetacea), Smithsonian Contr. Zool. 207, iv + 72 pp.

Miller, G. S., Jr. 1923, The telescoping of the cetacean skull. Smithsonian Misc. Coll. 76:1-71.

Norris, K. S., 1964, Some problems of echolocation in cetaceans, pp. 317-336 in: "Marine Bio-Acoustics", W. N. Tavolga, ed., Pergamon Press, Oxford.

Norris, K. S., Dormer, K. J., Pegg, J., and Liese, G. J. 1971, The mechanism of sound production and air recycling in porpoises: a preliminary report, Proc., Eighth Ann. Conf. Biological Sonar and Diving Mammals, 1971:113-129.

Norris, K., and Harvey, G. W. 1972, A theory for the function of the spermaceti organ of the sperm whale (Physeter catodon), pp. 397-417 in: "Animal Orientation and Navigation", S. R. Galler, K. Schmidt-Koening, G. J. Jacobs, and R. E. Belleville, eds., NASA, Washington, D.C.

Purves, P. E., and Pilleri, G. 1973, Observations on the ear, nose, throat and eye of Platanista indi, Invest. on Cetacea, 5:13-57.

Raven, H. C., and Gregory, W. K. 1933, The spermaceti organ and nasal passages of the sperm whale (Physeter catodon) and other odontocetes, Am. Mus. Novitates, 667:1-18.

Ridgway, S. H., 1971, Buoyancy regulation in deep diving whales, Nature, 232:133-134.

Ridgway, S. H., and Carter, D. A. 1988, Nasal pressure and sound production in an echolocating white whale, Delphinapterus leucas, pp. 53-60, in: "Animal Sonar: Processes and Performance", P. E. Nachtigall and P. W. B. Moore, eds., Plenum Press, New York.

Schenkkan, E. J., 1973, On the comparative anatomy and function of the nasal tract in odontocetes (Mammalia, Cetacea), Bijdragen tot de Dierkunde 43(2):127-159.

Schenkkan, E. J., and Purves, P. E. 1973, The comparative anatomy of the nasal tract and the function of the spermaceti organ in the Physeteridae (Mammalia, Odontoceti), Bijdragen tot de Dierkunde 43(1):93-112.

Van Valen, L., 1968, Monophyly or diphyly in the origin of whales, Evolution 22:37-41.

THREE-DIMENSIONAL RECONSTRUCTIONS OF THE DOLPHIN EAR

Darlene R. Ketten[1] and Douglas Wartzok[2]

[1]Department of Otology and Laryngology
Harvard Medical School, Boston, Massachusetts, USA
[2]Department of Biological Sciences
Purdue University, Fort Wayne, Indiana, USA

INTRODUCTION

The umwelt or perceptual world of odontocetes is largely defined by acoustic cues imperceptible to humans. Like bats, they use ultrasonic frequencies to echolocate. To penetrate this acoustic world, we must use indirect anatomical and psychophysical techniques. While bat research has incorporated anatomy and physiology to describe neural processing of echolocation signals, cetacean research, hampered by practical and legal restrictions, depends largely upon spectral and temporal analyses of emitted sounds coupled with behavioral observations. From these investigations, we have gained considerable information about the psycho-acoustics of dolphin echolocation, but we still know little about the receptor anatomy.

This study is based on an anatomical comparison of inner ear structure in twelve species. A comparative anatomy approach was chosen for two reasons. First, peripheral auditory structures are important components in determining hearing capacities (West, 1986; Stinson, 1983; Zwislocki, 1981; Iurato, 1962). Secondly, it is more feasible to obtain adequately preserved cochlea for most odontocetes than central nervous system tissues. In bats, ultrasonic vocalization spectra and auditory sensitivity are directly related and anatomical correlates for frequency ranges are well documented (Suga, 1983; Long, 1980; Neuweiler, 1980; Pollack, 1980; Bruns; 1976; Sales and Pye, 1974; Hinchcliffe and Pye, 1968; Grinnell, 1963). Electrophysiological recordings and neuroanatomical studies from Tursiops and Stenella show similar structural correlates for emitted signals (Ridgway, 1980; Bullock and Ridgway, 1972; McCormick et al., 1970; Bullock et al., 1968). All odontocetes recorded to date produce ultrasonics in species-specific frequency ranges and are assumed to echolocate (Watkins and Wartzok, 1985; Pilleri, 1983; Popper, 1980; Wood and Evans, 1980; Norris et al., 1961; Kellogg, 1959). Thus, we expect extreme differences in echolocation signals to reflect anatomical differences in the auditory periphery. Three-dimensional reconstructions in this study of odontocete cochlea show structural similarities between bat and odontocete inner ears related to ultrasonic perception and interspecific differences amongst odontocetes which correlate with echolocation ranges. These analyses also show specific adaptations of the dolphin ear for aquatic audition.

METHODS

Any investigation of cetacean sensory systems must consider practical and legal limitations. Live research implies the obvious difficulties of maintaining large aquatic animals and meeting Marine Mammal Protection Act strictures. Acute preparation studies are restricted and require exceptional skills and facilities (Ridgway et al., 1974; Ridgway and McCormick, 1967; Nagel et al., 1964). Access to animals in commercial facilities is limited and many captive animals have been treated with preventative antibiotic regimens which may include ototoxic agents (Montali and Migaki, 1980). Most morphometric studies of cetacean bullae use dehydrated or unpreserved tissues collected days to weeks post-mortem (Norris and Leatherwood, 1981; Fleischer, 1976; Kasuya, 1973; Fraser and Purves, 1960). An approach was needed that obviated the physiological, mechanical, and political problems of live animal research yet would yield more than standard morphometric detail. In contrast to the availability of live animals, substantial numbers of cetaceans are netted or stranded annually in fisheries and well-preserved material is archived worldwide. By developing collection and analysis techniques applicable to these tissues, the data base could be significantly increased in terms of both individuals and total species examined.

Key species were determined prior to collection for this research since distribution of specimens by species, functional diversity, and quality of tissue was crucial. Selection criteria included ultrasonic frequency spectra, taxonomic relationships, habitat, degrees of sociality, and feeding strategies. Sixty-three bullae were obtained through a specimen request survey, of which fifty-five were accepted for complete analyses (Table 1). Key species covered four odontocete families which represent different eras of collateral development and several degrees of specialization for the major evolutionary divisions of extant species (Kasuya, 1973). Habitats ranged from estuarine through sublittoral and pelagic to transitional bathypelagic for deep-diving species. Bullae from opportunistic (non-key) species of exceptional quality were analyzed by the same procedures as key specimens. Measurements from opportunistic specimens were used in family analyses, but, without conspecifics for comparison, were not considered adequate for species analyses. All tissues came from four sources: (1) bullae from fisheries or aboriginal hunts extracted and preserved in situ in buffered formalin; (2) bullae extracted and injected with buffered formalin or glutaraldehyde during necropsy; (3) whole heads or temporal blocks preserved on dry ice and thawed in buffered formalin; (4) perfused, archival animals analyzed only with radiography and returned to the lending facility.

Acoustic data for key and related species are listed in Table 2. A major concern in this study was to find a reliable measure of hearing for each species. Odontocetes have a wide functional range, but underwater auditory selectivity and sensitivity measures are available for very few species. Like most mammals, however, cetaceans produce sounds centered around frequencies at which their hearing is most acute. Lower frequency communication signals, as defined in Popper (1980), vary greatly with individuals and have little or no species-

Table 1. Specimen Distributions

Classification	Common name	Bullae	Left	Right	Male	Female
Suborder ODONTOCETI		57	31	26	19	15
Family DELPHINIDAE		39	21	18	9	13
Delphinus delphis	common dolphin	2	1	1	1	
Feresa attenuata	pygmy killer whale	2	1	1	1	
Globicephala macrorhynchus	short-finned pilot whale	2	1	1	1	
Grampus griseus†	Risso's dolphin	4	2	2	1	1
Lagenorhynchus albirostris†	White-beaked dolphin	3	2	1		3
Stenella attenuata†	spotted dolphin	16	8	8	1	7
Stenella coeruleoalba	striped dolphin	2	2		1	1
Stenella longirostris	long-beaked spinner	2	1	1	1	
Tursiops truncatus†	bottlenose dolphin	6	3	3	2	1
MONODONTIDAE						
Monodon monoceros	narwhal	2	1	1	1	
PHOCOENIDAE						
Phocoena phocoena†	harbour porpoise	8	4	4	2	2
PHYSETERIDAE						
Physeter catodon††	sperm whale	6	4	2	6	
PLATANISTIDAE						
Inia geoffrensis††	Amazonian boutu	2	1	1	1	

† Key species
†† Key specimens examined only with radiography

specifity. Echolocating animals vary pulse repetition rate, interpulse interval, intensity, and click spectra and it is known that captive animals selectively modify echolocation pulses in response to ambient noise (Moore, 1990; Supin and Popov, 1990; Thomas et al., 1988; Popper, 1980; Au et al., 1974; Norris, 1969; Schevill, 1964). Nevertheless, echolocation signals tend to be produced in species-specific frequency ranges. Echolocation signals, from which ultrasonic auditory ranges can be inferred, provide the most consistent comparative acoustic data for all species in this study. Species were categorized into two groups based on the peak spectra of their echolocation signal; i.e., the frequency of maximum energy in a typical, broadband echolocation click. Data from recordings of untrained animals in natural surroundings were used when available. Specimens from species with a peak signal energy located above 100 kHz, Phocoena phocoena and Inia geoffrensis, were designated Group I (Table 2). Those from species with ultrasonic peak spectra below 100 kHz comprised Group II. For some species, little or no acoustic data are available and they are subsequently categorized based on cochlear morphometry.

Table 2. Characteristics of Odontocete Sounds
(Adapted with permission from Popper, 1980)

Acoustic Group and Species	Type of sound	Frequency Range (kHz)	Maximum Energy (kHz)	References
I				
Inia geoffrensis	Click	25-200	100	Norris et al., 1972
			95-105	Kamminga et al., 1989
Phocoena phocoena	Pulse	100-160	110-150	Møhl and Andersen, 1973
II				
Delphinus delphis	Click		4-9	Busnel and Dziedzic, 1966
	Whistle	4-16		Gurevick in Evans, 1973
	Click	0.2-150	30-60	Gurevick in Evans, 1973
Lagenorhynchus obliquidens	Click	0.06-80		Evans, 1973
	Whistle	1-12		Caldwell and Caldwell, 1971
Stenella attenuata	Pulse	to 150		Diercks, 1972
	Whistle			Evans, 1967
Stenella longirostris	Click	.1-160	60	Ketten, 1984
	Pulse	1-160	5-60	Brownlee, 1983
	Whistle	1-20	8-12	Brownlee, 1983
Tursiops truncatus	Click	>octave	53[†]	Diercks et al., 1971
	Click	0.2-150	30-60	Diercks et al., 1971
	Bark	0.2-16		Evans, 1973
	Whistle	4-20		Evans and Prescott, 1962
	Whistle	2-20		Caldwell and Caldwell, 1967
Unknown[††]				
Physeter catodon	Coda	16-30		Watkins and Schevill, 1977
Grampus griseus	Whistle			Watkins and Wartzok, 1985

[†] Au et al. (1974) reported clicks by trained animals in noise with 100-130 kHz spectra.
[††] No wide band recordings are available for these species.

All specimens were screened by one or more of three radiographic techniques: single plane stereo-radiography, digital subtraction, and computerized tomography (CT scanning). The principal advantage of radiography for surveying cochlea is that it provides non-destructive techniques for viewing structures in situ (Fig. 1). CT also provides a numerical data base for quantitative analyses and, through multiplanar image reconstruction, a "dissection" of the cochlea without extraction or decalcification. CT scans were obtained with a Siemens Somatom DR3 in the Johns Hopkins Medical Institutions. The DR3 scanner generates parallel, contiguous transaxial slices of 1 to 8 mm thickness with an optimal resolution of 300μ. Like conventional radiography, CT measures tissue absorption of X-rays. Resolution depends upon the number of exposures, collimators, and detectors; the storage and manipulation capacity of associated computers; and the resolution of the image display. The mammalian cochlea, comprised of bone, soft tissue, and fluid, is an ideal subject for CT examination since optimal CT resolution occurs at interfaces of high and low density tissue (Maue-Dickson et al., 1983; Moran et al., 1983). Densest bone in normal humans measures <2000 H.U.[1], but in cetaceans, the temporal bone may exceed 3000 H.U.. This density allows exceptional cross-sectional imaging of both bony and residual soft tissues in odontocete bullae (Fig. 2). Scan parameters used in this survey were 90-100 kV accelerating voltage, 0.5-0.6 milliamp-seconds (MAS), 720 projections, 1-2 mm slices, 512 matrix high resolution imaging, and a 0.3 m image aperture. Data and images were stored on disk and magnetic tape as raw absorption data, cross-sectional images, and reconstructions.

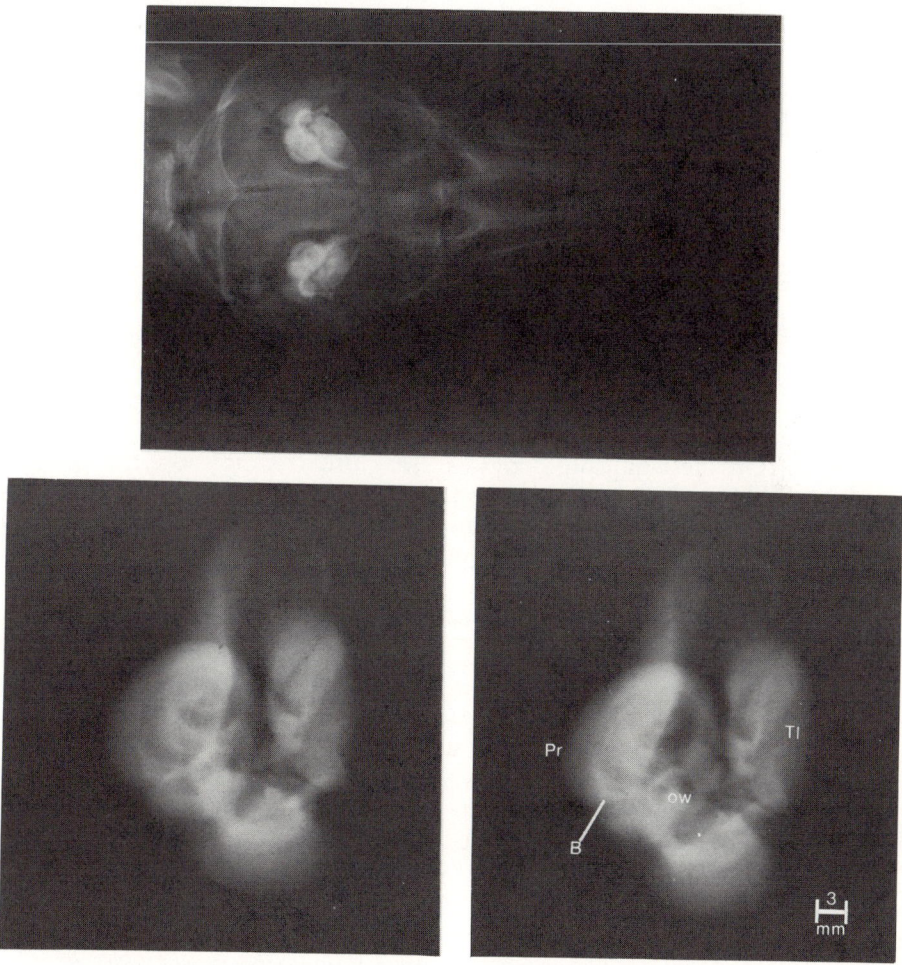

Fig. 1. Stereoradiography of Odontocete Ears. A single plane X-ray of a juvenile <u>Stenella attenuata</u> head and stereo-paired images of the right bulla show the cochlea in the promontorium (Pr) of the periotic medial to the less dense tympanic lobe (Tl). Radiographs image structures in a grayscale proportional to X-ray attenuation, from white for densest material to black for air. The high contrast of the bullae results from their exceptional tissue density. The basal turn of the cochlea (B) can be seen curving posteriorly and medially from the oval window (ow) in each ear.

---

[1]Hounsfield Units, derived from the linear attenuation coefficient of a substance normalized to water, provide a relative measure of tissue density or X-ray absorption characteristics in a range of -1000 (air) to <+4200 (metal).

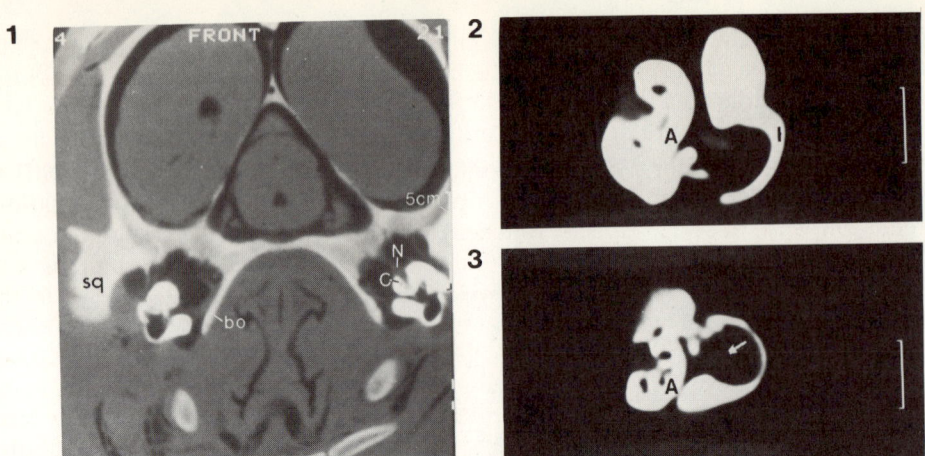

Fig. 2. CT Scans of Cochlear Anatomy. (2.1) A transaxial CT scan of an adult Tursiops truncatus head shows the relationship of the bulla to major cranial structures. The peribullar cavity is ventral to the cerebral hemispheres and is bordered by the ventrolateral process of the basioccipital (bo) and the squamosal (sq) bones. Suspensory ligaments in the cavity are gray as is the auditory nerve (N), which enters the periotic via the internal auditory canal and is flanked by cross-sections of the cochlear spiral (C). The hollow mandible (Ma) is ventral to the tympanic bulla. On the right, high contrast, magnified scans show the cochlea in cross-section in (2.2) Phocoena (1.5 turns) and (2.3) Stenella (2.5 turns). Gray bands of soft tissue in the tympanic cavity (arrow) are folds of the corpus cavernosum. Scale bars represent 1 cm. A apex; l lateral.

After scanning, bullae were extracted, cleaned, weighed, catalogued, and measured. For key species, the periotic was decalcified, embedded in paraffin or celloidin, and examined in 20 μ serial sections. Four methods of decalcification were attempted; a formic acid and formalin-based modification of Schmorl's solution provided the best balance of efficacy vs. distortion. Three to four fiducials were used to gauge processing artifacts and for alignment of reconstructions. Histological procedures are detailed in Ketten (1984).

Cochlear canal midpoints and anatomical contours were digitized from CT images and histological sections to obtain Cartesian triplets (X,Y,Z) for three-dimensional mapping, measurement, and reconstruction of the cochlear duct. Approximately 30 mid-canal triplets, progressing from the stapes to the helicotrema, were used to map the cochlea and to calculate spiral dimensions for each specimen. Contour coordinates were used to reconstruct cochlear duct and bullar components and to calculate exterior surface area and volume. The SAS statistical package was used for univariate and multivariate statistical analyses of spiral and bullar measurements. Left and right bullae were treated as individual entries to test for asymmetry. Statistical analyses were performed on both raw data and on values normalized by animal length for interspecific comparisons.

RESULTS

Bullar Morphometry

The bulla or temporal bone of odontocetes is distinctive and dense. It differs from terrestrial mammalian bullae in appearance, construction, location, orientation, and, in some aspects, function. It is not fused to the skull as in other mammals but is suspended by ligaments in a peribullar cavity with the long axis of the tympanic angled ventromedially (Figs. 1,2). The periotic is dorsal to the tympanic and the shorter, vertical bullar axis is rotated medially 15° to 20°. The acousto-vestibular (VIIIth) nerve projects inward from the dorsomedial edge of the periotic, traverses the retro-peribullar space, and enters a dense, bony canal. The periotic is relatively uniform in thickness, composed of compact bone, and encloses the cochlea, vestibule, and the residual components of the vestibular apparatus. The tympanic has a thickened posterior; a thin, friable body; and a narrow anterior process. The concha or tympanic shell is lined with a membranous corpus cavernosum and contains the ossicular chain and a partially ossified tympanic conus. In all species except Physeter, a band of fibrous tissue, analogous to the stylo-hyoid ligaments, joins the posterolateral edge of the bulla to the posterior margin of the mandibular ramus and stylo-basihyoid complex. In all whole heads, the right bulla was located anteriorly to the left.

Surface anatomy is resilient in dehydrated specimens and has been carefully assessed in other studies (Oelschlager, 1990, 1986; Kasuya, 1973; Reysenbach de Haan, 1956). All surface measurements in this survey (Table 3) are consistent with previous results and are strongly correlated with animal size ($r \geq 0.9$). Linear discriminant analyses redistributed surface data by species with the smallest squared distances amongst delphinids (1400-32000) and the largest between phocoenids and delphinids (110000-156000). T-tests on normalized data showed no significant differences between Groups I and II. Schematic surface reconstructions (Fig. 3) revealed no clear-cut group characteristics although there are visible differences amongst species in the solidity of the periotic-tympanic

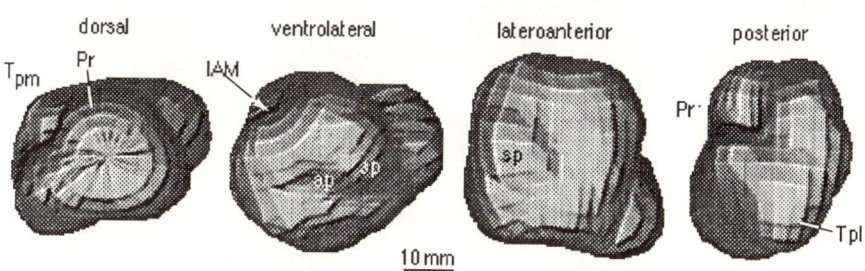

Fig. 3. Solid Surface Reconstructions. CT scans of the right bulla of Tursiops truncatus were digitized, reconstructed, and displayed in four rotations using Multiple Marker Analysis (Graves et al., 1984). Major surface features evident in the reconstruction include the cochlear bulge of the promontorium (Pr), the indentation of the internal auditory meatus (IAM), the petrotympanic aperture (ap) posterior to the sigmoid process (sp), and the lateral (Tpl) and medial (Tpm) posterior tympanic prominences.

Table 3. Bullar Surface Morphometry

Group and Species	Body length (cm)	Bullar Dimensions (mm)[1]	Periotic (mm)[1]	Tympanic (mm)[1]	VIIIth Nerve[2] (mm)	Bullar Weight (gm)	Surface Area[3] (mm²)	Bullar Volume (ml)[3]
I								
Phocoena phocoena	133	33.2 / 24.1	30.3 / 16.3	30.1 / 19.7	3.7	15.2	2341	9.600
Physeter catodon	1361	69.7 / 63.9	60.6 / 36.5	59.6 / 37.2	11.3	180.7		45.52 (periotic only)
II								
Grampus griseus	228	43.9 / 32.9	37.3 / 25.1	38.7 / 24	4.7	32.7	3190	14.708
Lagenorhynchus albirostris	207	42.2 / 32.6	30.7 / 19.0	36.3 / 22.5	4.6	23.7	2806	12.339
Stenella attenuata	185	33.3 / 24.7	28.5 / 19.3	30.9 / 18.7	4.9	13.3	2341	9.5585
Tursiops truncatus	259	44.0 / 28.8	31.6 / 19.8	34.0 / 21.4	5.5	25.2	2963.7	13.45

[1] Lengths of longest/shortest axes.
[2] Diameter of auditory nerve at the periotic aperture of the internal auditory canal.
[3] Calculated by MMAS from CT scans. There was no significant difference between calculated values and fluid displacement measurements for five test specimens.

suture, the proportions of the bullar divisions, and the complexity and relative position of surface convolutions. Thus, odontocete bullae have species-specific size and shape characteristics which are not correlated with ultrasonic frequency ranges but all are similarly constructed from exceptionally dense, compact bone and are completely isolated from the skull in a peribullar cavity (Fig. 4).

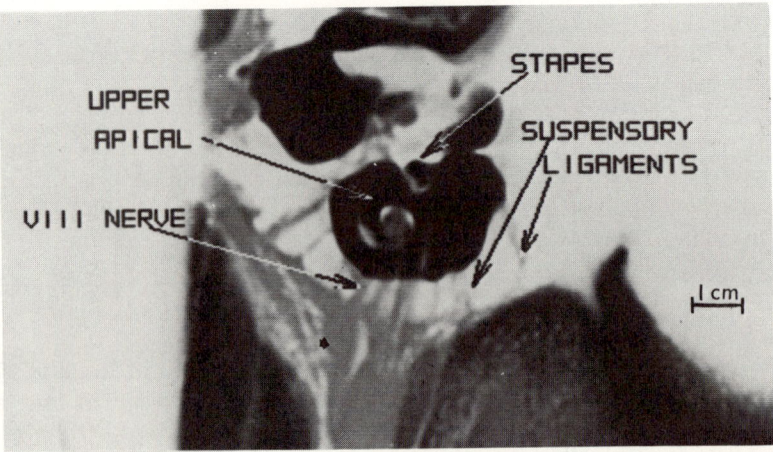

Fig. 4 Sperm Whale Periotic. A 2 mm CT section shows a Physeter catodon bulla from a dorsal view. Five sets of ligaments attach the tympanic and periotic to less dense lamellar bone surrounding the peribullar cavity. Most of the cochlea is visible as it curves for 1.5 turns from the stapes to the apex.

Cytoarchitecture of the Osseus Labyrinth

The dolphin cochlea has the prototypic mammalian divisions: scala media (cochlear duct), scala tympani, and scala vestibuli. The membranous labyrinth of scalae forms an inverted spiral inside the bony labyrinth of the cochlear canal (Fig. 5), which curves medially and ventrally in each periotic from the stapes to the helicotrema at the apex, around a core, the modiolus, containing the auditory branch of the acousto-vestibular nerve. The canal decreases in diameter from base to apex with cross-sectional area ratios ranging 1.6 (Phocoena) to 6.3 (Grampus). In all specimens, the vestibule is large but the semi-circular canals are reduced and

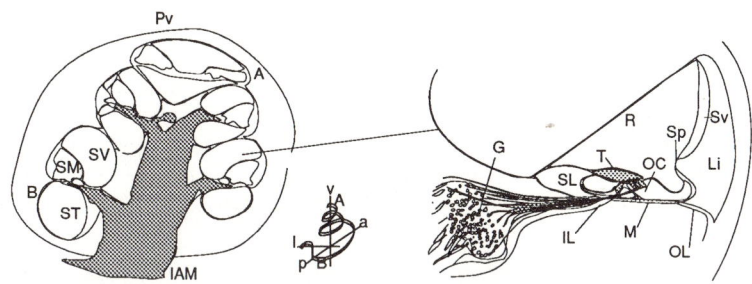

Fig. 5. The Cochlear Canal. Major cochlear structures are shown schematically for a right delphinid ear. On the left, the periotic is bisected along the neural axis in a mid-modiolar plane. It is drawn inverted from in vivo orientation (indicated on the small axes) for comparison with conventional mammalian representations which have the cochlear apex at the top of the image. The light micrographs in Figure 6 are shown in the same orientation. The right enlargement depicts scala media in the mid to upper basal turn, equivalent to the location of Figure 6.1. a anterior, p posterior, l lateral, v ventral; A apex; B basal turn; G spiral ganglia; IAM medial aperture of the internal auditory canal; IL inner spiral lamina; Li spiral ligament; M basilar membrane; OC organ of Corti; OL outer spiral lamina; Pv ventral edge of promontorium; R Reissner's membrane; SL spiral limbus; SM scala media; Sp spiral prominence; ST scala tympani; SV scala vestibuli; Sv stria vascularis; T tectorial membrane.

form incomplete channels; it is unclear whether all components of the vestibular system are functional. All odontocete cochleae examined in thin section differ significantly from other mammalian cochleae (Fig. 6) and structural differences in the basal turn separate odontocetes into two anatomically distinct groups. Three anatomical features of the inner ear which influence resonance characteristics and frequency perception are addressed in detail here: basilar membrane construction, osseous spiral laminae configurations, and spiral ganglion cell distributions.

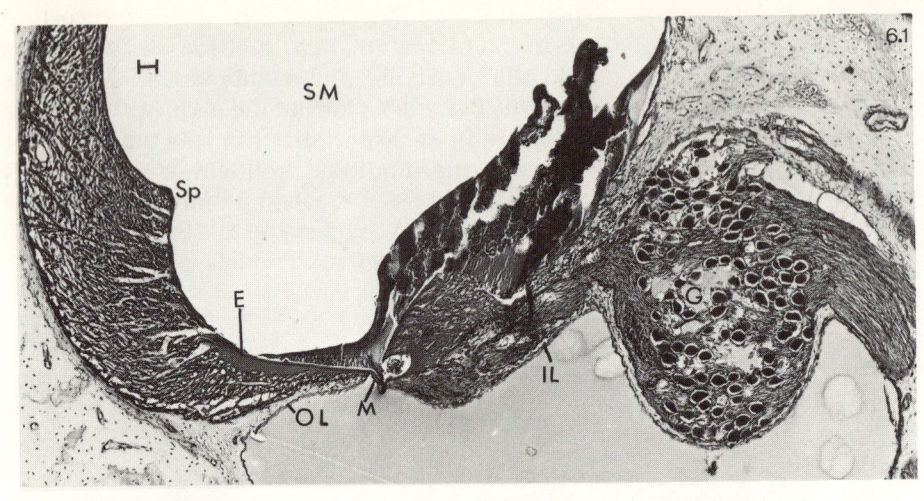

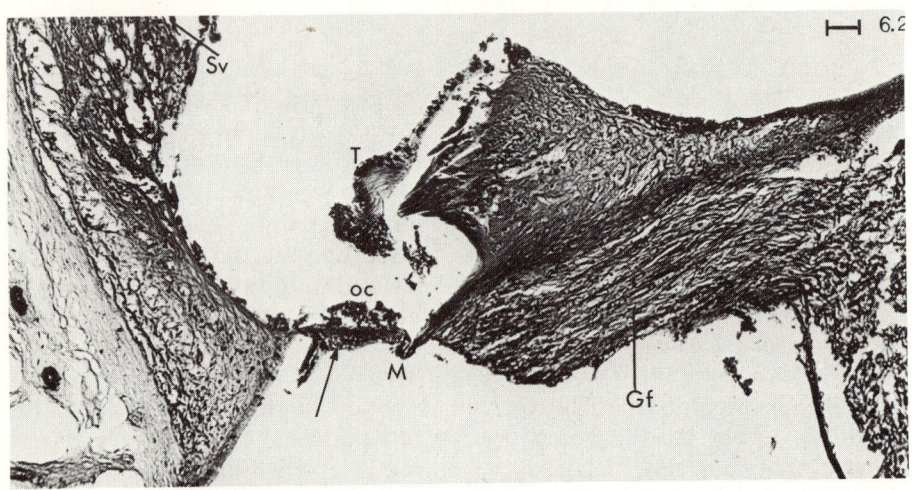

Fig. 6. Cochlear Duct Cytoarchitecture. Light micrographs of 20 μ mid-modiolar sections demonstrate structural characteristics of the odontocete cochlear duct. Descriptions below use conventional neurocentric orientations for the cochlea in which inner or medial mean towards the modiolus and outer or lateral refer to the anti-modiolar or abneural side of the cochlea. All tissues are from adult animals and represent average material preserved 5 hours to 4 days post-mortem by round window injection. They show preservation and processing artifacts similar to those of human temporal bones, including disruption and collapse of Reissner's membrane, absent or necrotic organ of Corti, acidophilic staining of the perilymph, and serous protein deposits in scala media (SM). Each scale bar represents 50μ .

(6.1.) The basilar membrane (M) of Phocoena phocoena in the upper basal turn, 7 mm from the oval window, measures 45μ x 20μ. It is stretched between inner (IL) and outer (OL) ossified spiral laminae. The outer lamina is is 30-40μ thick. There is heavy staining of the perilymph in scala tympani, but the endolymph of scala media (SM) is not contaminated, indicating the membrane is intact. Blood in scala media is the result of a concussion. A distinctive cellular layer (E) found only in the basal turn in odontocetes lines the lateral basilar membrane recess below the spiral prominence (Sp). Kolmer reportedly dubbed them "ersatzzellen" (Reysenbach de Haan, 1956), and although noted by several authors, these cells are unclassified and their function remains unclear. The large number of oblate spiral ganglion cells (G) clumped medially in a pocket of Rosenthal's canal which protrudes into scala tympani is a typical cross-section of the spiral ganglia in odontocetes.

(6.2) In the upper middle turn of Tursiops truncatus, 28 mm from the stapes, the basilar membrane (190μ x 10μ) is partly obscured by organ of Corti remnants and by mesothelial cells on the tympanic border (arrow). These cells are common in mammals and increase apically. Curvature of the membrane is a compression artifact. The stria vascularis (Sv) is characteristically dense and collagenous. The tectorial membrane (T) extends over the spiral limbus and broadens into a gelatinous flap over the basilar membrane. There is no fibrillar layer analogous to Hensen's stripe in humans. Large numbers of habenular fibers (Gf) are apparent between the inner spiral laminae.

(6.3) In an apical section 4 mm from the helicotrema in Phocoena, the membrane measures 200μ x 10μ. Only the spiral ligament (Li) supports the lateral edge of the basilar membrane at this point. Multiple cells of Huschke (H), the auditory teeth, are visible in the spiral limbus immediately below the limbal tectorial membrane (T).

Basilar membrane thickness and width vary inversely from base to apex in mammals. Highest frequencies are encoded in its narrow, basal region and progressively lower frequencies, towards the apex as the membrane broadens and thins. In all odontocetes, thickness varies uniformly from 25μ basally to 5μ apically with no apparent change in fibrillar density. Width increases 9-12 fold from a basal minimum of 30-40μ (Table 4). Delphinids have a greater increase in membrane width than phocoenids; however, Phocoena phocoena has the steepest rate of increase. Comparisons of bat and odontocete membrane thickness:width ratios (Fig. 7) show parallel slopes for all groups in the lower

Table 4. Basilar Membrane Dimensions

Group and Species	Membrane Length (mm)	Outer Osseous Lamina Length (mm)	Basal/Apical Width (μ)	Basal/Apical Thickness (μ)
I				
Phocoena phocoena	25.93	17.6	30/290	25/5
II				
Grampus griseus	40.5	-	40/420	20/5
Lagenorhynchus albirostris	34.9	8.5	30/360	20/5
Stenella attenuata	36.9	8.35	40/400	20/5
Tursiops truncatus	40.65	10.3	30/380	25/5

frequency apical regions of the cochlea and 2-3 fold higher ratios for odontocetes in the basal ultrasonic regions.

Inner and outer ossified spiral laminae are present throughout most of the basal turn in all species examined and are amongst the most striking features of the odontocete cochlea (Figs. 6,8). The internal osseous spiral laminae, tunneled by the foramina nervosa or nerve fiber tracts, form a bi-layered wedge which supports the medial margin (pars arcuata) of the basilar membrane. The thickness of the inner laminae varies inversely with distance from the stapes. In the lower basal turn, the inner laminar wedge averages 50μ at the membrane juncture. In the middle to upper basal turn it thins to 5μ, and, in delphinids, becomes a single shelf supporting the spiral limbus. The outer lamina in the basal turn in all odontocetes is 30-40 μ thick, heavily calcified, and functions as a housing for the spiral ligament and lateral attachment for the basilar membrane.

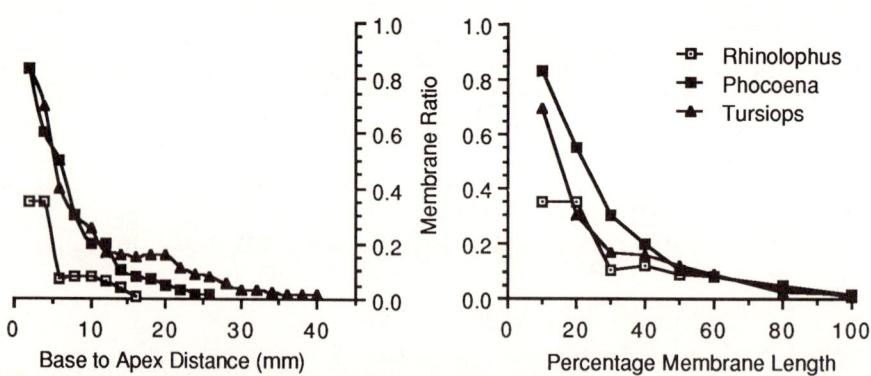

Fig. 7. Basilar Membrane Ratios. Thickness:width ratios for basilar membranes in the horseshoe bat, Rhinolophus ferrumequinum, and two odontocetes (Phocoena phocoena (I); Tursiops truncatus (II)) are plotted against location in the cochlea and as a percentage of cochlear length. The acute decrease in the bat ratio at 5 mm corresponds to a characteristic region of abrupt change in membrane configuration reported in some constant frequency bats associated with the specialized "foveal" membrane region encoding their echolocation signal (Camhi, 1984).

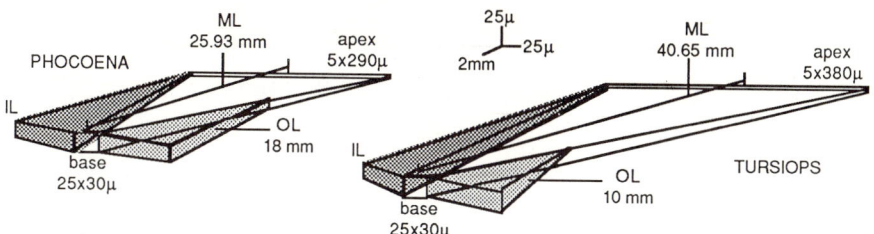

Fig. 8. Laminar Attachments of the Basilar Membrane. The basilar membrane (M) and inner (IL) and outer (OL) osseous spiral laminae are drawn in orthoscopic projection for representative Group I and Group II species. Different longitudinal and cross-sectional scales are used to permit a single reconstruction to show basal and apical basilar membrane configurations, membrane length (ML), and the proportions of inner and outer laminae.

Thus, in the extreme basal end, the basilar membrane is firmly anchored at both margins to a bony shelf. In Delphinidae, the outer lamina thins, paralleling the inner lamina, and the spiral ligament replaces it as the primary lateral membrane support; an ossified outer shelf is found only in the first 8 to 10 mm of the average delphinid duct (Fig. 8). In Phocoena, bony laminae are present medially and laterally for 17 to 18 mm, ending with no significant taper. The basilar membrane therefore is provided with substantial buttressing at both edges over 60% its length in Phocoena and 20 to 28% of its length in delphinids.

Total ganglion cell counts and ganglion cell densities were estimated for one specimen of Phocoena, Tursiops, and Stenella attenuata from 20 μ serial sections corrected for intersection partial cell duplications (Table 5). Data from human temporal bones, perfused Lagenorhynchus obliquidens, and horseshoe bats are listed for comparison. Normal human temporal bone data are useful comparisons because they have similar preservation hazards as fisheries and stranded animals. Surprisingly, odontocete data from previous studies for perfused animals and data from this study are similar. Ganglion cell densities for all odontocetes are more than twice those of bats and humans. They are also 30-50% greater than the highest densities reported in the basal, foveal regions in bats.

Cochlear Morphometrics and Topology

Most attempts at cochlear spiral measurements use two-dimensional interpolation techniques (Guild, 1921) or serial section plots (Schuknecht, 1953; Wever et al.,1971a). Although these methods provide reasonable approximations of spiral shapes, all orthographic projections have the same inherent disadvantage; i.e., flat plots are necessarily foreshortened. In shadow projections, tall or short spirals with equal interturn radii produce the same axial silhouette and they will appear to have the same length regardless of differences in height (Fig. 9). Useful allometric analyses require all three dimensions be taken into account.

Table 5. Ganglion Cell Density

Species	Total Ganglion Cells	Membrane Length (mm)	Average Density (cells/mm)
Phocoena phocoena	66933	24.31	2753.3
Lagenorhynchus obliquidens	70000[1]	34.90	2005.7
Stenella attenuata	82506	37.68	2189.6
Tursiops truncatus	105043	41.57	2526.9
Rhinolophus ferrumequinum	15953	16.10	1000/1750[2]
Homo sapiens	30500[3]	31.00	983.9

[1]Wever et al.(1972)
[2]Bruns and Schmieszek (1980); cochlear average/acoustic fovea densities.
[3]Schuknecht and Gulya (1986)

Three-dimensional cochlear spiral measurements for key species are compared with ultrasonic frequency ranges in Table 6. There is a strong negative correlation (-0.968<r<-0.791) for characteristic frequency and all spiral variables except scalae length and basal diameter, which have a positive correlation with animal length (0.84<r<0.92). Principal component analyses distribute the data into two frequency-weighted divisions with 91% of variability attributed to body length and spiral geometry. Excluding frequency, the data redistribute into three categories: short body length and low cochlear spiral parameters (Phocoena); long body length and low parameters (Physeter); and short to average length with

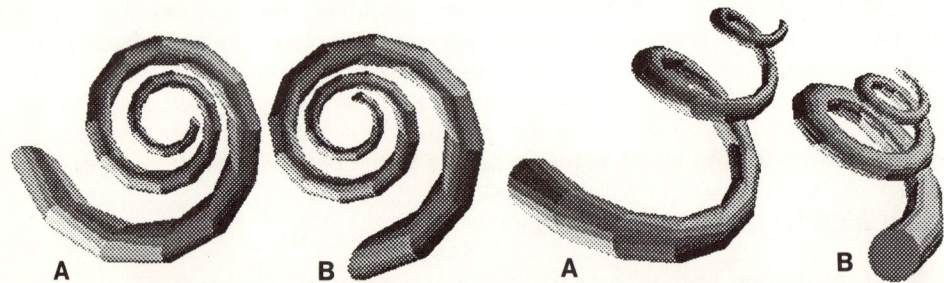

Fig. 9. Orthogonal vs. Three-dimensional Projections. The spiral pair on the right are 70° rotations of the left. An initial pair was generated by computing two spirals in which all variables are constant except vertical increment/rotation and final path length. These were then displayed at two rotations. In axial projections (left), their differences appear negligible and center-line plots of their paths would be indistinguishable. As is apparent in side views, spiral A is actually 15% longer with an axial height 135% that of B.

Table 6. Cochlear Canal Spiral Parameters

Group and Species	Turns	Scalae Length (mm)	Basal Diam. (mm)	Axial Height (mm)	Axial Pitch[1] (mm)	Basal Ratio[2]	Slope Ratio[3]	Echolocation Pulse Peak Frequency (kHz)
I								
Phocoena phocoena	1.5	25.93	5.25	1.47	0.982	0.280	.038	130
Physeter catodon	1.75	72.21	14.3	3.12	1.78	0.218	.025	unknown[5]
Inia geoffrensis[4]	1.5	38.2	8.5	-	-	-	-	200
II								
Grampus griseus	2.5	40.5	8.73	5.35	2.14	0.614	.053	unknown[5]
Lagenorhynchus albirostris	2.5	34.9	8.74	5.28	2.11	0.604	.061	40
Stenella attenuata	2.5	36.9	8.61	4.36	1.75	0.507	.047	60
Tursiops truncatus	2.25	40.65	9.45	5.03	2.24	0.532	.055	70

[1] $\frac{\text{axial height}}{\text{turns}}$ [2] $\frac{\text{axial height}}{\text{basal turn diameter}}$ [3] $\frac{[\text{axial height/scalae length}]}{\text{turns}}$

[4] All measurements for Inia are estimates from single plane X-rays.
[5] Echolocation has not been documented in these species.

high spiral parameters (Delphinidae). These data indicate two spiral morphometries, Type I and Type II, differentiated by turns, height, pitch, slope, and basal ratios. Species distributions for Type I and II spirals coincide with acoustic Groups I and II and with differences found in outer osseous laminar configuration. Although T-tests showed no significant differences between Groups I and II for bullar surface variables or standardized scalae length, spiral configuration data for the groups differ with significance levels beyond 0.1%.

Type I and Type II spirals are closely modelled by Archimedian (I) and equiangular (II) spirals (Fig. 10). In ideal forms, these two spirals represent (I) a constant interturn radius curve, like that of a tightly coiled rope, and (II) a gnomonic spiral with logarithmically increasing interturn radii; e.g., a chambered nautilus. The gnomonic spiral is common in nature and is the assumed configuration for mammalian cochlea. The archimedean curve is rare. The mathematical parameters for Type I and Type II are:

$$
\begin{array}{ll}
\text{I} & \text{II} \\
r = a\theta & r = e^{a\theta} \\
N < 2 & N > 2 \\
\left(\frac{r_n}{n}\right) \geq \left(\frac{r_{n+1}}{n+1}\right) & \left(\frac{r_n}{n}\right) \leq \left(\frac{r_{n+1}}{n+1}\right)
\end{array}
$$

where r = total radius at turn n; N = number of turns; θ = angular displacement (radians); and a = a spiral size constant. For odontocete cochlea, a is species-isometric and is calculated retroactively for the models from basal ratios (Figs. 10, 11).

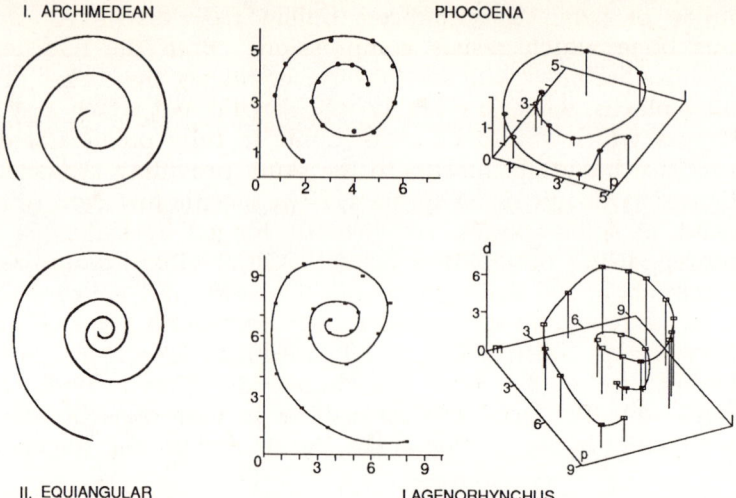

Fig. 10. Spiral Models and Species Cochlear Canal Plots. Ideal Type I and Type II spirals are shown for comparison with plots of cochlear canal midpoints in <u>Phocoena phocoena</u> and <u>Lagenorhynchus albirostris</u>. Two-dimensional projections are based on single-plane X-rays. Three-dimensional plots were obtained from 2 mm. CT scans of the same animal and include the cochlear hook which cannot be seen in a flat, axial image. Flattened contours appear in the three-dimensional curves where all points of a cochlear half-turn fell within one CT scan. Axes are scaled in millimeters. m medial, p posterior, l lateral, d dorsal.

CONCLUSIONS

The principal question posed in this study was whether three-dimensional comparative assessments of the auditory periphery could provide insights into the ability of odontocetes to echolocate in water. Anatomical analyses and three-dimensional reconstructions show a complex peripheral auditory architecture which is unique to odontocetes. Their temporal bone is adapted for both acoustic isolation and for pressures encountered in an aquatic environment, while the inner ear is clearly adapted for ultrasonic perception. Inner ear modifications related to ultrasonic audition found in all odontocetes included an exceptionally narrow basal basilar membrane, high spiral ganglion cell densities, and extensive, bony outer spiral lamina. Moreover, there are two cochlear spiral configurations which correlate with odontocete echolocation signal ranges. These configurations, when combined with species data on cochlear duct laminae distributions, can be used to predict classes of odontocete ultrasonic audition.

Comparisons of odontocete and bat bullae imply that temporal bone structure is strongly influenced by environmental factors. Unlike the fragile, inflated bullae of bats, all odontocete bullae are constructed of massive, porcelaneous bone which resists compression. Both the middle ear and peribullar cavities are lined with specialized membranes, the corpus cavernosum and peribullar plexus, which are highly vascularized and which may moderate volume changes which would be catastrophic in fully pneumatized cavities. Ligaments replace bony attachments to the skull, providing acoustic isolation. The periotic and tympanic differ in mass, have specific low density regions or windows, and, in some species, are virtually hinged by a flexible tympano-periotic suture, raising possibilities for differential vibration and conduction properties throughout the bulla. Although this study did not directly address alternatives to a tympanic or ossicular path for sound transmission to the cochlea, the position, construction, and ligamentous associations of the bulla support the "pan bone" theory of jaw conduction (Norris, 1969; Norris and Harvey, 1974), and, like facial ruff assymetries in barn owls (Knudsen, 1981), bilateral asymmetries in the location of the bullae may provide directional cues.

All bullae examined in situ were oriented with the periotic medial and dorsal to the tympanic. This orientation results in the cochlear apex ventral to the stapes, orthogonal to the terrestrial mammalian format. This placement, or displacement, of the ear may result from the spinal flexion and caudal brain case compression that occurred in odontocetes as they regressed to a fuselloid shape, but its utilitarian effect is a shorter, less angular pathway for the VIIIth nerve which crosses the peribullar cavity before entering the brain case. This "externalization" of the auditory nerve may be unique in odontocetes. It arises from the separation of the bulla from the skull, which is adaptive for aquatic echolocation, and provides a functional explanation for the dense fibrous sheath fully enclosing the VIIIth nerve.

Basilar membrane length varies with animal size and is not correlated with peak ultrasonic frequency. Membrane widths in this study are smaller than reports by Fleischer (1976) and Norris and Leatherwood (1981). Neither work listed animal lengths, which could account for 10% variability; however, differences in our results may be explained also by technique. Fleischer used a deep corrosive cast and estimated widths from interlaminar gaps. Norris and Leatherwood used an extremely corrosive, rapid decalcificant, trichloracetic acid, on salvaged tissues. Either method will distort or etch bony membrane supports and yield overestimates. The measurements in this study are comparable to those by Wever et al. (1971a; 1972) for perfused Tursiops and Lagenorhynchus.

Basilar membrane dimensions interact with its composition and support to determine resonance characteristics. Based upon dimensions alone, the odontocete membrane is a highly differentiated, anisotropic structure capable of an exceptionally wide frequency response. In the basal end, the basilar membrane in all odontocetes is nearly square with a 400-600μ^2 cross-sectional area and is tightly joined to double laminae. Apically, it thins to a 5μ x 300-450μ strip supported by ligaments. The basal construction is characteristic of echolocators (Hinchcliffe and Pye, 1969), but while bat and odontocete basilar membranes are similar, odontocete basal ratios are substantially greater. Membrane ratios plotted as a percentage of membrane length (Fig. 7) are significantly higher for the basal turn of odontocetes. Were other cochlear duct structures identical, the differences in the ratios, which reflect structural differences related to membrane stiffness, are sufficient to account for lower ultrasonic ranges in bats.

The diameter of the auditory nerve (Table 3), the volume of habenular nerve fibers in the osseous spiral lamina, and high ganglion cell counts are consistent with hypertrophy of the entire odontocete auditory system. For quantitative interspecific comparisons, ganglion cell density is a more effective measure than total cell population. Densities in this study ranged from 2000 cells/mm in Lagenorhynchus to 2700 cells/mm in Phocoena, which are higher than in any other mammal. Wever et al. (1971b, 1972) reported ganglion:hair cell ratios of 4:1 for Lagenorhynchus and 5:1 for Tursiops. Virtually all mammals average 100 inner hair cells/mm (N. Kiang, pers. comm.) with 2.5 to 4 rows of outer hair cells/inner hair cell. Wever's data imply a hair cell array of 1 inner and 4 outer rows in two odontocetes. Using this estimate with our data, we calculate a nearly 6:1 ratio for Phocoena phocoena, 5:1 for Tursiops truncatus, 4.4:1 for Stenella attenuata, and 4:1 for Lagenorhynchus obliquidens. The human ratio is 2.4; cats, 3; and bats range 3-5:1 (Firbas, 1972, Bruns and Schmieszek, 1980). Since 90-95% of all afferent spiral ganglion cells innervate inner hair cells, we estimate an average ganglion cell:inner hair cell ratio of 24 for odontocetes. This is more than twice the average density in bats and three-fold that of humans (Firbas, 1972). While data from three specimens are insufficient for a definitive analysis, ganglion cell:hair cell ratios appear to be proportional to frequency ranges in both bats and odontocetes and it is likely that higher afferent ratios in odontocetes are directly related to the extent or complexity of information extracted by neuronal processing of their echolocation signals.

The presence of an external bony lamina is virtually diagnostic for ultrasonic perception (Sales and Pye, 1974; Reysenbach de Haan, 1956). Differences in laminar structure amongst odontocetes are consistent with acoustic divisions and provide a simple but important mechanistic link for species differences in ultrasonic ranges. In Odontoceti, the extent of the ossified lateral spiral lamina is a species and group-specific character. Phocoenids have the highest frequency range and a substantial outer lamina over two-thirds of the cochlear duct. Delphinids have a characteristic outer lamina for 20-30% of the duct. Since the basal basilar membrane is similarly constructed in both groups, a longer outer lamina in Group I increases membrane stiffness, thereby increasing the resonant frequency for that membrane region, compared to equivalent, unsupported membrane locations in Group II.

Cochlear spiral measurements show a clear division of odontocete inner ears into two types which correlate with cochlear duct differences in laminae. The divisions are not determined by taxonomy but by complex spiral geometry. Categorizations of species by spiral format also coincide with high and low echolocation frequency groups:

I. Type I species have a nearly planar cochlear spiral with a slope ratio <0.04 and a constant radial increment. Axial height is less than 0.1% body length and the basal ratio ranges 0.2-0.3. There are less than two full turns. Outer spiral laminae buttress the basilar membrane for >60% its length. Peak energy of echolocation clicks for known species is located above 100 kHz.

II. Type II species have a more attenuated spiral with logarithmically increasing radii, an axial height more than 0.2% body length, and slope ratios >0.05. The basal ratio is >0.5. There are typically 2.5 turns with outer bony laminae present 25% of basilar membrane length. Type II species generally produce broad spectrum ultrasonic clicks with peak energy below 100 kHz.

Three-dimensional composite reconstructions graphically demonstrate the configurational differences between Type I and Type II odontocete cochlea (Fig. 11). The composites were produced by combining spiral model parameters, cochlear canal data, and cochlear duct contour measurements of Group I and Group II species. Contours of the basilar membrane, spiral ganglia, and inner and outer laminae were digitized, measured, standardized by animal length, and plotted in a computerized, three-dimensional framework to obtain a weighted average contour for each component. A principal spiral was generated by plotting a Type I or Type II spiral with constants derived from normalized species averages. Regular structures were produced centered on the spiral by superimposing the averaged contours on the spirals at the mid-modiolar plane and interpolating each component along the curve. In both Type I and Type II cochlea, the spiral ganglia are distributed in a continuous band for nearly 80% of cochlear duct length, but differences in cell densities are implied by the smaller volume of the ganglia in the Type II reconstruction. Differences in membrane buttressing between types are clear. The Type I cochlea has proportionately twice as much membrane supported by bony laminae as Type II. The basilar membrane, which normally stretches between the inner and outer spiral laminae or spiral ligament, is represented only by its outer edge to avoid blocking views of other structures. At the apex, the Type II membrane is broader, which suggests these species have a wider frequency range than Type I. This is likely to be true for lower frequencies, but differences in basal laminar support imply Type II cochlea have a lower ultrasonic capacity.

Type II spirals resemble the conventional, terrestrial cochlear format and include the delphinid species which have been most extensively investigated in the past. Type I represents a novel cochlear format with major deviations from conventional assumptions of cochlear modelling. The combination of Type I spiral configuration, more extensive laminar buttressing, and higher echolocation frequencies in Group I species argues strongly for an adaptive relationship to aquatic echolocation for this cochlear format. Since the data in this survey for Type I cochlea are dominated by Phocoena phocoena, it could be argued they represent a single, distinctive genus rather than a format for upper range ultrasonic audition. Evidence that there are at least two spiral formats can be adduced, however, from anatomical and behavioural studies in other species.

If we assume echolocation frequencies are correlated habitat, differences in species distributions for Type I and Type II spiral formats should correlate with environmental distributions as well. Fresh water and near shore species live in an information dense, structurally complex environment. Since wavelength is inversely related to frequency, using echolocation to differentiate the small structures typical of these waters requires exceptionally high frequencies. The same range of higher frequencies would be of little use to open ocean species which, by comparison, live in extremely low density environments and are primarily concerned with detection of larger, distant objects or communication

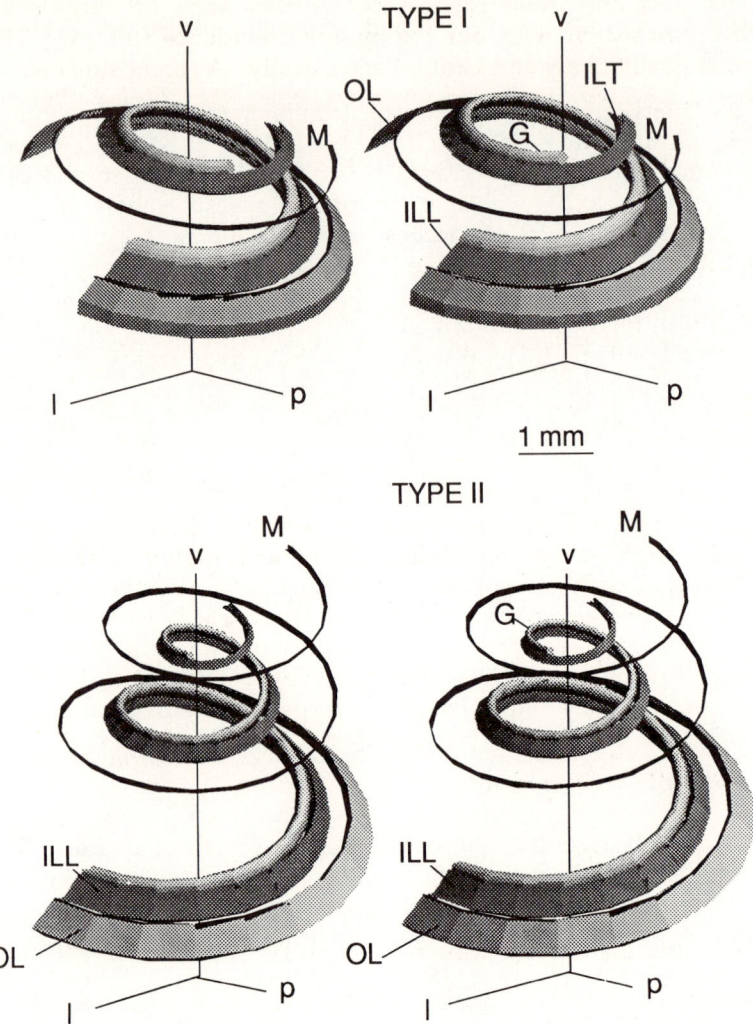

Fig. 11. Basilar Membrane, Spiral Lamina, and Neural Fiber Distributions in Odontocetes. Composite reconstructions, generated from standardized data, schematically represent major cochlear duct structural and neural components of Type I and Type II odontocetes. Principal features are described in the text. The images are reproduced as parallel stereo-pairs with an approximate viewing distance of 25 cm. Most conventional stereo-viewers may also be used. The spirals have been scaled to common axes to facilitate comparisons. The cochlea are inverted, as in Figures 6 and 7, from in vivo orientations. l lateral, p posterior, v ventral; G spiral ganglia; IL inner osseous spiral lamina (L limbal edge; T tympanal); M basilar membrane (lateral edge); OL outer osseous spiral lamina.

with conspecifics. With these assumptions, we would predict platanistid, riverine dolphins living in the Ganges or varzea lakes of the Amazon to be higher frequency, Type I species. In fact, echolocation signals of these species range to 200 kHz and illustrations show cochlea with 1.5 turns (Purves and Pilleri, 1983), consistent with our radiographic evidence for Inia. They fit the Type I format qualitatively and Group I acoustically. A recent study by Feng et al. (1986) shows 1.5 evenly distributed turns in Lipotes, the Chinese river dolphin, as well. These comparisons suggest that Type I spirals are not a unique adaptation of Phocoena. For two key species, Grampus griseus and Physeter catodon, we have no corroborative recordings. Grampus are off-shore animals which travel in pods and anecdotal reports indicate they whistle. The cochlear data for Grampus fall clearly within Type II parameters, implying an echolocation range below 100 kHz. In contrast, Physeter has a Type I format numerically, but there are several additional factors to consider. Unlike the earlier species presumed to be Type I, Physeter is pelagic and it has a substantially different bullar anatomy. There is no evidence to date for echolocation in Physeter. No structural analyses of any physeterid cochlear duct are yet available, but they are imperative for frequency estimates since spiral configuration alone cannot dictate inner ear resonance characteristics, and the immense size of Physeter catodon may radically affect cochlear structure. Our data indicate at least one alternative cochlear configuration exists in odontocetes which is coincident with extensive bony laminae, high basilar membrane ratios, and higher ultrasonic auditory ranges. Physterid cochleae appear, preliminarily, to resemble this format, but they may also represent a third alternative from which we may discover the limits which size alone can impose on the ability of even odontocetes to produce or perceive ultrasonics. While our data begin to reveal the diversity of odontocete cochlea, they impel us also to expand investigations to even more species if we are to understand the true range of the odontocete ear.

ACKNOWLEDGMENTS

This work was supported by the ARCS Foundation and NSF grant BNS-8118072. All specimens were collected under Permit no. 368, National Marine Fisheries Service, National Oceanic and Atmospheric Administration, Department of Commerce, in compliance with the Endangered Species Act and Marine Mammal Protection Act. The work could not have been completed without the encouragement and assistance of the staffs of the Departments of Anatomy, Experimental Radiology, and Neuroradiology of the Johns Hopkins Medical Institutions. Invaluable assistance with radiographic studies were provided by Frank Starr, III, James Anderson, and Arthur Rosenbaum. Alan Walker, Willard Graves, and George Carey provided advice and assistance with reconstructions. Key specimens were obtained through the efforts of Gregory Early, Joseph Geraci, James Gilpatrick, Richard Lammertson, Daniel Odell, William Perrin, James Mead, and Charles Potter. Nelson Kiang and Steven Rauch provided helpful reviews on the manuscript. Final reconstructions and graphics were produced with the cooperation of the Cochlear Implant Research Laboratory, the Eaton-Peabody Laboratory for Auditory Physiology, and the Department of Otolaryngology, Massachusetts Eye and Ear Infirmary.

Abbreviations in figures:

A apex
a anterior
ap petrotympanic aperture
B basal turn of the cochlea
bo basioccipital
C cochlear spiral
d dorsal
G spiral ganglia
Gf habenula perforata
H cells of Huschke
IAM internal auditory meatus
IL lamina spiralis ossea primaria (inner lamina)
l lateral
Li spiral ligament
M basilar membrane
m medial
Ma mandible
ML membrane length
N auditory nerve
OC organ of Corti
OL lamina spiralis ossea secundaria (outer lamina)
OW oval window (fenestra ovalis)
P periotic
p posterior
P_a anterior periotic edge
Pr promontorium
Pv ventral promontorium edge
R Reissner's membrane
SL spiral limbus
SM scala media
Sp spiral prominence
sp sigmoid process
sq squamosal
ST scala tympani
SV scala vestibuli
Sv stria vascularis
T tectorial membrane
Tl lateral tympanic lobe
Tpl lateral posterior tympanic prominence.
Tpm medial posterior tympanic prominence
v ventral

REFERENCES

Au, W.W.L., Floyd, R.W.,Penner, R.H. and Murchison, A.E., 1974, Measurement of echolocation signals of the Atlantic Bottle-nosed dolphin, Tursiops truncatus montagu, in open waters, J. Acoust. Soc. of Am., 56: 1280-1290.

Brownlee, S., 1983, Correlations between Sounds and Behavior in the Hawaiian Spinner Dolphin, Stenella longirostris, M.S. thesis, University of California, Santa Cruz.

Bruns, V., 1976, Peripheral auditory tuning in the Doppler shift compensating bat, Rhinolophus ferrumequinum: II. Frequency mapping in the cochlea, J. Comp. Physiol., 106: 77-86.

Bruns, V. and Schmieszek, E.T., 1980, Cochlear innervation in the greater horseshoe bat: Demonstration of an acoustic fovea, Hearing Res., 3:27-43.

Bullock, T.H., Grinnell, A.D., Ikezono, E., Kameda, K., Katsuki, Y., Nomoto, M., Sato, O., Suga, N., and Yanagisawa., K., 1968, Electrophysiological studies of central auditory mechanisms in cetaceans, Z. vergl. Physiol., 59: 117-156.

Bullock, T., and Ridgway,S., 1972, Evoked potentials in the central auditory system of alert porpoises to their own and artificial sounds, Jour. Neurobiol., 3: 79-99,

Busnel, R-G., and Dziedzic,A., 1966, Acoustic signals of the pilot whale Globicephala melaena and of the porpoises Delphinus delphis and Phocoena phocoena, in: "Whales, Dolphins, and Porpoises," K.S. Norris, ed., University of California Press, Berkeley.

Caldwell, M.C., and Caldwell, D.K., 1967, Intraspecific transfer of information via pulsed sound in captive odontocete cetaceans, in: "Animal Sonar

Systems: Biology and Bionics, II," R-G. Busnel, ed., Laboratoire de Physiologie Acoustique, Jouy-en-Josas.

Caldwell, M.C. and Caldwell, D. K., 1971, Statistical evidence for individual signature whistles in Pacific whitesided dolphins, Lagenorhynchus obliquidens, Cetology, 3: 1-9.

Camhi, J.M., 1984, "Neuroethology: Nerve Cells and the Natural Behavior of Animals," Sinauer Assoc., Inc., Sunderland.

Diercks, K.J., 1972, Biological sonar systems: A bionics survey, Applied Research Laboratories, ARL-TR-72-34, University of Texas.

Diercks, K.J., Trochta, R.T., Greenlaw, R.L., and Evans, W.E., 1971, Recording and analysis of dolphin echolocation signals, J. Acoust. Soc. Am., 49: 1729-1732.

Evans, W.E., 1967, Vocalizations among marine mammals, in: "Marine Bio-Acoustics," W.N. Tavolga, ed., Pergamon, New York.

Evans, W.E., 1973, Echolocation by marine delphinids and one species of fresh water dolphin, J. Acoust. Soc. Am. , 54: 191-199.

Evans, W.E., and Prescott, J.H., 1962, Observations of the sound production capabilities of the bottlenose porpoise: A study of whistles and clicks, Zoologica, 47: 121-128.

Feng, W., Liang, C, Wang, J., and Wang,X., 1986, Morphometric and Stereoscopic Studies on the Spiral and Vestibular Ganglia of Lipotes vexillifer,, (prepubl.).

Firbas, W., 1972, Über anatomische Anpassungen des Hörorgans an die Aufnahme hoher Frequenzen, Monatsschr. Ohr. Laryn.-Rhinol., 106:105-156

Fleischer, G., 1976, Hearing in extinct cetaceans as determined by cochlear structure, Jour. Paleon., 50: 133-152.

Fraser, F., and Purves.,P., 1960, Hearing in cetaceans: Evolution of the accessory air sacs in the structure and function of the outer and middle ear in Recent cetaceans, Bull. Brit. Mus. Nat. Hist., 7: 1-140.

Graves, W.L., Carey, G.A., Benac, S.L., and Cameron, L.W., 1984, Modeling and Graphic Display System for Cardiovascular Research Using Random 3-D Data, IEEE 1984 Int. Symp. on Medical Images and Icons, 304-308.

Grinnell, A.D., 1963, The neurophysiology of audition in bats: Intensity and frequency parameters, J. Physiol., 167: 38-66.

Guild, S.R., 1921, A graphic reconstruction method for the study of the organ of Corti, Anat. Rec., 22: 141-157.

Hinchcliffe, R., and Pye.,A, 1968, The cochlea in Chiroptera: A quantitative approach, Int. Audiol., 7: 259-266.

Hinchcliffe, R., and Pye., A., 1969, Variations in the middle ear of the Mammalia, J. Zool., 157: 277-288.

Iurato, S., 1962, Functional implications of the nature and submicroscopic structure of tectorial and basilar membranes, J. Acoust. Soc. of Am., 34: 1368-1395.

Kamminga, C.F., Engelsma, F.J., and Terry, R.P., 1989, Acoustic observations and comparison on wild, captive and open water Sotalia and Inia, Eighth Bienn. Conf. Biol. Mar. Mamm., 33.

Kasuya, T., 1973, Systematic consideration of recent toothed whales based on the morphology of tympano-periotic bone, Sci. Rep. Whale Res. Inst., 25: 1-103.

Kellogg, W.N., 1959, Auditory perception of submerged objects by porpoises, J. Acoust. Soc. Am., 31: 1-6.

Ketten, D.R., 1984, Correlations of Morphology with Frequency for Odontocete Cochlea: Systematics and Topology, Ph.D. thesis, The Johns Hopkins University, Baltimore.

Knudsen, E.I., 1981, The Hearing of the barn owl, Sci Am., 245(6): 113-125.

Long, G.R., 1980, Some psychophysical measurements of frequency in the greater horseshoe bat, in: "Psychophysical, Psychological, and Behavioural Studies in Hearing," G. van den Brink and F. Bilsen, eds., Delft University Press, Delft.

Maue-Dickson, W., Dickson, D.R., and Pullen, F.W., 1983, "Computed Tomographic Atlas of the Head and Neck," Little, Brown and Co., New York.

McCormick, J.G., Weaver, E.G., Palin, G., and Ridgway, S.H., 1970, Sound conduction in the dolphin ear, J. Acoust. Soc. Am., 48: 1418-1428.

Møhl, B., and Andersen, S., 1973, Echolocation: High-frequency component in the click of the harbor porpoise (Phocoena phocoena L.), J. Acoust. Soc. Am., 57: 1368-1372.

Montali, R.J., and Migaki, G., 1980, "The Comparative Pathology of Zoo Animals," Smithsonian Inst. Press, Wash., D.C.

Moore, P.W.B., 1990, Investigations on the control of echolocation pulses in the dolphin, (this volume).

Moran, P.R., Nickles, R.J., and Zagzebski, J.A., 1983, The physics of medical imaging, Phys. Today, July: 36-42.

Nagel, E.L., Morgane, P.J., and McFarland, W.L., 1964, Anesthesia for the bottlenose dolphin, Tursiops truncatus, Science, 146: 1591-1593.

Neuweiler, G., 1980, Auditory processing of echoes: Peripheral processing, in: "Animal Sonar Systems," R-G Busnel and J.F. Fish, eds., Plenum Press, New York.

Norris, J., and Leatherwood, K., 1981, Hearing in the Bowhead Whale, Balaena mysticetus, as estimated by cochlear morphology, Hubbs Sea World Rsch. Inst. Tech. Rpt. no. 81-132: 15.1-15.49.

Norris, K.S., 1969, The echolocation of marine mammals, in: "The Biology of Marine Mammals," H.J. Andersen, ed., Academic Press, New York.

Norris, K.S., and Harvey., G.W., 1974, Sound transmission in the porpoise head, J. Acoust. Soc. Am., 56: 659-664.

Norris, K.S., Harvey, G.W., Burzell, L.A., and Krishna Kartha, D.K., 1972, Sound production in the freshwater porpoise Sotalia cf. fluviatilis Gervais and Deville and Inia geoffrensis Blainville in the Rio Negro Brazil, in: "Investigations on Cetacea," G. Pilleri, ed., 4: 251-262, University of Berne, Berne.

Norris, K.S., Prescott, J.H., Asa-Dorian, P.V., and Perkins., P., 1961, An experimental demonstration of echolocation behavior in the porpoise, Tursiops truncatus, Montagu, Biol. Bull., 120: 163-176.

Oelschlager, H. A., 1990, Evolutionary morphology and acoustics in the dolphin skull, (this volume).

Oelschlager, H. A., 1986, Comparative morphology and evolution of the otic region in toothed whales, Am J. Anat., 177: 353-368.

Pilleri, G., 1983, The sonar system of the dolphins, Endeavour, 7: 59-64.

Pollack, G.D., 1980, Organizational and encoding features of single neurons in the inferior colliculus of bats, in: "Animal Sonar Systems," R-G Busnel and J.F. Fish, eds., Plenum Press, New York.

Popper, A.N., 1980, Sound emission and detection by delphinids, in: "Cetacean Behavior: Mechanisms and Functions," L.M. Herman, ed., John Wiley and Sons, New York.

Purves, P.E., and Pilleri, G.E., 1983, "Echolocation in Whales and Dolphins," Academic Press, Inc., Ltd., London.

Reysenbach de Haan, F.W., 1956, Hearing in whales, Acta Otolaryngol., Suppl., 134: 1-114.

Ridgway, S.H., 1980, Electrophysiological experiments on hearing in odontocetes, in: "Animal Sonar Systems," R-G. Busnel and J.F. Fish, eds., Plenum Press, New York.

Ridgway, S.H., and McCormick., J.G., 1967, Anesthetization of porpoises for major surgery, Science, 158: 510-512.

Ridgway, S.H., McCormick, J.G., and Wever, E.G., 1974, Surgical approach to the dolphin's ear, J. Expl. Zool., 188: 265-276.

Sales, G., and Pye, D., 1974, "Ultrasonic Communication by Animals," John Wiley and Sons, New York.

Schevill, W. E., 1964, Underwater sounds of cetaceans, in: "Marine Bio-Acoustics," W.N. Tavolga, ed., Pergamon Press, New York.

Schuknecht, H.F., 1953, Technique for study of cochlear function and pathology in experimental animals, Arch. Otolaryngol., 58: 377-397.

Schuknecht, H.F., and Gulya, A.J., 1986, Anatomy of the Temporal Bone with Surgical Implications. Lea and Feibiger, Philadelphia.

Stinson, M.R., 1983, Implication of ear canal geometry for various acoustical measurements, J. Acoust. Soc. Am., 74(S1): 8.

Suga, N., 1983, Neural representation of bisonar (sic) information in the auditory cortex of the mustached bat, J. Acoust. Soc. Am., 74(S1): 31.

Supin, A.Y. and Popov, V.V., 1990, Frequency selectivity of the auditory system of the bottlenosed dolphin Tursiops truncatus, (this volume).

Thomas, J., Chun, N., and Au, W., 1988, Underwater audiogram of a false killer whale (Pseudorca crassidens), J. Acoust. Soc. Am., 84: 936-940.

Watkins, W., and Schevill, W., 1977, Sperm whale codas, J. Acoust. Soc. Am., 62: 1485-1590.

Watkins, W.A., and Wartzok, D., 1985, Sensory biophysics of marine mammals, Mar. Mamm. Sci., 3: 219-230.

West, C. D., 1986, Cochlear length, spiral turns and hearing, 12th International Congress on Acoustics, 1: B-1.

Wever, E.G., McCormick, J.G., Palin, H., and Ridgway, S., 1971a, The cochlea of the dolphin, Tursiops truncatus: The basilar membrane, Proc. Nat. Acad. Sci., U.S.A., 68: 2708-2711.

Wever, E.G., McCormick, J.G., Palin, H., and Ridgway, S., 1971b, The cochlea of the dolphin, Tursiops truncatus: Hair cells and ganglion cells, Proc. Nat. Acad. Sci., U.S.A., 68: 2908-2912.

Wever, E.G., McCormick, J.G., Palin, H., and Ridgway, S., 1972, Cochlear structure in the dolphin, Lagenorhynchus obliquidens, Proc. Nat. Acad. Sci., U.S.A., 69: 657-661.

Wood, F.G., and Evans, W.E., 1980, Adaptiveness and ecology of echolocation in toothed whales, in: "Animal Sonar Systems," R-G Busnel and J.F. Fish, eds., Plenum Press, New York.

Zwislocki, J., 1981, Sound analyses in the ear: A history of discoveries, Amer. Sci., 69: 184-192.

SENSORY NEOCORTEX IN DOLPHIN BRAIN

Peter J. Morgane and Ilya I. Glezer[1]

Worcester Foundation for Experimental Biology, Shrewsbury, Mass. 01545, USA; [1]Department of Cell Biology and Anatomy, City University of New York Medical School, New York, N.Y. 10031, USA

ORIGINS AND ORGANIZATION OF CEREBRAL CORTEX

The fundamental plan and organization of the neocortex is the result of a slow evolutionary process which appears to have evolved through different transformations to eventually become the most intricate part of the nervous system. Its origins appear to have been at the reptile to mammal transition stage in the Triassic period of the Mesozoic era and continued for approximately 100 million years to reach a final prototype in the basal Insectivora in the late Cretaceous period of the Mesozoic era. Studies by Filimonoff (1949) showed that the great growth of the neocortex, and the complications of its structure as a whole, represent the principal characteristics of the evolution of the cerebrum in the course of phylogeny. In addition to the increase in the surface area of the neocortex, it also shows a qualitative progression and enrichment with differentiation of more specialized cellular elements. In this connection, the evolution of the cerebral cortex corresponds to a development and improvement of the sense organs, while the sequence of appearance of the cortical regions corresponds to the consecutive differentiation of these organs. In this regard, Pirlot (1987) argues against the view that brain structure, particularly neocortical organization, does not vary markedly across diverse species. This latter was based on concepts that selective forces may not act primarily on the brain, but rather on organs and systems which are in direct contact with the environment, i.e., the peripheral sensory-motor structures, particularly sensory receptors. We agree with Pirlot that brain features, especially those of the neocortex, are ecological variables. There certainly are significant structural diversities in neocortical organization across mammalian species, as well as differences in functional localization in the neocortex. This species diversity in cortical structure is present even though the fundamental structural plan of the neocortex is maintained across all mammalian species. This is important to keep in mind relative to many of the arguments presented in this discussion.

The mammalian neocortex shows basic similarities as regards the topographical relations of the main cytoarchitectonic formations throughout the comparative anatomical series extending from insectivores through primates. However, the developmental level and the relative dimensions of these formations are markedly different in the various mammalian orders (Brodmann, 1909; Filimonoff, 1949). These differences are due to both the phylogenetic features of a given order and to the ecological/functional loads placed on the sensory apparatus, the central parts of which form the main sensory fields of the cerebral cortex. In terms of sensory functions the physiological significance of the neocortex often has been interpreted by contrasting the development and variety of sensory capacities, which relates to the size of the primary sensory cortical fields, with the so-called associative capabilities which relate to the organization of the secondary and tertiary sensory cortical fields.

In considering its basic organizational plan, one of the most striking aspects of the cerebral neocortex is its lamination. In this regard, the individual neocortical layers have specific functions, as reflected in the afferent fibers they receive, the neuronal targets to which they project, and the receptive field properties of the cells within the various laminae. Brodmann (1909) first described and defined the basic principles of organization of the neocortex, based on its 6-layered structure in all mammals. This hexilaminar stratification pattern or "initial" type of organization for all mammals has formed the basis for the further evolution of the neocortex over the course of the past 100 million years. Over this period, the 6-layered neocortex has changed in each of the orders making-up Mammalia in an entirely independent fashion, both cytoarchitectonically, i.e., in the form and interrelationships of its constituent nerve cells, as well as quantitatively. Thus, a basic plan of neocortical organization is demonstrable throughout the mammalian series in which layers III and IV appear to be the major recipients of cortical afferents and from here nerve impulses are relayed mainly to layers II and III. Layer III is the source of long and short association fibers and intrinsic descending connections with layers V and VI, which contain a majority of cells projecting subcortically, as well as a number of intrinsic cells with ascending axons. This primordial "plan" suggests that the neocortex functionally is uniform in type at a rather fundamental level. However, the study of the varieties of cells, their synaptology and chemical identity, and the way they are interlinked by intrinsic wiring patterns in different mammals, clearly shows the existence of important variations, some of which may be relatively unique for a given species.

Relative to our views on the organizational plan of the convexity neocortex in cetaceans, we stress that the neocortex does not appear to be preprogrammed in terms of numbers and types of specific cellular elements, but rather has retained during evolution the capability to adapt or modify neuronal types to whatever form is most opportune for its specific function. This pertains particularly to the interneuron population which appear to be the best examples of neuronal adaptation as opposed to "hard-wired" pyramidal cells. In this regard, Jolicoeur et al. (1984) pointed-out that the neocortex, as a whole, is not preordained phylogenetically and that this feature enables animals to continuously modify behavior in response to novel events in the environment. Valverde (1986) has shown that in lower mammals, such as the hedgehog (Erinaceus europaeus), most of the neurons in the neocortex are pyramidal cells and also observed that most pyramidal cells in the neocortex of the hedgehog are atypical or transitional in type and appear to have no clearcut counterpart in other mammals. However, in previous studies (Morgane et al., 1985; 1986 a,b; 1990) we also observed this same feature in the cetacean visual neocortex (Fig. 1) and described the cytoarchitecture (Fig. 2) and neuronal types seen in convexity neocortex (Fig. 3) where these same types of modified, atypical pyramidal elements are the rule (Fig. 4). From Figure 1 it can be seen that the heterolaminar cortex (left hemisphere, small black squares) corresponds to the area in the depths and along the banks of the entolateral sulcus from which Sokolov et al. (1972), Ladygina et al. (1978) and Supin et al. (1978) obtained the shortest-latency visual evoked responses, while the homolaminar cortex shown by slanted lines in Figure 1 (left hemisphere) corresponds to the area from which they obtained the longest-latency visual evoked responses.

At this point it is important to summarize the major prevailing trends in progressive cortical differentiation since these form the basis of also assessing degrees of conservative traits in cerebral cortical evolution. In general, more advanced mammals show the following principal trends in neocortical evolution: (1) Stepwise appearance of granular cells superimposed on a basic, more hard-wired type of pyramidal organization. This disposition to granularization or stellarization, which also includes a strong tendency toward development of smaller perikarya, is a main progressive trend in higher neocortical evolution. There is emergence of an incipient inner granular layer IV via a dysgranular stage when granule cells still are intermingled with small pyramidal cells. Relative to the cortical growth ring concepts of Sanides (1970; 1972) discussed below, granular cells, including layer IV star types, make their first appearance in limbic and insular proisocortices; (2) Thickening of cortex, including particular layers, often with sublamination or splitting of individual layers; (3) Accentuation of lamination with clear delimitation of individual lamina; (4) Relative increase of lamina IIIc cells; (5) Dominance of development of the basal dendritic skirt of neurons (basal dendritic arborization); (6) In general, heightening of structural heterogeneity with greater varieties of neuronal forms, especially of the non-pyramidal types, which is characteristic of brains of higher mammalian forms, particularly primates; and (7) Some shift in emphasis, including increased laminar and cellular differentiation, from the inner cortical strata to the outer strata in most neocortical areas.

HYPOTHETICAL PRECURSORY TYPES OF CEREBRAL ORGANIZATION: THE "INITIAL" OR ARCHETYPIC BRAIN CONCEPT

Fossil evidence suggests that both the marsupial opossum and the placental insectivores existed in the Cretaceous pe-

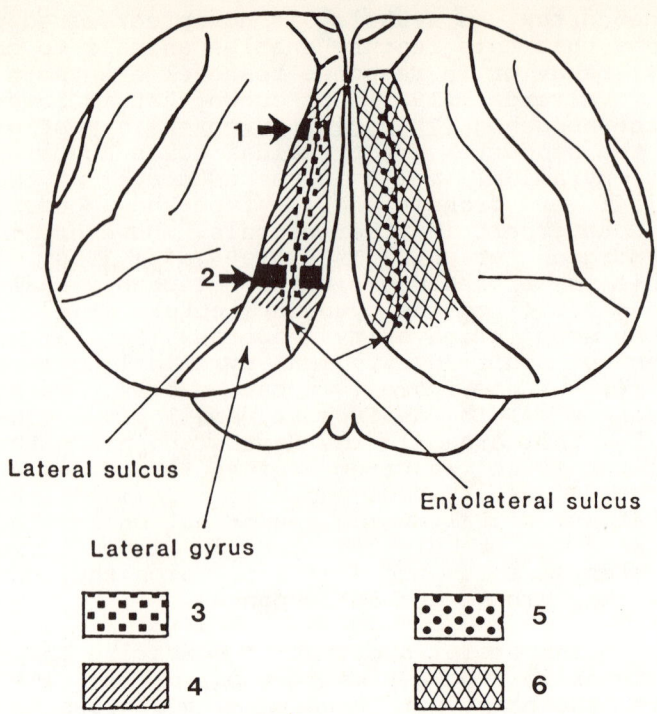

Fig. 1. Dorsal view of cerebral hemispheres in the dolphin showing the position, boundaries, and subdivisions of the lateral gyrus and surrounding sulci and gyri of the convexity cortex. The visual projection areas, as mapped physiologically by Sokolov et al. (1972), Ladygina et al. (1978) and Supin et al. (1978), are shown in the map of the right hemisphere as a cross-hatched area and black dots and the corresponding cytoarchitectonic areas are shown in the map of the left hemisphere as slanted lines and small black squares. Numbered sectors are: (1) sampled area (large black blocks) in anterior part of lateral gyrus representing the visual cortex medial and lateral to the entolateral sulcus, and (2) sampled areas (large black blocks) in the posterior part of the lateral gyrus representing the visual cortex medial and lateral to the entolateral sulcus. The blocks indicated by the heavy arrows 1 and 2 (left hemisphere) represent anterior and posterior areas of the lateral gyrus, including visual cortex medial and lateral to the entolateral sulcus. Boxes 3 and 4 represent cytoarchitectonic areas (indicated on left hemisphere) that we identified and they correspond to heterolaminar cortex (box 3) and homolaminar cortex (box 4). Box 5 (black circles, right hemisphere) represents the zone of the cortex shown in the Russian physiological studies to have the shortest-latency evoked potentials while box 6 (cross-hatched area, right hemisphere) represents the zone of the cortex shown in physiological studies to have the longest-latency evoked potentials (Magnification approximately 1/2 X).

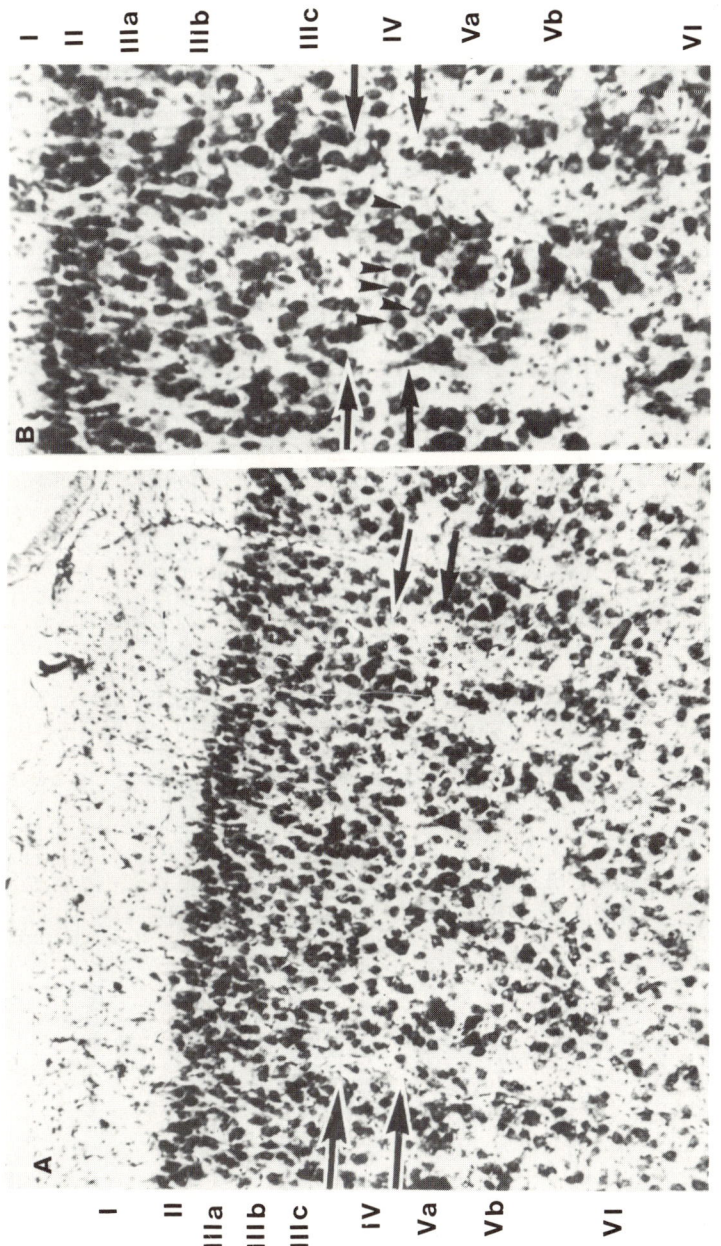

Fig. 2. Photomicrograph of cortical section from lateral bank of the entolateral sulcus showing cytoarchitectural arrangement of heterolaminar cortex of posterior part of the lateral gyrus in the dolphin (Tursiops truncatus). A. Note extremely wide layer I and marked accentuation of layer II. At this magnification it appears that a relatively acellular area occupies layer IV (defined by arrows). A few aspinous granule cells are present in layer IV (arrowheads in B). Layer V shows a well-expressed magnocellularity with pyramidal cells larger than those in layer III. Nissl stain. Magnification x 99. B. Higher magnification of A showing the presence of small granule cells (arrowheads) in incipient layer IV. Nissl stain. Magnification x 157.

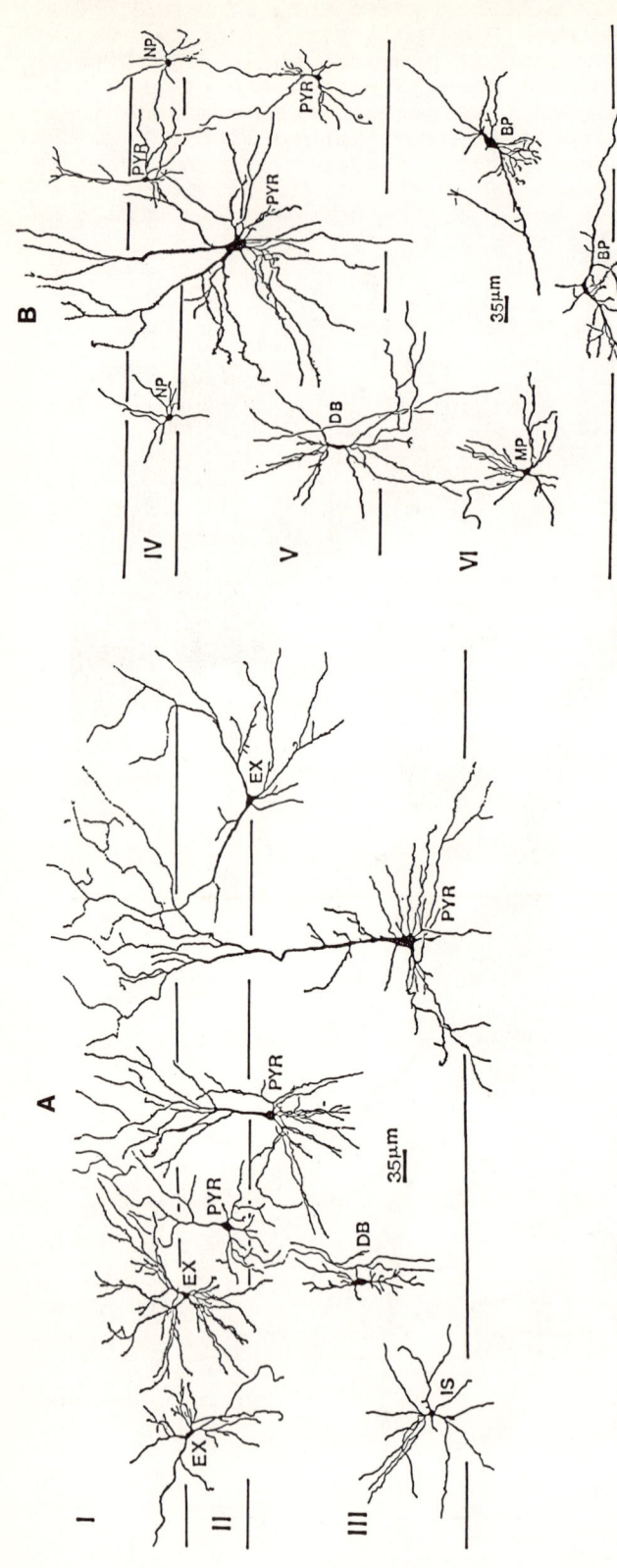

Fig. 3. Camera lucida drawings showing neuronal types in heterolaminar visual cortex of the dolphin (Stenella coeruleoalba). A. Composite drawing of neurons in layers II and III. In layer II extraverted neurons (EX) with wide-spreading apical dendrites are shown extending into layer I. An atypical, club-shaped pyramidal cell also is shown (PYR). Layer III is characterized by the presence of transitional, imprecise pyramidal neurons (PYR). Also, a nonpyramidal neuron of an isodendritic type (IS) is shown as well as a double bouquet (DB) cell. B. Neurons in layers IV, V and VI. Neurons in incipient layer IV are mostly aspinous nonpyramidal types (NP). Also, an atypical pyramidal cell is present (PYR). In layer V a large atypical pyramidal cell is shown (PYR) as is a smaller imprecisely-shaped pyramidal cell (PYR) and a double bouquet (DB) cell. In layer VI are neurons of the bipolar type (BP) and various pleomorphic, multipolar neurons (MP) which are difficult to classify precisely, though they are often of the isodendritic type. Rapid Golgi impregnation.

riod approximately 100 million years ago (Simpson, 1945). The North American opossum (<u>Didelphis</u> <u>virginiana</u>) is one of the least modified of marsupial mammals, while insectivores are the only known placental mammals which were present during the Cretaceous period. Paleontological evidence indicates that hedgehogs appear to have changed little from the ancestral prototype. Thus, there is direct evidence consisting of endocranial casts from the middle Oligocene epoch which suggests that the brain of the hedgehog has undergone

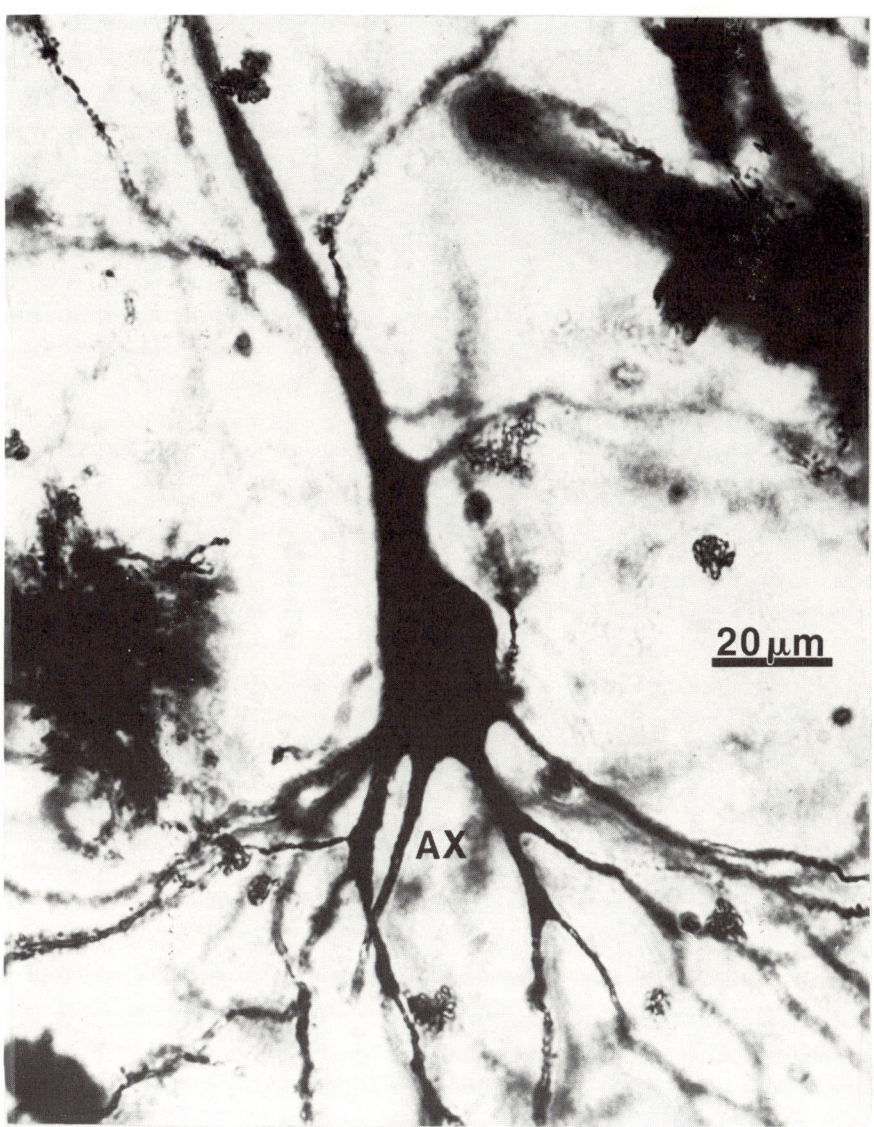

Fig. 4. Photomicrograph showing large atypical, club-shaped pyramidal cell in layer V of dolphin visual cortex. Initial segment of an axon (AX) also is shown. Rapid Golgi impregnation.

little change since the middle of the Tertiary period. The brains of extant basal Insectivora thus represent the best approximation of brain structure of extinct primitive Insectivora and depict models of the "initial" mammalian brain (Filimonoff, 1949; Diamond, 1967). The bat (<u>Myotis lucifugus</u>) brain also is thought to serve as an additional model of ancestral mammalian brains, as indicated by the comparative neocortical analyses of Sanides (1970, 1972). A comparison of forebrain organization in the opossum, hedgehog and bat is of special interest because these three animals are as close, perhaps, as any living forms to the archetypal mammals that first showed the multi-layered cellular pattern typical of mammalian neocortex. Comparison of forebrain organization and connections of the opossum, hedgehog and bat with that of reptiles, on the one hand, and with more specialized mammals on the other, has been thought to be a useful approach to clarifying the origin and development of neocortical structures. Consideration of the status of these brains as mammalian prototypes, or models of the "initial" brain, is important with reference to studies of the cetacean brain since the ancestors of modern whales returned to the water some 50-70 million years ago and, thus, are likely to reflect in their neocortical structure many of the features of these archetypal terrestrial mammalian brains before various specializations occurred. The phylogenetically early isolation of cetaceans on an independent evolutionary branch makes a study of their neocortices of particular interest.

It is obvious that it never can be established absolutely that any extant groups of mammals totally is representative of the ancestral organization of the brain. An exact evolutionary sequence of changes in any system, or even that such changes actually have occurred, can not be established definitively by studies of extant mammals. Thus, the absence of ancestral type brains prohibits complete answers on many aspects of brain evolution. The fossil record indeed is meager and never includes brains so that almost nothing on internal structure of the brain can be learned, for example, from endocranial casts. Hence, we instead must rely on indirect evidence from comparative embryology and comparative anatomy. Accordingly, various inferences from studies of extant brains need to be made in assessing hypothetical precursory types of cerebral organization. The use of recent material is possible since a number of archaic mammalian forms appear to have persisted almost unchanged (Simpson, 1945; Romer, 1966) and can thus be used profitably for the reconstruction of early phylogenetic stages of placental mammals. In this regard, hedgehogs especially are representative of an order closest to the source of a diverging evolutionary branch of mammals and for that reason can be considered as a baseline or reference against which to compare brain organization and development in other present-day mammals. There is considerable evidence that many, perhaps all, orders of placental mammals can be traced back to insectivore-like ancestors. Specifically, in considering a hypothetical, original type of cerebral cortex, the neocortical formations of basal insectivores also appear to be nearest to this original or "initial" type. The neocortex of the hedgehog is one of the most weakly-developed of all mammals. It also is characterized by a minimal size in relation to the archicortex and paleocortex. However, we emphasize the hedgehog brain as a model of the "initial" type of mammalian brain since it has fundamental archetypic organizational characteristics, including neocortical features, that are emphatic compared to those seen in other, more progressive, mammalian lines (Gould and Ebner, 1978; Gould et al., 1978). Relative to this, our studies, particularly in the past several years using Golgi and electron microscopic material for analysis of the cerebral cortex in several species of small whales (Tursiops truncatus, Phocoena phocoena, Stenella coeruleoalba, Globicephala melaena and Delphinapterus leucas), reveal many basic similarities between the micro-structure of dolphin neocortex and that of the hedgehog (Morgane et al. 1985; 1986 a,b; 1990; Glezer et al., 1988; Glezer and Morgane, 1990).

From an evolutionary standpoint and, especially in light of the types of neurons we see in dolphin neocortex, it also is essential to consider amphibian and reptilian cortices in relation to neuronal types seen in presumed evolutionary conservative mammalian forms. As is known, no separated cortex has been formed in amphibians and reptiles and the bulk of neurons is concentrated in the depth of the wall of the hemisphere and around the cavity of the brain ventricle (Herrick, 1948; Mazurskaya et al., 1966). As pointed out by Ramón y Cajal (1911, 1955) and Herrick (1948), their most characteristic feature is that the extension of the dendrites, which diverge from the bodies of the nerve cells in a fan- or cone-like manner, is oriented towards the surface of the hemisphere (Fig. 5). A likewise characteristic is the fact that the dendrites extend in two or three bundles from the angles of the cell and from that of its parts which face the surface of the hemisphere. The phylogenesis of neocortical pyramidal neurons in several vertebrate species, paralleled by ontogenesis of a pyramidal neuron in a mammal, is shown in Fig. 6.

At the lower stage of phylogenesis of mammals, as typified by basal insectivores, a prevalence of diverse transitional forms of mixed neurons is observed varying between pyramids, spindles and stars, as emphasized in the investigations of Zhukova (1953). These findings testify to a still feebly pronounced qualitative structural differentiation of the cortical neurons in insectivores in comparison with advanced mammals. In spite of this lack of differentiation, the principal structural distinctions between the efferent and internuncial (star) neurons, as represented in their typical forms, already manifest themselves quite distinctly in the neocortex of insectivores. It is also of interest in the context of our findings in the dolphin brain (see below) that in the hedgehog the neurons of the phylogenetically older, deep levels of the neocortex (layers V and VI) reach a considerably higher degree of development than the neurons of the superficial levels of the cortex. These latter, i.e., the complex of layers IV, III and II, are formed later in phylogenesis and are represented in the basal insectivores by a feebly differentiated complex. This calls to mind the organizational pattern of the hedgehog neocortex noted by Valverde (1983) and Valverde and Facal-Valverde (1986) in which they described a type of allocortical organization of the upper cortical layers superimposed on somewhat more typical neocortical deeper layers. As indicated below, we have described essentially this same pattern in the dolphin convexity neocortex (Fig. 3).

CETACEAN EVOLUTION AND NEOCORTICAL DEVELOPMENT

Cetaceans are thought to have descended back to the sea some 50-70 million years ago. They adapted themselves to the new aquatic conditions and appear to have preserved characteristic features of the original structure of the brain of primitive mammals extant at that time in greater measure than land animals. At the same time, the cetaceans were in a new environment conducive to develop specific features of brain adaptation not characteristic of mammals that remained on land. Given these evolutionary possibilities, there is little question that studies of the cetacean cortical structure thus may provide important information on prototypic mammalian brains and, in so doing, lead to answers of some important problems relating to the fundamental plan of the mammalian cerebral cortex and its evolutionary history.

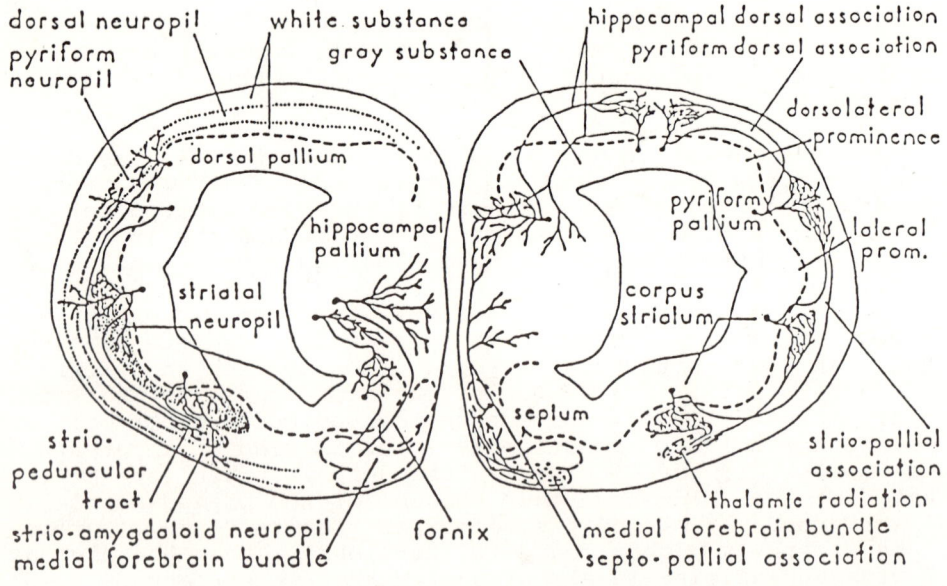

Fig. 5. Diagram of coronal sections through the end-brain of the amphibian <u>Necturus</u> depicting neurons with only subpial dendritic arborization (After Herrick, 1948). Note the cone-like extensions of dendrites from the angle of the neurons toward cortical surface (Magnification approximately 13.5 X).

In our studies aimed at unravelling the principles of cortical organization in the whale brain, we have taken the tack of attempting to reconstruct an "initial" or ancestral reference type of mammalian cerebral cortex, and postulated various pathways of derivation from this hypothetical "initial" basal prototype (Glezer et al., 1988; Morgane et al., 1990). A basic assumption is that cetaceans, in entering water, adapted to a medium that was unfamiliar to mammals and were able, to a greater extent than terrestrial forms, to preserve signs of the initial brain structure of terrestrial mammals extant at that time, as well as to develop specific additional adaptations to the aqueous environment. In this sense, whales are unique for the study of several crucial problems in mammalian evolution. They have a totally aquatic mode of life shared by no other order of mammals

Fig. 6. Diagram from Ramón y Cajal (1911) showing phylogenesis of cortical neurons (A-D), paralleled by ontogenesis of a pyramidal cell of a mouse (a-e). A, frog; B, lizard; C, mouse; D, human.

(the Sirenia being largely estuarine, herbivorous mammals). In this regard, we emphasize that the formation and development of the neocortex occurred in most mammals that are adapted largely to an aerial environment.

Our studies of the whale brain over the past 20 years indicate that the cortex of the dolphin is generalized to serve as a model of the status of mammalian brains of some 50-70 million years ago. The cetacean brain may, accordingly, represent a prototypic mammalian brain of special interest since whales returned to water before terrestrial forms specialized and, in particular, granularized their cerebral cortices. Thus, the return to water was before the rise of koniocortical foci of stellarization and hyperpyramidalization, these latter indicating, according to the concepts of Sanides (1970, 1972), the highest degree of sensory and motor cortical specialization seen in advanced terrestrial mammals. The assumption of similarity of the structural organization of the cetacean neocortex to the original architectonics of the neocortex of primitive mammals makes it possible to suggest that granularity of the neocortex of land mammals developed much later than at the time at which the ancestors of modern Cetacea descended into the sea. In considering the differences in the mode of structural evolution of the brain, as originally outlined by Zvorykin (1977), we hypothesize that in the cetacean brain there is a great increase in territory occupied by neocortex, i.e., there is an extremely vigorous quantitative expansion of neocortex, but without any substantial reorganization of the basic 6-layered stratification plan seen in the initial prototypic mammalian brain. It is of interest, relative to our arguments below, that this mode of cortical evolution is considered the most ancient among present-day mammals (Zvorykin, 1977).

FUNDAMENTAL PLAN OF CORTICAL ORGANIZATION IN DOLPHIN BRAIN

We will now construct some broader principles of organization of the convexity neocortices of dolphins, particularly the sensory neocortices. Our plan is to put this into an evolutionary and comparative neurological perspective to relate to the concepts outlined above. This includes more fully developing a conceptual plan of organization of the cetacean sensory neocortex and an integration of this with the cortical growth ring concepts of Sanides (1970; 1972) into an overall view of the phylogenetic level and status of the dolphin cerebral neocortex.

The ultimate goal in studying a portion of the central nervous system, such as the cerebral cortex, is to gain some insights into its function and the role it plays in the behavior of the animal. We can contribute to this goal by determining the neuronal composition of the structure, what kinds of connections the constituent neurons produce, how the assemblies of neurons are organized, and their chemical identity. In studies of the cerebral cortex in a variety of mammals, significant progress has been made in determining what types of neurons are present in various cortical areas, especially in particular layers of functionally defined cortices, and how they are related synaptically. However, we are still far from understanding how neurons in the cerebral neocortex are arranged to form functional groups, though the basic plan of vertical-integrating systems or columns is well-known and yielding to physiological analysis. Before we begin to understand how the neurons are organized in the dolphin neocortex and interact with each other, we need to know how many neurons of each type are present in a particular cortical area, what percentages of the neuronal types are contained within the population, how they apportioned to the different layers and the types of synaptic connections that are seen on particular neuronal types identified by their morphology and transmitter neurochemistry. In this regard, we have recently described a columnar type organization of the visual neocortex in the dolphin that differs considerably in terms of columnar size and number from that seen in other mammals (Morgane et al., 1988). We have also begun to identify local neuronal circuits in the neocortex (Morgane et al., 1990) and defined by immunocytochemical analyses several classes of chemically identifiable neurons (Glezer et al., this volume). We now need additional information on the thalamic input to particular neocortical layers and on the immunocytochemical characteristics of neuronal families in individual cortical laminae. The entrance of thalamic afferents into the cerebral cortex constitutes an important issue from a phylogenetic viewpoint. These afferents not only contribute to the differentiation of the developing cortex into a neocortical type, favoring the development of neocortical pyramidal cells, but, in addition, their maturation also appears to shape different patterns of intrinsic connectivity in the adult brain. Some studies of thalamo-cortical relations, using axon transport methods, are being carried-out by Garey and Revishchin (this volume). Previously, the only study of thalamocortical relations in dolphin brain were those of Krasnoshchekova and Figurina (1980) using Fink-Heimer degeneration techniques following medial geniculate lesions. Some of their findings, relatable to the organization of the dolphin brain that we propose, are summarized below.

To assess the neocortex in an evolutionary perspective, which we think is essential in order to draw conclusions about the type of neocortical organization present in the whale brain, it is important to first define its fundamental organizational plan and establish the nature of its deviations from the basic design. The primary characteristics of neocortical organization might be defined as: (1) the presence of so-called true neocortical-type pyramidal neurons; and (2) the presence of topographically organized and sensory-modulated thalamocortical afferents arborizing and terminating most consistently at middle cortical levels. On the other hand, the allocortical formations show no equivalent type pyramidal cells and their major afferent supply is formed through inputs to the first cortical layer. Thus, the essential characteristics of the various cortical formations, both phylogenetically new and old, are well-defined and can be used as bases for establishing the level of cortical development in different species. The dolphin convexity neocortex, as discussed below, essentially does not show the above typical features of neocortical organization, i.e., it contains neocortical pyramidal elements that are of the transitional type which are imprecisely formed and the inputs from the thalamus appear to ramify most strongly in Layer I and, to a lesser extent, in layer III (Krasnoshchekova and Figurina, 1980). We have pointed-out that the dolphin convexity neocortex demonstrates a type of hybrid organization with strong allocortical characteristics of its upper layers superimposed on lower cortical layers having features somewhat nearer to a more typical progressive neocortical plan (Morgane et al., 1985; 1986 a,b; 1990). These same characteristics have been described in the neocortex of the hedgehog by Valverde (1983) and Valverde and Facal-Valverde (1986).

We also can define more specifically the fundamental characteristics of cellular elements of sensory neocortex in mammals. Here, there are also two major morphological signs by which projectional sensory zones of cortex have been distinguished in terrestrial mammals: (1) the presence of layer IV granules which are predominantly small stellate type neurons (true granular elements); and (2) the presence of large numbers of small, primarily stellate, neurons which are disseminated throughout the cortical layers. The dolphin brain lacks this fundamental type of sensory cortical organization; there is only an incipient granularization with no hypergranular "cores", such as are seen in primary sensory cortex of most advanced mammals. Further, the granular cell layer IV itself is incipient or absent. The granule cells we find in dolphin neocortex are mostly large elements of the isodendritic type (Fig. 7), whereas the process of stellarization in evolution involves a strong trend toward smaller perikarya which also means smaller dendritic fields. In fact, the absence of layer IV can be considered as a sign of intermediate-type cortex (Filimonoff, 1949; Amunts et al., 1977). This feature defines the neocortex of the dolphin as showing transitional type cortical features and, therefore, not reaching the highest developmental stage of neocortical organization as defined by Sanides (1970; 1972).

SANIDES' CORTICAL GROWTH RING CONCEPTS

In defining the evolutionary status of the dolphin and other whale brains, the conceptual views of Sanides (1970, 1972) are especially useful, though some aspects of these concepts can only provisionally be accepted. Sanides' views of cortical evolution are thought-provoking and useful in serving as a basic starting-point for assessing the status of cortical formations across the phylogenetic scale. We have, accordingly, conceptualized the organization of the dolphin neocortical formations based, in part, on the cortical growth-ring analyses of Sanides. On the basis of extensive comparative cytoarchitectonic and myeloarchitectonic studies, Sanides elaborated a plan of neocortical evolution based on enumeration and analysis of the so-called growth rings of the neocortex. In this view, different cortical architectonic fields represent successive waves of circumferential differentiation outward from archicortex and paleocortex resulting in two moieties of neocortex. Sanides (1970; 1972) appropriately has termed these waves of cortical elaboration the "growth rings" of the neocortex. These concepts provide meaningful information on evolutional directions of cortical differentiation and may also shed light on

the level of cortical evolution that has been reached in a
given brain in the phylogenetic series. We have used these
same principles in recent years in our assessment of the cetacean brain and have "defined" and assessed the cetacean
cerebral cortex in this context (Morgane et al., 1985; 1986
a,b; 1990). One goal has been to trace sequences of cortical architectonic differentiation back to the primary territories of origin and outward toward putative hyperspecial-
ized "cores" of cortical differentiation. The premise is
that the gradations originating from phylogenetically older
cortices, in part, express the emerging makeup and assembly
of more recent cortices and, therefore, represent directions
of cortical differentiation during evolution.

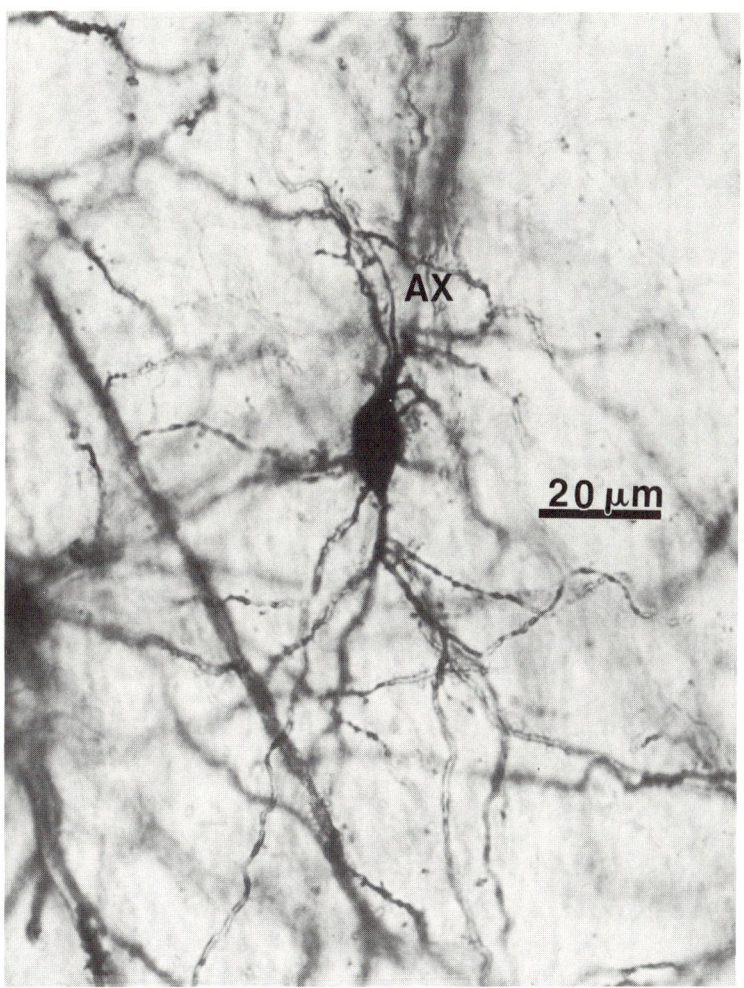

Fig. 7. Photomicrograph showing a giant stellate cell in
layer IIIb of visual cortex of the dolphin
(Stenella coeruleoalba) with its ascending axon
(AX). Note fusiform shape of the cell body and
long, weakly-spined dendrites.

We assume that in the convexity cortex waves of cortical growth and differentiation during evolution represent the adaptive responses to an ever-changing environment. The cortex evolves by successive waves of circumferential differentiation outward from the older cortical moieties represented by the paleocortical and archicortical formations. Together, these two ancestral cortices comprise what we term the allocortex. There is an explicit relationship between the cortical growth rings to thalamic differentiation and to differentiation of the affiliated peripheral sense organs. In these type of approaches, it is essential to examine and assess the successive intermediate structural steps in the cortical formations, i.e., periallocortex and proisocortex, which are intercalated between primitive allocortex and mature isocortex. Essentially, the dual nature of the neocortex means that there are two neocortical moieties, one differentiated in stages away from the hippocampus (archicortex) and the other in stages away from the piriform cortex (paleocortex). The cortical growth-ring concept can be summarized as follows: The first stage of a laminated cortex is the two-trata periarchicortex medially and peripaleocortex laterally. Periallocortex represents the first incipiently laminated cortex and forms the first growth ring of neocortex. The periallocortex is considered the first-step of neocortical evolution. As a second cortical growth ring, the proisocortical belt of the limbic lobe appears medially and insular cortex appears laterally. A third growth ring also is formed of two moieties, a paralimbic one medially, and parinsular one laterally. These are thought by Sanides (1970; 1972) to be sites of additional sensory and motor cortical representations (secondary and tertiary) which are not of the classically defined "primary" types. The paralimbic belt of fields thus may represent an intermediate stage between limbic proisocortex and the individual cortical fields of the convexity. These paralimbic fields are dysgranular or incipient granular (and in some cases agranular) and contain the supplementmentary motor and prokoniocortical representations as they do in higher primates, including humans (Sanides, 1970; 1972). The intermediate stage of the paralimbic belt in architectonic differentiation on the way from the limbic cortex is paralleled by a similar stage seen in the parinsular belt, composed of promotor and prosensory cortex, or so-called secondary sensory-motor areas on the way from insular proisocortex. In general, limbic features of the cortex undergo a diminution in the paralimbic and parinsular areas and new features are added. The last stage of cortical evolution is, in the case of sensory cortex, development of special cores of koniocortex that appear within the sensory regions. In the case of the primary motor cortex, the area gigantopyramidalis appears. It is noteworthy that the last stage of cortical evolution, i.e., koniocortex in the case of primary sensory cortical evolution, appears as a "core" packed with granule cells, rather than a ring within the cortical sensory regions. Relative to the cetacean neocortex, neither hypergranular cores of koniocortex nor giant pyramidal cells can be found and on these grounds we tentatively have put forward the view that the cetacean neocortex has not reached the final stage of cortical evolution as defined by Sanides (Morgane et al., 1985; 1986 a,b; 1990). The short-latency visual and auditory responses found in convexity neocortex of the dolphin by Sokolov et al.

(1972), Ladygina et al. (1978) and Supin et al. (1978) would certainly be termed "primary", not only due to their short latency, but also to the fact that each area responded to only one type of sensory input. In examining this cortex, we found no evidence of a koniocortical type of organization, but rather one with dominant paralimbic, parinsular type histological features. It was partly on this basis that we chose to describe this cortex as resembling the paralimbic-parinsular or third cortical growth ring stage of Sanides. This should not, however, be interpreted to mean that functionally there are no "primary" sensory cortices in the dolphin brain, but rather that these zones of short latency responses have not reached evolutionary maturity approximating those in more advanced terrestrial mammals.

In regard to the above growth ring concepts, hedgehogs (Erinaceus) can be considered "survivors from the Eocene" and are pertinent to that group of insectivores from which the primates arose. They have been considered unique among the recent Eutheria in keeping the paralimbic-parinsular stage as the highest leading level of cortical differentiation. Relative to this, the latest step in motor and sensory cortex evolution appears to have occurred in somewhat advanced mammals about 50 million years ago in the Eocene epoch of the Tertiary period. The hedgehogs, as survivors of the Paleocene epoch when archaic mammals were dominant thus did not reach the more advanced stage of sensory and motor cortex evolution (Sanides 1970; 1972). Likewise, we have pointed-out (Morgane et al., 1985; 1986 a,b; 1990) that the whales, which went to water in this same general period, did not develop granular-type cortices or large macropyramidal motor neurons and in the aquatic medium kept a pattern of highest cortical development approximately equivalent, as regards to histological organization, to the third growth ring stage, i.e., at the paralimbic and parinsular stage of cortical evolution.

CYTOARCHITECTONIC TRENDS IN PHYLOGENY

Classically, the sensory and motor regions of the neocortex were referred to as "primary" and the integration or "interpretive" cortices (formerly termed "association" cortices) were termed secondary and/or tertiary regions. These latter areas may well represent phylogenetically more ancient and primitive older levels of sensorimotor control than the classic, so-called "primary" representations as discussed in detail by Sanides (1970; 1972). In this view, the koniocortical areas or cores of hyperspecialization with intensive granularization and which form the heaviest myelinated cores of the sensory regions, would appear to represent the most recent stage in sensory cortex evolution. In a similar sense, the area gigantopyramidalis, which forms the heaviest myelinated core of the motor region, may represent the most recent stage in motor cortex evolution. These inferences also are suggested by the highest cytoarchitectonic differentiation of these primary areas and by their great degree of functional specialization. Sanides (1970; 1972) presents convincing evidence that the hypergranular sensory cortex and motor cortex containing giant pyramids represents the most advanced level of architectonic specializations as compared with the integration or so-called "association" cortices which remain more generalized in their organization. These concepts relate to the views of Sanides (1970; 1972) that secondary and tertiary sensory types of integration cortex may have evolved before primary cortices developed. This part of Sanides' view of cortical evolution is, of course, open to debate. In fact, Kaas (1987) argues against such a possibility and points-out that much of the massive development seen in cetacean neocortex may represent non-sensory areas and that the significance of these vast expansions is unknown. Obviously, many of these issues are unresolved, though we feel there is considerable conceptual usefulness in tentatively relating the cortical growth ring concepts of Sanides to our view of whale brain evolution and the status of their neocortical formations, particularly in terms of the organization of their sensory areas.

Sanides (1970; 1972) stresses that the paralimbic and parinsular belt of cortical fields represent an intermediate cortical evolutionary stage on the way from limbic and insular proisocortex to the individual so-called "primary" fields of the convexity. In advanced terrestrial mammals, higher stages of sensory-motor organization are flanked medially by paralimbic supplementary motor and ventro-laterally by parinsular secondary sensory-motor areas, and may be preceded in evolution by a stage where the paralimbic and parinsular representations still are contiguous (Sanides 1970; 1972). This latter is the situation we feel prevails, perhaps in somewhat modified form, in the whale neocortex.

CONSIDERATION OF NEURONAL TYPOLOGY IN CORTICAL EVOLUTION

Before discussing the neuronal typology of dolphin convexity sensory neocortex, we present a general classification of neurons of both the nonpyramidal- and pyramidal-type as a basis for understanding and interpreting the meaning of the neuronal elements in the dolphin neocortex. In general, nonpyramidal neurons are separated into three types of stellate cells: 1) short-radiators, 2) long-radiators, and 3) double bouquet cells, this latter being a special type of nonpyramidal cell which is not common in lower forms. Double bouquet cells have vertically-oriented axonal strands intimately related to pyramidal apical dendrites and are spine-free stellate cells. Nonpyramidal neurons also can be classified as multipolar, bipolar, and bitufted and as spinous or aspinous.

Ramón-Moliner (1975) pointed-out that there are two opposite groups of nerve cells from the viewpoint of their dendritic geometry, i.e., generalized and specialized. They are extremes of a spectrum made-out of various intermediary types. From this dendroarchitectonic point of view, a number of neuronal families have been characterized. Ramón-Moliner used this approach to reconstruct the probable phylogenetic history that led to the dendroarchitectonic neuronal families observed in the mammalian brain. Two main types of neurons were postulated to be characteristic of the primordial vertebrate brain: (1) nerve cells with subpial tufts, the dendritic configuration being reminiscent of that of the neurons of the dentate gyrus in the mammalian hippocampal formation and; (2) neurons with scarce and long dendrites, usually rectilinear and overlapping in type, very reminiscent of those types found in the reticular core and periventricular regions of the mammalian brainstem. These latter, so-called isodendritic cores are primogenial regions usually associated with limited input discrimination. The archaic cores of the brainstem can be considered as primordial entities containing a pool of pluripotential neurons which, in the course of evolution, remain diffusely distributed throughout the brainstem and which retain extremely generalized morphological features. When such undifferentiated correlation-type neurons are seen at the level of the neocortex, as we have described in the dolphin brain (Morgane et al. 1985; 1986 a,b; 1990), this is considered additional evidence of a more generalized cortical organization where neurons have not become highly specialized (Fig. 8). Thus, dominance of long-radiator neurons in dolphin neocortex over short-radiator types points to a more conservative type of cortical evolution.

Relative to considerations of neuronal typology in an evolutionary sense, various cellular types that exhibit generalized or primordial features are well-represented in the neocortices in the basal insectivores, bats, and, as we have pointed-out, in several species of small whales. In this regard, reticular formation cells represent the remnants of the presumed "original" type of neuron of the brain which have served as the matrix for various specific types of neurons that were differentiated eventually from it in the course of phylogeny. Leontovich and Zhukova (1963) describe, as a peculiarity in the hedgehog neocortex, the appearance of scattered "generalized neurons" which in other mammals are found only in neocortical layer VI. It is important to point-out that this is the same cell type which dominates the reticular core of the brainstem. This reticular neuron type has long rectilinear dendrites, hence the term "long-radiator", proceeding from the cell body in various directions without any definite orientation. These types of cells first were described by Ramón-Moliner and Nauta (1966) and designated as isodendritic neurons. In most advanced mammals scattered reticular type neurons are seen in small numbers only in the phylogenetically oldest layer VI of the neocortex, whereas in the cerebral cortex of the hedgehog they are quite numerous and more widespread across all the neocortical layers. We have described these same isodendritic, generalized-type, long-radiator neurons in several layers of the convexity neocortex of the dolphin (Morgane et al., 1985; 1986a,b; 1990) and this provides another basis for our considering that the whale neocortex shows an overall dominance of conservative features.

Relative to the short-radiator neurons among the stellate cells, it has been shown that, in addition to the fact that they are distinctly increased in number in the neocortex of rodents and rabbits, as compared to dolphins and hedgehogs, they also have on the average significantly smaller somas than in the insectivore and cetacean brains we have examined. This increase in short-radiator elements in more advanced mammals may be interpreted as another expression of the trend towards granularization of the stellate cells, which has culminated in the cerebral neocortices of higher primates. Their relatively low numbers in hedgehog and in dolphin, indicate a definite lack of a trend towards granularization of the stellate cells. In this regard, the presence of large stellate cells with smooth dendrites intercalated between input neurons and layer V pyramidal cell axons appears to be a basic circuit in both the hedgehog and mouse neocortex. Since we have demonstrated the presence of these large stellate cells as a major characteristic of the dolphin neocortex (Morgane et al., 1985; 1986 a,b; 1990; Glezer et al., 1990), we feel this same elementary circuit exists in dolphin neocortex. Our ongoing studies now are seeking to examine these local circuits in more detail in the neocortices of the dolphin, hedgehog and bat.

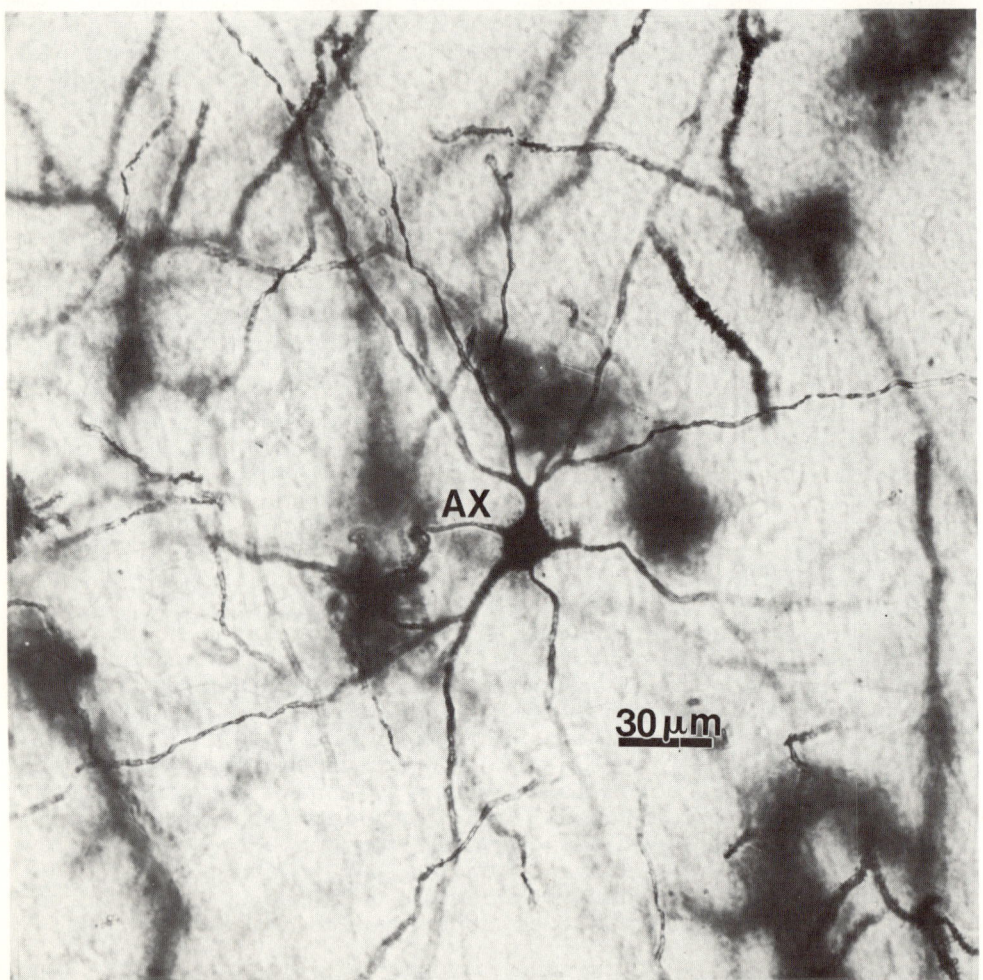

Fig. 8. Photomicrograph showing an isodendritic, relatively non-differentiated neuron in layer IIIa of visual cortex of the dolphin (Stenella coeruleoalba). Note long retilinear dendrites spreading in all directions from the cell body with minimal branching. Initial segment of an axon (AX) arises from the lateral side of the cell body and is tangential to the pial surface. Rapid Goldi impregnation.

We have found that the neocortices of the dolphin and various species of insectivores have a considerably greater share of long-radiator cells even when compared to the neocortex of rodents and the rabbit. In rodents and rabbit, the short-radiator cells are by far the prevailing stellate cell type and they appear to develop at the expense of long-radiator elements. These short-radiators correspond to the classic picture of star cells of granular layer IV described by Ramón y Cajal (1911) and Lorente de Nó (1938).

With regard to pyramidal-type neurons in dolphin convexity neocortex, there is clearly a predominance of modified, atypical pyramidal type forms over all other types of cells (Fig. 4). This feature has been enumerated many times in the literature by Kesarev, 1969; Kesarev et al., 1977 a,b; Krasnoshchekova, 1978; Burikova and Krasnoshchekova, 1983; and Morgane et al., 1985; 1986 a,b; 1990. In general, these modified or transitional-type pyramids are almost universally atypical and imprecise in shape in the dolphin neocortex, thus giving the appearance of immature, intermediate type neurons. These are often of transitional-type form, i.e., club-shaped, fusiform-shaped or of quadrangular and clavate shape with only a few showing typical triangular configuration. As to layer V pyramids, it would not in any case be appropriate, even in the dolphin motor cortex, to term these as "giant" pyramids since they are often three times smaller than the giant pyramidal neurons described by Betz in carnivore and primate motor cortex. Kesarev (1969) measured their mean maximum dimensions in dolphin motor cortex as 59 x 27 microns, whereas we found them to average 57 x 25 microns in several species of small whales. In general, we agree with Kesarev (1969) and Kesarev et al. (1977 a,b) that the abundance of transitional, imprecisely-formed pyramidal neurons, along with small numbers of stellate cells, serve as strong evidence of the low differentiation of the convexity neocortical formations in the dolphin brain.

Kesarev (1970) characterizes dolphin neocortex as being of monotonous character and showing incomplete splitting of the neocortical plate. He postulates that similarity in structure to the periarchicortex makes the name "neocortex" appear rather approximate and actually inappropriate in Cetacea. We do not agree with this extreme view but, as pointed-out below, the overall character of the neocortex in whales is one of markedly conservative evolution and one in which the final level of cortical differentiation, in terms of the growth rings of Sanides, does not appear to have been reached. Our surveys of the cerebral convexity neocortex of the dolphin reveal a remarkable similarity of structure with the intermediate cortical formations as well as, in the case of neocortical layers I and II, with the allocortex and periallocortex. Convexity neocortex in the dolphin clearly combines the features of both conservative and progressive brains in comparing the upper cortical laminae with the lower cortical laminae. The upper laminae are organized dominantly as intermediate cortex and allocortex, which appears superimposed on a more progressive type of neocortical organization of the lower layers. Thus, in several important features the neocortex of small whales corresponds to the "original architectonics" of the brain in the earliest mammals.

Granularity, which is markedly attenuated in the whale cortex, appears to be a later sign of differentiation of the neocortex of terrestrial mammals, developing some 50 million years ago after the ancestors of modern cetaceans had already made the transition from the terrestrial to the aquatic environment.

CONCEPT OF EXTRAVERTED NEURONS AND THEIR EVOLUTIONARY MEANING IN NEOCORTEX

Extraverted telencephalic neurons, as originally defined by Sanides (1970; 1972) and Sanides and Sanides (1972; 1974) in the European hedgehog (Erinaceus), Asian hedgehog (Hemiechinus) and African hedgehog (Aetechinus) and bat (Myotis lucifugus) brain, comprise all neurons which border the subpial zonal layer and preserve clear predominance of subpial dendrites over basal dendrites. Such extraverted neurons are not only found in the primitive allocortex (comprising paleocortex and archicortex) but also in the adjacent intermediate cortical formations of the periallocortex and, furthermore, even in the next cortical belt, the proisocortex (limbic and insular cortex). Sanides (1970; 1972) and Sanides and Sanides (1972, 1974) describe nerve cells, belonging to the extraverted cell family, over the entire neocortex in a group of primitive mammals, including the hedgehog and bat, and we have found similar cells over the entire neocortical convexity in the dolphin (Morgane et al., 1985; 1986 a,b; 1990; Glezer and Morgane, 1988). Figure 9 shows such an extraverted neuron from layer II of the dolphin visual cortex. This peculiar pyramidal cell type has been found to be a dominant form of neuron in cortices of all mammals showing a markedly accentuated layer II. Depending on the degree of extraversion, which varies considerably across conservatively organized neocortices in different mammalian species, they are strikingly similar to the superficial cell condensations characteristic of allocortex, periallocortex and, to a lesser extent, proisocortex.

In the hedgehog, layer II of paleocortex ("olfactory" cortex) contains large undifferentiated type pyramidal cells having ovoid or triangular (pyramidal) shaped bodies with two opposite bunches of dendrites richly endowed with spines. The ascending (pialward) dendritic bunch arborizes profusely in layer I where it receives a strong zonal, perhaps largely olfactory and thalamic, input. Such a type of dendritic polarization represents a stage of phylogenetic development more fully expressed at the amphibian level where one sees the most accentuated dendritic extraversion and the total absence of basal dendrites (Fig. 6). This latter characteristic is most typical of the archicortex and dramatically is illustrated by the granule cells of the dentate gyrus. We emphasize that in primitive mammals, such as hedgehogs and bats, this type of dendritic polarization is seen in layer II neurons over the entire convexity neocortex, whereas in advanced mammalian forms it is confined only to the older cortices. In these cortices, it represents an expression of the fact that the zonal layer still represents, particularly in allocortex and periallocortex, the main afferent and association plexus of the cortex. We have described a completely similar feature in the dolphin neocortex (Morgane et al., 1985, 1986 a,b, 1990) and this has been verified by the recent studies of Ferrer and Perera (1988).

Dendritic extraversion is a neuronal and architectonic feature of considerable evolutionary value. It is a conservative feature in neocortical evolution, being a protoneocortical mark indicating the originally prevailing layer I input of the axodendritic type. It marks, in addition to the allocortical formations, the periallocortical and proisocortical stage and, to a limited extent, the paralimbic/parin-

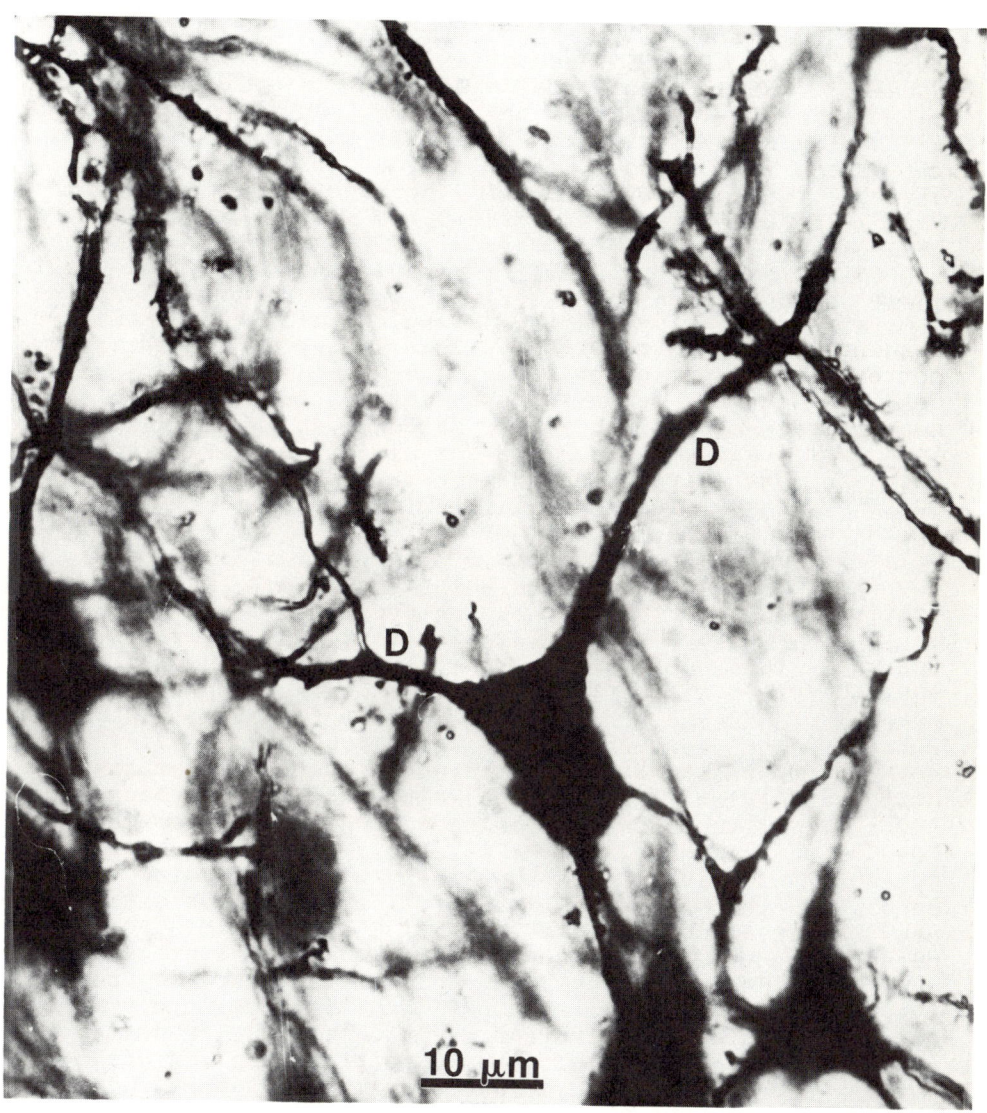

Fig. 9. Photomicrograph showing an extraverted neuron in layer II of visual cortex of the dolphin (Stenella coeruleoalba). Note quadrangular shape of the cell body and two apical dendrites (D) spreading widely into layer I. There is a relatively weak representation of the basal dendrites.

sular stage of neocortical evolution (Sanides 1970; 1972). It is a common characteristic of the ancient neocortical growth rings in the more recent placental mammals, whereas it actually prevails over the entire extent of the neocortex in some primitive mammals, including insectivores, bats, marsupials and monotremes. In advanced mammals accentuated layer II extends only to the level of insular and limbic proisocortex. Thus, mature neocortex in advanced mammals lacks the conspicuously accentuated layer II in strong contrast to its presence in the first and second cortical growth rings. In advanced mammalian forms layer II accentuation tends to fade-out by a stepwise decrease as progressive isocortical features appear. In these progressive mammalian forms we see in layer II of the neocortex a small pyramidal cell with either a short or no apical shaft and the same range of apical dendritic bouquet as basal skirt dendrites. The shapes of the layer II extraverted pyramidal cells and their dendritic pattern disclose their evolutionary origins from the oldest type of telencephalic neurons with subpial tufts, which are found in lower vertebrates where they precede formation of a superficial cortex. Extraverted neurons appear in Golgi material to be inverted cones with a prevailing zonal arborization of dendrites, or subpial panache, with a generally tufted character of these subpial dendrites. In the cetacean convexity sensory neocortex, we found that most of these layer II cells show a stronger dendritic arborization into the zonal layer than into the cortical plate by the basilar dendrites. Their presence in the convexity neocortex in dolphins is clearly a sign of conservative cortical development and this is another reason why we define cetacean neocortex as being of an intermediate and evolutionary conservative type.

From the evolutionary point of view, Ramón y Cajal (1911) adduced evidence that the pyramidal cells of the mammalian neocortex have developed from the amphibian stage of the endbrain neurons showing only a subpial dendritic arborization (Fig. 6). In primitive mammals, such as the North American opossum, the zonal layer, the original synaptic site of the cerebral cortex, still appears to play such a role with regard to the thalamic input, which further explains the significance of the extraverted layer II neurons. Layer II of the isocortex of hedgehog and bat possesses neurons with somewhat larger perikarya than those seen in the opossum and with much stronger and wider-spreading external dendrites than the basal ones. As noted, these extraverted neurons very much resemble those of the superficial cell condensations of the first and second cortical growth rings, i.e., the periallocortex and proisocortex, and they assuredly can be recognized as a phylogenetically ancient type of cortical neuron. The extreme example of the extraverted type of neuron goes back even to the amphibian level of brain evolution before any cerebral neocortex had developed. The extraversion of the neurons comprising layer II points to the greater significance of the zonal layer (layer I) in primitive mammals which is in-line with its ancient role in phylogenesis and ontogenesis prior to neocortical development. In allocortical formations the zonal layer is the principal collector of afferent fibers, as well as of ascending branches of the long axons of efferent neurons coming from other areas of the archicortex. An essential part of intercortical connections within archicortex is played by tangential connections in layer I. The presence of extraverted neurons over the entire convexity in the dolphin brain points-to the strong possibility that layer I also may represent the main, but certainly not exclusive, afferent and association plexus of the neocortex in cetaceans. Summarizing, we can say that the dendritic pattern of the extraverted neurons in the dolphin neocortex corresponds to a phylogenetically ancient functional pattern bearing many resemblances to the organizational plan seen in the neocortex of the hedgehog, bat, and opossum. This is in agreement with the observation in the hedgehog that terminations of thalamic afferents still reach the zonal layer (Ebner, 1969), where they make synaptic contact with the extensively overlapping external bouquets from layer II extraverted neurons. Such zonal thalamic terminations were also described in the periallocortex of the rat by Domesick (1969). In this regard, Krasnoshchekova and Figurina (1980), using the Fink-Heimer degeneration method following medial geniculate lesions, have described a strong thalamic input to the zonal layer of the neocortex in the dolphin. This direct evidence points to the importance of layer I as a major thalamic input stratum in the convexity neocortical formations of the brain of small whales.

SUMMARY

We will close with a summary of several key features that are characteristic of the dolphin convexity neocortex. An examination of these in toto gives clear indication of an extremely conservative development of the cetacean neocortex. The signs of conservative evolution we see in dolphin neocortex are: (1) Overall thin neocortex averaging 1.5 mm across all convexity cortical areas; (2) Poor transitional boundaries between cortical architectonic areas, giving a sense of overall monotony or homogeneity with a low degree of structural heterogeneity; (3) Only slight differentiation of cortical layers (weakly expressed stratification) and cells (poor neuronal differentiation); (4) Weak cellular specialization with predominance of simple, generalized neuronal forms over complicated forms, in other words a general absence of diverse collections of neuronal families; (5) The neocortex is highly pyramidalized with pyramidal elements being highly atypical and mostly of the transitional, imprecisely-shaped or intermediate and immature types; (6) The neocortex is largely agranular with few typical stellate neurons and with most stellate elements being large and thus non-granular in type. The absence or incipience of layer IV along with the pyramidal character of layer II strongly emphasizes the overall agranularity of the convexity neocortex; (7) Lack of hypergranular "cores" of specialization (koniocortex), as well as absence of giant pyramids (area gigantopyramidalis), indicates the neocortex of the dolphin may have only reached the paralimbic/parinsular level of cortical evolution (third neocortical growth ring of Sanides) and not the last (fourth ring) stage of neocortical evolution; (8) Presence of large numbers of isodendritic, reticular type long-radiator neurons throughout all cortical layers. These have relatively unbranched dendritic processes and resemble "non-specific", generalized type neurons characteristic of the brainstem reticular formation; (9) Presence of a darkly staining, highly accentuated layer II over entire convexity neocortex. This layer is made-up of modified pyramids showing widespread branching of apical dendrites into layer I, thus being similar to "extraverted" neurons described in the bat and hedgehog brain. This pattern in neocortex resembles features seen only in allocortex and periallocortex of advanced mammals; (10) Massive development of the phylogenetically older neocortical layers I and VI, representing in some areas almost half the thickness of the entire neocortex; (11) Presence of astriatal and supraradial pattern of fiber systems with weakly expressed vertical (radial) systems and with long horizontal fibers in layer I. There is strong presence of horizontal association fibers in layer I rather than in layers II, III, and IV as seen in progressive mammalian species; (12) Though the upper cortical layers are organized more conservatively and resemble an allocortical pattern, they are superimposed on more typically neocortical-type deeper layers creating an almost hybrid type of neocortical organization; (13) Prominant thalamic input to layer I from sensory specific thalamic nuclei.

Taken as a whole, the features summarized above indicate that the convexity neocortex of the dolphin brain is organized in many ways similar to that of the hedgehog and bat. Thus, in this sense the dolphin neocortex also may serve as a model of the "initial" neocortex of mammals. The fact that ancestors of modern whales returned to water when archaic mammals were dominant may, in part, relate to the findings of conservative neocortical evolution, including the lack of granularization seen in whale neocortex. The effect of the aquatic medium on the neocortex is not clear, given the many resemblances between the dolphin neocortex and that of the bat and hedgehog. In general, with an increase in overall brain and cortical size there usually is seen a parallel increase in differentiation. In the case of the brain and, especially, the neocortex, the two variables of a structure, i.e., size and differentiation, do not vary independently and increased size is usually paralleled by increased differentiation. The huge lateral expanse of neocortex of so-far unknown significance in whales will, as pointed-out by Kaas (1987), be of extreme interest for future analysis. Certainly the possibility exists that this expansion does not represent sensory neocortex. The significance of these vast physiologically unmapped cortical fields awaits further study. The differences between the dolphin neocortex and that of hedgehogs and bats is not so much in the cortices that are directly comparable, i.e., the known sensory and motor areas, but to the fact that, as opposed to the hedgehog and bat neocortex, the dolphin neocortex consists of vast cortical fields for which we have no ascribable functions. Keeping in mind the view of cetacean neocortical evolution in terms of the Sanides principles outlined above, could these cortical fields represent mainly additional paralimbic and parinsular cortices and thus types of protosensory and protomotor cortex such as supplementary motor, prokoniocortical and secondary sensory- motor areas? Only future work can give possible answers to these intriguing questions.

ACKNOWLEDGEMENTS

This research was supported by National Science Foundation grants BNS-84-14532 and 87-42032 and the New York Zoological Society.

REFERENCES

Amunts, V. V., Bogolepova, I. N. and Kesarev, V. S., 1977, Structural characteristics of the reticular formation of the brainstem, hippocampus and limbic region of the cortex, Zhurnal Neuropathologii Psikhiatrii S. S. Korsokova, 77:1766-1770.
Brodmann, K., 1909, "Vergleichende Lokalizationslehre der Grosshirnrinde", Johann Ambrosius Barth, Leipzig.
Burikova, N. V. and Krasnoshchekova, E. I., 1983, Cytoarchitectonics of the auditory cortex of the brain of animals with agranular type of neocortex, Vestn. Leningr. Univ. 9:64-70.
Diamond, I. T., 1967, The sensory neocortex, in: "Contributions to Sensory Physiology", Vol. 2, W. D. Neff, ed., Academic Press, New York, 51-100.
Domesick, V., 1969, Projections from the cingulate cortex in the rat, Brain Res., 12:296-320.
Ebner, F. F., 1969, A comparison of primitive forebrain organization in metatherian and eutherian mammals, Ann. N. Y. Acad. Sci., 167:241-257.
Ferrer, I. and Perera, M., 1988, Structure and nerve cell organization in the cerebral cortex of the dolphin Stenella coeruleoalba. A Golgi study, Anat. Embryol., 178:161-173.
Filimonoff, I. N., 1949, "Comparative Anatomy of the Cerebral Cortex of Mammals. Paleocortex, Archicortex, and Intermediate Cortex", Publication Academy Medical Sciences, Moscow.
Garey, L. J. and Revishchin, A. V., 1990, Structure and thalamocortical relations of the cetacean sensory cortex: histological, tracer and immunocytochemical studies, in: "Sensory Abilities of Cetaceans: Laboratory and Field Evidence", J. A. Thomas and R. A. Kastelein, eds., Plenum Press, New York.
Glezer, I. I., Jacobs, M. S. and Morgane, P. J., 1988, Implications of the "initial brain" concept for brain evolution in Cetacea, Behavioral and Brain Sciences, 11:75-116.
Glezer, I. I. and Morgane, P. J., 1990, Ultrastructure of synapses and Golgi analysis of neurons in the neocortex of the lateral gyrus (visual cortex) of the dolphin and pilot whale, Brain Res. Bulletin, 24:401-427.
Gould, H. J. and Ebner, F. F., 1978, Interlaminar connections of the visual cortex in the hedgehog (Paraechinus hypomelas), J. Comp. Neurol., 177:503-518.
Gould, H. J., Hall, W. C. and Ebner, F. F., 1978, Connections of the visual cortex in the hedgehog (Paraechinus hypomelas), J. Comp. Neurol., 177:445-472.
Herrick, C. J., 1948, "The Brain of the Tiger Salamander", Univ. of Chicago Press, Chicago.

Jolicoeur, P., Pirlot, P., Baron, G. and Stephan, H., 1984, Brain structure and correlation patterns in Insectivora, Chiroptera and Primates, Syst. Zool., 33, 14-29.

Kaas, J. H., 1987, The organization and evolution of neocortex, in: "Higher Brain Functions: Recent Explorations of Brains Emergent Properties", S. P. Wise, ed. Wiley, New York, 347-371.

Kesarev, V. S., 1969, Structural organization of the limbic cortex in dolphins, Arkhiv Anat. Gistol. Embriol., 56:28-35.

Kesarev, V. S., 1970, Certain data on neuronal organization of the neocortex in the dolphin brain, Arkhiv Anat. Gistol. Embriol., 59:71-77.

Kesarev, V. S., Malofeyeva, L. I. and Trykova, O. V., 1977a, Ecological specificity of cetacean neocortex, J. Hirnforsch., 18:447-460.

Kesarev, V. S., Malofeyeva, L. I. and Trykova, O. V., 1977b, Structural organization of the cerebral neocortex in cetaceans, Arkhiv Anat. Gistol. Embriol., 73:23-30.

Krasnoshchekova, E. I., 1978, Histologic study of the cortex of the temporal region of the dolphin brain (Phocoena phocoena), Nerv. Sist., 18:31-38.

Krasnoshchekova, E. I. and Figurina, I. I., 1980, The cortical projection of the medial geniculate body of the dolphin brain, Arkhiv Anat. Gistol. Embriol., 78:19-24.

Ladygina, T. F., Mass, A. M. and Supin, A. Ya., 1978, Multiple sensory projections in the dolphin cerebral cortex, Zh. Vyssh. Nerv. Deiat., 18:1047-1054.

Leontovich, T. A. and Zhukova, G. P., 1963, The specificity of the neuronal structure and topography of the reticular formation in the brain and spinal cord of carnivora, J. Comp. Neurol., 121:347-381.

Lorente De Nó, R., 1938, Architecture and structure of the cerebral cortex, in: "Physiology of the Nervous System", J. F. Fulton, ed., Oxford University Press, Oxford, 291-330.

Mazurskaya, P. Z., Davydova, T. V. and Smirnov, G. D., 1966, Functional organization of exteroceptive projections in the forebrain of the turtle, Fiziologicheskii Zhurnal SSSR imeni I. M. Sechenova, 52:1050-1059.

Morgane, P. J., Jacobs, M. S., and Galaburda, A., 1985, Conservative features of neocortical evolution in dolphin brain, Brain, Behavior and Evolution, 21:176-184.

Morgane, P. J., Jacobs, M. S. and Galaburda, A., 1986a, Evolutionary morphology of the dolphin brain, in: "Dolphin Cognition and Behavior. A Comparative Approach", R. J. Schusterman, J. A. Thomas and F. G. Wood, eds., Lawrence Erlbaum Associates, Hillsdale, 5-29.

Morgane, P. J., Jacobs, M. S. and Galaburda, A., 1986b, Evolutionary aspects of cortical organization in the dolphin brain, in: "Research on Dolphins", R. J. Harrison and M. Bryden, eds., Oxford University Press, Oxford, pp. 71-98.

Morgane, P. J., Glezer, I. I. and Jacobs, M. S., 1988, Visual cortex of the dolphin: An image analysis study, J. Comp. Neurol., 273:3-25.
Morgane, P. J., Glezer, I. I. and Jacobs, M. S., 1990, in press, Comparative and evolutionary anatomy of visual cortex of dolphin, in: "Cerebral Cortex, Vol. 8. Evolution and Comparative Anatomy of Cerebral Cortex." E. G. Jones and A. Peters, eds., Plenum Press, New York.
Pirlot, P., 1987, Contemporary brain morphology in ecological and ethological perspectives, J. Hirnforsch., 28: 145-211.
Poliakov, G. I., 1964, Development and complication of the cortical part of the coupling mechanism in the evolution of vertebrates, J. Hirnforsch., 7:253-273.
Ramón y Cajal, S., 1955, "Studies on the Cerebral Cortex", Year Book Publishers, Inc., Chicago, 179 pp.
Ramón y Cajal, S., 1911, "Histologie du Système Nerveux de L'Homme et des Vertébrés", Vol. II, Maloine, Paris.
Ramón-Moliner, E., 1975, Specialized and generalized dendritic patterns, in: "Golgi Centennial Symposium: Perspectives in Neurobiology", M. Santini, ed., Raven Press, New York, 87-100.
Ramón-Moliner, E. and Nauta, W. J. H., 1966, The isodendritic core of the brainstem, J. Comp. Neurol., 126:311-336.
Romer, A. S., 1966, "Vertebrate Paleontology", 3rd Ed., Univ. of Chicago Press, Chicago, 468 pp.
Sanides, F., 1969, Comparative architectonics of the neocortex of mammals and their evolutionary interpretation, Ann. New York Acad. Sci., 167:404-423.
Sanides, F., 1970, Functional architecture of motor and sensory cortices in primates in the light of a new concept of neocortex evolution, in: "The Primate Brain", C. R. Noback and W. Montagna, eds., Appleton-Century Crofts, New York, 137-208.
Sanides, F., 1972, Representation in the cerebral cortex and its areal lamination patterns, in: "The Structure and Function of Nervous Tissue", G. H. Bourne, ed., Academic Press, New York, 329-453.
Sanides, F. and Sanides, D., 1972, The "extraverted neurons" of the mammalian cerebral cortex, Zeit. Anat. Entwickl.-Gesch., 136:272-293.
Sanides, D. and Sanides, F., 1974, A comparative Golgi study of the neocortex in insectivores and rodents, Zeit. Mikrosk. Anat. Forsch., 88:957-977.
Simpson, G. G., 1945, The principles of classification and a classification of mammals, Bull. Amer. Mus. Nat. Hist., 85:1-350.
Sokolov, V. E., Ladygina, T. F. and Supin, A. Ya., 1972, Localization of sensory zones in the dolphin's cerebral cortex, Dokl. Akad. Nauk SSSR, 202:490-493.
Supin, A. Ya., Mukhametov, L., Ladygina, T., Popov, V., Mass, A. and Polyakova, I., 1978, Electrophysiologic study of the dolphin brain, Nauka Press, Moscow, 29-85.

Valverde, F., 1983, A comparative approach to neocortical organization based on the study of the brain of the hedgehog (Erinaceus europaeus), in: "Ramón y Cajal's Contribution to the Neurosciences", S. Grisolia, C. Guerri, F. Samson, S. Norton and F. Reinoso-Suàrez, eds., Elsevier, Amsterdam, 149-170.

Valverde, F. and Facal-Valverde, M.V., 1986, Neocortical layers I and II of the hedgehog (Erinaceus europaeus). I. Intrinsic organization, Anat. Embryol., 173:413-430.

Zhukova, G. P., 1953, On development of the cortical part of the motor analyzer, Arkhiv Anat. Gistol. Embriol., 30:32-38.

Zvorykin, V. P., 1977, Principles of structural organization of the cetacean neocortex, Arkhiv Anat. Gistol. Embriol., 72:5-22.

EVOLUTIONARY MORPHOLOGY AND ACOUSTICS IN THE DOLPHIN SKULL

Helmut A. Oelschläger

Dept. of Anatomy, J.W.Goethe-University
Theodor Stern-Kai 7
6000 Frankfurt am Main-70, FRG

INTRODUCTION

Cetaceans are the most fascinating creatures in the animal kingdom. They adapted so perfectly to their aquatic habitat that it is hard to imagine they originated from typical tetrapod land mammals. Naturally, this change to an extremely different habitat brought about profound modifications and specializations in the cetacean sensory world. Dolphins have a fairly effective visual system, which allows good sight above and below the water surface (Nachtigall, 1986; Dral, 1987). The peripheral olfactory system is reduced totally, but still occurs during early ontogenesis (Jacobs et al., 1971; Oelschläger and Buhl, 1985a, b) while the taste organ, in principle, is retained (Nachtigall and Hall, 1984). Cutaneous sensitivity seems to be fairly good in cetaceans (Herman and Tavolga, 1980) and there are indications that dolphins may possess a magnetic sense (Zoeger et al., 1981; Bauer et al., 1985; Kirschvink et al., 1986; Credle, 1988). However, it is the acoustic system, which is clearly dominant and optimized by a highly efficient ultrasound transmitter for echolocation. The latter may be connected functionally with the unusually well-developed terminalis system (Demski et al., 1985; Buhl and Oelschläger, 1986; Oelschläger et al., 1987; Ridgway et al., 1987; Oelschläger, 1989). Some toothed whales presumably can ..."emit sounds so intense that their prey is debilitated and capture made easier" (Norris and Møhl, 1983). This sonar system is tied to a variety of structural and functional (mechanical) prerequisites all over the head.

Because of sound conduction and reflection abilities of bone, the investigation of the hearing organ should include the skull as a whole. Although the ear has to be independent from the skull functionally, neither of them really can be understood without knowing the morphological and functional implications of the other. In the following, therefore, a survey is given as to the evolution of the cetacean skull as a whole with respect to acoustics, followed by discussions on the otic region, the ear bones, and the auditory ossicles.

SURVEY OF MORPHOLOGICAL TRENDS AND FUNCTIONAL IMPLICATIONS

Tetrapod Origin of Whales and Dolphins

As indicated by the fossil record (Szalay and Gould, 1966; van Valen,

Abbreviations and symbols		mx	maxillary
		n	nasal
Nomenclature after Kellogg (1936),		o	fenestra vestibuli
Reysenbach de Haan (1957), Fraser and		p	parietal
Purves (1960) and Nomina Anatomica (1983)		pbs	peribullary sac
		pl	palatine
a	tympanic aperture of facial canal	pe	periotic of generalized mammals
a'	porus acusticus internus	pe'	periotic of cetaceans
ac	external aperture of cochlear aqueduct	plf	posterior lacerate foramen
		pm	premaxillary
ae	bony porus acusticus externus of cetaceans	pp	paroccipital process
		ps	posterior sinus
ao	accessory ossicle (processus tubarius)	pt	pterygoid
		ptl	lateral lamina of pterygoid
at	auditory tube	ptm	medial lamina of pterygoid
av	external aperture of vestibular aqueduct	pts	pterygoid sinus
		r	fenestra cochleae
b	basioccipital	s	supraoccipital
bf	facet for processus tubarius of tympanic	sf	stylomastoid foramen
		sm	soft external auditory meatus
bp	lateral process of basioccipital	sq	squamosal
		st	stapes
co	occipital condyle	t	tympanic
d	groove for tensor tympani	th	tympanohyal
de	dental	to	tympanic orifice
e	elliptical foramen	v	vomer
et	ethmoid	1-4	premolar teeth
ex	exoccipital	*	bony nares
f	frontal	▲	orbita
fa	falciform process	◐	postorbital process of frontal
fh	fossa for head of malleus	◑	zygomatic process of squamosal
fi	fossa incudis	◆	temporalis fossa
fp	articulating facet of posterior process of periotic	◇	temporal fenestra
		◆	coronoid process
		○	articular process of dental
fs	fossa for stapedial muscle	●	postglenoid process of squamosal
ft	articulating facet of posterior process of tympanic		
		×	attachments of tympanic bone to surrounding elements except periotic
h	malleus		
i	incus		
j	jugal		
l	lacrimal	•—	external auditory meatus
li	ligament	◄ ►	anterior (left) and posterior (right) process of periotic
ll	lateral lobe of tympanic		
m	mastoid	◁ ▷	anterior (left) and posterior (right) process of tympanic
ma	malar		
ml	medial lobe of tympanic	■	sigmoid process
mp	mastoid process	⊙	facial canal
ms	middle sinus	✺	cochlear part of periotic

Fig. 1. Lateral view of skulls from selected cetacean groups. a) +Archaeoceti (+<u>Dorudon osiris</u>), after Stromer (1903, 1908) and Kellogg (1936); b) +Squalodontidae (+<u>Eosqualodon langewieschei</u>), after Rothausen (1968a, b), slightly modified and completed after van Beneden and Gervais (1868-1880), Brandt (1873), and Kellogg (1928); c) Pontoporiidae: La Plata dolphin (<u>Pontoporia blainvillei</u>), ZMA 16714 (Zoological Museum of Amsterdam); d) Phocoenidae: Harbor porpoise (<u>Phocoena phocoena</u>), (Coll. Dept. of Anatomy, Frankfurt am Main). Lower jaw translucent, osseous symphysis shown as hatched area. Scales: in a)-d) 10 cm. For abbreviations and symbols see above.

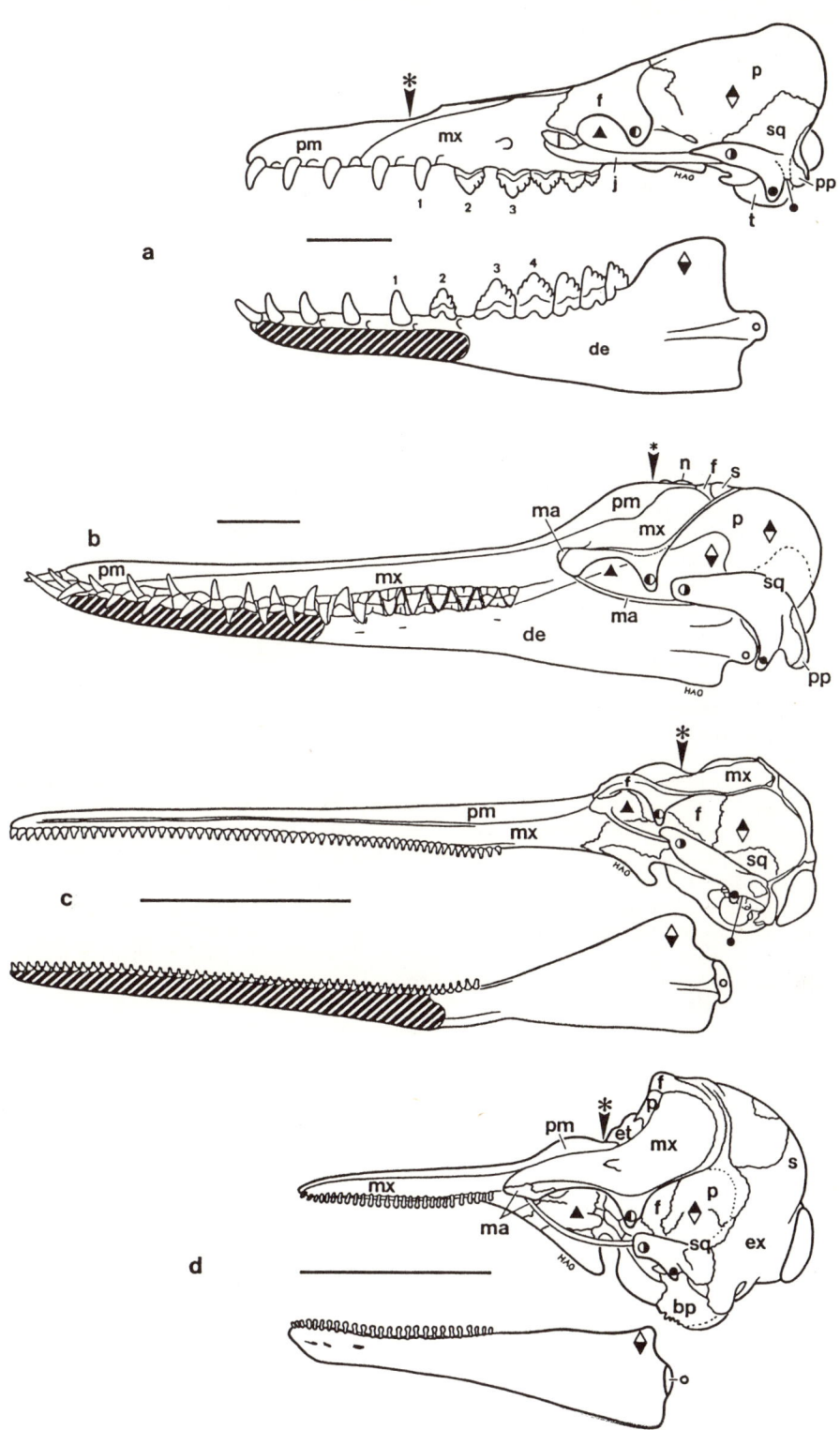

1966; Barnes, 1984), the ancestry of whales and dolphins goes back to
typical land mammals from the early Tertiary period (Paleocene), presumably
representatives of the order +Condylarthra (+Mesonychidae)[1]. These
carnivorous mammals obviously had close relationships to the ancestors of
the ungulate orders Artiodactyla and Perissodactyla, which are reflected in
similarities (if not homologies) among many organ systems in these groups
and in the Cetacea (Evans, 1987).

The mesonychid skull (+Mesonyx obtusidens, Middle Eocene) was compact
and must have had a biting force comparable to that of living carnivores.
The rostrum was relatively short and the jaws had a set of 10-11 teeth in
each ramus (incisors, canines, tritubercular premolars and molars). The
mandibles articulated with the squamosals in well-developed temporomandibular joints and were united rostrally in a solid osseous symphysis. The
three anterior premolar teeth in the opposing jaws constituted "scissors"
for cutting meat while the fourth premolar and the molar teeth may have
served for crushing and chewing food (Szalay and Gould, 1966). In accordance
with the firm masticatory part of skull, the neurocranium of the cetacean
ancestor was well-sculptured. The temporalis muscle had a vast area of
origin, including a high sagittal crest, and was attached to a strong coro-
noid process. Consequently, the horizontal "temporal fenestra" between the
lateral wall of skull and the zygomatic arch was wide. Bony pillars in the
walls and the roof of the skull together with the strong zygomatic arches
may have dissipated forces of the masticatory muscles over the neurocranium.
On the other hand, they should have been able to transform the contraction
energy of the neck and trunk muscles via the skull to the jaws (grasping and
holding, tearing of the prey).

The cranial vault was longer than wide and the brain rather small (low
degree of encephalization). Radinsky (1976) estimated the endocranial volume
of a 26 cm-skull of Mesonyx at about 80 cc. The skull roof was symmetrical
and showed the caudalward succession of bony elements typical for mammals.
The bony nares were situated on the tip of the upper jaw; the nose should
have been fairly well-developed, although the olfactory bulbs were
relatively small. The orbital cavity was moderately developed and the ear
region must have been adapted for the perception of air-borne sound (light
and morphologically complicated middle ear ossicles). Unfortunately, in
those members of the +Condylarthra, which obviously have the closest
relationships to the ancestor of the Cetacea (+Mesonychidae), the otic
region is poorly preserved (Scott, 1888). Presumably the periotic was firmly
integrated in the skull, the suspension of the tympanic characterized as
being similar to the situation in +Hyaenodon (Fraas, 1904; Oelschläger,
1986b). In another branch of the +Condylarthra (+Meniscotherium), the otic
region and other areas of the skull already show essential features of the
ungulate group (cf. Gazin, 1965).

Adaptation to the Aquatic Environment: Derived Characters

While the common ancestor of whales and dolphins had been a typical
land mammal, the successive adaptive radiations of cetaceans were characterized by an increasing number of specialized features correlated to life
in water. For instance, they had to adapt to aquatic food presumably consisting of invertebrates and fishes. Thus, already in the archeocetes at the
beginning of the Eocene (Fig. 1a), the jaws had become elongated (Fraas,
1904; Kellogg, 1936; Fordyce, 1985). In their rostral half, the teeth were
separate from each other and conical in shape. The rear part of the jaws had
highly compressed serrate premolar and molar teeth, linked together, but
interlocking in jaw closure, and was adapted to slicing prey, which may have

---

1) + = fossil taxon

been swallowed in pieces. Crushing or chewing of food obviously was no longer possible. However, archeocetes still had a deciduous and a permanent tooth generation (Kellogg, 1936).

Early odontocetes show a marked increase in tooth number (one tooth generation). Progressive squalodonts (Stromer 1911; Fig. 1b) already had long rostra with 15-16 teeth in each ramus, the rostral half of the dentition consisting of simple teeth, the caudal half comprising double-rooted, denticulated and laterally compressed teeth. In single, more advanced groups of dolphin-like forms (+Acrodelphidae; +Rhabdosteidae = +Eurhinodelphinidae; Pontoporiidae), the jaws constitute a long forceps equipped with hundreds of conical teeth (Kellogg, 1924; 1926; Barnes, 1984; Barnes et al., 1984; Fig. 1c). Presumably, such slender beaks extended the radius of action in early toothed whales, which may have been relatively slow and perhaps luring animals, and served to poke around in the sediment of estuarine and shallow coastal areas. The masticatory apparatus had changed from grasping and cutting the food to exclusively grasping smaller prey and swallowing it whole. In most of the living groups, the masticatory musculature drastically is reduced, together with the corresponding attachment sites of the skull and the lower jaw (grooves and crests, apophyses). Moreover, the mechanical coupling of the components in the temporomandibular joint became less intense. However, the bony symphysis between the rami of the lower jaw persisted and even was elongated in order to guarantee mechanical stability and precision in the function of this long rostrum. Consequently, in modern short-beaked toothed whales, the bony symphysis was lost (Fig. 1d). The zygomatic arch could be reduced because there was no need for a strong masseter muscle (Anthony, 1926). Furthermore it was no longer necessary to brace the rostral ("masticatory") part with the caudal (neural) part of skull laterally, because:

1) the intertemporal constriction between the zygomatic arches (temporal fenestrae), which had been extreme in archeocetes, gradually was compensated by the broadening of the skull roof in correlation with the expansion of the cranial vault, and

2) as the habit to cut down the prey was abandoned, a twisting of the two skull segments against each other (especially when slicing unilaterally) was no longer a problem.

But the reduction of the zygomatic arch still had another important effect; while in land mammals it transforms forces along the skull and allows bone conduction, its disintegration seems to have been a preadaptation for the acoustic isolation of the ear from the skull. In modern dolphins, the rostral part of the zygomatic arch is reduced to a flexible, thin bony rod that borders the orbital cavity below. Obviously, it had to persist because it holds the eye-ball in position. Interestingly enough, in modern toothed whales a "secondary zygomatic arch" has been established; it is composed of the postorbital process of frontal and the old part of the original arch, the zygomatic process of squamosal. However, both processes are separated by a gap filled with connective tissue. Obviously, this "secondary zygomatic arch" renders the stability needed free from bone conduction.

The development of a long rostrum in early odontocetes obviously was accompanied by the necessity to anchor it properly on the skull. Thus, the bones of the upper jaw (premaxillaries, maxillaries) have been extended far backward so that they in part cover the frontals and parietals to meet the supraoccipital bone (Fig. 1b-d). This phenomenon was called "telescoping process" by Miller (1923) and is most obvious in those fossil toothed whales that possessed the longest beaks ever known (rhabdosteids, +Zarhachis; Kellogg, 1924; 1926). In +Eurhinodelphis, the maxillary reaches even the level of the occipital condyles (Abel, 1901). By this telescoping process, flat bones come to lie upon each other, constituting a "sandwich structure"

(Fleischer, 1976). As a device for serial impedance mismatch, the latter obviously reflects sonar signals produced in a set of (secondary) accessory air sacs around the blowhole (Norris, 1964; 1968; 1980; Cranford, 1988; Ridgway et al., 1980; Ridgway and Carder, 1988) and thus, protects the ear against the animal's own sonar beam (acoustic shield). In addition, this new frontal area of skull is broadened and deepened in order to host the melon or spermaceti organ, a highly specialized adipose cushion (acoustic fat), which is supposed to focus the sonar signals for propagation (acoustic lens) with the help of the blowhole musculature. Finally, in modern dolphins, the skull roof becomes asymmetrical as the external bony nares are shifted to the left, one naris wider than the other and the whole narial region being rotated (Ness, 1967). There are some indications that this asymmetry is correlated with ultrasound generation and that the (asymmetrical) nasal tracts are specialized in different ways functionally (Mead, 1975;. Heyning, 1989).

The formation of the odontocete rostrum was connected with the backward shift of the nares together with the reduction of the olfactory system and the pertinent part of the nasal cavity (Fig. 1a-d). This caudalward migration of the nostrils can be observed in the early ontogenesis of toothed whales (Klima, 1987; Buhl and Oelschläger, 1988). Obviously, the terrestrial olfactory system (inherited from water-dwelling vertebrates) could not be adapted to the aquatic environment again (Oelschläger and Buhl, 1985a, b; Oelschläger, 1989). The nasal septum, on the other hand, was elongated together with the bones of the upper jaw; during ontogenesis it is the driving force for the development of the cetacean rostrum (Klima and van Bree, 1985; Klima, 1987). The remaining "respiratory tubes" of the nose are turned into a vertical position together with the cribriform plate, leading to a shortening of the cranial vault. Therefore, the brain of living cetaceans is wider than long. The various morphogenetic trends in the evolution of the skull were superimposed by the strong progression of the brain. Whereas, in the earliest cetaceans the brain case was relatively small, it is the dominating part in the skull of modern dolphins (Fig. 1d). Here the braincase has a remarkable capacity compared to the body weight (400-1500 cc; Schwerdtfeger et al., 1984).

During the evolution of the archeocetes and odontocetes, the periotic bone was uncoupled from the skull base, leaving part of the mastoid wedged between the squamosal and exoccipital, and rotated out of the cranial vault (Oelschläger, 1986a, b; 1987). This process is favored by the gradual extension of accessory air sinuses of the tympanic cavity, both trends leading to the acoustic isolation of the ear bones and allowing directional hearing. The tympanic bone was transformed into a large bulla which unites with the periotic in the tympano-periotic complex, one of the most remarkable structures in cetaceans. Surrounded by air- or foam-filled spaces and suspended by connective tissue only, the tympano-periotic complex falls from the skull following maceration after death and may be found on the sea-floor (Hyrtl, 1845; Yamada, 1953).

In dolphins, the vestigial external ear is located beneath the subcutaneous fat (blubber) while the soft external acoustic meatus is very narrow; their functional significance for hearing is a matter of debate (Yamada, 1953; Reysenbach de Haan, 1957; Purves, 1966; Norris, 1968, 1980). The bony porus acusticus externus is still formed by the tympanic bone, while the tympanic membrane has been stretched out into a ligament. Obviously, the perception of sound is achieved via the lower jaw and a large fat body in the wide alveolar canal, respectively (Norris, 1968, 1980; Norris and Evans, 1988), the composition of which is comparable to that of the melon and spermaceti organ. The alveolar canal ends in a wide medial foramen (acoustic window), where the fat body is in contact with the tympanic bulla. From there, transmission of sound to the fenestra vestibuli is accomplished via the auditory ossicles. In other words, the tympanic bulla has been added to the chain of ossicles (Fleischer, 1978; Norris and Evans, 1988).

MORPHOLOGY AND EVOLUTION OF THE OTIC REGION

Archeocetes

In 1983, Gingerich et al. reported cranial fragments from fluvial sediments of the early Eocene of Pakistan obviously belonging to a highly unspecialized archeocete (Fig. 2a). To conclude from the skull material presented to date +Pakicetus, in general, did not differ very much from other, more highly advanced (marine) archeocetes, especially +Protocetus (Fraas, 1904). While it is not obvious, whether or not the periotic already was separate from the mastoid (which seems to be true for other archeocetes), long anterior and posterior processes of this bone were wedged between neighboring skeletal elements. The ventralward and medialward rotation of the cochlea (cf. Boenninghaus, 1904) is not yet obvious; the fenestrae cochleae and vestibuli still are situated at the lateral margin of the pars cochlearis of periotic. The accessory air sinuses of the auditory tube and tympanic cavity are developed weakly in comparison with other archeocetes, perhaps with the exception of +Protocetus, which in this respect is not adequately known. Because there is no information as to whether the cetacean ancestor already possessed such air sinuses (also reported of hyracoids and perissodactyls), it cannot be determined whether these air sacs are a shared-derived feature within these groups or have developed independently several times and perhaps for different purposes. The tympanic bone of +Pakicetus seems to correspond to the typical auditory bulla of cetaceans, and Gingerich et al. (1983) suggested that the tympanic and malleus may have been fused partially. It is therefore probable that, apart from the tympanic, the chain of auditory ossicles (which have not yet been found) was largely adapted to aquatic hearing. However, the tympanic still was attached to neighboring bones of the skull base in four areas (Fig. 2a; cf. +Hyaenodon, Pompeckj, 1922) so that a typical tympano-periotic complex was not yet established.

In progressive archeocetes (Fig. 2b), the rotation of the pars cochlearis was in an advanced condition: here the fenestra cochleae is seen on the medial margin of the periotic. Simultaneously, the accessory air sacs of the auditory tube and tympanic cavity increased in size; they penetrate into adjacent skeletal elements (pterygoid, paroccipital process of exoccipital). The pterygoid sinus and auditory tube together form a cuneate space between bony laminae of the pterygoid, basioccipital, squamosal and exoccipital. In correlation with the extension of this cavity, the cochlear part of periotic now is completely free from the skull base while the posterior process still is wedged between squamosal and exoccipital. The peripheral attachments of the tympanic were lost or shifted onto the periotic (tympano-periotic complex) while the processes of the latter are considerably shorter. Thus, the mastoid bone, which has become separate from the periotic, is visible in the ventral aspect. The posterior process has a rugose surface which articulates with the posterior process of the tympanic bulla. The external auditory meatus in archeocetes, generally, is relatively wide, whereas the situation in +Pakicetus (narrow meatus) is exceptional and not understood.

Odontocetes

Very little is known about the morphology of the otic region in the earliest odontocetes (agorophiids, squalodontids). To conclude from the small amount of information in the literature (e.g. True, 1910; Abel, 1912; dal Piaz, 1916; Flynn, 1948; Rothausen, 1968a, b; Fordyce, 1981), they may have had attained a similar condition as the long-snouted rhabdosteid dolphins (Abel, 1901; Kellogg, 1924, 1926; Fordyce, 1983). Concerning the morphology of the ear region in living odontocetes (Fig. 2c), the Ganges river dolphin (Platanista sp.) retains a strongly plesiomorphous (Hennig, 1966) condition which resembles that in the archeocetes. The periotic still is integrated in the skull base; the squamosal bone is much extended ventrally and in part covers both processes of the periotic. Moreover, the squamosal hides the mastoid so that in the basal aspect only its lateral extremity is visible. However, the cochlear part of periotic has been rotated farther ventrally and medially, now fully exposing the fenestra vestibuli. As in progressive archeocetes (Kellogg, 1936; Kasuya, 1973), the tympanic bulla is attached to the rugose posterior process of periotic and the adjacent area of the squamosal. The external auditory meatus still is wide in comparison with most of the other living odontocetes. As in

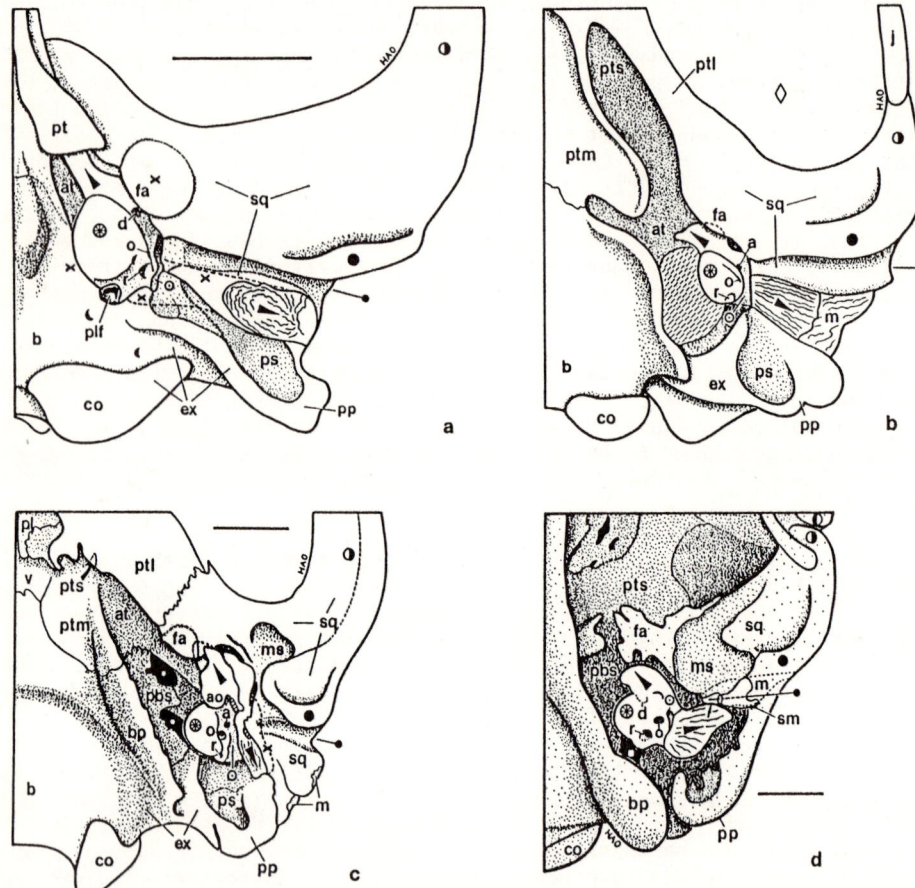

Fig. 2. Left otic region of archeocetes (a-b) and odontocetes (c-d) in ventral aspect. a) +Pakicetus inachus. Presumably concealed part of the posterior process of periotic (right arrowhead) indicated by dashed outline. After Gingerich et al. 1983. b) Advanced archeocetes, combined after figures of +Zygorhiza kochii and +Basilosaurus cetoides (Stromer, 1908; Kellogg, 1936). c) Ganges river dolphin (Platanista gangetica) ZMA 10465. Concealed parts of periotic processes (arrowheads) indicated by dashed outline. d) Atlantic white-sided dolphin (Lagenorhynchus acutus). Anterior margin of the posterior process of periotic (right arrowhead) touches the basal mastoid ridge (arrow). Full extension of mastoid (m) indicated by dashed outline. Bars in a), c) and d): 2 cm. After Oelschläger (1986a), modified.

archeocetes and fossil odontocetes (squalodontids, rhabdosteids), some of the accessory sinuses of the auditory tube/tympanic cavity are located between bony laminae (sheaths), the lateral one connecting the palatine roof with the zygomatic process of squamosal. Obviously, parallel to the formation and acoustic isolation of the tympano-periotic complex, which was achieved phylogenetically by the strong progression of air sinuses, as well as by the uncoupling of the periotic and its rotation out of the cranial vault, large gaps in the skull base occur ontogenetically (Boenninghaus, 1904; Yamada, 1953). These gaps reach their maximal size in the late fetal period when the air sinuses are beginning to evaginate from the main pterygoid sac. After birth, the gaps are closed by marginal growth of the bordering bones with the exception of the area above the periotic (vestibulocochlear nerve) and the foramina for other cranial nerves and blood vessels.

In the morphology of the lateral part of the skull base, the La Plata dolphin (Pontoporia) shows a somewhat ambivalent situation (Oelschläger, 1987). On the one hand, the accessory (tympanic) air sinuses still are partially ensheathed by bony laminae as is the case in progressive archeocetes, rhabdosteids and Platanista. However, the bony sheaths are rather thin and even perforated in some places; in modern living odontocetes they are largely reduced (see below). On the other hand, the periotic of Pontoporia is uncoupled from the skull base to the same degree as in modern dolphins (not shown). Moreover, as in many delphinid genera, the posterior process of the tympanic exceeds that of the periotic in length and makes contact with the skull base (squamosal). This reminds of the extreme situation in the beaked whales (ziphiids) and sperm whales (physeterids), where the periotic is suspended from the skull by a much elongated posterior process of tympanic; the latter may even be fused to the enlarged mastoid secondarily (Kogia). Because the latter bone only is integrated loosely in the skull base, complete maceration will deliberate here a tympano-periotico-mastoid complex (Oelschläger, 1986b). In Pontoporia, the mastoid is again largely covered by the squamosal ventrally and the bony external auditory meatus is wide.

As far as the otic region of skull is concerned, the Amazon river dolphin (Inia; not shown) is rather similar to modern toothed whales. The periotic is uncoupled ultimately from the skull and (as in Pontoporia) has very short processes while the osseous sheaths largely are reduced. However, the bony external auditory meatus is still relatively wide and in part formed by the mastoid. In delphinids like the Atlantic white-sided dolphin (Lagenorhynchus; Fig. 2d), the bony meatus has been transformed into a narrow cleft. The reduction of the meatus was achieved by a shortening of the otic region (Oelschläger, 1986a) which is also obvious in the ontogenesis of this species and of the bottlenose dolphin (Tursiops truncatus). The squamosal and exoccipital bones approach each other ventrally until the basal and medial extremity of the cuneiform mastoid is merely a crest (basal mastoid ridge) which bears the narrow soft external auditory meatus. The tympano-periotic complex stays in contact with the skull via the posterior process (Fig. 2d), which touches the mastoid ridge with a rounded anterior edge, the contact being stabilized by dense connective tissue (Fig. 3d). The bony sheaths of the air sinuses largely are reduced, isolating the palatal roof from the zygomatic arch. Residues persist in the double lamina of the pterygoid bone and in the falciform process of squamosal. In modern delphinids, the predominant part of the sinuses is bordered ventrally and laterally by the "pterygoid ligament" (Boenninghaus, 1904) which radiates from the palatal roof to the falciform and zygomatic processes of squamosal and to the lateral process of the basioccipital, incorporating the tympanic bulla and auditory tube.

TYMPANO-PERIOTIC COMPLEX

Because of the poor fossil record, there is very little direct evidence as to the morphology of the ear bones in the potential cetacean ancestor and the evolutionary processes leading to the formation of the tympano-perotic complex. However, the exceptional occurrence of tympanohyal bones in the bottlenose dolphin, a small element of the hyoid arch normally not present in cetaceans, facilitated the theoretical derivation of the dolphin ear from that of tetrapod mammals (Oelschläger, 1986b; Fig. 3). Comparison with the otic region of living carnivores and ungulates reveals a good general correspondence between cetaceans and ungulates, e.g., in the morphology and topographical relations of structures surrounding the facial canal as well as in their suspension from the skull. Observations in various other organ systems point in the same direction (Anthony, 1926; Kellogg, 1936; Flynn, 1948; Boyden and Gemeroy, 1950; Barnes, 1984; Ridgway and Wood, 1988).

The tympano-periotic complex evolved within the archeocete suborder. Extension of the auditory tube/tympanic cavity in the form of accessory sinuses, shortening of the periotic processes, as well as the deletion or shift of peripheral attachments of the tympanic bulla to the periotic, led to the formation of a morphologically unique structure which in most odontocetes is nearly totally separate from the skull (Figs. 3-5). Here both components contact each other in three areas: via the posterior processes, only very restricted at the sigmoid process, and by means of the accessory ossicle (Kellogg, 1936; Flynn, 1948; Kasuya, 1973; Pilleri, 1987). Whereas in archeocetes, squalodontids, Platanista (Fig. 2c), the ziphiids and physeterids the accessory ossicle may be separate from the tympanic and attached to the periotic, the situation in extant smaller (delphinoid) toothed whales is opposite. Thus, in rhabdosteids, kentriodontids, Pontoporia, the iniids, phocoenids and delphinids (Figs.4-5), the ossicle is fused completely to the tympanic and now called processus tubarius (Kellogg, 1927; Rensberger, 1969; Fleischer, 1973a; Fordyce, 1983; Oelschläger, 1986a). The origin of this ossicle within the mammalian bauplan, which also has been found in the fetal blue whale (ossiculum accessorium mallei; Ridewood, 1922) possibly goes back to the supra-angular of vertebrates (de Beer, 1937). Whether its fusion either to the periotic or to the tympanic bulla has any impact on the hearing process is not known.

In contrast to Platanista, which in this respect reminds one of the archeocetes, the periotic bones of the fossil squalodontids and rhabdosteids (dal Piaz, 1916; Kellogg, 1924; Fordyce, 1983; Cigala-Fulgosi and Pilleri, 1985) are very similar to those of modern smaller toothed whales (Figs. 4-5), perhaps with the exception of the plesiomorphous +Prosqualodon (Flynn, 1948). This is true for the position of the foramina, the facial canal, the grooves for the ear ossicles (malleus, incus), and the fossae of the middle ear muscles. In advanced forms, the posterior process is short and about as strong as the anterior one (rhabdosteids), or the anterior process is the slightly larger one (squalodontids). The posterior process wears a rather large, smooth and elongate triangular or round facet for the articulation with the posterior process of the tympanic bone. In Pontoporia, Inia, and in delphinid species like Tursiops (Figs. 2d, 3-5), the periotic processes are shorter than in most squalodontids and rhabdosteids . In this respect, Pontoporia, which has one of the smallest periotics known, shows an extreme situation (Figs. 4a-b, 5a). The posterior process has a narrow, nearly smooth facet and articulates with an often distinctly longer posterior process of tympanic (Fig. 5b). The periotic of Inia (Figs. 4c-d, 5c) is much larger, but similar to that of Pontoporia; in both species, the round window (fenestra cochleae) is relatively wide. Additionally, Inia has a strongly developed and spiny anterior process while the short posterior process bears a moderately large ovoid facet. In the periotic of Tursiops (Figs. 4e-f, 5e), the cochlear part has about the same size as in Inia while the anterior process is somewhat more slender, but not pointed. The posterior process of Tursiops is strong and bears a very large round, pleated facet which has about the same diameter as the whole cochlear part in Pontoporia. The medial margin of the facet is enlarged by a thin lamina; sometimes, it may show a co-ossified tympanohyal (Fig. 5e) and then covers the facial canal and the groove of the stapedial muscle as well. The internal auditory meatus of Tursiops is wide and sole-shaped.

The tympanic bulla (Figs. 3-5) is ovoid in archeocetes and more cuneiform in squalodontids and living odontocetes (Stromer, 1908; dal Piaz, 1916; Kellogg, 1924, 1936; Kasuya, 1973; Kaiya et al., 1979). In the caudal part, near the posterior process, the bulla is divided in two lobes by a longitudinal groove, the lateral lobe being rather prominent (Figs. 3b, 4b, d, f). The anterior process of the tympanic may be pointed as in squalodontids, rhabdosteids, Platanista and Inia, or more or less obtuse as in delphinoids.

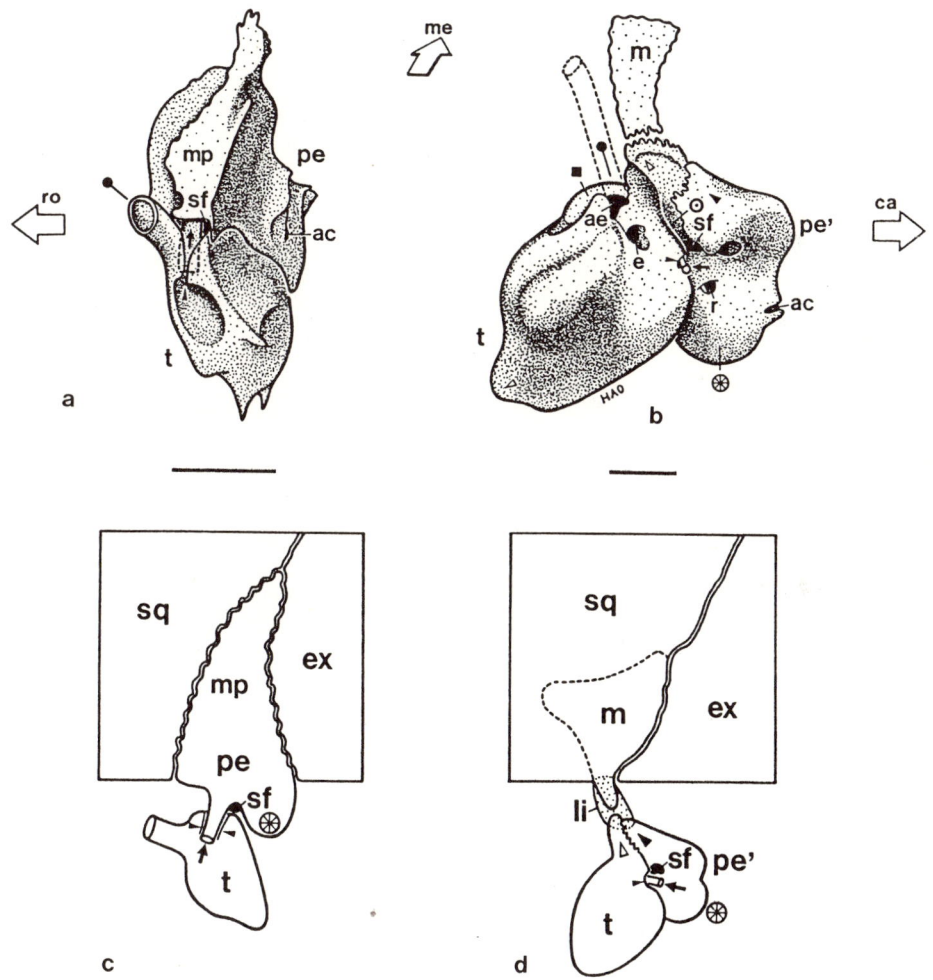

Fig. 3. Comparison of ear bones in an ungulate (a, c) and a delphinid toothed whale (b, d), with special reference to the tympanohyal-tympanic connection. a) and b) Original morphological results; c) and d) Formation of the tympano-periotic complex (scheme). ca = caudal, me = medial, ro = rostral. Arrow = tympanohyal, small arrowhead = groove of tympanic for tympanohyal. Bars in a) and b) 1 cm. After Oelschläger (1986b), modified.

The tympanic of Pontoporia to some degree is intermediate between those of Tursiops and Inia, but closer to Tursiops. Pontoporia corresponds with Tursiops in the general shape of the bulla (although it is much smaller) as well as in the morphology of processus tubarius and sigmoid process. However, the bulla of Tursiops is more elongate. In Inia, the anterior process is equipped with a spine and funnel-shaped as in Platanista while the body is more rounded and the posterior process very short. In Tursiops, the posterior process of tympanic forms an obtuse angle with the axis of the bulla; the same holds true for the posterior process of periotic. Both processes bear very large, pleated facets. Such firm and extended, complicated articulations between the posterior processes of periotic and tympanic are typical for advanced delphinids of rather different size as, e.g., Delphinus, Steno,

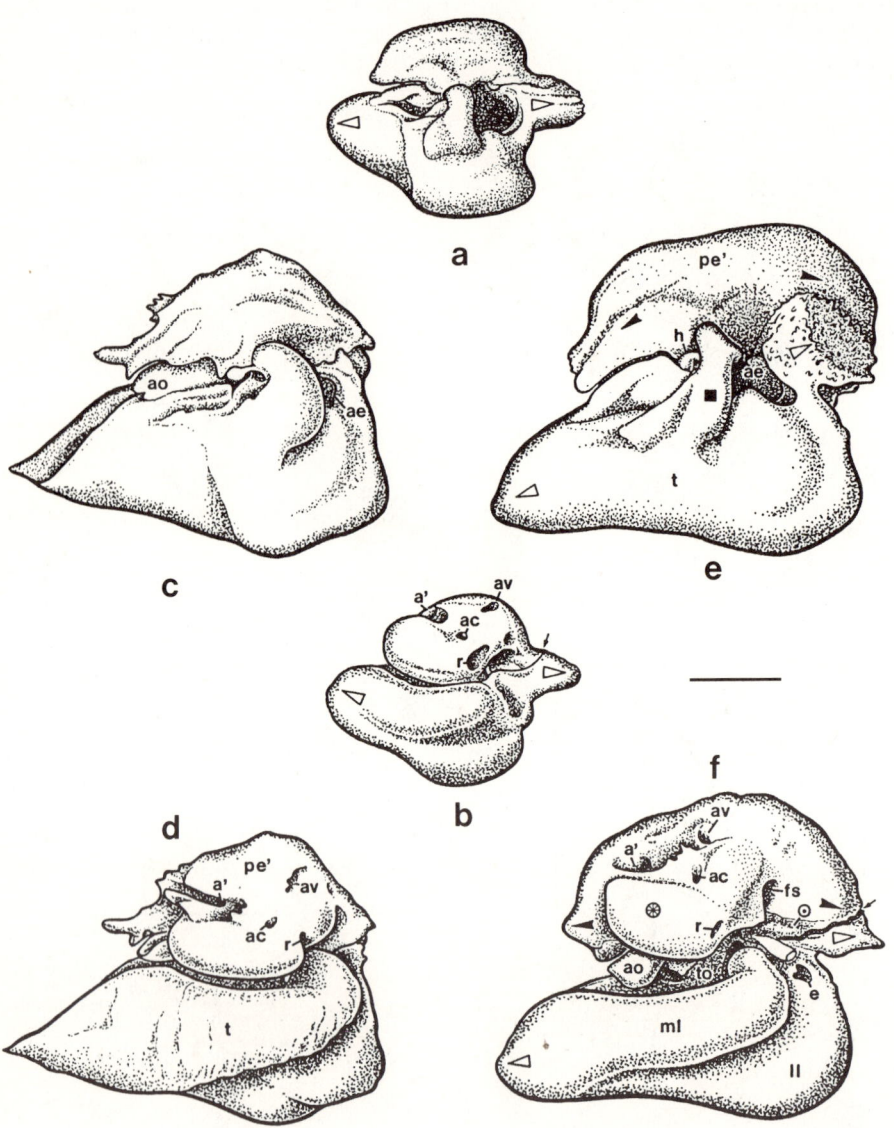

Fig. 4. Tympano-periotic complex in toothed whales. a) and b) La Plata dolphin (Pontoporia blainvillei), c) and d) Amazon river dolphin (Inia geoffrensis), e) and f) bottlenose dolphin (Tursiops truncatus). a),c),e) lateral aspect, b),d),f) medial aspect. Small arrows = posterior tympano-periotic articulation. For abbreviations and symbols see Fig.1. Bar = 1 cm.

Lagenorhynchus, Grampus, Stenella, Tursiops, Globicephala and Orcinus (Kasuya, 1973; Flower, 1874; Yamada, 1953; Flower and Lydekker, 1978). Presumably this type of connection (Tursiops-type) is needed for efficient mechanical coupling of the periotic and tympanic bones in the range of very low frequencies when the tympanic as a whole vibrates (Fleischer, 1973a).

AUDITORY OSSICLES

The middle ear ossicles of whales and dolphins, in general, are much stronger and more highly condensed than in tetrapod mammals. Obviously, this is primarily an adaptation to the much better propagation of sound in water

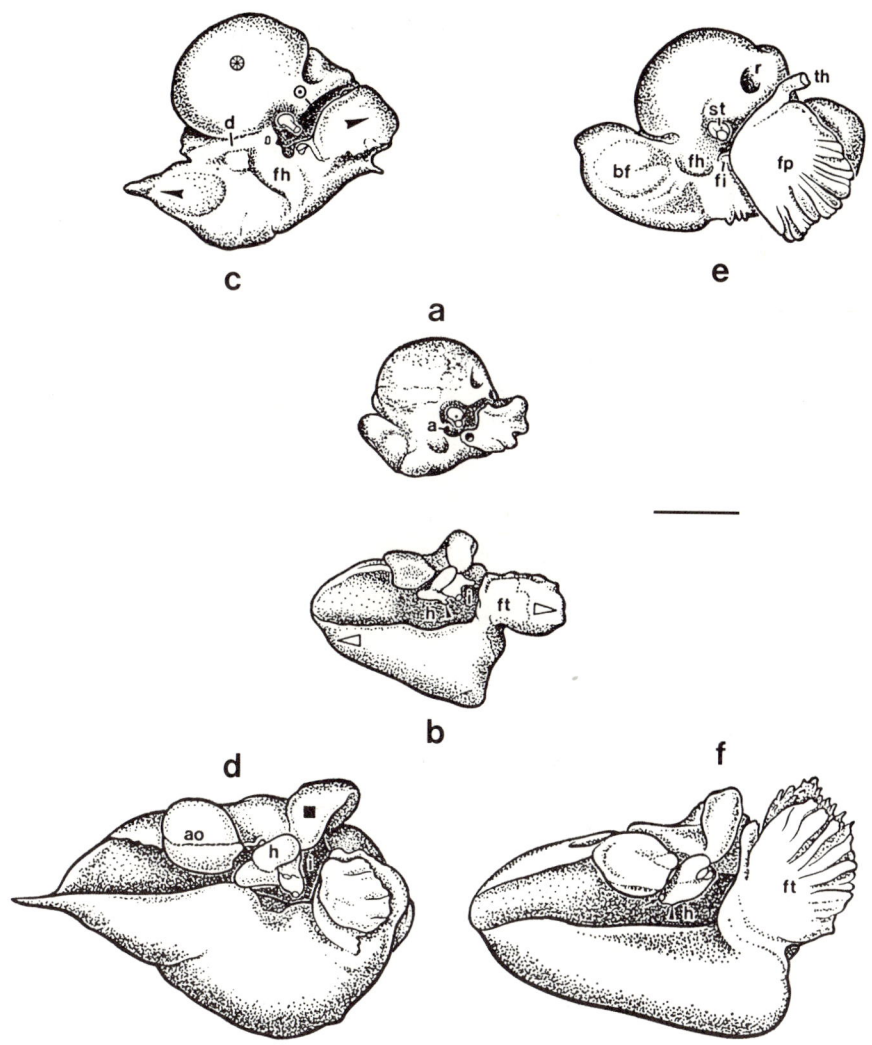

Fig. 5. Tympano-periotic complex opened; periotics above, in ventral aspect, tympanics below, dorsal aspect. a) and b) Pontoporia, c) and d) Inia, e) and f) Tursiops. Small arrow = attachment site of tensor tympani muscle, small arrowhead = manubrium mallei. Bar = 1 cm.

149

and thus higher sound pressure levels (increased mechanical stability), and secondarily to the conduction of ultrasound in odontocetes (Reysenbach de Haan, 1957; Fleischer, 1978). Consisting of very dense, ceramic-like bone, they are five times heavier in the harbor porpoise (Phocoena) than in the adult human with roughly the same body weight (Boenninghaus, 1904). In the series Pontoporia, Inia, and Tursiops, the ossicles become more compact. Archeocetes already had ossicles which were rather similar to those in extant toothed and baleen whales (Pompeckj, 1922; Kellogg, 1936) and were attached to or in contact with the surface of periotic in much the same way (fossa for head of malleus, fossa incudis, fenestra vestibuli). The malleus of +Dorudon was stout and fused with the tympanic bulla via the prearticular (gonial) element as in all other cetaceans, and there are strong indications that the caput mallei was perforated by the chorda tympani, too. Fossil squalodontids and other plesiomorphous odontocetes are nearly unknown with respect to ear ossicle morphology. The Miocene rhabdosteid dolphins had a malleus very similar to that of modern toothed whales (Fordyce, 1983). It was as large and stout as that of Tursiops and even had a thicker caput with large facets. However, there was still a short and curved manubrium and a pronounced apophysis for the tensor tympani muscle.

In Pontoporia (Fig. 5b), the head of malleus is slender and bears large articular surfaces for the incus, while the collum is well developed. The distinct manubrium reminds of the situation in +Dorudon and appears as a small denticle, which, however, is curved as in Globicephala (Reysenbach de Haan, 1953; Pilleri et al., 1987). The incus in Pontoporia is large and resembles that in the human and in artiodactyls; however both crura, not very different in strength, are somewhat shorter. In archeocetes (+Dorudon), the body and long process of the incus were very stout, with a stumpy conical process for the articulation with the stapes, the short process obviously slender as in modern toothed whales (Pompeckj, 1922).

Inia to some degree is similar to Pontoporia (Fig. 5b, d). However, the caput mallei of the former is more ovoid and the collum seems to be relatively larger. In both species, these main regions are distinctly separate by a circular furrow as was the case in +Dorudon. The manubrium is not visible in Figure 5d. The incus seems to be more compact than in Pontoporia.

Concerning the tendency to condense the middle ear ossicles, delphinoids like Phocoena, Tursiops, and Globicephala as well as the ziphioids and physeteroids are far-advanced, although various morphological/functional trends seem to be involved, complicating the situation (Beauregard, 1894; Boenninghaus, 1904; Yamada, 1953; Reysenbach de Haan, 1957). In this respect, Tursiops (Fig. 5e) and obviously Stenella (Kasuya, 1973) are among the most highly derived species . Here the malleus is extremely stout; the head and collum are not separate from each other while the facets for the incus are relatively small. The "collum" of Tursiops tapers in the direction of the manubrium which is nearly totally reduced. In Tursiops, the tympanic ligament is inserted into a tiny U-shaped protuberance which extends as a narrow sulcus in the direction of the caput mallei. The tensor tympani muscle inserts in a minute shallow ovoid fovea; of course, the reduction of the processus muscularis does not imply a reduction of the muscle proper. In the neonate bottlenose dolphin, head and collum mallei are still separated by a shallow sulcus while the collum is (absolutely) broader than in the adult. Therefore the distance between the hardly visible manubrium and the short but distinct muscular apophysis and its ovoid fossa is somewhat longer. The incus of modern smaller toothed whales is highly condensed. In Tursiops, the body is relatively long, nearly conical and wears a nipple-like crus longum and a short, rod-like crus breve.

The stapes was reported to be tuned to a certain frequency range mainly by increasing or decreasing its mass. Cetaceans depending on high-level ultrasound obviously possess light and short stirrups while their area of attachment (fenestra vestibuli) is relatively large (Fleischer, 1978). In archeocetes (+Dorudon), the stapes seems to have been only moderately condensed and the stapedial limbs still surrounded a large ovoid foramen (intercrural aperture; Kellogg, 1936). In squalodonts (+Prosqualodon; Flynn, 1948), the stirrup was slightly larger, but had a rather similar shape, a broad head and neck, relatively long limbs and a foramen of equal relative size as in +Dorudon. This situation, in principle, also is found in Lipotes, where the stapes is slightly smaller. In this species, the head and neck are narrower and the limbs diverge to join a relatively broad foot-plate; the large intercrural aperture is rounded triangular. Fossil long-beaked rhabdosteid dolphins still show a small intercrural foramen and a prominent tubercule for the insertion of the stapedial muscle (Fordyce, 1983). In Inia, the stapes has a long and slender neck, compared to the diameter of the foot-plate (Kasuya, 1973; Fig. 5c). The stirrup of Pontoporia (Fig. 5a) shows a somewhat shorter neck while the diameter of the base is slightly larger than that of Inia. In Phocoena and Delphinus (Beauregard, 1894; Boenninghaus, 1904), the stapes is short and resembles that of Tursiops (Fleischer, 1973b; Fig. 5e). Here the intercrural aperture is reduced to minute fossulae on both sides of the plate-like limb region; these foveolae are more distinct in neonate specimens. In Tursiops, the neck as well as the foot-plate have a distinctly larger diameter than in Inia and Pontoporia. All the toothed whales mentioned show an inconspicuous attachment site of the stapedial muscle.

Both the M. tensor tympani and the M. stapedius were well-developed in archeocetes; this is obvious from grooves and clefts on the periotic bone of +Dorudon (Pompeckj, 1922; Kellogg, 1936). In rhabdosteids and living odontocetes, the same situation is to be found (Fordyce, 1983; Fig. 5a, c, e) while squalodontids, in this respect, are not known sufficiently. In living toothed whales, the origin of the stapedius muscle is more or less hidden by a thin lamina of the posterior periotic process, which is part of its articular facet and roofs the facial canal. Somewhat exceptional are Pontoporia and Inia, where the posterior process of the periotic and the according facet are short and moderately developed (Fig. 5a, c). Modern smaller toothed whales (Phocoena and Tursiops) have very well-developed middle ear muscles (Boenninghaus, 1904; Yamada, 1953). Because of the phylogenetic rotation of the malleus-incus complex, the tensor tympani works against the tympanic ligament and can increase its tension, while the stapedial muscle pulls obliquely or even perpendicularly to the long axis of the foot-plate (Fleischer, 1978). The muscles are reported to tighten the articulations of the ossicular chain and to increase the stiffness of the whole system, respectively, and thus, to suppress the transmission of low frequencies and to shift the natural frequency of the stapes to higher frequencies (Reysenbach de Haan, 1957; Møller, 1974).

CONCLUSIONS

About 70 Million years ago, a group of rather plesiomorphous tetrapod mammals (order +Condylarthra) began to adapt to life in the water, under circumstances we can only speculate about. Comparison between typical living land mammals and cetaceans give an idea of how profound the alterations and modifications of the organ systems throughout the whole body may have been. Naturally, apart from locomotion and feeding, orientation and communication in the new environment must have been among the most challenging problems for these animals.

Looking back from the cetaceans of our time, the earliest whales known (archeocetes) already show a whole set of morphological features which, in living forms, are correlated to the perception and conduction of water-borne sound. Therefore, it appears that the acoustic underwater receiver was developed first. Such features are the heavy ear ossicles, which were in contact with the periotic as in extant whales; the cetacean tympanic bulla, which was added to the ossicular chain; and the large medial foramen in the proximal part of the mandible (acoustic window). The morphology of the tympanic bulla allows the conclusion that also soft parts as, e.g., the tympanic membrane, were adapted to underwater sound which should have been perceived via the lower jaw and conducted by its fat body in the wide alveolar canal (Norris, 1964). Archeocetes obviously were fairly well adapted to hearing in water; however, they supposedly were restricted in frequency range and their capability for directional hearing certainly was rather moderate in comparison with the highly effective living dolphins.

Another problem of early cetaceans may have been vocalization under water. Obviously, the devices for emission of ultrasound were evolved in a second stage. The majority of authors dealing with that question have found strong indications that the generation of ultrasound signals in toothed whales takes place in the region of the forehead, adjacent to or even within the so-called accessory air sacs of the nasal tract (Norris, 1968; Ridgway et al., 1980; Ridgway and Carder, 1988). If this is true, the early toothed whales shifted their phonation from the larynx to these air sacs, i.e., they had to develop the ability of sound-production a second time. Effective generation and emission of sound and ultrasound, at the same time avoiding the irritation of the own ears by these signals, however, made necessary a profound modification of the skull as a whole. In order to suppress bone conduction, the zygomatic arches were disintegrated. The anterior skull roof was transformed into an acoustic reflector by the "telescoping process" which led to a shift of the elements of the upper jaw over those of the frontal region. This telescoping effect may originally have been a device for increased mechanical stability of the elongated rostrum and, as a by-product, may have improved the reflection properties of the skull roof. The spermaceti organ or melon, a cushion of fat tissue located in front of the deepened acoustic shield and the nasal tracts, may have originated as a device for opening and closing the blowhole before and after respiration, respectively, effected through muscles innervated by the facial nerve. In some living odontocetes (Delphinapterus), the shape of the melon is constantly changed in the course of phonation and may act as a variable acoustic lens for focusing the sonar beam. Naturally, the differentiation and improvement of both the receiver and, secondarily, of the generator of ultrasound, too, had a strong impact on the evolution of the brain (see below).

One of the most fascinating phenomena in the natural history of cetaceans is the modification of the ear bone complex, which closely is related to the development of accessory air sinuses of the auditory tube and tympanic cavity, respectively. Obviously, the bony elements adjacent to the periotic receded to the same degree the sinuses expanded while the periotic processes were reduced in size. This detachment of the periotic from the skull base can be studied in early fetal stages of toothed whales (de Beer, 1937; Hombach, 1981; Kiessler, 1981). Boenninghaus (1904) described the phylogenetic rotation of the cetacean cochlea which leads to a vertical instead of a horizontal orientation of the cochlear axis found in primates and other mammals. However, as is shown by the position of the foramina, the whole pars cochlearis of periotic has been rotated during the evolution of the archeocetes; nevertheless, the semicircular canals of extant cetaceans seem to have the same orientation as in other mammals (Honigmann, 1917; Eales, 1950). During ontogenesis this rotation process is not obvious.

The tympanic bone also was uncoupled from the skull base and united with the periotic. This tympano-periotic complex is largely surrounded by the sinuses, isolated acoustically, and allows directional hearing. However, the sinuses also are correlated with diving mechanics. If an animal descends in water, the air volume in the sinuses will shrink very quickly, although it may be possible to transfer air from the lung before it collapses. In order to avoid hampering of the middle ear function, arterial blood is shifted into cavernous tissue inside the tympanic cavity, replacing the loss of air volume (Purves, 1966; Fleischer, 1978). Nevertheless, the air sinuses may be subject to more or less drastic volume changes. While originally the sinuses partially were ensheathed by bony laminae, the latter presumably have been reduced with the improvement of diving capability in modern toothed whales. Only a few plesiomorphous forms living in freshwater and shallow coastal habitats, respectively (Platanista and Pontoporia), retain these osseous sheaths.

A striking feature of cetaceans is the very small size of their vestibulum and semicircular canals, which is also known of archeocetes (Fleischer, 1976). While the cochlea of Phocoena is as large as in the horse but has only 1.5 to 1.75 turns, the volume of the vestibule is merely as large as in the rabbit, that of the semicircular canals as large as in the hamster (Claudius, 1858). The reason for the exceptional small dimension of these parts of the inner ear is not really understood; perhaps the size of the labyrinth is restricted by the necessity of a periotic as massive as possible with as little cavities as possible (Reysenbach de Haan, 1957), obviously with no loss of functional quality of this system. Thus, the vestibular nerve of dolphins seems to be well-developed, although it is dominated by the massive cochlear nerve (Morgane and Jacobs, 1972).

Already in archeocetes, the tympanic bone was fused to the periotic in two main areas. The anterior junction was achieved by means of the accessory ossicle which in extant toothed whales may be fused to the periotic or the tympanic bone. Apart from mediating this contact between the elements of the tympano-periotic complex, the accessory ossicle (processus tubarius) helps to support the head of malleus on the lateral lip of the tympanic bulla. The posterior contact of the two elements is via their posterior processes. The area size of the corresponding articular surfaces is rather different among the odontocete groups. They are small to very small in Platanista[2], Inia and Pontoporia, intermediate in Phocoena and closely related (phocoenid) forms, and maximally developed in delphinid species. Possibly in toothed whales a correlation exists between the type of this posterior articulation (area size of facets, pleating of the latter or not) and the frequency range of the emitted sound and of the audiogram, respectively. If this should hold true for toothed whales, in general, coarse interpretation of a species would be possible from the morphology of the ear bones. A very interesting hypothesis correlates the frequency range, the body size of toothed whales and the size of their prey (Evans, 1973; Watkins, 1980; Thomas et al., 1988); larger species feed on larger prey and thus, need far field (low-frequency) orientation. However, most of the toothed whales showing this firm tympano-periotic articulation (Tursiops-type) belong to the Delphinidae and comprise species as different in size as Delphinus and Orcinus.

The earliest whales certainly had some difficulties in hearing effectively under water. In their presumed transition from freshwater to estuarine, shallow coastal and then open sea habitats, the archeocetes and early odontocetes optimized their ability to produce and perceive waterborne soundwaves. However, within their respective ecological niches, they certainly were adapted to specific frequency ranges. In plesiomorphous forms

---

[2] Platanista shows a somewhat peculiar situation because the periotic is not really uncoupled from the skull base and because the tympanic also touches the squamosal as is the case in a number of other toothed whales (ziphiids, physeterids).

living in freshwater habitats and in shallow coastal waters (Platanista, Inia, Pontoporia?), a trend of specialization to medium- and high-level ultrasound frequency ranges seems to exist (Popper, 1980a, b; Nachtigall, 1986; Evans, 1987). Obviously Platanista and Inia ..."have little or no energy below 15 kHz" (Evans and Awbrey, 1988). Inia can emit sound up to 200 kHz and hear frequencies from 1-100 kHz, with a maximum sensitivity at 20-60 kHz (Nachtigall, 1986). The coastal Phocoena has a similar bandwidth; here the frequency range of emitted sound extends from 2-180 kHz, with a reported maximum auditory sensitivity between 100 and 140 kHz (!) within a frequency range from 1-160 kHz (Evans, 1987). Obviously these species are ..."sharply tuned to the dominant frequency of their echolocation pulses" (Purves and Pilleri, 1983; Voronov and Stosman, 1986; Evans and Awbrey, 1988). Dolphins living in the open sea (Delphinus, Steno, Tursiops), which have to echolocate over longer distances, but nevertheless need high resolution in the near field, are reported to emit very low to very high frequencies (Tursiops: 0.1-300 kHz) and to hear frequencies from 0.075-150 kHz, with a maximum sensitivity in the range from 40-70 kHz (Tursiops). In Orcinus, finally, the frequency range of the emitted sound, the maximum energy, as well as the maximum sensitivity, are rather low (0.1-80 kHz; 0.25-40 kHz; 16 kHz). Interestingly, in those forms with the medium- to high-frequency specialization (Platanista, Inia and perhaps Pontoporia), the caudal tympano-periotic articulation is fairly moderately developed; like the posterior processes, the facets are small and nearly smooth. In contrast, Orcinus, as well as other larger toothed whales (Globicephala, Physeter), have large pleated facets and obviously are specialized for relatively low frequencies (Evans, 1987; Watkins et al., 1988). In Orcinus, the highest absolute sensitivity among toothed whales (Popper, 1980b) and the rather low range of maximum sensitivity can be correlated with the large size of the prey and the necessity to echo-locate over large distances. Smaller and medium-sized delphinids (Delphinus, Steno, Lagenorhynchus, Stenella, Tursiops, Grampus) have also such large, complicated facets on the posterior processes of the tympano-periotic complex. Obviously, they form a third group of (evolutionary successful) toothed whales, capable of emission and perception at very low frequencies and extreme ultrasound.

Really fascinating is the idea that low frequencies (0.5-5 kHz) ... "fall into the range of hearing or labyrinth sensitivity of prey species"... so that odontocetes might be able to stun their prey with specific long duration, loud impulse sounds causing overloading of the unprotected fish ear (Marten et al., 1988). In contrast, ultrasonic sounds appear to have little effect on fish. It may be speculated that the firm posterior connection between periotic and tympanic increases the stiffness of the tympano-periotic complex leading to an increase of its fundamental natural frequency (Fleischer, 1978). On the other hand, the firm connection, which may change into a genuine synostosis of periotic and tympanic, makes ..."it also a receiver for low frequencies, although it is very likely not tuned to such extremely low frequencies as in mysticetes" (Fleischer, 1980). As the Tursiops-type of tympano-periotic articulation is also found in large species (Physeter and Orcinus) with a relatively low and narrow frequency range, the latter suggestion seems more important. Interestingly enough, just these species have been reported to emit loud low-frequency impulse sound, presumably for stunning prey. Thus, the Tursiops-type connection may generally increase the bandwidth to low and very low frequencies, whithout necessarily limiting the use of extreme ultrasound in many species.

In summary, whales and dolphins have adapted perfectly to a habitat which, in general, is hostile to most mammals. As far as is known today, one of the major steps in the evolution of cetaceans was the adaptation to water-borne sound reception. In the course of the Eocene and Oligocene, other systems (elements) of the head were integrated in a new framework that allowed not only hearing and discrimination of successively higher frequencies but also production of sound waves, the echo of which could give more and more detailed information about the "umwelt" of the animal. In modern smaller toothed whales, the whole head is modified distinctly as to sonar function; obviously, they are able to scan their surroundings acoustically and reconstruct a three-dimensional image of it, using a "channel of interactive information between organism and environment" (manipulatory feedback; Eglash, 1984). Some contributions to the sonar equipment came from other organ systems and/or were rendered possible by certain functional trends not primarily related to hearing. Thus, e.g., the telescoping effect as a prerequisite for the development of an "acoustic shield" obviously was correlated with a change in feeding strategy and caused by the necessity of a long rostrum. The development of the fat body in the alveolar canal and of the "acoustic window" were only possible because the masticatory apparatus was reduced, allowing the lower jaw to decrease in strength. This reduction tendency also favored the dampening of bone conduction and the disintegration of the zygomatic arches. Finally, the migration of the nostrils, correlated with the reduction of the olfactory system, was a prerequisite for the co-localization of sound-generator, melon, and acoustic shield.

The development of a single pre-adaptive feature may have been a rather limited factor in the evolution of early cetaceans. However, the coincidence of an increasing number of new morphological devices and physiological qualities in subsequent forms (odontocetes) presumably rendered possible the development of a complex and highly efficient echo-location system using extremely detailed and precise ultrasound signals and extreme speed in processing the echo (Ridgway, 1983, 1986b). These innovations not only facilitated the acquisition of food but may also have helped to compensate for the loss of the rostral olfactory system in odontocetes. The latter presumably had been a major information source in the land-living cetacean ancestor (Gazin, 1965; Radinsky, 1976) but in the course of perfect adaptation to the aquatic environment was gradually reduced and lost. Moreover, in correlation with a considerable number of advanced morphological and physiological features, the sonar system with its new possibilities may have had a fundamental impact on the social life of many odontocetes. As an ensemble, these new properties are likely to have produced the unexpectedly high organizational level of the odontocete brain. However, there is also an ecological component influencing encephalization. Thus, species living in freshwater habitats (river dolphins) and the coastal harbor porpoise ..."tend to have much less sophisticated (flexible) echolocation"... (Eglash, 1984) and ..." they are also much less social than their smaller marine relatives, having fewer communicative sounds (no whistles)". While in these rather plesiomorphous dolphins the size of the brain clearly surpasses that in most carnivores and ungulates, progressive smaller dolphins in this regard equal or even exceed simian monkeys (Schwerdtfeger et al., 1984; Ridgway, 1986a, b).

The dominant position of the acoustic system is obvious at every level of the dolphin brain. Therefore, although the number of hair cells in <u>Tursiops</u> is about the same as in the human, the number of cochlear ganglion cells is more than the double. The reason for this may lie in the "increased number of neural paths for transmission of high-frequency information to the brain "...and..." that more details about cochlear events are supplied to the auditory system" (Wever et al., 1971). Another reason may be that a larger amount of information processing already takes place in the cochlea. The cochlear nerve and the subcortical nuclei of dolphins are strongly to extremely well developed, and so are the primary projection fields in the neocortex (Morgane and Jacobs, 1972; Ladygina and Supin, 1978; Supin et al., 1978; Ridgway, 1983, 1986; Schulmeyer and Oelschläger, 1988). The neocortex, which is rather thin but has a very large surface, has been characterized as primitive because of organizational similarities with insectivore mammals (Morgane et al., 1985, 1986; Glezer et al., 1988). However, it should be kept in mind that cetaceans have followed a separate evolutionary line for about 70 million years and that they have adapted perfectly to a habitat totally different from that of nearly all other mammals. Because the aquatic environment was favorable for acoustic (high-frequency) orientation and communication, delivering large amounts of information (Worthy and Hickie, 1986), the dolphins may have attained a similar (analogous) developmental level of brain differentiation as primates. However, they followed another path leading into another sensual world. Therefore, it is very difficult to compare them with tetrapod mammals as to high-level neurophysiological capabilities of the brain.

ACKNOWLEDGMENTS

The author thanks Dr.Peter J.H.van Bree (Amsterdam, The Netherlands) for donating valuable cetacean material and for the possibility to investigate the rich collection of the Zoological Museum in Amsterdam. I am much obliged to Jutta S.Oelschläger for her conscientious technical help and to Horst L.Schneeberger for his skillful artwork. I am grateful to Dr.David Johnson for kindly editing the English manuscript.

REFERENCES

Abel, O., 1901, Les dauphins longirostres du Bolděrien (Miocène supěrieur) des environs d'Anvers, Měm. Mus. Roy. Hist. Nat. Belgique, 1:1-95, 10 pls.

Abel, O., 1912, Cetaceenstudien: Rekonstruktion des Schädels von Prosqualodon australe Lyd. aus dem Miozän Patagoniens, Sitz. Ber. Kaiserl. Akad. Wiss. Math.-Nat. Klasse, Wien, 121:57-74, 3 pls.

Anthony, R., 1926, Les affinités des cétacés, Ann. Inst. Océanogr., Paris, 3:93-134.

Barnes, L. G., 1984, Whales, dolphins and porpoises: Origin and evolution of the Cetacea, in: "Mammals". Notes for a short course organized by P. D. Gingerich and C. E. Badgley; T. W. Broadhead, ed., Univ. Tennessee Dept. Geol. Sci., Studies in Geology, 8:139-154.

Barnes, L. G., Domning, D. P, and Ray, C. E., 1984, Status of studies on fossil marine mammals, Marine Mammal Science, 1:15-53.

Bauer, G. B., Fuller, M., Perry, A., Dunn, J. R., and Zoeger, J., 1985, Magnetoreception and biomineralization of magnetite in cetaceans, in: "Magnetite Biomineralization and Magnetoreception in Animals: A New Biomagnetism", J. L. Kirschvink, D. S. Jones, B. J. McFadden, eds., Plenum Press, New York, pp. 489-507.

Beauregard, H., 1894, Recherches sur l'appareil auditif chez les mammifères, J. Anat. Physiol., Paris, 1894:366-413.

Boenninghaus, G., 1904, Das Ohr des Zahnwales, zugleich ein Beitrag zur Theorie der Schallleitung, Zool. Jb., 19:189-360.

Boyden, A., and Gemeroy, D., 1950, The relative position of the Cetacea among the orders of mammalia as indicated by precipitin tests, Zoologica, (New York), 35:145-151.

Brandt, J. F., 1873, Untersuchungen über die fossilen und subfossilen Cetaceen Europas, Měm. Acad. Imp. Sci. St.Petersbourg, Ser.7, 20:372 pp., 34 pls.

Buhl, E. H., and Oelschläger, H. A., 1986, Ontogenetic development of the nervus terminalis in toothed whales, Evidence for its non-olfactory nature, Anat. Embryol., 173:285-294.

Buhl, E. H., and Oelschläger, H. A., 1988, Morphogenesis of the brain in the harbour porpoise, J. Comp. Neurol., 277:109-125.

Cigala-Fulgosi, F. and Pilleri, G., 1985, The lower Serravallian cetacean fauna of Visiano (Northern Apennines, Parma, Italy), in: "Investigations on Cetacea", G. Pilleri, ed., Vol.17, Institute of Brain Anatomy, Berne (Switzerland), pp. 55-93, 10 pls.

Claudius, M., 1858, Physiologische Bemerkungen über das Gehörorgan der Cetaceen und das Labyrinth der Säugetiere, Kiel 1858.
Cranford, T. W., 1988, The anatomy of acoustic structures in the spinner dolphin forehead as shown by x-ray computed tomography and computer graphics, in: "Animal Sonar: Processes and Performance", P. E. Nachtigall and P. W. B. Moore, eds. (NATO ASI Series, Vol. 156), Plenum Press, New York, pp. 67-77.
Credle, V. R., 1988, Magnetite and magnetoreception in stranded dwarf and pygmy sperm whales, Kogia simus and Kogia breviceps, University of Miami thesis, 86 pp.
dal Piaz, G., 1916, Gli odontoceti del Miocene Bellunese. Parte seconda. Squalodon, Mem.Ist. Geol.Univ.Padova, 4:1-94, 11 pls.
de Beer, G., 1937, "The development of the vertebrate skull". Clarendon Press, Oxford.
Demski, L. S., Ridgway, S. H., Bullock, T. H., and Schwanzel-Fukuda, M., 1985, Terminal nerve of odontocete whales, Am. Zool., 25:107A (abstract).
Dral, A. D. G., 1987, On the optics of the dolphin eye, Aquatic Mammals, 13:61-64.
Eales, N. B., 1950, The skull of the foetal narwhal, Monodon monoceros L., Phil. Trans. Soc., London., B, 235:1-33.
Eglash, R., 1984, The cybernetics of Cetacea, in: "Investigations on Cetacea", G. Pilleri, ed., Vol.16, Institute of Brain Anatomy, Berne (Switzerland), pp. 151-197.
Evans, P. G. H., 1987, "The Natural History of Whales and Dolphins". Facts on File Publications, New York.
Evans, W. E., 1973, Echolocation by marine delphinids and one species of freshwater dolphin, J. Acoust. Soc. Am., 54:191-199.
Evans, W. E. and Awbrey, F. T., 1988, Natural history aspects of marine mammal echolocation: Feeding strategies and habitat, in: "Animal Sonar: Processes and Performance", P. E. Nachtigall and P. W. B. Moore, eds. (NATO ASI Series A, Vol. 156), Plenum Press, New York, pp. 521-534.
Fleischer, G., 1973a, Structural analysis of the tympanicum complex in the bottle-nosed dolphin (Tursiops truncatus), J. Auditory Res., 13:178-190.
Fleischer, G., 1973b, On structure and function of the middle ear in the bottle-nosed dolphin (Tursiops truncatus), Proc. 9th Annu. Conf. Biol. Sonar and Diving Mammals, Menlo Park, Cal.: Stanford Res. Institute, pp. 137-179.
Fleischer, G., 1976, Hearing in extinct cetaceans as determined by cochlear structure, J. Paleontol., 50:133-152.
Fleischer, G., 1978, Evolutionary principles of the mammalian middle ear. Adv. Anat. Embryol. Cell Biol., 55:1-70.
Fleischer, G., 1980, Low-frequency receiver of the middle ear in mysticetes and odontocetes, in: "Animal Sonar Systems", R.-G. Busnel and J. F. Fish, eds. (NATO ASI Series A, Vol. 28), Plenum Press, New York, pp. 891-893.
Flower, W. H., 1874, On Risso's dolphin, Grampus griseus, Trans. Zool. Soc. Lond., 8:1-21, 2 pls.
Flower, W. H., and Lydekker, R., 1978, An introduction to the study of mammals living and extinct, Arno Press New York (1978; reprint), pp.225-272.
Flynn, T. T., 1948, Description of Prosqualodon davidi Flynn, a fossil cetacean from Tasmania, Trans. Zool. Soc. Lond., 26:153-196.
Fordyce, R. E., 1981, Systematics of the odontocete whale Agorophius pygmaeus and the family Agorophiidae (Mammalia: Cetacea), J. Paleontol., 55:1028-1045.
Fordyce, R. E., 1983, Rhabdosteid dolphins (Mammalia: Cetacea) from the Middle Miocene, Lake Frome area, South Australia, Alcheringa, 7:27-40.
Fordyce, R. E., 1985, Late Eocene archeocete whale (Archaeoceti: Dorudontinae) from Waihao, South Canterbury, New Zealand, N. Z. J. Geol. Geophys., 28:351-357.

Fraas, E., 1904, Neue Zeuglodonten aus dem unteren Mitteleocän von Mokattam bei Cairo, Geol. Palaeont. Abh., Jena (N.F.), 6:1-24.

Fraser, F. C., and Purves, P. E., 1960, Hearing in cetaceans. Evolution of the accessory air sacs and the structure and function of the outer and middle ear in recent cetaceans, Bull. Br. Mus. (Nat.Hist.) Zool., 7:1-140.

Gazin, C. L., 1965, A study of Early Tertiary condylarthran mammal Meniscotherium, Smithson. Misc. Coll., 149:1-98.

Gingerich, P. D., Wells, N. A., Russell, D. E., and Shah, S. M. I., 1983, Origin of whales in epicontinental remnant seas: New evidence from the early Eocene of Pakistan, Science (New York), 220:403-406.

Glezer, I. I., Jacobs, M. S., and Morgane, P. J., 1988, Implications of the "initial brain" concept for brain evolution in Cetacea, Behav. Brain Sciences, 11:75-116.

Hennig, W., 1966, Phylogenetic Systematics, University of Illinois Press, Urbana, 263 pp.

Herman, L. M., and Tavolga, W. N., 1980, The communication systems of cetaceans, in: "Cetacean Behavior: Mechanisms and Functions", L. M. Herman, ed., Wiley, New York, pp. 149-209.

Heyning, J. E., 1989, Comparative facial anatomy of beaked whales (Ziphiidae) and a systematic revision among the families of extant Odontoceti, Contributions in Science, Natural History Museum of Los Angeles County, 405:1-64

Hombach, U., 1981, Morphogenese des Chondrocraniums von Physeter macrocephalus L. Mit besonderer Berücksichtigung der Ohrkapseln, Thesis, Faculty of Human Medicine, Frankfurt am Main, 115 pp., 43 figs.

Honigmann, H. L., 1917, Bau und Entwicklung des Knorpelschädels vom Buckelwal, Zoologica (Stuttgart), 69:1-87, 28 figs., 2 pls.

Hyrtl, J., 1845, Vergleichend-anatomische Untersuchungen über das innere Gehörorgan des Menschen und der Säugetiere, F. Ehrlich, Prag.

Jacobs, M. S., Morgane, P. J., and McFarland, W. L., 1971, The anatomy of the brain of the bottlenose dolphin (Tursiops truncatus). Rhinic lobe (rhinencephalon). I. The paleocortex, J. Comp. Neurol., 141:205-272.

Kaiya, Z., Weijuan, Q., and Yuemin, L., 1979, The osteology and the systematic position of the Baiji, Lipotes vexillifer. Acta Zool. Sin., 25:58-74, 4 pls.

Kasuya, T., 1973, Systematic consideration of recent toothed whales based on the morphology of tympano-periotic bone, Sci. Rep. Whales Res. Inst., Tokyo, 25:1-103.

Kellogg, R., 1924, A fossil porpoise from the Calvert formation of Maryland, Proc. U. S. Nat. Mus., 63:1-39, 18 pls.

Kellogg, R., 1926, Supplementary observations on the skull of the fossil porpoise Zarhachis flagellator Cope, Proc. U. S. Nat. Mus., 67:1-18, 5 pls.

Kellogg, R., 1927, Kentriodon pernix, a Miocene porpoise from Maryland, Proc. U. S. Nat. Mus., 69:1-55, 14 pls.

Kellogg, R., 1928, The history of whales - their adaptation to life in the water, Q. Rev. Biol., 3:29-76, 174-208.

Kellogg, R., 1936, A Review of the Archaeoceti, Carnegie Inst. Washington, Publication 482.

Kiessler, G., 1981, Die Morphogenese des Chondrocraniums von Phocoena phocoena mit besonderer Berücksichtigung der Ohrkapseln. Thesis, Faculty of Human Medicine, Frankfurt am Main, 122 pp., 29 figs.

Kirschvink, J. L., Dizon, A. E., and Westphal, J. A., 1986, Evidence from strandings for geomagnetic sensitivity in cetaceans, J. Exp. Biol., 120:1-24.

Klima, M., 1987, Morphogenesis of the nasal structures of the skull in toothed whales (Odontoceti), in: "Morphogenesis of the Mammalian Skull (Mammalia depicta)", H.-J. Kuhn and U. Zeller, eds., Parey, Berlin, pp. 105-121.

Klima, M., und van Bree, P. J. H., 1985, Überzählige Skeletelemente im Nasenschädel von Phocoena phocoena und die Entwicklung der Nasenregion bei den Zahnwalen, Gegenbaurs Morph. Jahrb., 131:131-178.

Ladygina, T. F., and Supin, A. Ya., 1978, On the homology of the different regions of the brain's cortex of Cetacea and other mammals, in: "Morskiye Mlekopita'yushchiye, Resul'taty i Metodi Issledovanii", V. E. Sokolov, ed., Izdatel'stvo Nauka, Moscow, pp. 55-64 (in Russian).

Marten, K., Norris, K. S., Moore, P. W. B., and Englund, K. A., 1988, Loud impulse sounds in odontocete predation and social behavior, in: "Animal Sonar: Processes and Performance", P. E. Nachtigall and P. W. B. Moore, eds. (NATO ASI Series A, Vol. 156), Plenum Press, New York, pp. 567-579.

Mead, J. G., 1975, Anatomy of the external nasal passages and facial complex in the Delphinidae (Mammalia: Cetacea), Smithson. Contr. Zool., 207: 1-72.

Miller, G. S., 1923, The telescoping of the cetacean skull. Smithson. Misc. Coll., 76:1-71.

Møller, A. R., 1974, The acoustic middle ear muscle reflex, in: "Handbook of Sensory Physiology", H. Autrum, ed., Vol. 5/1, Springer, New York, pp. 491-517.

Morgane, P. J., and Jacobs, M. S., 1972, Comparative anatomy of the cetacean nervous system, in: "Functional Anatomy of Marine Mammals", R. J. Harrison, ed., Vol. 1, Academic Press, London, pp. 118-244.

Morgane, P. J., Jacobs, M. S., and Galaburda, A., 1985, Conservative features of neocortical evolution in dolphin brain, Brain, Behav. Evol., 26:176-184.

Morgane, P. J., Jacobs, M. S., and Galaburda, A., 1986, Evolutionary morphology of the dolphin brain, in: "Dolphin Cognition and Behavior. A Comparative Approach", R. J. Schusterman, J. A. Thomas, F. G. Wood, eds., Lawrence Erlbaum Assoc., London, pp. 5-29.

Nachtigall, P. E., 1986, Vision, audition and chemoreception in dolphins and other marine mammals, in: "Dolphin Cognition and Behavior: A Comparative Approach", R. J. Schusterman, J. A. Thomas, F. G. Wood, eds., Lawrence Erlbaum Assoc., London, pp. 79-113.

Nachtigall, P. E., and Hall, R. W., 1984, Taste Reception in the bottlenosed dolphin, Acta Zool. Fenn., 172:147-148.

Ness, A. R., 1967, A measure of asymmetry of the skulls of odontocete whales, J. Zool., London, 153:209-221.

Nomina Anatomica, 1983, 11th Int. Congr. Anat. (Mexico City), 5th ed., Williams and Wilkins, Baltimore, pp. 1-86.

Norris, K. S., 1964, Some problems of echolocation in cetaceans, in: "Marine Bioacoustics", W. N. Tavolga, ed., Pergamon Press, New York, pp. 317-336.

Norris, K. S., 1968, The evolution of acoustic mechanisms in odontocete cetaceans, in: "Evolution and Environment", E. T. Drake, ed., Yale Univ. Press, New Haven, pp. 297-324.

Norris, K. S., 1980, Peripheral sound processing in odontocetes, in: "Animal Sonar Systems", R.-G. Busnel, and J. F. Fish, eds. (NATO ASI Series A, Vol. 28), Plenum Press, New York, pp. 495-509.

Norris, K. S., and Evans, E. C., 1988, On the evolution of acoustic communication systems in vertebrates. Part I: Historical aspects, in: "Animal Sonar: Processes and Performance", P. E. Nachtigall and P. W. B. Moore, eds. (NATO ASI Series A, Vol. 156), Plenum Press, New York, pp. 655-669.

Norris, K. S., and Møhl, B., 1983, Can odontocetes debilitate prey with sound?, Amer. Naturalist, 122:85-104.

Oelschläger, H. A., 1986a, Comparative morphology and evolution of the otic region in toothed whales (Cetacea, Mammalia), Amer. J. Anat. 177:353-368.

Oelschläger, H. A., 1986b, Tympanohyal bone in toothed whales and the formation of the tympano-periotic complex (Mammalia: Cetacea), J. Morphol., 188:157-165.

Oelschläger, H. A., 1987, Pakicetus inachus and the origin of whales and dolphins (Mammalia: Cetacea), Gegenbaurs Morph. Jahrb., Leipzig, 133:673-685.
Oelschläger, H. A., 1989, Early development of the olfactory and terminalis systems in baleen whales, Brain, Behav. Evol., 34:171-183.
Oelschläger, H. A., and Buhl, E. H., 1985a, Development and rudimentation of the peripheral olfactory system in the harbor porpoise, Phocoena phocoena (Mammalia: Cetacea), J. Morphol., 184:351-360.
Oelschläger, H. A., and Buhl, E. H., 1985b, Occurrence of an olfactory bulb in the early development of the harbor porpoise (Phocoena phocoena L.), in: "Functional Morphology in Vertebrates", H.-R. Duncker, and G. Fleischer, eds., G. Fischer, Stuttgart, pp. 695-698.
Oelschläger, H. A., Buhl, E. H., and Dann, J. F., 1987, Development of the nervus terminalis in mammals including toothed whales and humans, Ann. New York Acad. Sci., 519:447-464.
Pilleri, G., 1987, The Cetacea of the Italian Pliocene, Institute of Brain Anatomy, University of Berne (Switzerland), pp. 1-160, 63 pls.
Pilleri, G., Gihr, M. and Kraus, C., 1987, The organ of hearing in cetaceans. 1. Recent species, in: "Investigations on Cetacea", G. Pilleri, ed., Vol.20, Institute of Brain Anatomy, Berne (Switzerland), pp.43-125.
Pompeckj, J. F., 1922, Das Ohrskelett von Zeuglodon, Senckenbergiana, 4:43-100.
Popper, A. N., 1980a, Sound emission and detection by delphinids, in: "Cetacean Behavior: Mechanisms and Functions", L. M. Herman, ed., Wiley, New York, pp. 1-52.
Popper, A. N., 1980b, Behavioral measures of odontocete hearing, in: "Animal Sonar Systems", R.-G. Busnel, and J. F. Fish, eds. (NATO ASI Series A, Vol. 28), Plenum Press, New York, pp. 469-481.
Purves, P. E., 1966, Anatomy and physiology of the outer and middle ear in cetaceans, in: "Whales, Dolphins and Porpoises", Norris, K. S., ed., University of California Press, Berkeley, pp. 320-380.
Purves, P. E., and Pilleri, G., 1983, "Echolocation in Whales and Dolphins", Academic Press, New York.
Radinsky, L., 1976, The brain of Mesonyx, a middle Eocene mesonychid condylarth, Fieldiana (Geol.), 33:323-337.
Rensberger, J. M., 1969, A new iniid cetacean from the Miocene of California, Univ. Calif. Publ. Geol. Sci., 82:1-43.
Reysenbach de Haan, F. W., 1957, Hearing in Whales, Acta Otolaryngol., Stockh. (Suppl.), 134:1-114.
Ridewood, W. G., 1922, Observations on the skull in foetal specimens of whales of the genera Megaptera and Balaenoptera, Philos. Trans. Roy. Soc. London, (B) 211:209-272.
Ridgway, S. H., 1983, Dolphin hearing and sound production in health and illness, in: "Hearing and Other Senses: Presentations in Honor of E.G.Wever", R. R. Fay, G. Gourevitch, eds., The Amphora Press, Groton, CT, pp.247-296.
Ridgway, S. H., 1986a, Dolphin brain size, in: "Research on Dolphins", M. M. Bryden and R. Harrison, eds., Clarendon Press, Oxford, pp. 59-70.
Ridgway, S. H., 1986b, Physiological observations on dolphin brains, in: "Dolphin Cognition and Behavior: A comparative approach", R. J. Schusterman, J. A. Thomas, F. G. Wood, eds., Lawrence Erlbaum Assoc., London, pp. 31-59.
Ridgway, S. H., and Carder, D. A., 1988, Nasal pressure and sound production in an echolocating white whale, Delphinapterus leucas, in: "Animal Sonar: Processes and Performance", P. E. Nachtigall, and P. W. B. Moore eds. (NATO ASI Series A, Vol. 156), Plenum Press, New York, pp. 53-60.
Ridgway, S. H., and Wood, F. G., 1988, Cetacean brain evolution, Behav. Brain Sci., 11:99-100.

Ridgway, S. H., Demski, L. S., Bullock, T. H., and Schwanzel-Fukuda, M., 1987, The terminal nerve in odontocete cetaceans, Ann. New York Acad. Sci., 519:201-212.
Ridgway, S. H., Carder, D. A., Green, R. F., Gaunt, A. S., Gaunt, S. L. L., and Evans, W. E., 1980, Electromyographic and pressure events in the nasolaryngeal system of dolphins during sound production, in: "Animal Sonar Systems", R.-G. Busnel, J. F. Fish, eds. (NATO ASI Series A, Vol. 28), Plenum Press, New York, pp. 239-249.
Rothausen, K., 1968a, Die Squalodontidae (Odontoceti, Mamm.) im Oligozän und Miozän Italiens, Mem. Ist. Geol. Mineral. Univ. Padova, 26:1-18, 2 pls.
Rothausen, K., 1968b, Die systematische Stellung der europäischen Squalodontidae (Odontoceti, Mamm.), Paläontol. Z., 42:83-104, 2 pls.
Schulmeyer, F. J., und Oelschläger, H. A., 1990, Zellmorphologie des Trapezkörperkerns und der periolivären Kerne beim Delphin, Anat. Anz. Suppl., 166:533-534.
Schwerdtfeger, W. K., Oelschläger, H. A., and Stephan, H., 1984, Quantitative neuroanatomy of the brain of the La Plata dolphin, Pontoporia blainvillei. Anat. Embryol., 170:11-19.
Scott, W. B., 1888, On some new and little known creodonts. J. Acad. Nat. Sci. Philadelphia, 9:155-185.
Stromer, E., 1903, Zeuglodon-Reste aus dem oberen Mitteleozän des Fajum, Beitr. Paläont. Geol. Österr.-Ungarns u. d. Orients, Mitt. Geol. Paläont. Inst. Univ. Wien, 15:65-100.
Stromer, E., 1908, Die Archaeoceti des ägyptischen Eozäns, Beitr. Paläont. Geol. Österr.-Ungarns u. d. Orients, Mitt. Geol. Paläont. Inst. Univ. Wien, 21:106-177.
Stromer, E., 1911, Neue Forschungen über fossile lungenatmende Meeresbewohner, Fortschr. Naturwiss. Forsch., Berlin, 2:83-114.
Supin, A. Y., Mukhametov, L. M., Ladygina, T. F., Popov, V. V., Mass, A. M., and Poljakova, I. G., 1978, Electrophysiological studies of the dolphin's brain. V. E. Sokolov, ed., Izdatel'stvo Nauka, Moscow, pp. 7-85 (in Russian).
Szalay, F. E., and Gould, S. J., 1966, Asiatic Mesonychidae (Mammalia: Condylarthra), Bull. Amer. Mus. Nat. Hist., 132:127-174.
Thomas, J. A., Stoermer, M., Bowers, C., Anderson, L., and Garver, A., 1988, Detection abilities and signal characteristics of echolocating false killer whales (Pseudorca crassidens), in: "Animal Sonar: Processes and Performance", P. E. Nachtigall and P. W. B. Moore, eds. (NATO ASI Series A, Vol. 156), Plenum Press, New York, pp.323-328.
True, F. W., 1910, A new genus of fossil cetaceans from Santa Cruz territory, Patagonia; a redescription of a mandible and vertebrae of Prosqualodon, Smithson. Misc. Coll., 52:441-455, 3 pls.
van Beneden, P. J., and Gervais, P., 1868-1880, "Ostéographie des Cétacés, vivants et fossiles". Paris, Bertrand, 605 pp., 64 pls.
van Valen, L., 1966, Deltatheridia, a new order of mammals, Bull. Amer. Mus. Nat. Hist., 132:1-126.
Voronov, V. H., and Stosman, I. T., 1986, Electrical responses of the stem structures of the acoustic system of Phocoena phocoena to tonal stimuli, in: "The Electrophysiology of the Sensory Systems of Marine Mammals", V. E. Sokolov, ed., Nauk, Moscow.
Watkins, W. A., 1980, Click sounds from animals at sea, in: "Animal Sonar Systems", R.-G. Busnel and J. F. Fish, eds. (NATO ASI Series A, Vol.28), Plenum Press, New York, pp.291-297.
Watkins, W. A., Moore, K. E., Clark, C. W., and Dahlheim, M. E., 1988, The sounds of sperm whale calves, in: "Animal Sonar: Processes and Performance", P. E. Nachtigall and P. W. B. Moore, eds. (NATO ASI Series A, Vol.156), Plenum Press, New York, pp.99-107.
Wever, E. G., McCormick, J. G., Palin, J., and Ridgway, S. H., 1971, The cochlea of the dolphin Tursiops truncatus: Hair cells and ganglion cells, Proc. Natl. Acad. Sci., 68:2908-2912.

Worthy, G. A. J., and Hickie, J. P., 1986, Relative brain size in marine mammals, Amer. Naturalist, 128:445-459.
Yamada, M., 1953, Contribution to the anatomy of the organ of hearing of whales. Sci. Rep. Whales Res. Inst., Tokyo, 8:6-79.
Zoeger, J., Dunn, J. R., and Fuller, M., 1981, Magnetic material in the head of a common Pacific dolphin, Science, N. Y., 213:892-894.

TACTILE SENSITIVITY, SOMATOSENSORY RESPONSES, SKIN VIBRATIONS, AND THE SKIN SURFACE RIDGES OF THE BOTTLENOSE DOLPHIN, TURSIOPS TRUNCATUS

Sam H. Ridgway and Donald A. Carder

Biosciences Division
Naval Ocean Systems Center
San Diego, California, USA 92152

INTRODUCTION

Kramer (1960, 1977) was the first to suggest that dolphins have a compliant skin that enhances their hydrodynamic performance by damping incipient turbulence. Kramer also was the first to develop a synthetic vessel coating based upon dolphin skin. However, his coating contained no mechanism for active vibration or for other adjustments to changing boundary layer conditions. Lang (1966) reviewed the earlier work on dolphin hydrodynamics and evaluated the various theories. Concerning the idea that dolphins might actively change their skin surface to reduce hydrodynamic drag, he stated: "An alternate explanation for low drag with regard to cetaceans is that they actively adjust the flexibility and movement of their skin to damp out the microscopic disturbances in the laminar boundary layer. Betchov showed that the laminar flow might be extended indefinitely by this means."

Several authors have suggested that living dolphin skin makes adjustments that improve boundary-layer conditions to reduce drag during underwater swimming (Haider and Lindsley, 1964; Palmer and Weddell, 1964; Surkina, 1971; Khomenko and Khadzhinskiy, 1974; Kayan, 1974; Babenko and Nikishova 1976). Their evidence was based on skin anatomy, nerve structures in the skin, electrical potentials from the skin, or from microvibrations. These ideas can best be summarized by a quotation from Khomenko and Khadzhinskiy: "One of the reasons that dolphins are so hydrodynamically perfect is that they actively control (by reflexes) their skin, which contains specific receptors connected to the central nervous system."

Innervation of dolphin skin has been described by Palmer and Weddell (1964) and by Harrison and Thurley (1972 and 1974). The first authors were especially impressed:

> The presence of longitudinally disposed dermal ridges, the patterned arrangement of

the collagen and elastic fibres related to
them, together with the passage of preterminal nerves through tunnels in the base of
the epidermal ridges to serve large complex
terminals attached to papillary walls, all
suggest that the skin is a specialized
pressure-transducing mechanism. The number
and complex arrangement of other nerve
terminals in the skin further suggest that
the skin is instrumental in enabling the
dolphin to become 'aware' of its body image
in relation to the water around it; in other
words, that the skin has both tactile and
proprioceptive functions.

In support of this view can be cited the
fact that the environment in which it lives
virtually isolates it from the effect of
gravity and, moreover, its locomotor system
is covered with blubber. Further, all the
organized endings in the dermis are supplied
by two separate and distinct sets of nerve
fibres, one thick and the other fine. The
significance of this is not known, but by
analogy with muscle innervation it is tempting to regard the fine fibre components of
the organized endings as efferent and
capable of modulating the sensitivity of the
large fibre components.

If this is so, the skin is likely to be
able to detect even the slightest tendency
towards the onset of turbulence and, though
there is as yet no positive proof, a good
case can be made out for regarding the
specialised innervation of the skin in
<u>Tursiops truncatus</u> as part of a complicated
sensory-motor mechanism which permits the
maintenance of laminar flow... (Palmer and
Weddell, 1964).

Lende and Welker (1972) first studied dolphin skin
sensitivity. They recorded electrical potentials from an
area of the contralateral cerebral cortex to study the
representation of this area of somatosensory cortex on the
skin surface. Using stimuli such as tapping or lightly
touching or stroking of the skin or by allowing water droplets to fall on the skin, the investigators produced a map
of skin sensitivity based upon recordings from this area of
cortex. The greatest sensitivity was found in "a broad zone
extending below both eyes and ventrally around the neck...."

Kolchin and Bel'kovich (1973) used the galvanic skin
response (GSR) produced by a 0.3 mm weighted wire to make a
partial map of body skin sensitivity in the common dolphin
<u>Delphinus delphis</u>. Of the body portions studied, they found
the dolphin to be most sensitive (10 mg/mm^2) in separate
circular areas of about 5 cm diameter around the blowhole
and eyes. The snout, lower jaw, and melon were found to be
somewhat less sensitive (10 to 20 mg/mm^2) while still less
sensitivity (20 to 40 mg/mm^2) was observed along the back

in broad areas both anterior and posterior to the dorsal fin. The authors state that "from an ecological point of view the results we obtained are not unusual. The values for the threshold of sensitivity to touch in dolphins are 10--40 mg/mm^2 ; this is close to the values for a human being in the most sensitive skin areas, the tactile surfaces of the fingers, the skin of the eyelids, and the lips."

Microvibrations, minute tremor-like vibrations that occur in warm-blooded animals at all times over the entire body, have an amplitude of about 1 to 5 microns at frequencies of 7 to 13 Hz in relaxed humans. This "minor tremor" is thought to be important in maintaining body temperature (Rohracher, 1964). Shivering is believed to be a natural amplification of this continuously present tremor. Microvibrations in dolphin skin were studied by Haider and Lindsley (1964) who observed that the dolphin skin exhibited microvibrations three or four times the amplitude found in human skin and the peak frequencies they measured were 11 Hz for a human and 13 Hz for a dolphin subject.

One known mechanism for producing gross skin movements on the body surface is the rapid contraction of the extensive subcutaneous muscle. The panniculus carnosus is a sheet of muscle that lies beneath the skin in many animals that can move the skin. [Panniculus carnosus is the name used by anatomists of the 19th and early 20th century when most of the studies on gross cetacean anatomy were published; the modern term is musculus cutaneus. In the horse, musculus cutaneus is described in five parts, the largest of which is M. cutaneus trunci which covers much of the body (Sisson and Grossman, 1953).] In horses, a major component of this muscle is called by the common name "flyshaker." The panniculus carnosus is especially well-developed in dolphins and other small whales (Murie, 1873; Schulte and Smith, 1918). This thin sheet of muscle covers most of the dolphin's body except for the tail-stock, appendages, snout, melon and midback. This thin, broad muscle with numerous, apparently separate bundle groupings lies under the outer blubber layer (hypodermis), external to the subcutaneous depot fat (inner blubber layer) and in the bottlenose dolphin is 2 to 4 cm below the skin surface. The most anterior portion of the panniculus carnosus underlies the region of the eye and corner of mouth (angle of gape).

We replicated the experiments of Haider and Lindsley (1964) and extended the investigations begun by Lende and Welker (1972) to find out more about the sensitivity, reflexiveness, and responsiveness of the skin of the living dolphin. We sought answers to the following questions: Is the dolphin epidermis sensitive to small vibrations or pressure changes? Does the skin surface respond to such stimulation? Does the dolphin panniculus carnosus muscle, that underlies much of the skin, make rapid contractions in response to stimulation analogous to the rapid "flyshaker" movements of similar muscles in the horse?

Although the intracutaneous structure of dolphin skin has received considerable attention (Palmer and Weddell, 1964; Harrison and Thurley, 1974; Sokolov, 1982; Geraci et. al., 1986; Stromberg, 1989), the cutaneous surface has not been studied in detail. Because the cutaneous ridges appeared to be possibly an important feature of the skin we attempted to document their appearance and orientation. Although the cutaneous ridges are readily observed by careful inspection of the skin, they have not to our knowledge been documented in the literature, nor has their possible function been discussed. Geraci et al. (1986) mention the existence of the ridges, but other writers generally refer to the skin of dolphins and other small whales as "smooth." For example, Sokolov (1982) in his major review of mammal skin states that "in all Cetacea under investigation, the skin surface is relatively smooth."

MATERIALS AND METHODS

Physiological Studies

For each test, the dolphin subject (two adult females and one adult male T. truncatus) was placed in a padded tank in a relatively fixed position with its dorsal surface and blowhole above water. Exposed skin surfaces were kept wet throughout the test period by sponging or misting with water.

Two types of mechanical stimulation were used. Both types have been employed in human vibrotactile studies. A moving coil shaker (B&K 4810) presented stimuli with a great variety of frequencies and amplitudes. A piezoceramic bimorph (Sherrick, 1975) 20mm X 6 mm X 0.6mm was used to present low-amplitude tactile stimulation, and had the added advantage of low mass and low electromagnetic radiation. A function generator controlled stimulus rate and triggered the signal averaging computer. Stimulus rates varied from one per five s to forty per s for averaging. Stimuli were given at intervals as long as 30 s for observations of microvibration responses. Shaker stimulus parameters were gaged via accelerometer and force sensors in the shaker head. Stimuli were monitored on an oscilloscope. Fingertap and water-drop stimuli also were presented at low asynchronous rates and few repetitions per set.

Brain activity (electroencephalogram or EEG) was monitored from fine wires, the size of a human hair, placed subcutaneously adjacent to the skull near the vertex and left and right mastoid positions (a similar arrangement has been used for auditory evoked potentials in dolphins, Ridgway et al., 1980). For recording somatosensory evoked responses the vertex electrode position was moved slightly forward, and about two centimeters off the midline, over an area that has been identified as somatosensory cortex (Lende and Welker, 1972). For these recordings, a fine wire electrode in the snout also was used as a reference.

Microvibrations were sensed using piezoceramic bimorphs (Sherrick, 1975) and accelerometers (Model 9001, Vibra-Metrics Inc., Hamden, CT). The reed bender bimorphs employed as sensors ranged in size from 3 X 12 mm to 12 X 58 mm and 0.6 mm thick. The accelerometers were small, 6.6 mm high by 6.4 mm diameter, with a low mass of 0.75 g. The bimorphs were placed on the dorsal half of the body surface with petrolatum for adhesion, contact medium, and electrical insulation. Predetermined thicknesses of petrolatum were applied with templates to compare dolphin data with calibrations of the same bimorphs on a petrolatum-coated platform atop a shaker equipped with an impedance head. Accelerometers were attached with a cyanoacrylate bonding agent or petrolatum. Electrical leads were very low mass, mechanically-damped and isolated fine wires 0.1 mm in diameter.

Electrocardiogram sensors were 5 cm diameter silver-silver-chloride suction cup electrodes centered about 10 cm behind the posterior insertion of the pectoral fins. A ground connection was made to the water in the tank.

Electromyogram (EMG) was sensed from selected muscles with paired fine wire electrodes 0.1 mm in diameter bared 1 mm at the tip and placed in or near the muscles of interest through hypodermic needles. In the past, these fine wire sensors have been used successfully for recording EMG from many different muscles of dolphins (see Ridgway et al., 1980).

All signals were monitored on a polygraph (Grass 78D) and recorded FM on an instrumentation tape recorder at 15 ips, along with stimulus triggering pulses, shaker force, velocity, and phase, time code, dolphin phonations and respiration, and voice annotations.

Analysis Methods for Physiological Recordings

Somatosensory evoked potentials (SEP) in response to skin stimulation were analyzed on- and off-line with a signal averaging computer (Nicolet 1072). Averages of the EEG signal in the form of SEPs were stored for later analysis on digital magnetic tape and on X-Y plots. Latencies of components of the SEP were measured digitally, and amplitudes were calculated by digital measurement and compared with system calibrations. Sweep periods were 10, 50, 200, 500 ms and 1 and 5 s. Numbers of sweeps averaged ranged from 8 (5 s sweep periods) to 2048 (50 and 10 ms sweep periods). Artifact tests were conducted to assure that responses were not a result of extraneous stimulation, equipment fault, or electromagnetic radiation.

Microvibration signals were analyzed with a digital real-time spectrum analyzer (Spectral Dynamics SD-350-8), an oscilloscope, and the polygraph. (Settings of the spectrum analyzer were 100-Hz range, 256-transform size, 1-Hz resolution, linear frequency and amplitude, and usually 15-s averages--except during "blows" when all data was from the blow period). Spectral averages were taken from three dolphins at up to 25 sites on the dorsal half of the body from melon to tail and plotted with an X-Y plotter (HP 7035).

Polygraph charts containing up to eight channels of data (EEG, EKG, EMG, accelerometer and bimorph outputs, stimuli, and time code) were analyzed visually comparing time, phase, and amplitudes of stimuli and responses.

Documentation of Skin Surface Ridges

We observed the size and orientation of the cutaneous

ridges of the dolphin skin. We photographed the ridges, noted their orientation, and made histological sections. Photographs were made with a special 2 x close-up lens. Rough drawings were made of the orientation of the cutaneous ridges. Histological sections oriented with the long axis of the body were made for skin sections taken from one dolphin. The sections were taken 25 cm posterior to the blowhole and 10 cm to the left of the dorsal midline. During processing for histology (mounting in paraffin, sectioning on a microtome, and mounting on microscope slides), we took care that the sections were registered so that shrinkage or stretching could not distort the ridges which were visible readily when seen from the surface or from the lateral margins of the skin stion.

RESULTS

Dolphin Skin Sensitivity

We used the SEP to obtain a gross map of dolphin skin sensitivity. Figure 1 shows averaged potentials evoked by vibrating the dolphin skin compared to evoked potentials from the human wrist (Allison, 1962). The dolphin SEP was small (about one-fifth as large) compared to responses from the same leads evoked by sound stimulation. With electrode leads placed over the right side of the dolphin's brain, SEP's were recorded only to stimuli on the left side of the body. When the snout was stimulated on the left side near the tip, a robust SEP was observed, but when the stimulus was moved only 3 cm, over to the right side, the SEP vanished. Figure 2 shows a rough map of dolphin skin sensitivity based on the SEP.

On a dolphin's spindle-shaped body, transition from laminar to turbulent flow might be expected in the area below the dorsal fin. Good evidence for this has been obtained by observations of dolphins (Lagenorhynchus obliquidens) swimming through bioluminescent waters at night (Wood, 1973). The drawings in Figure 3 were done by L. E. McKinley who made these observations. We had expected that perhaps the skin of this transition area might be extra sensitive. However, our results did not show this.

Nerve bundles especially are prominent around the dolphin's eyes, blowhole, genital area and along the snout (Kolchin and Bel'kovich, 1973; Khomenko and Khadzhinskiy, 1974). One might anticipate more sensitivity in these regions and indeed our findings (Fig. 2) are consistent with this expectation. Our measurements can not be used to compare the magnitude of dolphin skin sensitivity with other species.

Microvibrations in Absence of Stimulation

In our dolphin subjects, microvibration amplitude peaks usually were evident in one or more frequency bands of 4 to 9 Hz, 13 to 20 Hz, 27 to 39 Hz, 45 to 64 Hz, and 75 to 85 Hz. The highest amplitude peak was never above the 45 to 64 Hz range. Microvibrations in the 75-85 Hz range were always less than one or more other peaks and absent when peak activity was in the 27-39 Hz range.

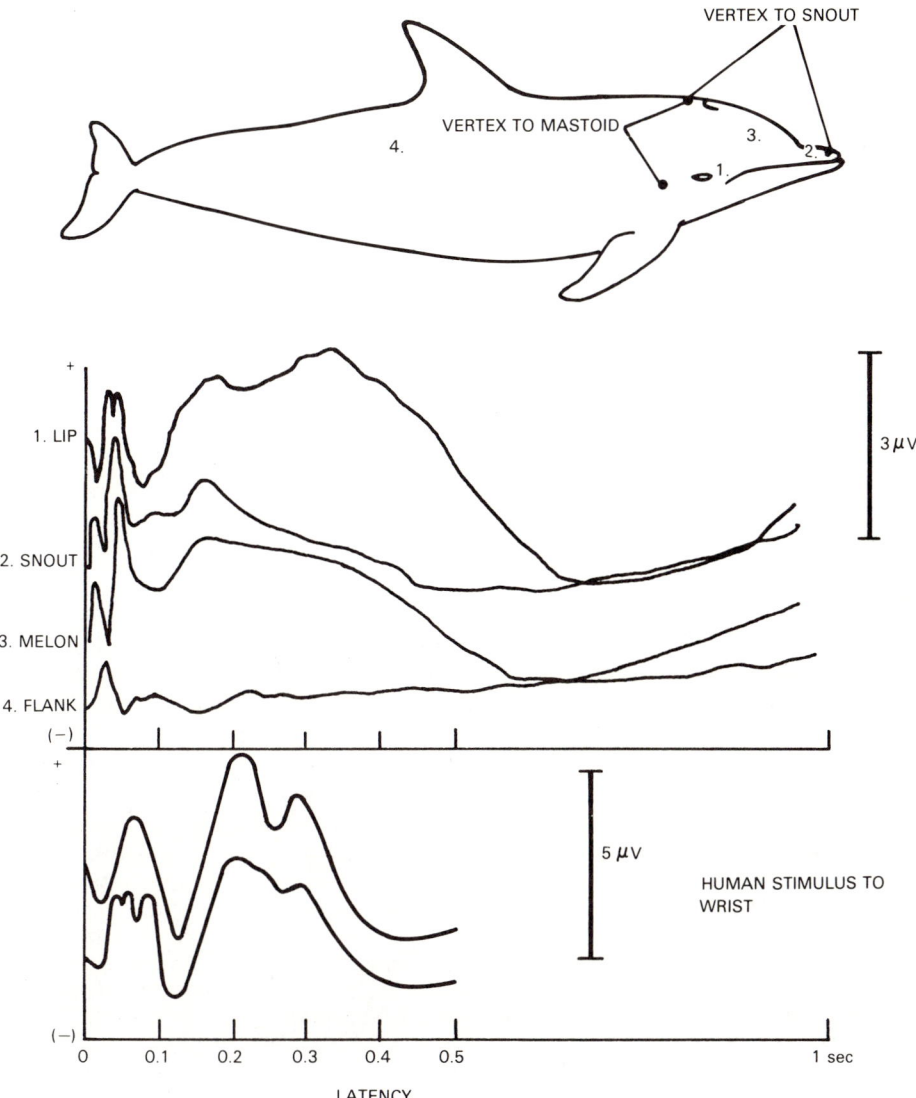

Fig. 1. Somatosensory evoked potentials from a dolphin compared to those from a human evoked by a stimulus to the wrist. Human evoked potentials redrawn from Allison (1962). Numbers indicate areas where stimuli were delivered to evoke the responses shown. Recording leads were on the dolphin's right side and stimulus sites were on the left.

Blows, the brief, forceful exhalation-inhalation cycle, produced microvibration peaks in the 4-8 Hz range with slightly smaller peaks in the 14-17 Hz range and little activity at other frequencies. During periods when large skeletal muscles directly under the sensor contracted, the frequency became predominately 4 Hz with a secondary peak at 6 or 8 Hz.

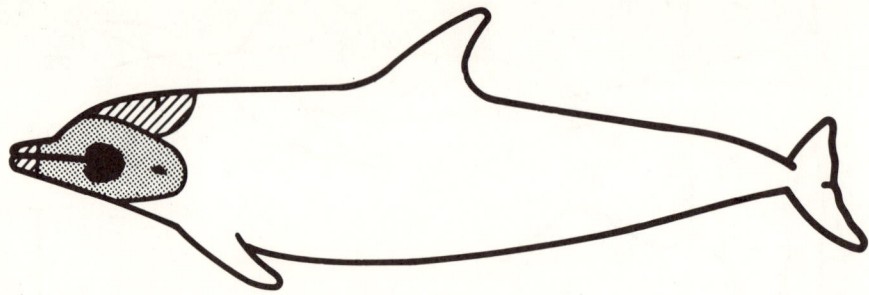

Fig. 2. Map of dolphin skin sensitivity based on somatosensory evoked potentials. The belly and genital area were not tested. 1. most sensitive, followed by 2., 3. and 4. in descending order.

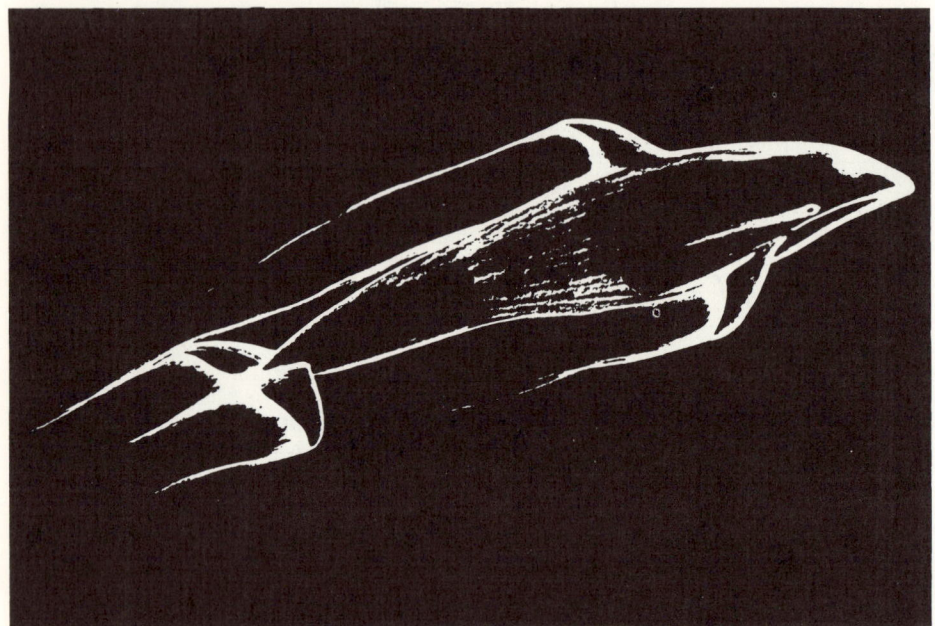

Fig. 3. Drawing of a dolphin swimming through Pacific Ocean waters containing bioluminescent organisms. Turbulence excites the organisms causing them to luminesce. The trunk between the eye and dorsal fin, where the cutaneous ridges run circumferentially forming lines perpendicular to the direction of water flow, is free of turbulence. Drawn by Mr. L. E. McKinley from an under water bubble on the R/V Sea See off Southern California in the 1960s (see also Wood, 1973).

If we accept that activity peaks recorded in the 4 to 8 Hz range were mainly due to blows, body movements, and large skeletal muscle contractions, which seems reasonable, then the major microvibration activity in our experiments was in the 13 to 20 Hz range (Fig. 4). Haider and Lindsley (1964) mentioned peak activity of 13 Hz in their dolphin subject.

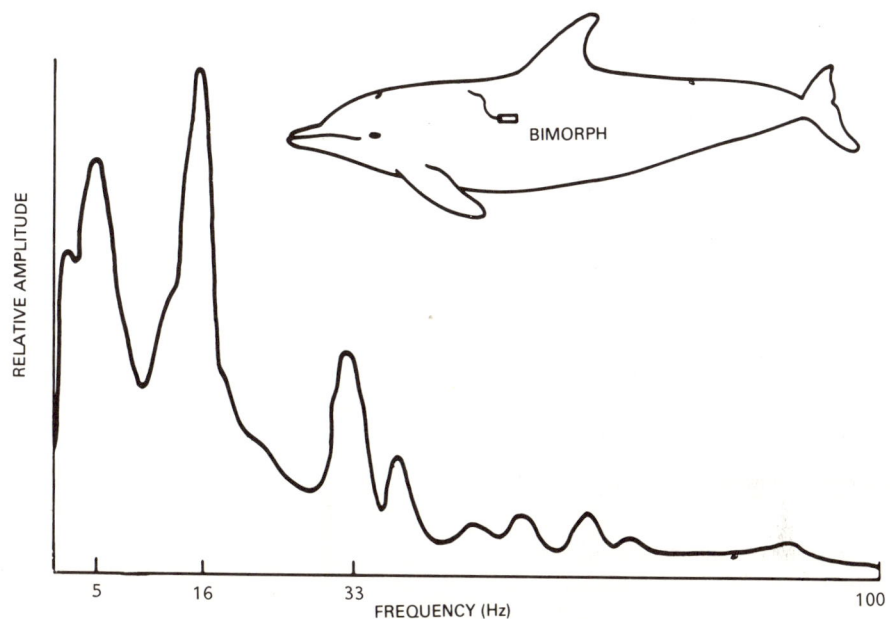

Fig. 4. Frequency analysis of microvibrations from the dolphin's skin. A 60-s epoch of signal from a bimorph, positioned as shown in the drawing, was averaged on the SD-350 spectrum analyzer to produce this plot. Major peaks in this period were at 5, 16, and 33 Hz. Significant peaks above 40 Hz were very rarely observed.

Microvibrations During Vibratory Stimulation of the Skin

Figure 5 shows the results of vibrating the dolphin's right side near the dorsal fin. The accelerometer was located on the skin about 10 cm posterior to the vibrating stimulus, and the bimorph 10 cm farther on. The stimuli were presented in a series about 10 s apart. In this instance, activity from both sensors continued after the stimulus was terminated. Low-level activity also was registered (arrows) just before the succeeding stimulus.

Occasionally, continuing activity (CA) was more pronounced, lasting over a s after a 25 Hz stimulus (Fig. 6). In Figure 7, CA is seen again with a 30-Hz stimulus and the low-level activity (arrow) is evident just before the stimulus as if the animal anticipated the vibration. This is quite possible since the stimuli were given in regular intervals of 5 to 20 s. In this case, the EKG shows prolonged heart-beat intervals around the beginning and ending of the vibration.

Figure 8 shows the response of dolphin skin to a stimulus of 50-Hz given 5 s after the termination of a 30-Hz stimulus. Note that CA, seen after the 30-Hz stimulus, was not observed after the 50-Hz vibration. CA was seen only after vibrations in the 20 to 45-Hz range.

Because the CA was at roughly the same frequency as the stimulus, we assumed that such activity was an artifact produced by the sensors; however, when we repeated the same stimuli with the same bimorphs and accelerometers using sorbathane or neoprene rubber rather than the living dolphin's skin, this ongoing activity was never found.

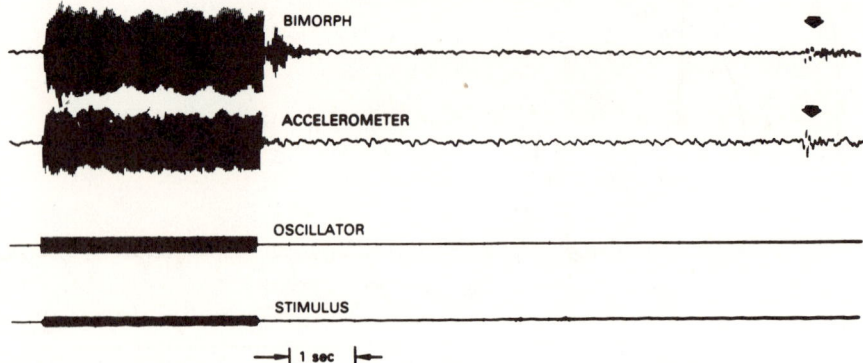

Fig. 5. Response of dolphin skin as recorded from an accelerometer and a bimorph affixed to the skin on the side near the dorsal fin. Activity from both of these sensors continues after the stimulus. Arrows indicate activity before the following stimulus in the series.

Since the CA was found only when we vibrated the skin with frequencies of 20 to 45 Hz, we next reasoned that CA may have been the result of a resonance phenomenon in the dolphin's elastic skin. However, since the CA was present less than 10% of the time, a purely physical phenomenon seemed unlikely. If the CA is indeed a resonance in the dolphin skin, the resonance must be effected by internal properties of the skin-- e.g. muscle tension or blood pressure in dermal arterioles.

It appeared that our dolphins, possibly, were capable of initiating vibrations in the 20 to 45-Hz range and perhaps of amplifying vibrations so that CA occurred. If swimming dolphin's skin is capable of responding to pressure at its surface by flexing, then drag may be reduced. In regions of turbulent-flow, the skin might be flexed away from high-pressure water and toward low-pressure water. In this case, the dolphin's skin would act in a similar way to what we observed during the CA following vibratory stimulation.

We can not assume that all changes would be aimed at reducing drag. In many cases, it might be advantageous for the animal to increase drag: for example, in changing directions, stopping, or when a mother "carries" an infant on her

pressure wave. In such cases, the response of the skin movement might simply be reversed in phase to achieve the desired effect.

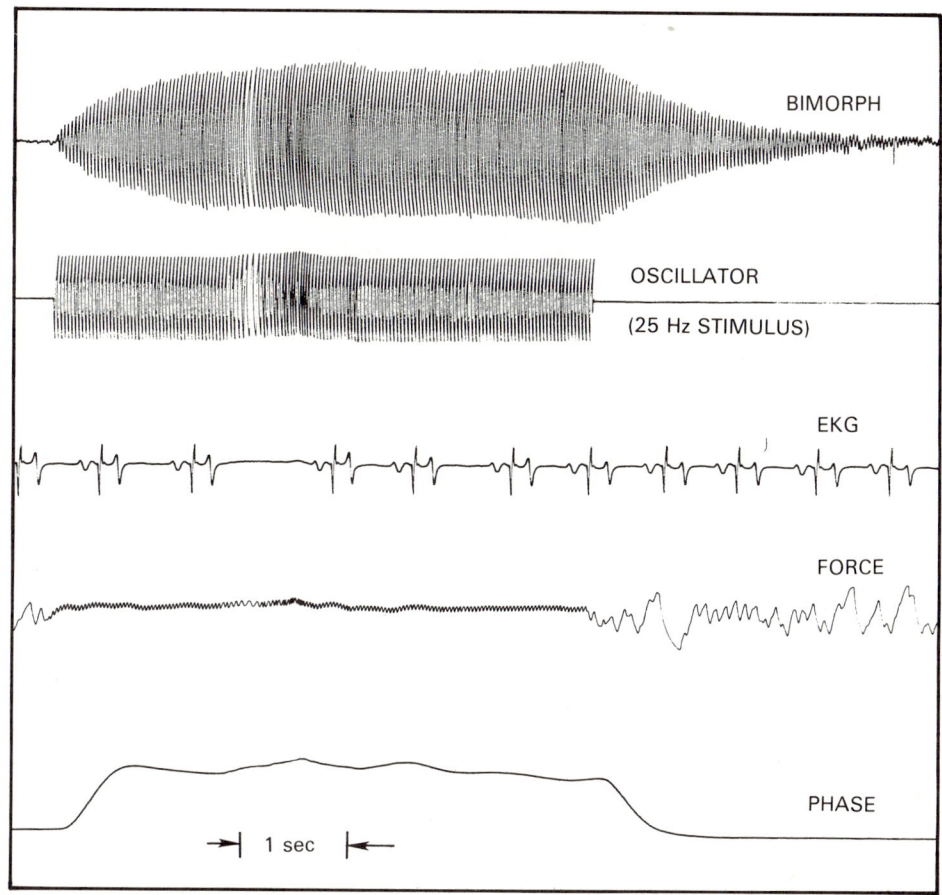

Fig. 6. Response of the dolphin skin as recorded from a bimorph affixed to the skin on the side just forward of the dorsal fin. Vibration continues for more than a s after the stimulus is off.

Electromyogram

In all of our experiments, we attempted to insert fine wire electrodes into subcutaneous muscle; however, we were not successful in recording EMG signals that correlated with the stimuli we presented or with the dolphin's responses (SEPs or CAs or microvibrations). Thus, we were not able to prove that the panniculus carnosus is, or is not, involved in the microscopic movements that we observed as microvibrations or as CA.

It is possible that the subcutaneous muscle is responsible for some skin movement, but that discrete areas of the muscle or bundle groupings move specific areas of skin.

Perhaps our electrodes never were in the right-place at the right-time. However, we never have seen dolphin's skin make the characteristic rapid gross movements that can be observed from the horse's "flyshaker."

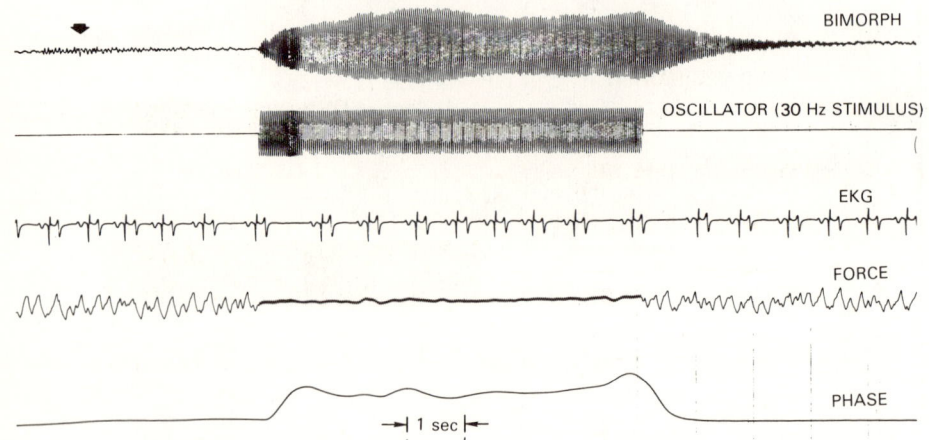

Fig. 7. Response to vibratory stimulation at 30 Hz. Arrow shows activity prior to the stimulus -- just one of a series. Activity recorded by the bimorph continues for more than one s after the stimulus is terminated.

Cutaneous Ridges

All dolphins we observed had small, regular cutaneous ridges over much of the surface of their bodies (Fig. 9). These ridges usually were faint at the surface of the skin and could not be seen from a distance; however, they nearly always were visible when one closely inspected the skin of the living animal. The ridges especially were easy to observe at an appropriate oblique angle or with a low-power magnifying lens.

Cutaneous ridges were not prominent on the snout, melon or lower jaw. They became prominent at the level of the blowhole and eyes. From about the blowhole back to the dorsal fin, their orientation was perpendicular to the long axis of the body. The ridges ran circumferentially around the occipital, cervical, and thoracic regions, forming lines perpendicular to the long axis of the body and at right angles to the direction of water-flow past the swimming dolphin. They ran circumferentially around the base of the dorsal fin, but were not prominent on the upper part of the dorsal fin. They were not observed on the flippers except near their insertion to the body. On a level with or posterior to the dorsal fin, the ridges usually were oriented obliquely or in some cases almost parallel to the body axis.

DISCUSSION

Our studies indicate that dolphin skin is sensitive to vibrations or small pressure changes on its surface. We have shown that the most sensitive areas are at the angle of gape, and around the eyes, snout, melon, and blowhole (Fig. 2). The exact magnitude of this sensitivity can not be determined from our data, but we tend to agree with Kolchin

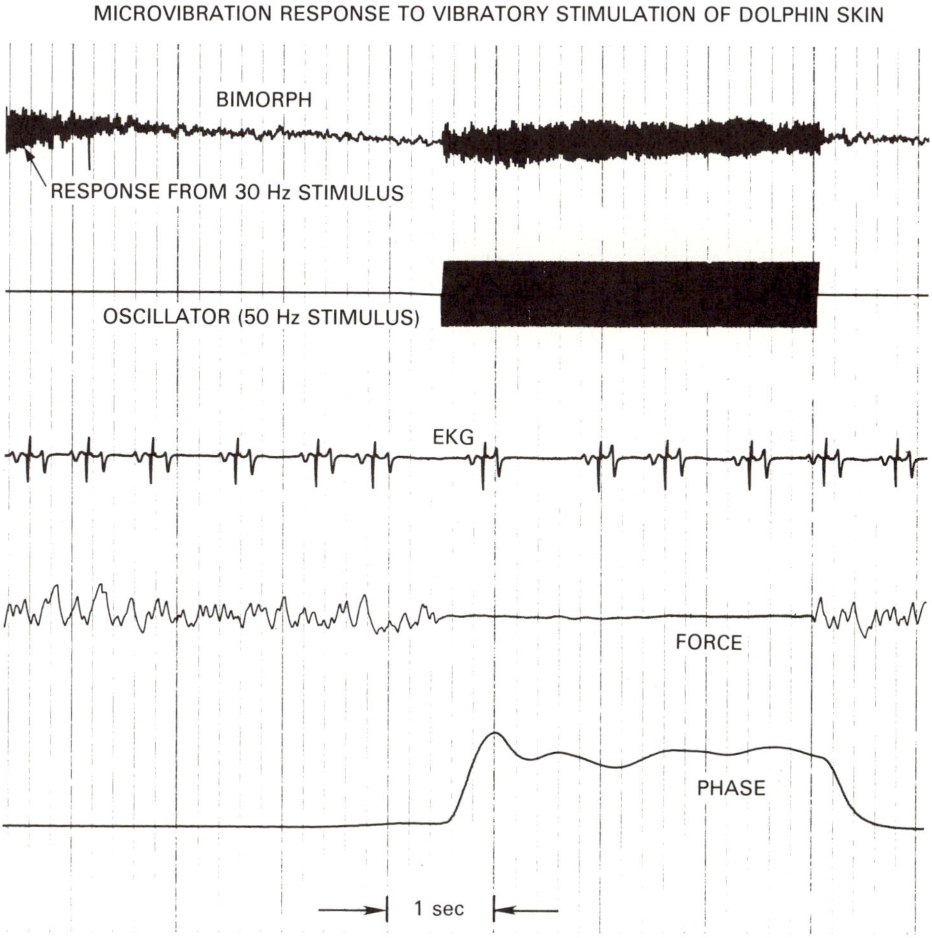

Fig. 8. Response of dolphin skin to a vibratory stimulus of 50 Hz delivered soon after a 30 Hz stimulus. The bimorph response has barely subsided when the 50 Hz stimulus is given. The response to the 50 Hz stimulus is small and terminates immediately when the stimulus terminates.

and Bel'kovich (1973) that the most sensitive areas of the dolphin skin are about as sensitive as the skin of the human lips and fingers. Since the whisker pits along the dolphin's snout are well-innervated, we expected to find them to be more sensitive than the surrounding skin of the snout and adjacent areas of the head. The stimuli we employed did not produce greater responses from the area of the snout containing the whisker pits. Neither our results nor those of other investigators show any great relative sensitivity in the region around the dorsal fin where a transition from laminar-to-turbulent-flow might be expected.

The dolphin's nervous system detects changes in pressure on its skin surface; however, our results only suggest that the dolphin's skin may reduce drag by moving synchronously with small vibrations impinging on its surface.

Our observations of CA and occasional amplifications of microvibrations suggest that the dolphin skin may be able to adjust to pressure changes by amplifying normal microvibrations or by producing microscopic vibrations with muscular contractions. Our studies suggest a mechanism by which the dolphin skin might move or vibrate to improve hydrodynamic performance as has been proposed (Haider and Lindsley, 1964; Palmer and Weddell, 1964; Khomenko and Khadzhinskiy, 1974; Kayan, 1974; Babenko and Nikishova, 1976). Probably, dolphin skin is sensitive enough to detect turbulent-flow. Drag may be decreased by decreasing the pressure gradient in the adjacent water layer. The skin may actively flex away from higher pressure and toward lower pressure.

We do not think it likely that the panniculus carnosus or other muscles connected to the skin are responsible for the "speed induced skin folds" observed by Essapian (1955) in the large tank at Marineland of Florida. Lang (1966) has commented that these skin folds are "undoubtedly drag-producing phenomena, and may occur only during acceleration, deceleration, or turning, as a by-product of loosely attached skin...". We observed such folds in small cetaceans at sea. They often occur during a short period just before a fast-swimming animal surfaces to breathe, and the folds disappear when the animal again is fully submerged. In this case, the phenomenon is probably a manifestation of surface drag which occurs within a body length of the surface.

Our experiments only suggest an active means whereby the dolphin skin may reduce drag. They do not prove swimming dolphins use this means of drag reduction. Further tests of skin motion and microvibrations must be done with instrumented swimming dolphins. Additional tests of skin sensitivity should include studies of motion detection and sensitivity to rapid changes in pressure along a larger area of skin surface. Anatomical and physiological studies should focus on determining the different types of skin receptors and the role of each in sensing motion, pressure, and temperature. Detailed studies are needed on the panniculus carnosus muscle, the extensive subdermal connective tissue sheath, and any other connections between muscle and skin. In the process of such studies, the muscle names should be modernized and homologized with those of a better studied mammals such as the horse. Electromyography of muscles associated with the skin should be studied and the portion of the skin moved by different muscle fiber bundles identified. A study of the metabolic fiber types in skin related muscle also might be instructive.

Cutaneous ridges were not documented in the past because they are not always clear in histological sections

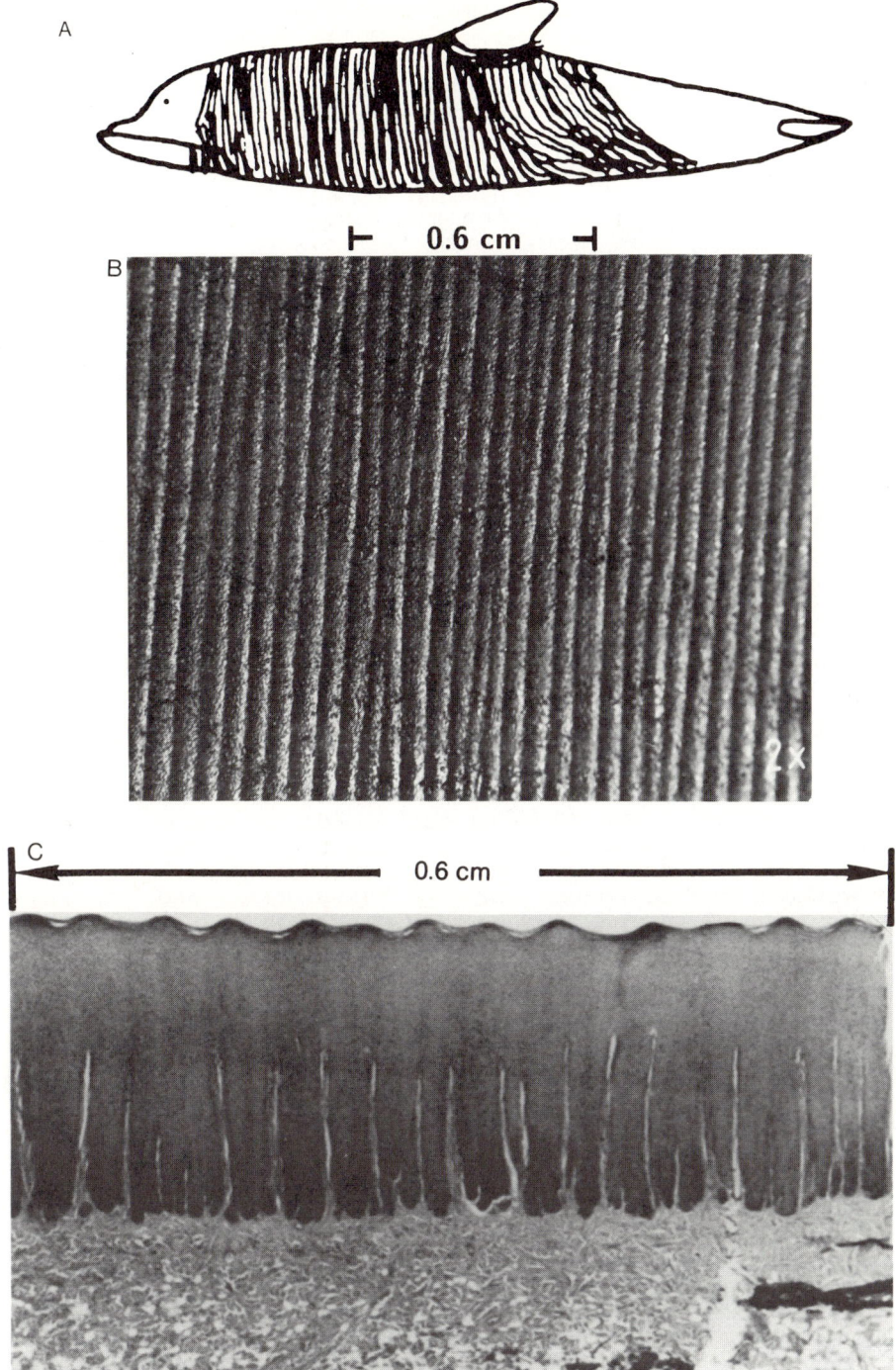

Fig. 9. A. Sketch showing orientation of cutaneous ridges on a bottlenose dolphin. B. 2X photograph of the skin about 25 cm posterior to the blowhole and about 10 cm lateral to the dorsal midline where ridges run perpendicular to long axis of body. C. A low-power photomicrograph of a section of skin taken parallel to the long axis.

of cetacean skin. The ridges may tend to relax or disappear from the surface of dead skin, particularly on a specimen which has been excised from a carcass. Past investigations of dolphin skin have concentrated on the deeper structures and sections may not have been taken in such a way as to preserve the surface ridges. However, when the ridges are preserved in histological sections, their spacing appears to be roughly the same as that of the underlying dermal papillae, although their association with the deeper epidermal and dermal structures needs better definition.

ACKNOWLEDGEMENTS

We thank Franklin Borkat and Richard Kataoka for technical assistance and Jeff Haun for his assistance in obtaining support for this work. We thank L. E. McKinley for making his drawings of dolphins swimming through bioluminescent waters available to us. F. G. Wood reviewed the manuscript and made many helpful suggestions.

REFERENCES

Agarkov, G. B. ,1970, Morphological investigations on hydrobionics. Vestnik. Akad. Nauk Ukrainskoy SSR. 8:58-68.
Allison, T. ,1962, Recovery functions of somatosensory evoked responses in man. Electroenceph. Clin. Neurophysiol. 14:331-343.
Babenko, V. B. and Nikishova O. D. ,1976, Some hydrodynamic patterns in structure of integument of marine animals. Bionika. 10:27-33.
Essapian, F. S. ,1955, Speed-induced skin folds in the bottle-nosed porpoise, Tursiops truncatus Breviora 43:1-14.
Geraci, J. R., St Aubin D. J. and Hicks B. D. ,1986, The epidermis of odontocetes: a view from within. in: "Research on Dolphins." Ed. by Bryden M. M. and Harrison, R. J., Oxford University Press. Oxford pp. 3-21.
Harrison, R. J. and Thurley, K. W. ,1972, Fine structural features of delphinid epidermis. J. Anat. 111:498-500.
Harrison, R. J. and Thurley, K. W. ,1974, Structure of the epidermis in Tursiops, Delphinus, Orcinus, and Phocoena. in: "Functional Anatomy of Marine Mammals." R. J. Harrison (ed.). Academic Press, London. pp. 45-71.
Haider, M. and Lindsley D. B. ,1964, Microvibrations in man and dolphin. Science. 146:1181-83.
Kayan, V. P. ,1974, Resistance coefficient of the dolphin. Bionika. 8:31-35.
Khomenko, B. G. and Khadzhinskiy, V. G. ,1974, Morphological and functional principles underlying cutaneous reception in dolphins. Bionika. 8:106-113.
Kramer, M. O. ,1960, Boundary layer stabilization by distributed damping. J. Amer. Soc. Nav. Engnrs. 72:25-33.
Kramer, M. O. ,1977, Boundary layer control by "artificial dolphin coating" Naval Engineers Journal. October, pp41-45.
Kolchin, S. and Bel'kovich, V. ,1973, Tactile sensitivity in Delphinus delphis. Zoologicheskiy Zhurnal. 52:620-622.
Lang, T. G. ,1966, Hydrodynamic analysis of cetacean performance, in: "Whales, Dolphins, and Porpoises" Ed. by K. S. Norris, University of California Press (Berkeley and Los Angeles) pp. 410-432.

Lende, R. A. and Welker, W. I. ,1972, An unusual sensory area in the cerebral neocortex of the bottlenose dolphin, Tursiops truncatus. Brain Research. 45:555-560.

Murie, J. ,1873, On the organization of the caaing whale, Globicephalus melas Trans. Zool. Soc. Lond. 8:235-301.

Palmer, E. and Weddell, G. ,1964, The relationship between structure, innervation, and function of the skin of the bottlenosed dolphin (Tursiops truncatus). Proc. Zool. Soc. Lond. 143:553-568.

Ridgway, S. H., Bullock, T. H., Carder, D. A., Seeley, R. L., Woods, D., and Galambos, R. ,1980, Auditory brain-stem response in dolphins. Proc. Natl. Acad. Sci. 78:1943-1947.

Rohracher, H. ,1964, Microvibration, permanent muscle-activity and constancy of body-temperature. Percept. Motor Skills 19:198.

Schulte, H. von W. and Smith, de F. M. ,1918, The external characters, skeletal muscles and peripheral nerves of Kogia breviceps (Blainville). Bull. Am. Mus. Nat. Hist. 38:7-72.

Sherrick, C. E. ,1975, The art of tactile communication. American Psychologist. 30:354-360.

Sisson, S and Grossman, J. D. ,1953, "The Anatomy of Domestic Animals." Saunders, Philadelphia. 972p.

Sokolov, V., 1982, "Mammal Skin." University of California Press, Berkeley.

Stromberg, M. W., 1989, Dermal-epidermal relations in the skin of the bottlenose dolphin (Tursiops truncatus) Anat. Histol. Embryol. 18:1-13.

Surkina, R. M. ,1971, Structure and function of the skin muscles of dolphins. Bionika. 5:81-87.

Wood, F. G. ,1973, "Marine Mammals and Man", R. B. Luce Inc., New York.

A POTENTIAL SYSTEM OF DELAY-LINES IN THE DOLPHIN AUDITORY BRAINSTEM

John M. Zook and Ralph A. DiCaprio

Department of Zoological & Biomedical Sciences and
College of Osteopathic Medicine, Ohio University
Athens, OH, 45701, USA

INTRODUCTION

   This is the second of an ongoing series of reports on the comparative cytoarchitecture of the dolphin auditory brainstem. The previous report (Zook et al., 1988) focused on unusually ordered cell arrangements within three auditory brainstem cell groups: the ventral cochlear nucleus (VCN), the medial nucleus of the trapezoid body (MNTB) and the ventral nucleus of the lateral lemniscus. Part of each cell group is distinguished by an orderly alignment of cells into straight rows or columns.

   This study focused on two of these cell groups, the VCN and the MNTB, which are well-developed in the dolphins, Tursiops truncatus and Stenella coeruleoalba. An intriguing aspect of these groups is the orientation of their cell rows relative to the course of fibers within the trapezoid body. The VCN has several cell rows at the entrance of the eighth nerve root, each of which runs perpendicularly to the reticulating fibers of the trapezoid body. Cells in the dorsal part of the MNTB are aligned in a single straight row, but this row is slanted at a consistent angle across the horizontally-running fibers of the trapezoid body. The sheer number of fibers present in the dolphin trapezoid body might be expected to confound any such alignment of cells, save perhaps a channeling of cells along the course of the fiber fasicles. Precisely aligned rows of cells, whether perpendicular or at a slant to the course of fibers, are not expected and their presence suggests some function.

   It is our working hypothesis that the VCN and MNTB cell arrangements may be two components of a larger functional network. This idea is based on the assumption that, as in other mammals, there is a direct projection between cells of the VCN and MNTB (Warr, 1972; Morest, 1968; Tolbert et al., 1982; Zook and DiCaprio, 1987, 1988). Specifically, we suggest that these perpendicular and slanting rows are elements of a system of varying temporal delay-lines. With the gradual shift in position of adjacent MNTB cells from ventral to dorsal in the row, there would be a concurrent incremental increase in axon length to the more dorsal MNTB cell in the row. Assuming a constant conduction velocity, information traveling via VCN axons would arrive first at cells in the ventromedial MNTB. Moving dorsomedially along the MNTB row, cells would receive information with gradually increasing delays.

To support this hypothesis, we would like to establish that the VCN and MNTB cell populations are linked directly in dolphins. However, we are limited by the fact that brainstem projections can not be traced directly in available tissue. Our approach is to relate aspects of dolphin cytoarchitecture to more detailed investigations of this pathway in other mammals, in particular several echolocating bat species. We will begin by establishing that the cell rows of the dolphin VCN and MNTB contain the cell types that are known to form the VCN-MNTB pathway in other mammals.

In its simplest form, the proposed delay-line system would exhibit a topographically ordered, one-to-one match between a given cell within a VCN row and a corresponding cell within the MNTB row. To illustrate the degree of specificity possible in dolphins, we have included some examples of VCN to MNTB projection patterns as shown by intracellularly labeled axons in the bat brainstem.

To estimate the range of delays possible in the dolphin brainstem, we measured the distances between the VCN and MNTB cell rows. Since conduction velocity varies with axon diameter, the distribution of fiber diameters within the trapezoid body also was determined. The estimated ranges of conduction velocity and distance were used to calculate the range of temporal delays. The proposed delay-line system is discussed in terms of its potential role within a coincidence or correlation system. Finally, we discuss the possible use of such analysis with regard to the auditory discrimination requirements of echolocating species.

METHODS

Materials

The _Tursiops_ brainstem used here was obtained from the collection of Drs. I. Glezer, P. Morgane and M. Jacobs. Additional data was obtained from extensive examination of their collection of brainstem slides of _Stenella_ and _Tursiops_. Bat brainstem material was prepared from the following species: _Pteronotus parnellii_, _Eptesicus fuscus_, and _Antrozous pallidus_.

Histology

Three brains from each bat species were fixed by transcardial perfusion of 4% buffered Formalin, embedded in celloidin and sectioned at 20 um in either horizontal, transverse or sagittal planes. Sections were stained with cresylecht violett or thionin for cells; alternate sections were stained by the Heidenhain method for fibers as described previously (Zook and Casseday, 1982). The _Tursiops_ brainstem was fixed by immersion in 10% Formalin. A portion of this brainstem was vibrotomed into 1 mm transverse blocks. A set of 40 um, transversely-cut, frozen sections were taken from each block for Nissl and fiber staining.

Semi-thin plastic sections were prepared for close examination of cellular morphology and for photomicrography. In each species, a series of 500 um square tissue samples was taken from each of the larger tissue blocks. These samples were embedded in EMbed 812 (EMS) and cut in 1 to 5 um transverse sections with an ultramicrotome (LKB). To measure axon diameters in the trapezoid body, additional plastic sections were cut in the sagittal plane at the level of the MNTB.

Intracellularly Filled Cells

Additional and supplemental descriptions of dendritic and axonal morphology were based on intracellularly dye-filled cells. Brainstem slices were prepared from the bats Pteronotus and Eptesicus. VCN and MNTB cells within physiologically active tissue slices of the brainstem were filled with an intracellular label, either horseradish peroxidase or the fluorescent dye, Lucifer Yellow. These slices were maintained in an in vitro recording chamber as previously described (Zook and DiCaprio, 1988). Intracellular dye injection in these tissue slices allowed us to examine details of axonal and dendritic morphology not found in Nissl stained material.

Tissue slices lightly fixed by immersion in aldehydes also were used for more limited dye labeling (Einstein, 1988). In this tissue, the extent of labeling was limited to the cell soma with occasional filling of proximal dendrites and axonal segments. Although incomplete, fixed tissue fills often were useful in revealing details of dendrites and axons. This technique was used mainly with bat brainstem tissue, however a few cells also were filled in brainstem slices of the dolphin, Tursiops.

Analysis

Nissl and intracellularly labeled cells were examined, drawn and photographed with a Zeiss standard microscope. The microscope was equipped with drawing tube, Nomarski and epifluorescence optics. Magnifications from 10X to 1200X were used.

RESULTS

Cell types in the dolphin ventral cochlear nucleus (VCN) and medial nucleus of the trapezoid body (MNTB) will be considered first and compared to earlier descriptions of these areas in the bat (Zook and Casseday, 1982; Schweizer, 1981) and the cat (Osen, 1969; Brawer et al., 1974). Secondly, axonal lengths and diameters in the dolphin trapezoid body will be quantified and compared to measurements in the cat (Spangler et al., 1985). Where applicable, intracellularly labeled cells will be described to supply additional details of axonal and dendritic morphology.

The Ventral Cochlear Nucleus

In both dolphin species, between two and four cell rows are visible in transverse sections through the VCN when viewed at low magnification. As described previously (Zook et al., 1988), the cell rows in the dolphin VCN are limited to the region where the eighth nerve enters between the anteroventral and posteroventral cochlear nuclei. From this region the rows spread rostrally into the caudal part of the anteroventral cochlear nucleus (AVCN). The rows varied in width from 30 to 60 um, in length from 0.5 mm to 4 mm, and were spaced from 0.3 mm to 0.6 mm apart. Each row contained an average density of 1 to 5 cells per 100 um of row length.

In dolphins, as in other mammals, the predominant cell types in this part of the VCN are a small to large multipolar cell and a large, round or ovoid cell-type often having an eccentrically placed nucleus (Fig. 1A). From their Nissl-stained appearance, cells of the latter type were originally referred to as globular cells (Harrison and Warr, 1962; Osen, 1969; Warr, 1972; Brawer et al., 1974). More recently, with reference to dendritic morphology, they have been termed globular bushy cells (Tolbert and Morest, 1982). As illustrated in Fig. 2A, these cells have a characteristic pattern of one to three primary dendrites, each having many secondary processes branching-off at right angles (Tolbert and Morest, 1982; Rouiller and Ryugo, 1984; Tolbert et al., 1982; Smith and Rhode, 1987; Wu and Oertel, 1984). The main difference between the dolphin and other mammals is the size of these globular and multipolar cells (Fig. 1A-B). These range from 35 to 45 um in diameter, while in the cat and bat they are usually no more than 25 to 30 um in diameter (Tolbert and Morest, 1982; Zook and Casseday, 1982). A few small round and elongate cells also appear in this region.

The cell rows of the dolphin VCN are made-up predominantly by globular bushy cells. A few multipolar cells also are present along with an occasional small elongate cell or medium multipolar cell. As a row is followed rostrally into the AVCN, the proportion of multipolar and other cell types increases.

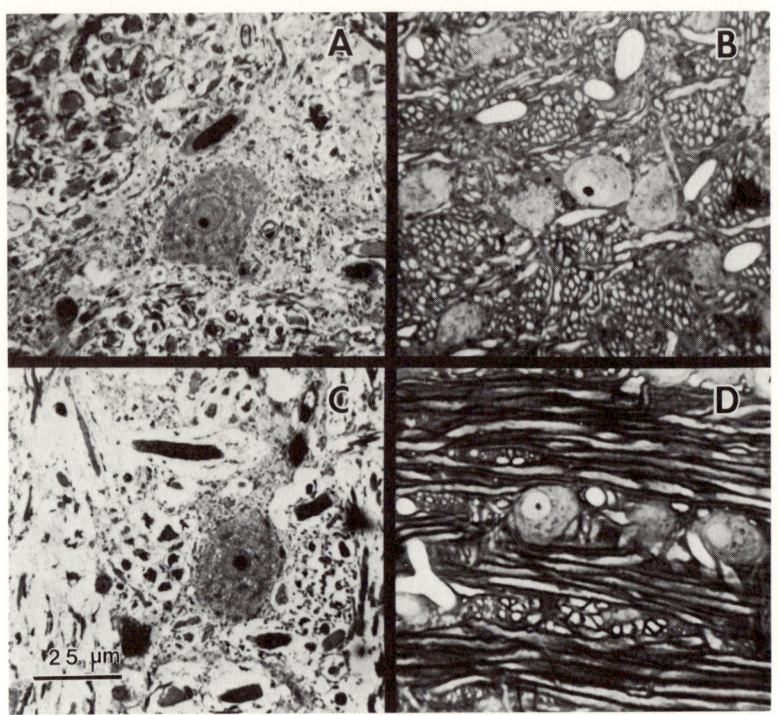

Fig. 1. Photomicrographs of 1 um plastic sections, Toluedine Blue stain. (A) Globular cell of the ventral cochlear nucleus (VCN) in the dolphin, <u>Tursiops</u>. (B) Globular cells of the VCN of the bat, <u>Eptesicus</u>. (C) A principal cell of the dolphin medial nucleus of the trapezoid body (MNTB). (D) Principal cells of the bat MNTB.

It has not been possible to trace single axons from the dolphin VCN into the trapezoid body. It was possible to follow axons for short distances in semi-thin plastic sections, as well as in labeled cells from fixed tissue sections. In these cases, the initial axon segments were found to vary in diameter from 2 to 12 um. These measurements may be misleading, because they reflect only the axon segment nearest the cell soma. The general pattern seen for intracellularly filled axons in bats was a small initial axon segment (1 to 3 um in diameter), which commonly increased to around 12 um in diameter at a distance of 30 um from the cell soma (see Fig. 2A). The axon usually maintained this diameter for most of its length. Axons commonly narrowed to a diameter of 1 to 3 um just before formation of the characteristic calyx of Held in the contralateral MNTB (see Fig. 2B; also Tolbert et al., 1982).

The Characteristics of Globular Bushy Cell Axons

It was possible to intracellularly fill and trace individual axons in the bats. These axons could be traced through the trapezoid body from the globular bushy cells of the VCN to the principal cells of the MNTB (Fig. 2). Similar intracellularly-filled axons have been traced in the rat (Friauf and Oswald, 1988). This pattern of projections was suggested in earlier studies which traced the retrograde transport of horseradish peroxidase (Tolbert et al., 1982; Zook and Casseday, 1985). Intracellular labeling experiments show that each axon contacts a single principal cell of the MNTB. This contact is in the form of a huge terminal

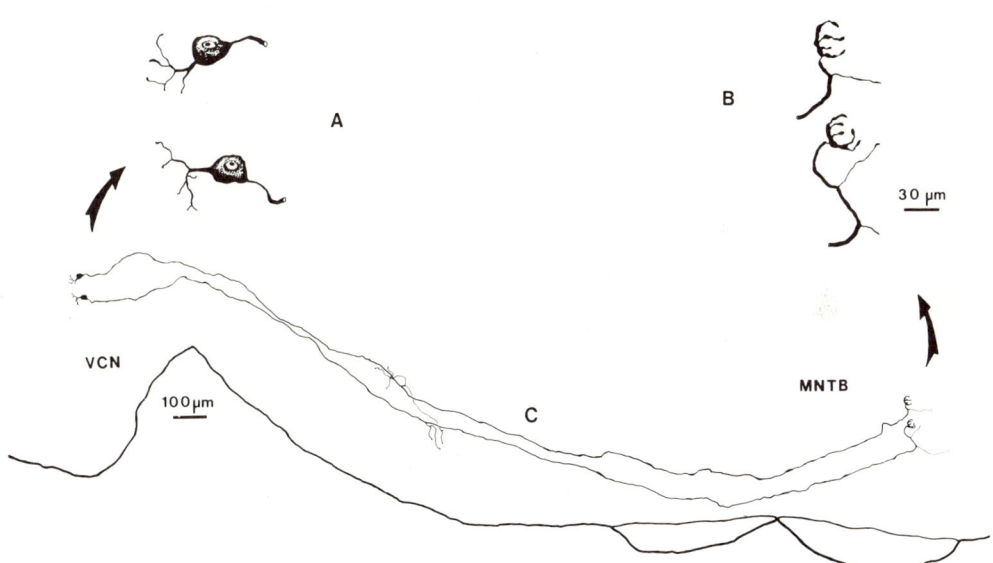

Fig. 2. Drawings of representative globular bushy cells in the ventral cochlear nucleus (VCN) intracellularly filled in the bat, Eptesicus. (A) High magnification drawing of the globular cell somata. (B) Same magnification drawing of the axon endings, the calyces of Held, large terminal specializations common to globular cell axons. Each calyx surrounds the soma of a single principal cell of the medial nucleus of the trapezoid body (MNTB). (C) Lower magnification drawing of the entire axon projection from VCN to MNTB. Note the relative positions of each cell soma within the VCN and calyciferous ending within MNTB.

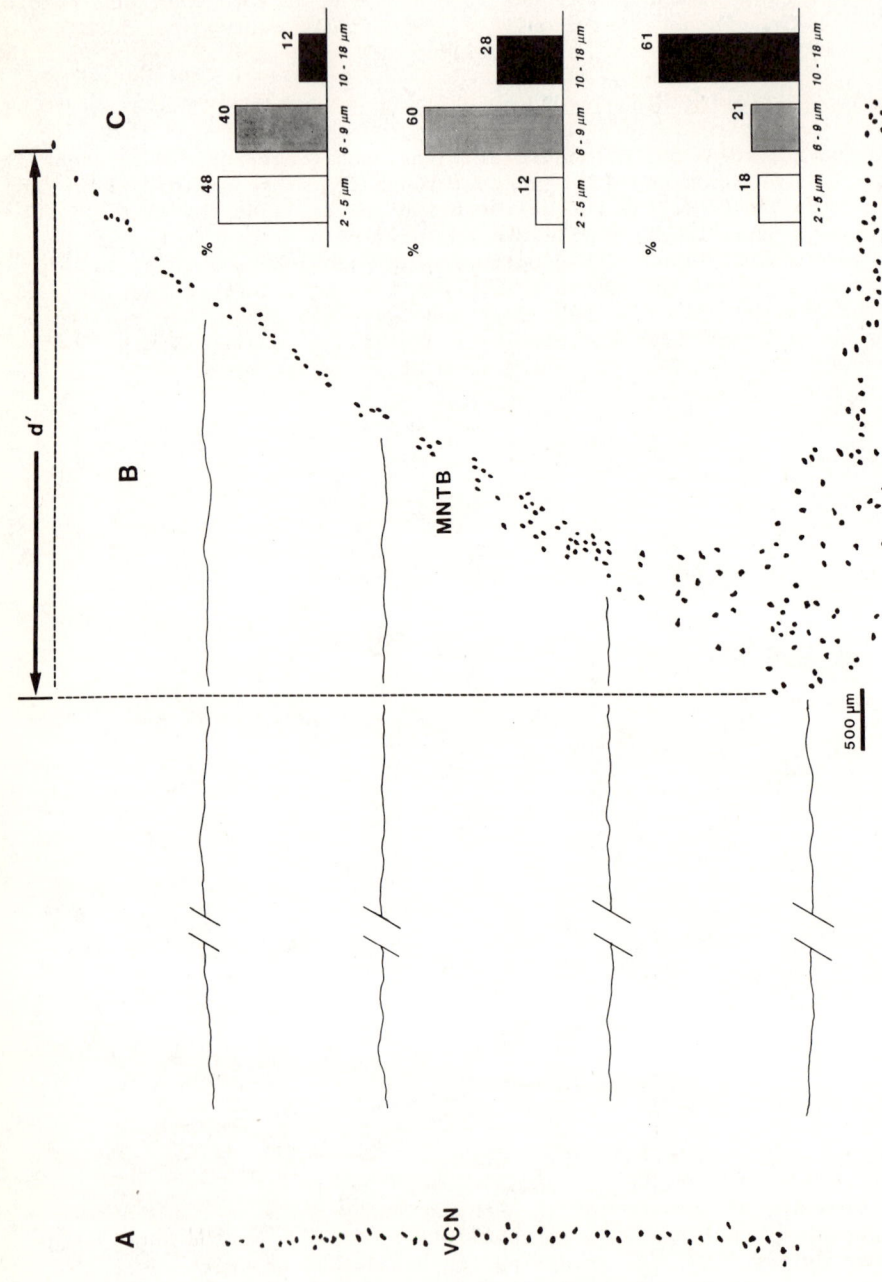

Fig 3. *Stenella* (A) A typical cell row in the ventral cochlear nucleus (VCN). Horizontal lines show the course of trapezoid body fibers. (B) A cell row in the medial nucleus of the trapezoid body (MNTB). Line d' is the difference in axon length to dorsal vs. ventral cells. (C) Bar graphs of axon diameters sampled in dorsal, middle and ventral trapezoid body.

specialization, the calyx of Held, each of which almost totally engulfs the soma of a single MNTB principal cell (Fig. 2B; see also Morest, 1968). These intracellularly filled axons suggest that there is a one to one relationship between each VCN globular bushy cell and MNTB principal cell. This relationship was found in all cases with intracellular labeling of calycine axons (Zook and DiCaprio, 1987; Kuwabara et al., 1989).

The course of intracellularly-labeled axons suggested that there can be considerable variation in the path an axon takes through the trapezoid body. However, there was often a close match between position of the labeled VCN cell and the position of the calyx within the MNTB. When multiple axons were labeled there was a distinct tendency for calyces to be ordered within the MNTB in the same way that the globular cell somata were arranged within the VCN (Fig. 2C).

The Medial Nucleus of the Trapezoid Body

As described in the previous paper (Zook et al., 1988), the dolphin MNTB forms a wedge-shape when viewed in transverse sections through the trapezoid body. The ventral arm of the wedge consists of a dense cluster of cells elongated mediolaterally within the lower part of the trapezoid body (Fig. 3B). The dorsal arm of the MNTB extends out from the medial end of the lower arm in a slender, slanted line. This line can run dorsolaterally as far as 7 mm. The slope of this line was consistent for all cases within each dolphin species, but varied between the two species, from around 1.6 in Stenella to 2.9 in Tursiops. This wedge-shape is most prominent in rostral parts of the superior olivary complex. It is visible throughout most of the rostrocaudal extent of the superior olivary complex, with the exception of the caudal-most pole, where the dorsal arm is not seen.

Where the MNTB has been examined, it appears to have a much more homogeneous cell population than the VCN (Morest, 1968; Spangler et al., 1985; Zook and DiCaprio, 1988). The predominant cell-type is the principal cell (Fig. 1C-D). In Nissl material, these distinctive cells have an oval or round, densely staining soma. Like the VCN globular cells, the dolphin principal cells are generally larger (Fig. 1C) than their counterparts in other mammals. Examples of the relatively small principal cells of the bat Eptesicus are shown in Fig. 1D. Although principal cells have a very uniform appearance in Nissl stained material, it is possible to divide them into distinct subclasses in the bat and mouse on the basis of dendritic and axonal morphology (Kuwabara et al., 1989). Unfortunately, there is insufficient data to subdivide the principal cell population in either dolphin species. As in other species, the dolphin MNTB has a small number of medium to large stellate and elongate cells.

Measurements of Trapezoid Body Axons

The trapezoid body is a massive structure in Tursiops. In sagittal sections cut near the midline, the trapezoid body forms a flattened oval roughly 6 X 8 mm. In transverse sections, the VCN and MNTB cell rows often are not found at the same level. The main body of the MNTB is displaced rostrally in the brainstem relative to the VCN cell rows. Due to this displacement, it was difficult to measure distances between VCN and MNTB rows with any precision. This distance was estimated to range from 1.0 to 2.5 cm. It was possible to measure the additional distance an axon must travel to the top of the MNTB cell row compared to the bottom of the row. As shown in Fig. 3, a line perpendicular to the trapezoid body was marked through the ventromedial cells of the MNTB row. A second line was drawn parallel to the trapezoid body through the top-most cell of the row. The distance between the intersection of the two lines and the top cell is the potential difference in axon distance between the top and bottom of the MNTB cell rows. This distance varied from 3.4 to 4.5 mm in Stenella and from 2.3 to 4 mm in Tursiops.

In the trapezoid body of the cat (Spangler et al., 1985) and of the bat species examined, there is a dorsoventral gradient in axon diameter. As conduction time between VCN and MNTB depends upon a particular axon's diameter, as well as axon length, it was important to consider the relative size and distribution of axons projecting to different parts of the MNTB in the dolphins.

In the trapezoid body of Tursiops, fiber diameters varied from 2 to 18 um. As seen in other mammals, there was a gradient of axon diameters dorsoventrally across the trapezoid body (Fig. 3C). In the ventral third of the trapezoid body, axon diameters varied between 8 and 18 um. In the middle third, axon diameters ranged from 6 to 12 um. In the dorsal third of the trapezoid body, axon diameters covered the widest range, from 3 to 9 um. Calycine axons tend to be the largest diameter fibers in the trapezoid body. In the cat, calycine axons were estimated to be from 7 to 16 um in diameter (Spangler et al., 1985). In the bats, a range of 4 to 14 um was measured directly from intracellularly filled calycine axons. It was estimated that calycine axons in the dolphins range from 4 to 18 um in diameter.

Conduction Velocity Estimates

Differences in conduction velocity and arrival time to the top and bottom of the MNTB rows were calculated on the basis of the patterns of axon diameter and axon length described above. In Stenella, the difference in conduction time to the top vs. the time to the bottom of MNTB cell row was calculated to be in the range of 0.68 msec to 1.7 msec. In Tursiops, assuming the same variation in axon diameter and distribution, the delay was calculated to be in the range of 0.21 msec to 0.35 msec.

DISCUSSION

We used cytoarchitectural descriptions to explore the possibility of a connection between the cell rows of the dolphin ventral cochlear nucleus (VCN) and medial nucleus of the trapezoid body (MNTB). To suggest the course and pattern of axons in this pathway, we examined the characteristic patterns of intracellularly labeled axons in several bat species. In the dolphin, we measured the range of axon lengths and axon diameters within the trapezoid body between the VCN and MNTB. These measurements were used to estimate the effective range of temporal delays between the cell rows of the VCN and MNTB.

We will discuss each of these elements: cytoarchitecture, axon characteristics and the range of temporal delays, in terms of their significance for the structure and nature of the proposed delay-line system. This will be followed by a consideration of the possible functional roles of a delay-line in mammalian audition and dolphin echolocation.

Cytoarchitecture

The existence of a projection between the cell rows in the VCN and MNTB is supported by the cytoarchitecture of both regions. The two cell-types, that in other mammals form the main projection between these nuclei, also were found in the dolphin: globular bushy cells in the VCN and principal cells in the MNTB. It is these cell-types that form most of the respective cell rows of the VCN and the MNTB. Although the presence and position of these cells in the dolphins does not prove the existence of a connecting pathway, it is consistent with this assumption. Experiments are in progress which may allow us to trace this projection with greater accuracy.

The cells in these rows represent a subset of the population of the globular cells in the VCN and principal cells of the MNTB. The location of these rows suggests a possible segregation by frequency response. In other mammals, cells in the brainstem auditory nuclei are grouped tonotopically, across each nucleus there is a systematic variation in the sound frequencies to which cells are most responsive. With some exceptions, the general tonotopic organization of brainstem auditory nuclei follows a general pattern that remains consistent across species. There is a tendency for middle to high frequencies to be represented around the entrance of the eighth nerve root in the VCN and toward the dorsomedial margin in the MNTB (Brownell, 1975; Guinan et al., 1972; Kiang et al., 1965). Thus, the cells in the dolphin VCN and MNTB rows might be expected to represent frequencies in the middle to high end of the spectrum.

Cell rows or other elaborations of the VCN or MNTB are not commonly found in the mammalian brainstem. The VCN of the harbor porpoise, Phocoena phocoena, was described in detail by Osen and Jansen (1965). No row arrangement in VCN or MNTB was reported. In the bat species examined there has been no more than a suggestion of ordered cell arrangements (Zook et al., 1988). The lack of spatially-ordered cell groups in the bat brainstem is not surprising. Staggered spatial arrangements would only be useful for establishing delay gradients in a large brain with pathways of sufficient length. Bats, as flying mammals, have relatively small brains and correspondingly short fiber tracts. There are, of course, other ways to establish delay-lines that do not require long fiber tracts. For example, a plausible system of temporal delays could be based upon a graded-shift in synaptic integration within a cell group such as the MNTB.

Axon characteristics

Although ordered cell arrangements are uncommon, there does appear to be some arrangement of axons in the mammalian VCN and MNTB. In species where axons have been traced, each globular bushy cell generally projects to one, and only one, MNTB principal cell (Friauf and Oswald, 1988; Zook and DiCaprio, 1987, 1988). Where more than one axon was filled and traced in the same brainstem section, a distinct topographic relation was found between the location of the labeled pair of globular cells in the VCN and the labeled pair of calyces in the MNTB (see Fig. 2C).

The features of the globular cell projection are well-suited for a temporal processing system. Studies in other mammals have suggested that the VCN and MNTB might play a role in the fast, temporally-secure transfer of information on the basis of the unusual size and conservative physiological characteristics of globular cell axons and their calycine terminals (Guinan et al., 1972; Morest, 1968).

Temporal Delay Calculations

The estimations of temporal delay based on path length and axon diameter by necessity are rough. For example, it is not clear if axon diameter remains constant over the distance between VCN and MNTB. As mentioned in the results, there is evidence for considerable variability at the beginning and end of each globular cell axon (see Fig. 2C). Furthermore, in the dolphins, it is not possible to match axon diameter with the position of cells in the rows. Thus, our calculations can give only a broad estimate of the maximum temporal delay.

Temporal Delay-lines in Auditory Signal Processing

Delay-lines are particularly useful as elements of larger circuits for coincidence or correlation analysis. Coincidence or correlation systems can take many forms. In neurological terms, a coincidence system could be based upon the convergence of two pathways, one direct and one indirect, carrying the same information to the same group of target cells. The main difference between the routes would be a range of temporal delays in the indirect pathway. As in the proposed VCN-MNTB system, the indirect path could incorporate different path lengths, axon diameters or levels of synaptic integration to generate a grade range of conduction-times. At the target cells, simultaneous activity in the direct and indirect pathways would be expected to increase the overall response. The coincidence system would come into effect whenever there was a regular rate of activity and the period of this activity matched a time-delay represented in the indirect route. In these cases, where the delayed input arrives one cycle (or an integral multiple of the cycle period) after the direct input, a temporal summation of inputs would occur, facilitating the response of a given target cell. If the indirect channel contained a range of possible delays, a temporal coincidence could occur over a range of periods and corresponding frequencies. This type of system would enhance detection of regularly occurring spike trains within a particular frequency range; the exact frequency range depending upon the range of time-delays in the indirect pathway.

In the proposed MNTB delay-line, our calculations suggest a propagation delay as long as 1.7 msec. in a coincidence system, MNTB principal cells would be expected to relay the delayed signal to a third cell group where the actual coincidence analysis would take place. The calycine terminal at the MNTB principal cell would represent a synapse in the indirect pathway that is not present in the direct pathway. We assume, therefore, an additional 0.5 msec synaptic delay in all delay-lines, based on the measurements of Li and Guinan (1971). The resulting maximal propagation delay of 2.2 msec would set a lower limit for rate of the axon firing that would be of value to the coincidence system. This lower frequency limit is estimated at around 450 Hz. As the time factor of the VCN-MNTB synaptic delay might reasonably be as long as 1.0 msec or more, this rate could be as low as 370 Hz.

This estimated bottom frequency is well within the range that neurons in the auditory periphery will readily phase lock to an auditory signal (Rose, et al, 1968). The upper range of frequencies, where auditory axons begin to lose the ability to phase lock, is around 1 to 3 kHz. This upper limit means that temporal delays less than 0.3 to 1.0 msec would not be useful in the coincidence analysis circuit. Thus, the maximal range of axon firing frequencies, or range of sound frequencies, which could be enhanced by the proposed system falls between 370 Hz and 3 kHz.

It is important to note that the frequency representation of the VCN-MNTB pathway probably extends well above 3 kHz in dolphins. Correlation analysis for low frequency spike trains still is possible within a high frequency system. The echolocation signals of dolphins and some bats often contain a constant frequency (CF) component, which effectively serves as a high frequency carrier for low frequency amplitude modulations. Such modulations would appear in the echo of a CF pulse when reflected off an irregularly shaped, oscillating target. Modulations of this type are commonly imposed upon bat echolocation signals by the wing beat patterns of flying moths (Schnitzler, 1987). In the brainstem auditory nuclei of CF bats, units with best frequencies around the CF-carrier commonly respond with bursts of activity locked to the period of the low frequency amplitude modulation (Pollak and Schuller, 1979; Suga et al.,1975). Judging from behavioral experiments, these bat species readily can discriminate between insect prey on the basis of wing beat patterns (Goldman and Henson, 1977; von der Emde and Menne, 1989). A correlation analysis system could effectively enhance such discriminations, particularly at low amplitudes or in noisy environments.

Exactly which part of the dolphins' natural environment might generate low frequency amplitude modulations in the range of 370 Hz to 3 kHz is not clear. For a dolphin, repeated fin movements or the tail flip escape responses of prey fish might impose a regular, amplitude modulation on a reflected dolphin signal. Unfortunately, these movements are not rapid enough for the proposed frequency range. The fastest movement, a tail flip, was measured in the range of 10 Hz. during the acceleration phase of the goldfish escape response (Eaton and Hackett, 1984).

Frequency range estimates used here are rough and can be refined. For example, it may be possible to sample the range of integration times for different MNTB synapses from intracellular recordings with the bat tissue slice preparation. Axon conduction velocities also could be measured directly and correlated with the projections of specific dye-filled axons.

ACKNOWLEDGMENTS

Special thanks to Dr. L. Ross for critical reading of the manuscript and to P. Toot, L. Owen and Dr. N. Kuwabara for reading and technical help. Supported by NIH Grants NS26304 and K04 DC00038 (J.M.Z.) and OUCOM.

REFERENCES

Brawer, J. R., Morest, D. K. and Kane,E. C., 1974, The neuronal architecture of the cochlear nucleus of the cat, J. Comp. Neurol., 155:251.
Brownell, W. E., 1975, Organization of the cat trapezoid body and the discharge characteristics of its fibers, Brain Res., 94:413.
Eaton, R. C. and Hackett, J. T., 1984, The role of the Mauthner cell in fast-starts involving escape in teleost fishes, in:"Neural Mechanisms of Startle Behavior", R.C. Eaton, ed., Plenum Press, New York.
Einstein, G., 1988, Intracellular injection of Lucifer Yellow into cortical neurons in lightly fixed sections and its application to human autopsy material, J. Neurosci. Methods, 26:95.
Friauf, E. and Oswald, J., 1988, Divergent projections of physiologically characterized rat ventral cochlear nucleus neurons as shown by intra-axonal injection of horseradish peroxidase, Exp. Brain Res., 73:263.
Goldman, L. J. and Henson, O. W., Jr, 1977, Prey recognition and selection by the constant frequency bat, Pteronotus p. parnellii, Behav. Ecol. Sociobiol., 2:411.
Guinan, J. J., Norris B. E., and Guinan, S. S., 1972, Single auditory units in the superior olivary complex. II. Locations of unit categories and tonotopic organization, Intern. J. Neurosci., 4:147.
Harrison, J. M. and Warr, W. B., 1962, A study of the cochlear nuclei and the ascending auditory pathways of the medulla, J. Comp. Neurol., 119:341.

Kiang, N. Y. S., Pfeiffer, R. R., Warr, W. B. and Backus, A. S., 1965, Stimulus coding in the cochlear nucleus, Ann. Otol. Rhinol. Laryngol., 74:463.

Kuwabara, N., DiCaprio, R. A. and Zook, J. M., 1989, Collateral axons of the medial nucleus of the trapezoid body, Soc. Neurosci., 15:745.

Li, R. Y-S. and Guinan, J. J., 1971, Antidromic and orthodromic stimulation of neurons receiving calyces of Held, MIT Q. Prog. Rep., 100:227.

Morest, D. K., 1968, The collateral system of the medial nucleus of the trapezoid body of the cat, its neuronal architecture and relation to the olivo-cochlear bundle, Brain Res., 9:288.

Osen, K. K., 1969, Cytoarchitecture of the cochlear nuclei in the cat, J. Comp. Neurol., 136:453.

Osen, K. K. and Jansen, J., 1965, The cochlear nuclei of the common porpoise, Phocaena phocaena, J. Comp. Neurol., 125:223.

Pollak, G. and Schuller, G., 1979, Disproportionate frequency representation in the inferior colliculus of horseshoe bats: Evidence for an "acoustic fovea", J. Comp. Physiol., 132:47.

Rose, J. E., Brugge, J. F. Anderson, D. J. and Hind, J. E., 1968, Patterns of activity in single auditory nerve fibres of the squirrel monkey, in:"Hearing Mechanisms in Vertebrates", A. V. S. deReuck and J. Knight, eds., Churchill, London.

Rouiller, E. M. and Ryugo, D. K., 1984, Intracellular marking of physiologically characterized neurons in the ventral cochlear nucleus of the cat, J. Com. Neurol., 225:167.

Schnitzler, H.-U., 1987, Echoes of fluttering insects: Information for echolocating bats, in: "Recent Advances in the Study of Bats", M. B. Fenton, P. A. Racey, I. M. V. Rayner, eds., Cambridge U. Press, Cambridge.

Schweizer, H., 1981, The connections of the inferior colliculus and the organization of the brainstem auditory system in the greater horseshoe bat (Rhinolophus ferrumequinum), J. Comp. Neurol., 201:25.

Smith, P. H. and Rhode, W. S., 1987, Characterization of HRP-labeled globular bushy cells in the cat anteroventral cochlear nucleus, J. Comp. Neurol., 266:360.

Spangler, K. M., Warr, W. B. and Henkel, C. K., 1985, The projections of principal cells of the medial nucleus of the trapezoid body in the cat, J. Comp. Neurol., 238:249.

Suga, N., Simmons, J. A. and Jen, P. H.-S., 1975, Peripheral specialization for fine analysis of doppler-shifted echoes in the auditory system of the CF-FM bat Pteronotus parnellii, J. Exp. Bio., 63:161.

Tolbert, L. P. and Morest, D. K., 1982, The neuronal architecture of the anteroventral cochlear nucleus of the cat in the region of the cochlear nerve root: Golgi and Nissl methods, Neurosci., 7: 3013.

Tolbert, L. P., Morest, D. K., and Yurgelun-Todd, D. A., 1982, Neuronal architecture of the anteroventral cochlear nucleus of the cat. Horseradish peroxidase labeling of identified cell types, Neurosci., 7:3031.

von der Emde, G. and Menne, D., 1989, Discrimination of insect wingbeat frequencies by the bat Rhinolophus ferrumequinum, J. Comp. Physiol., 164:663.

Warr, W. B., 1972, Fiber degeneration following lesions in the multipolar and globular cell areas in the ventral cochlear nucleus of the cat, Brain Res., 40:247.

Wu, S. H. and Oertel, D., 1984, Intracellular injections with horseradish peroxidase of physiologically characterized stellate and bushy cells in slices of mouse anteroventral cochlear nucleus, J. Neurosci., 4:1577.

Zook, J. M. and Casseday, J. H., 1982, Cytoarchitecture of the auditory system in the lower brainstem of the mustache bat, Pteronotus parnellii, J. Comp. Neurol., 207:1.

Zook, J. M. and Casseday, J. H., 1985, Projections from the cochlear nuclei in the mustache bat, *Pteronotus parnellii*, *J. Comp. Neurol.*, 237:307.

Zook, J. M., Jacobs, M. S., Glezer, I. and Morgane, P. J., 1988, Some comparative aspects of auditory brainstem cytoarchitecture in echolocating mammals: Speculations on the morphological basis of time-domain signal processing, *in*: "Animal Sonar: Processes and Performance", P.E. Nachtigall and P.W.B. Moore, eds., Plenum Press, New York.

Zook, J. M. and DiCaprio, R. A., 1987, A preparation for the *in vitro* study of efferent pathways in the auditory system, *Soc. Neurosci.*, 13:548.

Zook, J. M. and DiCaprio, R. A., 1988, Intracellular labeling of afferents to the lateral superior olive in the bat, *Eptesicus fuscus*, *Hearing Res.*, 34:141.

CONCLUDING COMMENTS ON SENSORY ANATOMY AND PHYSIOLOGY

Peter J. Morgane

Worcester Foundation for Experimental Biology
Shrewsbury, Mass., 01545, U.S.A.

Participants in the workshop on Sensory Anatomy and Physiology discussed many issues pertaining to aspects of continuing work on cetacean brain material. All agreed that among the major problems of carrying-out research in this area was one of inability to obtain adequately and properly prepared brains and other tissues. There especially is need for better communication and networking among scientists in the laboratory and the field relating to availability of material and special preservative and fixative techniques such as for electron microscopic studies or immunocytochemical analyses. Obviously, beached, dead specimens are of little use in specialized studies other then for the most gross observations. Brain material must be fresh and quickly perfused for serious microscopic studies. Perfusion must be carried-out to adequately fill the vast rete mirabile complex of the vascular network that is interposed between main blood vessels and the brain. Unfortunately, such perfusions may make the material less useful for many other types of studies such as bacteriological and various chemical analyses. In many cases investigators interested in the body as a whole are reluctant to allow a perfusion that enters the primary circulation and thereby makes many other types of studies impossible. It was made clear that the problem of competing for access to specimens continues to be a major one in this field. Thus, quite naturally each investigator has sought out his own material for his select lines of investigation and thereby wasted many body parts. The workshop discussions indicated that this is no simple or easily resolvable problem and that team-type research endeavors are needed so that scientists can go to distant sites, such as Japan or Brazil, where material is more readily available, with specific studies in mind, and with agents for perfusions and/or external fixation of material in hand. In most cases these types of approaches have been impractical as well as too time-consuming and expensive. They continue to remain as future alternatives and will require considerable cooperation among investigators. It was emphasized strongly that precise guidelines should be set-up for the collection and preservation of morphological material from stranded animals, from animals that die in captivity and from animals trapped in fishing gear.

In the discussions considerable emphasis was placed on attempting to combine physiological studies with behavioral observations, though the obvious difficulties of these types of approaches also was made clear. In this regard, it has not often been possible or practical, for example, to insert chronic brain electrodes in behaving small cetaceans to study many of the parameters examined in other species such as rats, monkeys, etc. Although the evoked potential mapping studies of cortical sensory areas by Lende and Welker (1972), Sokolov et al. (1972), Ladygina et al. (1978) and Supin et al. (1978) are noteworthy, most investigators have not been able to study brain physiology. For example, extended cortical mapping of the huge expanses of cortex, not explored by the Russian workers, is needed as are comparative studies involving examination of single unit activity in the giant reticular cells involved in REM sleep generation. It also was pointed-out that the cortical evoked potential studies will be extremely difficult to repeat and extend in the future due to protective policies now in force in most countries. It was stressed, however, that if and when such studies are done that they should involve extensive cooperation between behavioral and physiological laboratories. Such cooperative ventures in this field are not now being done to any extent. Each animal is extremely valuable and the risk factors in anesthesia and surgical procedures can be considerable.

On the morphological side, there is little doubt that approaches to the brain will continue to be productive and illuminating of many problems in comparative and evolutionary morphology. Such studies should be comparative in nature involving such prototypic mammalian species as basal insectivore and bats, among others. Thus, comparative, quantitative morphological studies across several species in terms of the acoustic system, such as carried out by Zook and his colleagues (1988), will continue to provide valuable information related to comparative aspects of signal processing. Similarly, our laboratory is involved extensively in comparative and evolutionary analyses of the neocortical formations as summarized in our two papers in this symposium. In this regard, it was emphasized that newer conceptual frameworks need to be developed to further assess the status and organization of the cetacean brain and that these should be, for example, based on studies similar to, as well as extensions of, the concepts of Sanides (1970, 1972) and with reference to evolutionary theories such as proposed by Zvorykin (1977). This boils-down to moving away from purely descriptive anatomy and seeing the larger picture, including development of ideas regarding the significance of this work, for example, such as shedding-light on the origins of the mammalian neocortex, and considering encephalization, particularly corticalization, as an adaptation to the demands of particular ecological conditions, and relating thalamo-cortical afferents to particular sensory adaptations and specializations. Mere collecting and cataloging, even in the case of rare and hard to obtain species, no longer suffices and is not likely to be a viable approach in the future. This includes such approaches as further surface analyses of brains, examination of sizes and weights of brain components and various allometric indices involving brain/body weight corrections which, though valuable, provide only limited advances and perspectives.

Relative to the brain mapping problem, it is important to note that a huge area of the cetacean neocortex has not been mapped due to difficulties of placing electrodes into the more lateral, medial and ventral parts of the cortical formations. These extensive cortical areas remain a mystery in terms of their function. In regard to the "initial" brain concepts outlined by Glezer et al. (1988) and reviewed by Morgane and Glezer in this symposium, this extended cortical mapping should have a high priority in future work. One major way the whale brain differs from that of the basal insectivores, as well as bats, is not only in sheer size but in the fact that, instead of the entire neocortex being sensory or motor in nature, there are extensive neocortical areas for which no ascribed function is yet known. Such cortical areas are not developed in the other known models of the "initial" brain such as is present in bats and basal insectivores. In this regard, one key to understanding cortical organization is to examine thalamo-cortical projections. Such studies, using axon transport methods, are underway by Garey and Revishchin (this volume). These will complement the earlier degeneration studies done by Krasnoshchekova and Figurina (1980). However, more details of the thalamo-cortical projections into neocortex need examination such as particular aspects of laminar patterns of thalamic projections. Relative to this type of issue, Garey and Revishchin (1988) examined the laminar distribution of cytochrome oxidase activity in the frontal and occipital areas of the dolphin neocortex and reported high levels of cytochrome oxidase in neocortical layers I and III. This is of particular interest because, in regard to the organization of the whale neocortex, Morgane et al. (1985; 1986a,b; 1989) note the likelihood of heavy inputs to layer I, including a heavy thalamic contingent. Since high levels of cytochrome oxidase coincide with high density of afferent terminals, this finding points to possibilities of dolphin neocortex receiving its heaviest inputs to layer I and III. Additionally, it was stressed in the workshop that the details of post-synaptic targets of thalamic inputs to the neocortex need to be worked-out. For example, so far we know only that the thalamus in cetaceans has many projections to the neocortex just as it does in all other mammals, though it remains to be seen what preferential layers of cortex are invaded by particular thalamic inputs. We need to extend these types of studies in small cetaceans to match those carried-out in other mammalian species. Thus, several specific types of laminar analyses need to be done on the thalamo-cortical systems in Cetacea. The evidence for channeling of different types of input to different laminae in the cerebral neocortex provides the rationale for closer examination of the neuronal constituents of these thalamic recipient regions and for attempts to trace the further projections of the neurons in the individual layers. The information about thalamic axon terminals in the neocortex suggests that in cetaceans the following specific questions need be addressed: (1) Is a single functional zone defined by using the thalamic input distribution to establish the boundaries of the layers? (2) What additional inputs to a given layer exist, other then the thalamic inputs, that can be helpful in defining the limits of the layer? (3) Do neurons lying in specific cortical laminae restrict their dendritic surface to a particular thalamic fiber termination zone or do the dendrites extend into other laminae? (4) Do thalamic afferents terminate on the most available perikaryal and dendritic surfaces in the layer or do they selectively establish contacts on particular neuronal types? (5) What are the patterns of distribution of the axons arising from different types of neurons in the layer, both within and outside of the lamina? (6) What types of synapses are formed by thalamic afferent fibers on particular neuronal types? It was made-clear that these are the types of thalamo-cortical studies needed to learn how the cetacean neocortical organization relates to that of other mammals. Merely noting rough cortical areas of inputs from the thalamus without these details will have only limited value in this field and can not help make fine distinctions between the organization of the cetacean neocortical inputs compared to those of other species. Additional studies on local circuit organization of the neocortex are needed, including further elaboration of the columnar nature of the dolphin neocortex described by Morgane et al. (1988). We must certainly understand more about the radially-organized modules or columns since these represent the basic processing units of the neocortex. The modular concept of cortical organization considers the neocortex as a mosaic of discrete vertical subunits which form the anatomical basis in a functional design. As such, their study deserves special emphasis. This has potential importance since, considering the "initial" brain concept in relation to organization of the neocortex in cetaceans, we may be dealing with models of the first columns in mammalian cortical evolution. These may tell us a great deal regarding evolution of the basic processing units of the neocortex. Examining how these progressed in terrestrial and semi-aquatic forms versus their character in cetaceans also may provide useful data on information processing capacities in cetaceans.

Thinking about the organization of cortical modules also led to further discussions of the functional meaning of cortical lamination in general. It is important to examine the afferent and intrinsic organization of laminated structures. Thus, questions arose as to what various lamination patterns imply regarding function. The peculiar hybrid type of neocortical organization in cetaceans, especially the allocortical type of organization seen in the upper layers of cetacean neocortex, which resembles the organization seen in bat and hedgehog convexity neocortex, needs to be examined further, including studies of these specific layers in the so-far unmapped expanses of cetacean neocortex. Such morphological analyses should shed some light on what the unmapped cortical formations represent, though eventual electrophysiological mapping will be essential. Obviously, until such studies are done there will remain lingering mysteries about the status and organization of cetacean neocortex. With regard to layering, it was observed that stratification permits more dependence on local interactions. In evolution there appears to be more confinement of the target of each type of axon to limited parts of the post-junctional neurons and there may be more differentiation of axon terminals, dendrites and output messages (Bullock, 1980). With the evolution of these forms of increased specification, a tendency to bring-together the corresponding parts of the cell type into one stratum would lead to several advantages. Stratification would provide far more connections per unit of volume and permit cells with short axons to become more influential and to reach more targets with their limited arbors. Even more important, it would permit relatively more dependence on local interactions. There is little question, therefore, in continuing comparative studies we must gain insight into the physiology of individual strata, of large and small cells and of the local circuit arrangements. With reference to cetaceans, we have reported in recent years and in this symposium, that there is little evidence of granularization in the dolphin neocortex with mainly large, almost giant stellate cells being the rule, though stellate cells are relatively few in number in the highly pyramidalized cetacean neocortex.

Several issues surfaced in the workshop pertaining to the concept of so-called "generalized" type of neocortex, such as we see in the convexity neocortex in the dolphin. This, in turn, brought-up the question of how to establish parameters of "specialization". Among the stellate neurons, we see a dominance of reticular type, isodendritic neurons scattered throughout all layers of the cetacean neocortex that closely resemble the generalized neurons of the phylogenetic old reticular core of the brainstem. Accordingly, we have postulated that a type of "extension" of the reticular core into the forebrain, such as is seen in the earliest mammals, is present in the dolphin so that "correlational" or "generalized" type neurons continue to predominate in these aquatic mammals. It is of interest in this regard that Bowsher (1976) made an important distinction between the reticular and lemniscal components extending into the diencephalon and telencephalon. Thus, the evolutionary trend towards a more evolved status in receiving and processing of sensory information would be directly related to the quantitative and qualitative progress of the lemniscal component in comparison with the reticular, i.e., between the specific and focussed input system compared to the generalized, more diffuse input system. This will be an important areas for future research in cetaceans compared to other mammalian lines. In evolution, as put forward by Ebbesson in his parcellation theory (1980), it was noted that most progressive change represents subdivision and differentiation with selective loss of formerly more generalized connections between major brain regions, rather than invention of new connections or new cell types. In whales it appears possible that the reticular, more generalized, correlation type connections may have been maintained as opposed to the situation that prevailed in most terrestrial mammals. Again, it was stressed that these types of concepts need to be assessed in future studies.

In regard to the concepts of generalized, reticular-organized neocortex, there is another questions to consider regarding cetaceans. Is it possible that they may use more than one principle for the operation of recognizing their ethologically important sensory stimuli with systems of neurons? Do cetaceans have some distinct form of input processing in cortical operations leading to recognition, i.e., appropriate responses? As pointed-out by Bullock (1983), a common distinction is made between two extremes: (1) A hierarchy of feature-sensitive neurons that achieves more and more specific recognition of some complex stimulus configuration by a limited number of highly specified neurons, and (2) a large population of variously sensitive but not so complex or specified neurons that recognizes by the particular spatio-temporal pattern of assemblage facilitated or inhibited by a stimulus configuration.

One modern approach to brain and cortical organization is that of immunocytochemical analysis to understand neurotransmitter-specific cells and how these are interlinked with other types of neurons. Transmitter or chemical neuroanatomy has not been done on the cetacean brain until recently by Garey and Revishchin (1989) and by Glezer et al. (this volume). It is still in its infancy in studies in cetaceans, but sure to provide valuable information on the organization of the cetacean neocortex. Studies of co-localization of classic neurotransmitters with neuropeptides also will be of considerable comparative and evolutionary value. These types of approaches will prove even more valuable when it can be determined, for example, where neuropeptide neurons lie with regard to cytochrome oxidase puffs in the neocortex such as examined recently in the primate brain. This again points-back to the problem of securing brains in a controlled fashion to assure proper perfusion and use of special fixatives needed to bring out the special chemical characteristics of neurons.

Regarding studies on the hearing apparatus of whales, which comprises a major defined area of research in this field, several of the problems and perspectives were discussed. The auditory region is the morphological domain that contains the organs of both hearing and balance, as well as the soft and hard tissues associated with them. Next to dentition the auditory region is the single most important and heavily sampled source of cranial taxonomic characters, particularly in primatology, as well as in many other fields of mammalian systematics. The taxonomic utility of the auditory region is matched by a correspondingly large literature on otic morphology in many mammalian groups. Such studies using modern techniques are now in progress on the cetacean otic system. In this regard, discussions were active regarding use of non-invasive procedures to study cetaceans, particularly with respect to the otic system. Computerized tomography (CT scanning) now has provided valuable information for three-dimensional reconstructions such as done on the cochlea by Ketten and Wartzok (this volume). Further, modelling of the auditory related neuronal networks will prove to be invaluable in providing essential basic information for later experimental analyses.

There is limited work, however, in one area, namely ontogeny, that has considerable potential relevance for otic character analysis and, ultimately, classification and phylogenetic reconstruction. In regard to ontogenetic approaches it is well-known how difficult it is to obtain a suitable ontogenetic series, though this, too, is being rectified and hopefully will provide important new information. Ontogenetic studies should comprise the results of a developmental and morphological investigation of certain otic structures found in Cetacea, as well as in certain primates and eutherian insectivores, particularly the basal insectivores. This comparative series is important enough historically and diverse enough structurally to permit meaningful character analyses. The features selected for examination should be those which are strongly emphasized, for example, in primate systematics and related fields and chiefly include, as opposed to an atlas of the entire auditory region, the soft and hard tissues of the middle ear. With regard to studies of the otic structures, it was again pointed-out that there are limiting factors regarding cetaceans which include the availability of appropriate specimens and the differential relevance of specific parts of the life cycle. There are two sources of accessible evidence that help to verify the continuum, namely the fossil record and the ontogeny of extant model mammals. Fossils are part of the substantive record of phylogenetic descent with modification and were in their time parts of living organisms and the products of separate ontogenies. It is, however, almost never possible to reconstruct precisely the entire continuum of ontogenies since individual fossils only represent isolated sampling points. Thus, since young stages of primitive mammals are rarely encountered, the mammalian fossil record principally is represented by adult forms. Ancient ontogeny, therefore, continues to be a mystery and we simply do not have adequate evidence for restructuring it in detail at this time. However, it was pointed-out that continuing work along such lines should be a consideration for the future. In any event, evolutionary sequences can be established legitimately without the complete record of the ontogenetic continuum. The fossil record, even though discontinuous, represents a true copy of the forms obtained by successive ontogenies in the past. Relative to the evolutionary modification of ontogenesis through history, the existence of fossils permits the reconstruction of the sequences in which these modifications occurred. Ontogenies of extant mammals do not permit such faithful reconstructions since they are not exact sequential records of evolutionary modification. Consideration of this distinction was discussed with regard to studies of Cetacea in terms of proper use of developmental data in evolutionary and systematic studies.

The reconstruction of developmental patterns in modern ontogeny relates to an additional matter relevant to the use of ontogenetic studies, namely the formulation and testing of various hypotheses about morphological change. It was noted that two main questions can be asked about form changes in an evolutionary lineage: (1) why they came about, which leads to hypotheses about selective factors that favored transformation along certain vectors; and (2) how they occurred, which leads to other hypotheses about the mechanisms of change along those same vectors. Obviously, embryology has a special role to play in questions of the second type given its concern with the nature of change in ontogenesis. It also was reemphasized in the workshop discussion that it does not follow that a shared structure that develops in a certain way in a descendant followed that exact same route in an ancestor. However, it was noted that examination of descendent ontogeny may reveal that there are definite preconditions for the development of a structure and that these almost assuredly also have to be present in the ancestor. Ontogenetic studies also go a long way in elaborating which among several competing hypotheses about a change is the most parsimonious, as measured by the scale of biological processes that must be inferred for the change to have occurred. For example, MacPhee (1979) observes that it is more parsimonious to assume that the bulla in primates is a derivative of the petrosal rather than a suppressed entotympanic because the former alternative does not require the occurrence of an unlikely ontogenetic event, i.e., primordial fusion. Ontogenetic studies, as pointed-out clearly by MacPhee (1977), also may reveal that very simple and potentially labile ontogenetic events can underlie the production of adult characteristics thought to be extremely conservative, such as the relationship of the ectotympanic to surrounding elements. Still, there are many reservations about ontogenetic mechanisms and it was emphasized that it is not possible to infer the primitive and derived states of a character simply on the basis of ontogenetic analyses.

All of these types of debates and discussions led to the view that we also need a large embryological series to be able to study sequences of the development of the brain and related sensory structures and, from these, make various phylogenetic inferences. Such a series does not exist and, therefore, these types of studies have been few and far between and almost always on randomly obtained fetal and embryonic material of unknown ages. It was pointed-out that the gaps in embryological studies in Cetacea are vast and that this is clearly an area needing future attention. It also was agreed in general that echo-locating and "language" capabilities of whales needs special consideration given that these are highly specialized features in Cetacea. Their representation in the brain, and especially in the cerebral cortex, needs to be better understood. In the case of the auditory system, which has received considerable attention in studies of cetaceans, we also need to characterize the descending auditory pathways in the brainstem and focus on cell groups which are especially well-developed in terms of size and differentiation in comparison with other echo-locating species. Analysis of vocal-auditory interactions is important in this field and needs to be considered as high priority studies in the future.

Overall, the discussants recognized many important areas for future research related to the sensory abilities of cetaceans. The need for good networking among the limited number of scientists active in this field was stressed. The animal material is scarce, protected by law, and is associated with special problems due to body size and difficulties in carrying out electrophysiological studies around salt water. All of these things have impeded progress in this area. However, there is little doubt that information gained about the cetacean nervous system, particularly the sensory system including its cortical representations, will provide valuable clues about the origins of the mammalian brain and the effects of powerful ecological forces in shaping the organization of the cortex. Understanding of the historical developments of the brain in a mammalian Order showing complete, secondary return to the aqueous medium in a geological era when primitive mammals were dominant, should provide many clues regarding the principles of organization of the mammalian brain.

REFERENCES

Bowsher, D., 1976, L'evolution des neurones chez les vertebres, La Recherche, 7:935-944.
Bullock, T. H., 1980, Comparative audition: where do we go from here?, in: "Comparative Studies of Hearing in Vertebrates", A. N. Popper and R. F. Fay, eds., Springer Verlag, New York, pp. 439-452.
Bullock, T. H., 1983, Why study fish brains? Some aims of comparative neurology today, in: "Fish Neurobiology", Vol. 2, R. E. Davis and R. G. Northcutt, eds., University of Michigan Press, Ann Arbor, pp. 361-368.
Ebbesson, S. O. E., 1980, The parcellation theory and its relation to interspecific variability in brain organization, evolutionary and ontogenetic development and neuronal plasticity, Cell Tissue Res., 213:179-212.
Garey, L. J. and Revishchin, A. V., 1988, Laminar distribution of cytochrome oxidase activity in porpoise neocortex, Doklady Akademii Nauk SSSR 302:1486-1489.
Garey, L. J. and Revishchin, A. V., 1989, Quantitative distribution of GABAimmunoreactive neurons in cetacean visual cortex is similar to that in land mammals, Brain Research, 485:278-284.

Glezer, I. I., Jacobs, M. S. and Morgane, P. J., 1988, Implications of the "initial brain" concept for brain evolution in Cetacea, Behavioral and Brain Sciences, 11:75-116.

Krasnoshchekova, E. I. and Figurina, I. I., 1980, The cortical projection of the medial geniculate body of the dolphin brain, Arkhiv Anat. Gistol. Embriol., 78:19-24.

Ladygina, T. F., Mass, A. M. and Supin, A. Ya., 1978, Multiple sensory projections in the dolphin cerebral cortex, Zh. Vyssh. Nerv. Deiat., 18:1047-1054.

Lende, R. A. and Welker, W. I., 1972, An unusual sensory area in the cerebral neocortex of the bottlenose dolphin, Tursiops truncatus, Brain Research, 45:555-560.

MacPhee, R. D. E., 1977, Ontogeny of the ectotympanic-petrosal plate relationship in strepsirhine prosimians, Folia Primatol., 27:245-283.

MacPhee, R. D. E., 1979, Entotympanics, ontogeny and primates, Folia Primatol., 31:23-47.

Morgane, P. J., Jacobs, M. S. and Galaburda, A., 1985, Conservative features of neocortical evolution in dolphin brain, Brain Behavior and Evolution, 21:176-184.

Morgane, P. J., Jacobs, M. S. and Galaburda, A., 1986a, Evolutionary morphology of the dolphin brain, in: "Dolphin Cognition and Behavior. A Comparative Approach", R. Schusterman, J. Thomas and F. Wood, eds., Lawrence Erlbaum Associates, Hillsdale, pp. 5-29.

Morgane, P. J., Jacobs, M. S. and Galaburda, A., 1986b, Evolutionary aspects of cortical organization in the dolphin brain, in: "Research on Dolphins", R.J. Harrison and M. Bryden, eds., Oxford University Press, Oxford, pp. 71-98.

Morgane, P. J., Glezer, I. I. and Jacobs, M. S., 1988, Visual cortex of the dolphin: An image analysis study, J. Comp. Neurol., 273:3-25.

Morgane, P. J., Glezer, I. I. and Jacobs, M. S., 1989, in press, Comparative and evolutionary anatomy of visual cortex of dolphin, in: "Cerebral Cortex, Vol. 8. Evolution and Comparative Anatomy of Cerebral Cortex." E. G. Jones and A. Peters, eds., Plenum Press, New York.

Sanides, F., 1970, Functional architecture of motor and sensory cortices in primates in the light of a new concept of neocortex evolution, in: "The Primate Brain", C. R. Noback and W. Montagna, eds., Appleton-Century Crofts, New York, pp. 137-208.

Sanides, F., 1972, Representation in the cerebral cortex and its areal lamination patterns, in: "The Structure and Function of Nervous Tissue", G. H. Bourne, ed., Academic Press, New York, pp. 329-453.

Sokolov, V. E., Ladygina, T. F. and Supin, A. Ya., 1972, Localization of sensory zones in the dolphin's cerebral cortex, Dokl. Akad. Nauk SSSR, 202:490-493.

Supin, A. Ya., Mukhametov, L., Ladygina, T., Popov, V., Mass, A. and Polyakova, I., 1978, Electrophysiologic study of the dolphin brain, Nauka Press, Moscow, pp. 7-85.

Zook, J. M., Jacobs, M. S., Glezer, I. I. and Morgane, P. J., 1988, Some comparative aspects of auditory brainstem cytoarchitecture in echolocating mammals: Speculations on the morphological basis of time-domain signal processing, in: "Animal Sonar: Processes and Performance", P. E. Nachtigall and P. W. B. Moore, eds., Plenum Press, New York, pp. 311-316.

Zvorykin, V. P., 1977, Principles of structural organization of the cetacean neocortex, Arkhiv Anat. Gistol. Embriol., 72:5-22.

TARGET DETECTION IN NOISE BY ECHOLOCATING DOLPHINS

Whitlow W. L. Au

Naval Ocean Systems Center
P. O. Box 997
Kailua, Hawaii 96734

It is well known that dolphins possess a sonar capability which allows them to project acoustic energy and analyze returning echoes in order to detect and recognize objects under water. The use of acoustic energy is probably the most effective way to probe an underwater environment for purposes of navigation, obstacle avoidance, prey and predator detection, and object localization and detection. Acoustic and other mechanical vibrational energy propagates in water more efficiently than any form of energy including electromagnetic, thermal and optical energy. Since the natural habitats of many dolphin species include shallow bays, inlets, coastal waters, swamps, marshlands, and rivers that are often so murky or turbid that vision is severely limited, these animals must rely heavily on their active and passive sonar capabilities for survival. Some of the sonar capabilities of dolphins have been described in review articles by Nachtigall (1980) and Au (1988).

In this paper, the target detection capability of the Atlantic bottlenose dolphin (_Tursiops_ _truncatus_) in the open waters of Kaneohe Bay, Oahu, Hawaii will be discussed and the dolphin's performance will be compared with an energy detector model. _Tursiops_ typically emit echolocation signals with peak frequencies between 110-130 kHz in Kaneohe Bay (Au, 1980). Kaneohe Bay has one of the noisiest "snapping shrimp" population in the world (Albers, 1965; Urick, 1984). An example of the ambient noise in the bay along with an example of the ambient noise in San Diego Bay, California and typical deep water noise spectral density for different sea states are shown in Fig. 1.

The target detection capability of any sonar system is limited by interfering noise and reverberation. The target detection sensitivity of a sonar can be measured by a variety of equivalent methods. The range of a target of known target strength can be increased until the target can no longer be detected. A fixed target range can be used and the size of the target can be reduced continuously until the target can no longer be detected. A fixed target range can be used and the echo signal-to-noise (E_e/N_o) ratio varied by either adjusting the amount of masking noise or by varying the target size, until the dolphin can no longer detect the target. Whatever method is used, certain important acoustic parameters must be measured for the detection experiment to be meaningful. The source level, target strength, and noise levels should be measured so that the E_e/N_o at the detection threshold can be determined.

BIOSONAR TARGET DETECTION CAPABILITY

A variety of biosonar experiments using the three equivalent methods mentioned in the preceding paragraph have been performed in Kaneohe Bay to determine the sonar detection capability of Tursiops truncatus. Murchison (1980) performed a maximum range detection experiment with two Tursiops, using a 2.54-cm diam. solid steel sphere and a 7.62-cm diam. stainless steel water-filled sphere as targets. The composite 50% correct detection threshold were at ranges of 72 and 77 m for the 2.54-cm and 7.62-cm spheres, respectively. However, a bottom ridge at approximately 73 m limited the animals' ability to detect the 7.62-cm target beyond 73 m. The animals' performance degraded rapidly when the target was in the vicinity of the ridge, suggesting that the dolphins were probably reverberation-limited with the 7.62-cm sphere.

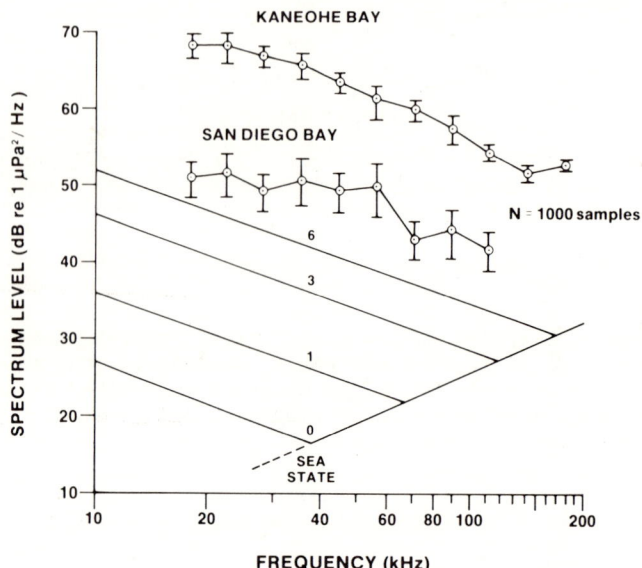

Fig. 1. Ambient noise of Kaneohe Bay measured in 1/3 octave bands. Deep water noise for different sea states are shown for comparison.

Au and Synder (1980) remeasured the maximum detection range in a different part of Kaneohe Bay using one of the same dolphins (Sven) and a 7.62-cm diam. sphere. Sven's target detection performances for the 2.54-cm sphere (Murchison, 1980) and the 7.62-cm sphere (Au and Synder, 1980) are plotted in Fig. 2 as a function of range. The 50% correct detection threshold for the 7.62-cm sphere occurred at 113 m, a considerably longer range than the 76.6 m reported by Murchison (1980).

The results shown in Fig. 2 are very specific to the ambient noise condition of Kaneohe Bay. In order to make the results more general and useful, the detection performance should be plotted as a function of the

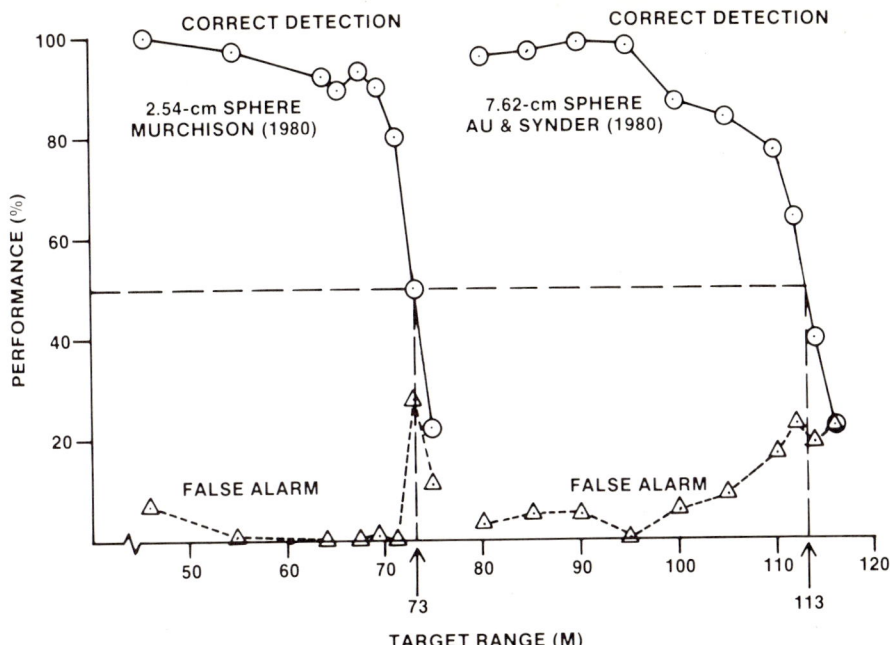

Fig. 2. Target detection performance of a <u>Tursiops truncatus</u> as a function of range for two different spherical targets (From Murchison, 1980; Au and Synder, 1980).

estimated received signal-to-noise ratio. The transient form of the sonar equation for a noise limited situation can be used to analyze the dolphin's performance shown in Fig. 2 in terms of the ratio of the energy in the received target echo to the noise spectral density. The transient form of the sonar equation expressed in dB can be written as (Au, 1988)

$$DT_E = SE - 2\ TL + TS_E - (NL - DI_R) \tag{1}$$

where: DT_E = detection threshold
SE = source energy flux density
TL = transmission loss
TS_E = target strength based on energy
NL = background noise level
DI_R = receiving directivity index

The detection threshold (DT_E) corresponds to the energy-to-noise ratio used in human psychophysics and is equal to $10\ \text{Log}(E_e/N_o)$, were E_e is the echo energy flux density and N_o is the noise spectral density level. During a sonar search dolphins typically vary the amplitude of their sonar signals over a 20 dB range making it difficult to estimate the detection threshold accurately. Au and Penner (1981) resorted to using the maximum source energy flux density per trial, which will lead to a conservative estimate of the detection threshold. Sven's sonar signals were measured in the study of Au et al. (1974) for target ranges of 59 to 77 m. The maximum peak-to-peak source level averaged over 12 trials at the 77 m range was 225 dB re 1 μPa and typical peak frequencies were approximately

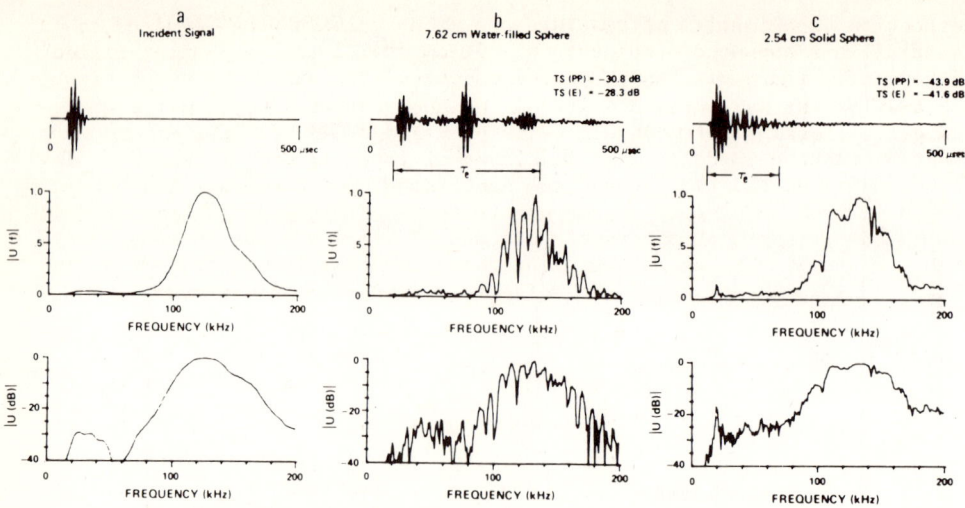

Fig. 3. Results of target strength measurements, (a) simulated dolphin click (incident signal), echoes from the (b) 7.62-cm and (c) 2.54-cm spheres (From Au and Synder, 1980).

120 kHz. Au (1980) showed that the energy flux density is approximately equal to the peak-to-peak SPL minus 58 dB for signals used by <u>Tursiops</u> in Kaneohe Bay, so that an SE of 166 dB re (1 µPa)^{2}s would be appropriate for use in the sonar equation. The target strength of the 2.54-cm and 7.62-cm spheres were measured by Au and Synder (1980) and their results are shown in Fig. 3. The target strength was -41.6 dB for the 2.54-cm sphere and -28.3 dB for the 7.62-cm sphere. From Fig. 1, the ambient noise level at 120 kHz is approximately 54 dB re 1 µPa^2/Hz. Au and Moore (1984) measured the receiving beam patterns of <u>Tursiops</u> at frequencies of 30, 60 and 120 kHz, and used the results to calculate the directivity index (DI_R). They found that the receiving directivity index can be described by the equation (Au, 1988)

$$DI_R(dB) = 16.9 \log f(kHz) - 14.9 \qquad (2)$$

For a peak frequency of 120 kHz, DI_R = 20.2 dB.

The dolphin's target detection results shown in Fig. 2 are replotted as a function of the echo signal-to-noise ratio in Fig. 4. The results indicate that the animal's performance was consistent for the two studies. The 75% correct thresholds were at 10.4 dB for the 2.54-cm sphere and 12.7 dB for the 7.62-cm. This difference of 2.3 dB is small considering the fact that the two studies were done approximately two years apart.

TARGET DETECTION IN NOISE

Au and Penner (1981) used the technique of fixing the target range and varying the level of a masking noise source to determine the target detection capabilities of two <u>Tursiops</u>. The animals were required to station in a hoop and echolocate a 7.62-cm stainless steel water-filled sphere at a range of 16.5 m. A noise source with a flat spectrum between 40 and 160 kHz was located 4 m from the hoop between the animal and the target. Masking noise levels between 67 and 87 dB re 1 $\mu Pa^2/Hz$ in 5-dB increments were randomly used in blocks of 10-trials for a 100-trial session. Turl et al. (1987) used the same technique to compare the detection capability of a <u>Tursiops truncatus</u> with a <u>Delphinapterus leucas</u>. A 7.62-cm sphere was used at ranges of 16.5 and 40 m and a 22.86-cm sphere at a range of 80 m. The experimental procedure was similar to that of Au and Penner (1981) except a smaller noise increment of 3 dB was used.

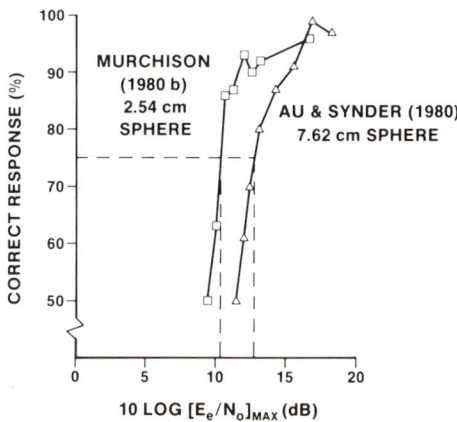

Fig. 4. Target detection performance of a <u>Tursiops</u> as function of the echo energy-to-noise ratio for the range detection data of Figure 2.

The dolphins' performance for both studies are plotted as a function of $(E_e/N_o)_{max}$ in Fig. 5. The average value of the maximum source energy flux density per trial was used in the calculations. The 75% correct response threshold occurred at $(E_e/N_o)_{max}$ of 7 and 12 dB in the study of Au and Penner (1981) and at approximately 10 dB in the study of Turl et al. (1987). The results shown in Fig. 5 indicate good agreement and consistency with only a small amount of inter-animal difference in target detection ability that was less than 5 dB. At the two highest noise level of the Au and Penner study, both dolphins began to "guess". One dolphin did not emit any detectable signals in 20% and 41% of the trials when the noise levels were 82 and 87 dB, respectively. The other dolphin did not emit any signals in 14% of the trials at the 87-dB noise level. Therefore, the average of the maximum signal per trial for the noise levels between 67 and 77 dB were used to calculate $(E_e/N_o)_{max}$. The animal

207

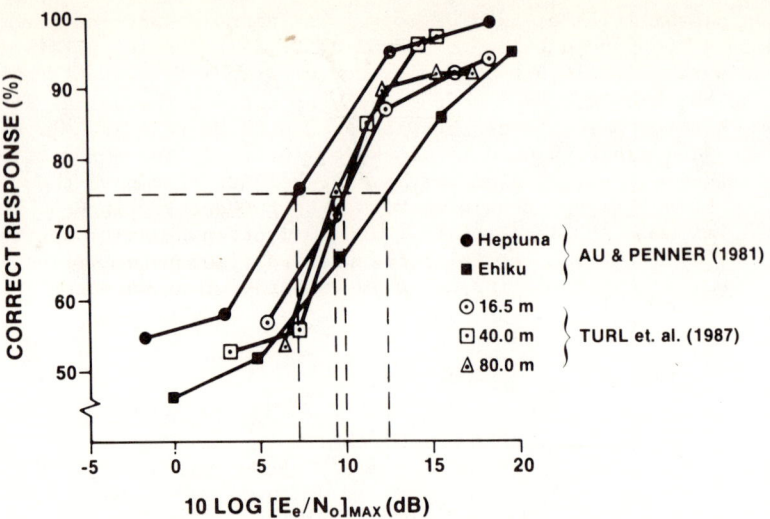

Fig. 5. Target detection in noise performance results as a function of the echo energy-to-noise ratio (from Au and Penner, 1981; Turl et al., 1987).

used by Turl et al. (1987) did not exhibit any "shut down" behavior probably because the highest noise level was 10 dB lower than that used by Au and Penner (1981).

The third technique of fixing the target range and reducing the target size was used by Au et al. (1988) to measure the target detection capability of a <u>Tursiops</u>. An electronic transponder system was used so that the effective echo strength could be varied by adjusting the level of the simulated echoes. A hydrophone located 1.9 m in front of a stationing hoop detected each projected signal which was digitized and stored in random access memory (RAM). The stored signal was then played back through a projector located 2.4 m from the hoop to simulate an echo from a target. Masking noise at a fixed level was also played to the animal and the intensity of the simulated echo was randomly varied in 10-trial blocks by increments of 2 dB. For each click emitted by the dolphin, two clicks separated by 200 μs were played back to the animal at a time delay corresponding to a 20 m target range. The dolphin was required to station in a hoop, echolocate and report if the phantom target was present or absent.

The phantom target results are shown in Fig. 6 along with the maximum range results (Fig. 3) and the noise masking results (Fig. 5). The results between the various studies agree extremely well considering the differences in animals, time periods and experimental procedures. Murchison (1980) and Au and Synder (1980) varied target range in small increments in terms of the resultant E_e/N_o. Au and Penner (1981), Turl et al. (1987) used a constant target range and randomly varied the masking noise levels. A relatively large increments of 5 dB was used by Au and Penner (1981) and a smaller 3-dB increment was used by Turl et al. (1987). Au et al. (1988) used a fixed phantom target range and noise

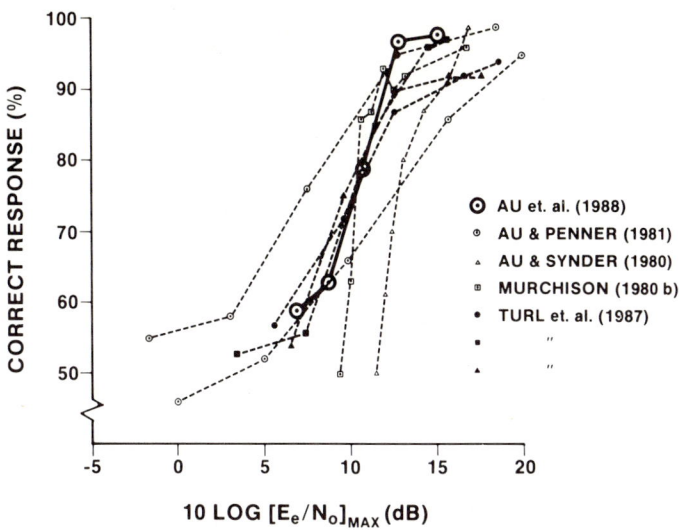

Fig 6. Phantom target detection performance of a <u>Tursiops</u> as function of the echo energy-to-noise ratio (from Au et al., 1988).

level and randomly varied the amplitude of the target echoes in 2-dB increments. The shallower slopes of the performance curves in the Au and Penner (1981) study were probably the result of using a large noise increment. The curves from the other studies have similar slopes.

DOLPHIN SONAR MODELED AS AN ENERGY DETECTOR

The auditory threshold as a function of signal duration and the critical ratio experiments of Johnson (1968 a,b) with <u>Tursiops truncatus</u> indicated that the dolphin's inner ear functions like the human inner ear and that the dolphin integrate acoustic energy in the same way as humans. Green and Swets (1966) showed that an energy detector is a good analogue of the human auditory detection process. Therefore, it seems reasonable to approach the dolphin auditory process as an energy detector. From Johnson's (1968a) auditory threshold data for pure tone signals of varying duration, the integration at 120 kHz should be approximately 2 ms. However, experiments with pulse sounds by Vel'min and Dubrovskiy (1975; 1976) indicated that dolphins may process short duration, broadband sonar signals differently than pure tone signals. Au et al. (1988) performed an experiment to measure a dolphin's integration time for sonar pulses using a phantom target with electronically simulated echoes that could be controlled with high precision. They first played back a single click for every click emitted by the dolphin and obtained a threshold in noise by progressively decreasing the amplitude of the single click echo. The threshold was obtained using a staircase psychophysical testing procedure. Next they played back two clicks for each click emitted by the dolphin and measured the dolphin's threshold. Various separation times between the first and second clicks of the double click echoes were used.

The results of the auditory integration-time experiment are shown in Fig. 7, with the echo energy-to-noise ratio in dB plotted against the separation time between the double click echo. The dolphin's threshold shifted approximately 3 dB when the stimulus changed from a single click to a double click. This shift is exactly what would be expected for an energy detector since there is 3 dB more energy in the double-click stimulus. As ΔT increased to 200 μs, the threshold remained essentially the same. As ΔT increased to 250 μs, the threshold began to shift towards

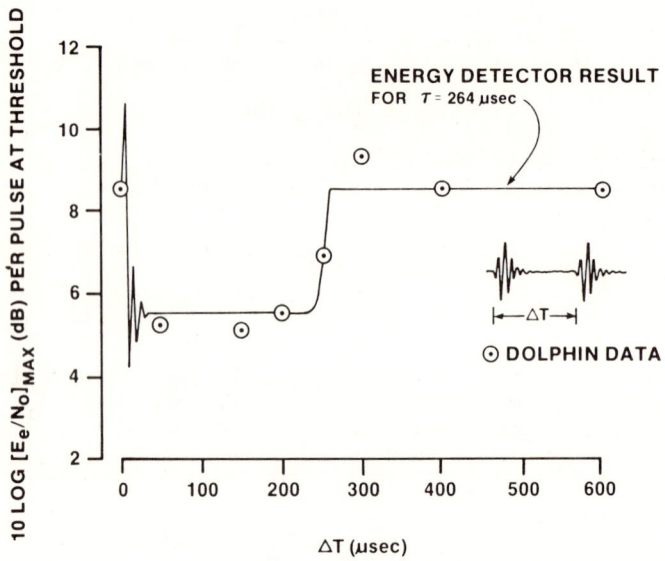

Fig. 7. Integration-time experiment results showing the average of the maximum E_e/N_o per pulse at threshold as a function of the separation time between pulses. Each echo at 0 μs consisted of a single click, while each echo at the other separation times consisted of double clicks. The solid curve is the response of an ideal energy detector that best fit the dolphin data (from Au et al., 1988).

the single-click threshold, reaching the single-click threshold at a ΔT of about 300 μs and greater. Therefore, the presence of a second click with ΔT greater than 300 μs did not help the dolphin in detecting the phantom target. The solid curve in Fig. 7 is the response of an ideal detector with an integration time of 264 μs. The curve associated with an integration time of 264 μs best fitted the dolphin's data with a minimum least-square error. The 264 μs integration time corresponded well with the 260 μs critical interval reported by Vel'min and Dubrovskiy (1975; 1976) for echolocating <u>Tursiops truncatus</u>. They defined the critical interval as a "critical time interval in which individual acoustic events merge into an acoustic whole," which may be another way of considering integration time. The integration time measured by Au et al. (1988) also agreed well with the backward masking threshold of 265 μs measured by Moore et al. (1984).

Energy detection processing by <u>Tursiops</u> was examined further by Au et al. (1988) using their electronic phantom target in another experiment. They played back echoes consisting of one, then two, and finally three replicas of each emitted click and measured the shift in the dolphin's threshold. All pulses were within the integration time of the dolphin's auditory system. Their results are presented in Fig. 8 along with an energy detector response curve. They found that the dolphin's sonar detection performance followed the response of an energy detector.

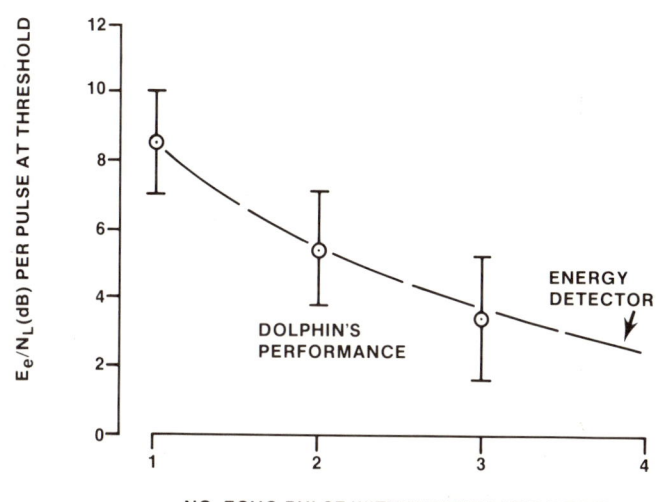

Fig. 8. Dolphin's performance as a function of the number of clicks in the echo with the animal's integration time (From Au et al., 1988).

The dolphin's performance data obtain in the presence of masking noise shown in Figs. 5 and 6 can be compared with a theoretical model of an energy detector. Urkowitz (1967) examined the detection of a deterministic signal in white Gaussian noise using an energy detector, and derived expressions for the correct detection and false alarm probabilities as a function of signal-to-noise ratio, an adjustable threshold level and the time-bandwidth product of the signal. The probability of a false alarm for a given threshold V_T is given by

$$P(FA) = 1 - Pr(V_T \leq \chi^2_{2TW}) \qquad (3)$$

where Pr is the area under the chi-square distribution curve with 2TW (time-bandwidth) degrees of freedom. For the same threshold level V_T, the probability of a correct detection is given by

$$P(D) = 1 - Pr(V_T/G \leq \chi^2_D) \tag{4}$$

where:
$$D = (2TW + E/N_o)^2/(2TW + 2E/N_o) \tag{5}$$

$$G = (2TW + 2E/N_o)/(2TW + E/N_o) \tag{6}$$

Pr is now the area under the noncentral chi-square distribution with a modified number of degrees of freedom, D, and a threshold divisor, G.

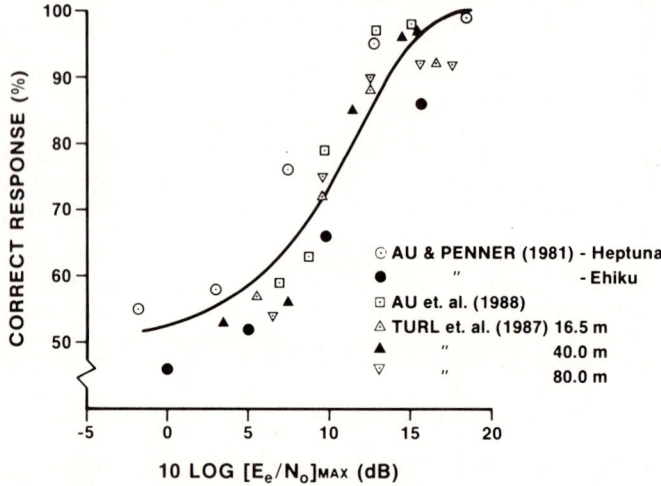

Fig. 9. Comparison between dolphin target detection in noise performance results and Urkowitz (1967) energy detection model for TW =11 (from Au, 1988).

These expressions derived by Urkowitz (1967) were applied to dolphin detection data by assuming an unbiased detector in determining the probability of a correct response P(C) from P(FA) and P(D) given in Eqs. (3) and (4). The calculation was done by first choosing desired values of P(FA) and 2TW and then determining V_T by an iterative procedure. Then with the iterated value of V_T, P(D) was calculated for different values of E/N_o. The procedure was continued for different 2TW degrees of freedom, until the values of P(C) were obtained as a function of E/N_o which best fitted the dolphin data. The performance data for <u>Tursiops</u> detecting targets in masking noise in three different studies are shown in Fig. 9 along with the results of Urkowitz energy detection model for 2TW = 22. Urkowitz's energy detection model agrees with the dolphins' data, further supporting the notion of the dolphin being an energy detector. Inserting an integration time of 264 μs into the TW product will result in a bandwidth of 42 kHz for the detector depicted in Fig. 8. Moore and Au (1983) measured a critical ratio of approximately 18 kHz at 120 kHz for Tursiops, which is in general agreement with the bandwidth for the energy detector model. The unbiased detector assumption used to derive P(C) is good for signal-to-noise conditions that correspond to performance at or above the 75% correct response threshold. Tursiops tend to be unbias for high signal-to-noise conditions (Au and Synder, 1980; Au and Penner, 1981)

COMPARISON WITH AN IDEAL RECEIVER

An ideal or optimal receiver is the best receiver in detecting a known signal in white Gaussian noise. Petersen et al. (1954) related the receiver-operating-characteristic (ROC) curves [P(D) versus P(FA)] to the signal-to-noise ratio at the receiver input required for detection of a signal in noise. They showed that the optimal receiver for the detection of a signal known exactly in white noise was a cross-correlator receiver, in which the input signal plus noise is correlated with a noise-free replica of the known signal. An equivalent receiver is a matched filter whose impulse response is the same as the waveform of the known signal reversed in time. Since the ideal receiver will detect a signal in noise better than any other receiver, the efficiency or effectiveness of any other receiver can be compared against an ideal receiver.

Au and Pawloski (1989) performed two experiments with an electronic phantom target to compare the target detection performance of an echolocating bottlenose dolphin with that of an ideal or optimal receiver. The first experiment was conducted to establish a more realistic method of estimating E_e/N_o at the dolphin's detection threshold. Two different types of echoes were used and the dolphin's threshold was determined by an up-down or staircase procedure. The first type of echoes consisted of two clicks, separated by 200 μs, which were replicas of each transmitted click. The amplitude of the echoes was directly proportional to the amplitude of the emitted clicks. With this echo, the $(E_e/N_o)_{max}$ at threshold was determined in a similar manner as for the results shown in Figs. 4-6. The second type of echoes consisted of a previously measured and digitized echolocation click from the animal which was stored in erasable programmable read-only memory (EPROM). The electronic target simulator was modified so that every time the dolphin emitted an echolocation signal, the EPROM was triggered to produce two pulses separated by 200 μs. The amplitude of the echo was fixed for each trial independent of the dolphin's signal level, resulting in a fixed E_e/N_o per trial, and an accurate estimate of E_e/N_o at threshold. The difference between $(E_e/N_o)_{max}$ and E_e/N_o was determined to be 2.9 dB, indicating that an accurate estimate of E_e/N_o can be obtained by subtracting 2.9 dB from $(E_e/N_o)_{max}$.

In the second experiment, Au and Pawloski (1989) obtained data that could be presented in an ROC format. The dolphin's response bias was manipulated varying the payoff matrix (number of pieces of fish reinforcement for correct responses). The payoff matrix was varied in terms of the ratio of correct detection to correct rejection in the following manner: 1:1, 1:4, 1:1, 4:1, 1:1, and 8:1. Six consecutive sessions were conducted at each payoff matrix, with the 1:1 payoff being the baseline. The results of the dolphin's target detection performance as its response bias was manipulated, are plotted in an ROC format in Fig. 10 for two different target strengths. The ideal isosensitivity curves associated with d' values of 2.2 and 1.6 for the strong and weak echoes, respectively, are included in Fig. 10. The detection sensitivity, d', represents the minimum value of E_e/N_o necessary to lead to the performance of an

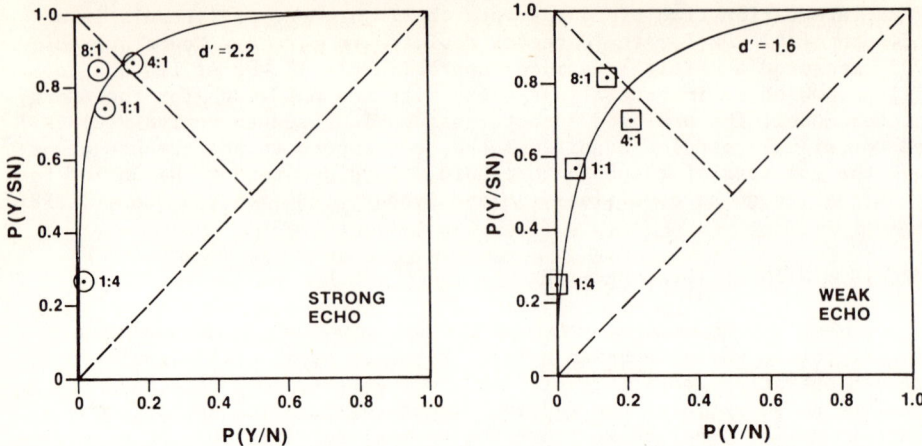

Fig. 10. Dolphin performance results plotted in an ROC format and iso-sensitivity curves that best matched the results. The ordinate is probability of detection, P(Y/SN) and the abscissa is probability of false alarm, P(Y/N) (from Au and Pawloski, 1989).

ideal receiver (Elliott, 1964). Each isosensitivity curve of Fig. 10 best matched the dolphin's performance with a minimum least-square error. From Fig. 10, the $(E_e/N_o)_{op}$ for an ideal detector to match the dolphin's performance can be determined and compared with the E_e/N_o for the dolphin. $(E_e/N_o)_{op}$ for an optimal receiver can be calculated from the definition of d' given in the equation

$$d' = \sqrt{2(E_e/N_o)_{op}} \qquad (7)$$

The echo energy-to-noise ratio in dB is

$$(E_e/N_o)_{op} = 10 \log(d'^2/2) \qquad (8)$$

Therefore, for an optimal receiver to approximate the performance of the dolphin, the following echo energy-to-noise ratios are needed

$$(E_e/N_o)_{op} = \begin{cases} 3.8 \text{ dB (strong echo)} \\ 1.1 \text{ dB (weak echo)} \end{cases}$$

The dolphin detection results were obtained with an E_e/N_o of 12.2 dB for the strong echo and 7.4 dB for the weak echo case. The difference in E_e/N_o between the dolphin and an optimal receiver can be expressed as

$$(E_e/N_o)_{dol} - (E_e/N_o)_{op} = \begin{cases} 8.4 \text{ dB (strong echo)} \\ 6.3 \text{ dB (weak echo)} \end{cases}$$

Averaging the differences for the strong and weak echoes, we conclude that an optimal receiver would outperform the dolphin by approximately 7.4 dB.

SUMMARY AND CONCLUSIONS

The sonar target detection sensitivity of *Tursiops truncatus* has been measured by determining: (a) The maximum detection range for two targets. (b) Target detection performance for a target at a fixed range in the presence of variable artificial masking noise. (c) Target detection performance for a variable sized target at a fixed range in the presence of artificial masking noise. The results of the various methods when considered in terms of E_e/N_o were very consistent with the detection threshold varying from 7.2 to 12.4 dB. The shape of the performance curves as a function of E_e/N_o was also similar except for the case in which the noise levels changed in 5-dB increments.

Target detection performance data suggest that *Tursiops* process sonar echoes like an energy detector with an integration time of approximately 264 μs. The data also suggest that dolphins may process short duration broadband signals in a different manner than long duration tonal signals. The integration time of 264 μs is smaller by a factor of 7.6 from the approximately 2 ms integration time for a 120 kHz tonal signal. Different processing mechanisms for short duration signals and long duration tonal signals have been suggested by Vel'min and Dubroskiy (1975; 1976) under the nomenclature of "active and passive hearing."

An accurate estimate for E_e/N_o at the detection threshold of a dolphin may be obtained by first calculating $(E_e/N_o)_{max}$ using the average of the largest energy flux density measured per trial, and subtracting a correction factor of 2.9 dB. An ideal or optimal receiver requires approximately 6 to 8 dB less energy to perform at the same level of accuracy as a *Tursiops truncatus* in detecting targets in noise.

REFERENCES

Albers, V. M., 1965, "Underwater Acoustic Handbook II", The Pennsylvania State University Press, University Park.
Au, W. W. L., Floyd, R. W., Penner, R. H., and Murchison, A. E., 1974, Measurement of Echolocation Signals of the Atlantic Bottlenose Dolphin, *Tursiops truncatus* Montagu, in Open Waters, J. Acoust. Soc. Am., 56: 1280-1290.
Au, W. W. L., and Snyder, K. J., 1980, Long-range Target Detection in Open Waters by an Echolocating Atlantic Bottlenose Dolphin (*Tursiops truncatus*), J. Acoustic. Soc. Am., 68: 1077-1084.
Au, W. W. L., 1980, Echolocation Signals of the Atlantic Bottlenose Dolphin *Tursiops truncatus*) in Open Waters. in: "Animal Sonar Systems", R. G. Busnel and J. F. Fish, eds., Plenum, New York, pp. 251-282.
Au, W. W. L. and Penner, R. H., 1981, Target Detection in Noise by Echolocating Atlantic Bottlenose Dolphins, J. Acoustic Soc. Am., 70, 687-693.
Au, W. W. L., 1988, Detection and Recognition Models of Dolphin Sonar Systems, in: "Animal Sonar: Processes and Performance", P. E. Nachtigall and P. W. B. Moore, ed., Plenum, New York, pp. 753-768.
Au, W. W. L., and Moore, P. W. B., 1984, Receiving Beam Patterns and

Directivity Indices of the Atlantic Bottlenose Dolphin <u>Tursiops truncatus</u>, <u>J. Acoust. Soc. Am.</u>, 75: 255-262.

Au, W. W. L., Moore, P. W. B., and Pawloski, D. A., 1988, Detection of Complex Echoes in Noise by an Echolocating Dolphin, <u>J. Acoust. Soc. Am.</u>, 83: 662-668.

Au, W. W. L., and Pawloski, D. A., 1989, A Comparison of Signal Detection Between an Echolocating Dolphin and An Optimal Receiver, <u>J. Comp. Physiol A</u>, 164: 451-458.

Elliott, P. B., 1964, Appendix 1-Tables of d', <u>in</u>: "Signal Detection and Recognition by Human Observers," J. Swet, ed., John Wiley, New York, pp. 651-684.

Green, D. M., and Swets, J. A., 1966, "Signal Detection and Psychophysics," Krieger, Huntington, N. Y..

Johnson, C. S., 1968a, Relation Between Absolute Threshold and Duration-of-Tone Pulses in the Bottlenosed Porpoise, <u>J. Acoust. Soc. Am.</u>, 43: 757-763.

Johnson, C. S., 1968b, Masked Tonal Thresholds in the Bottlenosed Porpoise, <u>J. Acoust. Soc. Am.</u>, 44: 965-967.

Moore, P. W. B., and Au, W. W. L., 1983, Critical Ratio and Bandwidth of the Atlantic Bottlenose Dolphin (<u>Tursiops truncatus</u>), <u>J. Acoust. Soc. Am.</u>, 74: (Suppl. 1): S73.

Moore, P. W. B., Hall, R. W., Friedl, W. A., and Nachtigall, P. E., 1984, The Critical Interval in Dolphin Echolocation: What is It?, <u>J. Acoust. Soc. Am.</u>, 76: 314-317.

Murchison, A. E., 1980, Maximum Detection Range and Range Resolution in Echolocating Bottlenose Porpoise (<u>Tursiops truncatus</u>), <u>in</u>: "Animal Sonar Systems," R. G. Busnel and J. F. Fish, eds., Plenum, New York, pp. 43-70.

Nachtigall, P. E., 1980, Odontocete Echolocation Performance on Object Size, Shape and Material, <u>in</u>: "Animal Sonar Systems", R. G. Busnel and J. F. Fish, eds., Plenum, New York, pp. 71-95.

Petersen, W. W., Birdsall, T. G., and Fox, W. C., 1954, The Theory of Signal Detectability, <u>Trans IRE, PGIT</u>, 4: 171-212.

Turl, C. W., Penner, R. H., and Au, W. W. L., 1987, Comparison of Target Detection Capabilities of the Beluga and Bottlenose Dolphin, <u>J. Acoust. Soc. Am.</u>, 82: 1487-1491.

Urick, R. J., 1984, "Ambient Noise in the Sea", U. S. Printing Office, Washington D. C.

Urkowitz, H., 1967, Energy Detection of Unknown Deterministic Signals, <u>Proc. I.E.E.E.</u>, 55: 523-531.

Vel'min, V. A., and Dubrovskiy, N. A., 1975, On the Analysis of Pulsed Sounds by Dolphins, <u>Dokl. Akad. Nauk. SSSR</u>, 225: 470-473.

Vel'min, V. A., and Dubrosvkiy, N. A., 1976, The Critical Interval of Active Hearing in Dolphins, <u>Sov. Phys. Acoust.</u>, 2: 351-352.

PRELIMINARY NOTES ON BEHAVIOUR OF A BLINDFOLDED FREE-SWIMMING DOLPHIN

PERFORMING A TARGET ECHOLOCATION TASK IN A POOL

Massimo Azzali and Gabriele Buracchi*

Research National Council, Institute of Marine Fishery Research
60100 Ancona, Italy; *Via Cigoli 2, 50042 Firenze, Italy

METHODS

A study has been carried-out since 1987 by the Research National Council (C.N.R.) in Ancona in collaboration with Adriatic Sea World (A.S.W.) in Riccione, to find-out how a blind-folded free-swimming Tursiops truncatus (Montagu) uses echolocation clicks for detecting a target, approaching it and making contact with its rostrum.

After a period of training (1987), the dolphin performed 72 sonar searches (February/ May 1988), that are the objects of discussion.

The experiments have been carried-out using a B&K type 8103 hydrophone arranged 5 cm above the target, an automatized underwater video camera, which an operator through a consolle can aim at the moving dolphin, equipments for recording and analysing both acoustic and video signals, properly synchronized (Figure 1). The target was a copper sphere (30 mm diameter), with target strength of 36 dB ± 0.5 at 120 KHz. It is the standard target used to calibrate the echo-sounders specialized to estimate size and weight of small pelagic fish usual prey of Tursiops.

The pool, the position of the target, the camera during the experiments are illustrated in Figure 2. The experiment was divided in two sessions. A session consisted of two sequences of six 3-trials blocks. In a sequence the target had fixed bearing and horizontal range but six different depths (see Table 1). For each depth a block of 3-trials was carried-out, but there were no evident of dependencies in the depths series. Then the bearing and the horizontal range were changed and a second sequence of trials repeated.

The experiment was conducted using an adult female Tursiops truncatus named Candy, but in the presence of two other adult dolphins. Initially, the three dolphins maintained the station by facing away from the target in front of the trainer. When Candy was blind-folded with eye cups, she turned and performed a sonar search of the target, while the other two animals mantained the station. In the correct trials, that were reinforced with fish, the dolphin acquired the target, approached it, made contact with its rostrum and at a recall signal of the trainer returned to the station. Incorrect trails were not reinforced and did not delay the commencement of the next trail.

Sensory Abilities of Cetaceans
Edited by J. Thomas and R. Kastelein
Plenum Press, New York, 1990

SURFACE UNIT

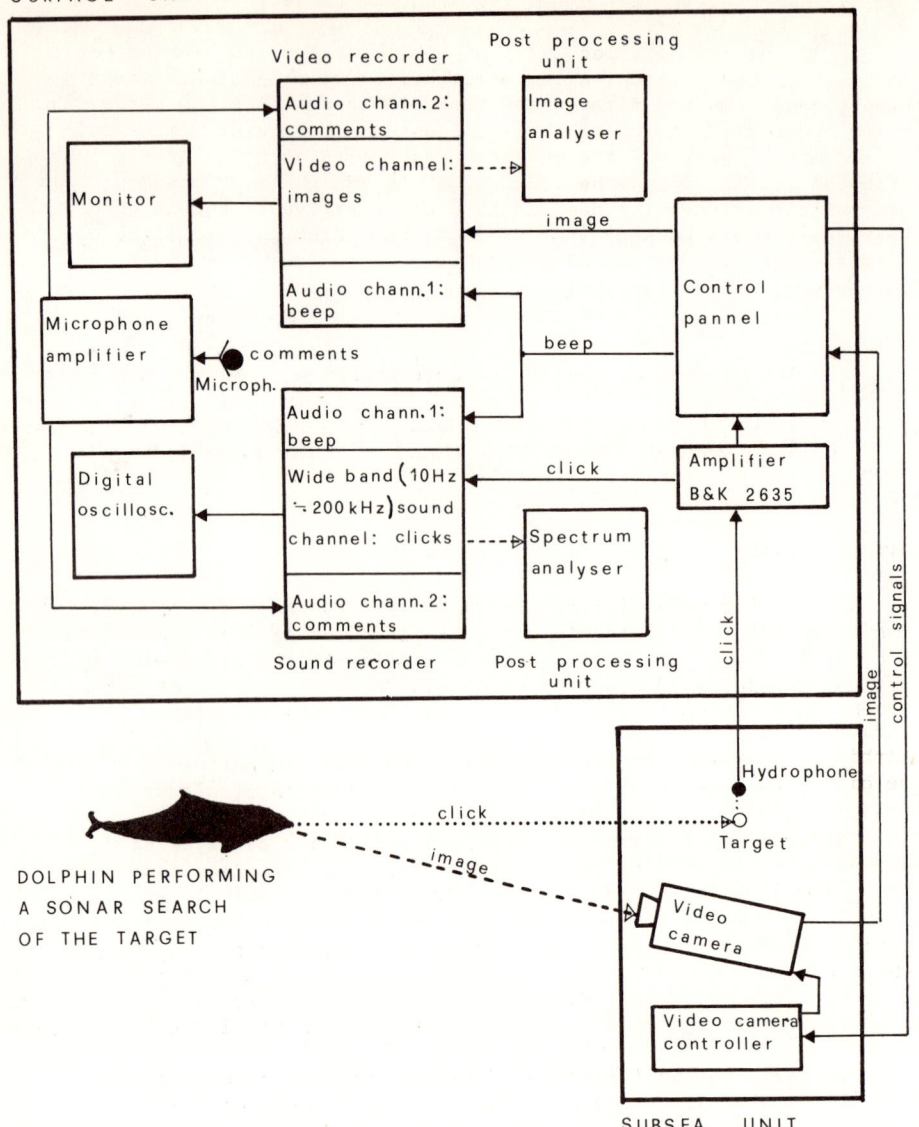

Fig. 1. Block diagram of tracking measuring and processing system. The outputs of the video camera and of the hydrophone are continuously recorded. A special circuit produces an audible signal "beep", whenever a train of clicks hits the hydrophone; it insures the operator that the dolphin is echolocating the target. The "beep" together with the comments of the esperimenter are used to synchronize the video and sound recorders.

SUBJECT OF DISCUSSION

Data collected in the experiment were analyzed to investigate the acoustical parameters of the emitted signals (click waveforms, spectra, click rate) and the tactical acoustical behaviour of the dolphin (method used in searching, tracking and reaching the target).

It is the second aspect of the problem that is the subject of discussion. The tactical acoustic behaviour of the dolphin comprises three phases. In the first phase the dolphin searches the target until it was located. This phase interests a restricted area of the pool up to 5 m around the station, the searching area of Figure 2. In the second phase the dolphin approaches the target, transducing acoustic informations in tracking decisions so as to arrive in the neighbouring of the target, the attacking area of Figure 2. From this area the dolphin attacks the target moving along spiral-like trajectories. However, in this last phase the behaviour of the dolphin strongly depends on the depth of target. This paper focuses mainly on the second phase.

FIRST PHASE: SEARCH FOR THE TARGET

The first problem that the dolphin has to solve is searching for the target. This operation that is performed around the release station, comprises three basic functions.

Target Detection (Figure 3A)

The dolphin must have the capability of scanning the entire tank volume and the ability to extract the "candidate targets" from noise and from reverberation of surface, bottom, sides and corners of the pool. It implies choosing locally appropriate thresholds. Figure 3A illustrates the attitude of the dolphin while he is detecting (and/or selecting) the candidate targets (the target). It is interesting to note that the dolphin in scanning the tank volume, behaved as though he had perceived the different probabilities of changing in the target bearing (a change every 18 trials) and in the target depth (a change every trial). In fact, the bearing angle at which the dolphin pointed the rostrum was dependent on the one immediately preceeding, whereas the entire azimuth sector was scanned in each successive trial.

Target Recognition

Upon completing the search of the acquisition space, the dolphin must decide which detection is most target-like and should be echolocated. Successful recognition of the target involves the unique ability of the dolphin to perform various tests on the detected signals to estimate shape, material, dimensions of the objects and to discriminate the true detection of the target against false detections of the other objects, such as the hydrophone, the cable, the video camera.

Table 1. Position of the Target Referred to a Cylindrical Coordinate System Centered at the Station

Coordinate	Target on the left side	Target on the right side
bearing angle	+ 19 degree	-42 degree
horizontal range	20 m	15.6 m
vertical range or depth	very close to bottom: in mid-water: very close to surface	4.8 m 4 m; 3 m; 2 m; 1 m 0.2 m

219

Target Location (Figure 3B)

In Figure 3B, taken two seconds after Figure 3A, the dolphin locates the target and estimates the "rostrum-to-target" line of sight vector. The "rostrum-to-target" position vector is perceived by the dolphin as the "error" that he must minimize, during the tracking process, to reach the target. Let $\underline{T}$ and $\underline{S}$ be the "station-to-target" and the "station-to-rostrum" position vectors respectively, expressed with reference to a fixed system of coordinate (Figure 4), then the "rostrum-to-target" vector, at time t_0, is the difference between the vector $\underline{T}$ (the "input" in the dolphin system) and the vector $\underline{S}$ (t_0) (the "output" in the dolphin system):

$$\underline{R}\,(t_0) = \underline{T} - \underline{S}\,(t_0)$$

However, since the rostrum is pointed to the target relative to dolphin body, it is most appropriate expressing $\underline{R}\,(t_0)$ in the dolphin body coordinate system as illustrated in the Figure 5. The form of the pointing vector (i.e. the rostrum-to-target vector) in the intrinsic coordinate system is given by

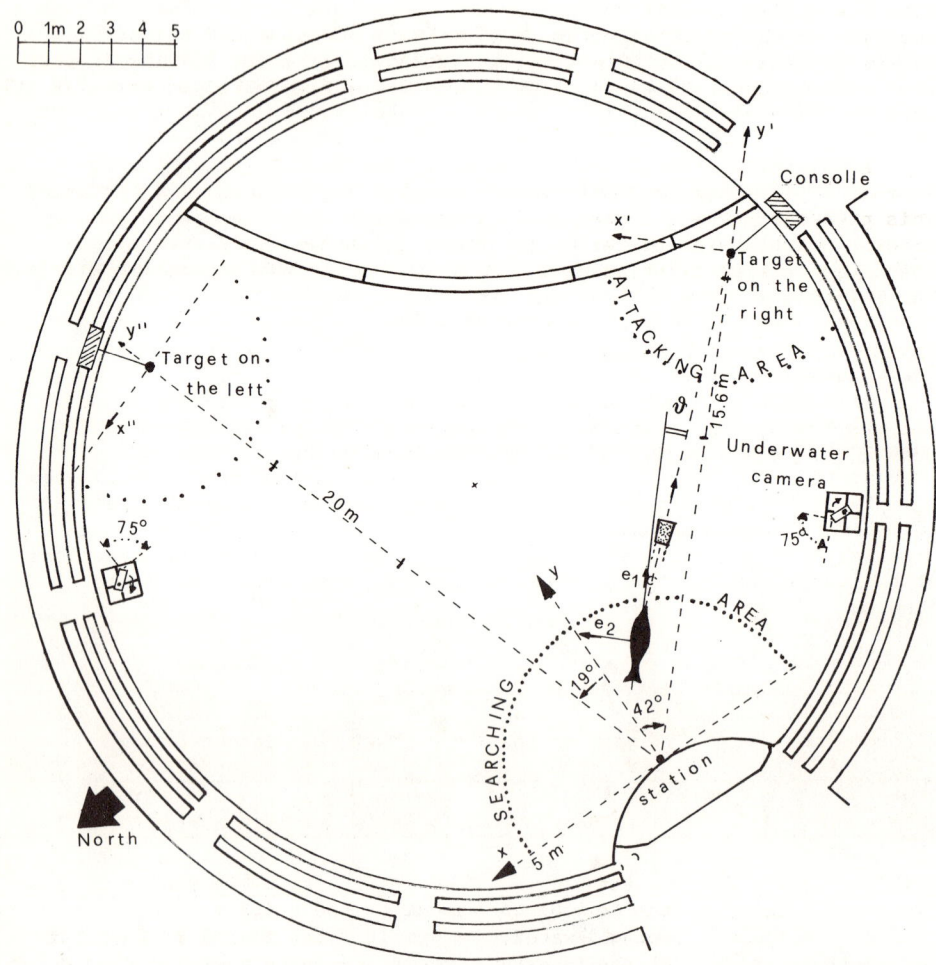

Fig. 2. Experimental set-up showing the two position of the target.

$$\underline{R}^b(t_0) = R(t_0) \cdot \underline{p}(t_0)$$

where:
$R(t)$ is the magnitude of the distance between the rostrum and the target
$\underline{p}(t)$ is the unit pointing vector; it measures the bearing θ and the azimuth ψ angular displacement of the rostrum, referred to the three primary axes of the dolphin body ($\underline{e}_1$, $\underline{e}_2$, $\underline{e}_3$):
$$\underline{p}(t) = \cos\theta * \cos\psi * \underline{e}_1 + \cos\psi * \sin\theta * \underline{e}_2 + \sin\psi * \underline{e}_3$$

The dynamic echolocation (i.e. echolocation closely tied to locomotion) requires that the dolphin is able to sense non only $R(t)$, through his sonar as in the static echolocation, but also the position, rotation and movement of his own body and head (i.e. the unit pointing vector $\underline{p}(t)$) through the vestibular canals and the appropriate proprioceptors located in the muscles, joints and tendons. All these cues, integrated by the brain, allow the dolphin to make inference on his body position referred to the target position and to the enviroment, at the time of the target location.

SECOND PHASE: THE TRACKING PROCESS

Once target location has occurred, the dolphin begins to transduce acoustic information in tracking decisions so as to guide its own body (the controlled system) to the target. This task is performed through a sequence of target echolocations, each being accomplished through a series of decisions (body movements, rotations of head) similar in that each is intended to minimize some functions of the perceived error ($\underline{R}(t)$ vector).

A characteristic of the tracking is that the dolphin can program and reprogram his "computations", while the tracking process is in progress. This programming is a consequence of decisions which depend on the tracking-displays characteristics in the dolphin's brain and on the adaptative characteristics of the dolphin. In general, tracking-displays can be categorized as:

Compensatory as in the Figure 6

The dolphin "sees'' only the magnitude of the difference $\underline{R}(t)$ between the system output $\underline{S}(t)$ and input $\underline{T}$. He has no way of determining the actual position of his own body/rostrum and of the

Fig. 3. (A) a typical attitude of Candy while she is detecting (and/or selecting) the "candidate targets" (the target). The other two dolphins, facing away the target in front of the trainer, are visible. (B) the dolphin locates the target (photo taken 2 seconds later than photo (A)).

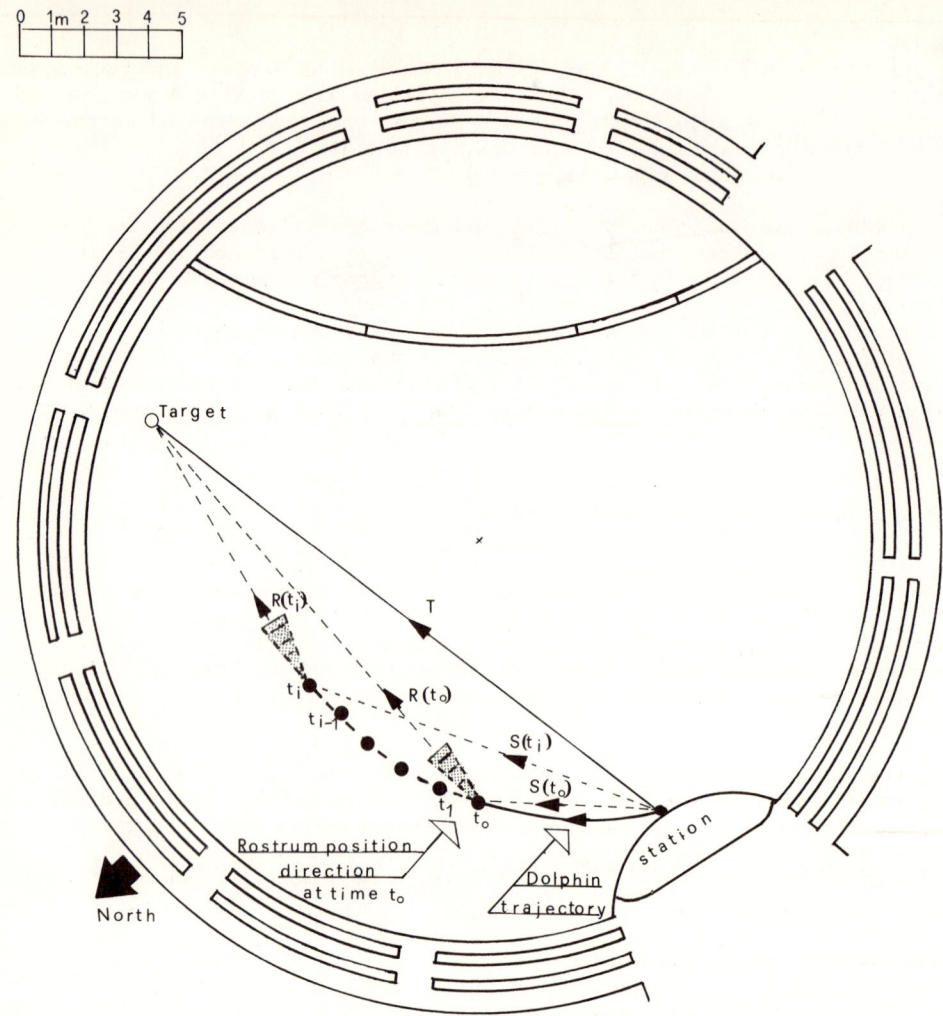

Fig. 4. A dolphin path described by the function: $\underline{R}(t) = \underline{T} - \underline{S}(t)$. The pointing vector $\underline{R}(t)$ is perceived by the dolphin as the "error" that must be minimized to reach the target.

target. The only involved sense in a compensatory tracking is echolocation. A method of tracking, that uses a compensatory display is the Track-While-Scan (T.W.S.), described here and largely used in artificial sonar and radar systems. The successive decisions in the T.W.S. are only governed by the criteria of minimizing the magnitude of the distance between the tracker and the target.

Pursuit as in Figure 7

The dolphin "sees" his own body together with rostrum position vector and the target position vector, both as a part of an environment within which the task is to be performed. It is evident that a pursuit tracking needs additional sensing of information outside the echolocation. The tracking methods, that use a pursuit display as the triangulation method described here, are characterized by significant changes of the criteria-error-function, in accord with particular situations and adaptations. The experiment indicated different types of

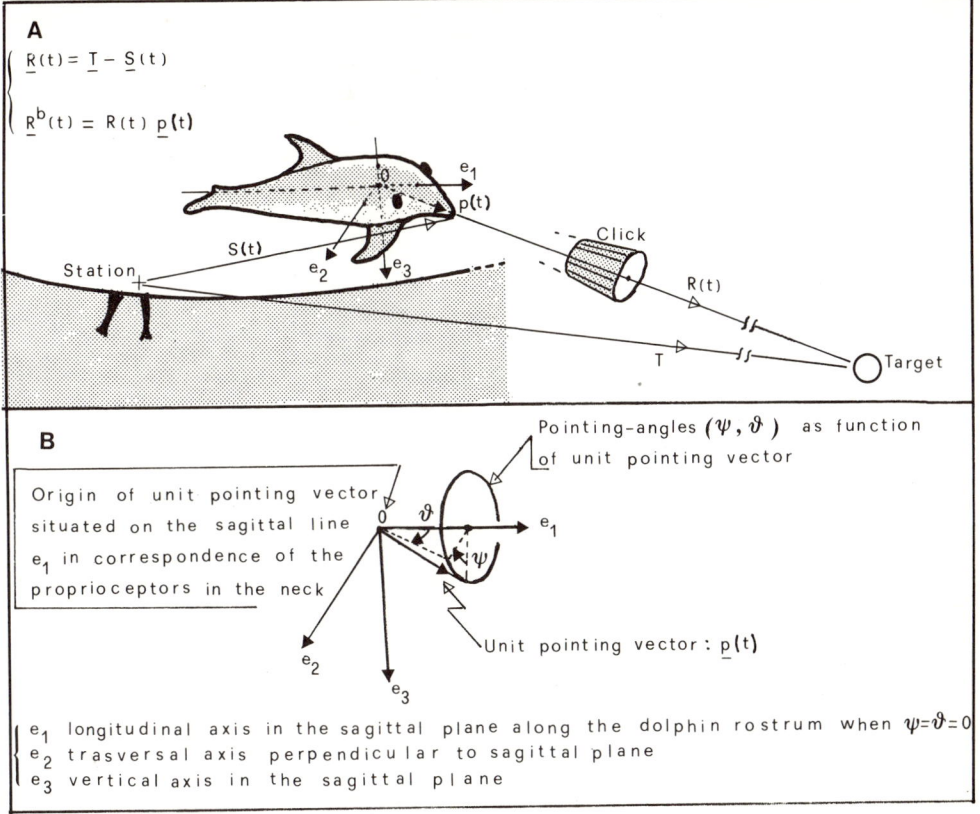

Fig. 5. (A) the pointing vector (or "error") formulated in a fixed coordinate system $\underline{R}$ (t) and in the dolphin body coordinate system $\underline{R}^b$ (t). (B) the unit pointing vector.

adaptation in the dolphin (such as adaptation to the target characteristics, to changes in procedures) and a capacity of developing skills based on past experience (learning).

FROM T.W.S. TO TRIANGULATION

At the beginning of the experiment, the dolphin approached the target using the method called Track-While-Scan (T.W.S.). The dolphin tried to retain the target in track (adjusting the position of the body and head, the click repetition), until the range R (t) (i.e. the perceived error) went to zero. In Table 2 is summarized the T.W.S. method, very well known in sonar/radar compensatory tracking systems. The number of clicks that illuminated the target was never less than 40. In some trials the dolphin seemed to drop the track after a series of K consecutive failures to update the target (K = drop-track threshold, unknown). Then the tactical behaviour of dolphin slowly evolved until in the last experiments (February/May 1988), he adopted an unusual strategy of target approaching. The target was ensonified by a number of clicks much lower than in the first trials (two or three clicks); the approaching path was not linear, as in the first experiments, but a poligonal-like curve; in the neighbouring of the target (the attacking area) the path assumed an arc of spiral shape. Table 3 and Figure 8 suggest a mathematical explanation of this behaviour based on successive triangulation method. In Figures 9, 10 and in Tables 4, 5 the results of

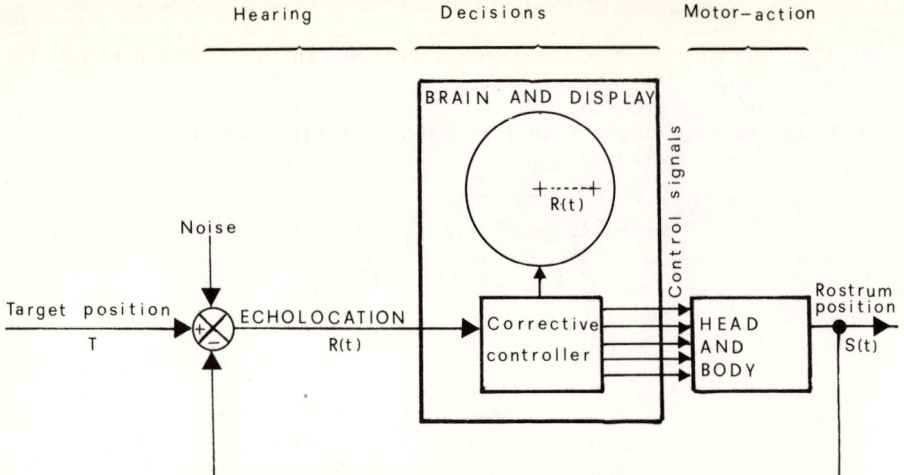

Fig. 6. Simple compensatory tracking system with display. The controller sees only the magnitude of the difference between the system output and input, namely the error R (t). The display representation is mono-dimensional.

a simulation of this method are illustrated. They show that it is possible to approach the target through one or two triangulations (two or three successul echolocations). Figure 11 shows some typical attitudes

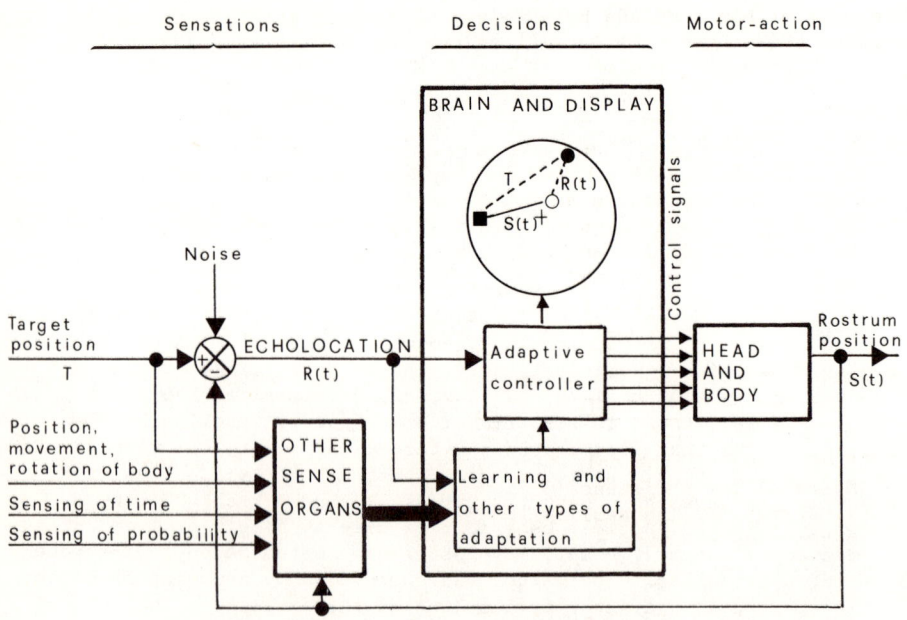

Fig. 7. Simple pursuit tracking system with display. The controller sees its own body with rostrum positions and the target positions, as well as their difference R (t). The display representation is in three-dimensional space.

Table 2. Track-While-Scan Method

a	At the time (t_0) of the first echolocation, the dolphin measures the initial range $R(t_0)$, the bearing $\theta(t_0)$ and the azimuth $\psi(t_0)$ angles.
b	Dolphin orients with the rostrum towards the target.
c	Dolphin swims along a line-like path, emitting clicks at time intervals Δt. The closing path is described by: $$\underline{P}(t) = \underline{S}(t_0) + \frac{(t-t_0)}{(t_f - t_0)} * (\underline{T} - \underline{S}(t_0))$$ where t_f is the time of impact
d	Range decrement per clicks is: $\Delta R = V*\Delta t$, so that the range $R(t_j)$, corresponding at the click j is: $$R(t_j) = R(t_0) - j*V*\Delta t; j=1,2,\ldots N; V=\text{speed}$$
e	Expected number of clicks emitted by the dolphin between the time of the first echolocation t_0 and the time of impact t_f is: $$N = ((1/\Delta t)*R(t_0))/V$$ For ex. if $(1/\Delta t) \geq 10s^{-1}$; $R(t_0)=20m$; $V \leq 5m/s$; then $N \geq 40$
f	Expected number of clicks that illuminate the target is: $$M = N*p(R); \quad p(R) = \text{average probability of updating the target in function of range R.}$$

of the dolphin in the tracking phase. The triangulation method assumes that:
(1) the target is "quasi-stationary" relatively to the dolphin.
(2) The dolphin has a precise perception of the position and angular movements of his body and head, viewed as a part of the environment, that includes the target. This implies the use of other sense organs in

Table 3. Triangulation Method

a*	At the time (t_0) of the first echolocation, the dolphin measures the initial range $R(t_0)$, the bearing $\theta(t_0)$ and azimuth $\psi(t_0)$ angles. The bearing is stored θ_0.
b*	The dolphin turns through a free angle $\Delta_1 < \theta(t_0)$ left (right), if the measured bearing angle is left (right).
c*	The dolphin swims along a line for a distance D_1 until he echolocates again the target at the same stored bearing angle θ_0. If $R(t_1)$ is the second measured range, at the time t_1 then $$\frac{D_1}{\sin \Delta_1} = \frac{R(t_0)}{\sin \theta_0} = \frac{R(t_1)}{\sin (\theta_0 - \Delta_1)}$$
d*	At the time t_1, the dolphin turns again through a free angle $\Delta_2 < \theta_0$ and swims for a distance D_2, until he echolocates the target at the stored bearing angle θ_0. If $R(t_2)$ is the third measured range at the time t_2 then: $$\frac{D_2}{\sin \Delta_2} = \frac{R(t_1)}{\sin \theta_0} = \frac{R(t_2)}{\sin (\theta_0 - \Delta_2)}$$
e*	The dolphin, arrived in an area around the target, attacks it using a large variety of strategies.

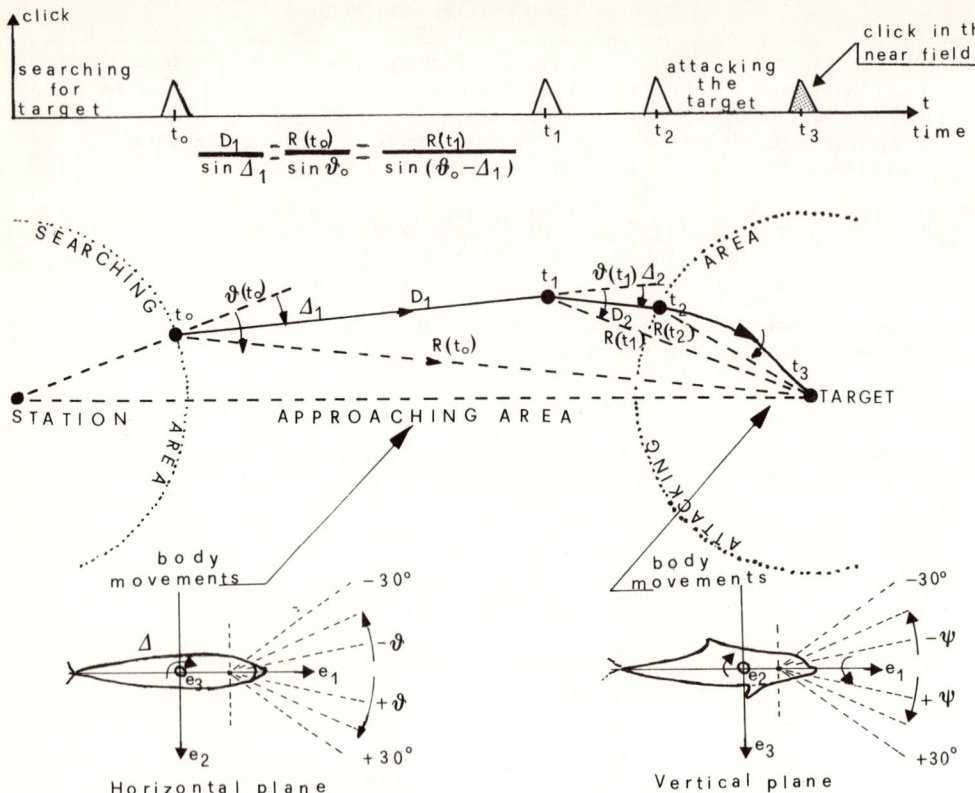

Fig. 8. Operational illustration of the triangulation method (Table 3 describes the relative algorithm). The triangulation is a "precognitive" method of tracking. The dolphin, choosing freely the turn angle Δ_i, can guess the future distance to be travelled (D_{i+1}) and the future range ($R(t_{i+1})$).

addition to echolocation.
(3) The dolphin has a versatile and complex mental display mechanism (pursuit tracking display). It allows the dolphin to have a "his body position image" that includes the target as well as the environment (see Figure 7).

The triangulation method needs a brain activity higher than the T.W.S. method, but offers many advantages.
(a) It is less sensitive to errors and it is stable. Therefore, the drop-track events are less frequent.
(b) The dolphin can program and reprogram the trajectory of approaching the target, while the tracking process is in progress.
(c) The triangulation method, differently from T.W.S. method, guides the dolphin smoothly to an area around the target, from where the dolphin can attack the target using a large variety of strategies.

ATTACKING STRATEGIES-THIRD PHASE

Arrived in the attacking area, the dolphin changes his tactical behaviour. He moves and rotates the body along spiral-like trajectories. Figures 12 and 13 illustrate typical behaviors of the dolphin, attacking the target. The trajectories depend strongly on the depth of the target and on the rostrum-to-target vector at the moment of last echolocation.

Table 4. Summary of Dolphin Orientation During Triangulation (see Fig. 9)

Number of triangulation	2 triangulations on the right			2 triangulations on the left			2 triangulations on the right		
Time (t_i) of echolocation	t_0	t_1	t_2	t_0	t_1	t_2	t_0	t_1	t_2
$R(t_i)$ in m	17.0	10.4	4.3	16.0	10.7	2.4	15.5	7.4	3.5
$\theta(t_i)$ in deg.	25	25		15	15		28	28	
Δ_i in deg.	10	15		5	10		15	15	
D_i in m	7.0	6.3		5.4	7.2		9.5	4.1	
Remarks	The dolphin arrives in the attack area with 2 triangulations (3 echolocations)								

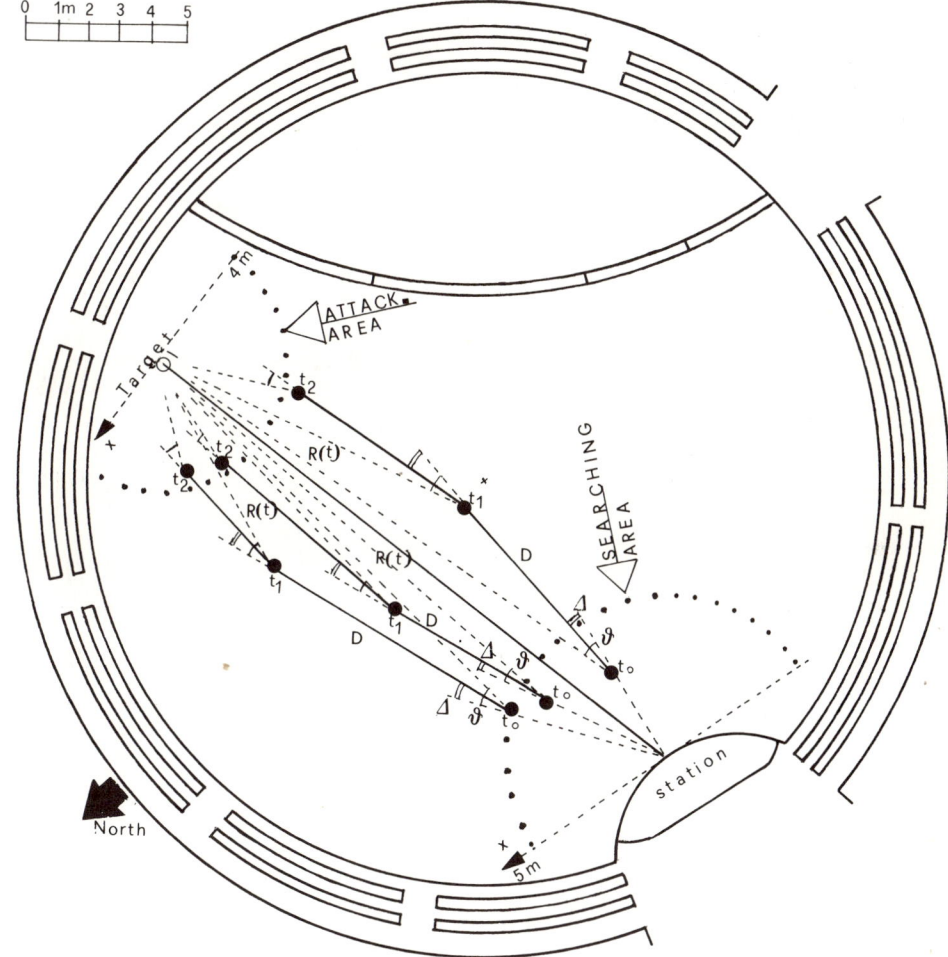

Fig. 9. Results of a simulation of the "triangulation method" (target on the left). They show that in general 2 triangulations are necessary to approach the target (confirmed by trials).

Table 5. Summary of Dolphin Orientation During Triangulation (see Fig. 10)

Number of triangulation	2 triangulations on the right			1 triangulations on the left		2 triangulations on the right		
Time (t_i) of echolocation	t_0	t_1	t_2	t_0	t_1	t_0	t_1	t_2
$R(t_i)$ in m	12.5	6.4	3.3	11.7	4.4	11	5.9	3.1
$\theta(t_i)$ in deg.	20	20		16		31	31	
Δ_i in deg.	10	10		10		15	15	
D_i in m	6.3	3.2		7.4		5.5	2.9	
Remarks	The dolphin arrives in the attack area with 2 or 1 triangulations (3 or 2 echolocations)							

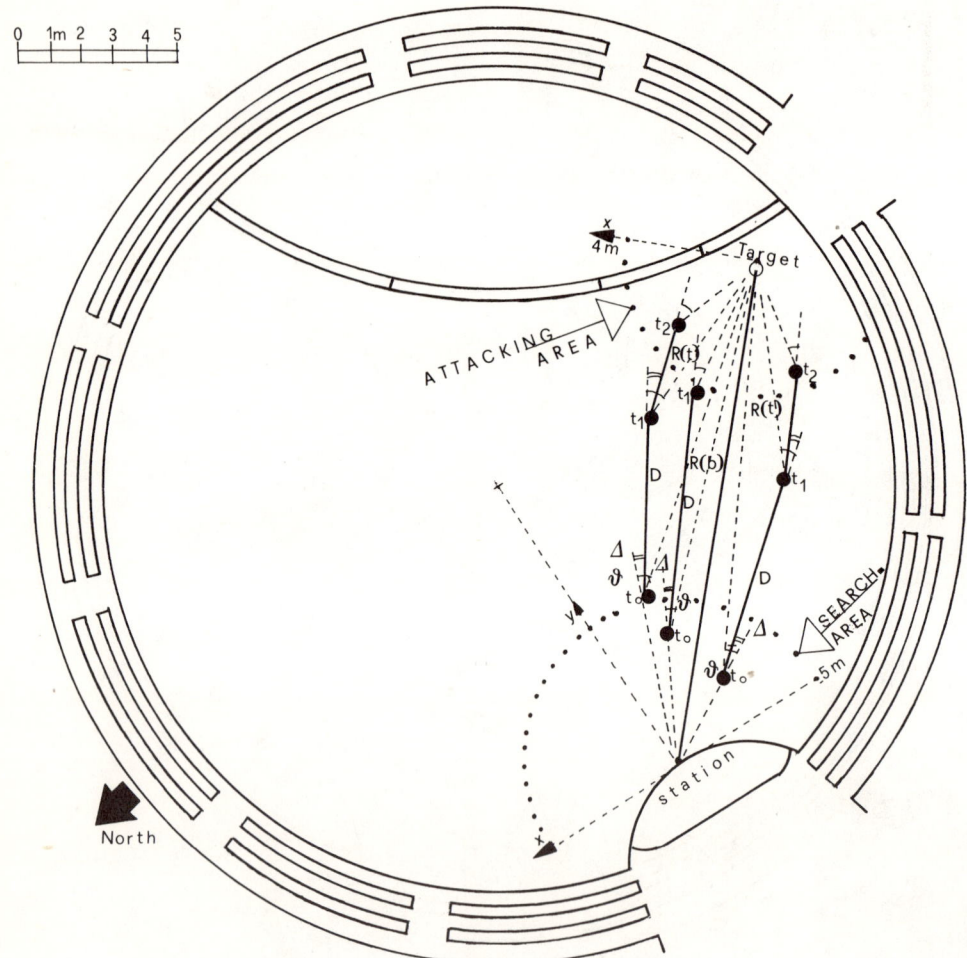

Fig. 10. Results of a simulation of the triangulation method (target on the right). They show that sometimes it is possible to approach the target through 1 triangulation (confirmed by trials).

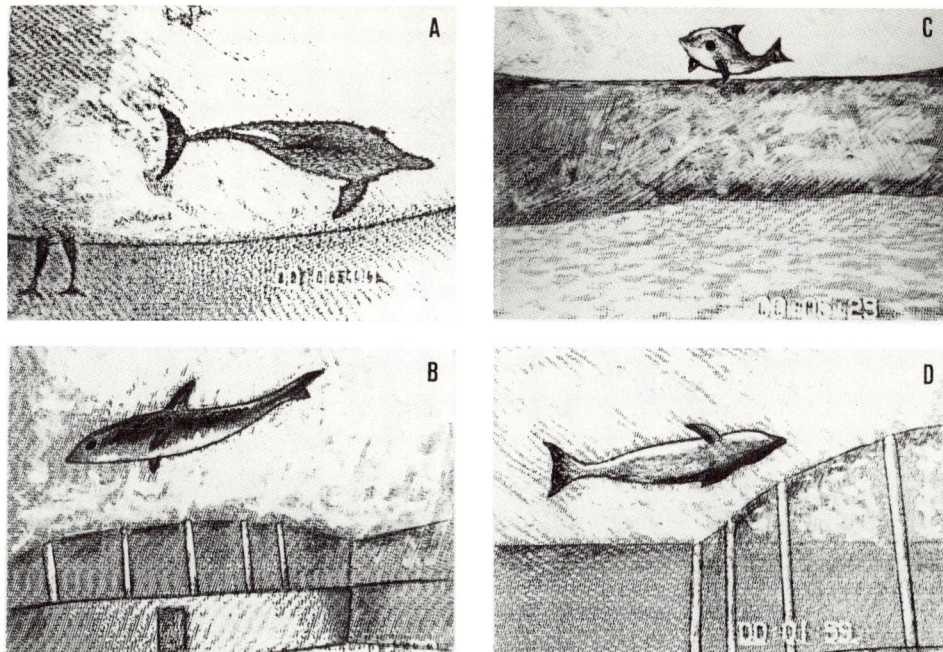

Fig. 11. Typical attitudes of the dolphin in the tracking phase. A, B: the target is on the left; C, D: the target is on the right.

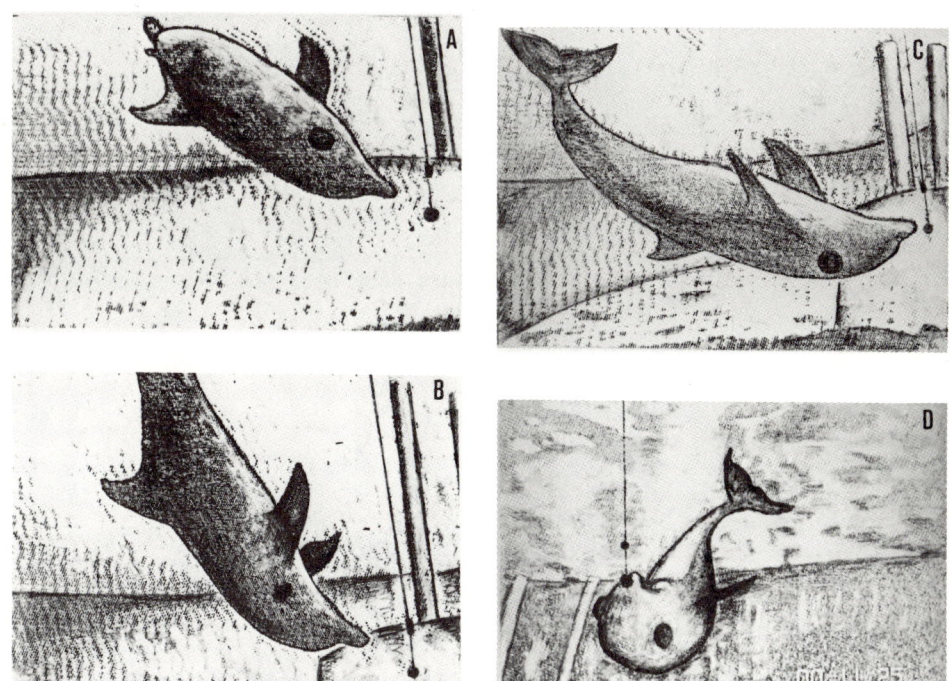

Fig. 12. The photos illustrate some typical attitudes of the dolphin in the final phase of tracking, when he echolocates the target in the near field.

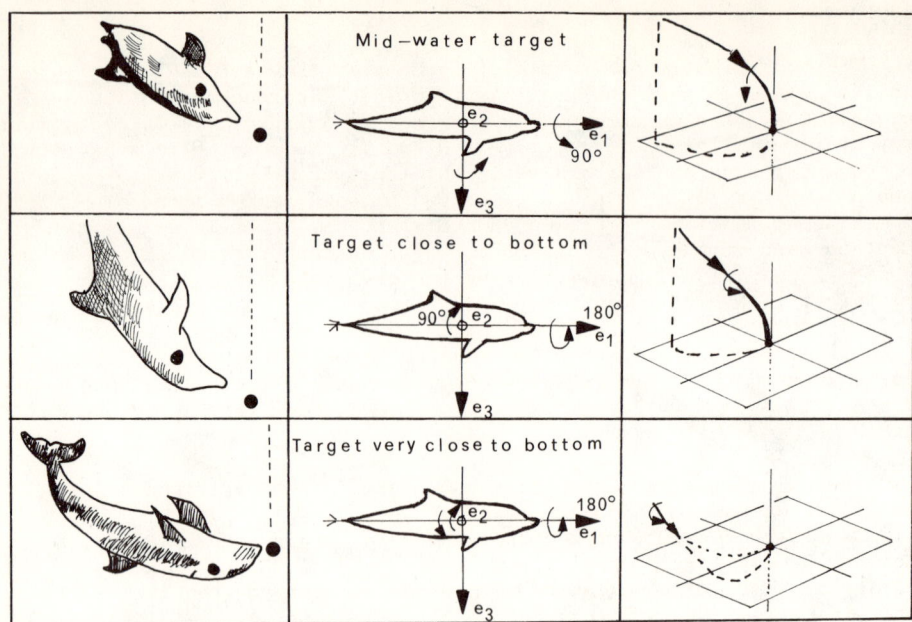

Fig. 13. Schematic representation of the dolphin attacking the target, in three tipical situations. If the target is in mid-water, the dolphin attacks it from up-to-down, rotating the body up to 90° around e_1 axis. If the target is close to bottom, the dolphin attacks it yet from up-to-down but rotating the body up to 180° around e_1 axis and up to 90° around e_3 axis. If the target is very close to bottom, the dolphin prefers to attack it from down-to-up, with the body completely reversed.

Roughly one can distinguish three basic situations.

Target close to surface. The dolphin attacks the target slightly from down-to-up. The dolphin, when arrives very near the target, echolocates it just before making contact with its rostrum (in the near field).

Mid-water target. The dolphin attacks the target from up-to-down, travelling along a spiral curve. Before making contact with its rostrum, the dolphin echolocates the target (see Figures 12 A, B and 13).

Target very close to bottom. The dolphin makes a nose-dive against the target, rotating the body so that he can echolocate and hit the target with the rostrum from down-to-up (see Figures 12 C, D and 13).

It is interesting to note that the dolphin always echolocates the target immediately before making contact with its rostrum and this last click, in the near-field, has structural characteristics very different from the clicks emitted in the far-field, during the phases of searching and approaching the target.

CONCLUSIONS

The study, based on 72 trials, focuses on the tactical behaviour of a blind-folded dolphin that searches, locates, approaches and attacks the target. The dolphin improved his tactical behaviour performances during the experiment. In the training period and in the first trials he seemed to concentrate on the "error" as it is displayed in a T.W.S. (or compensatory) tracking system. Then the dolphin seemed to become capable of making appropriate inference on the position, rotation of his own body/head and on the position and acoustic characteristics of the target. He used these cues and, probably, some stored information, to approach the target in a very unusual way. It seemed that the dolphin was tracking "precognitively" and acting as a "pursuit tracking system". It is tried to interpretate those results assuming that the dolphin after having learned the task, is able to approach the target using successive triangulations. This method offers many advantages respect to the T.W.S. method.

The hypothesis suggested here requires further verification and, if confirmed, opens many problems as:
(1) are the acoustical strategies related to single individuals and/or to individual temporary dispositions?
(2) How much do characteristics of the environment/the captivity condition affect the tactical behaviour of a dolphin?
(3) How will the dolphin react to a moving (or live) target?

AKNOWLEDGMENTS

The participation of the Adriatic Sea World in the Research National Council program has provided the circumstances for developing the work. In particular, the authors would like to thank Dott. Cariglia G. and Dott. Stanziani L. of the A.S.W., who trained the dolphins and have been very helpful in many ways, Valsecchi E., student of Pavia University, and Dott. Tangerini P. for their assistance in the experiments.

REFERENCES

Bar-Shalom, Y., 1978, Tracking methods in a multitarget environment,
    IEEE Trans Automat. Contr., Vol.AC-23, pp. 618-626.
Ferrel, J. L., 1981, Retention probability in a Track-While-Scan radar,
    IEEE Trans.on Aerospace and Electronic Systems, AES-17, pp. 134-144.
Marc, Megel, 1981, Three bearing method for passive systems with unknown deterministic bias,
    IEEE Trans on Aerospace and Electronic Systems, AES-17, pp. 814-818.
Au, W. W. L. and Moore, P. W. B., 1986, Echolocation transmitting beam of the Atlantic bottlenose dolphin,
    J.Acoust.Soc.Am. 80(2), pp. 689-691.

ON THE TWO AUDITORY SUBSYSTEMS IN DOLPHINS

Nikolai A. Dubrovskiy

N. N. Andreyev Acoustical Institute
Academy of Sciences of the USSR
Moscow, 117036 USSR

INTRODUCTION

   The task of the auditory system as a biological analyzer is to determine the sound source characteristics. The auditory system in echolocating animals and, in particular, in dolphins solves that task both in a passive regime, when the animals are interested in objects which produce sounds (I will call them ambient sounds) and in an active regime, when the objects are ensonified by a dolphin's outgoing (echolocation) signals, thus becoming the echo sources.

   The auditory system in these two regimes essentially functions differently. To begin with, ambient sounds and echo-signals differ considerably in amount of aprior information about its source. In the first case, the animal is not aware of bearing or distance to the source, sound intensity, time of arrival and temporal or spectral characteristics. Consequently, for effective reception of ambient sounds the auditory system must be "panoramic", that is, it must be ready to receive signals from any possible bearings, distances, at any intensities, times of arrival and temporal or spectral characteristics. In the second case, bearing of the echo's source is always known, for it practically always coincides with the direction of echolocation pulse. The distance to the source, echo intensity and the time of arrival are not known only at the very beginning of echolocation. When the first echoes are received, these parameters become known. The frequency range of echo coincides, as a rule, with the frequency range of the echolocation click and then, also is known by the animal. Thus, the main uncertainty in the echo characteristics lies in its fine temporal and spectral structure.

   Because of the discussed differences in the auditory system functioning in active and passive regimes are very important, it should be natural to assume, that in the process of dolphins evolution (and, possibly, of other echolocating animals) there could be developed two subsystems of the auditory system with specific function, which we called passive and active hearing (Bel'kovich and Dubrovskiy, 1976; Vel'min and Dubrovskiy, 1975).

For echolocating cetaceans this assumption proves to be true indeed since echolocation has appeared in the process of secondary adaptation to life in water of their terrestrial predecessors, which possessed well-developed passive hearing.

Different functional conditions of passive and active hearing were supposed to lead to a number of important differences in its characteristics. The necessity to receive a signal from the unknown direction demands that passive hearing be omni-directional. Since the directional characteristics mainly is attributed to the ratio of a wavelength, λ, to the interaural base, d, (or, in a more general case, to the typical head or body dimension of the animal), then this demand may be satisfied for the frequency range with the upper limit, $f_{max} = C/\lambda_{min}$, (C is the sound velocity, λ_{min} is the minimum wavelength) being defined from the condition $\lambda_{min} \approx d$. Because the base d for the adult bottlenose dolphin is approximately equal to 15-20 cm, then f_{max} is 7.5-10.0 kHz.

The fact that passive hearing has high sensitivity in the low frequency band enables dolphins to get the necessary detection range and the recognition of sources of ambient sounds since sound attenuation in water at low frequencies is small. The most important sound sources for dolphins are low frequency sounds from fish; this fact also predetermines passive hearing to be sensitive to low- frequency signals.

On the contrary, for echolocation analysis the high directivity of a radiation field and reception is necessary, that may be obtained by increase of radiation frequency. To have a forward directed radiation specific dimensions of the dolphin's radiator, defining its directivity, will coincide in order of magnitude with head size (d $\approx$ 0.2 m), that is hwy the average frequency range, which may be effectively used, for example, by a bottlenose dolphin for generating a narrow beampattern must be defined from the relationship, $\lambda \approx 0.1$ d, which may be achieved at $f \approx 10$ C/d or $f \approx 100$ kHz. The same specific separation distance of two sounds receivers (d ≈ 0.2 m) allows to ensure directional reception at frequencies in the order of 100 kHz. Thus, considering how a dolphin recognizes the bearing of the sound source one may conclude that the frequency range of passive hearing is likely to be within low-frequency range (about 1-10 kHz) and that of active hearing within high-frequency range (around 100 kHz). A dolphin may determine the distance to the sound source in a passive case only due to the <u>apriori</u> knowledge of the source intensity and spectral and temporal features of their signal under various acoustic conditions (Be'ke'sy, 1960). The active auditory subsystem must be able to measure the time interval between click's radiation and the moment of echo return for assessing the distance to the source. Consequently, the way of the signal processing necessary for distance to a sound or an echo source estimation also is quite different for passive and active hearing.

Processes of recognition of ambient sound and echo sources for passive and active hearing also are different. In the first case, there exists practically an infinite variety of sounds, which differ by their spectral and temporal structures; in the second case, the echo frequency range coincides, as a rule, with the frequency range of the echolocation signal. Echoes from the variety of objects present a sequence of pulses similar in waveform with the echolocation click.

Because maximum dimensions of the objects of main interest for small cetaceans (like Tursiops truncatus) are about 20 cm, the maximum time-delay between highlights of echoes must be less than 500 μs. It is just in this time interval where the auditory echo analysis must be especially thorough, as it is important for features extraction and estimation. Here the presence in active hearing of specific mechanisms of echo analysis lying in the range of echo duration less 500 μs, should be expected.

Passive and active hearing distinctly differ also in the way of overcoming effects of noise. In the passive case, the main interference is of low-frequency additive noises. In the active case, the main interference is reverberation, to overcome which it is necessary to use short broadband pulses (clicks) with high spacial (time) resolution. As compared to the passive hearing the active hearing must function in close interaction with the radiation system.

Proceeding from the above mentioned acousto-ecologic considerations it may be concluded, that the hypothesis of a specific echo analysis subsystem (active hearing) seems quite probable. By the moment of publication of this hypothesis (Bel'kovich and Dubrovskiy, 1976; Vel'min and Dubrovskiy, 1975; Vel'min and Dubrovskiy, 1978), a number of experimental facts proving functional peculiarities of passive and active hearing already have been accumulated. We point-out the most important of them: (1) the similarity of click's spectrum with absolute hearing sensitivity characteristics (Vel'min and Dubrovskiy, 1976); (2) the presence of minimum difference limens in frequency in the range of lower frequencies as compared to the range of maximum hearing sensitivity, as it occurs in non-echolocating animals and humans (Jacobs, 1972); (3) selectivity of evoked electrical potentials to short clicks at the level of the inferior colliculi (Bullock and Ridgway, 1972); (4) the ability of dolphins to produce simultaneously at least two signals different in purpose and in frequency band (one is echolocation of higher frequency, and the other is communicative of lower frequency), which makes sense only if there is a possibility of simultaneous and parallel reception of these signals, that is echoes and communication signals from other individuals (Markov, 1977); (5) failure of frequency differentiation when passing from long lasting tones to pulses, which points-out the essential differences in reception of durable signals and short pulses (Irapetjants and Konstantinov, 1974). Additional experimental data (both my own and by other authors) have been gathered to date in favor of a duplex theory of hearing in dolphins. The discussion of these data is the main purpose of this paper.

EXPERIMENTAL TECHNIQUE

We first examine our experimental procedures used in the investigation of hearing in dolphins (Bel'kovich and Dubrovskiy, 1976; Vel'min and Dubrovskiy, 1975; Vel'min and Dubrovskiy, 1978). The tests were conducted in concrete tanks with dimensions of 20 X 15 X 3 m^3 or 10 X 6 X 4 m^3. In training and data collecting periods, we used conditioning with a food reinforcement. Along the pool (Fig. 1) a kapron net was installed, enabling the operator to set a minimum distance, from which the dolphin had to choose a location side of positive (reinforced) stimulus (from the left or the right of the net). At signal detection task, the positive stimulus was the signal itself, and at signal discrimination task the positive stimulus was that one, on which the reflectory reaction of the dolphin had been formed (the approach to the radiator of the positive stimulus). The side of the net where the positive stimulus was produced was changed at random. Before the beginning of each trial, the dolphin 5 was located at the end of the net at the distance of 7-8 m from the transducer. In the case of correct choice of the positive stimulus the dolphin got fish. The signal was switched-off when the dolphin made its choice. The overall duration of the signal (a sequence of pulses) was about 8 s. If the signal was not present, the dolphin went back to the starting position. The level of the signal at which the dolphin detected or discriminated the signal in 75% of trials was considered as threshold one. The pulses were formed in accordance with the set-up shown in Figure 1. The pulse generator 7 was used to start the generator 9. The pulses from the output of the latter being delayed in block 8 exited the signal source.

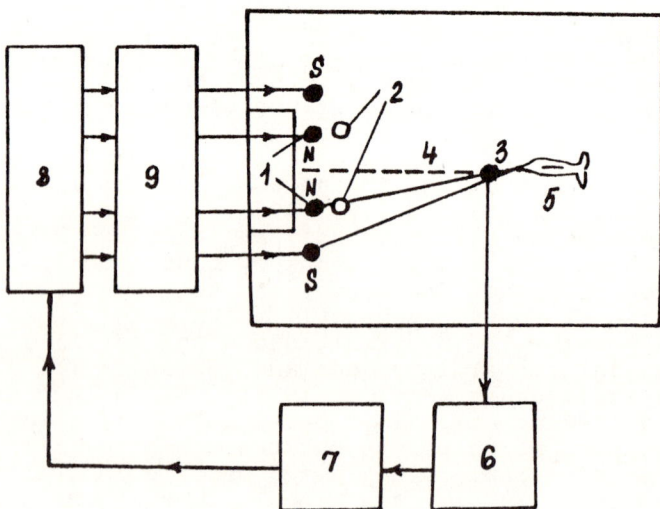

Fig. 1. Diagram of test tank: 1. projector, 2. positions of the targets, 3. receiving hydrophone, 4. dividing net, 5. dolphin at the starting position, 6. amplifier, 7. pulse generator, 8. delay-line, and 9. pulse generator.

The pulses, simulating those of the dolphins were produced by shock excitation of spherical piezoceramic transducers 9 (with the diameters from 5 to 30 mm) by unipolar electrical pulses with the duration from 2 to 100 μs from the generator. Thus, the signal presented a response of the transducer to the initial rise and the end fall of an electric video pulse. Varying the duration of a video pulse and making it multiple to the period of resonant vibrations of the transducer, it was possible to a certain degree to control duration and a waveform of the radiated acoustic pulse. The sound field was measured by a omni-directional hydrophone with sensitivity of 40 V/Pa. After the signal generated by the hydrophone was amplified (Block 6) it was photographed from the oscilloscope screen the scanning of which was started synchronously with the signal radiation into water. The threshold for parasite clicks perception due to the functioning of electronic devices had been determined beforehand in order to keep them as low as possible.

A sound pressure level of clicks was measured by a broadband tract with a flat frequency response up to 170 kHz (Bezrukov et al., 1973). The length of the dividing net in different series of tests varied from 6.35 to 7.35 m.

Now, we pass to the facts in favor of duplex theory of hearing in dolphins and, in particular in favor of functional specificity of active hearing.

Sensitivity of Active Hearing

In this case, active hearing specificity should manifest itself in high sensitivity to single clicks and in complete absence of temporal summation outside the interval 500 μs. Thresholds for perception of the click's sequence versus the pulse repetition rate are presented in Figure 2.

The absolute sensitivity at rate 3 pulses per second (P.P.S.) was equal for two Tursiops; $4.4 \pm 0.4 \times 10^3$ μPa or 73 ± 0.8 dB relative to 1 μPa and $5.6 \pm 0.5 \times 10^3$ μPa or 75 ± 0.7 dB, respectively. At 30 P.P.S. and click duration close to 15 μs (Fig. 2) the sensitivity for these dolphins appeared to be equal ($3.0 \pm 0.3 \times 10^3$ μPa or 70.5 ± 0.9 dB. It should be noted that these thresholds are very close to those estimated through evoked potentials at the auditory cortex of the bottlenose dolphin when broadband clicks were used as stimuli (Supin et al., 1978). Minimum value of a threshold was found to be $3 \cdot 10^3$ μPa or 70 dB.

Repetition rate increases from 3 to 100 P.P.S. decrease the auditory threshold due to auditory temporal summation in average up to 1.2 dB per doubling of repetition rate, F. This summation is considerably less than energy summation (3 dB per doubling of pulse repetition rate) observed in humans (Zwislocki, 1963).

If clicks summation occurs with time constant τ, signals intensity I_F will increase as (Zwislocki, 1963):

$$I_F = I_1 \{1-\exp(-1/F\tau)\}^{-1}$$

where: I_1 is intensity at 1 P.P.S. At $\tau F \gg 1$ the value $I_F \approx I_1/\tau F$. Assuming that the signal is detected when its

intensity exceeds some threshold Io, then in order to keep
the value IF constant and equal to the threshold I_o, the
summation in the auditory system should be compensated. To
achieve it, we should take (I) in a form:

$$I_F = I_o; \quad I_1 = I_o (1-\exp(-1/\tau F),$$

when $\tau F \gg 1$, $I_1 = I_o/\tau F$. That is, the intensity of a single
click must be reduced at 3 dB per doubling of F.

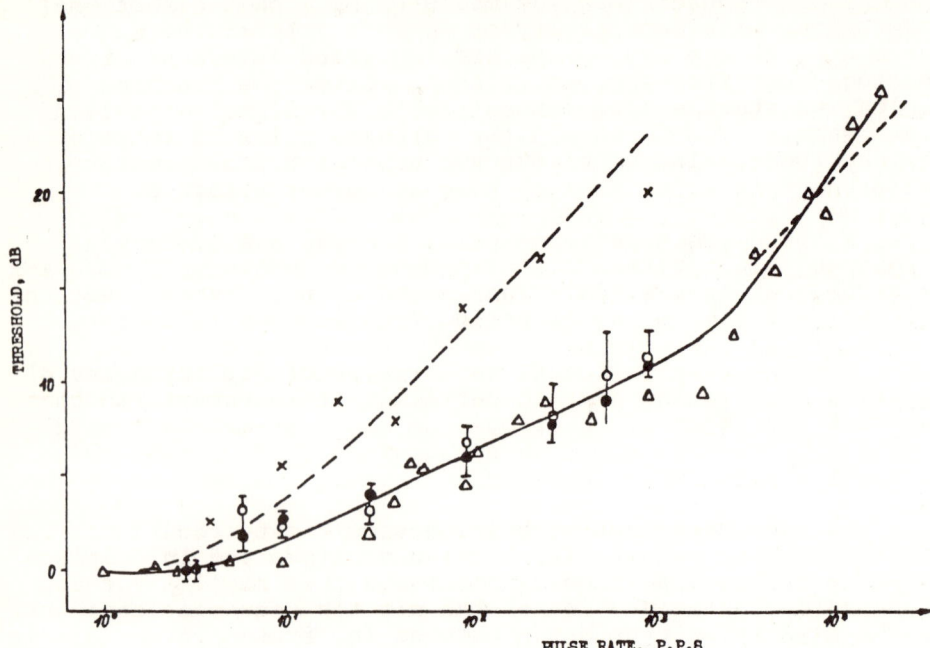

Fig. 2. Variation of the auditory thresholds of synthesized
pulses in dB versus their pulse rate; unnoticed and
noticeable circles denote our own data for two
bottlenose dolphins, triangles denote the data, obtained
by Zanin et al, 1977, crosses denote human data by
Zwislocki, 1963.

In Figure 2 a relation (2) is shown in dashed line at $=\tau$
200 ms (in logarithmic coordinates). From data presented in
Fig. 2 it follows that in the most widely used range of
dolphin click repetition rate from 12 to 60 P.P.S. the
temporal summation can not exceed 4 dB and, evidently can not
effect significantly the sensitivity of active hearing and
its resistance to noise (Babkin and Dubrovskiy, 1971).

Figure 2 also shows data from Zanin et al. (1977)
received by the same technique on one of our test dolphins,
but using broader range of pulse repetition rate (from 1 to
20,000 P.P.S.). The comparison of these data with our
results shows that first, in coinciding parts of pulse
repetition rate (from 3 to 100 P.P.S.) both groups of data
are in good agreement. Besides, from data by Zanin et al.
(1977) it is obvious that when click rate exceeds $3-4 \times 10^3$
P.P.S. summation effect rapidly grows and becomes close to
energy summation.

The temporal summation curve with time constant 250 us is shown in Fig. 2 in dashed-line. Au et al. (1986; 1988) found that auditory integration of click pairs occurred within time interval of 264 µs; that is in good agreement with our results.

The time summation constant of 250 us obtained in our experiments (Bel'kovich and Dubrovskiy, 1976; Vel'min and Dubrovskiy, 1975) and in experiments of Au et al. (1986) considerably differ from Johnson's estimate (1968) on summation of tonal bursts; the estimated integration time constant to be dozens of milliseconds. This shows the difference of auditory processing of short broadband clicks and of narrowband tonal stimuli. Zaslavskiy and Zanin (1978) studied the temporal summation of clicks in the porpoise. Either impulses of standard form with duration of 40 µs or bursts of noise in the 80-180 kHz band with the duration of 60 µs were used. As in a bottlenose dolphin (Fig. 2) the curve of temporal summation of pulse succession in the porpoise is divided in two parts: from 10 P.P.S. to 10^3 P.P.S. and from 2×10^3 P.P.S. farther on. In the first part of the curve the threshold decrease due to click summation was at 1.2 ± 1.3 dB per doubling of pulse repetition rate. In the second part of the curve the threshold decrease reaches 5 and 4 dB per doubling correspondingly for noise bursts and pulses of a standard form. It is remarkable that the rate threshold decreasing (5 dB and 4 dB) was definitely higher than that at the energy summation (3 dB per doubling of pulse repetition rate). As the stimulus summation in the auditory system usually lessens with the growth of its level over the auditory threshold, it should be expected the complete absence of the suprathreshold summation in dolphins in pulse repetition range 1-1000 P.P.S. It may imply that the auditory detection of an echo may take for a dolphin just one ongoing pulse. Measured by means of evoked potentials in the brainstem (Popov and Supin, 1986) the temporal summation was observed in the duration range from 200 to 300 µs, which is well consistent with the temporal integration estimation in the auditory system of a bottlenose dolphin, which has been gained in the behavioral experiments.

Discrimination of Synthesized Clicks on Intensity and Temporal Intervals

The specific features of active hearing when the difference limen on intensity (DLI) and the difference limen on temporal intervals (DLT) are measured must be exhibited in small threshold values of these quantities.

When DLI were measured signals to be compared were radiated simultaneously at both sides of the dividing net (Fig. 1) 30 times per second. The more intensive pulse at random was radiated from the left or from the right sides relative to the separative net. The task of the animal was to define the side of more intensive click radiation. The stronger to weaker signal ratio (in dB) was considered as DLi in case of 75% of correct reactions. The measurements were taken at several levels of the signal above its sensitivity threshold. The results of the measurements are given in Table 1.

Table 1. Average DLI and 90% Confidence Intervals in dB

No. of Dolphins	Signal Level Above Sensitivity Threshold, dB			
	(5)	(20)	(36)	(45)
1	2.5± 0.5	2.1± 0.8	2.1± 0.8	1.± 30.6
2	1.7± 0.4	0.8± 0.1	0.7± 0.1	-------

The values of DLI clicks are close to 2 dB near the sensitivity threshold.; with the increase of the signal level above the hearing threshold up to 40 dB, DLI lessens twice or more, reaching the average value of 1 dB.

From the above data it follows that DLI in bottlenose dolphin clicks approximately coincide with DLI for tone and noise signals; that proves a high intensity resolution in "active hearing". In this respect, our data agree very well with DLI estimations made by means of evoked potentials at the inferior colliculi (Bullock et al., 1968) and the auditory cortex (Supin et al., 1978). Secondly, DLI satisfactory agrees with the estimations, obtained from tone masking curves (1.0, 0.35, and 2.0 dB) by Johnson (1970). Thirdly, DLI also well coincides with the relative target strength ($\approx$1dB), if echo-signal contains only primary echo (Bel'kovich and Dubrovskiy, 1976). Thus, these results prove good adaptation of active hearing to intensity discrimination of short pulses.

As the dolphin uses clicks for echolocation, an echo from many objects consists of two and more pulses, which may differ in time intervals between their components. Therefore, the difference limen for intervals (DLT) measurements may characterize the active hearing and dolphins ability to discriminate targets.

When DLT were measured, pairs of clicks were radiated with 30 pairs per second simultaneously on both sides of the net. The dolphin's task was to choose the side of the net with bigger interclick intervals. Standard intervals were 50, 100, 200, and 500 μs. DLT in μs was defined as a difference of intervals to be compared (in case of 75% correct reactions); the DLT in dB or percent was defined as a ratio of DLT used compared to standard interval. The DLTs at standard intervals of 50 and 100 μs for one of the dolphins were the same and equal to approximately 0.4 dB (Table 2).

When the interval was 200 μs the DLTs were equal to 21.5 us and 25.5 μs, that corresponds to 10.7 and 12.7 percent (0.4 and 0.5 dB). At the standard interval of 500 μs the dolphin failed to discriminate intervals despite long training. Only once the estimation of DLT 34 μs was received, that corresponded to 6.5% (0.3 dB).

The ability of a dolphin to discriminate time intervals were measured in another bottlenose dolphin in the interval range of 50-100 μs with a step of 10 μs and further at 150,

180 and 200 µs (Dubrovskiy et al., 1978). Average DLT values were nearly constant and varied from 5% to 8%, that is slightly lower than the values presented in Table 2. Minimum DLT values were close to 1-2 µs, that is in good agreement with the data from echolocation experiments. Those experiments showed that dolphin required to recognize targets had to discriminate the intervals between pulses differing only by 1.8 µs (Bel'kovich and Dubrovskiy, 1976). At intervals in 180 and especially in 200 µs the discrimination was unstable. At the interval in 180 µs the threshold values were obtained in four cases out of six. Twice the thresholds were obtained not by a usual way, that is when a greater test interval was approached to a standard (smaller) one but on the contrary by fixing the standard (greater) interval and approximating to it a test (smaller) interval. In spite of the difficulties in getting the threshold values, they were still low of the standard interval in 180 µs. At the standard interval in 200 µs we succeeded in getting the threshold value in 3 cases out of six; however, the values of these DLTs were already very high (25-32 µs).

Table 2. Average DLT and 90% Confidence Interval for Each Standard Interval

Standard Interval (µs)	50	100	200
DLT, µs	4.8± 0.8	9.4± 1.7	11.5± 2.4
DLT, %	9.6± 1.6	9.4± 1.7	5.7± 1.2
DLT, dB	0.4± 0.1	0.4± 0.1	0.2± 0.1

Data given in Table 3 characterize the ability of one dolphin to discriminate the time intervals exceeding 300 µs.

As Table 3 shows the discrimination of intervals higher than 300 µs is absent. The distinction approximates 50% in different series of trials. Similar observations were obtained with anther bottlenose dolphin (Dubrovskiy et al., 1978).

Table 3. Discrimination of Click Intervals Exceeding 300 µs

Intervals to be Compared (µs)										
300	500	500	500	500	500	1000	1000	1000	1800	2000
400	500	556	600	700	1200	1100	1200	1500	2000	2000
Correct Discriminations (%)										
56	50	53	64	46	60	36	50	45	57	43

An attempt was made to teach a dolphin to discriminate the pairs of noise bursts 50 us in duration and repetition rate 2 per second (Irapetjants and Konstantinov, 1974). The

intervals between pulses in the compared pairs were either
100 and 1100 μs or 200 and 250 μs. The dolphin failed to
discriminate these intervals.

According to these results, we may assume that a dolphin
perceives the sequence of echo-pulses as an acoustic whole if
it ranges within a time interval not higher than 200-300 μs.
This critical time T_{cr} corresponds also to the maximum size
of an object ≈ 20 cm, its echo is perceived by dolphins as a
merged echolocation image. This time corresponds to the
duration of the interval where effective summation of click
succession occurs.

Masking in Active Hearing

The phenomenon of masking of one acoustic signal by
another is used widely in humans and animal hearing
investigations when the interaction of two stimuli; noise and
signal are in question. We used masking to identify the
characteristics of active hearing in the process of masking
of one click by another and of masking of the real echo-
signal by a phantom synchronized with a dolphin's ongoing
pulse.

When studying masking phenomenon in active hearing in
dolphins, they were exposed to two sequences of clicks
simultaneously produced by both sides of the separative net.
Sequence of single pulses 30 times per second was generated
by one transducer, the sequence of two pulses was generated
by the second transducer, where the first pulse in a pair
coincided in time with a single pulse meanwhile the amplitude
and the delay-time of the second pulse in a pair varied
during the experiment. The single click and first click in a

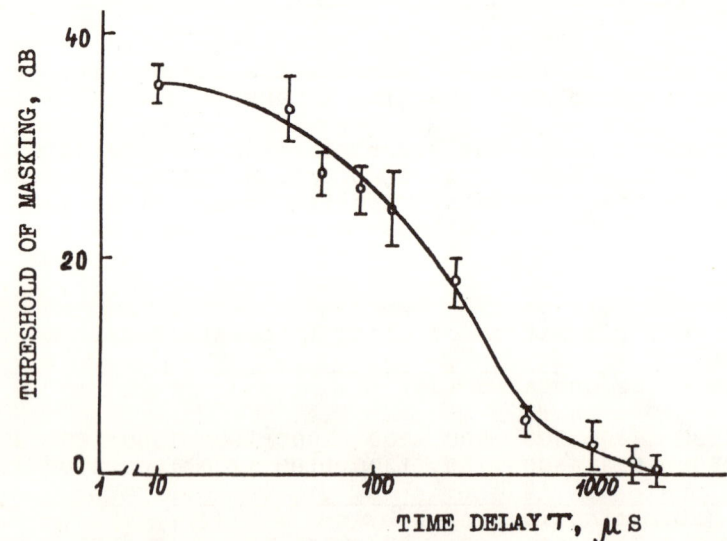

Fig. 3. Thresholds of backward temporal masking for one
dolphin versus time-delay T between a signal click and
an interference click. Stimulus repetition rate is 30
per second. Level of an interference click is 46 dB
above its threshold. Vertical lines represent 90%
confidence intervals.

pair were radiated simultaneously through transducers located on both sides of the net. Choice of a transducer to produce a pair of clicks or a single click was done at random. The dolphin's task was to detect the presence of the second pulse in a pair. The time interval between clicks in a pair could vary in different series of trials from 0 to 10 ms. The error of T-measurements was not higher than 3 μs. The error of measurement of sound pressure level was bout 2 dB.

The results of the experiments for one of the dolphins are given in Figure 3. At T of 500 μs the thresholds of masking approximate the absolute threshold of a single pulse within the limits of measurement error. It follows that auditory events do not interact when they are separated by 500 μs. On the contrary, at T $\leq$300 μs strong interaction of stimuli is observed.

Our data are in good agreement with the results on forward temporal masking obtained by means of evoked potentials in brain cortex of the bottlenose dolphin (Popov and Supin, 1986; Bibkov et al., 1986) when an interference click comes before a signal click. It corresponds to dolphin echolocation of the objects at small distances when dolphin's ongoing click becomes an interference. From this point, our data on backward masking are related to the case when the weak echo click is masked by a stronger one, delayed relative to the first.

Evoked potentials are separated at delays between an interference and a signal clicks at 2-5 ms (Popov and Supin, 1986; Bibikov et al., 1986). When an ongoing click acts as interference itself then the time separation of evoked potentials caused by both clicks was observed even at shorter time-delays near 1 ms (Popov and Supin, 1986). The first peak restoration of the evoked potential took place in 100% of the cases when time-delay was only 500 μs (Popov and Supin, 1986). Minimum response to a signal when signal and masker amplitudes coincided was observed at interval 200 μs (Popov and Supin, 1986).

Similar threshold values of the interval of 200 to 300 μs, when response to the second click in a pair can be identified with certainty, have been obtained in the process of electrophysiological experiments with the Azov Sea Harbor porpoise (Dubrovskiy et al., 1978). Thus, if we take the experiment with backward temporal masking as an illustration, we again come back to the important fact that mechanisms of auditory echo-pulse processing differ depending on whether highlights of the stimulus fall within the time interval of 300 μs or beyond. Evidently, the specific features of active hearing in a bottlenose dolphin should be revealed fully in perception of echo from an object. Observations of echolocation ability in dolphins under strong reverberation (detection and identification of objects near a wall or sea bottom) prove that dolphin hearing is well-protected from reverberation. Considering these observations, we can admit that cetaceans have a specific mechanisms of protection from reverberation pulses or from their own ongoing pulses (Bel'kovich and Dubrovskiy, 1976). For example, the gating mechanism when the auditory system is blocked at the moment of intense power click arrival, but its sensitivity grows when weak echoes are received.

To find-out active hearing features in the presence of dolphin's ongoing clicks the experiments were carried-out where bottlenose dolphins detected a steel sphere which had a diameter of 40 mm on the background of a phantom echo. The latter imitates in waveform and spectrum dolphin ongoing pulses. Time-locking of a phantom echo to dolphin echolocation clicks and to echo from the steel sphere was obtained by triggering the phantom click with dolphin ongoing pulses. By changing a delay there was possible to control spacial and temporal relations of an echo from a sphere and a phantom echo to ongoing clicks from a dolphin.

The percentage of correct discrimination of a sphere versus a delay between the echo from a sphere and a phantom echo is presented in Fig. 4. The ratio echo from a sphere to phantom echo was (-50.3± 3.5 dB). At T<-300 µs and T>200-500 µs detection of a sphere was quite reliable. It implies that the phantom click does not mask an echo from the sphere.

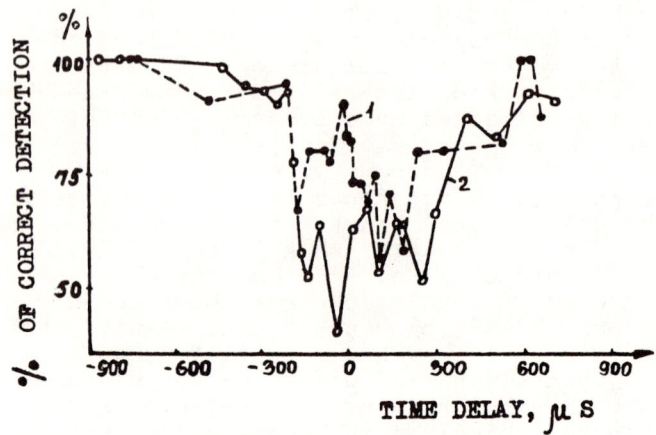

Fig. 4. Active hearing resistance to masker clicks, triggered by an ongoing dolphin click. Percentage of a sphere correct detection versus time-delay between a masker click and echo from a sphere.

Time-resolution values, obtained in this experiment approximately coincide with that in the experiments in the absence of phantom pulse (backward temporal masking). Proceeding from these facts, we may assume that high resistance to interference is achieved rather by a high temporal resolution of active hearing than by a gating mechanism.

In Figure 4, a steeper slope is observed when the echo from the sphere comes earlier than the phantom pulse and smoother when the phantom pulse comes first. Secondly, there is a sharp rising of the masking curve (for one of the dolphins) and an abrupt fall shown on the curve for another dolphin near a zero delay between the echo and the masker. In my opinion, this curve behaviour is attributed to the dolphin transition from the detection of echo from a sphere to the discrimination of a single phantom pulse from a pair of "phantom pulses plus echo". A similar phenomenon also was observed in the experiments with backward masking (Bel'kovich and Dubrovksiy, 1976). Moore et al. (1984) have determined the values of the critical interval in the experiment with backward temporal masking when a masker was triggered by an ongoing click of a dolphin. The results of this experiment have confirmed our data on backward temporal masking for an echolocating dolphin (Fig. 4); the critical interval in this paper was estimated at 265 μs. In this experiment, we again come across the critical interval in the natural conditions of active hearing (in the process of echolocation).

On the previous experiments, we observed masking when time separation of a signal and masker took place. In the following experiment masking of a click simulating an echolocation click spaced by masker was investigated (Vel'min and Dubvoskiy, 1977). We begin with the description of experimental procedures. In the sea tank a separative net (Fig. 1) was stretched from planked footways along both sides of which two pairs of spherical piezoceramic transducers N and S were installed. A pair of transducers N placed at a constant distance of 0.3 m from the net 4, was used for the masker click radiation, and another pair of transducers S was used for signal click radiation at random from the left side or from the right side of the net. The transducers S were positioned symmetrically respective to the net though their distance to the net could be changed. The masker click level over the threshold of their audibility at the nearest to the dolphin edge of the net attained 114 dB relative to 1 μPa. The masker clicks were simultaneously produced from both sides of the net (synchronously with the signal clicks radiation by one of the transducers). The repetition rate of all pulses was equal to 30 pulses per second. the click duration was close to 10 μs and maximum of its spectrum approximated to 80 kHz. A dolphin being at the beginning of the net ought to detect the signal click and swim to that side of the net where the transducer S was positioned in this trial. At a given distance to the transducer from the net the signal detection threshold was estimated according to 75% correct choice.

Decreasing of masking at different azimuth separation of S-S and N-N is shown in Figure 5 (curve 1). These results demonstrate high resistance to masker clicks in active hearing when the signal and masker sources are spaced. Masking achieves 43.6 dB when signal and masker are superposed at $\theta = 0$. Separation of the signal source only for $\pm 3.2°$ aside from the masker source reduces masking by 8-10 dB and separation by approximately $\pm 15°$ decreases masking on an average by 30 dB. The specific features of active hearing are revealed in this case in the fact that withstanding masking proves to be much higher for clicks simulating echolocation ongoing signals than for the high-frequency tonal signals (Fig. 5).

So at the level of 0.4 from a maximum, the width of curves 1-3 corresponds respectively to 6°, 23°, and 40° though the central frequency for the signal click (80)kHz coincides with the carrier-frequency of one of the tonal signals. The curves 1-3 characterize spatial resolution of hearing and in this respect we can compare them with a directivity pattern of hearing that describes spatial anisotrpy of auditory sensitivity. As it was shown

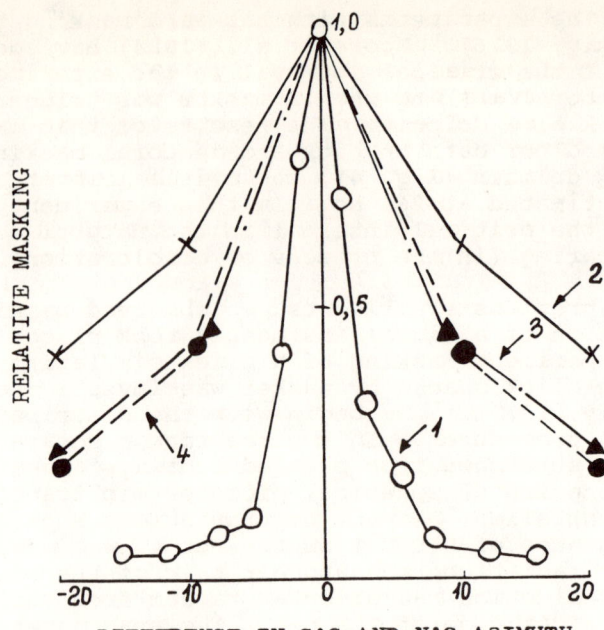

Fig. 5. Masking of a signal click by an azimuthally spaced masker click versus angular separation of the signal and masker transducers; 1-data given in (Vel'min and Dubrovskiy, 1977); 2-data for tone 80 kHz (Zajtseva et al., 1975); 3-data for tone 120 kHz (Zajtseva et al., 1975); 4-directivity pattern for tone pulse with the frequency of 80 kHz and duration of 40 µs (Akopian et al., 1977).

previously (Zatseva et al., 1975; Akopian et al., 5). the directivity pattern at reception narrows when the signal of the same frequency shortens (curve on Fig.

Thus, a functional specificity of active hearing in this experiment displays itself in the fact that the synthesized pulse simulating the dolphin echolocation pulse seems to be the most effective stimulus enabling active hearing to produce a high spatial selection of the signals on the background of masker clicks that is, the masker coincides with the signal in its spectral-space structure.

Critical Interval of Active Hearing

The peculiarities of click auditory analysis, revealed in considerable difference of auditory processing mechanisms of click pairs when single clicks in a pair is separated by more than 500 µs or less than 200 µs made it necessary to take more precise measurements of this critical interval, T_{cr}, (Bel'kovich and Dubrovskiy, 1976; Vel'min and Dubrovskiy, 1975; Dubrovskiy et al., 1978; Vel'min and Dubrovskiy, 1976).

Measuring the critical interval duration T_{cr}, the task of the experimental dolphin was to discriminate between the standard pair of pulses, separated by 100 μs and the test pair, being intervals between pulses in this pair varying from 150 to 400 μs. The sound pressure level of each click exceeded the threshold level by 40 dB. The standard and test intervals were radiated at random from the left side or from the right side of the separative net 7 m long. In this experiment, we used so-called emergency techniques of click pair presentations, separated by the test interval, on the background of two "standard" click pairs presentation with intervals of 100 μs and 150 μs. In this experiment, the task of the animals was to choose that side of the net where the couple of clicks with a greater interval was produced.

The results of this experiment are demonstrated in Fig. 6. As the Figure shows the percentage of correct choices made by the dolphin which varies from 100% at 150 μs < T_{test} < 200 μs up to 0 % at T_{test}>325 μs. Zero percentage of discriminations means that if the test interval occurs beyond the critical one the dolphin chooses as a positive stimulus not a pair of clicks separated by the test (greater) interval, but a pair of clicks separated by a standard (shorter) interval. In our opinion, it occurs due to disintegration of a merged acoustic image corresponding to the test pair in two independent acoustic images corresponding to each separate click in a pair when the test interval increases beyond the limits of critical values. In this case a pair of clicks with standard interval of 100 μs turned to be more "positive" stimulus than each independent click of the test pair. A complete lack of discrimination occurs at T_{test} = 260± 25 μs, and the threshold (75%) discrimination takes place at T_{test} = 230±40 μs. Thus, in this experiment the value of the critical interval has been determined within which a merging of acoustic images corresponding to the single click into an acoustic whole relative to a click pairs occurs. The significance of this experiment is expressed rather in definition of the basic content of the critical interval concept as a characteristic time interval within which pulse interaction occurs and two acoustic events merge into an acoustic whole than in an exact measurement of T_{cr}.

One can observe, how perception of a click pair is changing with interclick interval rise. It is evident that transition from merging of a pair of clicks to complete independence of auditory images of separate clicks will be gradual and will depend on a number of such physical characteristics as click sound pressure level, their amplitude ratio and so on. So, it may be expected that at T <100 μs auditory image of a click pair will be merged completely, but at 100 < T < 1000 μs its gradual disintegration will develop. In other words, the beginning of disintegration of a merged image may be already discovered at T <100 μs however, on the other hand, some partial merging of auditory images of separate pulses may be still observed even at T = 500 μs. Thus, one can obtain one or another value of T_{cr} in given experiments depending on what dolphin discrimination is based on; features pertaining to merging of stimuli or features characteristic of stimuli separation. A geometric mean value of extreme limits equals approximately 300 μs and it may be taken as some average value of T_{cr}.

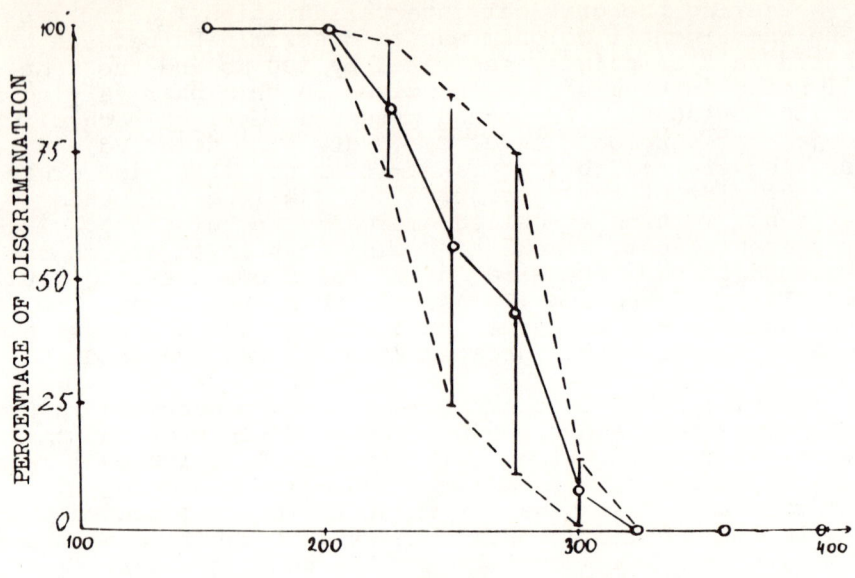

Fig. 6. Percentage of correct discrimination of test and standard click pairs versus duration of a test interclick interval. Vertical bars represent 20% confidence intervals.

The meaning of "critical interval" concept includes also the idea that mechanisms of auditory analysis of echo pulses falling within this interval prove to be sufficiently different from mechanisms of auditory analysis of echo-pulses that do not fall within this interval. In fact, considerable increase in threshold temporal summation observed when the interval between two adjacent clicks T equaled 300 μs (Fig. 2). Estimation of T_{cr} according to these data give the values T_{cr} lying with the range 200 μs to 500 μs.

Afterwards, however, one of the dolphins has been trained successfully to discriminate pairs of clicks with interclick intervals greater than critical one, but it was obvious that in this case difference in time intervals between pulses in compared pairs served as a feature for discrimination. Transition from differentiation of a pair of clicks with intervals more than critical to differentiation of pairs with intervals less than critical has always been accompanied by the criteria change and an increase in scattering of the experimental data, that is typical for the change in the discrimination criteria (Vel'min and Dubrovskiy, 1975). Thirdly, forward and backward temporal masking of one click by another takes place within the critical interval. Estimations of limits for T_{cr} according to masking curves lead to the following result: 100 μs < T_{cr} < 1000 μs. In the fourth place, forward and backward masking of echo-pulse by a pulse synchronized with an ongoing one for two dolphins are effective only within the critical interval. Estimates obtained in this experiment yield 200 μs < T_{cr} < 500 μs. In the fifth place, the direct measurement of the critical interval as an interval of merging of auditory images of separate clicks into an acoustic whole allowed us to receive the following estimation for T_{cr} limits: 200 µs < $T_{c}r$ < 350 µs.

M. N. Sukhoruchenko (Dubrovskiy and Sukhoruchenko, 1986) developed the hypothesis that has been tested in a behavioral experiment with a bottlenose dolphin by the method of selective adaptation. Hypothesis implies that the auditory system of dolphins has specific channels for estimation of time intervals.

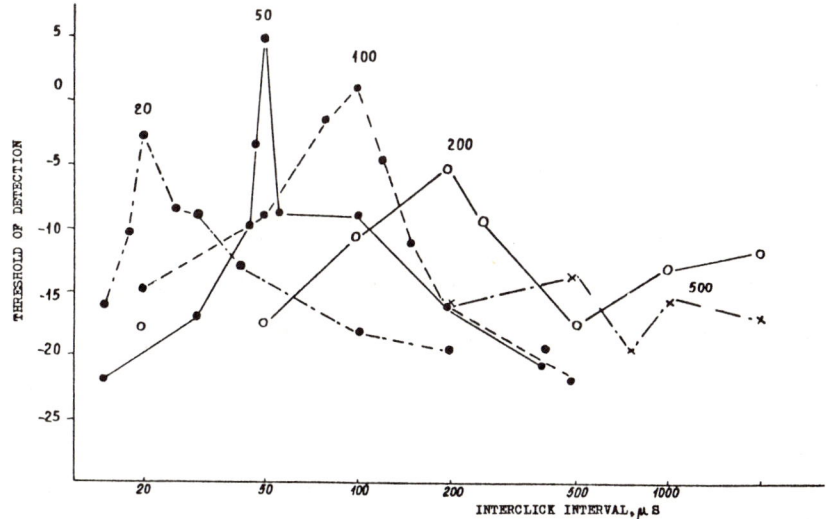

Fig. 7. Thresholds for detection of the second click in a signal pair of clicks after application of the adapting pairs of clicks. Horizontal axis presents interclick interval, T, in the adapting pairs. Different threshold curves correspond to different interclick intervals in the signal pair.

The threshold curves for the second pulse detection in a test pair versus the interpulse interval in a pair used as an adaptive stimuli have been plotted on Fig. 7. For test paris with the interclick interval in a pair up to 200 µs, the curve reveals the peak of adaptation at coinciding values of test and adaptive intervals. For the 500 µs test interval dependence of the threshold value on the adaptive interval is not observed (Fig. 7).

Summing the above data we may conclude:

(1) the auditory system of the dolphin is well-adapted to the analysis of acoustic stimuli consisting of one or two clicks simulating ongoing pulses or echo-signals received from underwater objects. This adaptation displays itself (I) in a high sensitivity to pulse signals (≈70 dB relative to 1 µPa);

(2) in a low threshold summation of clicks in a range of their repetition rates from 10 to 60 P.P.S. (reduction 1.2 dB at doubling of repetition rate), that speaks for mono-pulse character of auditory detection, that is one single echo-signal was enough for detection;

(3) in a low difference limens on intensity (≈1 dB); the values of DLs for short pulses coincide with those for long lasting noise and tonal signals;

(4) in essential differences of auditory processing mechanisms of clicks sequence falling within or outside the critical interval that is the interval incorporates echo components from ecologically important underwater objects.

Hence, the proposed hypothesis about two functionally (and possibly anatomically) specific auditory subsystems-active and passive hearing-is confirmed by large amounts of experimental facts.

Mechanisms of Auditory Pulse Analysis within the Critical Interval

As it has been pointed-out the basic feature of a processing mechanism with T_{cr} is the merging of acoustic events in an acoustic whole. Specific manifestations of this feature can be quite different; it can be displayed in an increase of the temporal summation effectiveness, the sensitivity to spectral and phase structures of stimuli. The limits to which such features are exhibited also will depend on what inherent quality of merged auditory image is investigated in a specific experiment.

In this respect, processing mechanism used by the active hearing inside T_{cr} is similar to that of information processing mechanism used by passive hearing inside critical band: the merge of auditory images, corresponding to single spectral components into an acoustic whole. Evidently, such similarity is not quite superficial. It should be expected that the auditory analysis mechanism inside T_{cr} is not less complicated than the mechanisms inside a critical band. The processing mechanisms inside T_{cr} may prove to consist of a number of mechanisms each having its own interval less than T_{cr}: a "matreshka" principle. As it will be demonstrated below, the obtained experimental data make it possible to assume the existence of two such subintervals at least. We first examine the results of this experiments set to reveal auditory processing mechanisms inside T_{cr} (Dubrovskiy et al., 1978).

The first problem that we attempted to consider was as follows: do dolphins analyze an echo-signal profile within the critical interval, in other words, does it use all information available in echo-signal or does it analyze signals based on reduced description, for example, on echo-signal energy spectrum. In the first experiment a dolphin was given a pair of clicks of a following kind:

$$S_1(t) = aS(t) + bS(t-T) \quad \text{and}$$

$$S_2(t) = bS(t) + aS(t-T)$$

where, s(t) is a profile of a single pulse, and a and b are positive numbers (a>b), T is a time-delay between clicks. Pairs of clicks $S^1(t)$ and $S_2(t)$ are characterized by identical energy spectra. In fact, if amplitude spectrum S(t) equals G(ω), then spectrum of a pair will be equal to:

$$G(\omega) = \{a + b \exp(i\omega T)\}$$

and its energy spectrum is

$$|G(\omega)|^2 = (a^2 + b^2 + 2ab \cos \omega T)$$

By analogy spectrum of the pair $S_2(t)$ equals

$$G(\omega) = \{b + a \exp(I\omega T)\}$$

and modulus square $|G(w)|^2$ equals

$$|G(\omega)|^2 (a^2 + b^2\ 2ab \cos \omega T)$$

Differentiation or non-differentiation by a dolphin of a pair of clicks $S_1(t)$ and $S_2(t)$ will testify that a dolphin uses either only energy spectrum of a click pair or also a phase spectrum.

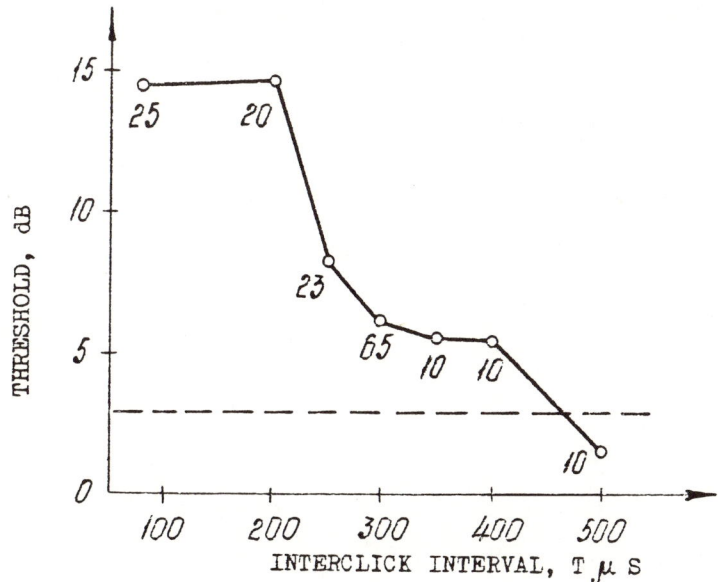

Fig. 8. Results from one experimental set on discrimination of click pairs $S_1(t)$ and $S_2(t)$. The threshold amplitude ratio 20 log a/b versus interval T us between clicks in a pair. A click sound pressure level above auditory sensitivity threshold is 40-45 dB. Numbers indicated near experimental points show a number of trials, used for estimation of the threshold. Dashed-line denotes the DL for intensity, measured in this test.

The value 20 log a/b, dB versus an interval T between clicks in microseconds is plotted in Fig. 8. Circles denote threshold values 20 log (a/b), corresponding to 75% of correct stimuli discrimination. The only exception is a point at T = 500 μs, that corresponds to 100%. Thus, a part of the first quadrant between the axes of coordinates lower than the experimental curve corresponds to nondiscrimination.

As it is seen in Fig. 8 discrimination within T_{cr} occurs at amplitude ratio 20 log a/b more 14.5 - 15.0 dB. When T rises up to 250 µs this ratio quickly falls down to 1.6 dB at T equals 500 µs. We are of the opinion that the dolphin discriminated stimuli by using the amplitude difference between the first and the second click pairs, that is, in the case when two clicks combined in a pair do not merge in a whole auditory image. On the contrary, at $T < T_{cr}$ non-differentiation of click pairs at 20 log a/b < 14 dB testifies that a dolphin could not make-out stimuli through their time profiles. From the spectral point of view, it means that the dolphin was not able to discriminate stimuli within the critical interval when their phase spectra were different. This result implies that either discrimination of the click pair is based on exclusively on energy spectra, that is a pure spectral processing mechanism with T_{cr}, or it could also mean that a dolphin could not use in our tests corresponding temporal or phase characteristics for discrimination. That is why the experiment for differentiation of click pairs $S_1(t)$ and $S_2(t)$ was continued at constant ratio of 20 log a/b (10 dB) and at varied T. In Fig. 9 the percent of correct differentiation versus the interval between clicks T is presented. At the beginning of the experiment complete non-discrimination of stimuli occurred at T = 10, 20, 40, and 200 µs (corresponding number of trials 30, 20, 20, 60, respectively). Later on, however, the dolphin was able to differentiate click pairs at intervals T equal 20 (88% of 36 tests) 30 and 60 µs; discrimination at these values of the interval confirms that the dolphin succeeded in finding-out time profile features or phase spectrum features. It should be marked that T-dependent ratio curve has more or less pronounced oscillations with a period of T = 15 µs.

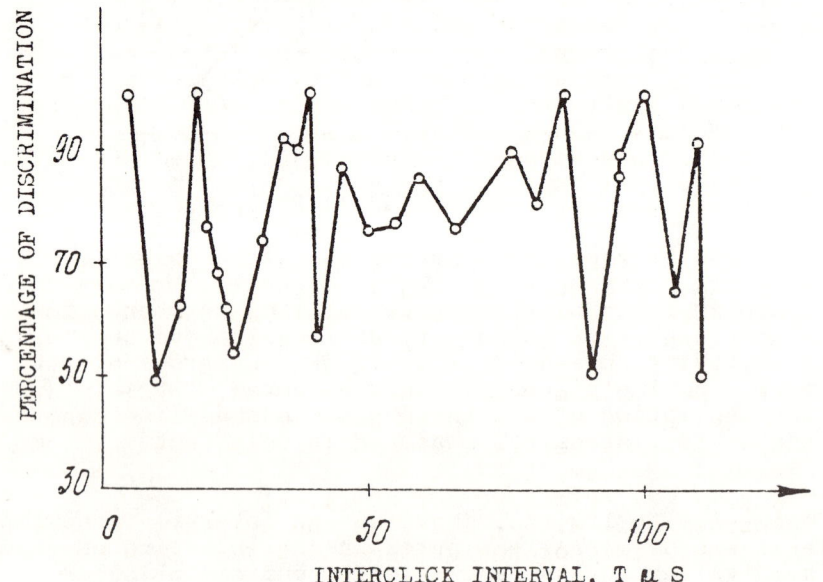

Fig. 9. Discrimination of pairs of the kind $S_1(t)$ and $S_2(t)$. Percentage of correct discrimination versus interclick interval,T. Average number of trials equals 50. Ratio 20 log a/b was constant and equalled 10 dB.

As far as the experiments with these stimuli went on the percentage discrimination of pairs grew to 100. That means this study undoubtedly has proven the possibility of temporal stimuli processing within the range of $T_c r$. Thus, pure spectral (based on energy spectrum) as well as time processing is quite valid within T_{cr}.

REFERENCES

Akopian, A. I., Zaitseva, K. A., Morozov, V. P., and Titov, A. A., 1977, Spatial directivity of the dolphins' auditory system in perception of varied frequencies signals in noise. 9th All-union Acoust. Conf. Moscow.

Au, W. W. L., Moore, P. W. B., and Pawloski D. A., 1986, The perception of complex echoes by an echolocating dolphin. J.Acoust. Soc. Am., S1, 80: S107.

Au, W. W. L., Moore, P. W. B. and Pawloski, D. A., 1988, Detection of complex echoes in noise by an echolocating dolphin. J. Acoust. Soc. Am. 83: 662-668.

Babkin, V. P. and Dubrovskiy, N. A. ,1971, On a range and noise resistance of the dolphin echolocation system in detection of different targets. Trudy Akust. Inst.Moscow, 17: 90-98.

Bel'kovich, V. M., and Dubrovskiy, N. A., 1976, Sensory basis of cetacean orientation, Leningrad, Nauka, 204 pp.

Be'ke'sy, G., 1960. "Experiments in Hearing", New York.

Bezrukov, V. K., Gassko, R. E., Dubrovskiy, N. A., and Zaslavskiy, G. L., 1973, The multichannel measurement system for registration of pulsed hydroacoustic signals 8th All-union Acoust. Conf. Acoustical Institute, Moscow, 145-147.

Bibikov, N. G., Rimskaya-Korsakova, L. K., Zanin, A. V., and Dubrovskiy, N. A., 1986, Investigation and modelling of auditory evoked potentials in the brain stem of Phocoena phocoena. in: "Electrophysiology of sensory systems in marine mammals". V. Ye. Sokolov, ed. A. N. Severtsov Inst. Evol. Morphol. Anim. Ecol. Akad. Nauk, SSSR. Moscow, Nauka, 56-84.

Bullock, T. H., and Ridgway, S. H., 1972, Evoked potentials in the central auditory systems of alert porpoises to their own and artificial sounds. J. Neurobiol. 3: 79-99.

Bullock, T. H., Grinnell, A. D., Ikezono, E., Kameda, K., Katsuki, Y., Nomoto, M., Sato, O., Suga, N., and Yanigasawa, K., 1968, Electrophysiological studies of central auditory mechanisms in cetaceans. Z. Vergl. Physiol., 59: 117-150.

Dubrovskiy, N. A., Rimskaya-Korsakova, L. K., and Sukhoruchenko, M. N., 1978, Discrimination of time intervals in a bottlenose dolphin. in: "Marine Mammals". Moscow, 113-114.

Dubrovskiy, N. A., Rimskaya-Korsakova, L. K., and Sukhoruchenko, M. N., 1978, On the criterion change near the upper limit of the critical interval in a bottlenose dolphin. in: "Marine Mammals", Moscow, 114-115.

Dubrovskiy, N. A., and Sukhoruchenko, M. N., 1986, Dolphin discrimination of paired and single clicks on background of paired clicks masker. in: "Marine Mammals'. Arkhangelsk, 127-128.

Dubrovskiy, N. A., Krasnov, P. S., and Titov, A. A., 1978, Auditory discrimination of acoustic stimuli with different phase structures in a bottlenose dolphin. in" "Marine Mammals", Moscow, 114-115.

Irapetjants, E., Sh. and Konstantinov, A. I., 1974, Echolocation in nature. Leningrad, Nauka, 512 pp.

Jacobs, D. N., 1972, Auditory frequency discrimination in the atlantic bottlenose dolphin, Tursiops truncatus Montague. J. Acoust. Soc. Am. 52: 692-698.

Johnson, C. S., 1968, The relation between absolute threshold and duration-of-tone pulses in the bottlenosed porpoise. J. Acoust.Soc. Am. 43: 757-763.

Johnson, C. S., 1970, Auditory masking of one pure tone by another in the bottlenose porpoise. J. Acoust. Soc.Am., 49: 1317-1318.

Markov, V. J., 1977, On bottlenose dolphin signals generated simultaneously by its three sound generators. 9th All-union Acoust. Conf. Moscow.

Moore, P. W. B., Hall, R. W., Friedl, W. A., and Nachtigall, P. E., 1984, The Critical Interval in Dolphin Echolocation: What is it? J. Acoust. Soc. Am. 76: 314-317.

Popov, V. V., and Supin, A. Ya., 1986, Measurements of the auditory characteristics in a bottlenose dolphin based on evoked potentials of brain stem. in: "Electrophysiology of the sensory systems of marine mammals" V. Ye. Sokolov, ed., A. N. Severtsov Inst. Evol. Morphol. Animal Ecol. Akad. Nauka SSSR. Moscow, Nauka, 85-106.

Supin, A. Ya., Mukhametov, L. M., and Ladygina, T. F.,Popov, V. V., Mass, A. M., and Polyakova, I. G., 1978, Electrophysiological investigations of the dolphin brain, Moscow, Nauka, 215 pp.

Vel'min, V. A., and Dubrovksiy, N. A., 1975, On the analysis of pulsed sounds by dolphins. Dokl. Eakad. Nauk SSSR, 225:470-473.

Vel'min, V. A., and Dubrovskiy, N. A., 1976. The critical interval of active hearing in dolphins. Sov. Phys.Acoust., 2: 351-352.

Vel'min, V. A., and Dubrovskiy, N. A., 1977, Spatial selectivity of active hearing in a bottlenose dolphin. 9th All-union Acoust. Conf. Moscow, 5-8.

Vel'min, V. A., and Dubrovskiy, N. A., 1978, Auditory perception by bottlenosed dolphin of pulsed signals in: "Marine Mammals: results and methods of study" V. Ye. Sokolov, ed. A. N. Severtsov Inst. Evol. Morphol. Anim. Ecol, Akad. Nauk SSSR, pp. 90-98.

Zanin, A. V., Zaslavskiy, G. L., and Titov, A. A., 1977, Temporal summation of pulses in the auditory system of the bottlenose dolphin. 9th All-union Acoust. Conf., Moscow, 21-23.

Zaslavskiy, G. L., and Zanin, A. V., 1978, Temporal summation in the auditory system of Azov Sea Harbor porpoise. in: "Marine Mammals" Moscow, 128-129.

Zajtseva. K. A., Akopian, A. I., and Morozov, V. P., 1975, Noise resistance of the auditory analyzer in dolphins as a function of noise presentation angle. Biofizika, 20: 519.

Zwislocki, J., 1963, Analysis of some auditory characteristics. in: "Handbook of Math. Psych.', Chapter 17, New York.

A PROPOSED ECHOLOCATION RECEPTOR FOR THE BOTTLENOSE DOLPHIN

(Tursiops truncatus): MODELLING THE RECEIVE DIRECTIVITY FROM TOOTH

AND LOWER JAW GEOMETRY

A. David Goodson

Sonar and Signal Processing Research Group
Electronic & Electrical Engineering Department
University of Technology, Loughborough. Leicestershire. LE11 3TU. U.K.

Margaret Klinowska

Research Group in Mammalian Ecology and Reproduction
Physiological Laboratory, University of Cambridge
Downing Street, Cambridge. CB2 3EG. U.K.

ABSTRACT

Perception and production of underwater sound in Odontocetes is known to extend over a wide band of frequencies but the transmit and receive pathways employed for echolocation appear unconventional. The lower jaw has been demonstrated to be an important component in the echolocation receive mechanism, but the mode of signal coupling to the auditory sense organs is still unclear. The resolving power of sonar signals is severely restricted by physical acoustics and to explain the excellent performance of the dolphin sonar during target tracking and food capture manoeuvres presupposes the presence of an efficient matched receiver fully exploiting the highest frequency components within the pulse transmissions.

The lower jaw is examined as an acoustic construct using the tooth spacing and jaw geometry to compare various detection enhancement hypotheses. Frequency/sensitivity receive 'beam patterns' have been computed to test for similarities with the transmitted signals. Of several hypotheses examined, one simple and apparently efficient model has been isolated which implicates the tooth/mandibular nerve structure as part of a high frequency echo pulse receptor which can accurately match the transmitted signal parameters. The model would also provide an explanation for the evolution of homodonty and polydonty, for the variability of tooth numbers between individuals in a given species, and for the alternate spacing of the teeth in opposite sides of the mandible.

This paper examines the component parts of this hypothetical echo receptor and collates the supporting evidence noted so far. The theory suggests that discriminatory sensing of a target echo range and bearing, particularly during the final hunting phase leading to food capture, exploits a dedicated high frequency receptor operating in parallel with the conventional wide band auditory sense organ.

INTRODUCTION

The echolocation emissions of Odontocetes represent an evolutionary peak in the development of underwater bio-sonar. In particular, the many studies of the bottlenose dolphin (Tursiops truncatus) while performing echolocating tasks, illustrate the superior target discrimination and detection ability of this dolphin. Our knowledge of this observed sonar performance is not yet matched by a clear understanding of all the bio-acoustic mechanisms employed for signal generation and detection. However the deceptively simple acoustic signals involved can be seen to function as part of a sophisticated, adaptive, multi-tasking sonar[1]. The wide ranging sonar capabilities displayed seem to imply that an additional dedicated sensory mechanism may be needed to achieve optimum performance in the high resolution modes.

All active sonar systems represent design compromises between conflicting acoustic, oceanographic and technological parameters, and as a result practical sonar systems are limited in application. Man-made designs usually originate from specific target size/range detection requirements combined with some knowledge of the anticipated target's echo strength. Choice of operating frequency and transducer array size are selected for target detection range and resolution, as are the transmitted pulse characteristics. Manufactured sonars are therefore normally limited to quite narrow specialised applications and several different systems may be needed simultaneously when attempting to investigate the wider underwater environment. Odontocete systems have evolved to solve similar acoustic problems and the signal variations between species may be assumed to reflect optimised performance matching environmental factors and preferred food type/behaviour.

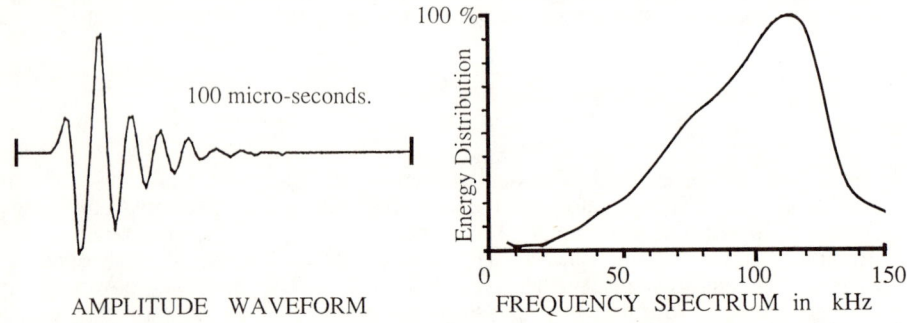

Fig.1. A typical click waveform and spectrogram from a bottlenose dolphin.

THE TRANSMITTED PULSE

Studies of the dolphin pulse emissions, characterised by the typical 'click' waveform in Fig.1, demonstrate a spectral energy distribution across the projected beam pattern which contains a highly directional high intensity component, peaking near 120 kHz, which is projected ahead of the animal. The changing spectral distribution recorded by sampling hydrophones positioned progressively off the transmit axis[2], supports the theory that the melon can be conceived as an acoustic beam former, analogous at the high frequencies to a 'Luneberg' lens antenna[3]. An effective acoustic lens requires a graded or 'onion' layered shell construction with slow sound propagation material in the centre and progressively increased velocities towards the outer layers. A graded velocity index has been observed in the bottlenose dolphin[4]. A directional beam will be projected from a signal source located at a focal point near, or in contact, with the rear surface of the lens. The non-spherical shape of the melon would indicate that the Luneberg lens is an oversimplification, but the concept illustrates its application as an acoustic beam former with provision for multiple sources projecting independent beams. Various structures in the Odontocete facial complex have been suggested as the sources of echolocation signals[5,6]. If a number of structures can be involved as sound injection points for the melon, then the concept can also explain the apparently anomalous high source levels detected by contact hydrophones at the tip of the rostrum[7].

The generation of highly directional acoustic signals for target localisation requires a beam forming technique and, regardless of whether the method employed is a lens, a curved mirror or a transducer array, there will exist a Fresnel zone close to the source within which the beam pattern is ill defined. This 'near-field' distance is a function of the beam forming aperture at the frequency of operation, i.e. defined by the acoustic cross-section of the dolphin melon. Our estimated maximum horizontal aperture for a typical adult Tursiops melon dimension (10 λ at 120 kHz) would imply that the associated near-field effects could extend to over 1.2 m (A_h^2/λ). Calculations suggest that beyond this distance, the horizontal beam width projected in the 'far-field' may approach 6 degrees, (measured at the -3 dB points for the 120 kHz component) and this would represent a theoretical limit to the angular resolution of the transmit beam former. If this postulated acoustic lens is under some muscular control and adjustable in focus, as the anatomy of the surrounding musculature suggests[5], then the near-field effects can be reduced and the transmitted far-field beam pattern could be established at a shorter range. This would permit some angular target discrimination, down to the ranges where eyesight would normally be employed during the final phase of target capture. The apparent rigidity of the Tursiops melon would seem to discount a major focus changing capability but the ellipsoid shape should offer different focus options depending on the position of the feed point. If the melon can be shifted with respect to the sound injection point or if the nasal diverticulae function[5, 6] as multiple sound sources into the melon, an astigmatic focus shift in the vertical plane could be achieved (Fig.2). That a close focus mechanism exists would seem to be evidenced by experimental measurement of the near/far-field transition[7], which has been determined to occur in practice as close as 0.4 m in front of the rostrum tip and at slightly increased ranges when measured on the tilted up major transmission axis, with beamwidths measured at 10 degrees in the far-field. This elementary transmission model ignores the effects of the rostrum and associated jaw tissue which must cause some shading effects especially at very close ranges.

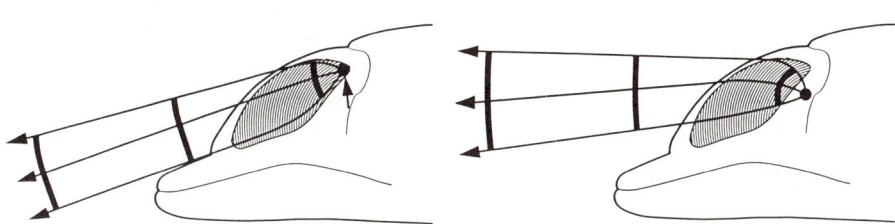

Fig.2. The melon interpreted as a bi-focal beam former.

ECHOLOCATION RECEPTION

For an active sonar system to be effective, both the transmitted pulse and the receive mechanism need to be closely matched to achieve optimum performance. It is clearly more difficult to establish the reception pathways in the dolphin because little is available to be observed. Traditionally, most observers have deduced that the echolocation sense, being based on acoustic emissions, will naturally use the conventional auditory sense mechanisms. It must be recognised that the typical mammalian ear evolved to exploit much lower frequencies than 120 kHz and has been optimised to accept complex wide band signals encoded with varying frequency and amplitude information. The excellent hearing sense to airborne sound[8] and to the sophisticated signature/communication whistles employed underwater would indicate that dolphin perception of this important band of sounds has not been compromised by the highly developed pulsed sonar. There is also evidence that whistling and clicking can occur simultaneously[5] indicating a multi-task capability that is difficult to appreciate if the mechanisms of either transmission or reception are common.

The studies by Brill et al[9] clearly demonstrate that the lower jaw is essential to the reception of the echolocation signals, however the assumption that the received sound is intercepted by the thin pan bones of the lower jaw and transferred by acoustic conduction through the mandibular fatty tissue merits closer examination. The reception of lower frequency sounds via this route seems possible as sounds that can be classified as signature/communication emissions require similar processing to their airborne equivalents and the conventionally proposed[10] fatty tissue pathway should conduct these acoustic signals from the water to the auditory system effectively.

CLOSE RANGE MOVING TARGETS

A different detection strategy seems necessary to localise and track a moving target's position with precision, especially if the target is a fish attempting to escape. Continuous information about the direction and range to the target needs to be sensed from the echo returns to achieve the catch. The time delay between transmitted click and the echo provides precise range information if the transmitted pulse duration is short. A short pulse also improves discrimination between the direct and multipath echoes. The time/bandwidth product of the Tursiops 'click' waveform seems to have been optimised for the temporal sensing of range and the detection of small targets. To sense unambiguously the changes in angular direction as a target manoeuvres implies paired reception channels for phase comparison. At close range, if sight is ignored, the angular resolving power of the sonar needs to be very good if the final stage of food capture is to be efficient. The 'click' signal is well suited to excite the impulse response from a target and the echo spectrum may provide considerable additional information about the target. The limited power contained in the single 'click' pulses would indicate a mode of operation that has been optimised for short range performance.

If the target structure resonates, or re-radiates internal reflections, when excited by the incident impulsive wavefront, then the returning echoes will contain a reverberation signature which can be interpreted spectrally to give information about the target's size and character. A strongly resonant structure, such as the swim bladder in some fish may be excited by matching the transmit repetition rate to the natural resonant frequency. Once frequency 'locked' onto such a target the very strong returns elicited would tend to mask any alternative weak echoes. i.e. from nets! The function can be seen to work very selectively as only a targeted fish from a shoal is chased to its muscle exhaustion/paralysis point.

A specular target echo returned from a position not directly ahead of the animal will contain some indication of the off-axis error derived from the lower frequency components of the ensonifying pulse and, by a head scanning movement, the dolphin can centre its direction onto the target. The leading edge of the on-axis echoes should contain a significant 120 kHz component which will be effectively absent in the returns from targets much more than 5 degrees off-axis. The echo amplitude and any spectral 'coloration' of the reverberant component when taken in conjunction with the time interval permits the target to be assessed for size and provides clues to discriminate its material and construction.

To be able to exploit the information contained in the returning echoes, an efficient matched receiving system must exist. With 'jaw echo-sensing' now a firmly established concept a detailed appraisal of the jaw tissues as an acoustic construct specific to echo location is necessary. If the conventional hearing sense is extended in frequency range to include the 120 kHz component present in the 'click' signals, then the extreme bandwidth implies a very poor signal/noise ratio which will impair the target detection performance, unless significant band limiting or matched filtering is introduced. A simple two element wideband receiver based on the normal auditory sense mechanism will also exhibit little forward directivity and, in conventional sonar terms, appears to fail as a matched system on a number of counts.

MULTIPLE SENSOR ARRAYS

Where high directivity is required in a manufactured sonar system, advantage is usually taken of multiple sensors arrayed together to reinforce detection in a specified direction. Band limiting techniques are usually applied as close to the transducer sources as possible, to maximise the signal/noise performance. Processing the outputs from multiple sensors also allows flexibility in beam forming and the direction of maximum sensitivity of many arrays can be 'steered' by introducing phase shifts in the data channels. Sensors arrayed for sonar applications are commonly employed in the 'broadside' mode where the summed outputs of a linear array[11] naturally beam form a directivity pattern normal to the line of the array. By choosing appropriate spacings and/or phasings, it is also possible to exploit an 'endfire' mode in which the peak sensitivity occurs in the same direction as the line of elements. The endfire array exhibits a significant advantage over most other beam formers when applied at short range. The sensitivity and beam width of the major lobe, formed along the line of the array, is maintained within the near-field and exhibits slightly improved beamwidth[12] as contact with the end of the array is approached!

Acoustically the most interesting structure in the Tursiops lower jaw, which must have been noted by many observers, are the straight rows of unusually regular and remarkably well matched teeth set in each side of the jaw. Indeed so uniform are the mid-row teeth that their length[13] in one case, and width[14] in another, have proved to be a powerful means to differentiate stocks in spinner (Stenella longirostris) and bottlenose dolphins respectively, although curiously, the number of teeth in each tooth row varies between individuals in almost all Odontocetes. If the teeth are modelled mathematically as an array of point sensing transducers in a continuous wave acoustic field, several interesting directivity patterns can be predicted, taking one side of the jaw as an isolated structure. The simplest model of the series exhibiting the best angular resolution (Fig.3), predicts that at one wavelength (λ) spacing a very sharp directivity lobe will form perpendicular to the array (about 3 degrees wide) and a wider lobe in line with the end of the array (25 degrees). Both of these main lobes exhibit equal sensitivity and offer signal/noise improvements some 20 times that of a single omni-directional element. Given the typical 11.4 mm tooth spacing[15] of adult Tursiops, this pattern would occur at 130 kHz, i.e. close to the maximum frequency of interest. At frequencies above this, ambiguous 'grating' lobes will start to form. A serious disadvantage of this model is the implicit assumption that the ensonifying source is a

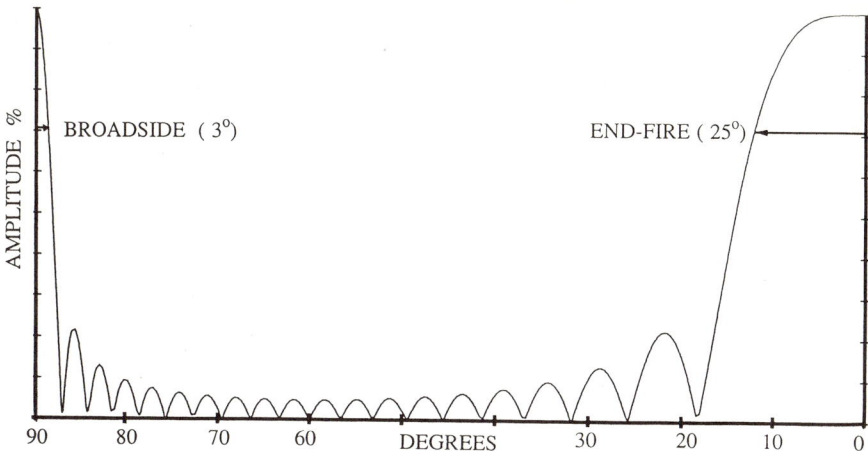

Fig.3. Point source line array pattern, predicted for 20 elements spaced at λ pitch.

continuous wave and that all the elements are contributing simultaneously, this clearly is not the case with a 'click' type signal.

PASSIVE RESONANT ELEMENTS

A model postulating the teeth as passive resonant elements (as in a TV antenna) matched to the peak frequency in the 'click' was then considered. If the teeth are sharply resonant structures, spaced at 1 λ pitch, and excited by a single broadband pulse then beamforming effects can occur but only if quite narrow bandwidths are assumed. If the 'Q' factor is low then signal/noise ratio is poor and beamwidths wide. (The 'Q' or quality factor is the ratio of centre frequency to bandwidth or approximately the number of oscillations needed to excite the maximum resonant amplitude.) If the 'Q' factor is very large the angular resolution factors improve but the sensitivity (to an impulse) becomes very low as the excitation period is too small to excite a significant tuned response. This model suggests that a compromise occurs when the 'Q' approximates 15-25, i.e each tooth continues to vibrate for sufficient time for the reradiated energy to interact with that from the remainder in the array. The resultant signal to be coupled acoustically to the auditory sense organ would comprise a relatively low frequency envelope response with both end fire and broadside directivity peaks at 125 kHz. The overall efficiency would seem to be rather poor. The short duration echo pulse and the precise tooth spacings noted (0.95 to 1 λ at 125 kHz in seawater) tends to suggest that the best angular resolution at endfire will occur near the 120 kHz spectral peak but the endfire lobe is reduced in sensitivity. The very narrow broadside response would be appreciably more sensitive. Re-modelling at 0.5 λ pitch, i.e. 62.5 kHz, promises more sensitive endfire performance, as the 'Q' can be lower, but the acoustic length of the tooth array in λ is halved and the predicted beam width will be doubled. If bone conduction coupling into the jaw is assumed in order to carry the enhanced sounds to the auditory sense then inter-element coupling effects via the bone will certainly introduce additional phasing effects which may further impair these first order predictions.

Several other models, based on passive resonant teeth interacting, were examined but although forward directivity improvements to aid angular target discrimination can be obtained, all of the modes examined seem inefficient and are regarded as unsatisfactory mechanisms if the detection sensing is assumed to be via an acoustically coupled mechanism along the lower jaw.

THE TOOTH NERVES AS PRESSURE SENSORS

Postulating the involvement of the tooth nerves as pressure sensors changes the model very significantly and at the same time the concept becomes generally applicable to all homodont and polydont cetaceans.

First, the propagation velocity in the mandibular nerves is very slow when compared to the velocity of sound in seawater. From human physiology[16] we know that the maximum conduction speed in nerves is of the order of 120 m/s. Conduction velocities of intradental nerves, i.e. within the tooth, have been studied in cats and are estimated for the faster fibres to be of the order of 30 to 45 m/s[17]. This is very slow when compared to the 1500 m/s velocity of sound in seawater. This slow propagation implies that a delayed response related to the tooth position in the jaw must exist. Since all the tooth nerves eventually proceed in parallel as part of the mandibular (and eventually trigeminal) nerve, it can be seen that the signal arrival times at the pons, for onward processing in the higher brain centres, are dependent on the individual nerve lengths. Random length 'delay line' coupled responses of individual sensors distributed within the ensonifying field can be represented as the inputs to a cross-correlation beam former[18]. In an idealised structure, the non-random lengths of these signal delay lines (individual tooth nerve paths) will translate the responses received from the teeth along one side of the jaw as they are sequentially ensonified by an on-axis sound pressure wave into a single time coincident event (Fig.4).

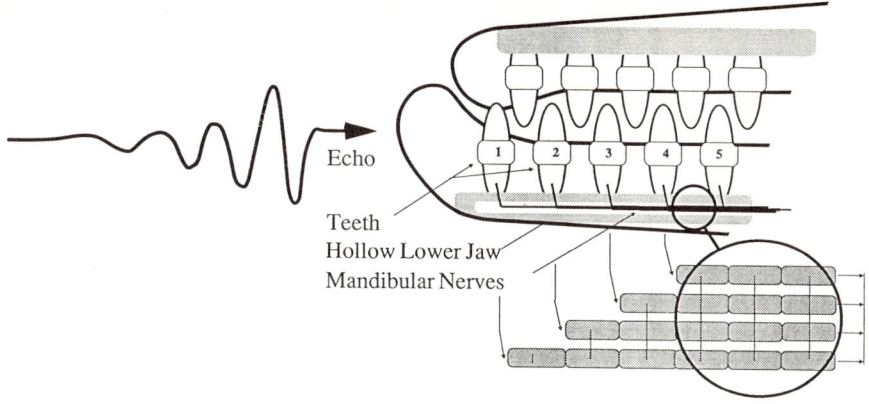

Fig.4. The signals propagating along the mandibular nerves can arrive simultaneously at the central nervous system.

Making this assumption and using the actual tooth spacing dimensions, we find that the peak response now exists only in the wanted 'endfire' direction and the unwanted 'broadside' responses are effectively nulled (time stretched by the progressive delays into low amplitude 'buzz').

The fixed timing 'advances' implied by 'delay line' beamforming also suggests that coarse angular information may be extracted from lower frequency off-axis signal components. A returning wavefront along an off-axis vector will ensonify all of the teeth in the array in less time than an on-axis signal. The resulting spread in the nerve propagated signal arrival times at the summing point may advantageously match the longer wave spectral component of these signals. Exploitation of this secondary effect, which relies on the transmitted frequency distribution predicted for the simple acoustic lens model, may be questioned on bandwidth grounds and would need additional frequency selective sensing in each tooth. Such off-axis perception is inherently of low angular resolution and the signals derived from the two jaw sides would need to be compared to localise the target position unambiguously.

Suggesting that the teeth are potential pressure transducers should not be too surprising an idea. Some human teeth are known to be acutely sensitive to ultrasound even when coupled in air, and dentine is a much better acoustic match to water than to air. 'Squeaky chalk' and similar sounds which "set one's teeth on edge" and more specifically the pain experienced by some dental patients exposed to ultrasonic cleaning techniques (typically 25 kHz), all illustrate that sensitivity to acoustic pressure effects exists in human at least in a rudimentary form. Further, the sensory trigeminal ganglion has a somatotopic organisation.[19]

TOOTH INNERVATION

Myelinated and unmyelinated sensory nerve fibres from the trigeminal ganglion traverse the pulp longitudinally giving branches to ramify in the cell-rich parietal zone. Here the fibres lose their myelin sheaths and continue into the odontoblast layer, some entering the dentinal tubules. Nerve fibres have not been found further that 100 μm into human dentine, but in rats they have been observed to run as far as the enamel-dentine border. Stimulation of dentine by thermal, mechanical or osmotic stimuli evokes a diffuse pain response in humans[16, 17]. The mechanism of stimulus transduction is unknown, but is unlikely to involve direct stimulation of the nerve endings in dentine because newly erupted teeth are equally sensitive, do not have a plexus of pulpar nerves, nor do pain-producing chemicals and local anaesthetics have much ability to stimulate or anaesthetise exposed dentine. The current hypothesis suggests that stimuli generate displacement of intra-extra cellular fluid along the dentinal tubules, causing in turn a local distortion of the pulp, sensed by the free nerve endings in the plexus. Recent evidence suggests that odontoblasts are joined by continuous tight junctions, and thus may be involved directly in relaying intratubular fluid movements to the nerve endings[17].

Detailed studies of the dental tissues of Odontocetes have been carried out by Boyde[20] which indicate that the dentine structure differs qualitatively from that of other mammals. In particular, there are large diameter bundles of parallel collagen fibres (von Korff fibres) parallel to the tubule axis, which may extend through the thickness of the dentine. Such radial fibres may function to strengthen the tissues and define their elastic properties. Another interesting feature is the high proportion of lateral and anastomotic branches of the dentine tubules. Such branches would serve to equilibrate the pressure within adjacent tubules and, together with the radial fibres, seem to indicate that the Odontocete tooth may have indeed evolved as a frequency selective acoustic pressure transducer.

In the examples of Tursiops examined so far, the teeth exhibit a maximum diameter in a band just below the gum line and this dimension in all the teeth (ignoring any anomalous specimens at the jaw tip) measured is remarkably constant and near circular. The actual cross-section appears slightly elliptical and the major/minor diameter ratio seems constant at about 8%. This might be interpreted as indicating that tubules on the minor axis are intended to match selectively nerve ends to the highest frequency on-axis component in the pulse envelope and the major axis of the tooth diameter might favour the slightly lower frequency off-axis echoes. More probably this may imply the bandwidth of a tooth receptor that has evolved to fully exploit the angular resolution of the system.

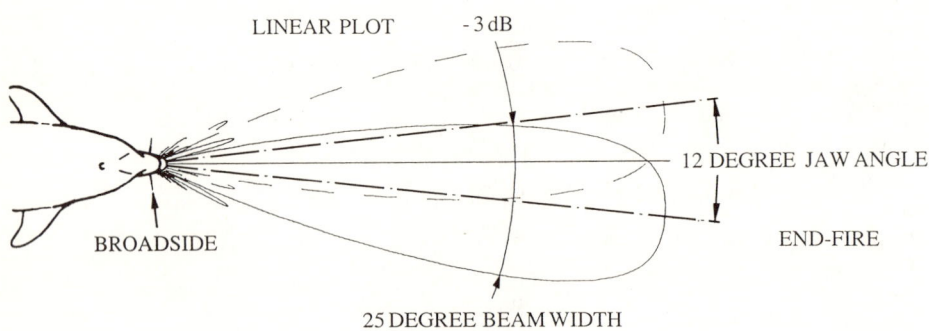

Fig.5. Computed sensitivity beam patterns at 120 kHz, overlayed for each side of the jaw.

THE DUPLICATE CHANNEL

Taking the second row of teeth in the lower jaw as a duplicate transducer array, the resulting directivity patterns cross over at the tip of the beak and produce a combined sensitivity in a forward capture field about 40 degrees wide symmetrical about the jaw (Fig.5). Each array in isolation will detect targets within its 30 degree lobe, but the two resulting signals are available for phase comparison and the resulting sensing of the target's angular position and direction of movement in the overlayed area can be made with much greater precision. Since both arrays are endfire constructs the degrading effects of the near-field zone, which are very significant in the broadside direction, are almost non-existent at endfire and the system should function unambiguously until the teeth contact the target.

Geometric data from a number of Tursiops skulls have been recorded, and while some positional errors may exist due to tooth movement after gum tissue removal, the overall repeatability of dimensions is remarkable. Three live 8 year old female specimens of Tursiops have now been persuaded to give dental imprints[15] and the plaster casts from these indicate a more precise spacing than in the museum examples. A further interesting detail that became evident when examining these casts was that the asymmetry of the jaw placed the left hand teeth exactly 180 degrees out of step with those on the right when receiving a signal on the central swim axis. i.e. at 6 degrees off axis the response from the two sides will be one tooth increment out of step. We speculate that neurological phase comparison decisions are simplified as a result of the time separation of signals received from the left and right tooth arrays. If these arrays were symmetrically placed, the phase shifts resulting from a target moving off-axis would be ambiguous, assuming that time separation is the criterion. The asymmetry of the arrays ensures that a target moving to the left or right of the centre axis is sensed as an increasing or decreasing interval between the two channel responses; the included angle between the two tooth arrays enabling target placement with 'vernier' precision in this central 12 degree sector.

Click trains recorded when approaching a target have been analysed by many researchers and the correlation between target range and click rate has been the subject of detailed study[2]. The processing time constant involved in attention to a specific target range has been quoted as a range independent, within limits, and reasonably consistent factor between 15-25 milliseconds[21]. More recently latency figures of 7-9 milliseconds have been determined[22] in conditions where the animal has acquired prior knowledge with which to anticipate the target range. It would therefore be interesting to investigate the propagation time required for the ensonified first tooth response to reach the central nervous system via the trigeminal nerve. The length of the endfire tooth array can be seen to function as a temporal sampling energy integrator. The time for an acoustic wave to transit the length of the array is determined by the sound velocity in the water and the minimum integration period, ignoring the time stretching effect due to the mechanical 'Q' of the tooth structure, will therefore be in the order of 143-173 microsecs (19-23 teeth). Critical interval times determined by Au et al[23] from experiments using synthesised multiple echoes indicate that an interval of 264 microseconds exists in practice. These figures are not in conflict with our hypothesis but can be seen to impose an acceptable upper limit to the 'Q' of the proposed tooth detector.

The shorter path followed by the tooth nerves along the upper jaw (via the maxillary division of the trigeminal nerve) can now be considered as a reference signal path. Triggered directly by the next transmission the responses will arrive after a smaller delay than from the lower jaw. Assuming that the simultaneous arrival of target echo and reference pulse is required for replica correlation processing, it can be seen that the next transmission is effectively timed to desense the receiving sensors after they have acquired the echo data. This effect will 'lock out' over-range information 'clutter'. In a conventional commercial sonar deliberate desensing of the receiver, timed to the transmission, is employed to initiate a time varying gain (TVG) function to compensate for square law spreading and acoustic absorption losses. The detection of over-range target echoes and the sensing of echo reverberation 'quality' seems likely to remain a function of the 'normal' hearing mode, i.e. separated from range/bearing discrimination.

FEATURES OF THE PREFERRED RECEPTOR MODEL

1)   The teeth of the lower jaw comprise two independent end-fire arrays of pressure transducers, primarily sensitive to the high frequency component in the 'click' waveform. This type of array works well without loss of bearing resolution within the near-field.

2)   The slow signal propagation along the mandibular nerves of each tooth's response provides the incremental 'delay lines' necessary to be able to sum each contribution in phase. This type of beam forming is required to exploit the transient nature of the 'click' echo.

3) The unusual structure of the dentine in the Odontocetes indicates significant differences from other mammals which we believe supports the concept that the teeth have evolved as sensitive pressure sensors matched to the acoustic pulse. The sensing element is assumed to be matched mechanically to the high frequency signal by the dentine structure and protected by it from out of band acoustic interference and mechanical damage.

4) Each end-fire array has a nominal directivity index of some 18 dB and an enhanced sensitivity in the zone ahead of each array some 20 times better than to either side. The removal or severe displacement of individual elements will cause very little impairment to this performance (i.e. loss of teeth and/or variation in their number). Beam widths at the -3 dB points, for each 20 tooth array (at 120 kHz), are estimated to be 25-30 degrees, but the overlayed 'crossed pair' pattern caused by the tapered jaw widens the overall sensitive sector to about 40 degrees (Fig.5).

5) The overlayed beam patterns permit excellent 'stereo' comparisons of the target echo phase which ensures that a target attempting evasive manoeuvres within the central 12 degree overlap sector can be tracked very accurately in angle and range down to the capture point without significant 'near-field' resolution problems. Note. In the far-field the measured 10° transmit beam-width at 120 kHz efficiently insonifies this predicted receive pattern over-lap zone.

6) As the teeth of the upper jaw receive direct excitation from the transmitted signal, the inclusion of the next transmitted impulse from this route can provide the reference signals for a correlation processing function. The shorter nerve pathways from the teeth of the upper jaw permit a replica of the next transmission to arrive at the central nervous system simultaneously with the received signals from the lower jaw for correlation comparisons whilst ensuring that the new transmission with its consequent desensing of the receiver occurs after the echo has been received. The propagation times for the lower jaw signals also need consideration when examining repetition rate/range processing time discrepancies.

The hypothetical sonar receptor proposed here is a specific construct optimised to exploit the high frequency, high energy component present in the echolocation 'click' spectrum. The time/bandwidth product of the waveform seems ideally suited to the precise determination of range and bearing, but the accepted auditory pathways do not seem optimised to exploit them. Other echo qualities induced by the target's structure may well be discriminated by the conventional auditory sense.

Additional clues implicating the tooth/jaw structure as a sound receptor exist. Our observations of <u>Tursiops</u> while examining unfamiliar targets reveals that sonar emissions are frequently employed at very short ranges, and occasionally while contacting the object with the tip of the lower jaw. Some dolphinarium animals have been observed to 'acoustically test/taste' dead fish gripped between their teeth before swallowing, which might suggest that acoustic attenuation through the fish tissue is being exploited. Deliberate excitation of strong mechanical resonances in some sample structures has now been observed on several occasions, e.g.[1]. Spectrograms of these very short range transmissions indicate a target interrogation technique involving sweeping the click repetition rate (occasionally to over 1 kHz) until strong target excitation occurs, followed by a short 'locked on' period while sustaining the precise frequency specific to the target.

The benefits to increased efficiency in the sonar system implied by this model would serve to explain why homodonty and polydonty have been so strongly favoured by evolution in many Odontocetes. The size and shape of the individual teeth have become so uniform in some groups that these characteristics alone can serve to differentiate between stocks of the same species. Further, given this uniformity, it would also explain the apparent anomaly that individual animals of the same species may have varying numbers of teeth in each row. It is known that elderly individuals, and those suffering accidents and illness during life may lose teeth but remain apparently otherwise well nourished; such tooth losses are acceptable if the precise number of teeth is not a critical parameter affecting the sonar performance. Loss of individual mid-row teeth will increase the amplitude of the sidelobes before the primary beam widens. A missing element which shortens the array will only increase the beam width by a factor of 1/n (where n is the number of sensors in the array). Thus polydonty provides a degree of redundancy to the structure. Loss of individual sensors will degrade the signal/noise ratio but this should not impair performance at short range. Tooth wear should not degrade performance, provided that the maximum diameter of the tooth (just below gum level) is not affected. A practical test of this model could be made using an edentulous animal. Such an animal would be expected to have significant difficulties in capturing a live fish whilst blindfolded. The ability to perceive the presence or absence of static test-targets should not be appreciably degraded. It would be expected that when tracking a moving target, more exaggerated head 'scanning' movements would be required to resolve left/right ambiguities. Training edentulous animals to accept 'eye-cup' blindfolds may prove difficult, because of the animal's greater dependency on eyesight for precision spatial coordination.

CONCLUSIONS

The high frequency component of the *Tursiops* sonar emission must be assumed to be the dominant factor when attempting to explain the target discrimination performance and acoustic resolving power of this bio-sonar.

The additional high frequency receptor mechanism proposed helps to explain a number of observed features of *Tursiops* sonar performance, particularly those involving efficient feeding behaviour on acoustically small, agile targets. It also serves to explain a number of otherwise curious features of Odontocete evolution.

The tooth array receptor described implies an additional sensory ability, optimised for small target range/bearing discrimination, which complements the auditory mechanisms described to date. As the trigeminal pathway is not primarily associated with hearing or vision, echo perception of a target via this receptor may equate to the acoustic equivalent of a tactile sense. Conventional auditory processing of returning echoes is also still assumed to occur as the spectral quality of the reverberation signature provides essential clues to target texture and structure.

The complete echolocation system may be modelled as a closed loop 'servo-system' in which the 'click' firing rate of a pneumatically driven 'relaxation oscillator' source is maintained in step with the echo return timing. Many of the sonar behaviour patterns observed during target acquisition and tracking are explained by this analogy. However at very close ranges the frequency/range relationship is observed to break down and it seems likely that the maximum pulse repetition rate is limited to little more than 1 kHz. At very short range the need for correlation processing is unnecessary and the echoranging time interval can be sensed directly.

The proposed reception mechanism has good directivity and is efficiently matched to the transmission source. The concept provides spatial perception of targets throughout the full detection range but it functions uniquely well at very short ranges, i.e inside the near-field of the transmission mechanism.

ACKNOWLEDGEMENTS

M. Wood and G. Gouvin, Theater of the Sea, Islamorada, USA.; M. Jacobs and M. Saunders, BBC(Bristol); P. Bloom and C.Wright, Flamingoland Dolphinarium (UK); P. A. Kelley, Cambridge University Dental Service; W. Wood; R. Beitsma; S. Rudland;.
Financial support from the Commission of the European Communities and from Isobel Goldsmith is gratefully acknowledged.

REFERENCES

1. A. D. Goodson, M. Klinowska and R. Morris, "Interpreting the Acoustic Pulse Emissions of a Wild Bottlenose Dolphin Tursiops truncatus", Aquatic Mammals, 14(1): 7-12 (1988).
2. W. W. L. Au, "Echolocation Signals in Open Waters", 260-269, in: Animal Sonar Systems, R. G. Busnel & J. F. Fish, eds., Plenum Press. New York (1980).
3. M. I. Skolnik, "Antennas", 252-254, in: Introduction to Radar Systems. 2nd Ed., McGraw Hill, New York (1980).
4. K. S. Norris and G. W. Harvey, "Sound transmission in the porpoise head", J.Acoust.Soc.Am., 56(2): 659-664 (1974).
5. J. G. Mead, "Anatomy of the External Nasal Passages and Facial Complex in Delphinidae (Mammalia: Cetacea)", Smithsonian Contributions to Zoology, 207: 1-72 (1975).
6. S. H. Ridgway, D. A. Carder, R. F. Green, A. S. Gaunt, S. L. L. Gaunt and W. E. Evans, "Electromyographic and Pressure Events in the Nasolaryngeal System of Dolphins during Sound Production", 239-249, in: Animal Sonar Systems, R. G. Busnel & J. F. Fish, eds., Plenum Press. New York (1980).
7. W. W. L. Au, R. W. Floyd and J. E. Haun, "Propagation of Atlantic Bottlenose Dolphin Echolocation Signals.", J.Acoust.Soc.Am. 64(2): 411-421 (1978).
8. T. H. Bullock, A. D. Grinnell, E. Ikezono, K. Kameda, Y. Katsuki, M. Nomoto, O. Sato, N. Suga and K. Yanagisawa, "Electrophysiological studies of central auditory mechanisms in cetaceans, Z. fur vergl. Phys. 59: 117-156 (1968).
9. R. L. Brill, M. L. Sevenich, T. J. Sullivan, J. D. Sustman and R. E. Witt, "Behavioural Evidence for Hearing Through the Lower Jaw by an Echolocating Dolphin (Tursiops truncatus)", Marine Mammal Science, 4(3): 223-230 (1988).
10. K. S. Norris, " The Evolution of Acoustic Mechanisms in Cetaceans", 297-324, in: Evolution and Environment, E. T. Drake, ed. Yale University Press. (1968).
11. R. C. Hansen, "Linear Arrays", in: The Handbook of Antenna Design, A. W. Rudge, K. Milne, A. D. Olver and P. Knight, eds., Inst.Elec.Eng. P.Perigrinus, London. (1986).
12. H. O. Berktay and J. A. Shooter, "Nearfield Effects in End-Fire Line Arrays" J.Acoust.Soc.Am. 53(2): 550-556 (1973).
13. P. A. Akin, "Geographic Variation in Tooth Morphology and Dentinal Patterns in the Spinner Dolphin Stenella longirostris", Marine Mammal Science, 4(2): 132-140 (1988)
14. W. A. Walker, "Geographic Variations in Morphology and Biology of Bottlenose Dolphins (Tursiops) in the Eastern North Pacific", South West Fisheries Center Administrative Report, LJ-81-03c.54 (1981).
15. P. Bloom, A . D. Goodson, M. Klinowska and S. Rudland. "Tooth and Jaw Geometry in some Odontocetes.", (In prep.)
16. P. L. Williams, R. Warwick, M. Dyson, L. H. Bannister, eds., in: "Gray's Anatomy", Churchill Livingstone, London. (1989).
17. D. J. Anderson, A. G. Hannam and B. Matthews, "Sensory Mechanisms in Mammalian Teeth and Their Supporting Structures", Physiol. Rev. 50: 171-195 (1970).
18. P. N. Denbigh and P. A. Tollman, "Beamforming by Cross-Correlation of Received Spectra", 439-446 in: Adaptive Methods in Underwater Acoustics. H. G. Urban, ed., Reidel, Dordrecht. (1985)
19. J. M. Gregg and A. D. Dixon, "Somatotopic Organisation of the Trigeminal Ganglion in the Rat", Archs. oral. Biol. 18: 487-498 (1973).

20. A. Boyde "Histological Studies of Dental Tissues in Odontocetes." Rep.Int.Whal.Commn., Special Issue.3, 65-87 (1980).

21. S. H. Ridgway, T. H. Bullock, D. A. Carder, R. L. Seeley, D. Woods and R. Galambos, "Auditory Brainstem Response in Dolphins", Proc.Nat.Acad.Sci. 78(3): 1943-1947 (1981).

22. P. W. B. Moore, "Dolphin Echolocation and Audition", 161-168, in: Animal Sonar Processes and Performance, P. E. Nachtigall and P. W. B. Moore, eds., Plenum Press, New York (1988).

23. W. W. L. Au and P. W. B. Moore, "The Perception of Complex Echoes by an Echolocating Dolphin", 295-299, in: Animal Sonar Processes and Performance, P. E. Nachtigall and P. W. B. Moore, eds., Plenum Press, New York (1988).

STUDIES ON ECHOLOCATION OF PORPOISES TAKEN IN SALMON GILLNET FISHERIES

Yoshimi Hatakeyama and Hideo Soeda[1]

National Research Institute of Fisheries Engineering
5-5-1,Kachidoki,Chuo-ku,Tokyo,104 Japan
[1]College of Agriculture and Veterinary Medicine,Nihon University
3-34-1,Shimouma,Setagaya-ku,Tokyo,154 Japan

INTRODUCTION

In the Bering Sea, marine mammals, particularly Dall's porpoises, Phocoenoides dalli, get incidentally entangled in gillnets of the Japanese mothership salmon fishery. The maximum total number of Dall's porpoises to be caught in a year was fixed by U.S. law. In order to develop techniques and devices to reduce the incidental catch in the gillnets, it is necessary to learn why and how they get entangled.

Until recently, only four studies on clicks emitted by Dall's porpoise were known. Three of them reported only low frequency clicks with peak energy below 10 kHz.[1-3] However Awbrey et al.[4] could not find low frequency components. They described detailed characteristics about the echolocation high frequency clicks with peak energy between 120-160 kHz, but the source level of the clicks was not measured. They also estimated the Dall's porpoise's auditory capability from cochlear morphology. Based on an estimation of net target strength and the porpoise's ability to hear resulting echoes, they doubted that Dall's porpoises could detect monofilament gillnets. In order to determine whether Dall's porpoises can detect and avoid the gillnets, we recorded their clicks and observed their reactions to the gillnets in the open sea.

It is difficult to catch Dall's porpoises alive and maintain them in captivity for a long time. However, a closely related species the Harbor porpoises (Phocoena phocoena) are easily caught alive in the Japanese coastal seas and are kept in several aquariums. So we attempted to examine the echolocation ability of Dall's porpoise through experiments using Harbor porpoises which belong to the same family as the Dall's porpoise. The Harbor porpoise emitted clicks of high frequency (more than 100 kHz) to detect objects.[5-7] Detection of fine wire[8,9] and discrimination of the height of a metal cylinder[10] were studied to estimate the echolocation ability of the Harbor porpoise. The audiogram[11], transmitting directivity[7] and minimum audible angle[12,13] of the Harbor porpoise were reported. Thus, there is considerably more information on clicks and echolocation abilities of Harbor porpoises, but their reactions to gillnets have not yet been observed. Therefore, we observed their reactions to gillnets in a pool and estimated their detectable ranges for the gillnet from waveform characteristics of their clicks, reflectivities of the gillnet materials and their hearing ability.

This report is a summary of parts of six documents submitted to the meeting of the Scientific Subcommittee of the Ad Hoc Committee on Marine Mammals of the International North Pacific Fisheries Commission(INPFC).[14-19]

MATERIALS AND METHODS

Materials

Dall's Porpoises in the Bering Sea. On 1982 July 2, about 10 Dall's porpoises approached a salmon research vessel while it was stationary in the Bering Sea after retrieving the gillnet. Two to four individuals swam around the vessel at the speed of 2-3 m/s, as if they were interested in the hydrophone hanging on the port side. They approached the hydrophone within a distance of 1 to 2 m repeatedly. Recordings were made for about 10 minutes. Because the weather was fine and the sea was calm, the condition was perfect for recording.

Dall's Porpoise in a Pool. On 1984 May 7, two purse seiners(14.5 GRT each) discovered a group of 20 Dall's porpoises in an area 10 miles off Hitachi city, Ibaraki Prefecture, and caught three Dall's porpoises with a mackerel purse seine. Two were dead on retrieval, but the remaining one (body length 160 cm, weight 77 kg, male) was alive and brought quickly to a pool(7x5x3 m) of the Oarai Aquarium about three hours after capture. After acoustic studies on May 9, the porpoise was transported to a pool(12x8x3 m) of the Kamogawa Sea World because of inadequate facilities at Oarai. The transportation time was about four hours. On May 10, acoustic studies were conducted at the Kamogawa Sea World.

We continued to provide food such as live sardine and live squid up to May 17 but, as he did not eat them, forced feeding was also conducted each day from May 10. From May 11, antibiotics and hyperadrenocortical hormones were administered each day. On May 18, the twelfth day after arival in human care, the porpoise suddenly had trouble swimming and died. Although Ridgway[1] succeeded in holding a Dall's porpoise using mackerel and jack mackerel as food for a period of 18 months, the feeding in the present experiment was not successful and the porpoise died within 12 days.

Harbor Porpoises in a Pool. Five Harbor porpoises were caught in all at a set-net fixed near Hakodate city, Hokkaido, from 1985 April to 1986 May. As a result four of them were used for the following two experiments. Three of them were male and one was female. Their body lengths ranged from 136 to 147 cm and weights from 41 to 43 kg. Three of them were kept in a pool(17x12x3.5 m, Fig.2) of the Kamogawa Sea World to record their clicks. Horse mackerel and sillaginoid, *Sillago sihama*,(10-15 cm in body length) were thrown into the pool and clicks were recorded while they were approaching and echolocating the baits. The distance between baits and the hydrophone was set at about 3 m. Two of them were used for the experiment on the observation of their reactions to a gillnet in the same pool.

Recording and Analyzing System

Systems for recording and analyzing clicks are shown in Fig.1. The upper system in Fig.1 was used for Dall's porpoise's clicks. The lower system was used for Harbor porpoise's clicks. The hydrophone of each system was B&K 8103. The amplifiers of the systems were B&K 2650 and 2635, respectively. Total frequency ranges of two systems were up to 200 kHz. Signals were passed through 2 kHz high-pass filter to reduce ambient sea noises and aim for the recording of high frequency clicks. The tape recorder of the both systems was a TEAC R410, four channels data recorder with a frequency range of 300 Hz to 200 kHz at 152 cm/s.

Parameters to be analyzed for clicks were sound pressure, pulse width, peak spectrum frequency, band width, interval and total number of clicks in a series. Pulse width was a time width in which an envelope of the click declined to half of its peak value and band width was a frequency width in which a spectrum of the click decreased by 3 dB from its peak value.

In the system (a) Fig.1, an output from the tape recorder was added to a storage oscilloscope(Sony-Tectronics 564B) and waveform analyses were made from polaroid photographs. Frequency was calculated from an average period of 3 to 5 cycles around the peak amplitude. Other parameters, such as the total number and the highest sound pressure related to the source level at 1 m, were measured from the whole photograph or record of a series of clicks.

In the (b) system, clicks were analyzed by a FFT analyzer(National VS-3310A) and a high speed graphic recorder(Rion LR-50). In order to conduct an effective analysis of clicks, 32 consecutive clicks were analyzed with an automatic continuous analysis memory function(Auto Store) of the FFT analyzer. The tape speed was finally reduced to 1/100 by conducting twice operations where the tape speed was reduced to 1/10 so that we could facilitate analysis. The waveform and power spectrum of the click were recorded with the internal video printer. The frequency range of the FFT analyzer was 10 Hz to 40 kHz, the number of sampling and spectral lines were 1,024 and 400, respectively and the capacity of data memory was 64,000 words. In this

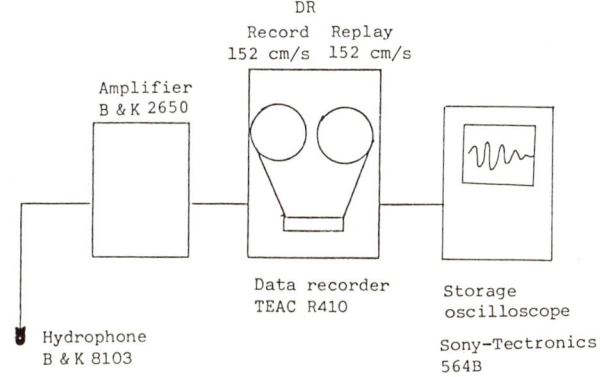

(a) System for Dall's porpoise's clicks

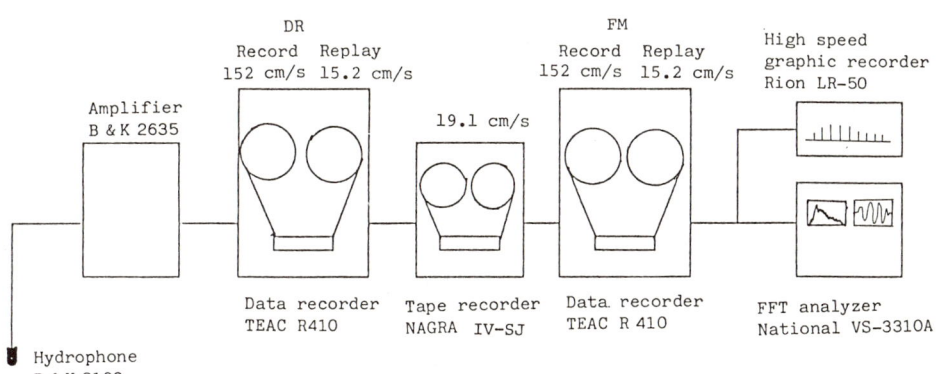

(b) System for Harbor porpoise's clicks

Fig.1 Blockdiagram of the measuring and analyzing system.

study, the equivalent frequency resolution power was 1 kHz. The tape recorder(NAGRA IV-SJ) was used to reduce the tape speed and its frequency range was 25 Hz to 20kHz at the tape speed of 19.1 cm/s. In order to study the total number of clicks, interval and source level, a series of clicks were recorded at a time constant of 0.01 s with the high speed graphic recorder. Its frequency range was 1 Hz to 20 kHz and its resolution power was 0.5 dB.

System for Observing Harbor Porpoise's Reaction to the Gillnet

Since the salmon gillnets in the fishery in the Bering Sea are set at sea during darkness from evening to the following morning, it is desirable to observe reaction to gillnets by the porpoise in a darkened pool. All mercury lamps over the experimental pool were turned off. However two fluorescent lights on the path beside the pool and three mercury lamps on the

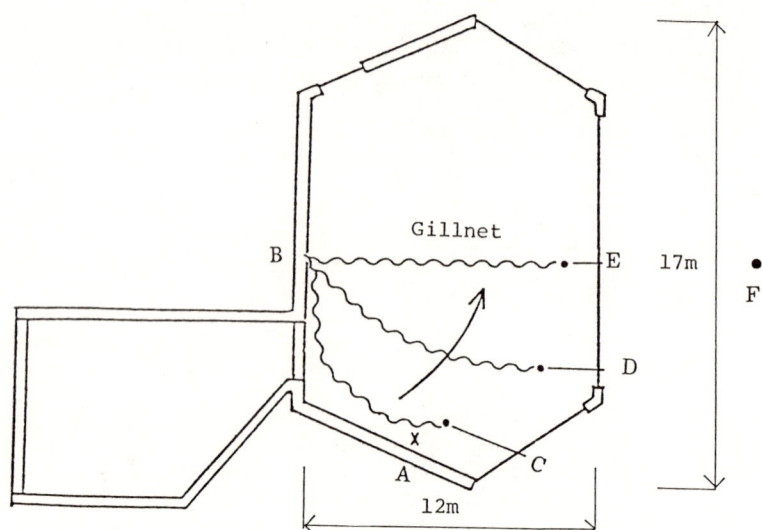

Fig. 2 Top view of the net in the experimental pool. The underwater irradiance was measured at position A. One end of the float-line was fixed at position B and another end was moved from position C to position E. The nightscope was set at position F.

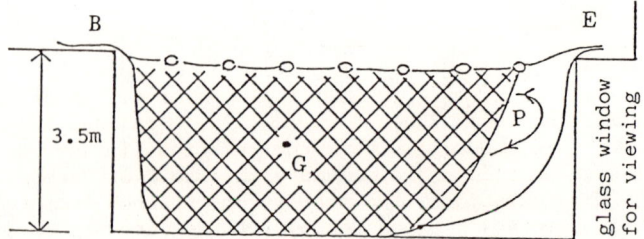

Fig. 3 Side view of the net in the experimental pool. B and E were ends of the float-line. The Harbor porpoise got entangled at position G. The Harbor porpoises could pass between the rope and net at position P.

272

Table 1. Waveform Characteristics of Clicks.

Porpoise	Peak Frequency (kHz)	Source Level (dB)[*]	Pulse Width (μs)	Interclick Interval (ms)	Total Number of clicks in a series	References
Dall's porpoise in the open sea	135-149	165-170	50-60	8-150	9-47	this paper
"	120-160		50->1000	13-143	9-40	Awbrey et al.(1979)
Dall's porpoise in the pool	90-115	154-157	15-60	9-48	64-176	this paper
Harbor porpoise in the pool	125-140 (130)▲	158-162 (160)	29-83 (43)	10-123	4-23	"
"	110-150	132-149 (140)	(100) 70 ▲▲	20	14	Møhl and Andersen (1973)

▲( )--- mean value, ▲▲ ___---calculated by us from photographs in their paper.
[*] --- dB re 1 μPa at 1 m.

reserve pool remained lit for potential emergency measures in case of entanglement. The underwater irradiance at position A in Fig.2 was 1.4 lux. vertically and 0.8 lux. horizontally and conditions were such that nearby netting at a distance of about 2 m could be seen dimly by the naked human eye accustomed to darkness.

Salmon gillnet was cut and remodeled to a small-sized gillnet(4 m in height, 15 m in length) and the net was set from position B in Fig.2. The net was placed in order in position C, D and E along the pool and finally stretched from position B to position E in the center of the pool. At position E, the net was adjusted by shortening the lead-line, but some space was produced between the rope and net, as shown in Fig.3 so that the porpoise could pass through the space.

Reactions to the gillnet by two Harbor porpoises were observed with a nightscope(FUGINON, Nightscope FNS-P101) at position F in Fig.2. Its amplification factor of brightness was 50,000 and sightable distance was about 1,000 m when sighting a human in the moonlight(0.01 to 0.1 lux.).

Method for Evaluation of Echolocation Ability

In the experiment on the ability to avoid a fine wire by a blindfolded Harbor porpoise, the porpoise recognized the wire of 0.5 mm diameter from a distance of 0.5 m by echolocation and avoided it. Møhl and Andersen[6] attempted to explain this phenomenon from the relationship between the waveform characteristics of clicks and auditory threshold of Harbor porpoise. According to their calculations, the echo level obtained from the source level and the back-scattering cross section of the wire was 88 dB re 1 μPa, and the auditory threshold corrected for small pulse width, was 92 dB. Both values were of the same order of magnitude. We can evaluate the echolocation ability on the condition that the detection threshold(DT) is 4 dB lower than the corrected auditory threshold. A detectable range is an intersection point on the graph which indicates the echo level as a function of the distance and the detection threshold.

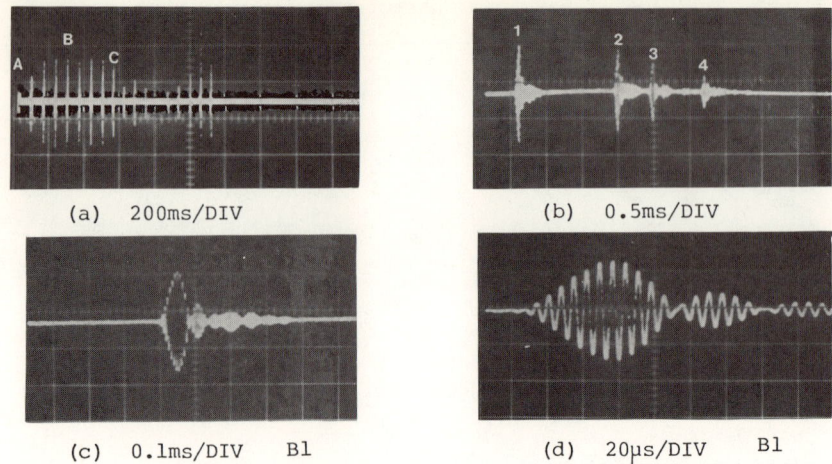

(a) 200ms/DIV   (b) 0.5ms/DIV
(c) 0.1ms/DIV B1   (d) 20µs/DIV B1

Fig. 4 Waveforms of clicks emitted by Dall's porpoise.

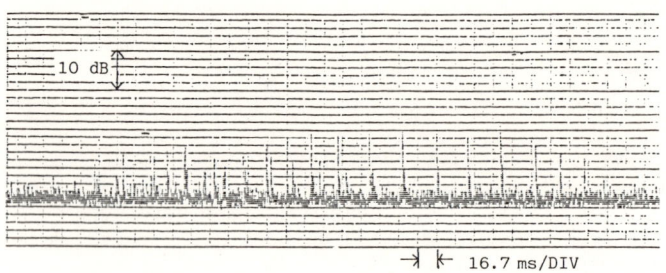

Fig. 5 Change of level and interval of clicks recorded with high speed graphic recorder (Harbor porpoise).

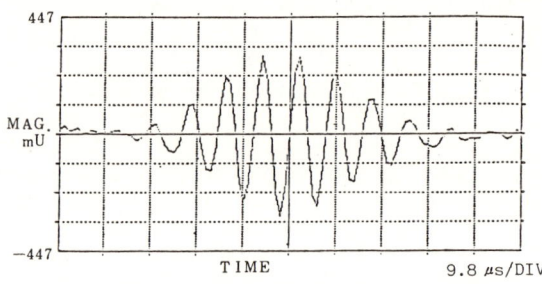

Fig. 6 Waveform analysis with FFT analyzer (Harbor porpoise).

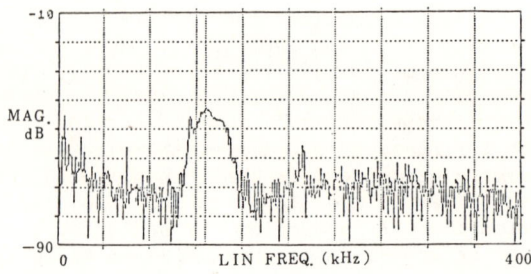

Fig. 7 Spectrum analysis with FFT analyzer (Harbor porpoise).

RESULTS AND DISCUSSION

Waveform Characteristics of Clicks

   Dall's Porpoises in the Bering Sea. From the 10 minute recording, seven portions with clear clicks were selected for detailed analysis. The photographs of a typical echolocating click of one portion are shown in Fig.4. Waveform characteristics of these clicks are listed in Table 1. Low frequency component was not found. Data reported by Awbrey et al.[4] also are listed in Table 1. When comparing these data, two total numbers almost agree. However, the longest pulse width of the latter is remarkably long. Dall's porpoises described in this paper emitted clicks with a narrow frequency range. No data on the source level were reported in the latter. The two maximum intervals are nearly equal but our minimum interval is smaller than that of the latter. Detailed observation indicated that each click consisted of one or a series of 2 to 4 pulses. However, Awbrey et al. reported that the signals of Dall's porpoises were single or double pulses of constant frequency.

   Dall's Porpoise in a Pool. Total recording time was 72 min. and total number of click emissions was 33. Three representative series(A,B,C) of clicks were analyzed. Series A was recorded right after a rope on the water surface was raised to 2 m above in the Oarai Aquarium. Series B was recorded immediately after the Dall's porpoise was caught for a physical check up in the same aquarium. Series C was recorded during his swimming in the Kamogawa Sea World.

   Waveform characteristics of his clicks are listed in Table 1. We could not find any low frequency component. Compared with wild Dall's porpoise's clicks, the frequency is lower by about 30 to 50 kHz, the number of clicks in a series is larger, the source level is lower by about 10 dB and many clicks with shorter pulse width and interval were found. Clicks of the porpoise in captivity might be influenced by a shock of being captured and the limited environment of the pool.

   Harbor Porpoises in a Pool. Four series of clicks with considerably high sound pressure were selected and 48 clicks were analyzed in total. An example of the record taken with high speed graphic recorder is shown in Fig.5. A typical waveform and spectrum of the click obtained with the FFT analyzer are shown in Figs.6 and 7. Waveform characteristics of their clicks are listed in Table 1. The band width of clicks ranged from 9 cycles to 33 kHz with mean value of 21 kHz. The clicks included about 9 cycles of narrow band sine waves which became gradually larger and mostly reached a maximum at 4th cycle.

   Data reported by Møhl and Andersen are also listed in Table 1. The source level in this paper is about 20 dB higher than observed by them. It could be considered as the reason for high source level that the pool was large and that the three porpoises competed with one another in catching food. The source level of clicks emitted by wild Dall's porpoises is slightly higher than that of Harbor porpoises in this paper. In Møhl and Andersen's paper, there was no description on the total number and interval of clicks but these values could be calculated to be 14 and 20 ms, respectively, from their photograph. These are in the range of the values in this paper. They also reported that a click was a gated sine wave of 7 to more than 11 cycles and the maximum frequency and amplitude were usually found within the first 2 or 3 cycles. The frequency range of major energy in this paper is narrower than that reported by them but the mean values are equal.

   In comparison with the Dall's porpoise, the frequency of clicks emitted by the Harbor porpoise is 12 kHz lower, the pulse width is 11 µs shorter and its click is a single pulse. The mean peak spectrum frequency of 130 kHz is close to the upper hearing limit of the Harbor porpoise[11]. We may say that the Harbor porpoise lays more stress on the reflectivity, and the distance and angle resolution than on the auditory sensitivity.

Observation on the Reaction to the Gillnet

Reactions of Dall's Porpoises to the Gillnet in the Open Seas. In July 1983, we observed behaviors of Dall's porpoises around the vessel in relation to the gillnets which were being retrieved in the Bering Sea. Two Dall's porpoises out of three in a group dived and passed under the gillnet and reappeared on the other side. However the third one, following them, got entangled in the vertical mid portion of the net. In addition, it was twice observed that a Dall's porpoise passed through a damaged hole(1.5 m wide, 1.0 m high) of the netting in the upper portion of the net without changing its swimming speed(3 to 4 m/s).

In August 1983, capture experiments were conducted in the coastal area of east Hokkaido. After discovering a school of Dall's porpoises and determining their direction of movement, we set gillnets(1,300 m long, 6 m deep) and chased them toward the gillnets with 4 boats. In general, the porpoises changed their swimming direction in front of the net and then swam along the net or dived and passed under it. However, in one case, two porpoises out of the group of 3 dived suddenly at about 4 to 5 m in front of the net and surfaced about 10 m on the other side of the net, but the third one rushed into the net, broke through it and escaped.

These facts show that Dall's porpoises have a high resolution echolocation ability by which they can detect the net and distinguish even minute irregularities of the net, such as a hole in the net. They can avoid getting entangled in the net. It appears that when a number of Dall's porpoises swim together, those which follow other porpoises do not use their echolocation, but merely follow the porpoises ahead and, as a result, they can become entangled in the net. It still remains to be studied in detail why and how they get entangled in the net in spite of their good echolocation ability.

Reactions of Harbor Porpoises to the Gillnet in the Pool. Reactions to the gillnet by the Harbor porpoises were sorted by patterns as shown in Fig.8 and listed on Table 2. The Harbor porpoises passed between the rope and the net(see P in Fig.3) eight times, and made nine U-turns immediately in front of the net. One porpoise approached the net at a right angle 4 minutes and 22 seconds after setting the net and in the act of turning counter-clock-wise immediately in front of the net, the caudal fin of the porpoise was entangled in the middle of the net(G in Fig.3). The porpoise got entangled in a complicated manner, the caudal fin was injured and there were net marks on the pectoral fin and head. The approximate swimming speeds were calculated from

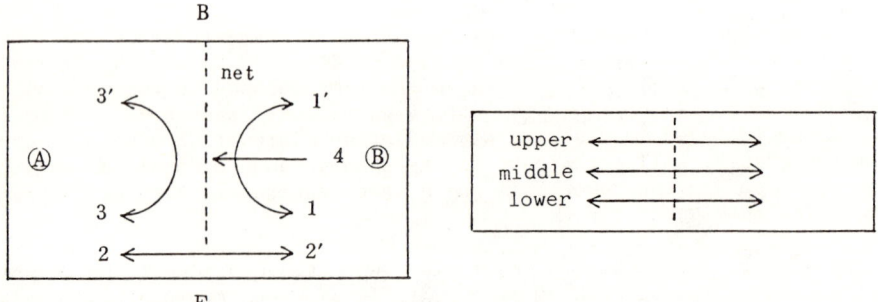

Fig. 8 Patterns of Harbor porpoise's behaviors near gillnet.

Table 2 Harbor Porpoise's Behaviors near Gillnet.

Time (h:m:s)	Pattern of behavior		Swimming speed(cm/s)	Time (h:m:s)	Pattern of behavior		Swimming speed(cm/s)
18:47:00 [1]				18:59:59	3'	lower	
57:48	2'	middle	121	19:00:16	3'	middle	
58:03	2	"	97	29	2'	lower	106
10 [2]				41	2	"	166
37	3'	lower		53	2'	"	146,225
56	3'	"		1:07	1	upper	
59:09	2'	upper	132	26	1	lower	
25	2	lower	94	40	2	middle	
35	2'	"	224	2:03	2'	lower	
37	3'	middle		22	1	middle	
49	3'	upper		32 [3]	4	"	85

1) start time, 2) The net was completely set in the pool.
3) One porpoise got entangled.

pictures of the nightscope and shown in Table 2. The swimming speed prior to entanglement was a maximum of 225 cm/s and a minimum of 94 cm/s (146 cm/s on average). The swimming speed at time of entanglement was 85 cm/s, which was the lowest one reported.

It appeared that at first they recognized the float-line and lead-line by sight and or echolocation and swam carefully. However, when they became accustomed to such circumstances, they approached the net and one got entangled because their echolocation ability was not good.

Evaluation of Echolocation Ability of the Harbor Porpoise

We calculated an echo level which a porpoise would receive by taking into account a propagating supersonic pulse wave with a very short pulse width and a very narrow beam angle.

Occurrences of reflected wave are considered on the assumption that the echolocation beam incidences on the lead-line or on the midpoint of the depth of the net. Supposing that the supersonic wave with a pulse width of τ hits the net, reflected waves which return to the location of the porpoise at the same time are composed of the waves reflected by the net within a range from R to $(R+c\tau/2)$, where c is velocity of sound in the water and R is distance between the receiving point and closest reflecting portion of object. Figure 9 illustrates a top view, where O and XY denote the location of the porpoise and the length of the net, respectively and $OP_1=R$ and $P_1P_2=P_2P_3=P_3P_4=\cdots=c\tau/2$.

When the supersonic pulse wave is propagating towards the net and the pulse wave front reaches P_2, the portion of the net included in a circle with a diameter of Q_2R_2 will contribute to the simultaneous composition of one reflected wave. Successively, when the wave front reaches P_3, the portion of the net included in the ring with a inside diameter of Q_2R_2 and outside diameter of Q_3R_3 will contribute to the simultaneous composition of the reflected wave. As the supersonic pulse wave propagates, this ring spreads toward the outside of the net.

Figure 10 illustrates a front view from the location of the porpoise towards the gillnet. EFGH denotes a gillnet. When the inner and outside diameters of the ring become larger than the depth of the net, the portion of the net contributing simultaneously to the reflection is limited by the depth of the net to the shadowed area of the ring. As the supersonic pulse wave propagates, the portion of the lead-line contributing to the simultaneous composition of one reflected wave moves from inner to outer rings consecutively; in this order: Q_2R_2, $Q_2Q_3+R_2R_3$, $Q_3Q_4+R_3R_4$, -------

S and L denote an area and length of the portion, respectively, related to the simultaneous composition of the reflected wave, as explained above. We assume that the porpoise can uniformly receive the reflected wave from each portion of the net on which the propagating pulse wave of the click incidences successively. On this assumption the reflection factor increases in proportion to the square root of S or L.

In general, when the porpoise echolocates on a small object, the echo level(EL) which the porpoise receives is given by:

$$EL = SL - 40\log R - 2\alpha R + TS \qquad (1)$$

where SL is source level(dB re 1μPa), R is distance between the porpoise and the object, α is absorption coefficient(dB/m) and TS is target strength(dB).

In case of a small object, the reflection factor does not change with the distance. However, in case of a long or large object, the reflection factor changes with the distance because S and L change with the distance as explained above. The TS of the large object is given by:

$$TS = TS_1 + 10\log(Sr/S_1) \qquad (2)$$

where TS_1 is the target strength measured at 1 m, S_1 and Sr are maxima of areas(Ss) at 1 m and R m, respectively. Replacing S in (2) with L, the TS of the long object is obtained.

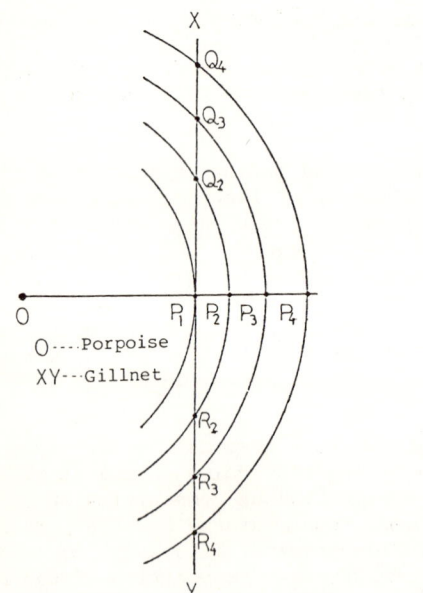

Fig. 9 Successive positions of a supersonic pulse wave front and reflecting portions. (Top view)

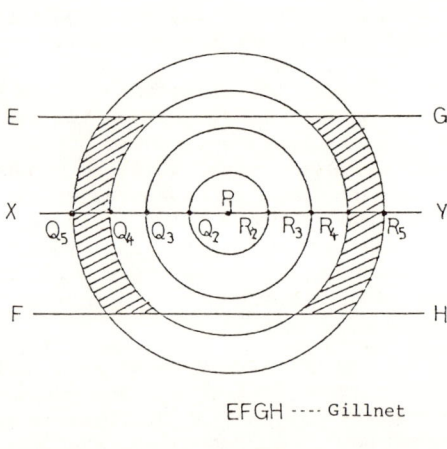

Fig. 10 Successive spreads of reflecting portions (Front view)

The waveform characteristics of clicks of Harbor porpoise, frequency, pulse width and source level were assumed to be 130 kHz, 43 μs and 160 dB, respectively. Directivity when the porpoise emits clicks and receives echoes is closely related to its echolocation ability. Pilleri et al.[7] reported that its transmitting directivity was estimated to be 9° horizontally and 18° vertically from anatomical data of the head of the Harbor porpoise. However, their suggestions have not yet been verified by actual observations. Therefore, in the calculation conducted here, both directivities were assumed to be 10°. The reflection directivities of lead-line and netting in the gillnet were determined to be 10° and 6°, respectively, according to Pence[20].

In the previous study[16], the target strength(TSs) of lead-line and netting were -33 dB and -55 dB, respectively. It is necessary to make a correction for the difference between the supersonic waves of measuring system and the porpoise's echolocation clicks. Frequency(f), beam angle(ϕ) and pulse width of the former were 143 kHz, 16° and 100 μs, respectively. The TSs were corrected using the equation of $\phi^{1/2}$ for the line, ϕ for the netting and $f^{3/2}$ for both materials[21]. Influence of the pulse width was not corrected for because it was negligible for the short distances. The TSs of the line and netting were fixed at -36 and -60 dB, respectively. The absorption coefficient of horizontal propagation of 130 kHz also was fixed at 0.033 dB/m.

The auditory threshold of the Harbor porpoise at 130 kHz was 68 dB and this was the threshold for a sound pulse of 1.5 s. It is necessary to correct this threshold at the time of pulse width(43 μs) of the click emitted by the Harbor porpoise, but no data are available for the Harbor porpoise at this time. If we use data for bottlenose dolphin[22], the correction value is 30 dB. Consequently, detection threshold(DT) was 94 dB.

The echo level(EL) was obtained as shown in Fig.11 by taking into account the distance-related variation of S and L. The detectable ranges of Harbor porpoise for lead-line and netting were 9 m and 2 m, respectively. Judging from these values, Harbor porpoises have to approach closely before detecting the netting by echolocation. According to data on reflection directivity by Pence[20], if the incident angle toward the netting inclines by 3° to 4° from 90° at the frequency of 150 and 170 kHz, TS decreases by about 10 dB. When TS decreased by 10 dB, the detectable range was 60 cm. In such a case Harbor

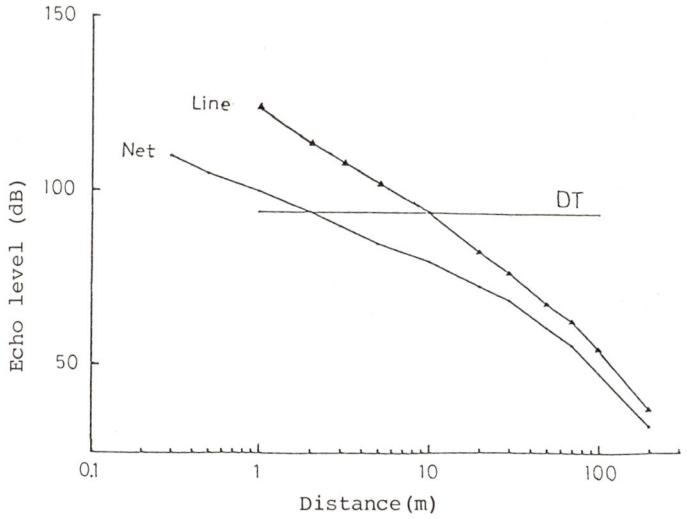

Fig.11 Echo level(EL) as a function of distance and detection threshold(DT).

porpoises will be entangled if they are not swimming slowly and carefully and not using their vision to some extent.

In this study, the detectable ranges were calculated based on simplified models of the gillnet and reflection mechanism also by using data on auditory characteristics of a dolphin belonging to another family. The gillnets actually set in the sea complicatedly change their configurations three-dimension-wise. Therefore, further studies are required on the reflection of supersonic pulse waves and the porpoise's ability to detect the gillnet. It is expected that the echolocation ability of the Dall's porpoise is better than that of Harbor porpoise because the source level, pulse width and frequency of clicks emitted by the former are more suitable for this type of echolocation than those of the latter.

CONCLUSIONS

Both Dall's and Harbor porpoises emitted echolocation clicks with a high frequency component, short pulse width and typical source level. The Dall's porpoise's reactions to the gillnet in the open seas show that they have high resolution echolocation ability and normally can avoid getting entangled in the net. Although the Harbor porpoise seemed to be able to detect the netting at a short distance, some finally got entangled to the gillnet in the darkened pool. The rough estimations of the Harbor porpoise's ranges for the detection of netting and lead-line were 2 m and 9 m, respectively. Judging from the waveform characteristics of clicks, the echolocation ability of Dall's porpoise seems to be better than that of Harbor porpoise for detection of nets. It still remains to be studied further why and how Dall's porpoises get entangled in gillnets in spite of their excellent echolocation ability.

ACKNOWLEDGMENT

The authors would like to thank Mr. T.Shimamura of Nihon University, Mr. K.Ishii of National Research Institute of Fisheries Engineering and Dr. T. Tobayama of Kamogawa Sea World for their helpful discussion and cooperation.

REFERENCES

1. S.H.Ridgway,1966, Dall porpoise, Phocoenoides dalli(True):Observations in captivity and at sea, Norsk, Hvalfagst-Tidende,5:97-110.
2. W.E.Schevill, W.A.Watkins and C.Ray, 1969, Click structure in the porpoise,Phocoena phocoena, J.Mammal.,50:721-728.
3. J.S.Leatherwood and D.K.Ljungblad, 1979, Background research in support of a proposed method for reducing mortality of Dall's porpoises, Phocoenoides dalli in the Japanese Pacific high seas fishery for salmon, Contract Report to Marine Mammal Div., National Marine Fisheries Service.
4. F.T.Awbrey, J.C.Norris, A.B.Hubbard and W.E.Evans, 1979, The bioacoustics of the Dall's porpoise-salmon drift net interaction, H/SWRI Technical Report, pp.79-120.
5. N.A.Dubrovskii, P.S.Krasnov and A.A.Titov, 1971, On the Emission of Echolocation Signals by the Azov Sea Harbor Porpoise, Soviet Physics-Acoustics, 16(4):444-447.
6. B.Møhl and S.Andersen, 1971, Echolocation:high frequency component in the click of the Harbor Porpoise(Phocoena ph. L.), J.Acoust.Soc.Am., 54(5):1368-1372.
7. G.Pilleri, K.Zbinden and C.Kraus, 1980, Characteristics of the Sonar System of Cetaceans with Pterygoschisis, Directional Properties of the Sonar Clicks of Neophocaena phocaenoides and Phocoena phocoena (Phocoenidae), in:"Investigations on Cetacea," G.Pilleri ed., Hirnanatomishes Institut, Bern, Vol.XI, pp.157-188.
8. R.G.Busnel, A.Dziedzic and S.Andersen, 1965, Seuils de Perception du

Systéme Sonar du Marsouin Phocoena Phocoena L., en Function du Diamétre dún Obstacle Filiforme, C.R.Acad.Sc., Paris,260:295-297.
9. R.G.Busnel and A.Dziedzic, 1967, Résultats Mestroloqiques Expérimentaux de Ĺecholocation Chez le Phocoena Phocoena, et Leur Comparison Avec Ceux de Certaines Chauve-Souris, in:"Animal Sonar Systems," R.G. Busnel ed., INRA-CNRS, Jouy-en-Josa; France, pp.307-336.
10. G.L.Zaslavskii, A.A.Titov and V.M.Lekomtsev, 1969, Investigation of the Underwater Echolocation Capabilities of the Azov Sea Harbor Porpoise, Report of the Karadag Section of the Institute of Biology of the Southern Seas(in Russian).
11. S.Andersen, 1970, Auditry Sensitivity of the Harbor Porpoise Phocoena phocoena,in:"Investigations on Cetacea," G.Pilleri ed., Hirnanatomisches Institut, Bern, Vol.II, pp.255-259.
12. S.Andersen, 1970, Directional Hearing in the Harbor Porpoise Phocoena phocoena, ibid., pp.260-263.
13. W.Dudock v.Heel, 1960, Sound and Cetacea, Diss, Netherlands, J.Sea.Res. 1:4.
14. Y.Hatakeyama, 1983, Study of the Dall's Porpoise's Echolocating Pulses and Specification of the Sound Generators, Document submitted to the meeting of the Scientific Subcommittee of the Ad Hoc Committee on Marine Mammals, INPFC, Tokyo, Japan, 14pp.
15. Y.Hatakeyama and T.Shimamura, 1984, Acoustic Studies on Dall's Porpoise in the Bering Sea, ibid., 9pp.
16. Y.Hatakeyama, 1984, On Reflection Loss of gillnet and Maximum Detectable Range for Dall's Porpoise, ibid., 14pp.
17. H.Taketomi, 1984, Experiments in capturing Dall's porpoise on the coast of Hokkaido in the sea of Okhotsk and experiments on Dall's porpoise's behavior in response to sounds of killer whales, ibid., 7pp.
18. Y.Hatakeyama and H.Shimizu, 1985, Feeding Trial and Acoustic Studies on Dall's Porpoise Captured Alive, ibid., 14pp.
19. Y.Hatakeyama, K.Ishii, H.Soeda, T.Shimamura and T.Tobayama, 1988, Observation of Harbor porpoise's behavior to salmon gillnet, ibid., 17pp.
20. A.E.Pence, 1986, Monofilament Gill Net Acoustic Study, Applied Physics Laboratory, University of Washington, 13pp.
21. V.G.Welsby and G.C.Goddard, 1973, Underwater acoustic target strength of nets and thin plastic sheets, J.Sound Vib., 28(1):139-149.
22. C.S.Johnson, 1967, Relation between absolute threshold and duration-of-tone pulses in the bottlenosed porpoise, J.Acoust.Soc.Am., 43(4): 757-763.

VERY-HIGH-FREQUENCY ACOUSTIC EMISSIONS FROM THE WHITE-BEAKED DOLPHIN

(LAGENORHYNCHUS ALBIROSTRIS)

RONALD B. MITSON

Ministry of Agriculture, Fisheries and
Food, Fisheries Laboratory, Lowestoft
Suffolk NR33 OHT, UK

ABSTRACT

Reception by sonar of the signals on board the Ministry of Agriculture, Fisheries and Food's (MAFF's) research vessel, CLIONE, from a school of white-beaked dolphins (Lagenorhynchus albirostris), shows that there is significant energy in their acoustic emissions to at least 325 kHz. A sector scanning sonar of 0.33º bearing resolution and 100 µs time resolution was used in active and passive modes when the dolphins approached the ship. Examples of the recorded pulse signals are shown in photographic sequences. Some of the analysed pulse rates varied from 100 to 750 pulses per second when averaged over 20 ms periods. Crude estimates of the source level of the signals are between 123 to greater than 207 dB re 1 µPa at 1 m.

INTRODUCTION

This paper is based on a series of recordings made on board MAFF's research vessel, CLIONE, in the Wellbank Flat area of the southern North Sea. The vessel was using an electronic sector scanning sonar device to observe sandeels moving from the sea bed into midwater when a school of dolphins (later identified as Lagenorhynchus albirostris) appeared on 11 June 1970. Dolphins again were close to the ship in the same area on the morning of 13 June. With the sonar in active mode, echoes were received from the bodies of the dolphins as they swam in the beam. At the same time, the dolphins emitted signals which were clearly seen on the sonar displays. The sonar transmitter was switched-off for a period and recordings made of the dolphin signals using the scanned receiving beam. The first brief report of these observations was made by Mitson and Morris (1988).

SONAR SYSTEM

Voglis (1972) described the design of the sonar in detail and Mitson and Cook (1971) its installation, performance and potential applications relative to fisheries research. For the past 20 years it has been used,

© British Crown Copyright, 1990

mainly in combination with acoustic transponders, for observations of the behaviour of unrestricted, identified, individual fish in the open sea. The technique has been applied (Arnold et al., 1990) to the following uses:

(a) describing the reactions of plaice (Pleuronectes platessa) to a Granton otter trawl and measuring the efficiency of the gear;

(b) following the movements of migratory fish in the southern North Sea relative to tidal streams; and

(c) telemetering the compass-heading from free-ranging fish.

One of the notable discoveries was that of selective tidal stream transport used by a variety of species for migration.

The most important characteristics of the sonar for the purpose of this paper (which make it a unique analyser) are the very narrow transducer beam of 0.33º (which is electronically scanned over a sector of 30º at a rate of 10,000 times per second, see Fig. 1) and the time resolution of 100 µs. At 90º to the scanned plane, the beam angle is 10º. The sonar sector can be steered manually through ± 270º with respect to the ship's head and the transducer is capable of being tilted from 0º to 90º. In addition, the transducer can be rotated so that its long axis is in the vertical plane and, in this mode, the beam is scanned in elevation, thus making it possible to obtain the depth of targets. The transducer also can be tilted from 0º to 90º in this mode.

In the active mode, the transmitting transducer of the sonar ensonifies the full 30º sector in the scanning plane and an angle of 5º in the non-scanning plane. A 75-element linear array, having a length of 0.75 m, forms the receiving transducer. At the time of these observations, it had a frequency response centered on 305 kHz with a bandwidth of 40 kHz. Thus, the received signals fell within the band of 285 to 325 kHz. Sensitivity over this band was -204 dB re 1V/1 µPa. The first sidelobes were -12 dB relative to the main lobe.

In this sonar, the display system comprises two electro-magnetically deflected, rectangular cathode ray tubes (CRT's) that display bearing horizontally and time vertically (in active mode, time represents range). One display has a total timescale of 242 ms, while the other shows a 74 ms portion of this timescale which is selected by a manually controlled electronic gate and expanded to occupy the full tube height. Only 26º of the 30º sector actually appears on the displays due to the flyback time of the timebase.

Each CRT has a raster scan display repeated at a rate of 4 Hz. The tubes have a blue flash with an orange afterglow from the long persistence phosphor. After the initiation of each scan, lines are generated every 100 µs, a total of 2420 for the 242 ms timescale. The resolution of the CRT's is limited to about 400 lines, with the result that several lines are integrated, thereby reducing the resolution on the time axis of the main timebase display, to around 0.6 ms (about 6 lines). On the expanded display, the resolution is about 0.2 ms (2 lines). There is a further small, but variable and unquantified, loss due to the recording film and subsequent copying and printing. Recording is by means of single-shot 16 mm cameras using Ilford FP4 film.

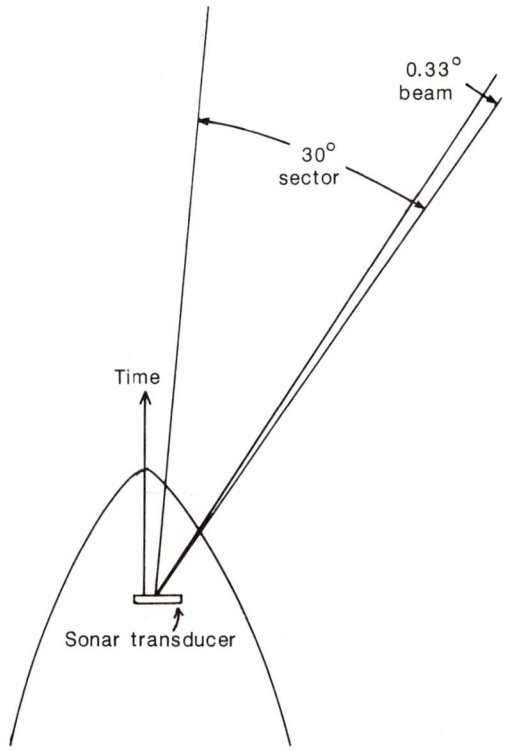

Fig. 1. Configuration of the sonar beam and the sector which it sweeps over at a rate of 10,000 times per second.

Fig. 1 shows a plan view of the sonar sector when scanning in the horizontal plane. For each sweep of the 0.33º beam, a 100 µs line is generated on the main (242 ms) timebase display and any signals occurring during the time of the sweep appear at the angle at which they are received by the transducer. In Fig. 2, there is an illustration of the principle on which this sonar works. It is based on deducing, from a given pressure distribution along the transducer face, the angular position of acoustic sources. The pressure distribution on the elements of the transducer will vary with the angle at which the signal originates relative to the long axis of the transducer.

At the time of the observations reported in this paper, the sea temperature was about 12ºC and the salinity 35 ppt which resulted in an acoustic propagation speed of 1493 m per second. Using the recent formula due to Francois and Garrison (1982), the absorption loss would be about 78 dB per km.

The receiving system of the sonar has a high sensitivity and is capable of detecting signals down to thermal noise level. At 305 kHz with a bandwidth of 40 kHz, this corresponds to an acoustic pressure level at the transducer face of about 80 dB re 1 µPa, producing an input terminal voltage of about 0.7 µV. The minimum detectable coherent CW signal at each input should be $75^{0.5}$ smaller, i.e. 0.08 µV. At the other extreme, the signal level which overloads the receiver is about 165 dB re 1 µPa, producing 10 mV at the input terminals. Levels of this order and above manifest themselves by horizontal lines across the display.

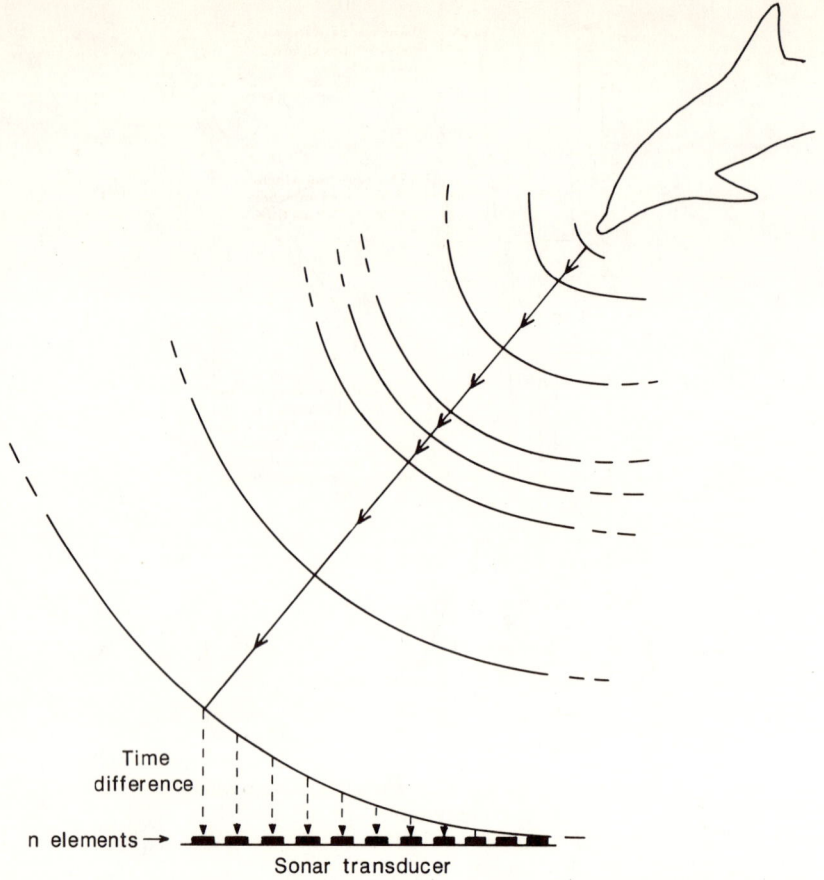

Fig. 2. Principle of directional reception of a succession of dolphin signals transmitted at different time intervals.

The signals recorded from this sonar are viewed best by watching a projection of the film at the original speed of 4 Hz, but there is an advantage in speeding the projection to 16 or even 24 Hz. By these means, it is possible to convey the swimming movements of the dolphins and the dynamics of the signals, which cannot be achieved by the study of still-frames.

DOLPHIN SIGNALS

Active sonar

Fig. 3 shows the coordinates and details of the displays which can be used to interpret Figs. 4 and 5.

In Fig. 4, the row of 20 frames (a) (from 1983 to 2002) illustrate successive periods of 242 ms, separated only by 3 ms (the film 'drawdown' period in the camera). Thus, from the top of the signal train in frame 1983 to the bottom of the signal train in 1984 there is no more than a 3 ms gap. This means that, in the sequence from 1983 to 2002, signals are appearing for a total period of 4897 ms (4.987 s) with 3 ms gaps between frames.

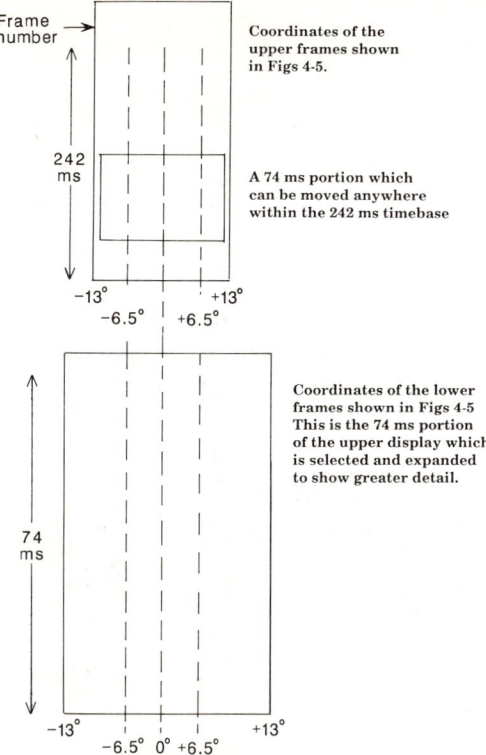

Fig. 3. An illustration of the display coordinates.

Because of the intensity and duration of the signals, very little detail is visible in row (a) until frame number 1985, when the bars are separated and an echo from the body of a dolphin can be seen at 73 m range on a bearing of -6°. This image is clearly seen on row (b) from the first frame (1983). The range of the dolphin appears to be reducing slightly during the sequence from 1983 to 2002, but this only can be determined accurately if measurements are taken directly from the original film; it should be noted that the actual start of the timebase on the photographic prints cannot be identified positively on all frames.

Detail can be seen in some of the signal 'bars' on row (b), which appears to be amplitude modulation (fluctuations in the intensity and/or range extent of the bars across the display) but the resolution cannot exceed 100 µs and several scanning lines are integrated. This effect is most noticeable in frames 1983 to 1987 (b). The number of bars decreases considerably from frame 1985, when certain groupings and spacings become clearer.

The dolphin, whose echo is first seen in frame 1983 (b) on a bearing of -6.5°, is still visible in frame 1999 at a bearing of about -4°. It is at a range of about 80 m in frame 1983 (a) but detail is lost in the reproduction and it cannot be seen clearly in the (a) frames until 1985. Initially, it seems to be heading at a negative angle (away from the centre bearing) but turns back in 1987, appears 'head on' in 1988, then is angled towards 0° in 1998, but appears to be moving out of the beam on

Fig. 4 Echoes and signals emitted from dolphins, displayed by the sector scanning sonar device when used in the active mode (see text for explanation).

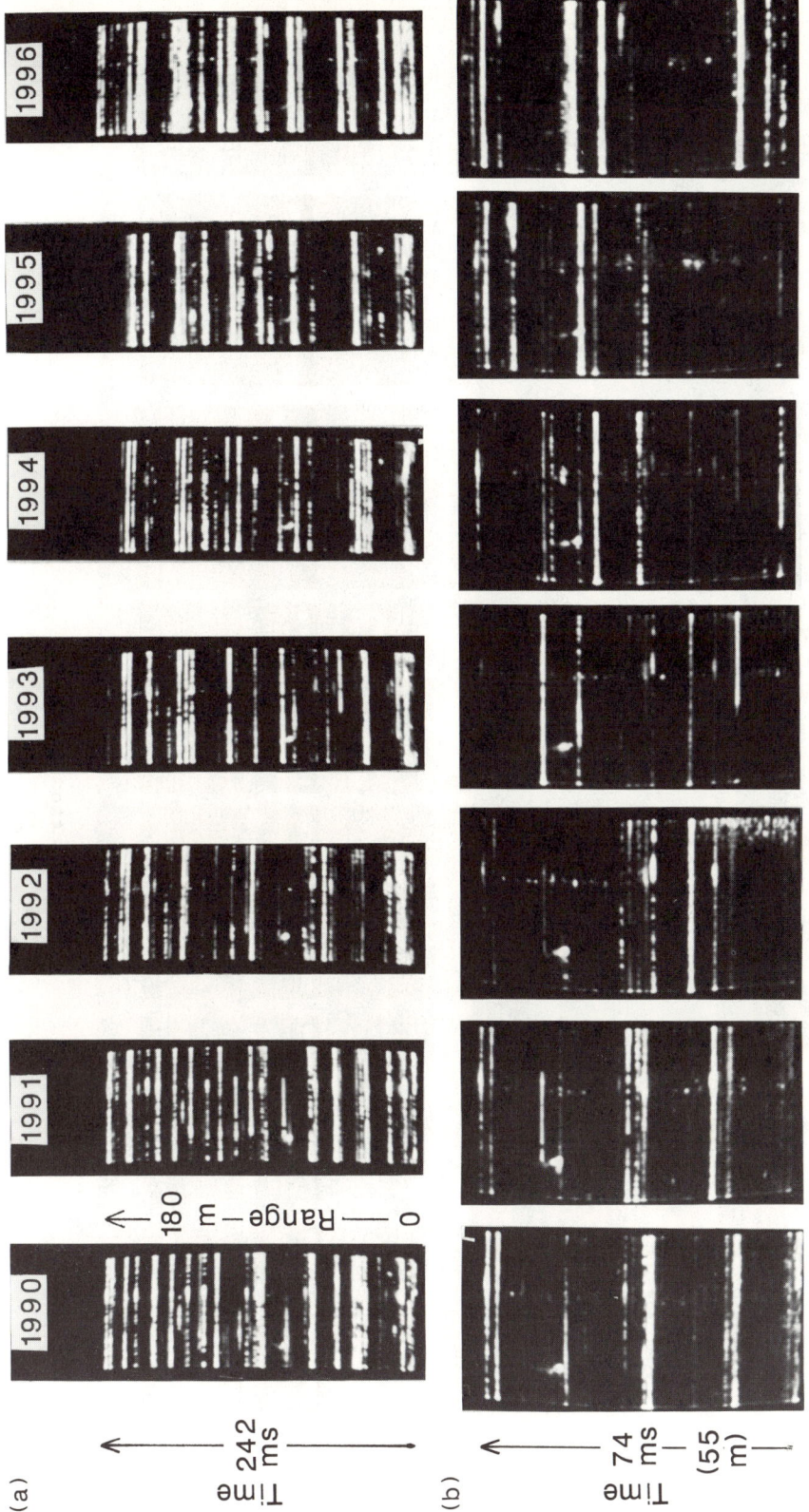

Fig. 4 Continued.

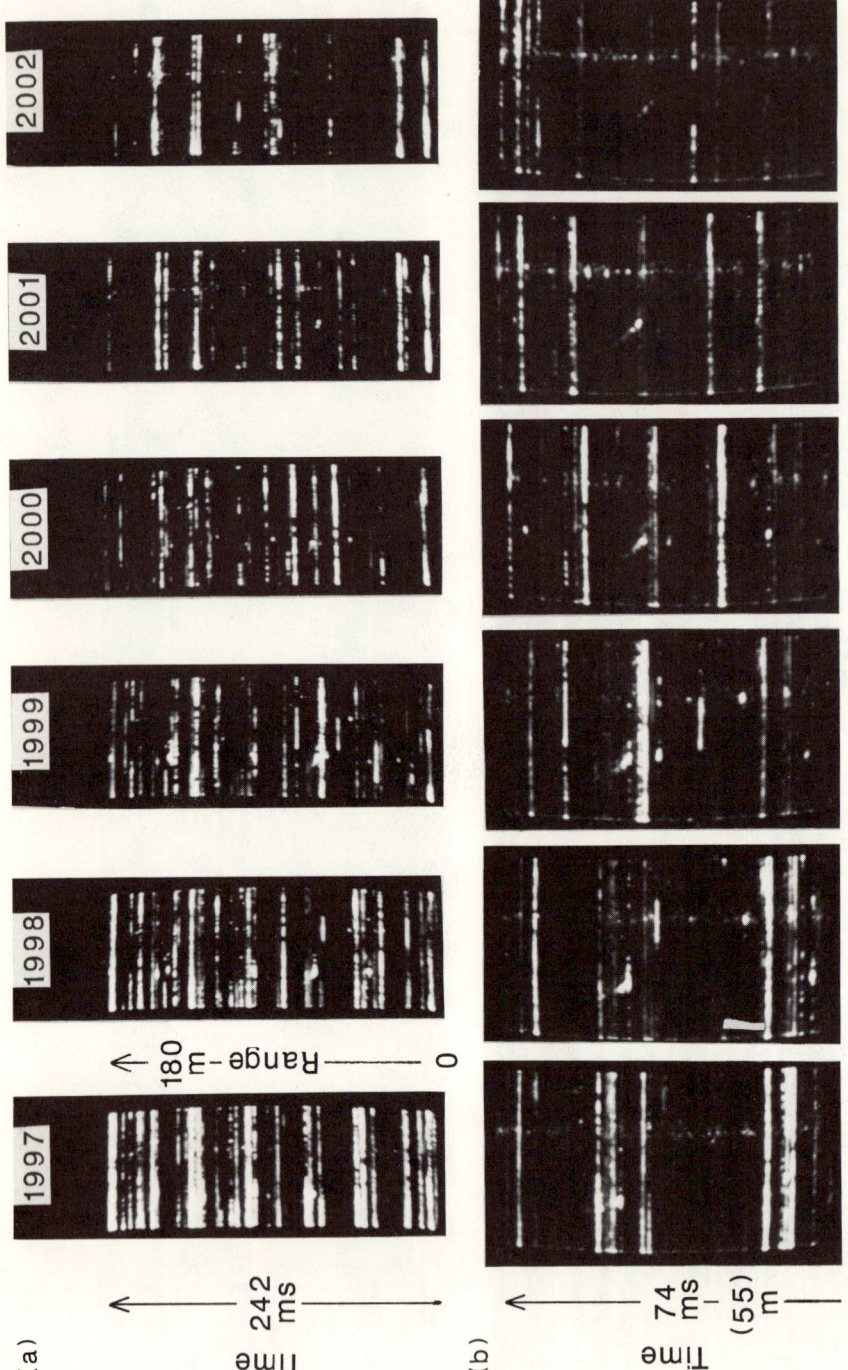

Fig. 4 Continued.

the following three frames. There is no firm evidence to link the other signals on the display to this dolphin although there are pulses that are coincident with the leading edge of the echo on frames 1986, 1987 and 1988, and similarly on 1990, 1995 and 1996.

Passive sonar

In Fig. 5, (frames 2288-2300) there are no sonar-generated echoes, because at film frame 2030 the sonar transmitter was switched-off. Therefore, all visible signals are those transmitted by the dolphins. In frame 2288, the pattern on displays (a) and (b) shows a line of dots spaced about 35 ms on a bearing of -3º which are in an apparent bearing gap left by the other signals. On 2288 (a) there is a train of pulses at -9º with a spacing of about 28 ms. Two of these pulses can be seen on 2288 (b).

Frame 2289 (a) has a train of pulses on the lower half of the display at about -4º which are difficult to resolve but they are seen clearly on 2289(b), where the first five are spaced at 4 ms intervals, followed by two intervals of 6 ms, with around 8 more pulses at 4 ms intervals. It is possible that the intensity of signals in frame 2289 has resulted in only their afterglow being visible in 2290, because the positions are identical, but the pulses much fainter (mostly not visible on the reproduction) and there is no sign of them in 2291. This frame is followed by a gap of 7 frames (about 1.7 s). In frame 2299, very quiet conditions prevail; the marker for the start of the expanded display can be seen and there is just a suggestion of a pulse train (short vertical signal at -5º in 2299 (a) only). In 2299 (b) one horizontal set of dots is seen for which there is no obvious explanation. In 2300(a), an intermittent series of signals occurs at about -4º to -5º, a portion of which can be resolved in frame (b), where a train of 23 pulses occurs over the full 74 ms, fairly uniformly spaced at about 3.2 ms.

Fig. 6 illustrates the changing rate of pulses during one 242 ms scan. The rates were estimated over consecutive 20 ms periods with data taken from frame 2801 which is not reproduced in this paper.

ESTIMATION OF SOURCE LEVEL

From the details of the sonar given in the previous section, it is possible to make crude estimates of the source level (SL) of signals received and recorded from these dolphins. Where there is a clear train of pulses not greater than about 3º wide, the pressure level (PL) at the transducer must have been below the overload level of 165 dB re 1 µPa.

There is evidence in frame 1999(a) (Fig. 4) of a dolphin swimming at about 70 m range from the sonar at a bearing of -5º. It is visible clearly on the expanded display (frame 1999(b)) where it appears to be heading at an angle of 45º towards the centre bearing. Assuming that it could be responsible for the full-width signals on the display, the minimum SL of these signals can be calculated from the overload threshold of the sonar, plus the propagation loss due to spherical spreading and absorption, i.e. 165 + 20 log 70 + 78 x 70/1000 = 207 dB re 1 µPa at 1 m.

However, if these signals emanated from a dolphin close to the transducer, the level would not have been reduced by the propagation loss and the SL might have been as low as 165 dB re 1 µPa at 1 m. For a dolphin at 70 m range, if the signals are not overloading the receiver, the

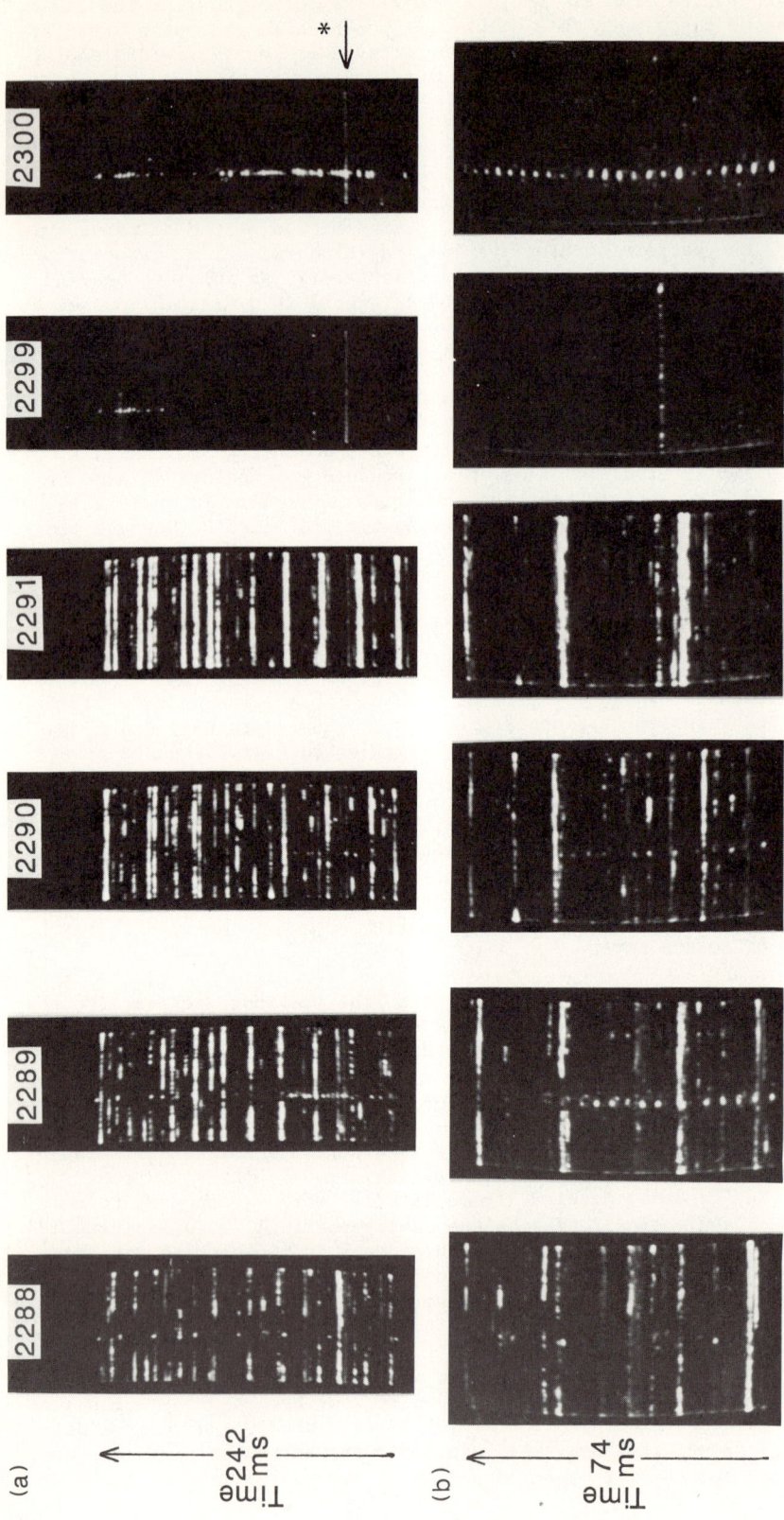

Fig. 5 Pulse signals of varying types from dolphins, displayed by the sector scanning sonar device when used in the passive mode (see text for explanation)

* Marker for expanded display

Individual frames (each ±13°)

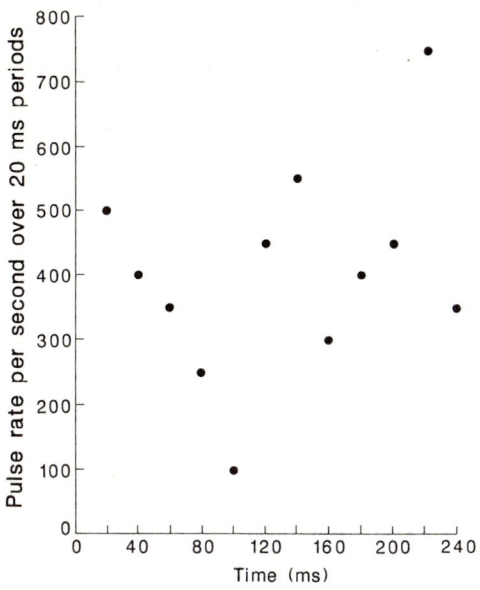

Fig. 6. Rapid change of pulse rate over one 242 ms scan - frame 2801.

SL would have been less than 207 dB by at least the amount of the propagation loss of 42 dB, i.e. SL = 207-42 = 165 dB re 1 µPa at 1 m. If the signals had emanated from very close range, the SL might have been as low as 165-42 = 123 dB re 1 µPa at 1 m.

The above figures of SL are consistent with the wide range of source levels reported in the literature. For example, Au et al. (1978) measured peak energies of 220 dB re 1 µPa at 1 yard (0.914 m), from Atlantic bottlenose dolphins (Tursiops truncatus) at frequencies between 120 and 130 kHz.

There is no firm evidence to link the signals on the displays to the dolphin at 70 m range, although two pulses, spaced by 36 ms (27 m), are on the same bearing as the leading edge of the echo. The closest of these pulses to the echo is also at precisely the same range as two other pulses, one of which appears at +5º and seems to be part of a pulse train on that bearing with pulse spacings of 10 ms (8 m) to 7.3 ms (5.5 m). The other, (at +7º) appears in isolation apart from its range coincidence.

There is a great deal of evidence to prove that many species of dolphin emit frequencies of around 150 kHz, but no previous reports of reception as high as 300 kHz. It is feasible to speculate that the signals recorded from the scanning sonar were due to non-linear effects in the water, because of high source levels transmitted by the dolphins.

SUMMARY

The 52 individual frames (26 full-scan, 242 ms and 26 expanded-scan, 74 ms) reproduced in this paper show some of the variability in types of signal and transmission rates recorded on film from the sonar displays. It must be emphasised that the echoes from the dolphin bodies and the signals transmitted by the dolphins can only be appreciated when viewed dynamically, i.e. by projecting the films. A few copies of the 16 mm cine films, each containing many thousands of frames recording these dolphin signals, are available for study purposes by application to the author.

ACKNOWLEDGEMENTS

Thanks are due to Dr J. C. Cook and Mr A. D. Goodson for corrections and helpful comment.

REFERENCES

Arnold, G. P. Greer Walker, M. and Holford, B. H., 1990, Fish behaviour: achievements and potential of high-resolution sector-scanning sonar. Rapp. P.-v Réun. Cons. int Explor. Mer, XXX: in press

Au, W. W. L., Floyd, R. W. and Haun, J. E., 1978, Propagation of Atlantic bottlenose dolphin echolocation signals. J. Acoust. Soc. Am., 64(2): 411-422

Francois, R. E. and Garrison, G. R., 1982, Sound absorption based on ocean measurements: Part II. Boric acid contribution and equation for total absorption. J. Acoust. Soc. Am., 72(6): 1879-1890

Mitson, R. B. and Cook J. C., 1971, Shipboard installation and trials of an electronic sector scanning sonar. Rad. Electron. Eng., 41(8): 339-350

Mitson, R. B. and Morris, J. W., 1988, Evidence of high-frequency acoustic emissions from the white-beaked dolphin (Lagenorhynchus albirostris). J. Acoust. Soc. Am., 83(2): 825-826

Voglis, G. M., 1972, Design features of advanced scanning sonars based on modulation scanning. Parts 1 & 2. Ultrasonics, 10(1) and (3): 16-25 and 103-113

Disclaimer: The reference to proprietary products in this paper should not be construed as an official endorsement of these products, nor is any criticism implied of similar products which have not been mentioned.

HIGH INTENSITY NARWHAL CLICKS

Bertel Møhl, Annemarie Surlykke[1] and Lee A. Miller[1]

Department of Zoophysiology
University of Aarhus
DK-8000 Aarhus C
Denmark
[1]Biological Institute
Odense University
DK-5230 Odense M
Denmark

INTRODUCTION

The hypothesis that some odontocetes use their sonar not only to find prey, but also to debilitate it (Norris and Møhl, 1983) requires that odontocetes produce sound pressures in excess of 230 dB re. 1 µPa (Zaegaesky, 1987; Hubbs and Rechnitzer, 1972). While maximum source levels[1] (SL) of clicks recorded from trained <u>Tursiops</u> (Au et al., 1974) and <u>Delphinapterus</u> (Au et al., 1987) are only a few dB short of this value, there is a gap of 50 to 120 dB between the debilitation threshold and the SL's reported for odontocete clicks in nature (Levenson, 1974; Watkins and Schevill, 1974; Watkins, 1980a,b).

Is this dB gap real? Or, is it simply an indicator of the difficulties inherent in SL measurements of free-swimming animals? On face value, the latter explanation seems unlikely; it implies millions of extremely powerful sound pulses being emitted each day and in all oceans, but without having been noticed.

Requirements for measuring SL's of foraging odontocetes include: reliable, non-interfering access to the animals, a favourable acoustic environment, suitable recording equipment, and methods to derive source positions relative to the hydrophone. As an attempt to meet these requirements we brought

---

[1]SL is defined as the intensity of the sound, referred to 1 m from the source and measured in the direct acoustical axis of the emitted beam. The acoustical axis specification is important due to the high directionality of odontocete clicks (Au et al., 1986; Au et al., 1987).

linear, vertical hydrophone arrays and instrumentation recorders to Inglefield Bay (Thule district, NW-Greenland (77°N, 67°W), where narwhals (<u>Monodon monoceros</u> enter during the month of August to feed on halibut and polar cod. Our findings show that foraging narwhals produce clicks as powerful as those emitted by trained, captive odontocetes. For the narwhal, at least, the dB gap is not real.

MATERIALS AND METHODS

Site

Inglefield Bay is attractive for acoustic studies of narwhals due to its depth (in excess of 600 m), and low background noise (Thiele, 1983). The latter is in part due to restrictions on motor boats, which are banned in certain areas by local regulations as a measure not to frighten the narwhals that are important to the prevailing subsistence hunting economy.

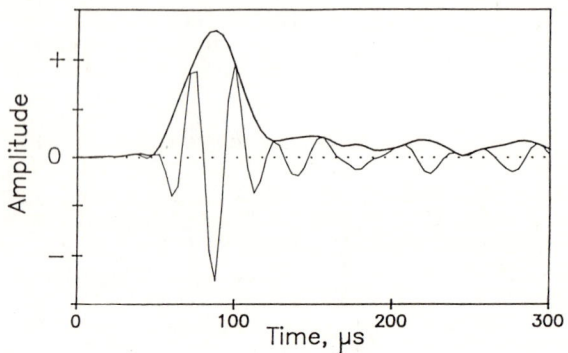

Fig. 1. Waveform and envelope of a narwhal click.

Narwhals are wary of any unnatural sounds. We had to be on station with the arrays deployed before the pods arrived. Only once were they observed at close range. Between 13 and 19 August 1987, we had a total of 3 hours of recording time and about 20 hours of monitoring. Often we recorded clicks when narwhals were not seen. No other cetaceans were observed in the area within this period.

Instrumentation

Two identical arrays, each consisting of 3 hydrophones, were lowered to depths of ca. 5, 35, and 65 m from two, 5 m dinghies. The two deeper hydrophones were model B & K 8101, while the shallow ones were uncalibrated, surplus sonobuoy types. The sonobuoy hydrophones were operated above their normal frequency range at resonances at 24, 41 and 56 kHz; the data were used for time-of-arrival measurements only. Two instrumentation taperecorders were used, one with an upper 3-dB limit at 60 kHz (B&K 7005, 4-channel), and the other with a bandwidth extended to 100 kHz (B&K 7006, 4-channel). A B&K 4223 hydrophone calibrator served to verify calibration of the entire recording chain at the recording sites with an accuracy better than 1 dB. The frequency response of the chains were governed largely by the recorders. Accuracy of time difference measurements was limited by recorder flutter, specified to be below 0.5 %. The depths of the hydrophones were calculated using pingers at the surface. The sound velocity profile (SVP) was not measured, but an average velocity of 1.455 m/ms was assumed, based on data from nearby Baffin Bay (Mellen et al., 1975).

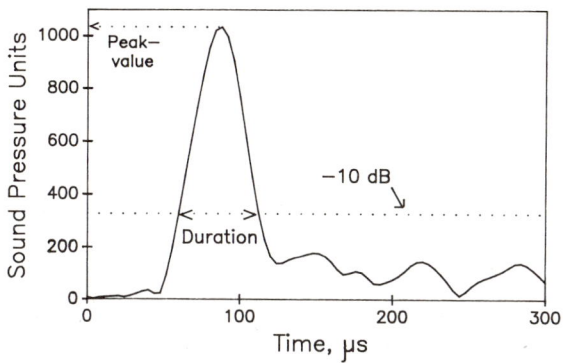

Fig. 2. Envelope of a narwhal click. Conventions for reading duration and peak value are shown.

Analysis

Based on criteria of non-saturation of electronics and sufficient number of time differences for positioning, we analysed 62 clicks from 8 distinct positions and probably representing 8 different animals. The 4-channel recordings were anti-aliasing filtered, multiplexed and digitized at an effective rate of 250 kHz, using 12-bit resolution. Time domain measurements (time of occurrence, amplitude and duration) are based on the click envelope, generated from the Hilbert transformed signal (Thrane, 1984, cf. Fig. 1). Amplitudes are given in dB referenced to the rms value of a continuous sine wave signal having the same amplitude as the transient (Stapells et al., 1982). Durations are given at the -10 dB level on the envelope function (cf. Fig. 2). In the frequency domain, clicks are characterised by the lower and upper -10 dB cut-off of the spectrum (Fig. 3).

Calculations of positions were done assuming the hydrophones were on a straight, vertical line. Surface reflected signals were treated as if recorded by virtual hydrophones above the surface. Thus, each click is represented by a set of up to 6 time series (the click set). Two algorithms were used. One calculated and plotted the hyperbolas for each set of "time-of-arrival-differences" for all possible pairs of hydrophones (Fig. 5). The other algorithm used a least-square method to estimate positions. Source levels are derived from the latter method. When the number of hydrophones with usable signals changed within a click series, positions were calculated using the lowest common set of usable hydrophones (usually 5).

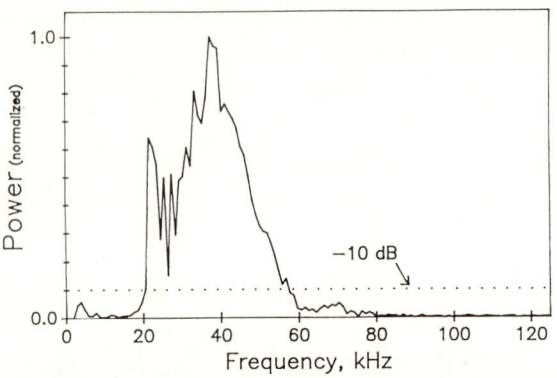

Fig. 3. Power spectrum (128 points) of a narwhal click. The oscillations are caused by the low-amplitude, trailing oscillations (see Fig. 1).

RESULTS

The waveform of narwhal clicks is a simple, 1.5-cycle oscillation that lasts for about 50 µs. The waveform resembles that reported for <u>Tursiops</u> by Au and co-workers (e.g. Au et al., 1974), using "state-of-the-art" equipment. The spectrum, however is distinguished by having a sharp low frequency (LF) cut-off at 20 kHz, and a more variable high frequency (HF) cut-off in the 50 kHz region. While the latter appears to be influenced by directionality, the former displays a remarkable invariance throughout our recordings. Quantitatively, the spectrum resemble that of clicks from a captive beluga in San Diego Bay (Au et al, 1985). A low-frequency component at 1 to 5 kHz is often observed in low-level recordings, but never above the -10 dB level of the ultrasonic component.

The maximum SL observed for each source position is plotted in Figure 4. Within each set of click-representations, the SL varies considerably, the maximum observed difference was 28 dB. The highest SL found was 218 dB re 1 µPa. It occurred within a series of 17 clicks with a repetition rate of about 4 pps, and with a mean level of 214 ± 3 dB re 1 µPa. Within some click series, maximum sonification was observed to change from one hydrophone to another, indicating a moving, directional source.

On some occasions, we recorded clicks with a high repetition rate. In a typical recording where there was no interference from other whales we recorded several clicks from a whale swimming close to the surface around 300 m from the hydrophone array (Fig. 4). The repetition rate varied around 2 Hz and the SL was 180-190 dB re 1 µPa. Suddenly, the whale emitted a series of about 30 clicks at a rate of 200 Hz, (resembling the buzz phase of the emissions from a bat pursuing an insect). The duration and spectrum of such a click was like that of the clicks previously described, but the amplitude was much lower (-20 dB relative to the preceding clicks). The high repetition rate and reduced amplitude was typical for all recorded buzz sequences.

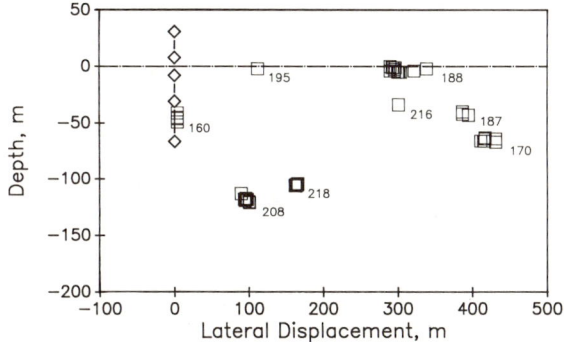

Fig. 4. Composite plot of positions of 66 click sources, probably representing 8 animals. Open diamonds: hydrophones. Open squares: source positions. Maximum measures SL in dB re 1 µPa is given for each position.

DISCUSSION

The accuracy of our determinations of the 2-dimensional positioning of click sources is unknown. Many poorly known factors are involved in these determinations like the sound velocity profile, deviations from strict verticality (skewness) of the array, and the geometry of the source in relation to the skewed array. However, the use of surface-reflected clicks effectively increased the number of sensors above the minimum of 3 required for calculation of the source position. Besides adding infor-mation about the clicks, the redundant sensors carry information about the magnitude and sign of the skewness of the array, see Fig. 5. From this information we estimate the ranging error to be less than 30%, except for sources near the axis of the array. Positions reported here are not corrected for skewness of array, as the resulting error from this source in the SL estimates is 3 dB or less.

The SL measurements are also subject to errors deriving from the non-uniform directivity of the hydrophones in the XY-plane. This error, which can be as high as 60 dB, can only reduce the value of the estimated SL.

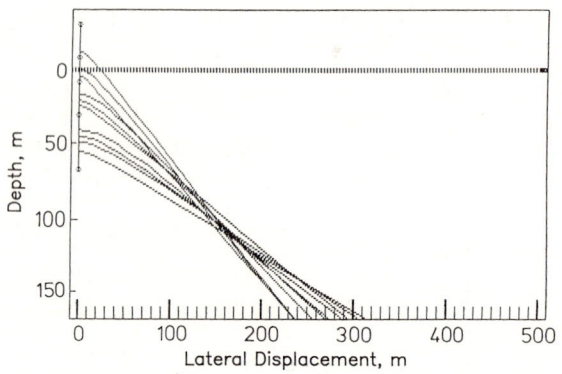

Fig. 5. Family of hyperbolas, generated from differences in time-of-arrival of a narwhal click to the hydrophones of an array. The two hydrophones above the surface symbolise virtual hydrophones, introduced to process surface reflected clicks. The de-focussing effect is caused mainly by deviation from ideal linearity of the array.

Using the envelope for time domain measures was adopted in some of the analyses because it incorporates the entire waveform, rather than just a few samples of the analog waveform, as in conventional pp-measurements. While the advantage of using the envelope function (Fig. 2) is obvious with regard to measurements of durations (and often also in time difference measurements where phase changes may cause ambiguity), it is less obvious with regard to amplitude measures. The method returns peak values, not peak-to-peak values, as traditionally used. This difference, however, is inconsequential as long as the nature of the reference is stated. However, for clicks that are asymmetric around zero, values obtained by the envelope method are about 1 or 2 dB larger than those obtained by traditional pp-measures. In our opinion, the method is useful by eliminating a number of minor error sources, while still retaining a general compatibility with peak measuring methods. Further, the envelope method presents the salient temporal and amplitude data in a simpler manner relative to that of the 'raw' waveform.

The vertical hydrophone arrays proved possible to operate from small boats, but methods to estimate skewness should be used when greater accuracies are required. Surface reflections were found valuable to estimate the size and sign of the errors in position determinations.

The spectra and waveforms of the narwhal clicks in our recordings differ from previously reported values (Watkins et al., 1971; Ford and Fisher, 1978) by having all significant energy confined to the ultrasonic range and by having durations of about 50 µs. The differences seem well-explained by differences in instrumentation, which in the previous works was limited to audio frequencies. Energy may well be present in the narwhal clicks above our limit of 100 kHz.

With the conventions adopted here, the time-bandwidth-product of the narwhal pulse is 1.75 (50 µS * 35 kHz). If the -3 dB values are used, the product reduces to about 1. The low value for the time·bandwidth-product indicates a signal achieving a maximum concentration in the time/frequency space. Theoretically, such signals yield the best possible signal-to-noise ratio for echoes to be detected by a sonar limited by broad band noise (Wiersma, 1988). Also, increased resolution in time difference measurements using cross-correlation is not possible with this kind of signal.

From the observed depths, we assume that some of the animals were feeding, rather than just travelling. Our admittedly limited data set contains no indications of a depth-related effect on the waveform of the clicks. Still, the observation of very intense SL's at depths of about 100 m may have some relevance with regard to speculations about the sound generation mechanism. Here, the ambient pressure is increased by a factor of 10. If a cavitation mechanism is hypothesized, it should include means of dealing with large variations in hydrostatic pressure.

We want to emphasize that a SL of 218 dB re 1 µPa represents an acoustic intensity of about $\frac{1}{2}$ W/cm², corresponding to a level in air of 156 dB SPL. To our knowledge, this is the most intense sound recorded so far from an animal in nature. And yet, our conditions do not insure that we have sampled the maximum capacity of the narwhal. However, the maximum SL's observed are similar to the most intense ones measured experimentally on trained, captive dolphins. While our results thus meet the objection against the hypothesis of acoustical debilitation by odontocetes, namely that high intensity pulses have not been observed in nature, they certainly are not evidence that the clicks served such purposes, neither do other parameters in the analysed sequences suggest such a function. Rather, our results imply that no major evolutionary step is required to bridge the gap between the source levels observed during feeding and those necessary for prey debilitation.

ACKNOWLEDGEMENTS

We thank E. Born (Greenland Home rule), for bringing the potentials of Inglefield Bay to our attention, Jørgen Daorana (Qaanaaq), for logistic support, N. Kristiansen (University of Aarhus), M-P. Heide-Jørgensen (Greenland Fisheries Research Institute) and S. Leatherwood (Hubbs-Sea World Research Institute) for participation in the field work, and E. Thue Poulsen (Mathematical Institute, University of Aarhus) for the source positioning algorithm. Greenland Environmental Research Institute provided major financial, logistic and field support. Danish Natural Science Research Council is acknowledged for instrumentation grants (no. 11-5997, 511-7149).

REFERENCES

Au, W. W. L., Floyd, R. W., Penner, R. H., and Murchison, A. E., 1974, Measurement of echolocation signals of the Atlantic bottlenose dolphin, Tursiops truncatus Montagu, in open waters. J. Acoust. Soc. Am., 56(4):1280-1290.
Au, W. W. L., Carder, D. A., Penner, R. H., and Scronce, B. L., 1985, Demonstration of adaptation in beluga whale echolocation signals, J. Acoust. Soc. Am., 77(2):726-730.
Au, W. W. L., Moore, P. W. B., and Pawloski, D., 1986, Echolocation transmitting beam of the Atlantic bottlenose dolphin. J. Acoust. Soc. Am., 80(2):688-691.
Au, W. W. L., Penner, R. H., and Turl, C. W., 1987, Propagation of beluga echolocation signals. J. Acoust. Soc. Am., 82(3): 807-813.
Ford, J. K. B., and Fisher, H. D., 1978, Underwater acoustic signals of the narwhal (Monodon monoceros). Can. J. Zool., 56: 552-560.
Hubbs, C. L., and Rechnitzer, A. B., 1972, Report on experiments designed to determine effects of underwater explosions on fish life. Calif. Fish Game, 38:33-366.
Levenson, C., 1974, Source level and bistatic target strength of the sperm whale (Physeter catodon) measured from an oceanographic aircraft. J. Acoust. Soc. Am., 55(5):1100-1103.
Mellen, R. H., Browning, D. G., Ross, J. M., and Merklinger, H. M., 1975, Low frequency sound attenuation in Baffin Bay. J. Acoust. Soc. Am., 57:1201-1202.
Norris, K. S., and Møhl, B., 1983, Can odontocetes debilitate prey with sound? Am. Nat., 122:85-104.
Stapells, D. R., Picton, T. W., and Smith, A. D., 1982, Normal hearing thresholds for clicks. J. Acoust. Soc. Am., 72:74-79.
Thiele, L., 1983, Ambient noise in the sea off Thule. North Greenland. Rep. 83.88, Ødegaard & Dannskjold-Samsøe, Copenhagen.
Thrane, N., 1984, The Hilbert transform, Brüel & Kjær Technical Review 3:3-22.
Watkins, W. A., Schevill, W. E., and Ray, C., 1971, Underwater

sounds of Monodon (Narwhal). J. Acoust. Soc. Am., 49(2): 595-599.

Watkins, W. A., and Schevill, W. E., 1974, Listening to Hawaiian Spinner Porpoises, Stenella CF. longirostris, with a three-dimensional hydrophone array. J. of Mammalogy, 55(2): 319-328.

Watkins, W. A., 1980a. Click sounds from animals at sea, in: "Animal Sonar Systems", R-G. Busnel & J. F. Fish, eds., pp 291-297, Plenum Press, New York.

Watkins, W. A., 1980b, Acoustics and behaviour of sperm whales, in: "Animal Sonar Systems", R-G. Busnel & J. F. Fish, eds., pp 283-290, Plenum Press, New York.

Wiersma, H., 1988, The short-time-duration narrow-bandwidth character of odontocete echolocation signals, in: "Animal Sonar", P. E. Nachtigall & P. W. B. Moore, eds., pp 129-145. Plenum Press, New York.

Zaegaesky, M., 1987, Some observations on the prey stunning hypothesis, Marine Mammal Science, 3(3):275-279.

INVESTIGATIONS ON THE CONTROL OF ECHOLOCATION PULSES IN THE DOLPHIN

(TURSIOPS TRUNCATUS)

Patrick W.B. Moore and [1]Deborah A. Pawloski

Naval Ocean Systems Center, P.O. Box 997
Kailua, Hawaii 96734 U S A
[1] SEACO, A Division of Science Applications International
Corp. 146 Hekili St., Kailua, Hawaii 96734 U S A

INTRODUCTION

Schusterman and Kersting (1980) first demonstrated control over dolphin echolocation emissions in a binary (on/off) condition. The dolphin performed a detection task and learned to echolocate during the presence of a tone and to remain silent if no tone was given. Binary control was confirmed by wideband, low level, recordings that verified stimulus control had been attained over dolphin click emission.

Mackay (1981) used two Atlantic bottlenose dolphins (Tursiops truncatus) to determine the dolphin's capability to control whistle emissions in the 5 to 16 kHz range. Using operant conditioning techniques and automatic feeders activated by specific frequency sounds, Mackay (1981) showed that dolphins can control their whistles.

Au (1980) reported differences in dolphin echolocation signals from animals kept in San Diego Bay (a biologically quiet environment) as compared with animals in Kaneohe Bay (a biologically noisy surrounding). Echolocation pulses shifted in peak frequency from low to high, as did the pulse's sound pressure level (SPL). Au et al. (1985) also studied the beluga (Delphinapterus leucas) and found this animal also shifted both the peak frequency and SPL of its emitted clicks, between San Diego and Kaneohe Bay, presumably to compensate for noise in the environment.

Dolphin clicks are short duration (10 to 100 μs) wideband transients. During echolocation the pulse amplitude is increased or decreased depending on the target and task. Although previous studies have shown that dolphins control the repetition rate of emitted clicks as a function of target range (Morozov, et al., 1972; Au et al., 1982; Kadane and Penner, 1983; Penner, 1988), it has not been demonstrated that dolphins are capable of independent control of the emitted frequency content and SPL of their clicks.

Recent research revealed behavioral control can be obtained over the SPL of the echolocating dolphin. Moore and Patterson (1983) used operant procedures to develop control over the dolphin's emitted SPL. Detection performance during SPL control was maintained at 90% correct levels, while average SPL's for "high" and "low" amplitude clicks showed differences of 23 dB. Pawloski and Moore (1987) reported various training procedures used to control both SPL and frequency content of a dolphin's echolocation clicks. These observations lead to the interesting question of how, and at what cognitive level does the control of echolocation signals occur? Does the animal have voluntary and direct control over the SPL and frequency content of its echolocation click? Voluntary control implies a learning mechanism allowing the animal to regulate SPL and frequency for a particular echolocation task. This mechanism would permit the animal to learn to change its echolocation emissions to help recognize various objects in its environment. This paper presents a description of the procedures and methods used to train and test a dolphin's control over the amplitude and frequency of its echolocation pulses.

METHOD

Subject

The subject of this study, an adult male Atlantic bottlenose dolphin (Tursiops truncatus) Tt-018, was previously trained to control the SPL of its echolocation click. The dolphin was 20 years old and weighed 211.8 Kg with a body length of 277.8 cm by the end of the three year study.

Apparatus

The dolphin was housed in floating pens located in Kaneohe Bay, Hawaii. The testing pen measures 6.1 m by 6.1 m and the adjoining holding pen was 9.1 m by 6.1 m. Located on the deck of the test pen was a small instrument shelter which housed the electronic equipment used to measure the dolphin clicks and control the experimental procedures. The dolphin was trained to station at a fixed point facing the experimenter in the instrument shelter between trials. From this location the animal was sent by hand signal to a bite-plate/tail-rest assembly located 1.0 m below the water surface. This position was the designated station for the trial and the position from which all click measurements were made.

The use of the bite-plate/tail-rest assembly insured accurate click evaluation by positioning the dolphin's melon region directly in-line with the collecting hydrophone. The bite-plate consisted of acoustically transparent polystyrene material designed to fit the dolphin mouth. A 60 cm by 106 cm aluminum screen, centered 1.0 m below the water surface, separated the dolphin from the Bruel & Kjaer (8103) collecting hydrophone during inter-trial intervals. A diagram of the testing facility is shown in Figure 1.

The control and the reinforcer tones were generated by an Apple IIE computer system. Realistic (SA-150) stereo amplifiers were used to drive University Sound (UW-30) underwater loudspeakers. An interrupter device produced the frequency control cue by introducing two distinct interruption rates on the high and low amplitude control tones.

The collection hydrophone output was channeled through an Ithaco (4120) electronic filter/amplifier, set to pass a frequency band of 5 to 165 kHz, to eliminate background noise. An eight-channel peak-hold filter (Ceruti and Au, 1983) separated the echolocation click's relative energy output into eight frequency bins with center frequencies between 30 and 135 kHz in 15 kHz bands. The computer simultaneously measured and analyzed the emitted clicks. The averaged results were immediately available to the trainer for each trial.

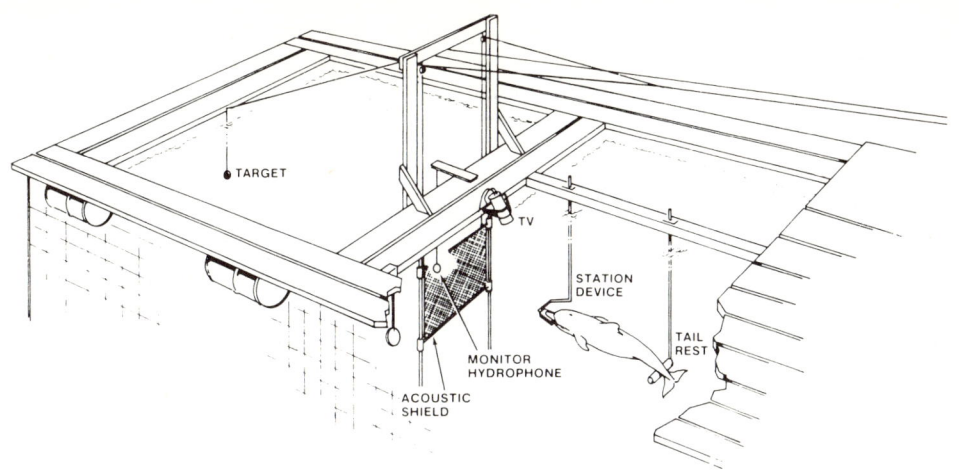

Fig. 1 Diagram of the pen showing the dolphin at station on the bite-plate/tail-rest device. The response paddle for target detection is at the left of the acoustic screen.

The computer was programed to track the dolphin's click output in real time and to reinforced the first correct click (or series of clicks) to shape and subsequently train the dolpnin's behavior. Other computer programs were used which averaged the amplitude for the first 35 clicks produced in a trial and compared the average to a previously entered amplitude or frequency criterion to shape the behavior.

A trial began when the dolphin received a hand signal to move into position on the bite-plate/tail-rest. The trainer initiated amplitude control tones with the superimposed frequency interruption rate while the dolphin settled into a straight, stationary position. The screen was then lowered, signaling the dolphin to emit clicks for up to three seconds. After an incorrect response, the stimulus tone ended and the screen raised into position. The dolphin was then required to return to the starting position facing the trainer. When the dolphin emitted clicks of correct frequency or amplitude, the reinforcer tone immediately sounded under water. The dolphin reacted by surfacing for the fish reward. Sessions consisted of 50 trials and averaged 45 minutes in duration. Behavioral control was maintained, even during the most difficult stages of training, with simple "time-outs" from training.

TRAINING FOR AMPLITUDE CONTROL

The detection task was simple for the dolphin. The target was a 12.5 mm diameter steel ball-bearing, always situated 6.0 m directly in front of the dolphin. Before any training for SPL or frequency control the animal was allowed to stabilize outgoing click level during the detection task. Successive sessions of detection trials were used to measure the animal's average emitted SPL. When the average SPL stabilized the last 10 sessions were pooled to determine the animal's preferred level for detection. An example of previous amplitude data for two animals trained for this detection task by Moore and Patterson (1983) is shown in Fig.2

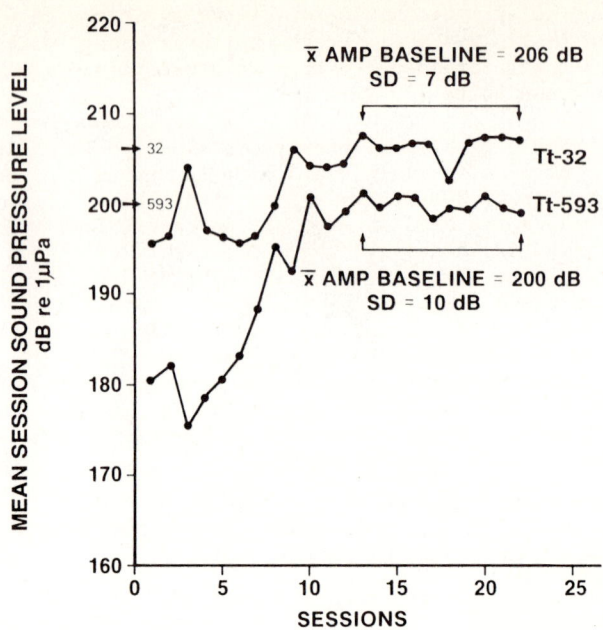

Fig.2 Average SPL as a function of sessions for a simple target detection task (from Moore and Patterson, 1983). Both dolphins have stabilized the emission SPL by session 15. The last 10 sessions were averaged to determine each animal's preferred SPL for the detection task.

After detection performance was at 90% or better, training to control SPL began. At the start of a trial the computer turned on a tone from one of two underwater speakers. The right speaker cuing the dolphin to increase the SPL of its clicks and the left to decrease level. As the animal emitted clicks, a comparison was made between the preset criterion level and the emitted level. If a click was above (or below) the criterion, the computer interrupted the trial by sounding the secondary reinforcer tone while the animal still emitted clicks. Slowly adjusting the criterion higher or lower, in accordance with the right or left tone, trained the animal to use the tones to produce high or low level clicks.

Bimodal frequency output is the simultaneous occurrence of energy peaks in frequency bins at opposite ends of the frequency range. Bimodal frequency output was observed in high amplitude trials. During trials which averaged over 197 dB SPL but lower than about 210 dB, the spectra of the dolphin's clicks could be either broadband (ie, relativity flat across the frequency bins of 35 to 120 kHz), predominantly low frequency (major energy in the 30 to 60 kHz bins), or high frequency (major energy in the 105 to 135 kHz bins). For trials with amplitude levels above 205 to 210 dB, the relative energy output could be elevated in both low and high frequency ranges (bimodal) with less energy output in the median range (75 to 90 kHz bins).

Wideband analog recordings using a Racal (store 4 DS) recorder connected to the monitor hydrophone were made after the initial detection portion of this study and before any frequency training. The click train in Fig. 3A is typical of the energy distribution in a click train with no frequency constraints. The SPL criteria was 205 db or greater. The fist 23 clicks have peak energy at the lower frequencies centered at about 45 kHz and SPLs below 210 dB. As the SPL increases, bimodal energy appears with peaks at 45 kHz and 120 kHz. Immediately after the first appearance of bimodal clicks the energy becomes wideband across frequencies from 35 to

110 kHz. Around click #68 six bimodal-clicks again emerge with energy ranging from 30 to 110 kHz and a second narrower energy peak at about 130 kHz. The click train terminates with a narrower high-frequency energy distribution between 80 to 100 kHz.

A second example of bimodal energy distribution is shown in Fig 3B. The figure is typical of a click train emitted during a detection trial with a SPL criterion of 205 dB or greater with target present. As the click train develops the major energy of the clicks is bimodal with -3 dB bandwidths centered at 45 and 120 kHz. The first few clicks are not bimodal and their SPLs fall below 208 dB. The SPLs for clicks in the remainder of the click train was above 209 dB (the difference in the number of clicks between the two trial types is a function of the target reporting paradigm -- for a target present the animal must leave the station to report -- for absent the animal remains at station). Bimodal output as a function of SPL has been reported previously for this particular training problem (Ceruti et al., 1983). Even though bimodal clicks were noted, we disregarded them at this stage of training.

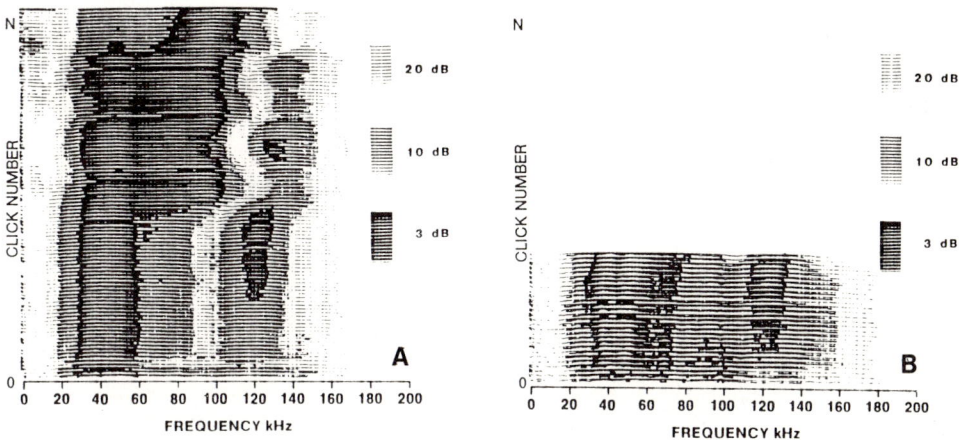

Fig. 3 Waterfall displays of the distribution of energy of clicks in a click train across the frequency spectrum. Each click is plotted vertically as a series of horizontal spectra (click 1 at the bottom of the plot and the last click plotted at the top of the vertical axis). Three shades of gray represent the 20, 10, and 3 dB energy bandwidths. Figure 3A displays a click train of 101 clicks emitted during a detection trial with the target absent. The SPL criteria was 205 db or greater. Figure 3B shows a train of 29 clicks emitted during a target present trial with the SPL criterion at 205 dB or greater.

The ability of the dolphin to control the amplitude of their echolocation clicks has already been documented by Moore and Patterson (1983). They required a dolphin to produce high and low amplitude clicks in blocks of five trials all with the same amplitude criterion. Amplitude control was demonstrated by analyzing the animal's emitted level on only the first trial of each random amplitude control block, their results are shown in Figure 4.

In this study the subject was trained to control the amplitude of the clicks in a random presentation. Early training incorporated several separate sessions of simple target detection or amplitude training. Continued training sessions imposed amplitude criteria on the dolphin that grew in difficulty as the animal improved.

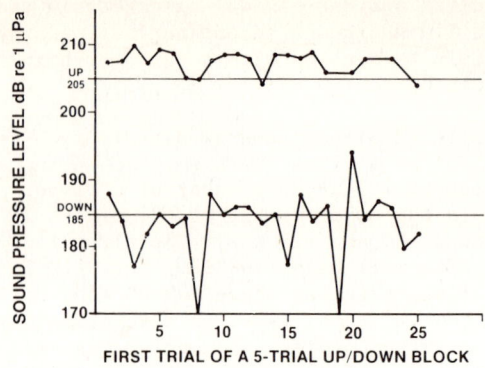

Fig. 4 Averaged SPL for clicks in each first-trial of a randomly cued high or low amplitude five trial block. Each point represents the averaged level of the clicks emitted during the first (reversal) trial. The criterion level is indicated by the two solid lines (Moore and Patterson, 1983).

Once the animal demonstrated 95% correct amplitude control and detection performance, these separate chores were combined over seven sessions (350 trials) into a single task and designated "baseline" sessions. A high and low emitted level criterion was set for this animal. High level was 205 dB and low was 195 dB. This animal was tested over seven sessions of randomly cued high and low criterion trials and required to perform a target detection task. The dolphin maintained 90% or better performance, meeting the combined amplitude and detection criteria when presented with both tasks randomly. As shown in Figure 5, our animal had a mean of 209 dB for the high level criterion and 189 dB for the low level criterion. Target detection for all seven sessions averaged 98% and the animal's score for meeting the amplitude criterion was 95%. Averaging across sessions, the animal maintained a 20 dB difference between the seven test sessions.

After we obtained SPL control we trained the dolphin to respond to only one speaker, which produced both tones, to control the animal's emitted level. Before moving into frequency training, the tones were reversed in location and gradually moved together until a single speaker, producing both tones, controlled amplitude. This action forced the controlling cue for amplitude to be the frequency of the tone and not it's location.

TRAINING FREQUENCY CONTROL

Confusion occurred between the amplitude and the frequency tasks in early training. To differentiate the frequency task from the amplitude task the underwater speaker was relocated behind the dolphin during frequency training. Then, as the subject improved in the separate tasks, the speaker was moved gradually toward the final location near the front of the dolphin. Training concentrated on the frequency task and detection was eliminated from the training sessions. Amplitude control/detection sessions, or "baseline" sessions, were conducted intermittently to maintain this separate combination of behaviors.

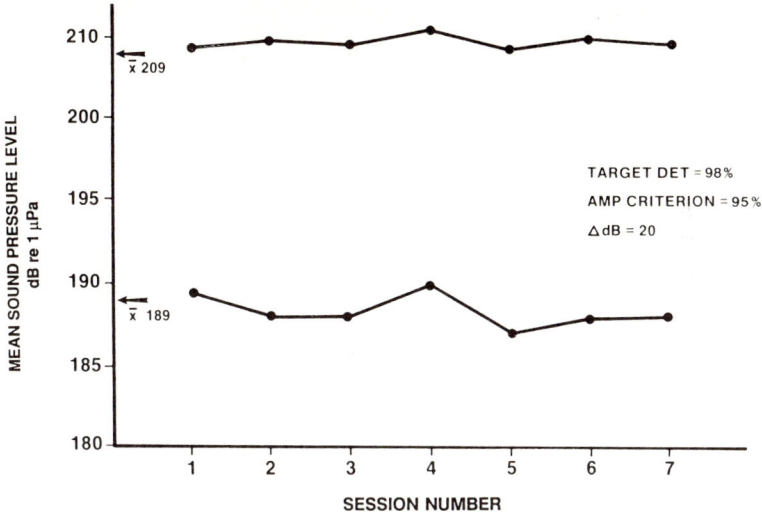

Fig. 5 Results from the seven test sessions when the dolphin was required to perform a random detection task with a random amplitude criterion cue. Each point represents the averaged level of all trials of either the high or low criterion for that session.

During early frequency training the dolphin continued to shift its amplitude in response to the intended cue for frequency change. Frequency training required a computer program change from averaging the frequency bins of the emitted clicks in a click train, to a real time analysis of the energy in each frequency bin of each emitted click. The real time program reinforced the dolphin on the first correct click which met the low (60 kHz or below) or high (105 kHz or above) frequency criterion, without considering the large number of previous <u>incorrect</u> clicks. This design worked for the generalized idea of amplitude training, because the subject gradually moved its emitted level (high or low) to reach the correct amplitude level. Frequency training, we discovered, required reinforcement of a <u>consecutive</u> number of correct clicks for the dolphin To master the task. We could select 8, 16, 32 or 64 consecutive correct high or low frequency clicks for the dolphin to receive reinforcement. After separate training of high and then low frequency emission we tested the animal's ability to switch when randomly cued. The normalized energy distribution for 200 randomly cued high and low frequency trials are presented in Fig 6.

Note that for the low frequency condition the maximum energy occurs in the 60 kHz - bin but substantial energy still exists in both the 120 and 135 kHz - bins. For the high frequency condition major energy is in the 129 kHz - bin, but is still well distributed in the lower bins (45, 60, 75 and 90).

CONTROL OVER AMPLITUDE AND FREQUENCY

Typically, the dolphin used clicks of 200 dB for the detection task before any training in amplitude control. Since the amplitude measuring range of the apparatus was centered in the dolphin's normal output range the animal learned to maintain levels within 185-195 dB for low amplitude trials and 205-215 dB for high amplitude trials.

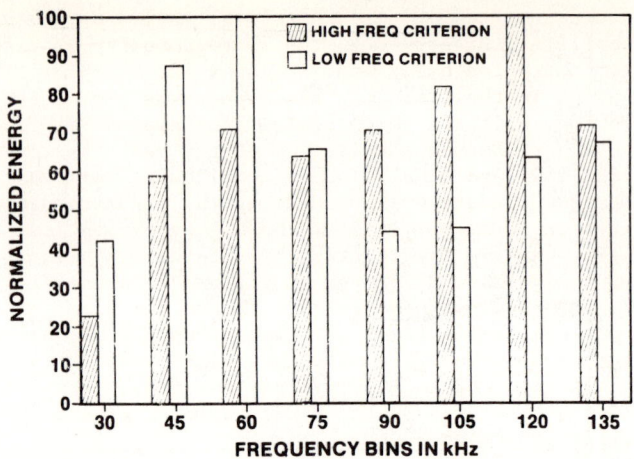

Fig. 6 Normalized energy across frequency bins for 200 randomly cued frequency control trials. Both conditions display a bimodal distribution of energy. The average number of clicks to meet the frequency criterion was 36 for the high frequency condition and 62 for the low frequency criterion.

When faced with the frequency control task, the dolphin continued to shift its amplitude when the correct response should have been a change in click frequency content. Because high amplitude clicks were inherently bimodal in frequency, we incorporated an amplitude criterion of greater that 195 dB on the frequency training trials. Then we trained the animal to shift its emitted bimodal frequency peaks based on the frequency cue of repetition rate superimposed on the amplitude cue tone. Frequency training in the high amplitude range (>195 dB) required a frequency criterion which limited the amount of bimodality in the emitted clicks.

The computer programs were modified to compare the low frequency bins (60 kHz or below) to the high frequency bins (105 kHz or above) and to restrict the energy between them. The computer checked the frequency bin with the greatest output in the opposing (incorrect) frequency range and computed the ratio of the energy emitted in the highest bin of the incorrect range to the energy emitted in the highest bin of the correct range. By allowing some bimodal frequency output (a ratio of .75) stimulus control was maintained in a very difficult task. The dolphin responded to the ratio criteria by simply reducing it's bimodal frequency output to conform to the energy ratio criteria imposed.

In the next stage of training, the bimodal allowance in the ratio criterion was reduced each time the dolphin reached 80% performance levels of frequency control. The energy ratio criteria levels used were 75%, 62%, 50%, and 35% bimodal output allowance. The computer program tracked the dolphin's emitted clicks with an eight click rolling average, so the number of clicks required to reach the criterion indicated the dolphin's learning. When the number of clicks required to reach frequency criterion stabilized, the dolphin was assumed to be at its peak performance for that criterion setting.

The high frequency criterion required clicks to have major energy at 105 kHz or above, and the low frequency criterion required clicks to have major energy at 60 kHz or below. The initial average number of click to reach the frequency criterion at either level varied by about 55 clicks. After additional training sessions, the average number of clicks required to reach either criterion had dropped to only 10.

At this point in the training the dolphin was given progressively more difficult frequency criteria which limited the amount of bimodal output allowed. At 80% performance levels, the energy ratio criterion was reduced from .75 to .62, .50, and finally .35 bimodal output allowance. This procedure allowed secondary energy peaks at the opposite (incorrect) frequency criterion to be only .35 of the highest energy output in the correct frequency range. The restrictions of the .35 bimodal output proved impossible for the animal and caused an extreme response bias that strongly influenced the animal's performance. To remove the bias the subject was regressed to the easiest level of .75 bimodal output allowance. As 80% performance standards were again met, the dolphin was stepped-through the energy ratio criteria (.62 and .50) in six sessions. The .35 bimodal output allowance was omitted in the final training because of the animal's apparent inability to solve the task.

During frequency training with energy ratios of .62 and .50, the dolphin's frequency error response was to change its emission amplitude. This developed into two distinctly different amplitude levels for high versus low frequency output during training. The subject remained above 85% performance for frequency control, but distinctly varied its amplitude output for high versus low frequency trials.

Our final test combined the high amplitude tone cue with the frequency cue of interruption rate. A limit of no more than 20 dB difference in the emitted amplitude level was added to prevent large SPL fluctuations during subsequent frequency control training. The dolphin's performance, measured by the number of clicks required to reach frequency criterion, stabilized after about 200 trials.

Testing of frequency control continued for an additional 550 trials to investigate the ability of the animal to learn to control frequency. Curiously, the dolphin continued to maintain an amplitude difference of 13 dB between high and low frequency output. After 750 trials, low frequency clicks averaged 197 dB, while high frequency clicks averaged 209 dB. However, the average number of clicks required to reach the high or low frequency criteria dropped to only five.

Training data showed that approximately 20% of the frequency trials had reversed amplitude outputs. During these trials, the dolphin switched the amplitude it normally associated with the separate frequency criteria. In these trials, the average energy spectra for low frequency was a perfectly declining sweep in energy in frequencies from low to high, with peak energy at 45 kHz. High frequency, however remained slightly bimodal, with peak energy at 135 kHz and a small secondary peak at 45 kHz. Although the dolphin continued to produce bimodal output its progress indicated task acquisition.

The dolphin's bimodal frequency output seemed related to the emitted level. During trials with amplitude levels above 203 dB, measured frequencies almost always were bimodal. We initially thought the bimodality arose because the dolphin confused the two tasks of frequency control and amplitude control. After extensive training (four months) we could not suppress bimodal emissions, and we now believe there is a physical reason for bimodal frequency output with high amplitude in echolocating dolphins. This same finding was reported for the Moore and Patterson study (1983), by Ceruti et al. (1983) and by Pawloski and Moore (1987). Peak energy in the emitted signal may be limited to one or two vibrational modes due to the size and shape of the production device.

A final stage of testing was conducted to investigate the affect of bite-plate composition on the ability to control emitted frequency. Neoprene rubber, 1/8" thick, was attached to the top- and bottom-faces of the bite-plate, and used in the frequency control tests. The assumption was that the air-filled neoprene would tend to isolate the upper head, the presumed sight of click production, from the lower jaw, the presumed sight of click reception. This isolation could possibly reduce acoustic coupling between the click source and the lower jaw, via the bite-plate, thereby reducing the perceived intensity of the animal's emissions. Possible interference to the lower jaw from the click source located above the jaw would be blocked by the reflective quality of the neoprene.

Random, 30-trial sessions with either the polystyrene or the neoprene covered bite-plate were conducted. The same criteria of high (>105 kHz) and low (<60 kHz) frequency with amplitude level at >195 dB and an allowable amplitude range of 20 dB was used to test the affect of blocking possible reflections of clicks to the lower jaw.

Eight sessions with random use of the different bite-plates were completed. The results are shown in Figure 7 and the energy in the frequency bins are normalized to the maximum bin value.

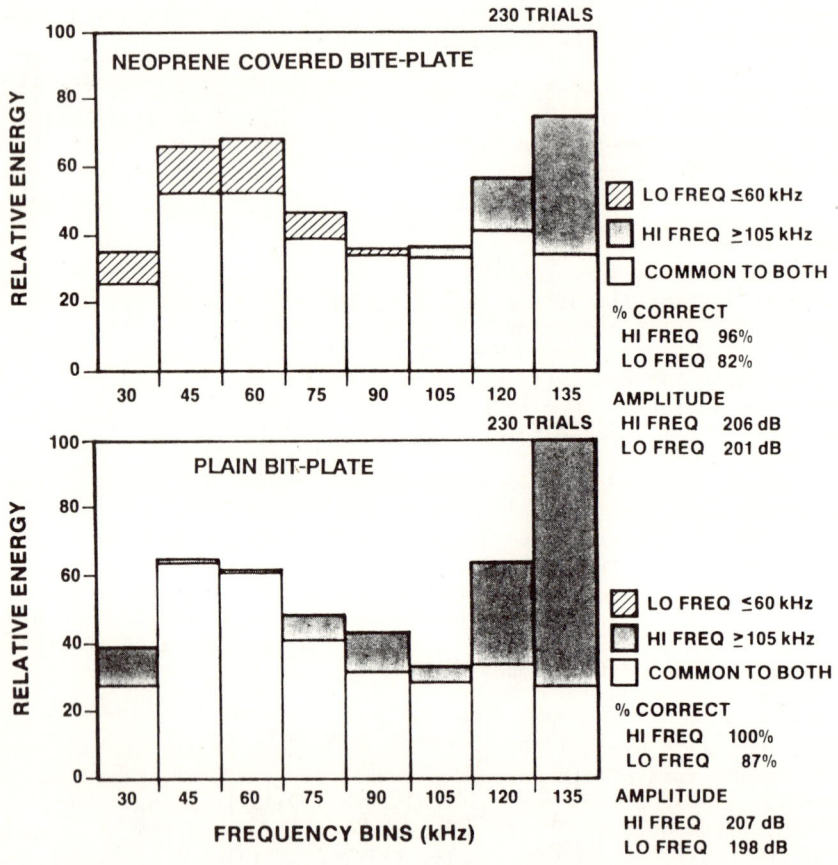

Fig. 7 Top panel - Relative energy across frequency bins for sessions of trials using the neoprene covered polystyrene bite-plate. Bottom panel - relative energy across frequency bins for sessions of trials using the plain polystyrene bite-plate.

The top panel indicates that with the neoprene covered bite plate, high frequency cued trials had major energy distributed in the high frequency bins (105 to 135 kHz), low frequency trial had energy distributed in the low frequency bins (30 to 60 kHz) as well as the 75 and 90 kHz bins. The overall amplitude levels of both high and low frequency trials are similar ($\pm$ 5 dB) as compared with the tests using the plain polystyrene bite-plate (bottom panel) which showed a 13 dB difference. Additionally, with the neoprene covered bite-plate both types of trials maintained relativity separate spectra. With the plain bite-plate (bottom panel) high frequency trials produced energy across the frequency bins and the SPL difference between the two conditions is 9.0 dB. It appears that with the plain bite-plate the animal simply added high energy for the high frequency trials especially in the 120 and 135 kHz bins. These data demonstrated relatively stable amplitude levels for both high and low frequency output using the neoprene covered bite-plate as compared with the plain polystyrene bite-plate. Using the neoprene covered bite-plate we conducted seven additional sessions of combined amplitude and frequency cuing and demonstrated combined amplitude/frequency control in the dolphin's echolocation emissions.

CONCLUSIONS

The dolphin exhibits the ability to change the frequency content of its clicks in a click-by-click fashion and produced echolocation signals with dual energy peaks in widely separate frequency bins (bimodal clicks). Dolphins also are capable of controlling both the frequency and amplitude of their echolocation clicks, within the bounds of our criteria, in response to operantly conditioned cues for such behavior. Generally, as the echolocation SPL moves above 210 dB emitted clicks can become bimodal, having two separate energy peaks in the low frequency and high frequency regions, as we defined them.

The training of frequency control is very difficult, taking three- to five-times as longer than training of amplitude control. This difference in difficulty probably is due to the limitations imposed by the physical structure of the dolphin's signal generating mechanism. This difficulty was apparent with the observation of bimodal clicks, and the production of at least three types of clicks: (1) predominantly low frequency (30 - 60 kHz); (2) wideband bimodal clicks (30 - 100 kHz); and (3) high frequency clicks (100 - 130 kHz). These observations suggest that the echolocation signal generation mechanism is constrained somewhat in the frequency domain by the level of the clicks.

REFERENCES

Au, W. W. L. 1980. Echolocation signals of the Atlantic bottlenose dolphin (Tursiops truncatus) in open waters. in: "Animal Sonar Systems," R.G. Busnel and J. Fish, eds. Plenum Press, New York, pp251-282.

Au, W. W. L., Penner, R.H., and Kadane, J., 1982, Acoustic behavior of echolocating Atlantic bottlenose dolphins, J. Acoust Soc. Am., 71: 1269.

Au, W. W. L., Carder, D. A., Penner, R. A., and Scronce, B. L., 1985, Demonstration of adaptation in beluga whale echolocation signals, J. Acoust. Soc. Am.., 77: 726-730

Ceruti, M. G. and Au, W. W. L., 1983, Microprocessor-based system for monitoring a dolphin's echolocation pulse parameters. J. Acoust. Soc. Am., 73: 1390.

Ceruti, M. G., Moore, P. W. B. and Patterson, S. A., 1983, Peak sound pressure levels and spectral frequency distributions in echolocation pulses of the Atlantic bottlenose dolphin, (Tursiops truncatus), J. Acoust. Soc. Am., 74 (S1): S73.

Kadane, J. A. and Penner, R. A., 1983, Range ambiguity and pulse interval jitter in the bottlenose dolphin, J. Acoust. Soc. Am., 74: 1050-1061

Mackay, R. S., 1981, Dolphin interaction with acoustically controlled systems: aspects of frequency control, learning, and non-food reward, Cetology, 41.

Moore, P. W. B., and Patterson S. A., 1983. Behavior control of echolocation source level in the dolphin (Tursiops truncatus), in: "Proceedings of the Fifth Biennial conference on the Biology of Marine Mammals", Boston, MA.

Morozov, V. P., Akopian, A. I., Burdin, V. I., Zaytseva, K. A., and Sokovykh, Y. A., 1972, Tracking frequency of the location signals of dolphins as a function of distance to the target. Biofizika, 17: 139

Pawloski, D. A. and Moore, P. W. B., 1987, Combined stimulus control of peak frequency and source level in the echolocation emission of the dolphin (Tursiops truncatus), in: "Proceedings of the 15th annual conference of the International Marine Amimal Trainers Association," New Orleans, Oct.26-30, pp. 3-9

Penner, R. H., 1988, Attention and detection in dolphin echolocation, in: "Animal Sonar: Processes and Performance," Nachtigall, P. E. and Moore, P. W. B., eds., Plenum Press, New York.

Schusterman, R. J. and Kersting, D. A., 1980, Stimulus control of echolocation pulses in Tursiops truncatus, in: "Animal Sonar Systems," R.G. Busnel and J. Fish, eds. Plenum Press, New York, pp.981-982.

ACKNOWLEDGEMENTS

The original computer programs were written by Mr. Dana Pieterson. Modifications to the programs were made with the help of Dr. Richard A. Johnson while at the Naval Ocean Systems Center on a ASEE fellowship. Thanks to Dr. Marion Ceruti, Ms. Sue Patterson and Mr. Jim Richards for their assistance training the animals. Dr. Whit Au designed the echolocation click measuring device. The idea for this study was conceived by Dr. Paul Nachtigall. This study spanned several years and involved many trainers, technicians and other psychologists, we would like to thank them all. Thanks also to Dr. Jeanette Thomas for helpful advice on the manuscript.

PURPOSEFUL CHANGES IN THE STRUCTURE OF ECHOLOCATION PULSES IN TURSIOPS TRUNCATUS

Evgeniy V.Romanenko

A. N. Severtsov Institute of Evolutionary
Animal Morphology and Ecology, USSR Academy of Sciences
Leninsky prosp. 33, 117071, Moscow, USSR.

In echolocation of obstacles and fish, dolphins normally rely on stereotyped pulses (Romanenko, 1974; Belkovitch and Dubrovsky, 1976). These pulses consist of 1-1.5 waves and their shape and frequency spectra are resistant to environmental changes. Even a considerable increase in the noise level in the ambient environment by means of sources distant from the dolphin causes no noticeable changes in pulse shape and frequency spectra.

Nevertheless, there are grounds to believe that a dolphin can manipulate the emitted pulse shape and frequency spectra in case noise sources are located in the immediate vicinity of the organ of hearing. To test this hypothesis, special experiments were carried-out; during these experiments intensive noise was generated alternately in the immediate vicinity of the sites on the dolphin's head of the possible reception of acoustic information (meatus acusticus externus, mandible and melon) (Romanenko, 1978). The noise level at these sites in OK frequency range made-up 120+6dB in re 0.02 mPa. The noise spectrum was constant in the frequency range from 5 to 30 kHz and it dropped by 6 - 7 dB per octave at higher frequencies. Two, 30 - mm diameter spheres of piezoelectric ceramics served as noise sources; they were fixed alternately with rubber suckers at the above mentioned sites (one sphere at each acoustic duct, on the right and left side of the mandible and both spheres near each other on the melon); they were fed by the same noise generator.

Echolocation pulses were received by miniaturized hydrophones, one of which was fixed on the anterior part of the melon (Fig 1), the second on the right or left side of the melon at equal distance between the first hydrophone and the blowhole. A third hydrophone was placed near one of the noise sources to monitor the noise level. Figure 1 shows a scheme of hydrophones and noise sources disposition. These are hydrophones: number one, two and three and noise sources: number one and two, which are placed near the acoustic duct.

The echolocation pulses were registered by a three-channel tape recorder, fixed on the dorsal fin of a dolphin together with a noise generator, a radiotelemetric system to turn on all the equipment. A light indicator made it possible to observe dolphin's movement in complete darkness. (The method of receiving the pulses on a dolphin's head was first proposed by Dierks and others, 1971).

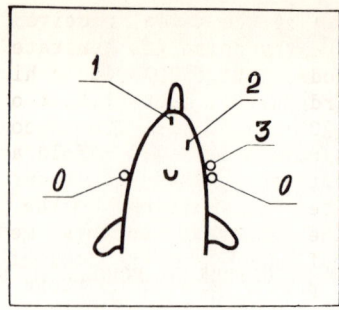

Fig.1. A scheme of noise sources (0) and hydrophones (1-3) positions.

The experiments were performed in a pool measuring 12.5 x 6 x 1.2 (depth). Initially the dolphin was placed at the starting position at one side of the pool. A fish was dipped with a splash on the opposite side of the pool. The fish was attached to a cotton thread and was moved noiselessly 2.5 - 3 m sideways. When hearing the splash, the dolphin would start to swim towards the fish, emitting echolocation pulses. A noise generator was switched on by radio when the dolphin was half way to the fish. The dolphin went on moving, echolocating to invariably locate the fish.

The dolphin's response to the switching-on of the noise generator in two ways: firstly: visual observation of changes of the luminous indicator trajectory, and secondly: by changes in echolocation activity.

When noise sources were placed near the external auditory meatus, a clear moving response was observed as indicated by a change of luminous trajectory. The dolphin gave a start when the noise generator was switched-on. Sometimes such experiments were carried-out in broad day light and one could clearly see that dolphin sharply moved his head when hearing the noise, trying to escape it, but soon calmed down.

When noise sources were placed near mandible and melon no moving activity changes were detected.

Analysis of the tape with the recorded signals showed the following: when noise sources were placed near the external auditory meatus, the echolocation pulses, which were received by all the hydrophones, were of stereotyped nature before exposure to noise (Fig.2, the first and the second pulse). After switching on the noise generator, the hydrophone

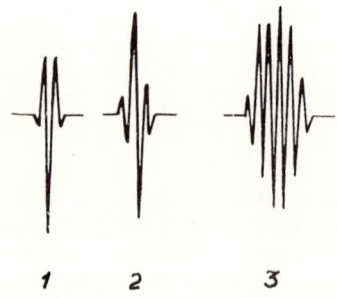

Fig.2. The stereotyped (1 and 2) and oscillatory (3) pulses.

fixed on the anterior part of the melon, received pulses of the same amplitude as the ones before the noise was generated, but they were oscillatory, with 4 - 5 periods, and displayed a higher-frequency, narrow spectra (Fig. 2, the third pulse). Transition from stereotyped pulses to oscillatory pulses took 200 - 300 ms. The second hydrophone received mostly stereotyped pulses with a 2.5 to 3 - fold amplitude. Occasionally, alongside with the stereotyped pulses, it recorded oscillatory pulses which were somewhat shifted in relation to the stereotyped pulses. The hydrophone, fixed near the external auditory meatus received emitted pulses as well as those reflected from the pool bottom, the water surface and the fish. The direct pulse, as well as those reflected from the pool bottom and water surface, were invariably stereotyped. By contrast, the pulses reflected from the fish were stereotyped prior to noise generation and became oscillatory when the noise started. In the case when noise sources were fixed on the mandible and melon no changes were recorded in the echolocation pulses after noise generator was started.

There is one other interesting fact. The pulses erradiated by dolphins when the noise generator is on, sometimes have a so called "forerunner" mentioned earlier by Zaslavsky (1974) and Dubrovsky (1975). We call "forerunners" pulses preceding each echolocation pulse. The intensity of "forerunner" changes in wide ranges and sometimes can run-up to the intensity of the main pulse: the pulse becomes double. Nevertheless, the duration of a "forerunner" remains shorter than the main pulse.

The results of our experiments suggest that: (1) the dolphin probably can change the echolocation pulses spectra as the main means of overcoming the wide-range noise. (2) Stereotyped and oscillatory pulses can be emitted by different sources. (3) The acoustic information enters the internal parts of auditory system via the external acoustical meatus.

REFERENCES

Belkovitch, V.M. and Dubrovsky, N.A., 1976, "Sensor Bases of Cetacea Orientation", Science, Leningrad.
Diercks, K.J., Trochta, R.T., Greenlow, C.F. and Evans, W.E., 1971, Recording and analysis of dolphin echolocation signals, J. Acoust. Soc. Am. 49 (6): pt.1: 1729.
Dubrovsky, N.A, 1975, "Echolocation in dolphins (Review)" TsNII Rumb Publ., Leningrad.
Romanenko, E.V. 1974, "Bioacoustic Physics Bases", Science, Moscow.
Romanenko, E.V. 1978, Some results dolphin's acoustic investigations, in: "Marine Mammals" V.E.Sokolov ed., Science, Moscow.
Zaslavsky, G.L., 1974, Experimental study spatial and temporal pattern of dolphin's echolocation signals, Candidate thesis dissertation, Karadag.

ECHOLOCATION CHARACTERISTICS AND RANGE DETECTION THRESHOLD OF

A FALSE KILLER WHALE (PSEUDORCA CRASSIDENS)

Jeanette A. Thomas and Charles W. Turl[1]

Western Illinois University, Macomb, Illinois
61455; [1]Naval Ocean Systems Center, Kailua, Hawaii
96734

INTRODUCTION

False killer whales (Pseudorca crassidens) are deep-diving, pelagic, toothed whales that inhabit temperate and tropical oceans. Pseudorca prey on squid, large fish, and perhaps even dolphin (Tomich, 1986). They produce many types of sounds, including whistles, frequency-modulated sweeps, and pulses (Busnel and Dziedzic, 1968; Watkins, 1980).

A preliminary study by Thomas et al. (1988b) showed that when the vision of a false killer whale in a concrete pool was obstructed it could detect a 7.62 cm diameter, steel sphere 4 m away. The peak frequency of echolocation pulses used by this Pseudorca in a concrete pool ranged from 20 to 65 kHz, had a bandwidth (at -3 dB) of 5 to 16 kHz, and peak-to-peak source levels of 145 to 150 dB re 1 μPa (Thomas, et al., 1988a). The later study suggests that the low peak frequency and amplitude of these pulses (compared with those produced by the bottlenose dolphin, Tursiops truncatus) probably are caused by the reverberant concrete pool. Thomas et al. (1988a) documented the underwater hearing sensitivity of a captive false killer whale. Pseudorca hearing is most sensitive (that is, within -10 dB of the maximum sensitivity) in the range from 16 to 64 kHz, but the whale has good sensitivity (i.e., within -40 dB) from 8 to 105 kHz. The broadband, U-shaped hearing curve of the false killer whale is characteristic of echolocating odontocetes.

Thomas et al. (1990, in press) documented low critical ratios (17-42 dB) in a false killer whale relative to terrestrial mammals and humans. The masked hearing ability of Pseudorca is similar to, perhaps slightly better than, that reported for the bottlenose dolphin, Tursiops truncatus, (Johnson, 1967;1968) and the beluga, Delphinapterus leucas, (Johnson et al., 1989).

The long-range detection of a 7.62 cm steel sphere by a bottlenose dolphin initially was studied in Kaneohe Bay, Hawaii by Murchison and Penner (1975) and Murchison (1980a; 1980b). They found that irregularities in the bottom topography produced reverberation-limited range detection thresholds. Au and Snyder

(1980) measured range detection by a bottlenose dolphin at another location in Kaneohe Bay and reported that correct detections fell to chance at 113 m.

The goals of our study were to: 1) document the characteristics of echolocation pulses from a false killer whale in open-water during a target detection task, and 2) measure the range detection threshold for a false killer whale.

METHODS

Subject

A subadult male false killer whale (Pc 739M) was the subject. The whale weighed 410 kg, was 3.25 m long, and about 5 years old. Although this was the animal's first study, he learned the task in 6 months. The whale received his daily ration of approximately 18 kg smelt, 5 kg mackerel, and 5 kg herring during two daily sessions.

Skyhook Range

We conducted the experiment in the same floating pen complex and "skyhook range" in Kaneohe Bay, Oahu, Hawaii used by Au and Snyder (1980) to measure the range detection threshold in a bottlenose dolphin. Au and Snyder (1980) provide a detailed descriptions of the "skyhook range". A hut on the pen housed the researcher and acoustic monitoring equipment. Figure 1 shows the skyhook range, which is a set of monofilament lines and pulleys suspended between two poles over the whale's pen. A set of five, 7.62 cm diameter, water-filled stainless steel spheres (TS_{p-p} = -30.8 dB) were suspended from monofilament lines to the water. The distance between the whale and the first target was adjusted from a reel located on each end pole of the range. The researcher lowered or raised individual targets into the water using one of five reels inside the hut. The depth of each target was adjusted, based on tides, to 1 m below the water surface before each session.

Procedures

The whale started each trial, facing the researcher, with its rostrum on a stationing post. The researcher gave a 3 kHz tone that sent the whale to station at an underwater hoop (Figure 2) that was placed at 0 m on the skyhook range and centered 1 m below the surface. An acoustic screen, made of aluminum plate and neoprene rubber, was in front of the hoop station to prevent the whale from echolocating the target. Before each trial, a target was either gently lowered into the water or left suspended in the air. Lowering the acoustic screen allowed a clear path to the target and cued the whale to echolocate. The whale ensonified the range for as long as it desired, backed from the hoop, and responded by touching either a paddle on the right (to indicate target present) or a paddle on the left (to indicate target absent). The whale received a fish reward for proper responses. False alarms (reporting a target as present when it was not) or a misses (reporting a target as absent when it was present) were not reinforced. The animal maintained an inter-trial interval of about 1 min.

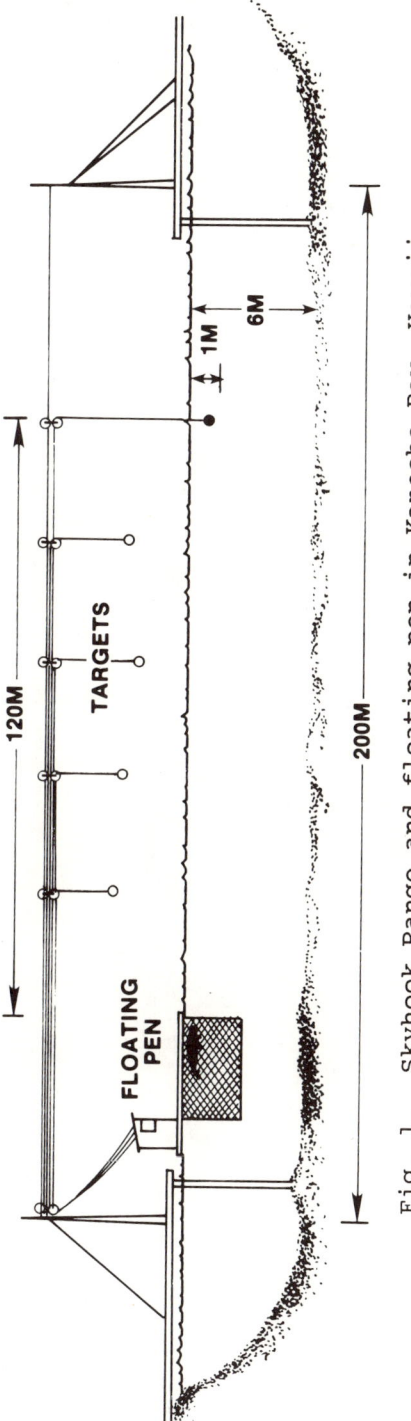

Fig. 1. Skyhook Range and floating pen in Kaneohe Bay, Hawaii.

Data were collected during a morning and an afternoon session five days per week from September to December 1988. A session consisted of 50 trials divided into five blocks of 10 trials. Each block was assigned a different target distance. Equal numbers of target-present and target-absent trials were distributed randomly in a block based on modified Gellerman (1933) tables.

To bracket the whale's range detection threshold, we initially ran 20 sessions with 5 targets spaced 20 m apart, with the first target 40 meters from the whale. After determining that the threshold was near 100

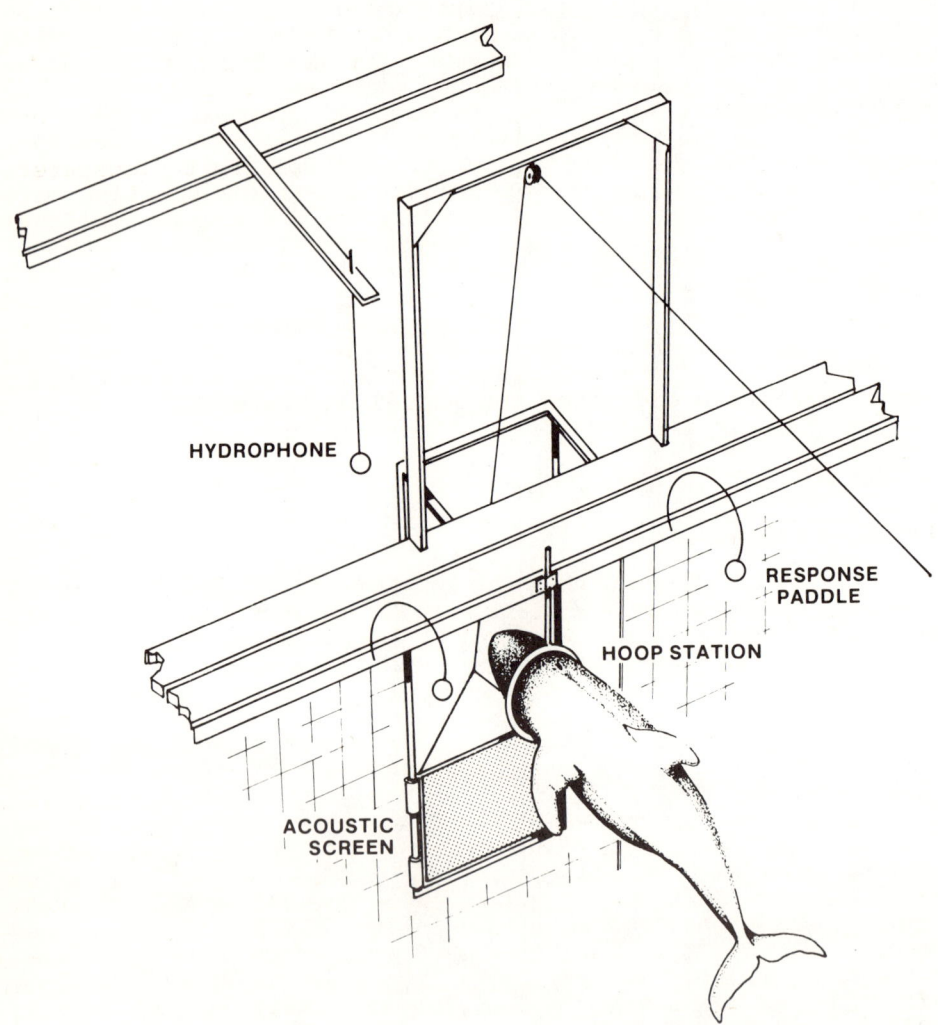

Fig. 2. Experimental configuration showing the floating pen, hoop station, acoustic screen, response paddles, and skyhook suspension system.

m, we reconfigured the range with 5 targets spaced 5 m apart. Twenty sessions were conducted for each of three range configurations, which started at 80 m, then 90 m, and then 100 m from the whale, respectively.

To address whether experience at a specific distance influenced the whale's performance, we presented the targets in two ways for each range configuration (80-100, 90-110, and 100-120 m). First, we collected data from 10 sessions with targets presented in order from near-to-far. We then conducted 10 sessions with five target distances presented randomly.

Acoustic Monitoring Equipment

Each echolocation pulse produced by the whale was received using a Bru"el & Kjaer 8103 hydrophone (2 m from the whale) and a MacIntosh Plus personal computer equipped with an analog-to-digital converter. To prevent the system from triggering on snapping shrimp noise (ambient levels from 45 to 65 dB re 1 $\mu Pa/Hz^{1/2}$), signals from the hydrophone were amplified, high-pass filtered at 20 kHz and low-pass filtered at 200kHz. This filtering may have caused slight changes in the frequency structure of the recorded pulse, but other pulse characteristics (amplitude, number of clicks, interclick interval) should not be affected. At the beginning of a trial, the researcher started data collection with a switch controlling the computer. For each click, 128 points of data at a 500 kHz sampling rate were collected each time a pulse triggered a preset amplitude level. The date, time, trial number, target distance, and the animal's response were cataloged with each pulse train and stored on a diskette along with the number of clicks, interclick intervals in msec, and the peak-to-peak source level for each click in dB re 1 μPa. During each block of 10 trials, the entire pulse train from one target-present trial and one target-absent trial were digitized and stored on diskette.

RESULTS AND DISCUSSION

Number of Clicks per Trial

Figure 3A shows that the number of clicks per trial increased as a function of distance between the animal and target. In trials where the animal responded with a miss or a false alarm, the number of clicks was more variable by distance, which may reflect the whale's uncertainty in the detection.

The mean number of clicks per trial was lowest for trials when the whale responded properly. Trials with misses or false alarms had consistently longer click trains. A Paired Sign-test ($p < .05$) showed significant differences between the mean number of clicks in these comparisons: miss versus false alarm, correct rejection versus miss, and correct rejection versus false alarm. This may suggest that the false killer whale required additional sampling, i.e., more clicks, when the task was difficult. Several studies have documented that the number of clicks produced by *Tursiops truncatus* increase with the difficulty of the task (Au and Penner, 1981; Au and Kadane, 1982; Au and Turl, 1983; and Au et al., 1987).

Interclick Interval

Figure 3B shows the interclick interval increased as a

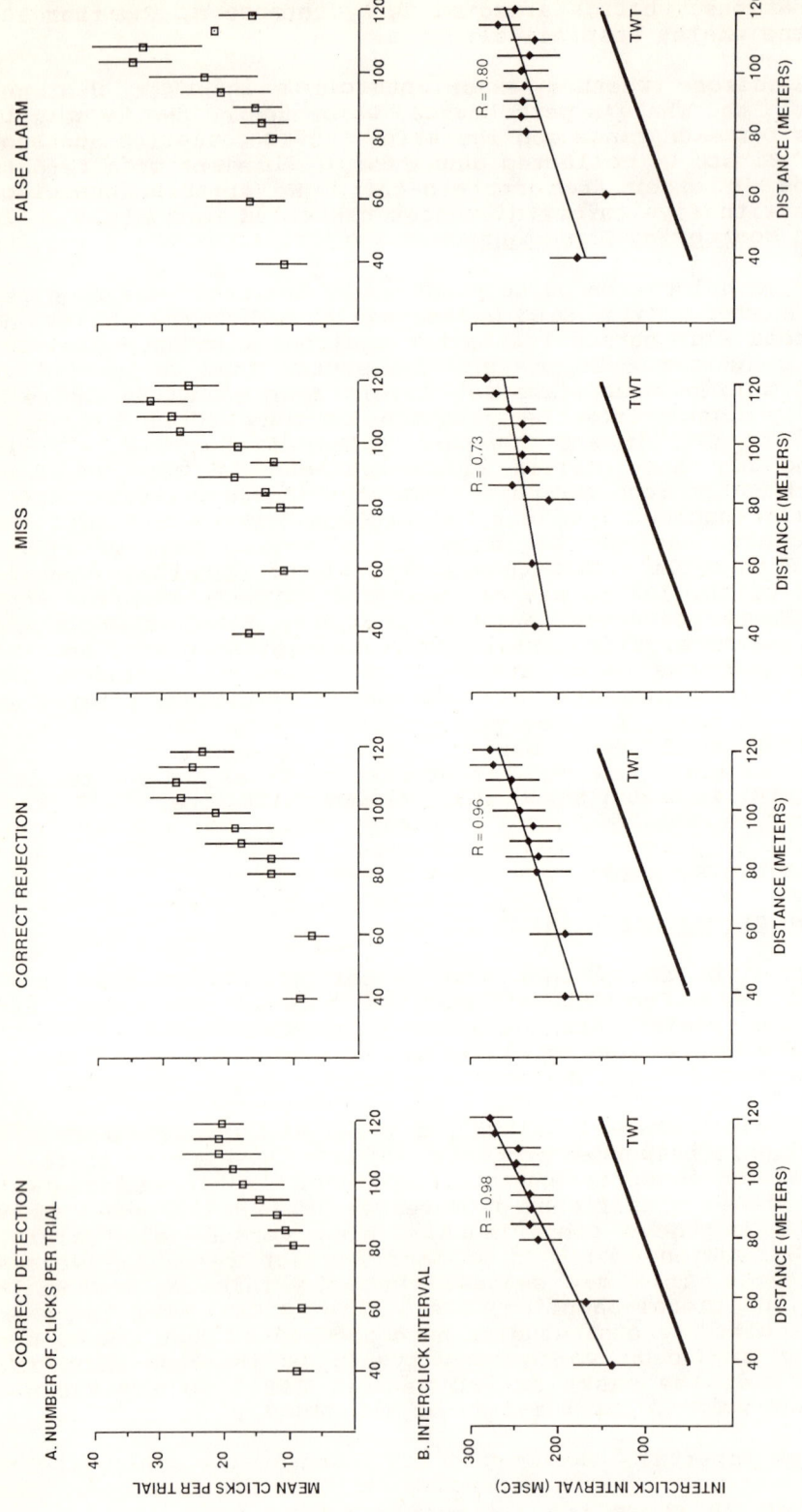

Fig. 3. A. Mean clicks per trial (±st. dev.) by target distance and type of response. B. Mean interclick interval (±st. dev.) by target distance and type of response. R is linear regression correlation coefficient; TWT is the the two-way travel time from the animal to the target.

function of distance between the target and the animal. Linear regression analysis showed significant correlations ($p < .05$) between the interclick interval and distance to the target, regardless of the animal's response.

All click trains had an interclick interval that was greater than the two-way travel time (TWT) that would be expected between the animal and target. For correct responses, the interclick interval was about 80 to 110 msec greater than TWT. Miss or false alarm trials had more variable interclick intervals, indicating that when the task was difficult the whale may have searched for the target at several locations along the range.

Bottlenose dolphins and false killer whales emit clicks at a rate that eliminates overlap between the outgoing signal and the echo. Morozov et al. (1972) first correlated interclick interval with distance to the target and reported that the interclick interval for <u>Tursiops</u> was 3 to 20 msec over the TWT. Au et al. (1974) report that for targets 60 to 80 m away the average interclick interval from a bottlenose dolphin is 30 to 50 msec greater than the TWT. Penner and Kadane (1980) report that the interclick interval for <u>Tursiops</u> was 20 msec over the TWT for targets 40 to 120 m away. Recently, Penner (1988) reported that the interclick interval of <u>Tursiops</u> also is affected by the animal's expectation of target distance. When targets were presented repeatedly at the same distance the interclick interval indicated the dolphin scanned only to that distance. However, when targets were presented at random, the dolphin scanned over the entire range of distances.

Peak Frequency and Bandwidth

Figure 4 illustrates that clicks within a train were consistent in peak frequency, waveform, amplitude, and duration. Echolocation pulses lasted approximately 50 to 70 usec. Head movements by the animal may have caused some clicks to be off-axis from the hydrophone and therefore, may account for some variation in amplitude among pulses.

The distance to the target did not affect the peak frequency or bandwidth of the echolocation pulses. The echolocation pulses from the false killer whale during the range detection task (Figure 5A) had a higher peak frequency (105-110 kHz) than reported by Thomas et al. (1988b) for a <u>Pseudorca</u> (20-65 kHz) in a concrete pool. We found no evidence of low-frequency components (5-12 kHz) as reported from a <u>Pseudorca</u> by Thomas et al. (1988b); however, because we filtered below 20 kHz these components were obscured. The bandwidth of echolocation pulses (Figure 5B) was broader (20-25 kHz) in the open-water task than reported for a false killer whale (5 to 16 kHz) in a pool. Differences in peak frequency and bandwidth could be attributed to the age or sex differences of the subjects or to the fact that pulses were directed towards a fish versus a metal sphere. However, we believe, as Au et al. (1985) found for <u>Delphinapterus</u>, that characteristics of pulses from <u>Pseudorca</u> probably are related to differences in the reverberation and background noise of open-water versus a pool.

Au et al. (1985) documented that odontocetes can adapt the peak frequency of clicks depending on the ambient noise. The same beluga whale produced lower frequency pulses in San Diego

Bay than in Kaneohe Bay, which has a lot of snapping shrimp noise. The peak frequency of pulses from the false killer whale produced during this study were similar to those reported by Au and Snyder (1980) for Tursiops (121 kHz) during the same range detection task. We believe that Pseudorca also has the flexibility to adapt pulse characteristics based on acoustic conditions.

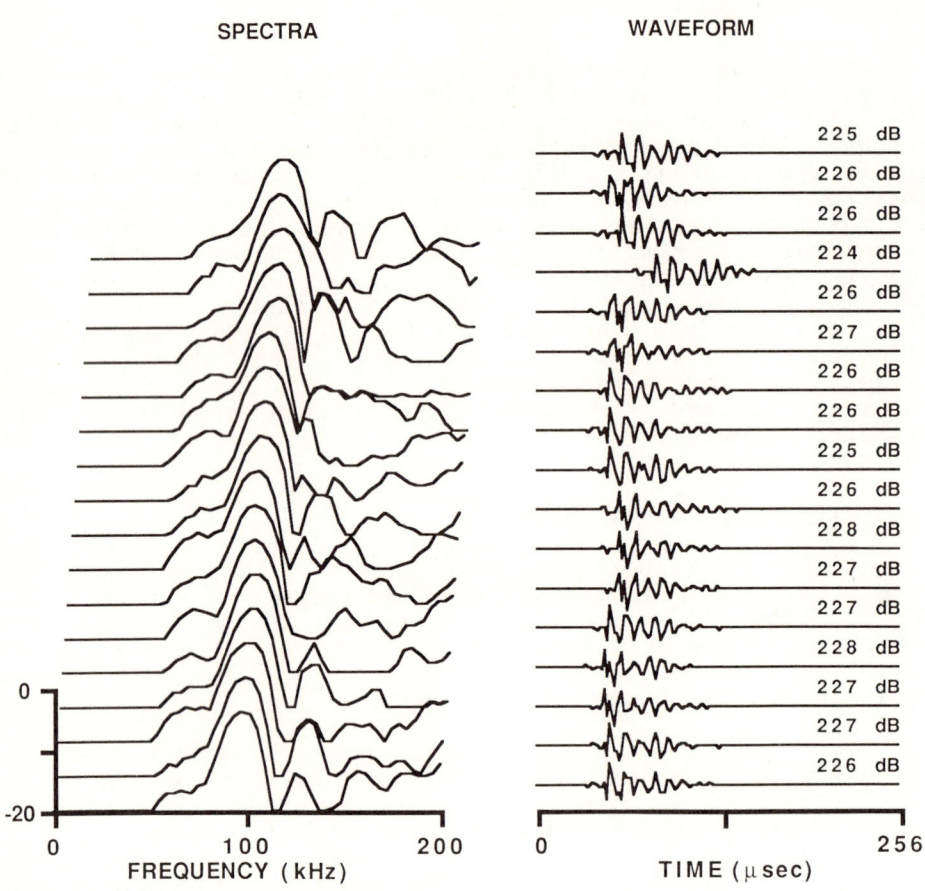

Fig. 4. Waterfall displays of spectra, waveforms, and source levels from pulses in a click train during the 100 m target detection.

Source Level

Figure 5C shows that the source level of echolocation pulses most commonly was between 220 and 225 dB re 1 μPa. There was a gradual increase in the mean amplitude of pulses with distance to the target (at 40 m = 218 dB, at 120 m = 221 dB). Small variations in the source level may be caused by pulses received slightly off-axis relative to the hydrophone.

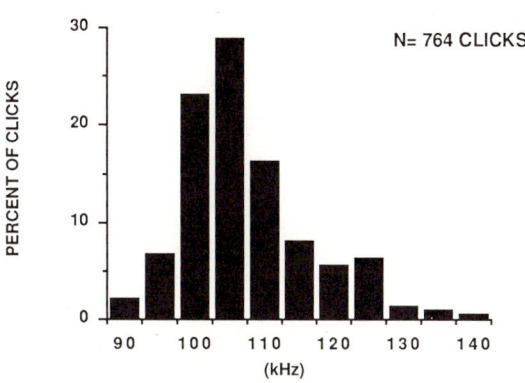

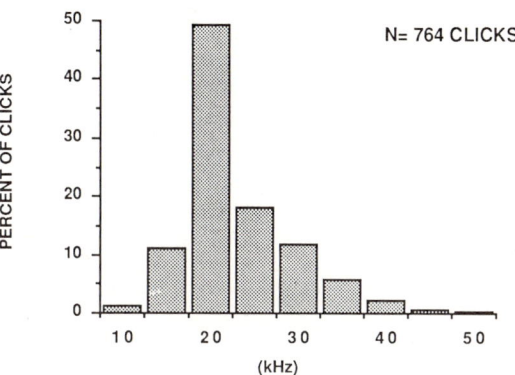

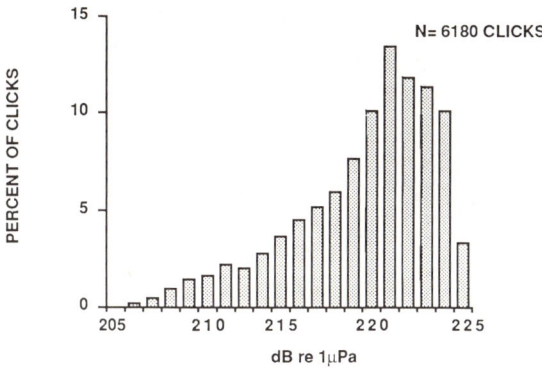

Fig. 5. Mean pulse characteristics by target distance during correct detection trials. A. peak frequency, B. bandwidth, C. source level. A and B data from 2 of 10 trials per block, C data from all trials.

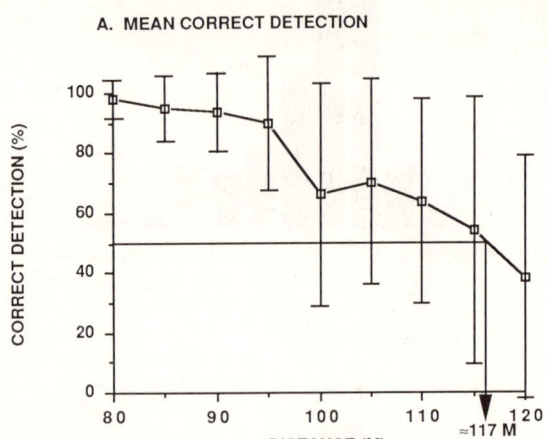

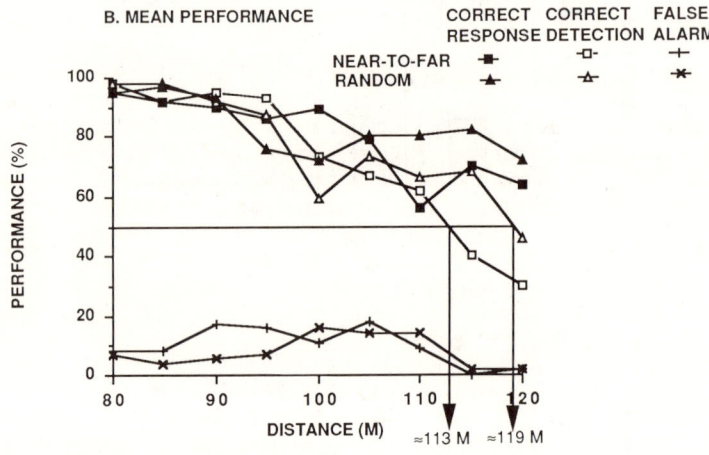

Fig. 6. A. False killer whale's average range detection performance (+ 1 st. dev.) for targets 40 to 120 m away. B. Mean performance for near-to-far versus random target presentations. Detection thresholds are marked by arrows and were estimated at the 50% performance level for correct detections.

The source level of pulses during the open-water task were higher than reported by Thomas et al. (1988b) from a captive *Pseudorca* (150-165 dB). Differences in source levels most likely result from different acoustic environments during the two studies. *Tursiops* pulses had an average source level of 220 dB during the same range detection task (Au and Snyder, 1980), which is similar to source levels in *Pseudorca*.

Performance

Figure 6A shows the average percent correct detections ($\pm$ 1 st. dev.) by the false killer whale for targets from 40 to 120 m away. The range detection threshold for these data was 117 m, as estimated by the distance at which performance was 50% for correct detections. This estimated range detection threshold for the false killer whale is similar to that measured for the bottlenose dolphin (113 m) by Au and Snyder (1980).

Figure 6B shows the false killer whale's average performance at each distance for targets presented in a near-to-far sequence versus targets presented at random. We estimated the threshold detection range (50% correct detection) as 113 m for near-to-far target presentations and as 119 m for random target presentation. Performance for near-to-far versus random target presentations were compared for each distance in each range configuration using a Student's t-test ($p < .05$). Except at 100 m in the 100-120 m configuration, there was no significant difference between the whale's performance in near-to-far versus random presentations. Range detection thresholds in our study were collected in a shallow water bay. We do not know how the threshold might change in open, deep water.

Figure 7 illustrates the probability of detection, $P(Y/SN)$, versus the probability of false alarm, $P(Y/N)$. As Schusterman (1974) reports, a low false alarm rate suggests the whale was conservative in reporting the presence of a target. The whale's false alarm rate was similar to that reported for the bottlenosed dolphin conducting the same task (Au and Snyder, 1980). Both species showed a slight increase in false alarm rates near threshold distances.

The false killer whale's performance beyond 90 m became more variable. Performance during one session might be 100% and drop to chance during the other daily session. We could not identify any particular pattern in change of performance. We do not know how oceanographic conditions, currents, schools of fish, target movement, or the animal's attentiveness might affect the whale's performance near threshold distances.

CONCLUSIONS

In most aspects, the false killer whale produces pulses similar to those from a bottlenose dolphin echolocating on the same targets on the same range. Small differences in peak frequency, bandwidth, and source level might be accounted for partially by differences in sound-generating anatomy, size or type of prey, or life history strategies. Both species were conservative in detecting targets and obtained similar thresholds on the skyhook range.

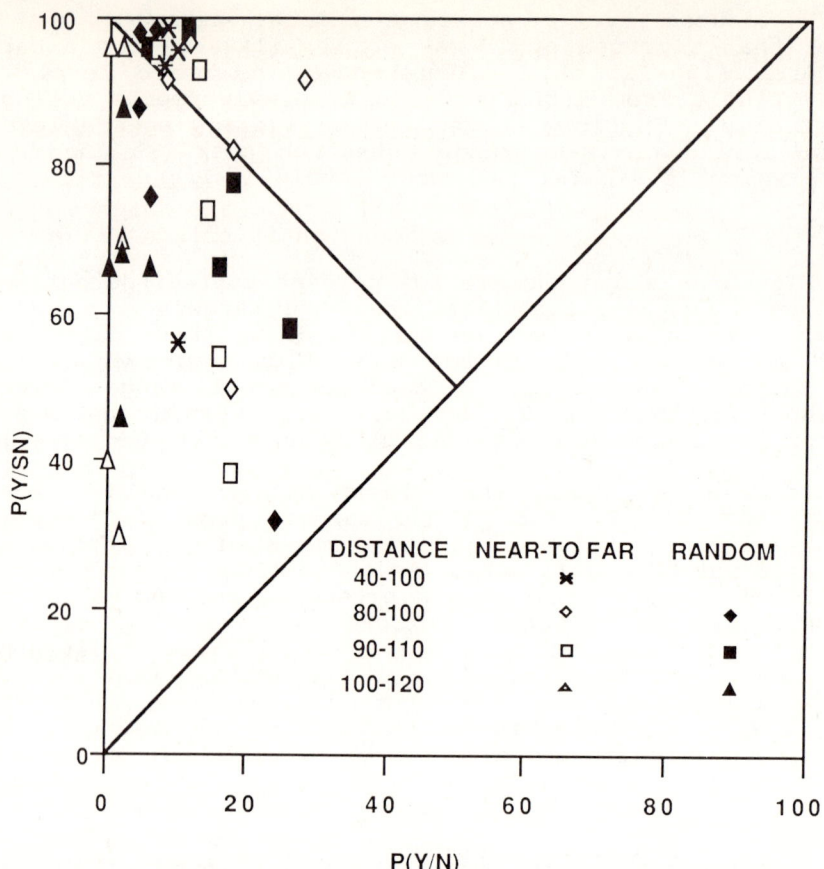

Fig. 7. False killer whale's performance with the probability of detection, P(Y/SN), plotted against the probability of false alarm, P(Y/N).

ACKNOWLEDGMENTS

We thank Rick Kahikina and his crew of NOSC employees for the construction of the floating pen complex and design of the acoustic screen. Jeff Pawloski provided many helpful suggestions about training. Whitlow Au, Bill Friedl, and Patrick Moore gave helpful reviews of the manuscript.

REFERENCES

Au, W. W. L., Floyd, R. W., Penner, R. H., and Murchison, A. E., 1974, Propagation measurements of echolocation signals of the Atlantic bottlenose dolphin, Tursiops truncatus Montagu, in open water, J. Acoust. Soc. Am. 56: 1180-1290.

Au, W. W. L. and Snyder, K. J., 1980, Long-range target detection in open waters by an echolocating Atlantic Bottlenose dolphin (Tursiops truncatus), J. Acoust. Soc. Am. 68(4): 1077-1084.

Au, W. W. L. and Penner, R. H., 1981, Target detection in noise by Atlantic bottlenose dolphins, J. Acoust. Soc. Am. 70: 687-693.

Au, W. W. L. and Kadane, J. ,1982, Acoustic behavior of echolocating Atlantic bottlenose dolphins, J. Acoust. Soc. Am. 71: 1269-1275.

Au, W. W. L. and Turl, C. W., 1983, Target detection in reverberation by an echolocating Atlantic bottlenose dolphin (Tursiops truncatus), J. Acoust. Soc. Am. 73: 1676-1681.

Au, W. W. L, Carder, D. A., Penner, R. H., and Scronce, B. L., 1985, Demonstration of adaptation in beluga whale echolocation signals, J. Acoust. Soc. AM. 77(2): 726-730.

Au, W. W. L., Penner, R. H., and Turl, C. W., 1987, Propagation of beluga echolocation signals, J. Acoust. Soc. Am. 82(3): 807-813.

Busnel, R.-G. and Dziedzic, A. ,1968, Caracteristiques physiques des signaux acoustiques de Pseudorca crassidens OWEN (Cetacea Odontocete), Mammalia 32(1): 1-5.

Gellerman, L. W., 1933, Chance orders of alternating stimuli in visual discrimination experiments, J. of Genetic Psychology. 42: 206-208.

Johnson, C. S., 1967, Sound detection thresholds in marine mammals, in: "Marine Bio-acoustics," W. N. Tavolga, ed. Pergamon, New York, Vol. 2, pp. 247-260.

Johnson, C. S., 1968, Masked tonal thresholds in the bottlenosed porpoise, J. Acoust. Soc. Am. 44: 965-967.

Johnson, C. S., McManus, M. W. and D. Skaar, 1989, Masked tonal threshold in the belukha whale, J. Acoust. Soc. Am. 85(6): 2651-2654.

Morozov, V. P., Akopian, A. I., Burdin, V. I., Zaytseva, K. A. and Yu. A. Sokovykh, 1972, Repetition rate of ranging signals of dolphins as a function of distance to target, Biofisika 17(1): 49-55.

Murchison, A. E., 1980a, Detection Range and Range Resolution in Echolocating bottlenose porpoise (Tursiops truncatus), in: "Animal Sonar Systems", R. -G. Busnel and J. F. Fish, eds. Plenum Press, New York.

Murchison, A. E. and Penner, R. H., 1975, Open water echolocation in the bottlenose dolphin (Tursiops truncatus):metallic sphere detection threshold as a function of distance, Conference on the Biology and Conservation of Marine Mammals, University of California, Santa Cruz, 4-7 December.

Murchison, A. E., 1980b, Maximum detection range and range resolution in echolocating bottlenose porpoises Tursiops truncatus (Montagu), Ph.D. dissertation, University of California, Santa Cruz.

Penner, R. H. and Kadane, J., 1980, Biosonar interpulse interval as an indicator of attending distance in Tursiops truncatus, J. Acoust. Soc. Am. S68 (Suppl. 1).

Penner, R. H., 1988, Attention and Detection in Dolphin Echolocation, in: "Animal Sonar: Processes and Performance," P. E. Nachtigall and P. W. B. Moore, eds. Plenum Press, New York.

Schusterman, R. J., 1974, Low false-alarm rates in signal detection by marine mammals, J. Acoust. Soc. Am. 55: 845-848.

Thomas, J., Chun, N., Au, W., and Pugh, K., 1988a, Underwater audiogram of a false killer whale (Pseudorca crassidens), J. Acoust. Soc. Am. 83 (3): 936-960.

Thomas, J., Stoermer, M., Bower, C., Anderson, L. and Garver, A.,1988b, Detection abilities and signal characteristics of echolocating false killer whale (Pseudorca crassidens), in: "Animal Sonar: Processes and Performance," P. E. Nachtigall and P. W. B. Moore, eds. Plenum Press, New York.

Thomas, J., Pawloski, J., and Au, W. (submitted). Masked hearing abilities in a false killer whale (Pseudorca crassidens), J. Acoust. Soc. Am.

Tomich, P. Q., 1986, "Mammals in Hawai'i". Bishop Museum Press, Honolulu, Hawaii.

Watkins, W., 1980, Click sounds from animals at sea, in: "Animal Sonar Systems," R. -G. Busnel and J. Fish, eds. Plenum, New York, pp. 291-297.

PRELIMINARY HEARING STUDY ON GRAY WHALES

(ESCHRICHTIUS ROBUSTUS) IN THE FIELD

Marilyn E. Dahlheim and Donald K. Ljungblad[1]

National Marine Mammal Laboratory, 7600 Sand Point
Way, Seattle, Wa. 98115; [1]Naval Ocean Systems
Center, Code 514, San Diego, Ca. 92151 USA

INTRODUCTION

Hearing capabilities of several odontocetes have been investigated through behavioral response techniques (Johnson, 1967; Anderson, 1970; Hall and Johnson, 1971; Jacobs and Hall, 1972; Ljungblad et al., 1982; Thomas et al., 1988), cortical evoked potential techniques (Seeley et al., 1976), and electrophysiological methods (Bullock et al., 1968; Ridgway et al., 1981). Past studies have been limited to toothed whales and dolphins held in captivity. Because of their large size, it is unlikely that mysticete whales will be kept in captivity for long-term behavioral studies. Similarly, it also is unlikely that a baleen whale could be restrained temporarily for a physiological hearing test. Alternative, non-invasive techniques will be necessary to test hearing in baleen whales. Hearing data on baleen whales are important to judge potential impacts from underwater man-made noise on these animals.

In 1984, concurrent with a study to determine the effect of man-made noise on communication among gray whales (Eschrichtius robustus) in Laguna San Ignacio, Baja California Sur, Mexico (Dahlheim et al., 1984; Dahlheim, 1987), a study was completed to determine the feasibility of conducting hearing sensitivity measurements on free-ranging mysticete whales. Gray whale behavioral responses to underwater signals produced at several different frequencies and amplitude levels were documented. This paper describes the field techniques used to collect this information, reports on our preliminary results, and provides recommendations for future hearing studies on free-ranging cetaceans.

METHODS

Study Area

Laguna San Ignacio, Baja California Sur, Mexico, located along the Pacific coast of Baja California, is one of four major calving areas for gray whales. Between January and

March each year approximately 300 to 450 whales representing adults of both sexes, juveniles and calves, occupy this lagoon (Jones et al., 1987).

Laguna San Ignacio is a system of narrow, relatively deep channels surrounded by extensive intertidal flats, extending

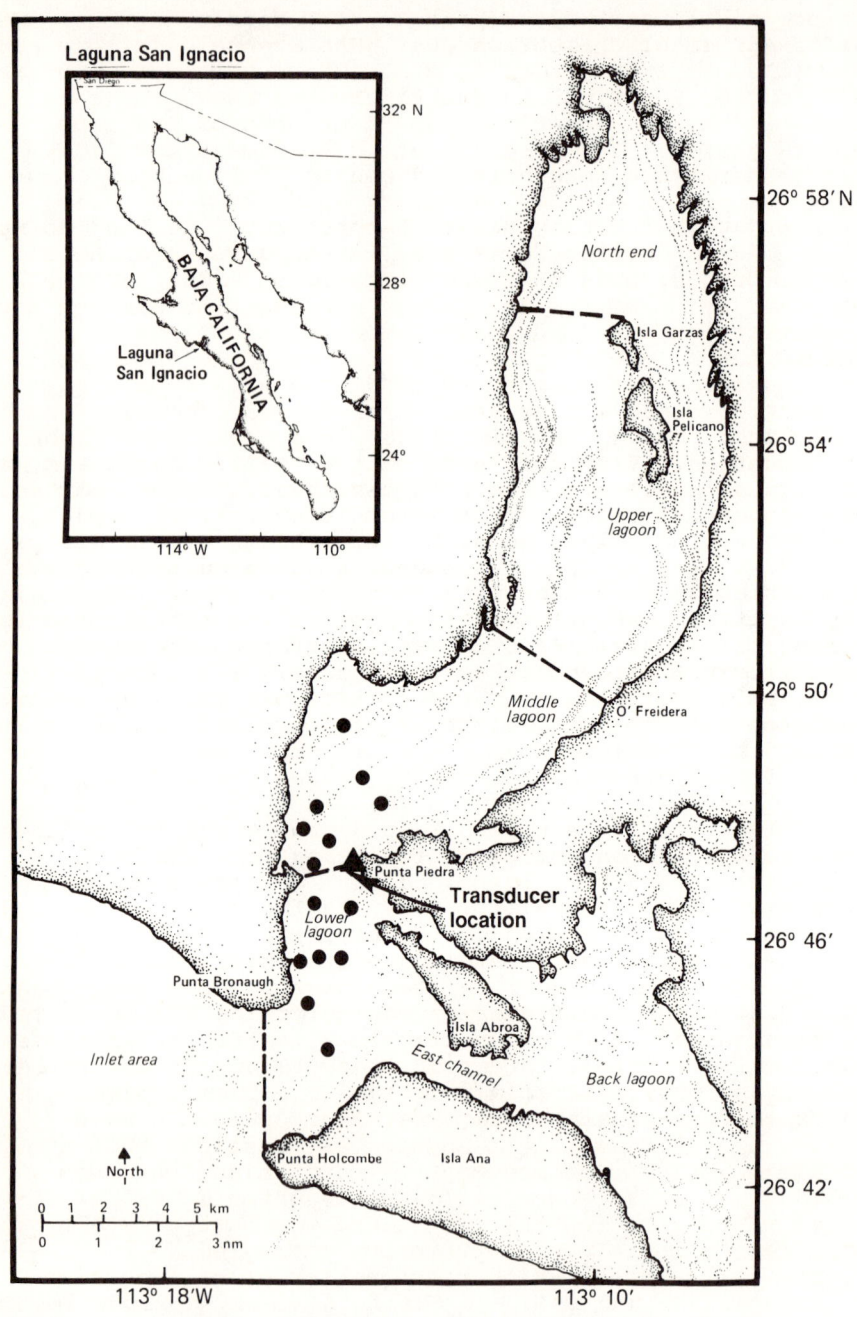

Fig. 1. Study area--Laguna San Ignacio. Ambient noise measurements (•); Transducer location (▲).

inland approximately 32 km, ranging in width from 1.8 to 6.5 km (Fig. 1). Jones and Swartz (1984) investigated the bathymetry and sedimentology. In the Lower lagoon, they report a steep-walled channel extending from the 3-km wide inlet to a constriction 1.8-km wide at Punta Piedra. The widest channel in the Lower lagoon has a maximum depth of 25.9 m. The deepest regions of this channel are covered with hard-packed, poorly sorted fine- to coarse-grained sand. West of Punta Piedra, ridges of medium-grained sand are interspersed with occasional rock outcroppings. The Middle lagoon is made up of a system of three channels. The channels in this area range in depth from 7.6 to 21.3 m. There the sediments are fine- to coarse-grained sand with crushed shell and some rocky areas. Extensive sand bars support strands of eelgrass (Zostera marina); the channels are devoid of plant life. Water temperatures range from 17 to 19°C and salinity values average 32 ppt (Dahlheim, 1987). A comparison of these parameters with depth indicated a homogenous, well-mixed water column. Depth, bottom substrate, temperature, and salinity have profound effects on sound propagation. The physical characterization of the lagoon was necessary to examine sound propagation pathways and predict transmission losses. Absolute sound levels and frequencies reaching the animal during experimentation must be estimated with some accuracy to ensure hearing was not masked.

Recognizing the importance of the calving lagoons to the reproductive success and continued recovery of the gray whale, Mexico enacted a series of legal and administrative provisions which (1) established a total ban on whaling within its territorial waters, (2) created national refuges for gray whales, and (3) controlled access to these areas (Swartz, 1986). Thus in 1979, Laguna San Ignacio was established as a national refuge area for gray whales and human activities, such as tourist use and commercial fisheries within the lagoon, were regulated. Vessel permits are required to enter the lagoon and tourist traffic is not permitted north of Punta Piedra.

Field Equipment

A Lubell Acoustic Transducer[1] (Model 98) was bottom-mounted (Fig. 2) at a distance of 75 m off the farthest promontory available near the Punta Piedra shore-based station (Fig. 1). The measured depth of the transducer at high-tide was 8 m, within the depth specifications required by the transducer manufacturer. The semi-diurnal tides ranged from 0.9 to 2.4 m. To ensure proper depth, bottom-placement and equipment safety, the transducer was attached in the center of a 1.5 cube of PVC pipe. Several holes were drilled into this pipe, allowing this structure to sink. A diver guided the transducer cage to the lagoon bottom and placed a sand-filled sack over each lower leg to secure its location; this ensured that there would be no movement with the tidal currents (7 to 9 knots). The transducer cable was weighted at 5-6 m intervals and ran from the cube along the bottom contour to ensure that a whale could not become entangled.

The shore-based equipment consisted of an Acoustics

---

[1]Reference to trade names does not imply endorsement by the National Marine Fisheries Service, NOAA.

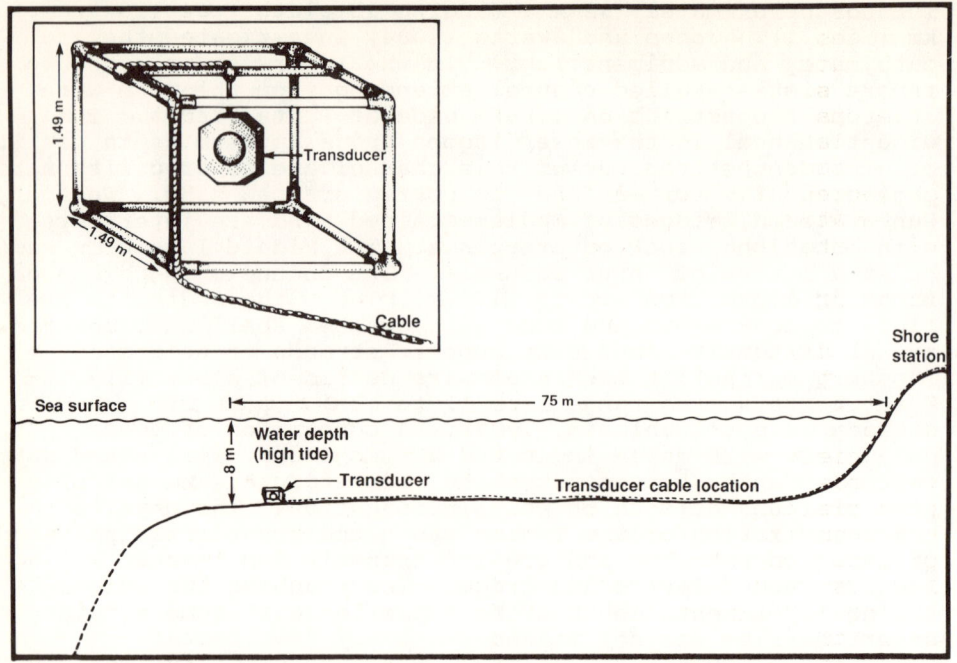

Fig. 2. Bottom-mounted transducer in Laguna San Ignacio.

Systems, Inc., Power Amplifier (Model TS 107). The input terminals of the amplifier were connected to an oscillator (Wavetek) capable of producing tones of various frequencies. Our initial study focused on the projection of low frequency sounds (between 200 and 2500 Hz); gray whales emit sounds in this range. Projected levels depended on the frequency characteristics of the transducer and ranged from 70 to 145 dB referenced to 1 μPa. A quiet switch was added to the system between the amplifier and oscillator to ensure that the whale heard a test tone and not the electronic switching noise. The projection system was driven by a 12-volt battery.

Calibrations

Studies on ambient noise levels and experiments to document sound propagation near the transducer site were completed concurrently in Laguna San Ignacio (Dahlheim, 1987). All acoustic equipment was calibrated before and after the field season. In addition to the numerous ambient noise measurements made throughout the lagoon from a skiff (deploying a hydrophone over the side of the platform), a bottom-mounted KSP hydrophone also was positioned 5 m from the transducer. The hydrophone cable ran along the bottom contour of the lagoon, terminating at the shore-based station. There it was connected to a SJS IV Nagra reel-to-reel tape recorder. Ambient noise samples and all signals projected during hearing studies were recorded.

To calculate noise levels from biological sources, one-third octave average sound pressure levels were computed for each station using a Spectral Dynamics 345 Spectrum Analyzer set for 0-20 kHz bandpass and for either 32 or 64 linear averages. The reported average sound pressure levels were calculated by averaging as many linear averages as possible from a single station. Using this method, the levels for the stations represent a range of actual sampling times equivalent to a flat-weighted, averaged sound pressure level taken at 1 meter sampling from 1.28 to 14.0 s. Sound pressure levels of projected signals were profiled using a Nicolet Scientific Corporation FFT Spectrum Analyzer, Model 446. Displays of the calculated spectra were made on an X-Y plotter.

Behavioral Data Collection

A difficult problem arises when studying hearing in an animal that has not been trained to inform the investigator that it heard a sound. The basic assumption in the present hearing study is that if a gray whale hears a sound of sufficient source level, it will respond to it, and that if it responds, an observer can detect the response. However, it is possible that some of the whales could have heard the sound but found it so nonthreatening that it did not elicit a visible response. To reduce or eliminate other contributing factors (i.e., factors that made whales appear to respond to the projected noise when in fact something else made them alter their behavior), it was necessary that certain experimental conditions existed. A response by the whale had to occur within 5 s of the projected tone. If a reaction occurred in this time frame, it was listed as a positive response. Whales were required to be traveling through the area and not milling or engaged in other activities. Projections were not made when boats were in the area. A 1-hour interval was required between tests to eliminate the possibility of whales signaling each other.

Maximum decibel levels for all frequencies were projected at random schedules in the hope to elicit a startle response (an abrupt behavioral change immediately following the projected tone, indicating surprise or fright). Other surface behaviors also were investigated. As part of the ongoing work at Laguna San Ignacio, four researchers tracked whales from shore-based locations directly overlooking the transducer site (tracking methodology described in Dahlheim, 1987). During the control experiments (n = 30), information was collected on the whale's identity, group composition (mothers with calves, single adults), respiration rates, tracklines through the lagoon relative to the transducer, and speed of transit. Average dive times for individual whales and all whales were calculated. Respiration rates of cow and calf pairs and single whales were statistically compared. Comparisons of whale diving behavior or surface movements were made between control (no signal projected) and experimental (exposure to signal) periods.

Whales moving south and north through the study area commonly passed over the transducer site. When a whale passed close to the transducer site (within 20 m), a pre-selected tone, lasting 1 s, was projected. The tracking team was not informed when projections occurred. Whale behavior before, during, and immediately after exposure to the tone was recorded by the tracking team and also by the researcher projecting the signal; thus two independent assessments were made of the whale's response. Tracklines of whales relative to transducer location were investigated between control and experimental periods by comparing distances offshore, precise movements of whales, and transit speeds.

RESULTS

Ambient noise was recorded at 14 stations in the lagoon (Fig. 1). Noise spectra varied among stations, with levels below 2 kHz increasing to high levels between 2 and 5 kHz and then declining gradually through 20 kHz. The average ambient noise levels attributable to biological sources ranged from 94 to 110 dB re 1 μPa. Numerous biological organisms were identified and deemed responsible for the high levels of ambient noise documented in the lagoon (see Dahlheim, 1987 for details). An inspection of synoptic ambient samples (24-hr recordings) from the shore-based station revealed no significant differences in levels or frequencies over a 24-hr period ($t = 1.17$, $p > 0.05$). The only sources of man-made

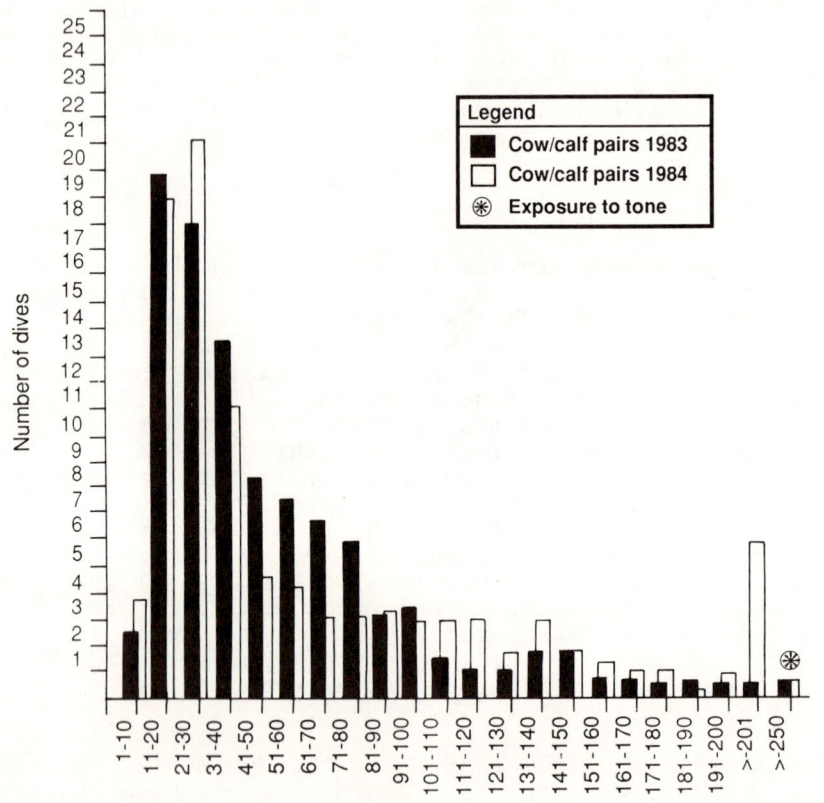

Fig. 3. 1983/1984 frequency distribution of dive durations on cow and calf pairs in Laguna San Ignacio. * Dive duration obtained during exposure to a 1000 Hz tone.

noise, documented during the studies, were vessel and skiff traffic. Dahlheim (1987) gives profiles of sounds emanating from these sources.

A total of 61 propagation measurements, collected at various angles and depths were made within a 1 km radius of the transducer site. Measurements were made at high-slack water during Beaufort 1 or 2 sea states. A sound profile was made for each of the 61 samples and comparisons yielded only minor fluctuations in predicted sound propagation pathways (Dahlheim, 1987). Areas of sound enhancement or reduction near the transducer site were not found. Therefore, we are confident that projection levels were relatively consistent over time.

Average dive duration of all whales during the 1984 season was 92.5 sec ± s.d. 58.1 s (total dives = 1287). Average dive time of cow and calf pairs was 70.9 ± s.d. 30.7 s, statistically different (t = 4.57; p < 0.05) from those obtained for single whales (107.7 ± s.d. 55.5 s, n = 67 whales, total dives = 519). The adult females may shorten their dive time to compensate for the newborn calf. Figure 3 represents the frequency distribution of dives for cow and calf pairs collected during the 1983 and 1984 seasons. Six dives greater than 201 s have been recorded. Prior to our hearing studies, the longest dive noted for a cow and calf pair was 242 s long.

Whales were observed to travel directly through the study area with little evidence of milling. The trackline depicted

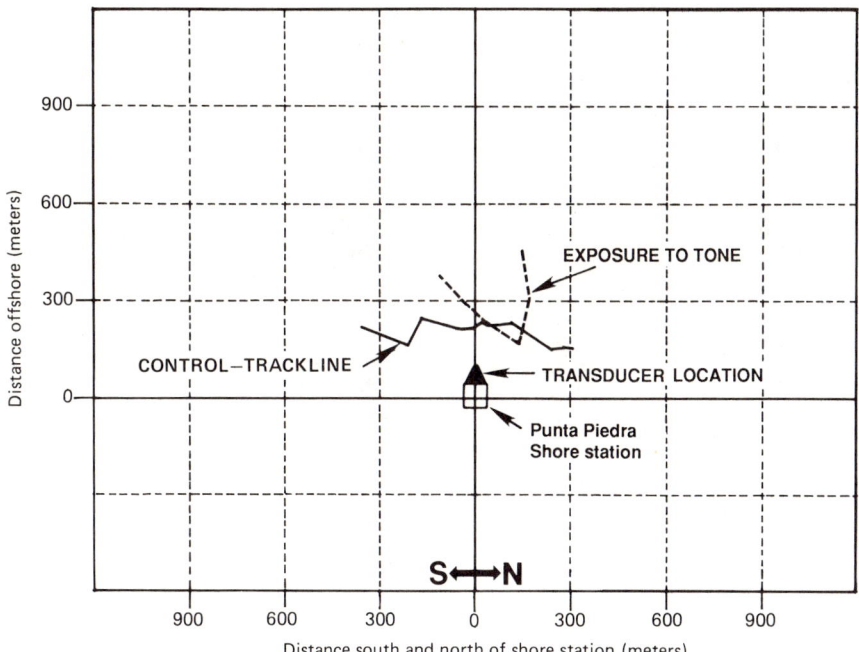

Fig. 4. Trackline of a cow and calf pair during control period (——) and during exposure to tone (---). Whale movement is from left to right.

in Figure 4 demonstrates a typical movement pattern past the shore-station during control periods (no intrusion of noise). The average perpendicular distance the whales traveled offshore was 230.4 ± 69.4 m. On average, observers spent 6.5 ± s.d. 4.0 min observing each whale.

Twenty-four tones, covering nine different frequencies, were projected to gray whales during February 1984 (Table 1). Whale responses varied: (1) a startle reaction; (2) change in course, transit speed or respiration; or (3) no detectable reactions. Whales did not exhibit behavioral changes during the projection of tones at 200 Hz produced at 70 dB (n = 4). Whale responses to 800 Hz tones projected at 95 dB (n = 2) resulted in either a rapid change in direction (Fig. 4) or no no reaction. Three tones were projected at 900 Hz at a level of 100 dB. In one of these instances, a startle response occurred in which the whales' behavior abruptly changed. A considerable amount of water disturbance was observed. The whale then began tail-lobbing and quickly left the study site. During the other two tests with 900 Hz, the whales exhibited no detectable response. After the projection of tones of 1000 Hz at 108 dB (n = 4) and 1500 Hz at 135 dB (n = 2), a measurable response occurred in all cases, ranging from a change in respiration (a cow and calf dive time of 362 s

Table 1. Responses of Gray Whales to Tones of Various Frequencies and Levels.

Frequency (Hz)	Decibel Level (re 1 μPa)*	Social Combination	Response
200	70	Single	None
200	70	Pair	None
200	70	Single	None
200	70	Single	None
800	95	Cow/calf	CC **
800	95	Cow/calf	None
900	100	Cow/calf	Startle
900	100	Single	None
900	100	Cow/calf	None
1000	108	Cow/calf	RC **
1000	108	Single	Lost
1000	108	Single	Startle
1000	108	Single	Startle
1500	135	Cow/calf	Startle
1500	135	Single	Startle
1800	142	Cow/calf	Lost
1800	142	Single	None
1900	145	Cow/calf	None
1900	145	Cow/calf	None
1900	145	Cow/calf	None
2000	145	Cow/calf	None
2000	145	Pair	None
2000	145	Cow/calf	None
2500	145	Single	None

* Decibel level at projector. All whales within 20 m of transducer when exposed to test signal. For each doubling in distance from the source, a 3 dB loss would result. At 20 m, a loss of 13 dB would occur.
** CC = Course change; RC = Respiration change.

immediately after the projected tone, Fig. 3), disappearance of a whale from the transducer area (the tracking team was unable to locate the animal after the tone was projected), and four startle responses (water disturbance, tail-lobbing, and avoidance of area). In one trial of a tone of 1800 Hz at 142 dB, (n = 2), the trackers were unable to relocate the whale. However, in another trial of 1800 Hz, no apparent reaction was documented. No measurable change occurred when whales were exposed to signals produced at 1900 Hz or higher all of which were projected at maximum transducer levels of 145 dB: 1900 Hz (n = 3); 2000 Hz (n = 3), and 2500 Hz (n = 1). In-water, low frequency hearing sensitivities of gray whales based on this preliminary study are given in Figure 5.

DISCUSSION

Laguna San Ignacio afforded us a unique opportunity for conducting this type of investigation. A dense concentration of animals, representing both sexes and all age classes, occurred in a small area ensuring an adequate sample size. Tracklines and respiration rates of whales were relatively easy to obtain since gray whales can be individually recognized. Working from a shore-based station provided many advantages. In particular, sensitive electronic equipment was protected against salt-water damage. Since whales frequently travel close to shore, the shore-based location provided a good vantage point for observations of behavior. Shore-based work eliminated the possible effect of the survey platform

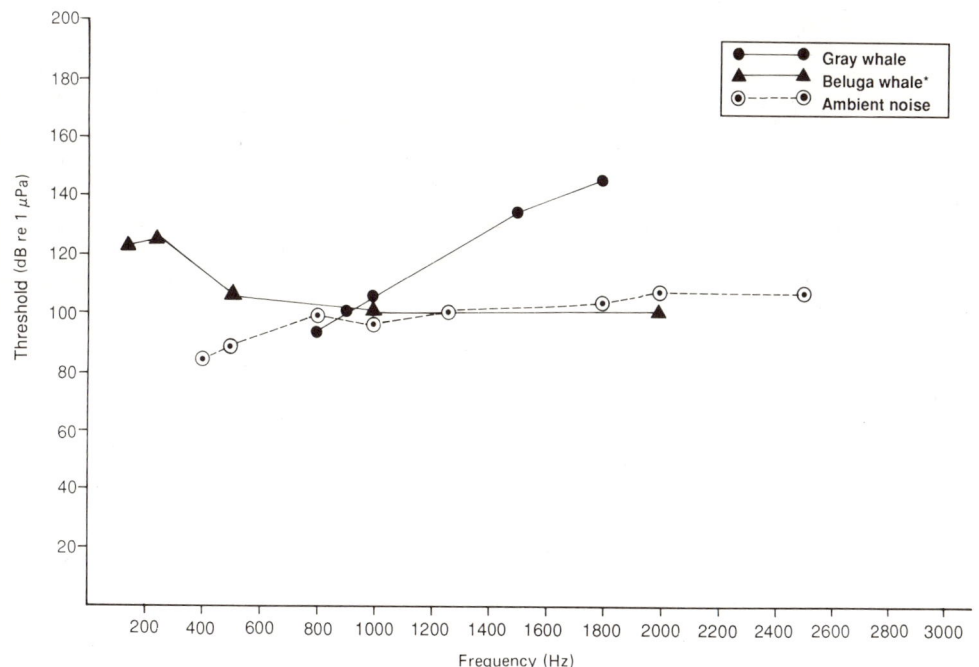

Fig. 5. In-water low frequency auditory sensitivities of gray whales (this study) and beluga whales, <u>Delphinapterus</u> <u>leucas</u> (Awbrey et al., 1988).

altering the whale's behavior during experimental trials. Restricted human activities within certain areas also minimized interference.

A number of environmental factors need to be considered when conducting hearing studies on free-ranging cetaceans. Measurements must be collected on ambient noise and on sound propagation. Absolute sound levels and frequencies reaching the whale must be known. Physical properties influencing sound propagation also should be described (e.g., depth, substrate, temperature, and salinity). High-level ambient noise environments like Laguna San Ignacio may preclude obtaining a species' sensitivity curve to various frequencies as some signals may be masked by environmental noise. In addition to environmental influences, biological factors need to be addressed when conducting hearing studies. Hearing may vary among individuals, dependent upon their age or sex (Awbrey et al., 1988). The whale's behavioral state also may influence its response (Ljungblad et al., 1988). Single adult whales may respond differently than pairs or groups of whales. Females with calves may be particularly sensitive to unfamiliar stimuli in the environment. If the same whale is exposed repeatedly to a projected sound, habituation may occur.

Whale behavioral responses to tones were examined in a number of ways during the current investigation. A startle response is a good indication that an animal heard the tone; however, quantification of this type of response can be difficult. To obtain startle responses, a signal of rapid onset and significant level must be projected. Data on whale movement and speed can be compared easily during control and experimental conditions. We were most encouraged by the results obtained during tracking experiments. Although dive duration varied in response to a projected tone, the variation in most cetaceans' dive profiles could make cause and effect relationships difficult to assess. Other surface or dive profiles of whales in response to sound should be investigated. Examples of this include the number of blows per surfacing, the length of time at the surface (surface interval) or the blow rate (the number of blows divided by the combined length of the surface interval and subsequent dive; see Ljungblad et al., 1988). In some cases, videotaping may prove to be a useful tool. Data obtained through telemetry devices (e.g., time/depth recorders or equipment designed to monitor the animal's physiological state) may be worth investigating.

We have not attempted to fully document the hearing capabilities of gray whales during this investigation. Our sample size is too small to draw any final conclusions from the research. We can, however, make some preliminary interpretations of our data. We assume that the whales' hearing is most sensitive in the frequency range where they produce their own sounds. Gray whales produce broadband signals ranging from 100 Hz to 2000 Hz, with an high-amplitude band between 300 and 825 Hz (Dahlheim et al., 1984). In most mammalian species, the hearing range is broader than the animal's sound range. Gray whales reacted to tones from 800 (95 db re 1 μPa) to 1800 Hz (142 dB re 1 μPa), with most responses documented between 1000 and 1500 Hz. Gray whales responded to tones that were well within their sound production range. Whales did not respond to the four tones produced at 200 Hz. However, the projected level at this frequency was only 70 dB re 1 µPa and may have been masked by the ambient conditions (Fig. 5). Whales also did not respond to frequencies greater than 1900 kHz produced at the highest projected levels of the study (145 dB re 1 µPa) during several trials (n = 7). It is possible that they either did not hear the sound or heard it but did not react to it. The shape of the sensitivity curve in figure 5 for gray whales suggests a decrease in sensitivity as frequency is increased, similar to odontocetes. Additional studies are needed to accurately determine gray whale hearing capabilities.

The primary purpose of the 1984 study was to determine the feasibility of conducting a hearing study on free-ranging gray whales, the first such study for any Mysticete. We are encouraged by the results obtained during this preliminary study and are optimistic about future investigations. As humans increase their use of the sea, a corresponding increase will occur in ambient noise levels. Knowledge on hearing in free-ranging mysticete whales remains a critical gap in our understanding of how man-made noise may impact cetaceans.

ACKNOWLEDGMENTS

The authors are grateful to the Mexican government (Deparatamento de Pesca), which issued the necessary permits to conduct this research. We thank Edith Polanco Jaime, Jorge Carranza Frasier, Alfonso Yanez, Alonso Lopez, Walter O'Campo, Pedro Mercado Sanchez, and Luis Fleischer. Field assistance and support were provided by M. Bursk, T. Crawford, R. Dahlheim, Jr., and J. Essley. Our gratitude is extended to S. Murphy (Applied Physics Laboratory, Seattle, Washington) for access to the analysis equipment. T. Loughlin, D. Rugh, and A. York reviewed the manuscript. K. Zecca provided the needed graphics. A special thanks is given to J. Thomas for her support and encouragement in the final stages of this publication.

REFERENCES

Anderson, S., 1970, Auditory sensitivity of the harbour porpoise, Phocoena phocoena, Dept. of Bioacoustics, Univ. Oolense, Strip, Denmark, "Investigations Cetacea" 2, 255-259.
Awbrey, F. T., Thomas, J. A., and Kastelein, R. A., 1988, Low-frequency underwater hearing sensitivity in belugas, Delphinapterus leucas, J. Acoust. Soc. Am., 84(6):2273-2275.
Bullock, T. H., Brimmell, A. D., Ikezomo, E., Kameda, K., Katsuki, Y., Nomoto, M., Sato, O., Suga, N., and Yamagisawa, K., 1968, Electro-physiological studies of central auditory mechanisms in cetaceans, Z. Vgl. Physiol., 59: 117-156.
Dahlheim, M. E., 1987, Bio-acoustics of the gray whale (Eschrichtius robustus), PhD. Dissertation, Univ. of Brit. Columbia, Dept. of Zool., Vancouver, B.C., Canada, 311 pp.

Dahlheim, M. E., Fisher, H. D., and Schempp, J., 1984, Sound production by the gray whale and ambient noise levels in Laguna San Ignacio, Baja California Sur, Mexico, in "The Gray Whale", Jones, M. L., Swartz, S. L., and Leatherwood, S., eds. Academic Press, New York, 511-541.

Hall, J. D., and Johnson, C. S., 1971, Auditory thresholds of a killer whale, Orcinus orca, J. Acoust. Soc. Am., 51: 515-517.

Jacobs, D. W., and Hall, J. D., 1972, Auditory thresholds of a fresh water dolphin, Inia geoffrensis, J. Acoust. Soc. Am., 51: 530-533.

Johnson, C. S., 1967, Sound detection threshhold in marine mammals, in: "Marine Bio-Acoustics", Vol. II, W. N. Tavolga, ed., Pergamon, New York, 247-260.

Jones, M. L. and Swartz, S. L., 1984, Demography and phenology of gray whales and evaluation of whale-watching activities in Laguna San Ignacio, Baja California Sur, Mexico, in: "The Gray Whale", Jones, M. L., Swartz, S. L., and Leatherwood, S., eds., Academic Press, New York, 309-374.

Jones, M. L., Swartz, S. L., and Dahlheim, M. E., 1987, Gray whale abundance and distribution in San Ignacio Lagoon during 1985, with a comparison to demographic trends from 1978-1982, Final Rept. to the Marine Mammal Commission, Cont. MM 2911023-0, 28 pp.

Ljungblad, D. K., Scoggins, P. D., and Gilmartin, W. G., 1982, Auditory thresholds of a captive Eastern Pacific bottle-nosed dolphin, Tursiops spp., J. Acoust. Soc. Am., 72(6): 1726-1729.

Ljungblad, D. K., Würsig, B., Swartz, S., and Keene, J. M., 1988, Observations on the behavioral responses of bowhead whales (Balaena mysticetus) to active geophysical vessels in the Alaskan Beaufort Sea, Arctic, 41(3):183-194.

Ridgway, S. H., Bullock, T. H., Carder, D. A., Seeley, R. L., Woods, D., and Galambos, R., 1981, Auditory brainstem response in dolphins, Proc. Natl. Acad. Sci. USA, 78(3):1943-1947.

Seeley, R. L., Flanigan, W. F., and Ridgway, S. H., 1976, A technique for rapidly assessing the hearing of the bottlenose porpoise, Tursiops truncatus, Naval Undersea Center, TP 522, San Diego, Ca.

Swartz, A. L., 1986, Demography, migration, and behavior of gray whales, Eschrichtius robustus, (Lilljeborg, 1861) in San Ignacio Lagoon, Baja California Sur, Mexico and in their winter range, Ph.D. Dissertation, Univ. of California, Santa Cruz, Dept. of Biology, Santa Cruz, California, 58 pp.

Thomas, J. A., Chun, N. K. W., Au, W. W. L., and Pugh, K, 1988, Underwater audiogram of a false killer whale (Pseudorca crassidens), J. Acoust. Soc. Am., 84(3):936-940.

INFERENCES ABOUT PERCEPTION IN LARGE CETACEANS, ESPECIALLY HUMPBACK WHALES, FROM INCIDENTAL CATCHES IN FIXED FISHING GEAR, ENHANCEMENT OF NETS BY "ALARM" DEVICES, AND THE ACOUSTICS OF FISHING GEAR

Jon Lien[1], Sean Todd[2], and Jacques Guigne[3]

(1) Ocean Sciences Centre and Department of Psychology
(2) Department of Psychology, Biopsychology Programme
(3) Centre for Cold Ocean Engineering
Memorial University of Newfoundland
St. John's, Newfoundland, Canada

INTRODUCTION

The incidental entrapment of non-target species in fishing gear is a problem of increasing concern for the fishermen and fishery managers of Newfoundland. By-catches of marine mammals however is not a problem unique to Canadian waters. Worldwide, thousands of cetaceans or pinnipeds collide with gear and are caught every year (Perrin, 1988; Fowler, 1988; Bonner, 1982). The problem causes considerable losses to fishing crews and such by-catch may seriously effect populations (Lien et al., 1989a). In addition, for many such losses are ethically and socially undesirable (Lien, 1989). In spite of these concerns, very little is known about the factors which result in cetacean collisions with fishing gear, the whale's ability to detect gear or the factors which impede detection.

We have monitored large whale collisions with fixed fishing gear in Newfoundlands' inshore fishery for the past decade (Lien, 1989). There are many confounded and confusing factors which limit the usefulness of such field data in making inferences about perceptual abilities of whales that fail to detect fishing gear. But data has provided us with the incentive and some opportunities to evaluate perceptual factors which may contribute to collisions.

Figure 1. A humpback whale entrapped in a Newfoundland codtrap.

CETACEAN COLLISIONS WITH FISHING GEAR IN NEWFOUNDLAND

In 1978 Memorial University was contacted by fishermen in Trinity Bay for assistance in retrieving fleets of groundfish gillnets which held two humpback whales (Megaptera novaeangliae). The whales had been entrapped for three months. The fishermen had abandoned efforts to release the animals and get the nets back themselves, and had exhausted contacts with agencies in looking for help. These whales were released by students and staff of Memorial University of Newfoundland with help from the fishermen.

This exposure to the problem of whale entrapment in inshore fishing gear was the basis for the organization of a whale research group at the University (Lien and Merdsoy, 1980). Since 1978, in cooperation with the Department of Fisheries and Oceans, the Newfoundland and Labrador Department of Fisheries and the Newfoundland Fishermens' Union, we have offered a program of assistance to fishermen who inadvertantly catch large whales and have studied factors which produce collisions.

Whale and shark collisions with inshore fishing gear are not a new problem for inshore fishermen in Newfoundland and Labrador. There is much anecdotal and historical evidence indicating that inshore gear damage due to large whales and sharks has always occurred at a low irregular level. Increased effort in the inshore fishery during the early 1970's was accompanied by more complaints of whale damage from fishermen. Whales and large sharks also were more commonly entrapped in fishing gear, held by nets made of much stronger modern synthetic materials. During the mid-seventies, there was a substantial increase in the amount of damage reported (Lien, 1980). Fishermen indicated whale problems had become very serious.

The number of whales, particularly humpbacks, sighted in inshore waters off Newfoundland and Labrador has been related negatively to the immature capelin (Mallotus villosus) biomass offshore (Whitehead and Carscadden, 1985). A collapse of immature capelin offshore in the mid-1970's caused humpback whales to move inshore in record numbers to feed on spawning capelin in near shore fishing areas (Whitehead and Lien, 1982). There were negligible increases in the inshore abundance of other common large whales including minkes (Balaenoptera acutorostrata) and finback (Balaenoptera physalus) (Lien and Whitehead, 1983). There are typically high, positive correlations between humpback whale sightings inshore and the number of collisions and entrapments which occur in an area (Lien, 1980).

With the recovery of collapsed capelin stocks in 1981, collision damages due to whales declined dramatically as the distribution of humpbacks returned to normal. Losses to fishermen from whale

Figure 2. Capelin, the humpbacks' primary food.

collisions in 1979-1980, the peak of the problem, were estimated at about two million dollars per year. Damages declined to several hundred thousand dollars per year in the early 1980's. Since 1985, fishermens' losses due to whales have cost about one-half million dollars per year (Lien, 1989). Numbers of entrapped humpbacks remains substantial ranging from a low of 31 animals in 1982 (Lien et al., 1982) to a high of 70 in 1989 (Lien et al., 1989c). The continued level of losses probably reflects increased effort in the inshore fishery and a larger population of humpbacks in waters off Newfoundland and Labrador (Lien et al., 1989a).

Humpback whale collisions with fishing gear appear to be due to a failure of the whale to detect the presence of the net or, at least, detect it in time to avoid it (Lien, 1980). The whale is not attempting to steal fish from the net. Humpbacks feed almost exclusively on capelin (Whitehead, 1987). They could be enticed to forage near fishing gear by the presence of capelin that tend to be abundant in good fishing areas; inshore fishermen also typically set their nets in areas where bait is plentiful. It also is possible that a dense school of capelin beside a net might visually or acoustically obscure it and make detection difficult for the whale. Although capelin can pass easily through the mesh of nets set for cod (Gadus morhua) and other larger species, they typically congregate beside net barriers (Lien, 1989).

Other species of whales and sharks also are responsible for net collisions. It has been estimated that about 25% of the collision damage experienced by fishermen in the early 1980's was due to basking sharks (Cetorhinus maximus) (Lien, 1980). Most of this damage occurred to salmon (Salmo salar) gillnets. Accidentally-caught basking sharks are reported each year in Newfoundland, although numbers vary widely (from 17-147). By-catch has declined recently because of closure of much of the salmon fishery (Lien et al., 1986).

While fishermen often do not distinguish large sharks from whales, shark damage to nets is unique due to their rough skins and the slimy material which they deposit on the net. Basking shark occurrence in inshore Newfoundland waters also is closely tied to surface water temperatures ranging from 8-12 degrees Celsius (Lien and Fawcett,

Figure 3. Humpback whale being released from a codtrap.

1986). Thus, damage due to sharks easily can be distinguished from that due to whales (Lien and Aldrich, 1982).

Other cetacean species such as minke whales (from 7-13/year), fin whales (from 0-3/year), occasional right whales (Eubalaena glacialis) and numerous smaller whales also collide with inshore fishing gear and are taken incidentally each year. Presumably not all cetaceans that collide with nets are entrapped. Hence, if they are not entrapped there is no way to document the species involved in the collision. Entrapped animals are the only means of estimating which species of whales are failing to detect the fishing gear and colliding with it. This means that in inshore Newfoundland waters, the best estimate of collision damages to fishing gear are primarily associated with humpback whales (Lien et al., 1989a). Abundance data also support clearly this conclusion (Whitehead and Lien, 1982).

There are several varieties of inshore fishing gear that are effected by whale collisions in Newfoundland and Labrador. Groundfish, salmon and lumpfish (Cyclopterus lumpus) gillnets as well as all types of codtraps (Figure 4). Because of the value of codtraps (from $5,000-$12,000), and their economic importance in total fish yield, our investigations have tended to concentrate on them.

FACTORS THAT INFLUENCE PROBABILITY OF COLLISIONS

In 1979-1980, fishermen that used a total of 85 codtraps volunteered to participate in a study of whale damage to traps. Prior to the fishing season they were interviewed about the characteristics of their codtraps and fishing berths. During the course of the fishing season, they reported damage to their gear; 20 of their traps were damaged by whales and 65 received no damage. Interviews and on-site investigations revealed a confusing picture (Lien, 1980).

Collisions most often occur during periods when capelin were spawning near shore which extends from mid-June through July. In 75% of the cases where the traps were damaged, capelin were plentiful in the vicinity of the gear at the time of the accident. Damaged codtraps were typically the most productive ones, usually producing more than 1800 kg of codfish per day. The best fishing berths in an area were the ones most likely to be hit by whales. Length of trap leader, size

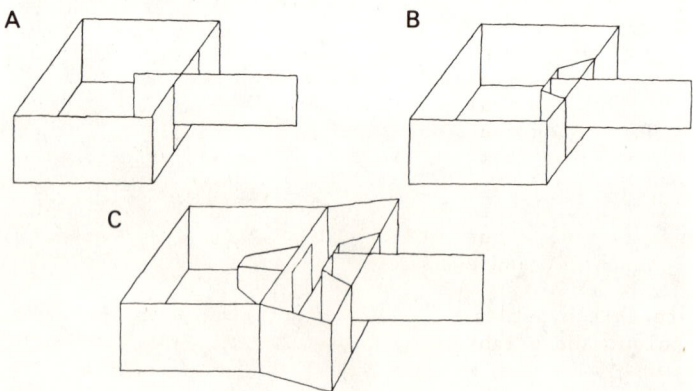

Figure 4. Types of codtraps; (a) Newfoundland (b) Modified Newfoundland and (c) Japanese codtrap.

of trap, distance from shore, depth and a variety of other gear and
environmental characteristics were not significantly related to
damages (Lien, 1980).

The leader of the trap incurred damage in 60% of collisions. The box
of the trap was damaged in 37% of collisions. Doors and "wings" are
the most likely damaged parts of the box; sides are less likely
damaged, and the back of the trap is the least likely to be damaged.

Collisions occur most frequently at night (72%) and less frequently
during the day (28%). Collisions are not related clearly to visibility
in the water column as measured by a Secci disc. Visibility is
generally poorer in areas where fishing is best. However, in good
fishing areas, we found there is no difference in water turbidity
between traps that were and were not damaged. There is some
relationship with discovery of net damage and weather because
fishermen can work their gear only in mild and moderate conditions.
However, weather or surface water conditions do not appear to be
associated with collisions (Lien, 1980).

The highest probability of a collision occurs on the first day a trap
is placed on its berth (9%); 25% of collisions with traps occur in the
first 96 hours and 50% of collisions occur in the first 14 days the
trap is in the water. This may indicate that the location of gear
placed in an area is learned (Lien, 1980). An alternative explanation
is that while nets are in the water they become fouled with biological
growth. Clean fishing gear may be visually or acoustically less
detectable than fouled gear.

Fishermen were very accurate in predicting where whale collisions
would occur. They based these predictions on the frequency of whale
collisions which occurred at a berth in previous years. Collisions
occurred with codtraps in the same berths year after year. We have yet
to discover a set of features which clearly characterize these high
risk berths.

VISUAL DETECTION OF FISHING GEAR

It is interesting that most collisions with codtraps occur at night.
There can be several explanations for this. One is that whales exhibit
periodicity in near shore activity and more activity occurs near shore
at night. It is reasonable to assume that their activity is, in part,
controlled by bait movements and diurnal variation in capelin has been
observed by many investigators (Dragesund and Monstad, 1973; Piatt et.
al, 1989). Perhaps the whales avoid boats which are abundant in the
nearshore areas during the day. A second possibility is that the gear
is more difficult for them to detect at night. If this is true it
might suggest the animals were using visual cues, at least to some
degree. Since visibility is generally poorer in the best fishing areas
where most collisions occur, this would suggest that humpbacks may use
visual cues to detect and avoid nets.

To evaluate this hypothesis, we placed fishing gear components which
varied in colour and brightness, which might be important either to
the whales or target fish species, under water to ascertain their
relative detectability. The gear was rigged under water in a standard
manner at 10 m depth. For each test, a diver clipped a surveyors chain
to the mount which held the test gear and swam away. He recorded the

distance at which the test object was no longer visible. During each te¬st, light meter readings were taken laterally, as well as up and down in the water column. Tests were repeated under "excellent" (> 10 m), "average" (> 3 m) and "poor" (< 3 m) water conditions. A total of 27 different gear components were tested in this manner (Lien, 1980).

Some variation in the disappearing distances of different gear was recorded, but it is difficult to assign a value of relative visibility to any one colour or brightness. When water visibility was good, determinates of visibility seemed to be size and brightness. When the water was turbid, all colors disappeared in very short distances and the only distinguishing feature of the gear was its shadow contrast. Small bright objects could disappear as easily as dark or colored objects, since their high reflectance in murky water made them blend well into the background of diffuse light.

However, the visual characteristics of the gear itself contributed relatively little variance to detectability by human divers. The most detectable gear in the best visibility condition would give divers a 30-40 % increase in the distance at which it could be detected over the least detectable gear. At poor visabilities, that distance advantage was only 10-20 %. Far more important than the visual properties of the gear was water turbidity. The types of gear hardest to detect under the best visual circumstances could be seen at 3-4 times the distance of the most detectable gear under poor visibility conditions. In the changing ocean environment, quality and quantity of light is extremely variable especially in inshore locations. No one particular type or characteristic of gear is substantially more or less visible than others. Under the relatively poor visibility conditions (4-5 m) which characterize productive fishing zones in inshore Newfoundland waters, vision is a risky modality to use in detecting nets with confidence or for more biologically normal tasks, such as finding bait. Variance in visibility due to water conditions encountered in the inshore waters is far more important than gear characteristics for human divers to detect.

There are suggestions that cetaceans rely on detecting the contrast of an object against its background, or alternatively, use optical pigments that specifically respond to those wavelengths which are most effectively transmitted through the water column (Watkins and Wartzok, 1985). However, given that the limitations illustrated by the above experiment are more a function of the physics of light in water rather than the optics of the human eye, presumably similar constraints would apply to cetacean vision. For much of their inshore activity, vision would be an unreliable guide to humpbacks in avoiding obstacles or finding bait.

It is still intriguing that humpback collisions with fishing gear occur most commonly at night and occur in fishing zones with low visibility. If vision was essential to avoiding gear, one would expect other circumstances or water characteristics which limit visual function to be related to the frequency of gear collision. However, that has not been found clearly, although we have investigated visibility at traps that had and did not have collision damage. Although we would like more data, our working conclusion, based on the limited information available to us, is that humpbacks are not orienting using visual cues during their inshore fishing activities. It is more likely that acoustical orientation is the primary modality that guides the animals in inshore waters.

ACOUSTIC ENHANCEMENT OF FISHING GEAR

There are several possible cues which humpbacks may use to detect and avoid nets that are acoustical in nature. One possibility is that nets produce sounds as water moves through them. Fish, held by the nets, may also produce sounds. A third possibility is that nets attenuate ambient background sounds in the near vicinity, by acting as an acoustic 'shield'. A final acoustical cue to the nets presence could be generated by echos resulting from sounds of the whale itself. Given the variability in water visibility, the fact that humpbacks are successful most of the time in avoiding nets at night with extremely low light levels and in water with poor visibility, it seems likely that net cues are, at least in part, acoustic in nature. Because they can move around netted areas at night without actively emitting any sounds (Lien, 1980), it seems unlikely that some form of echolocation is involved.

If humpbacks feed inshore using acoustical searching images, enhancing the acoustics of fishing nets may facilitate net detection and prevent collisions. We decided to enhance the acoustical cues associated with codtraps by adding sound devices which might aid the whale in detecting and locating the net. Sound devices which generated signals above the relatively low frequency range of codfish hearing (Brukle 1967; 1969) were chosen. Initially, a variety of high frequency electronic devices and mechanical sound devices which produced lower frequency sounds were chosen and tested to observe the humpbacks reaction to them. Each device tested emitted signals loud enough that they were easily detected by standard recording equipment (Uher 4400 recorder; Gould 17UT hydrophone) at 100m.

We presented a wide variety of these sounds to humpbacks that had habituated to the presence of our boat. Sounds were generated by devices which we believed might serve as "alarms" to facilitate detection of fishing nets. Initially, we observed the whales activities and recorded their proximity to our boat during a baseline period. This was followed by a stimulation period during which sounds were presented, and finally, another baseline period. Changes in the whales activity pattern and proximity to our boat were used to indicate if the whale detected the sounds presented.

Humpbacks responded to all the devices that we tested but the pattern of response was highly variable. Reactions can be summarized as: (1) approach to the sound source, (2) withdrawal from the sound source and (3) no reaction; no change in activity, direction of movement or proximity to the boat. On 45 % of presentations of all devices there were no detectable differences in the activities of the whales. In these cases it is not known if the animal failed to detect the sound or simply 'ignored' it. About half of the remaining whales approached; about half withdrew from the sound.

As humpbacks responded to many different sounds, we decided to select devices which were inexpensive, reliable and compatible with fishing methods and fishermens views. These included: (1) High frequency "pingers" (27 to 50 kHz), (2) Low frequency electronic "beepers" (3.5kHz) and (3) Low frequency mechanical "clangers" (500 Hz to 1 kHz). These devices are illustrated in Figure 5.

Such novel, artificial sounds have no clear biological meaning to the humpback. The sounds did produce 'curiosity' or 'fear' reactions and

the whales were able to localize the sound source as they either
approached or withdrew. Thus, these novel sounds might through
experience come to indicate to the whales the presence of a visually
and acoustically cryptic net.

A pilot test of devices was made in 1979. Following modifications, a
second test was conducted in 1980. Meetings were held with fishermen
in 24 villages in St. Mary's Bay, the Southern Shore and on the Avalon
Peninsula in eastern Newfoundland during early winter in 1980. As
humpbacks would have to learn that the sounds produced by the devices
meant the presence of nets, it was felt that the test area should be
somewhat concentrated to insure that whales might repeatedly encounter
the devices. Also this area was one which traditionally had the
highest incidence of whale collisions in Newfoundland (Lien, 1980).

A total of 191 fishing crews agreed to test "alarms" on a total of
560 codtraps. During the meetings we enlisted volunteers and asked the
fishermen to report fishing gear damages they had experienced during
the previous five years. Fishermen that volunteered frequently
reported gear damage: over 50% reported whale damage in the previous
year; only 30% of the volunteers reported gear damage due to ice or
storms in the past five years.

Fishermen were asked to indicate how many codtraps they fished, the
history of whale damages at each berth and on which traps they wished
to use "alarms". Fishermen always indicated that some berths were
"high-risk" berths and wished to use alarms on these. Alarms were
placed on some "high-risk" berths; other "high-risk" berths were
designated as controls. Berths that fishermen did not designate for
alarm use were included in the study as "low-risk/no alarm" controls.

Field crews instructed fishermen on placement of the "alarms" and
assisted in their installation. In many cases the actual installation
of the devices was left to fishermen but it was suggested that one
electronic device be placed mid-leader between the shore and the box

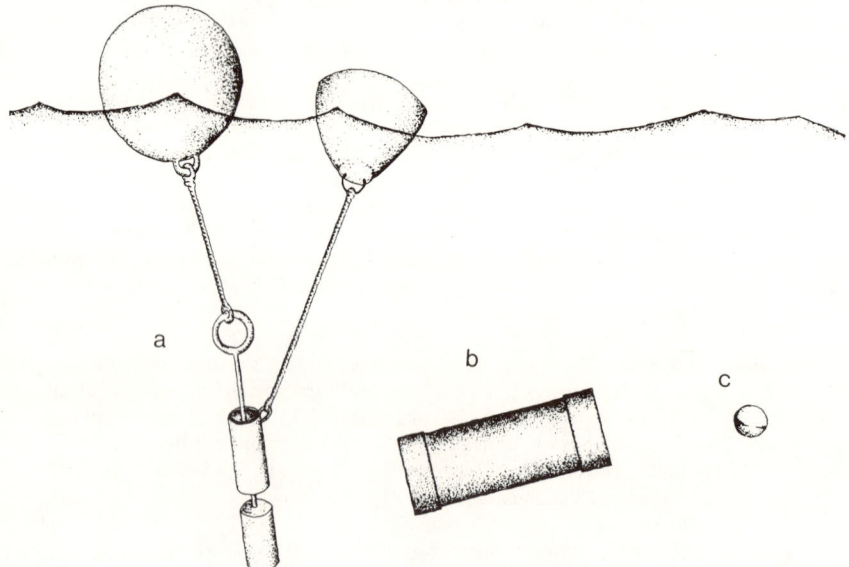

Figure 5. Acoustical "alarms" placed on codtraps; (a) "Clangers" (b) "Beepers" and (c) "Pingers".

Table 1. Results of adding acoustical enhancement to codtraps. Total number of codtraps = 560; total days of fishing effort = 26,040.

Type of Device	No. Codtraps	Days Fished	No. Collisions	P of Collision Per Trap Day	Mean Cost (in $)
Low-Hz "clanger"	81	2,721	35	.012	511.
Hi-Hz "pinger"	24	950	20	.019	255.
Low-Hz "beeper"	34	1,251	10	.007	238.
High-risk/ No device controls	97	2,604	32	.012	620.
Low-risk/ No device controls	265	13,592	93	.007	621.

of the trap, one at the entrance to the box and one on the rear of the box. Installation of the "low frequency clangers" differed in that the devices were operated by wave action on surface buoys and had to be installed on each corner of the box of the trap and one mid-leader, where side moorings were attached. Thus codtraps that used "clangers" had five devices (Figure 6).

Fishermen participating in the experiment were interviewed on a weekly basis to determine whale sightings, fishing effort, yields, and whale damage. If an accident was reported, the fisherman was questioned about details. Estimates of gear damage were based on the fishermans' opinion of the extent of repair work required. Results of the experiment are presented in Table 1, for a total N of codtraps of 560, and a total number of days of fishing effort of 26,040.

Fishermen were very successful in selecting berths at which they would have trouble with whales. Collisions at "high-risk" berths occurred with a probability of .012/trap day; "low-risk" berths had a significantly lower probability of whale collisions (.007/trap day). There was no difference in the costs of accidents at the two types of berths.

Mechanical low frequency "clangers" required constant tuning to insure that they were working. During inspections, we frequently found that fishermen had improperly adjusted buoys which operated them and

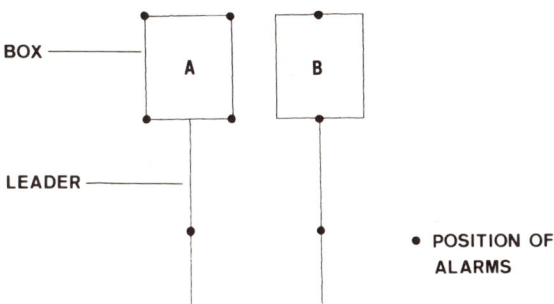

Figure 6. Locations where acoustical "alarms" were placed on codtraps for; (a) "Beepers", "Pingers" and (b) "Clangers".

they were not producing sounds. A second major problem with "clangers" was that, during extreme tidal currents, they occasionally became entangled in netting. Perhaps because of these difficulties, "clangers" did not significantly reduce the probability of a whale collision or the cost of a collision.

High frequency "pingers" did not lower the probability of an accident, but significantly reduced the cost of an accident. This may be explained by details of accidents provided by fishermen. "Pinger" accidents often appeared to be of the sort where the whale was maneuvering to avoid the gear, but did so unsuccessfully. This may mean that these devices were too quiet or difficult to localize and that the whale detected them too late to completely avoid the net, or that the alarms were of so high a frequency that they occurred at an insensitive part of the whale's hearing curve.

Low frequency "beepers" lowered the probability of a collision significantly as well as the cost of a collision. We had the clear impression with all "alarm" devices that fishermen were more likely to find and report any damage to their nets; minor damages to control nets were less frequently detected and reported. This tendency may account for part of the general reduction in cost of collisions experienced with "alarm" devices.

Fishermen have continued to use low frequency "beepers" extensively in the past decade. We have developed instructions for building them from easily accessible components and now case them in PVC sewer pipes. We presently are redesigning both mechanical and electronic "alarms" for minimizing the frequency of inshore whale collisions (Lien et al., 1989b). The new devices will be louder and randomize both sound characteristics and pattern in which they are emitted.

ACOUSTICS OF INSHORE NETS

While acoustical add-ons may offer help in alleviating collisions, basic redesigns of fishing gear may offer better possibilities to enhance detectability of fishing gear by whales. We have conducted experiments which radically modify leaders of codtraps by replacing netting with "bubble" leaders or limiting the vertical dimensions of trap leaders (Knight, 1980; Kingsley, 1982). However, fishermen are skeptical of such radical departures from tradition.

Therefore, we have begun investigations designed to acoustically measure the inshore environment where accidents occur. Certain types of fishing gear, and characteristics of capelin which the primary food of humpbacks are the primary focus of the study (Todd, 1989). A field program was started in the summer of 1989. By these experiments we hope to develop methods to modify major types of inshore fishing gear used in Newfoundland and Labrador.

Initially, to investigate the acoustics of nets, a mooring system was designed and installed in 10 m depth of water, 150 m offshore from St. Phillip's, Conception Bay, Newfoundland. It consisted of a main riser buoyed with a large float which remained approximately 2 m below the surface at low tide. This prevented any excessive noise of the float interacting with surface water, yet permitted a taut frame (Figure 7).

A highly sensitive hydrophone (Bruel and Kjaer Type 8101) was clamped to the main riser, 5 m from the seafloor and the hydrophone cable was laid along the seafloor to the shore. On shore, the hydrophone was

Table 2. Summary of physical characteristics of fishing gear tested for acoustical properties.

Type	Mass/ Unit Area (gm/m²)	No. of Twines	Mesh Size (m)	Diameter of Thread (mm)	Surface Area of Net Exposed /m² (cm²)
Cod trap, (C1)	74	3	0.21	1.78	1060
Cod trap, (C2)	54	3	0.23	1.26	690
Cod trap, (C3)	46	3	0.16	1.38	1080
Cod trap, (C4)	64	4	0.10	1.23	1550
Herring net, (H1)	7	3	0.175	0.60	430
Gill net, (G1)	1	1	0.245	0.58	290
Capelin trap (CT1)	180	3	0.020	0.97	6070

connected to an 8 channel multiplexer (Bruel and Kjaer Type 2811), charge amplifier (Bruel and Kjaer Type 2635), and finally led to a flat response instrumentation recorder (Hewlett Packard 3964A) and a precision digitizer (Data Lab DL 1200).

The net hanging system consisted of a hoop of sealed PVC plastic piping 16 m in circumference. The hoop was positioned and fixed directly over the hydrophone. From the hoop, at just over 1 m intervals, were hung fifteen lengths of hanging twine, each 2.5 m long. These were used to attach to the top rope of the net being sampled. Samples of net were cut into shape so that when laid they would form a cylinder of 16 m in circumference and 3.5 m in depth. Once laced into place, the sample was positioned 2 m below the surface. Therefore the hydrophone was situated just below the net system. Table 2 presents the types of gear tested and its physical characteristics. Acoustic characteristics of different nets are shown in Figure 8.

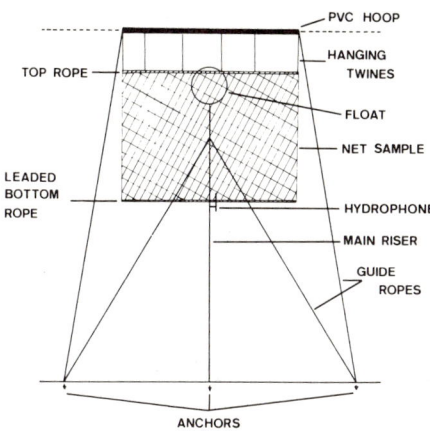

Figure 7. Diagram of the hydrophone assembly to assess net acoustics.

From our results, it is clear that nets vary substantially in their acoustics. This may be attributed to two factors. First, the nets differ in their material composition, and thus may respond differential as sound conductors or attenuators. While the net itself might not actively produce sound, it acts to redistribute or modify ambient noises present in the water column. A second, but dominant factor, involves the drag coefficient of the net. Nets set in the near shore zone are subject to tidal currents. Currents flowing through the mesh of the net may produce a 'streaming' effect. Such streaming or vibration produces acoustic energy in a strong band limited white noise. Drag can be linked by a simple equation to the surface area of a net (Baranov, 1975). In turn, the resultant acoustical emissions can be proportionally linked to drag.

However, these relationships also are dependent on the texture of the netting. Texture can vary from relatively smooth in a gill net constructed from singular nylon fibers, to relatively coarse in the twined nets such as cod traps, capelin traps, and herring nets (each twine is a core of 3-4 smaller twines). Therefore, gillnets should be more acoustically quiet based on the texture of the filaments since smooth fibers tend to produce low turbulences. Capelin traps, with their composite fibre construction should produce much more flow noise. Composite, twisted mixtures of fibers would induce a stronger backwash velocity gradient and therefore have a more acoustically pronounced signature.

Whether the net is acting as a conductor of sound or as an area of high drag actively producing sound, the process of biological fouling of the net will make the net more noisy. Biological growth will act to both change the resonant properties of the net and also to increase its drag. This might explain the common observation by fishermen that humpback whales mainly collide with new, clean traps. Fishermen often believe this is a particularly evil twist of character in humpbacks. It is more likely that clean traps are simply less likely to be detected by cetaceans than fouled ones because they are acoustically less obvious.

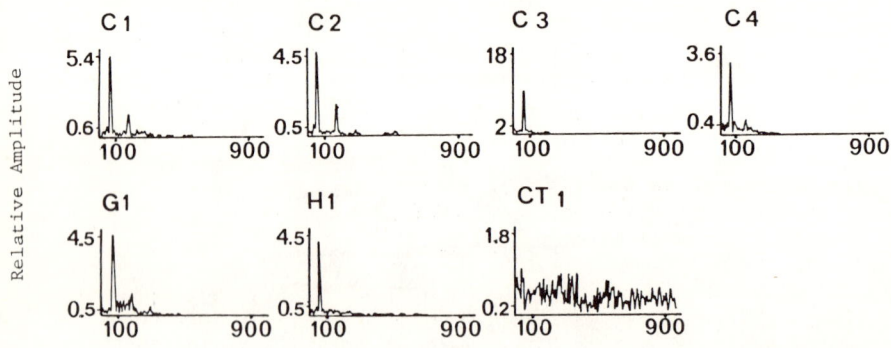

Figure 8. Acoustic characteristics of net samples. See Table 2 for details of net samples. Peaks at 60 Hz and at subsequent harmonics are due to a submerged power cable in the near vicinity. Note: vertical scale varies in magnitude.

The most acoustically obvious net was the sample of capelin trap, CT1, while the samples of cod trap - C1, C2, C3, and C4 - were certainly less so (Figure 8). The smaller mesh-size of capelin trap material results in relatively high surface areas and, therefore, is highly susceptible to the streaming effect. This streaming would translate in high broad spectrum noise levels. This finding is interesting in the context of collisions in the inshore fishery.

Unfortunately, there are no accurate estimates of fishing effort by capelin and codtraps in Newfoundlands inshore fishery from 1979-1989, the period during which careful records of whale collisions with fishing gear have been maintained (Lien et al., 1989a). In 1978, the last year in which numbers of codtraps fished were recorded, there were 7,516 codtraps in use (Lien, 1980) and it is believed that this effort has remained approximately level throughout the 1980's. Estimates of the number of capelin traps fished also are inadequate and vary by year (B. Nakashima, Fisheries and Oceans, personal communication). In 1983 it was estimated that about 827 capelin traps were fished; in 1988 about 3,125 traps were used. Average fishing days for capelin traps ranges from 11.0-17.3 days (Mean = 14.8 days). Duration of codtrap use is more variable but typically is from 30-40 days in length. Lien (1980) found mean duration of codtrap use in 1979-1980 to be 37.4 days.

Thus, the ratio of codtrap effort in days to capelin trap effort would be roughly 10-15:1. Although this is a rough approximation of effort, the ratio is informative. Based on monitoring from 1979-1989, the ratio of humpback entrapments in codtraps to capelin traps is 146:1. Capelin traps may indeed be much more detectable than codtraps. We know that capelin netting has a radically different acoustic signature and this may be the basis for their enhanced detectability.

FURTHER WORK

Our work on the problem of humpback whale collisions with fishing gear has focused on acoustics, for sounds are of obvious importance to most cetaceans. Other modalities are of less utility and can provide less reliable guidance for orienting and navigational tasks the animals must constantly face. There are a number of recent reviews on these capacities and the importance of hearing in the orientation of marine mammals (Popper, 1980; Forbes and Smock, 1981; Watkins and Wartzok, 1985).

While it is likely that marine mammals, unlike fish, typically detect nets by sound, there are few studies on the sounds of nets. We have begun studies on the sounds of nets by mapping their ambient pressure signatures (Todd, 1989). Our next step is to quantify the true radiated power with is emitted within the target structure. Underwater dynamic acoustic intensity scanning (DAIS) (Guigne et al., 1987a; 1987b; 1988; 1989) will be attempted.

The DAIS technique deals with vectorial intensity quantities when measuring any flow of energy radiating from targets or structures (Guigne and Williams, 1989). Since acoustic intensity is captured by multiplying the velocity of the flow with pressure with respect to time (watts/meter square), reactive energy can be separated from activeness in a sound field. Distinct directional signatures and information can therefore be isolated which are unique to a net in a particular environment. Background environmental noise that tends to be stationary in behaviour (reactive) normally dominates conventional pressure measurements. When intensity is used, the active flow is tracked with an overall gain in signal to noise of 20 db. Intensity is the context of all biological hearing involving directional capacity and identifying sources of sound. Therefore, use of intensity will highlight the sound flows which radiate from schools of fish and fishing gear environments as would be sensed by the hearing apparatus of marine mammals.

Other factors, besides the net itself, will effect the reliability with which humpbacks can detect fishing gear. While inshore in Newfoundland waters, humpbacks almost feed exclusively on capelin which are found commonly in areas where fishing gear is plentiful. In addition to the DAIS approach, we currently are analyzing the acoustical characteristics of capelin to better understand the acoustical searching image which humpbacks are using as well as to determine how capelin could mask nets (Todd, 1989).

Finally it is not clear how precisely humpbacks, and other large whales, are able to localize sounds. To understand what information is available to accomplish this, we have begun to analyze differences in sounds as they arrive at the two ears of the whale. Ear structures in the severed head of a minke whale, that was killed incidentally in fishing gear, have been replaced with sensitive hydrophones. With assistance of a sterotaxic technique based on an X-ray study of the anatomy of the head, specific points on the submerged head are then stimulated with a sound projector. Sounds as they arrive at the two ears are analyzed through a real time, dual channel, digital filtering analyzer. Intensity spectrums are calculated. The actual directional characteristics of the head are plotted as cosine characters. By measuring intensity directly through the head, we hope to begin to understand the whale's basis for localizing sounds as well as to find the internal roles that anatomical structures have on hearing.

The problem of cetacean by-catch in fishing gear is a difficult one with social, cultural, economic as well as biological complications. We hope, through our studies of perceptual aspects of the problem, that we can devise solutions and implement a program in Newfoundland and Labrador that will minimize the fishing gear collision problem both for the whales and the fisherman.

ACKNOWLEDGMENTS

Support for our work came from the Department of Fisheries and Oceans and the Newfoundland and Labrador Department of Fisheries. Without the cooperation we have received from the inshore fishermen of Newfoundland and Labrador, our work would not be possible. We are extremely grateful for their help.

We also acknowledge support for Sean Todd both from the School of Graduate Studies at Memorial University, and from private sponsors. Sue Johnson, Sharon Grey, Bora Merdsoy, Ben Davis and others of the Whale Research Group all assisted with "alarm" experiments. Peter Hunt of C-CORE provided technical assistance with field acoustical studies, and Dawn Nelson assisted with illustrations. We thank them all.

REFERENCES

Baranov, G. I., 1976, Selected works on fishing gear. in: Volume 1: "Commercial Fishing Techniques". Translated from Russian by E. Vilim, P. Greenberg, ed., Israel Program for Scientific Translations, Jerusaleum, Wiley, New York, 631 pp.

Bonner, W. N., 1982, Seals and man: A study of interactions. University of Washington Press, Seattle, 170 pp.

Buerkle, U., 1967, An audiogram of the Atlantic cod Gadus Morhua. Journal of the Fisheries Research Board of Canada 24: 2309-2319.

Buerkle, U., 1969, Auditory masking and the critical ban in the Atlantic cod Gadus morhua. Journal of the Fisheries Research Board of Canada 26: 1113-1119.

Dragesund, O. and T. Monstad, 1973, Observations on capelin, Mallotus villosus, in Newfoundland waters. ICNAF Research Document 73/33, 9 pp.

Forbes, J. L. and C. C. Smock, 1981, Sensory capacities of marine mammals. Psychological Bulletin 89: 288-307.

Fowler, C. W., 1988, A review of seal and sealion entanglement in marine fishing debris. in: "Proceedings of the Pacific Rim Fishermen's Conference on Marine Debris", in press.

Guigne, J. Y., P. G. Williams and V. H. Chin, 1987a, A concept for the detection of fatigue cracks in welded steel nodes. Marine Technology 18 (4).

Guigne, J. Y., P. G. Williams and V. H. Chin, 1987b, Analysis of the deformation in a partially cracked welded T - plate. Marine Technology 18 (4).

Guigne, J. Y., P. G. Williams and D. K. Mak, 1988, Acoustical imaging using a DAIS technique for non-destructive testing. in: "Proceedings of 17th International Symposium on Acoustical Imaging", Plenum Press, New York.

Guigne, J. Y., P. G. Williams and J. G. Adams, 1989, A note on the directionality of intensity measurements made in a cylindrical tank. Presented at the 12th World Conference on Non-Destructive Testing, Amsterdam.

Guigne, J. Y. and P. G. Williams, 1989, Dynamic acoustic intensity for the detection of fatigue in offshore structures. Contract report for the Department of Energy, Mines and Resources, C-CORE Contract No. 89-C4, Memorial University of Newfoundland, St. John's, Newfoundland.

Kingsley, R., 1982, Tests of modifications to the Newfoundland codtrap. Unpublished report. Newfoundland and Labrador Department of Fisheries, St. John's, Newfoundland, 14 pp.

Knight, K., 1980, Use of compressed air leaders in codtraps to minimize whale damages. Honors Thesis, Memorial University of Newfoundland, St. John's, Newfoundland, 33 pp.

Lien, J., 1980, Whale collisions with fishing gear in Newfoundland. Report to Fisheries and Oceans, Canada: Newfoundland region. 31 December 1980. 316 pp.

Lien, J. and B. Merdsoy, 1980, The humpback is not over the hump. Natural History, June issue: 46-49.

Lien, J., J. Dong, L. Baraff, J. Harvey and K. Chu, 1982, Whale entrapments in inshore fishing gear during 1982: A preliminary report to Fisheries and Oceans Canada, St. John's, Newfoundland, 20 September, 36 pp.

Lien, J. and D. Aldrich, 1982, The basking shark (Cetorhinus maximus) in Newfoundland. Report to the Department of Fisheries, Government of Newfoundland and Labrador, St. John's, Newfoundland, 186 pp.

Lien, J. and H. Whitehead, 1983, Changes in humpback (Megaptera novaeangliae) abundance off NE Newfoundland related to the status of capelin (Mallotus villosus) stocks. in: "Proceedings of the Fifth Biennial Conference on the Biology of Marine Mammals", 60.

Lien, J. and L. Fawcett, 1986, Distribution of Basking Sharks, Cetorhinus maximus, Incidentally Caught in Inshore Fishing Gear in Newfoundland. Canadian Field-Naturalist, 100 (2): 246-252.

Lien, J., K. Breeck, D. Pinsent, and H. Walter, 1986, Whale and shark

Lien, J., 1988, Whale and shark entrapments in inshore fishing gear during 1986: A preliminary report to Fisheries and Oceans Canada, St. John's, Newfoundland, 33 pp.

Lien, J., 1989, Problems of Newfoundland fishermen with large whales and sharks during 1987 and a review of incidental entrapment in fishing gear during the past decade. The Osprey 19(1): 30-38, and 19(2): 65-71.

Lien, J., Stenson, G. B., and Ni, I-H., 1989, A review of incidental entrapments of seabirds, seals and whales in inshore fishing gear in Newfoundland and Labrador: a problem for fishermen and fishing gear designers. in: "Proceedings of the World Symposium on Fishing Vessel and Fishing gear design", J. Huntington, ed.

Lien, J., J. Guigne and F. Chopin, 1989, Development of acoustic protection for fixed fishing gear to minimize incidental catches of marine mammals. Fisheries Innovation Industrial Support Program, St. John's, Newfoundland, 24 pp.

Lien, J., W. Ledwell and J. Huntington, 1989, Whale and shark entrapments in inshore fishing gear in Newfoundland and Labrador. Report to the Newfoundland and Labrador Department of Fisheries and the Department of Fisheries and Oceans.

Perrin, W.F., 1988, Dolphins, porpoises and whales: An action plan for the conservation of biological diversity: 1988-1992. IUCN/SSC Cetacean Specialist Group and U.S. National Marine Fisheries Service, NOAA, 28 pp.

Piatt, J. F., D. A. Methven, A. E. Burger, R. L. McLagan, V. Mercer and E. Creelman, 1989, Baleen whales and their prey in a coastal environment. Canadian Journal of Zoology 67 (6): 1523-1530.

Popper, A. N., 1980, Sound emission and detection by delphinids. in: "Cetacean Behaviour: Mechanisms and Functions", L.M. Herman, ed., John Wiley and Sons, New York, 1-52.

Todd, S. K., 1989, An investigation of the acoustics of humpback whale collisions with fishing gear. M.Sc. Thesis, Memorial University of Newfoundland, St. John's, Newfoundland, (in preparation).

Watkins, W. A. and D. Wartzok, 1985, Sensory biophysics of marine mammals. Marine Mammal Science, 1 (3): 219-260.

Whitehead, H. and J. Lien, 1982, Changes in the abundance of whales, and whale damage, along the Newfoundland coast 1973-1981. CAFSAC WP/34/01, 8 pp.

Whitehead, H. and J. E. Carscadden, 1985, Predicting inshore whale abundance - whales and capelin off the Newfoundland coast. Canadian Journal of Fisheries and Aquatic Science, 42 (5): 976-981.

Whitehead, H., 1987, Updated status of the Humpback Whale, Megaptera novaeangliae, in Canada. Canadian Field-Naturalist 101 (2): 284-294.

FORMATION OF AN ADAPTIVE STRUCTURE OF THE PERIPHERAL PART OF THE AUDITORY

ANALYZER IN AQUATIC, ECHO-LOCATING MAMMALS DURING ONTOGENESIS

Galina N. Solntseva

N. K. Koltzov Institute of Developmental Biology U.S.S.R.
Academy of Sciences, 26, Vavilov str., Moscow, 117808,
U.S.S.R.

INTRODUCTION

The structure of the peripheral auditory system of cetaceans and pinnipeds in post-natal ontogenisis is discussed in a number of detailed studies (Bogoslovskaya and Solntseva, 1979; Fraser, Purves, 1954; Fleischer, 1971, 1973; Pye, 1966; Ramprashad et. al,1973; Ramprashad, 1975; Reysenbach de Haan, 1957; Yamada, 1953) of structures that evolved as an adaptive response for life in water and the use of echo-location. The lack of data in the available literature on the development of the peripheral auditory system of aquatic mammals in post-natal ontogenesis made it impossible to find answers to many important questions, in particular, the regularities of evolutionary transformations in the peripheral auditory system of mammals that belong to different ecological groups.

Under these circumstances, we during set out to explore the development of this system in aquatic animals in ontogenesis to identify the stages of organogenesis at which adaptive structures, specific to the ear of aquatic and echo-locating animals, appear.

To solve the problem, we investigated embryos from representative cetaceans (Odontoceti- Stenella attenuata, Tursiops truncatus, Delphinus delphis, Delphinapterus leucas; Mysticeti- Balaenoptera acutorostrata) and pinnipeds (Phocidae- Erignathus barbatus, Phoca hispida; Otariidae- Eumetopias ubatus; Odobenidae- Odobenus rosmarus divergens) compared auditory development in terrestrial mammals fetuses (Insectivora, Chiroptera, Rodentia, Artiodactyla) from 20 to 150 mm in sinciput - to - tail length (Solntseva, 1983-1988). We compared development all stages of the full number of turns of the cochlea, which is as the least changing structure in the evolution of most mammals.In spite of the uniform basic type of the peripheral auditory system in mammals, however in the limits of this type demonstrates substantial morphological modifications of all components, which seems to be related to adaptation to different habitats and development of different ways of direction finding and communication.

Let us examine the adaptive structures and their development in the outer, middle and inner ear of cetaceans and pinnipeds.

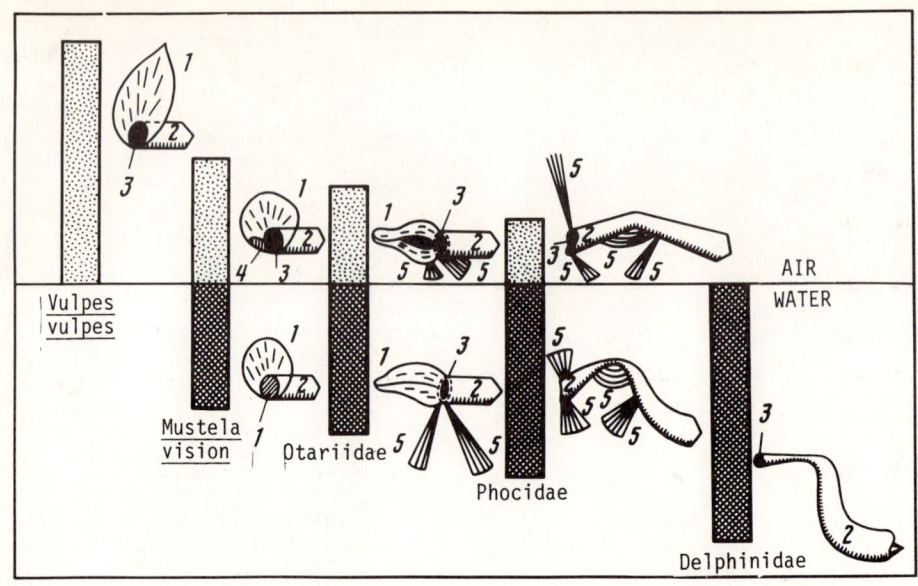

Fig. 1. Drawing of the structure of the outer ear of terrestrial, semi-aquatic and aquatic mammals. Mechanisms that open the meatus in air and close it under water. Vertical columns are average specific duration of stay in air and under water: 1- pinnal helix; 2- auditory meatus; 3- orifice; 4- skin valve; 5- ear muscles.

TABLE OF FIGURE TERMS

A- apical turn;
a- meatus acusticus externus;
B- basal turn;
ca- cartilago auriculae;
cs- plexus cavernous;
ch- cochlea;
CT- cavum tympani;
CR- cerebrum;
FO- fenestra ovalis;
gc- glandulae ceruminosae;
gr- processus gracilis;
i - incus
LF- ligamentum fibrosum;
M - medial turn;
m - malleus;

MT- membrana tympani;
Mtt- musculus tensor tympani;
MS- musculus stapedius
PL- processus lenticularis;
nCh- nervus vestibulo-cochlearis;
s- stapes;
si- sinus venosus;
st- scala tympani;
sw- scala vestibuli;
Y- organ vestibulum.

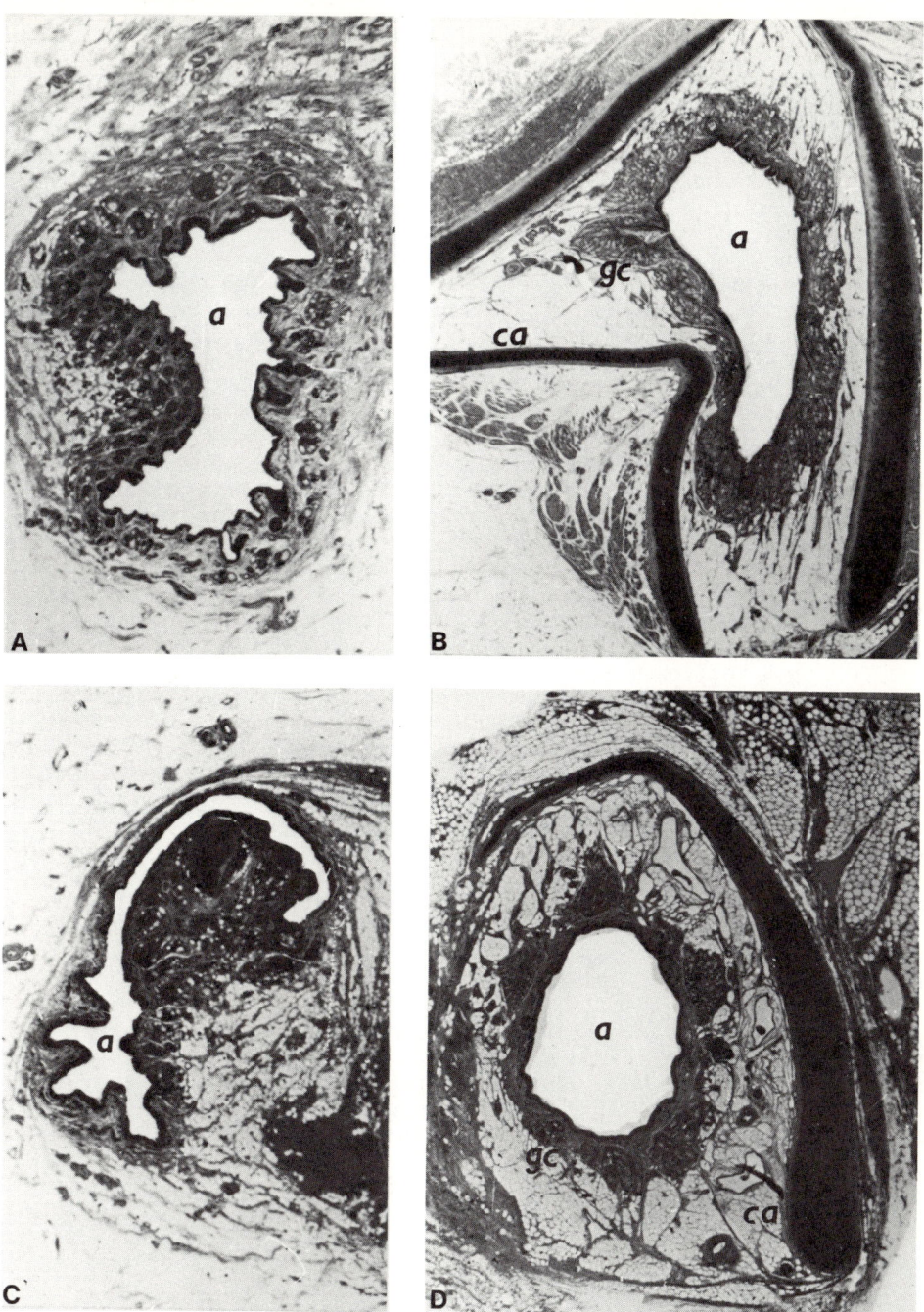

Fig. 2. Cross-sections the external auditory meatus of caspian seal (Phoca caspica): A- orifice; B- L- shaped bend in auditory meatus; C- expanded lumen in medial part of auditory meatus; D- end section of cartilaginous part of auditory meatus.

THE OUTER EAR

Post-natal ontogenetic studies conducted by the above mentioned authors showed that different species of aquatic mammals have substantially different structures of the outer ear. Isolation of the auditory meatus from the environment is a mophological adaptation of the outer ear or sound transmission under water. This is achieved by different structural features of the pinna and the meatus (Fig.1). Thus, the meatus of the eared seal and sea otter Enhydra lutris is closed by the twisting of the pinna. Because the pinna of the true seal is reduced, the meatus itself seals this cavity under water. The meatus distends considerably, its diameter and the form of the lumen change (Fig.2). Under water the external auditory meatus close by the cartilaginous spiral laminae as a result of its tension by means of the musculus antitragicus (Ramprashad et al., 1971). Cetaceans do not possess a mechanism for periodic opening and sealing of the meatus. The meatus of toothed whales is completely sealed in the distal section (Fig.3) (Clarke, 1948; Solntseva, 1971; Yamada, 1953). In baleen whales it is sealed tightly by a wax plug (Purves, 1955). The overgrown meatus of the Odontoceti gives rise to the assumption that sound could reach the inner ear by other pathways, getting round the outer and middle ear (Norris, 1964) and that there were other systems of sound transmission and perception unrelated to the auditory organ (Agarkov et al.,1971; Reznikov, 1972). It was suggested, as far back as 1957 by Reysenbach de Haan, that sound could reach the whales cochlea through the closed meatus.

The hypothesis was confirmed by calculations for the closed meatus of dolphins that showed that sound travels through the proximal section of the meatus, i.e. the section from the overgrown locus to the tympanic membrane (Lipatov and Solnseva, 1974). It was established that the threshold acoustic pressure on the tympanic membrane of Tursiops truncatus and Delphinus delphis at the frequency of the highest auditory sensitivity (70 khz) is only half the pressure on the tympanic membrane of humans at a frequency of 1 kHz. The findings of the simulation model make is possible to assume that normal sound reception under water, i.e. without a loss of auditory sensitivity, is possible for semi-aquatic and other aquatic mammals that possess a mechanism for closing the meatus under water, only when the meatus is closed (Solntseva, 1972, 1973).

As is known, what in early organogenesis the outer meatus of all mammals is blocked by epithelial cells. Embryos of cetaceans and pinnipeds begin to form the meatus when the developing fetus 30 to 40 mm long (Fig.4). At this stage it looks like a short tube. The formation of species-specific features of the outer ear was observed on 40 to 60 mm long embryos of toothed whales and true seals as a noticeable constriction of the meatus. Further on, the meatus distends, bends and acquires an S-shaped form characteristic of definitive forms (Fig.5). The meatus expands in the proximal section of embryos that are 70 to 100 mm long. It was established that at later stages the absolute dimensions of the meatus increase in direct proportion to the growth of an embryo.

THE MIDDLE EAR

Different groups of mammals have all the components of the sound transmitting mechanism of the middle ear, though these vary greatly in terms of structure, ossicular couplings and junction of the ossicles in the tympanic cavity. Additional structures related to the aquatic environment and reception of broad-band frequencies, including ultrasound, develop in the middle ear of aquatic mammals.

Ossicles of different mammal species demonstrate a wide range of structural variations (Fleischer, 1973). Adaptation to the aquatic environment includes a trend toward the thickening and shortening of the handle of the malleus (Enhydra lutris, Phocidae, Otariidae) and its complete reduction in cetaceans, as well as the lengthening and thickening of the thin process of the malleus. A striking feature of the malleus is the lengthening and thickening of the long process. The intercrural space in the stapes diminishes sharply, and disappears completely in toothed whales. Ossicles of echo-locating dolphins are smaller; coupled at a right angle in the area of the malleus-incus articulation, they are suited for transmission of broad-band frequencies. Toothed whales have more solid ossicles and the thin process of the malleus fuses with the tympanic ring in non-echo-locating animals the process is elastically connected to the wall of the tympanic cavity by a cartilage. Middle ear ossicles of toothed whales are suited for transmission of vibrations to the inner ear from the aquatic environment, with the reduced by 60 times amplitude and with the increased intensity (Reysenbach de Haan, 1957).

To maximize transmission of incoming sound energy, the specific acoustic impedance of the environment should match the auditory receiver. The matching is ensured by diversity of transmission structures of the middle ear, in accordance with the sound transmission medium. For optimal transmission of sound signals in the aquatic environment the auditory receiver should have a high elasticity, which is ensured by a more rigid articulation, i.e. at a right angle, of the malleus and the incus. Moreover, the peculiar location of ossicles in the middle ear cavity of dolphins increases pressure on the tympanic membrane two to three fold thanks to the "propeller effect" generated by the inclination of the rotation axis of the malleus and the incus transmits the sound reception to the malleus like a "jack hammer" (Lipatov, 1972). Extension of the range of auditory sensitivity to the ultrasound region is related to a higher resonance frequency of vibrations of the ossicles, and a greater elasticity of the tympanic membrane. This transformation substantial losses of sound energy in cartilages and articulations of the middle ear and makes the system more rigid, which creates optimal conditions for transmission of ultrasound. The rocking ossicle chain of terrestrial animals was replaced by an elastic vibrating system, which is capable to restore its initial state following a sound wave (Simkin, 1977).

Most mammals have a very thin roundish tympanic membrane which is slightly conical in shape. Aquatic animals, such as pinnipeds, have a much thicker tympanic membrane (Fig.6). The membrane of dolphins is not connected directly to the handle of the malleus, in contrast to most mammals. They are connected by a triangular ligament, which is asymmetrically joined to a roundish and very thick membrane. The tympanic membrane of toothed whales consists of two parts: a hairless "glove protrusion" which enters the meatus cavity, which a connecting fibrous ligament and the cylindrical handle of the malleus. The fibrous ligament of baleen whales is analogous to the triangular ligament of toothed whales (Fraser, Purves, 1960). Echo-locating animals (toothed whales and bats) have a smaller tympanic membrane.

Echo-locating forms display a tendency to isolate the tympanic bulla from bones of the cranium. Unlike all other aquatic mammals, dolphins have their tympanic connected to the cranium by connective tissue, which reduces sound transmission through bones; both ears become autonomous receivers and can receive directed acoustic signals (Reysenbach de Haan, 1957). Because of this isolation sound can reach the inner ear only via the ossicle system.

As is known, the middle ear is an adjunct of the labyrinth and is formed by protrusion of the first pharyngeal sac whose endoderm is transformed into a general tube-tympanic protrusion. All elements of the

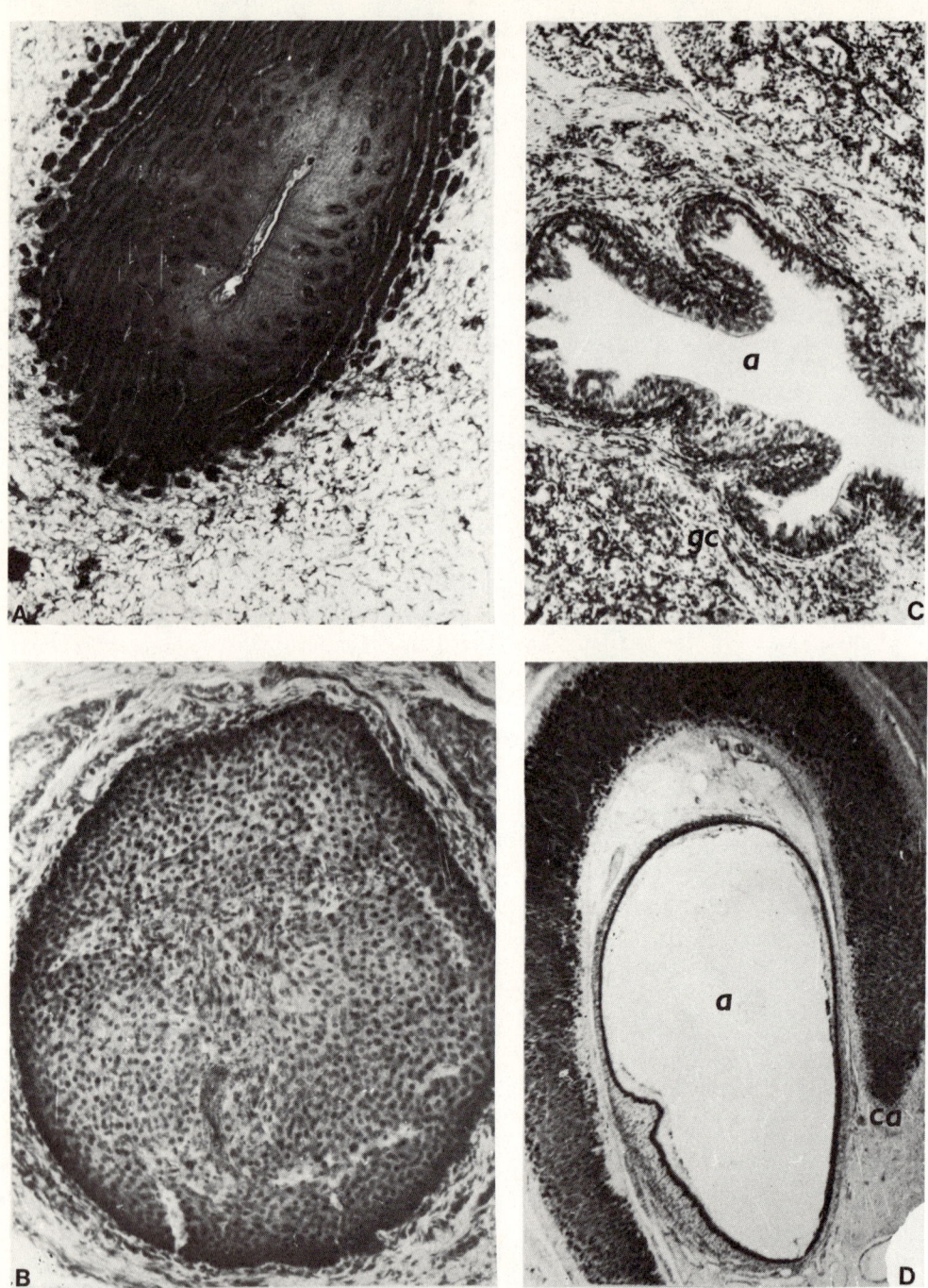

Fig. 3. Cross-sections the external auditory meatus of bottle-nose dolphin (Tursiops truncatus): A- orifice; B- blocking of lumen in auditory meatus; C- medial part of auditory meatus. D- end section of cartilaginous part of auditory meatus.

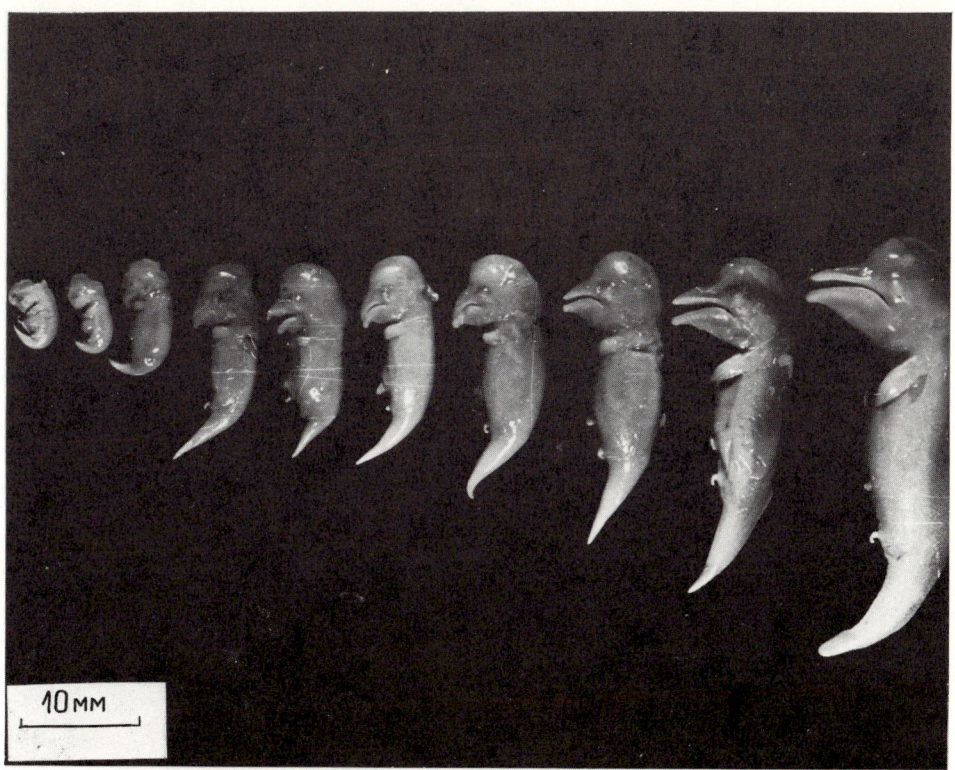

Fig. 4. Comparative view of embryos of <u>Stenella attenuata</u> at early stages of pre-natal development; 20 to 150 mm in sinciput-to-tail length.

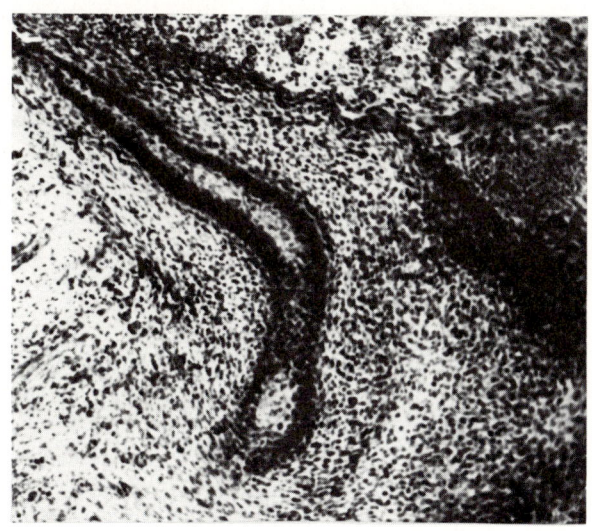

Fig. 5. Longtitudinal section of the auditory meatus of <u>Stenella attenuata</u>, 62 mm long. The meatus is much narrower in the distal part.

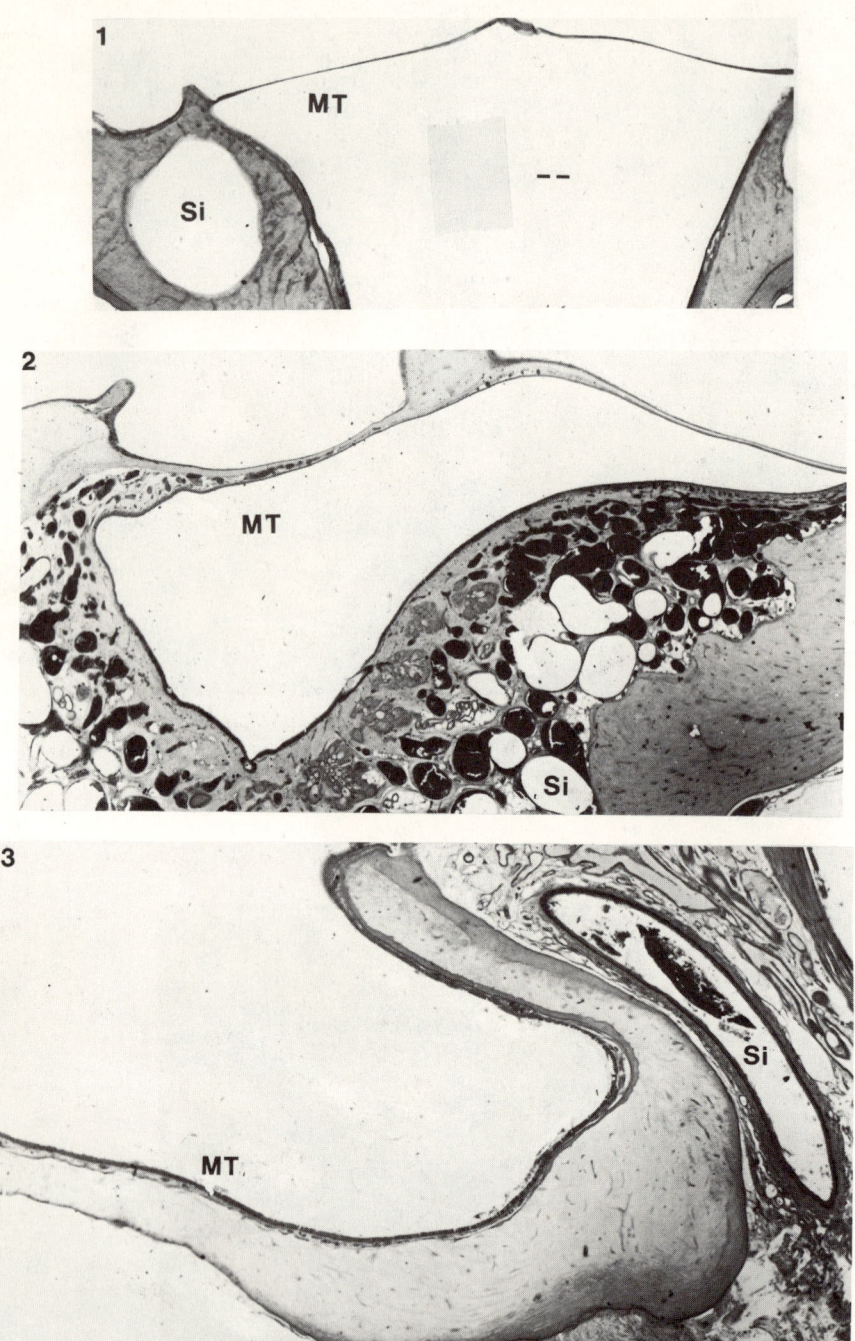

Fig. 6. Cross-sections of tympanic membranes of different species of mammals: rodent (Myocastor coypus); 2- harp seal (Pagophilus groenlandicus); 3- bottlenose dolphin (Tursiops truncatus).

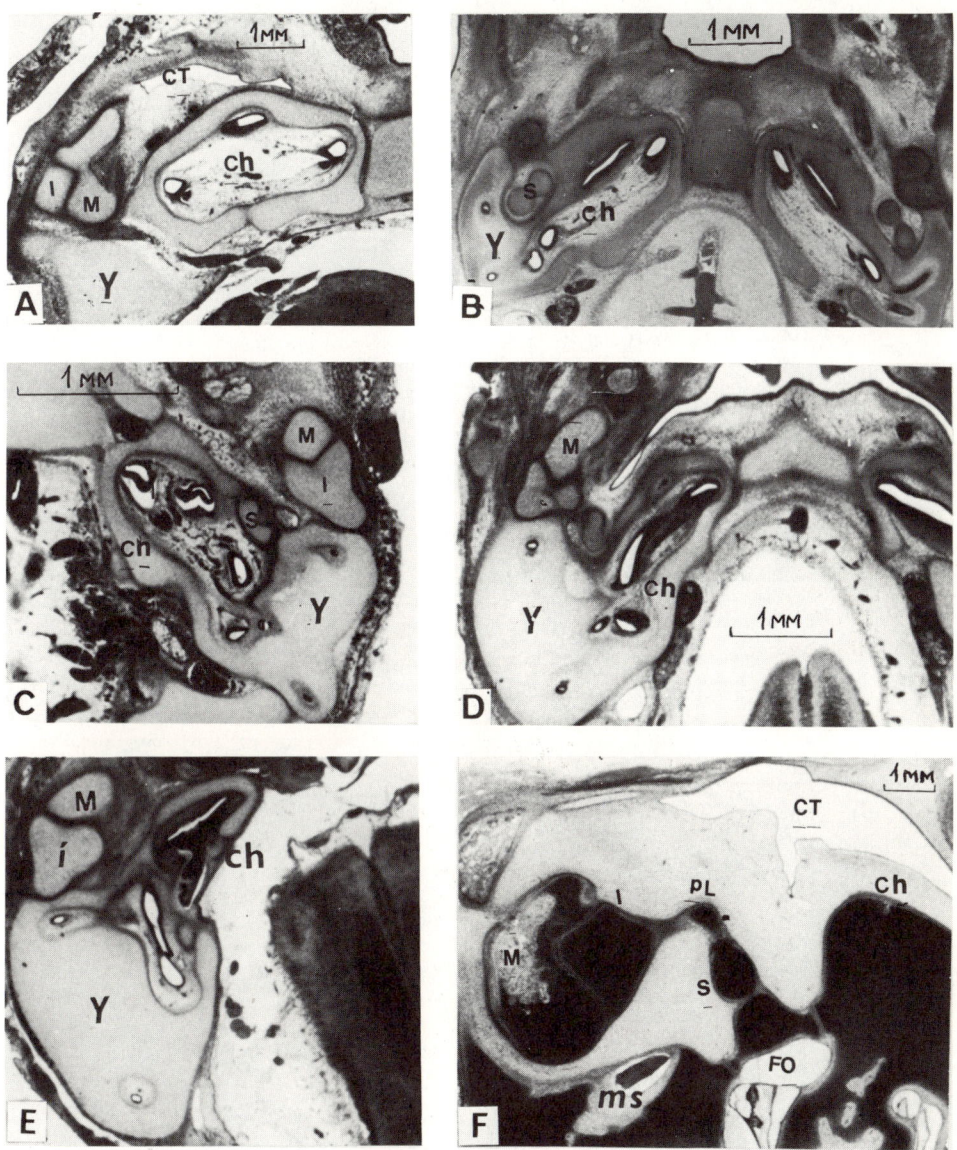

Fig. 7. Dorsal-ventral sections of the head the peripheral part of the auditory analyzer of aquatic mammals: A- <u>Stenella attenuata</u>, 62 mm long; B- <u>Balaenoptera acutorostrata</u>, 40 mm long; C- <u>Erignathus barbatus</u>, 25 mm long; D- <u>Eumetopias ubatus</u>, 35 mm long; E- <u>Odobenus rosmarus</u>, 35 mm long; F- <u>Delphinapterus leucas</u>, 250 mm long. The malleus-incus articulation of different aquatic species at different stages of development. Separate elements of ossicles are formed in early organogenesis.

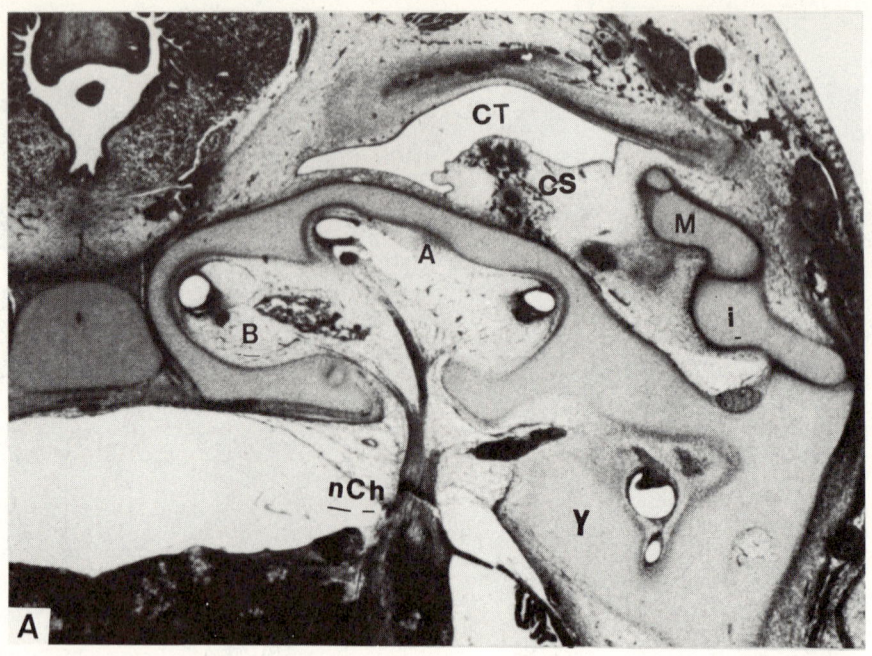

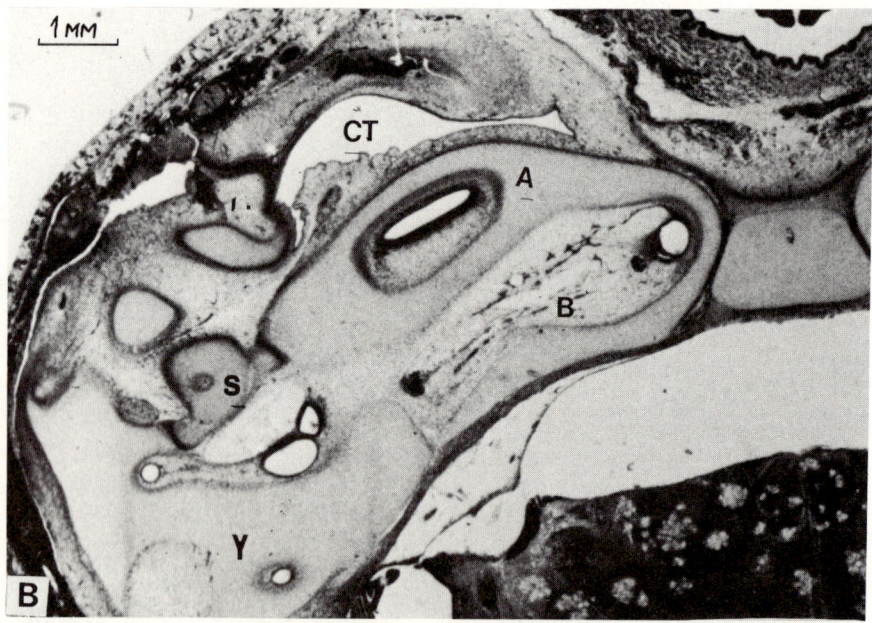

Fig. 8. The peripheral part of the auditory analyzer of a 90 mm long Stenella attenuata in dorsal-ventral sections of the head: A- development stages of cavernous plexus; B- end of formation of the tympanic membrane and additional ligament of middle ear.

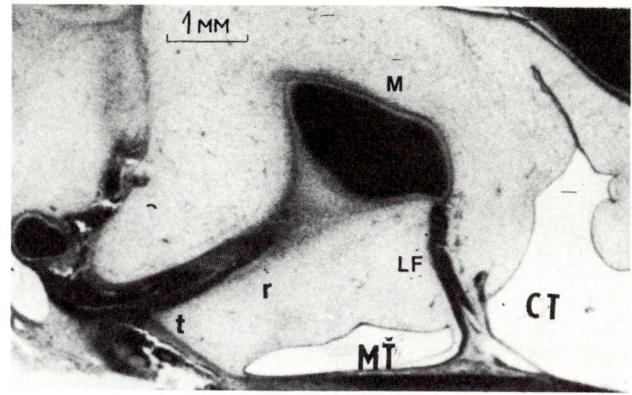

Fig. 9. Cross-sections of elements of the middle ear of a 250 mm long Delphinapterus leucas embryo. Rigid insertion of the long process of the malleus in the tympanicum, and connection of the head of the malleus with the tympanic membrane by an additional ligament.

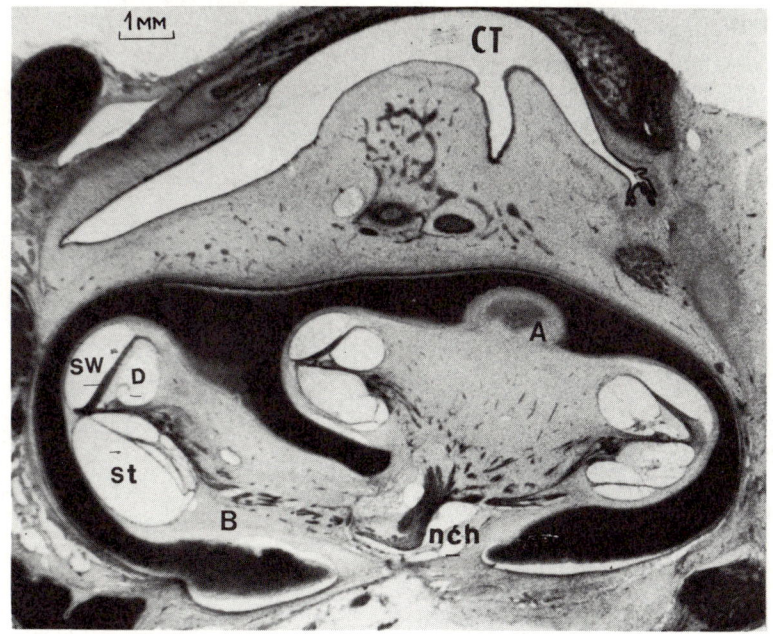

Fig. 10. The cochlea of a 250 mm long embryo of Delphinapterus leucas. The basal turn of the cochlea is widened, like in other echo-locating species. Differentiation of elements of the cochleate canal.

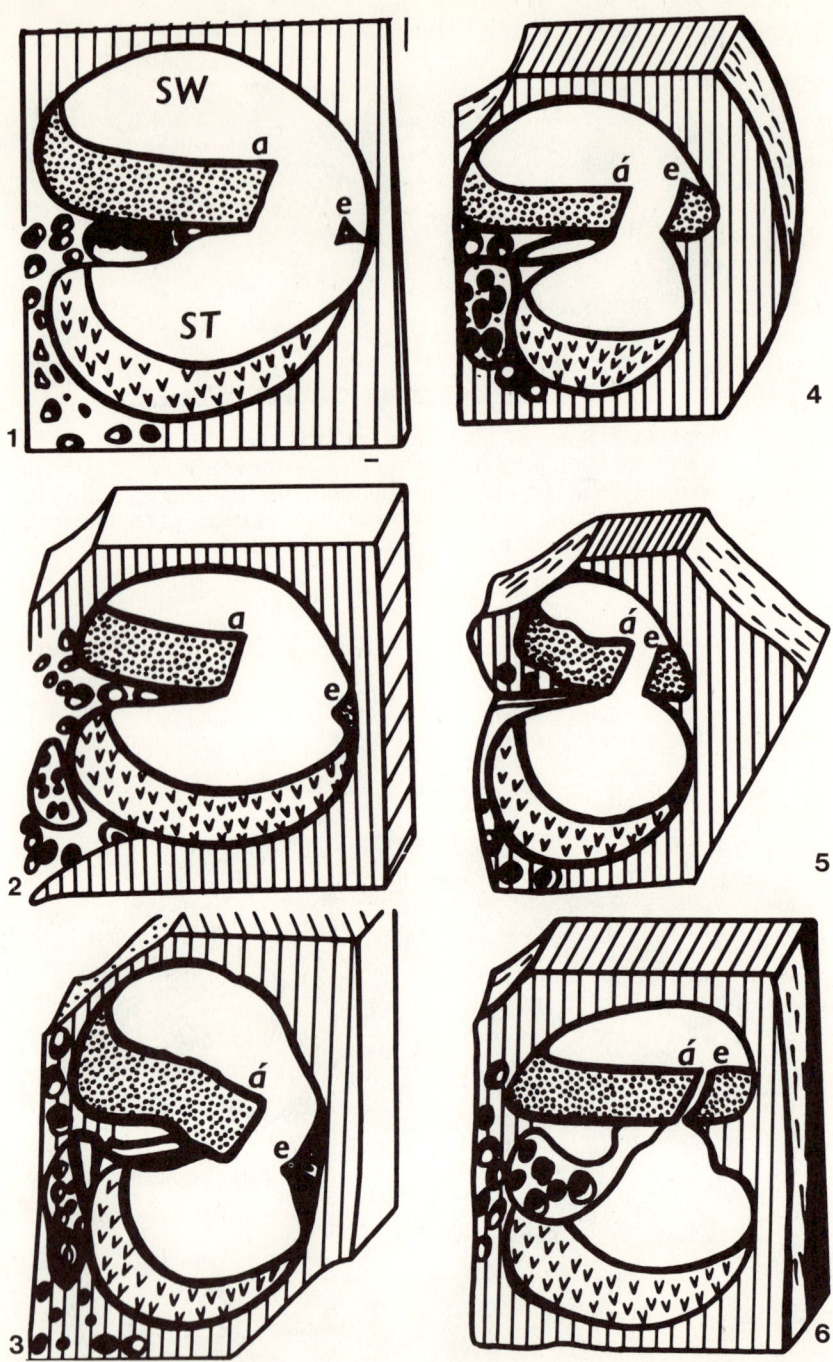

Fig. 11. Cross-sections of the cochleate canal of echo-locating and non-echo-locating mammals: 1- baleen whales (Balaena mysticetus); 2- human (Homo sapiens); 3- dog fruit bat (Rousettus); 4- shrew (Crocidura russula); 5- bat (Rhinolophus simulator); 6- harbor porpoise (Phocoena phocoena) a- primary osseus spiral lamina; b- secondary osseus spiral lamina. According to Fleischer (1973).

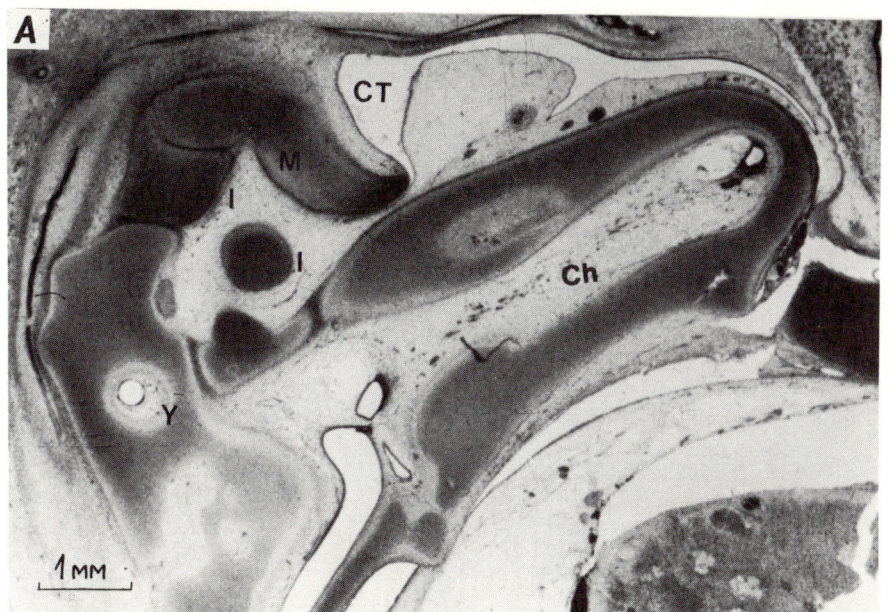

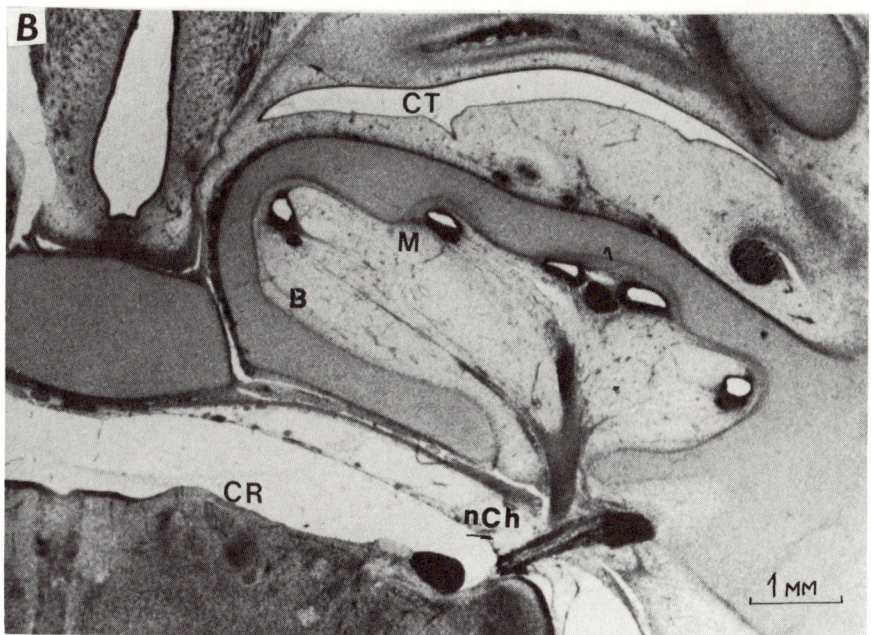

Fig. 12. The peripheral part of the auditory analyzer of a 86 mm long Balaenoptera acutorostrata in dorsal-ventral sections of the head showing: A- rigid coupling of malleus and incus and insertion of the long process of the malleus on the tympanicum; B- the cochleate canal on different stages of cellular differentiation.

middle ear develop from the mesenchyme and mesodermal elements. Ossicles develop from dorsal ends of cartilages of the visceral arches. In 20-30 mm long cetacean and pinniped embryos, the anlage of ossicles lie separately from the anlage of the tympanic cavity. Insertion of ossicles into the posterior tympanic space begins in embryos that are 30-40 mm long. The ossicles are connected continuously at earlier stages of development. Articulations begin to form in 30-50 mm long embryos (Fig.7). The ossicles are separated by perichondrium, which consists of fine flat chondroblasts. The most clear-cut borders were observed in 40-50 mm long embryos.The tympanic cavity begins to form in embryos of pinnipeds and cetaceans that are 30 to 40 mm long, cells are torn in the locus of the tympanic cavity, with the latter turning into a small canal.

The specific location of the ossicles and their differentiation into constituent elements were observed in 50-60 mm long embryos of cetaceans and pinnipeds. The malleus and the incus are located at a right angle to each other, like in mature mammals. This is particularly true for cetaceans. The stapes is differentiated into arches in toothed whales and pinnipeds. In toothed whales, the stapes form a column.

Venous sinuses in the osseus part of the meatus of pinnipeds and the cavernous tissue in the middle ear cavity of toothed whales are additional structures exclusive to aquatic animals. The structures ensure pressure equilibration in the middle ear during diving. The emergence of the cavernous tissue was observed in embryos of dolphins that are 50-60 mm long (Fig.8 A). Venous sinuses in embryos of pinnipeds are formed during replacement of cartilaginous tissue by osseus tissue, i.e. in 100-150 mm long embryos .

A species-specific feature of the middle ear of toothed whales is the triangular ligament that fixes the tympanic membrane to the handle of the malleus. The ligament performs an important function for it is an additional lever of the middle ear (Lipatov and Solntseva, 1972).

In dolphins, the tympanic membrane-ligament begins to appear when the embryo is 40 to 50 mm long. The process of formation is completed when it is 90 to 100 mm long (Figs.8B,9). At this stage, the connection between the tympanic membrane-ligament and the shortened handle of the malleus becomes evident. The tube-tympanic membrane protrusion begins to distend considerably and divides into the tympanic cavity and the Eustachian tube. The formation of the middle ear muscles was observed in 40-80 mm long embryos, and completed in embryos with length of 90 to 100 mm. No separation of the tympanic bulla from the cranium was observed in early organogenesis. Specific peribullar sinuses that separate the tympanic bulla from the cranium form at later stages, i.e. in 200-250 mm long embryos of the beluga (Delphinapterus leucas) (Fig.10). Cartilaginous tissue is replaced by osseus tissue in 90-100 mm long embryos, with separate spots of ossification developing in cranial bones. Ossification of ossicles was observed in embryos that were up to 250 mm long in the form of massive areas on the head of the malleus.

THE COCHLEA

The structure of the cochlea in the inner ear is similar in almost all mammals. The number of the turns varies from species to species. An increase in the number of turns is due to its morphological progress (Fleischer, 1973). While in echo-locating mammals, such as the dolphin, the cochlea makes 1.75-2 turns, the number of turns in echo- locating bats is 3.5. This diversity in the number of turns of the cochlea, does not affect the range of perceived frequencies. The cochlea of echo-locating species (dolphins) has the following important adaptive anatomical transformations: an increase in the basal turn of the cochlea, which perceives both low and high frequencies; the emergence of a well-developed secondary osseus spiral lamina; a substantial reduction of the distance between the primary and secondary spiral laminae in the cochleate canal; and lastly, thinner and wider of the basal membrane to which cells of the Corti's organ are fixed. In non-locating species the secondary lamina is developed only in the basal turn of the cochlea, while in echo-locating animals it occupies the entire cochleate canal, from the base of the cochlea to its apex (Fleischer, 1973)(Fig.11). It turns-out that the narrower the basal membrane the more developed is the secondary spiral lamina. The secondary osseus spiral lamina is underdeveloped in pinnipeds and baleen whales, which is evident from the great distance between the two laminae. The distance corresponds to the width of the basal membrane; about 0.37 mm in true seals and only 0.06 mm in dolphins.

A number of authors (Wever et al., 1971; Ramprashad et al., 1976) showed that the number of receptor cells and their distribution along the basal membrane are similar in most mammals, regardless of the frequency sensitivity of their auditory systems. Some researchers however, note specific features in the structure of accessory elements of the dolphins organ of Corti with these compactly located cells being of a bigger size (Wever et al., 1972). Furthermore, a number of authors noted a considerable increase in the number of cells of the spiral ganglion of echo-locating species (3 times) compared to humans, which testifies to their great capacity for processing acoustic information, starting with the peripheral auditory system (Firbas and Welleschik, 1973). The authors who studied the spiral ganglion, employed different methods to determine the specific density of neurons in the cochleate canal. They calculated the average number of cells per half of the cochlea's turn (Guild et al., 1931), per unit of length of the cochlea (Wever, 1949; Wever et al., 1971, 1972) per unit of volume of the Rosenthal canal (Schuknecht, 1953), and per unit of area (Ramprashad, 1976). The methods revealed the following general regularity: nervous cells are packed most densely in the second half of the basal turn and the first half of the medial turn, smallest the least density was observed in the apical turn, which correlates well with the neuron receptor ratio (4- 5.5 : 1 in dolphins; 3 : 1 in seals and cats; and 2 : 1 in humans). For the bats of the vespertilionidae family the ratio is 6 : 1 (Firbas and Welleschik, 1973). Therefore, most ganglion cells are located in the cochleate area which perceives high and medium frequencies.Interestingly, the harp seal (Pagophilus groenlandicus), which has different frequency characteristics in air and under water, shows two peaks of high neuron density, at a distance of 1-1.5 mm and 20 mm from the basal end of the cochlea (Ramprashad, 1976).

The inner ear is known to appear much earlier than the middle and outer ears. It forms of thickened ectoderm whose cells originate from the anlage of the labyrinth membraneous.

When embryos of cetaceans and pinnipeds are 20 mm long, the inner ear is differentiated into cochleate and vestibular structures. In cetaceans, the cochleate structure is much bigger than the vestibular structure, which also is characteristic of adult mammals. Like in most mammals, the vestibular structure is bigger the than the cochleate structure in pinnipeds.

In 20-30 mm long embryos, the cochleate canal looks like a tube whose base is formed by multilayered epithelium, and the top is formed by three layers of cubic epithelium cells. The differentiation of the cochlea into basal and apical turns was observed in 30-40 mm long embryos of cetaceans and pinnipeds. Species-specific features of the structure of the cochlea, such as the widening basal turn, begin to form in embryos that are 30-40

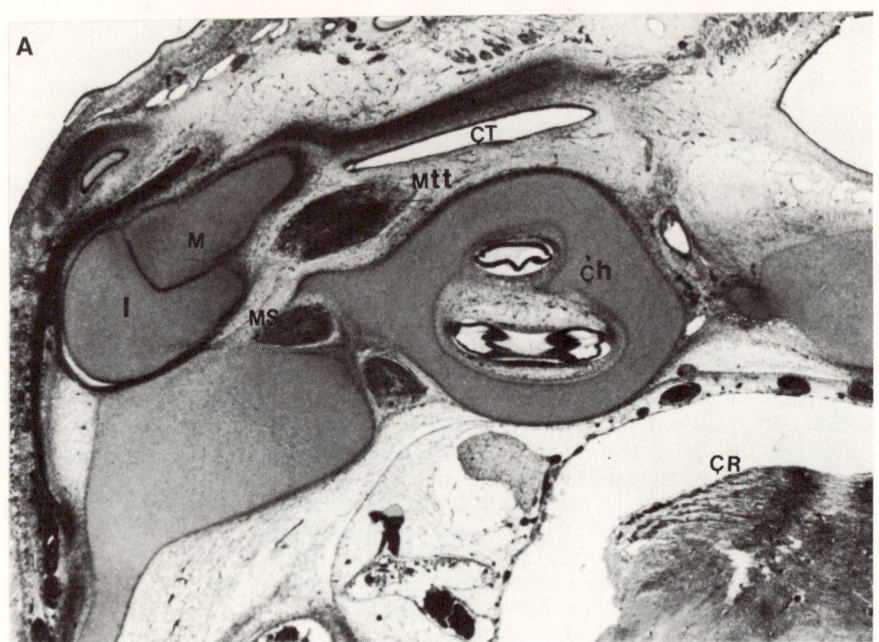
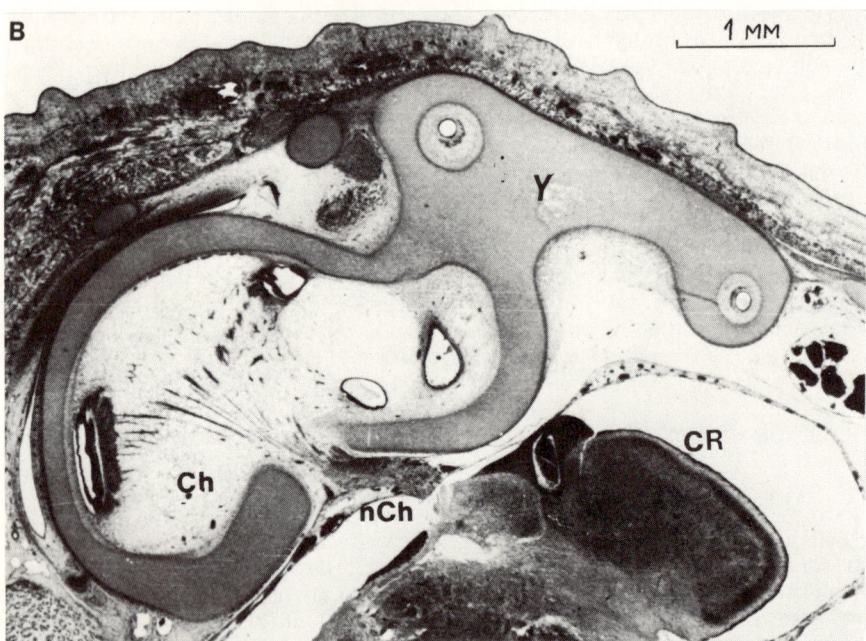

Fig. 13. The peripheral part of the auditory analyzer of a 90 mm long Erignathus barbatus in dorsal-ventral sections of the head: A- incus shown magnified in relation to malleus; B- the basic process of the cellular differentiation of the cochleate canal is completed.

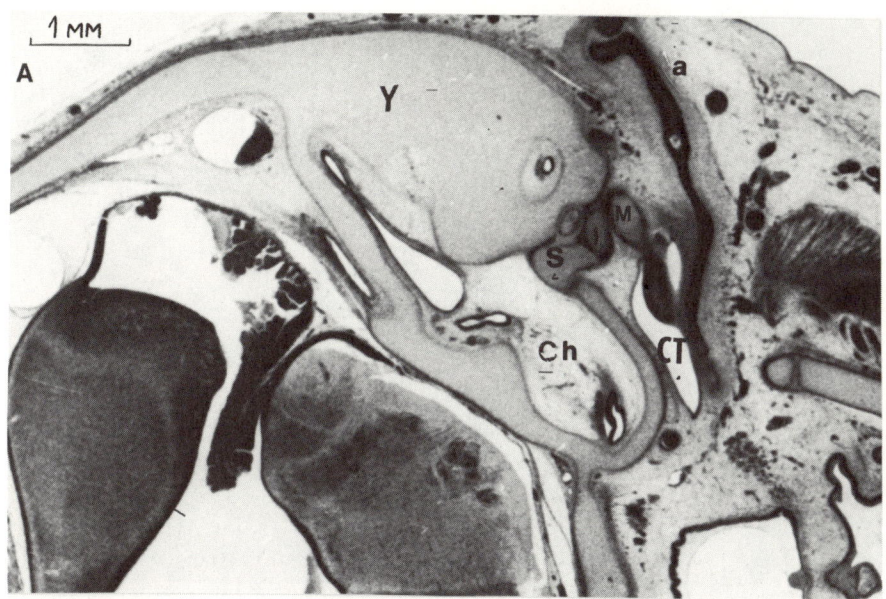

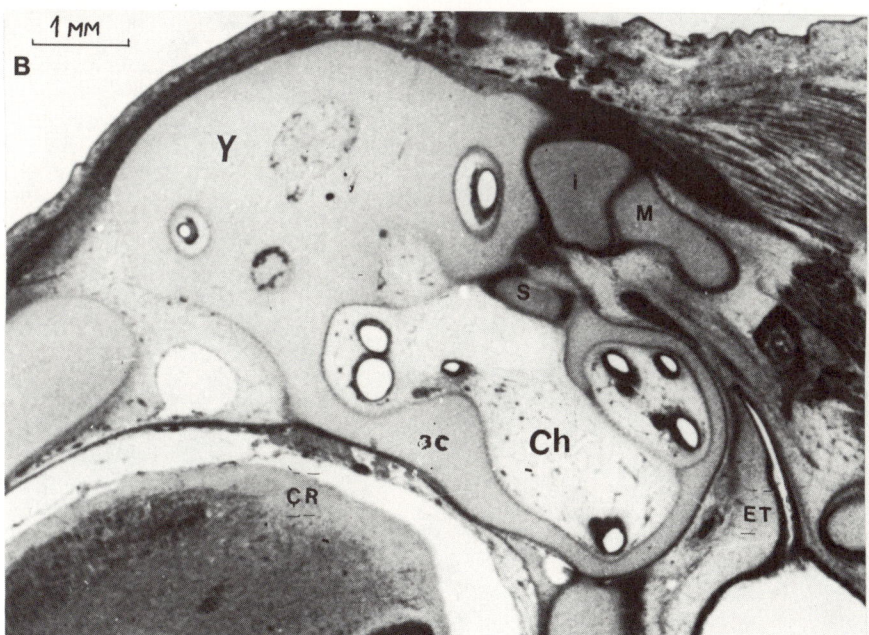

Fig. 14. Peripheral part of the auditory analyzer of a 90 mm long <u>Eumetopias ubatus</u> in dorsal-ventral sections of the head: A- the size of the vestibular structure is bigger than the size of the cochleate structure; B- in formed cochleate canal the process of cellular differentiation is continued.

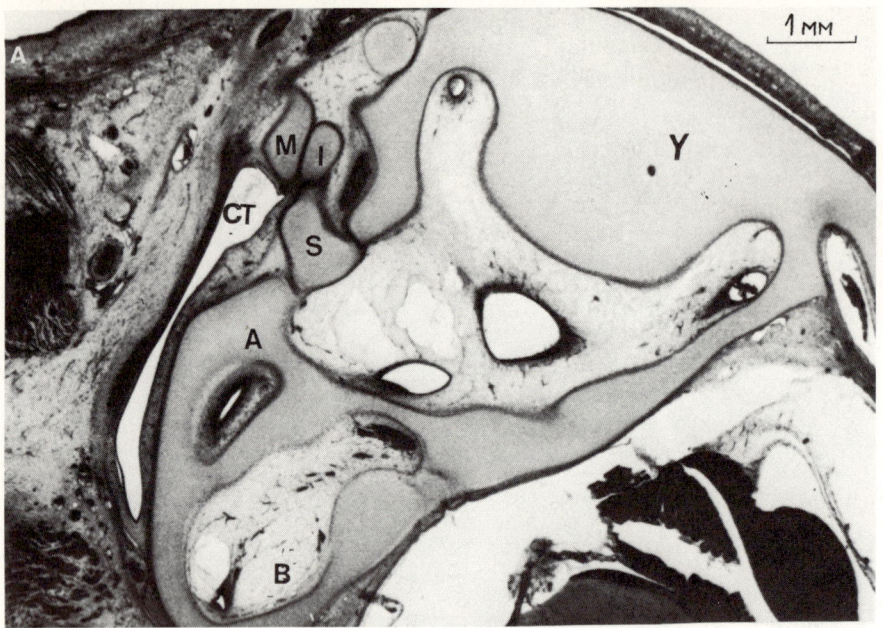

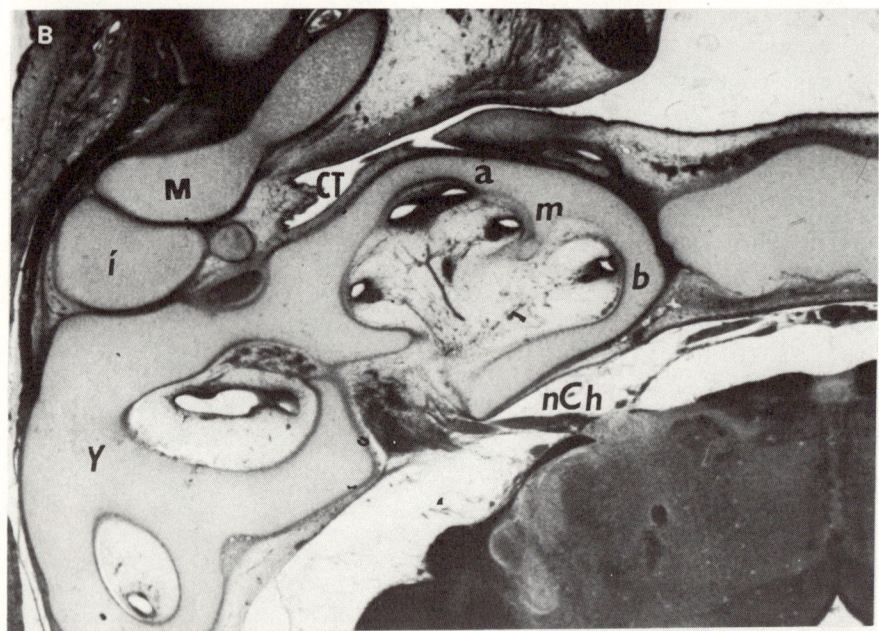

Fig. 15. The peripheral part of the auditory analyzer of a 100 mm long <u>Odobenus rosmarus</u> in dorsal-ventral sections of the head: A- end of differentiation of elements of the cochleate canal in the basal turn of the cochlea; B- differentiation of elements of the cochleate canal continues in the medial turn; the process has just started in the apical turn.

mm long. Formation of the secondary osseus spiral lamina in the cochleate canal and the development of the basal membrane were observed in 100-150 mm long dolphin embryos. Embryos that are 60-80 mm long clearly display the cochleate and vestibular branch. The vestibular structure are form somewhat earlier, i.e. in embryos that are 40 to 60 mm long.

Cellular differentiation of the organ of Corti is similar to that in terrestrial mammals. It begins after complete formation of the turns of the cochlea and elements of the middle ear, i.e. when embryos of cetaceans and pinnipeds are 60-70 mm long. The process starts with differentiation in cellular elements of the basal turn and ends much later at the apex of the cochlea. The main process of cellular differentiation, in the organ of Corti and elements of the cochleate canal of cetaceans and pinnipeds, ends when embryos reach a length of 100-150 mm (Figs.12-15).

Premature mammals are born with differentiated receptor structures, but accessory structures develop only by the 10th day of post-natal life. Mature newborn dolphin have good direction-finding abilities in the ocean, being capable of receiving and producing broad-band sounds. Hearing in young pinniped is less perfected, since pups can not locate their mother by sound; but they emit high-frequency sounds and by which their mother finds them. In this context, the development of accessory structures in dolphins goes somewhat more advanced evolutionarily than in seals.

CONCLUSION

As is known in ontogenesis, all mammals first develop the inner ear as a core structure of the peripheral auditory system. As it continues to develop, it is surrounded by additional peripheral structures of a different evolutionary age. Such additional structures as the outer and middle ear, expand the range of perceived frequencies, improve their analysis and consolidate contacts with cerebral structures, primarily motor centers.

Morphological progress of evolutionary change in the peripheral auditory system of mammals manifests itself in polymorphism and the emergence of new structures. Polymorphism is demonstrated by the diversity in the organization of components of the outer, middle and inner ear. New structures, specific only to aquatic mammals, developed according to the principle of complementarity, i.e. the old components did not disappear.

Early embryogenesis of aquatic mammals thus demonstrates original features of the development of the peripheral auditory system that are related to the aquatic habitat and perception of broad-band frequencies. Unlike pinnipeds , cetaceans do not repeat the development of these initial features of terrestrial species.Features of mammals that are an adaptive to different environments appear at the embryogenetic stage, including the earliest stages, regardless of the fact that the embryo is not affected immediately by the environment.

By comparing the formation of the auditory system of aquatic and terrestrial animals, we identified general regularities of the development of the peripheral auditory system of all representatives of this class:

1. Irrespective of the species and ecological specificity at early stages of development, most mammals demonstrate basic common features of the formation of the peripheral auditory system.

2. Species-specific features of the structural organization of the peripheral auditory system are formed in early organogenesis, depending on

the ecology of a given species and the frequency tuning of its auditory system.

3. Species-specific features that are formed in the beginning of organogenesis are preserved in late organogenesis and throughout the entire period of postnatal development.

The findings of our comparative study of embryogenesis of the peripheral auditory system show that each of the studied groups of mammals followed its own path of evolutionary transformations for this section of the auditory system, depending on a group's phylogenesis and ecological features.

REFERENCES

Agarkov, G. B., Solucha, B. V., Homenko, B. G., 1971, On the capacity of dolphins to echolocation, in: "Bionics". Kiev, 5: 32-57 (In Russian).
Bogoslovskaya, L. S. and Solntseva, G. N., 1979, The auditory system of mammals, Moscow, Nauka Publishers, 238 p. (In Russian).
Clarke, R., 1948, Hearing in Cetacea, Nature, London, 161: 979-980.
Firbas, W., and Welleschik, B., 1973, A quantitative study on the spiral ganglion of the Chiroptera. Period. biologorum, 75 (1): 67-70.
Fleischer, G., 1973, On structure and function of the middle ear in the bottle-nosed dolphin (Tursiops truncatus), Proc. 9th Ann. Conf. Biol. "Sonar and Diving Mammals", Standford Research Inst. Press., 137-179.
Fleischer, G., 1971, Uber Schwingungsmessungen am Skelett des Mittelohres von Halicore (Sirenia), Z. Saugetierk., 36: 350-36.0
Fleischer, G., 1973, Studien am Skelett des Gehororgans der Saugetiere, einschlisslich des Menschen, Saugetierk. Mitt., 21 (2-3): 131-239.
Fraser, F. C., and Purves, P. E., 1954, Hearing in Cetaceans, Bul. Brit. Mus. Natur. Hist. Zool., 2 (5): 101-114.
Fraser, F. C., and Purves, P. E., 1960, Hearing in Cetaceans. Bul. Brit. Mus. Natur. Hist. Zool., 7 (1): 1-140.
Guild, S. R., Crowe, S. J., Bunch, C. C., and Polvogt, L. M., 1931, Correlations of differences in the density of innervation of the organ of Corti with differences in the acuity of hearing, including evidence, as to the location in the human cochlea of the receptors of certain tones. Acta Otolaryngol., 15 : 269-308.
Lipatov, N. V. 1972, Functional principles of underwater hearing of mammals, Synopsis of the paper presented at the 5th All-Union Conf. on the study of aquatic mammals, Makhachkala, 2: 137-140 (In Russian).
Lipatov, N. V., and Solntseva, G. N., 1972, Some features of the biomechanics of the middle ear of dolphins, Ibid.: 140-143 (In Russian).
Lipatov, N. V., and Solntseva, G. N., 1974, Morpho-functional features of the external auditory meatus of Delphinus delphis and Tursiops truncatus, in: "Bionics", 8: 113-117 (In Russian).
Norris, K. S., 1964, " Some problems of echolocation in Cetacea", in:" Marine bioacoustics" (Ed) W. N. Tavolga N. Y.: Pergamon Press, 317-336.
Purves, P. E., 1955, The wax plug in the external auditory meatus of the Mysticeti, Discovery Repts., 27 : 293-302.
Purves, P. E., and Utrecht, W. L., 1963, The anatomy and function of the ear of the bottle-nosed dolphin, Tursiops truncatus, Amsterdam: Beaufortia, 111: 241-256.
Pye, A., 1966, The structure of the cochlea in Chiroptera, I. Microchiroptera, J. Morphol., 119: 101-109.
Ramprashad, F., Corey, S., and Ronald, K., 1971, The harp seal, Pagophilus groenlandicus (Erxleben, 1777), XIII. The gross microscopic structure of the auditory meatus. Can. J. Zool., 49(2) : 241-248.

Ramprashad, F., Corey, S., and Ronald, K., 1973, The harp seal, Pagophilus groenlandicus (Erxleben, 1777).XIV. The gross and microscopic structure of the middle ear, Canad. J. Zool., 51 (6): 589-600.

Ramprashad, F., 1975, Aquatic adaptations in the ear of the harp seal, Pagophilus groenlandicus (Erxleben, 1977), Rapp. et proc. verb. reun. Cons. intern. explor. mer., 169 : 102-111.

Ramprashad, F., Ronald, K., Geraci, J., and Smith, T. G., 1976, A comparative study of surface preparations of the organ of Corti of the seal, Pagophilus groenlandicus (Erxleben, 1777) and the ringed seal (Pusa hispida), I. Sensory cell population and density. Canad. J. Zool., 54 (1): 1-9.

Ramprashad, F., 1976, Population and density of the bipolar ganglion cells in the cochlea of the harp seal, Pagophilus groenlandicus,(Erxleben, 1777), Canad. J. Zool., 54 (11) : 1918-1926.

Reysenbach de Haan, F. W., 1957, Hearing in Whales, Acta Otolaryngol., suppl., 134 : 101-114.

Reznikov, A. E., 1972, About sound seeing in dolphins, Abstracts 23th scientific-technical conf. LIAP, Leningrad (In Russian).

Schuknecht, H. F., 1953, Techniques for study of cochlear function and pathology in experimental animals, Arch. Otolaryngol., 58: 377-397.

Simkin, G. N., 1975, "Acoustic direction-finding of mammals", in: "Bioacoustics", Moscow, Vysshaya shkola Publishers, 156-162 (In Russian).

Solntseva, G. N., 1971," Comparative anatomical and hystological features of the structure of the outer and inner ear of some dolphins", in: Research on marine mammals, Kaliningrad, 237-244 (In Russian).

Solntseva, G. N., 1973, Features of biomechanics of the middle ear of terrestrial, semi-aquatic and aquatic mammals, Reports of the 8 th All-Union Acoustic Conf.,29-32 (In Russian).

Solntseva, G. N., 1973, Morpho-functional features of the outer ear of terrestrial, semi-aquatic and aquatic mammals, Ibid.:. 25-28 (In Russian).

Solntseva, G. N., 1983, Early embryogenesis of the Peripheral Part of the Auditory Analyzer of a representative of the toothed whales, Stenella attenuata, Ontogenesis, 14 (3): 312-318 (In Russian).

Solntseva, G. N., 1985, Early Embryogenesis of the Peripheral Part of the Auditory Analyzer of Baleen whales, Balaenoptera acutorostrata, Papers of the USSR Academy of Sciences, 280 (6): 1428-1432 (In Russian).

Solntseva, G. N., 1985, Formation of the Peripheral Part of the Auditory Analyzer of a representative of true seals, Erignathus barbatus, Papers of the USSR Academy of Sciences, 285 (6): 1504-1508 (In Russian).

Solntseva, G. N., 1986, Early Embryogenesis of the Peripheral Part of the Auditory Analyzer of the walrus, Odobenus rosmarus divergens, Papers of the USSR Academy of Sciences, 288 (4): 984-988 (In Russian).

Solntseva, G. N., 1987, A Direction of evolutionary transformations of the Peripheral Part of the Auditory Analyzer of mammals of different ecology, J. General Biology, XLVIII (3): 403-410 (In Russian).

Solntseva, G. N., 1988, Formation of the Peripheral Part of the Auditory Analyzer of a representative of the true seals, Pusa hispida, Papers of the USSR Academy of Sciences, 302 (6) : 1489-1493 (In Russian).

Wever, E. G., 1949, Theory of hearing. N.Y. J. Wiley and Sons, 291-293.

Wever, E. G., McCormick, J. G., Palin, J., and Ridgway, S. H., 1971, The cochlea of the dolphin, Tursiops truncatus: hair cells and ganglion cells. Proc. Nat. Acad. Sci. USA, 68 :2908-2912.

Wever, E. G., Mc.Cormick, J. G., Palin, J., and Ridgway, S. H., 1972, Cochlear structure in the dolphin, Lagenorhynchus obliquidens, Proc. Nat. Acad. Sci. USA, 69: 657-661.

Yamada, M., 1953, Contribution to the anatomy of the organ of hearing of Whales. Sci. Repts. Whales Res. Inst., 8: 1-79.

FREQUENCY-SELECTIVITY OF THE AUDITORY SYSTEM IN THE BOTTLENOSE DOLPHIN, TURSIOPS TRUNCATUS

Alexander Supin and Vladimir Popov

Severtsov Institute of Evolutionary Morphology
and Ecology of Animals,
USSR Academy of Sciences
Leninsky prosp. 33, 117071, Moscow, USSR

INTRODUCTION

The auditory system of Cetaceans is characterized by a number of unique features. One important characteristic of the auditory system, used for fine discrimination of sound images, is the frequency selectivity of the auditory channels (filters). In the past years, attempts were made to measure the pass bands of the auditory filters (critical bands) in dolphins. These measurement were carried-out by the tone-tone and noise-tone masking (Johnson, 1968; Moore and Au, 1983). These studies were conducted using behavioral methods with specially trained animals. The complexity and time-consuming task of the behavioral method limits the number of subjects that can be measured, so accessible data are limited.

In this study, the frequency selectivity of the bottlenose dolphin's hearing was studied using the electrophysiological method. We recorded evoked potentials (EP) from the surface of the animal's head, thereby examining the perception of sound signals.

The possibility to record EP from the skull and body surface of dolphins has been shown in a number of papers (Ridgway et al., 1981; Popov and Supin, 1985; Supin and Popov, 1985). These were so-called auditory brainstem responses (ABR) with short latency and duration. However, short ABR duration limits the effective duration of stimuli, and short stimuli can not have a sufficiently narrow spectrum. This is a disadvantage when using ABR to measure frequency selectivity of hearing.

Therefore, we proposed to use a longer type of evoked potential. This EP also can be recorded from the surface of the dolphin's head (Popov and Supin, 1986; Supin and Popov, 1986a). The similarity between the EP thus obtained, and that recorded directly from the auditory cortex (Ladygina and Supin, 1970; Popov and Supin, 1976; Popov et al.,1986), implies that this EP is of cortical origin. It will be referred to as the auditory cortical response (ACR).

METHODS

The experiments were carried-out on 4 bottlenose dolphins, Tursiops truncatus. During the experiments, animal was placed in a bath with sea water and was supported by a stretcher so that the dorsal part of the head with the blowhole remained above the water surface.

Needle electrodes were used to record EP from the body surface. The electrodes were inserted 2-3 mm deep into the skin. This procedure was practically painless for the animal and required neither anaesthesia nor curarization. The EP were amplified in the 0.3-3000 Hz frequency band and 300-3000 realizations were averaged.

Acoustic signals were emitted into the water through a piezoceramic transducer. All the characteristics of signals were monitored by a hydrophone with a pass band of 200 kHz, located near the animal's head.

In the experiments with tonal masking, tone-bursts 25 t0 100 kHz were used as stimuli. These bursts had a linear rise/fall with the following characteristics, 1 ms rise-time, 1 ms plateau, and 1 ms fall-time. The intensity of the stimuli is indicated as dB re 1 mP of plateau effective sound pressure (1 mPa is a convenient reference level for designation of sounds intensity in water, because this level is close to the auditory thresholds of aquatic animals). Continuous tones 20 to 160 kHz were used as masking signals.

Another method of measuring frequency selectivity involved the stimulation by so-called "comb-filtered noise" with a rippled spectrum. In these experiments a continuous noise of 60 dB intensity served as signal. The noise spectrum had a frequency band of 1 octave with 1/2 octave steps: 8-16, 11-22, 16-32 kHz etc., except for the last band of 128-180 kHz (1/2 octave). Within the given band the spectrum had a rippled structure: it contained maxima and minima, alternating at equal frequency intervals (Fig.1). Such a spectrum may be characterized by the lower and upper boundaries of the frequency band, as well as by ripple density. The ripple density can be expressed in absolute and relative units. Absolute density is the number of ripples per a certain frequency interval, for instance, per 1 kHz: $D = 1/dF$, where dF is the frequency interval between the consecutive ripples. The relative density is $R = F/dF$, where F is the frequency at which measurements are performed (see Fig.1).

With this comb-filtered noise, the reversal of a spectrum ripple phase served as a stimulus. At the moment of stimulation, the positions of spectral maxima and minima at the frequency scale were interchanged (compare the upper and lower parts of Fig.1). Such a change can evoke an EP if the auditory system is capable of recognizing the spectral structure (Fig.1, left part). If the ripple density is too high and the ripples are fused, the change can not be recognized and a response is not evoked (right part of Fig.1).

RESULTS

Form and Conditions of Recording ACR

ACR recorded from the dorsal surface of the bottlenose dolphin's head had a latency of about 10 ms and a duration of about 30 ms. The maximum response amplitude was recorded from points on the body midline, about 20 cm caudally from the blowhole (Fig.2). When the recording point was displaced from this focus point, the ACR amplitude fell; recordings from rostral sites showed reversal of the response.

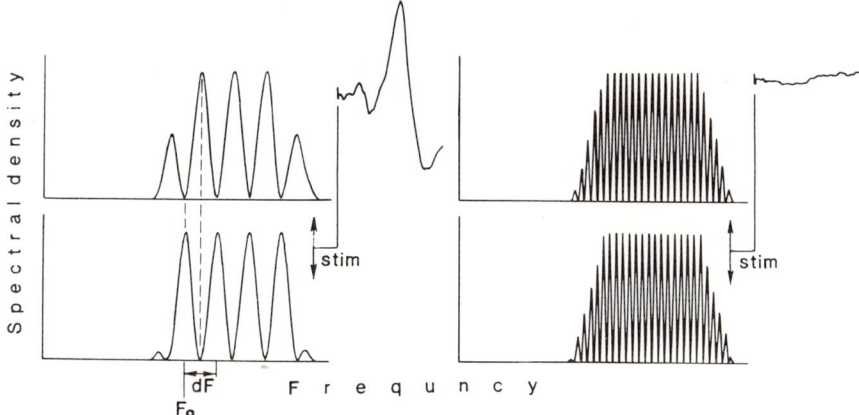

Fig.1. Frequency spectra of signals used for measuring FRP (scheme). The upper and the lower spectra differ in their ripple phase; the vertical interrupted lines show mutual position on the frequency scale of maxima and minima of two spectra. On the left and on the right are spectra with different ripple densities.

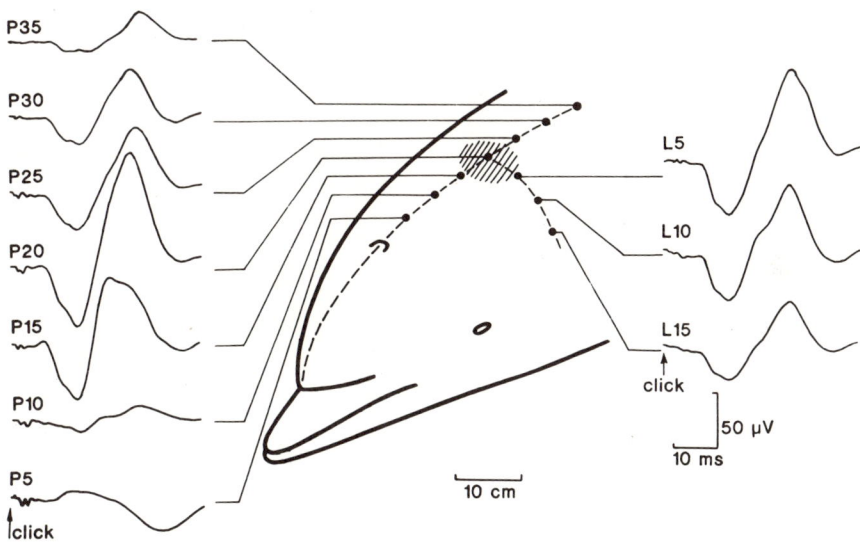

Fig.2. ACR recording from the body surface. P5-P35 - recording points on the midline, 5 to 35 cm posteriorly from the blowhole, L5-L15 - points 5 to 15 cm laterally from the midline at the P20 level. The area of maximal ACR amplitudes is shaded. Stimulus - a click of 60 dB intensity, repetition rate is 10/s, each oscillogram is the averaging of 1000 responses, negativity of the active electrode upward.

ACR recorded in the region of its greatest amplitude consisted of a positive and a negative waves. Some additional ripples can be seen on the both waves. The response amplitude could amount to several dozens of microvolts.

Studies of Tonal Masking

In these experiments, the ACR thresholds without masking initially were determined. For this purpose the tone-burst stimulus intensity was changed in 2-10 dB steps and the ACR threshold was determined with an accuracy of 2 dB. Then the masking tone of constant intensity (60 dB) was switched on, and the ACR threshold was measured again. The test frequency was kept constant, and the masking frequency was varied; the dependence of the threshold on the masker frequency, i.e. masking curve, was determined. Similar measurements were repeated for various test frequencies. This gave a set of masking curves, covering a considerable part of the frequency range of the dolphin's hearing.

Figure 3 illustrates the ACR changes under the tonal masking. Without masking (A) the stimulus of 100 kHz, 20 dB evoked a response of considerable amplitude, the stimulus of 10 dB evoked a small, near-threshold, response, and the stimulus of 0 dB evoked no response. Thus, the ACR threshold was between 0 and 10 dB. More precise measurement (omitted in the Figure) yielded a threshold of 4 dB. At a background of the masking frequency of 100 kHz the threshold was much higher: between 30 and 40 dB (B); according to a more detailed measurement, it was 36 dB. If the masking frequency was considerably differed from that of the test tone (70 kHz), the ACR threshold was close to that without masking: nearly 10 dB (C).

The given example shows ACR at two masking frequencies. Each full masking curve was obtained by measuring thresholds at 10-20 masking frequencies. A set of masking curves thus obtained is shown in Fig.4. This set covers a frequency range of 20 to 160 kHz.

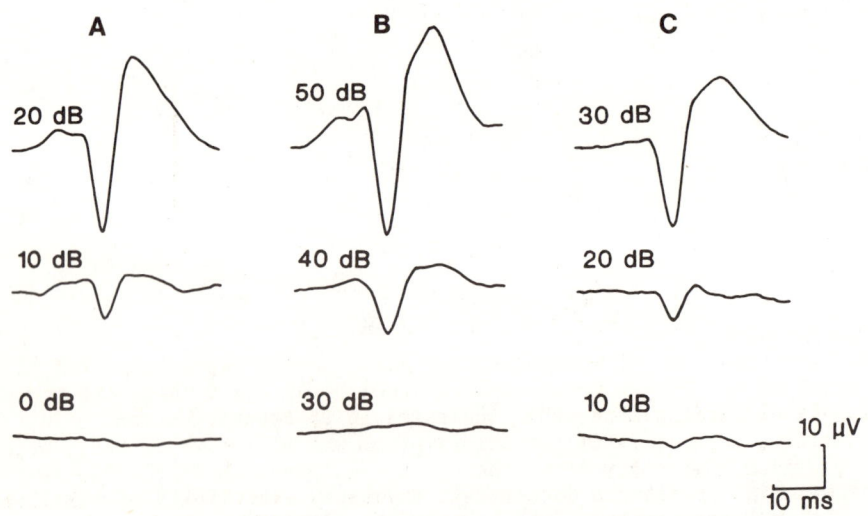

Fig.3. ACR to a test-signal of 100 kHz without masking (A) and with masking of 100 kHz at 60 dB (B) and of 70 kHz at 60 dB (C). Intensity of test-signal is indicated near the oscillograms.

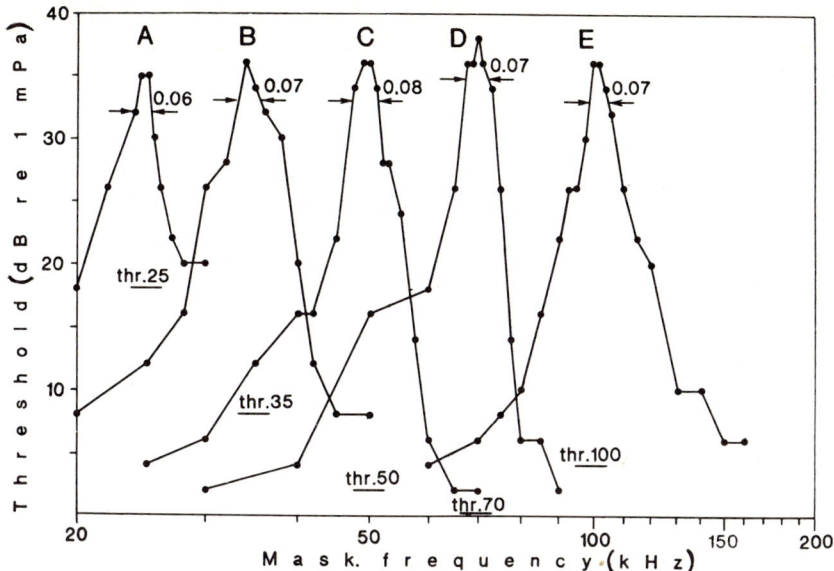

Fig.4. Masking curves for test-signals of 25 kHz (A), 35 kHz (B), 50 kHz (C), 70 kHz (D), and 100 kHz (E). Arrows indicate the width of the curves at a level of -3 dB below the maximum. Thr.25, thr.35 ... thr.100 are thresholds of response to test-signals of 25, 35 ... 100 kHz without masker.

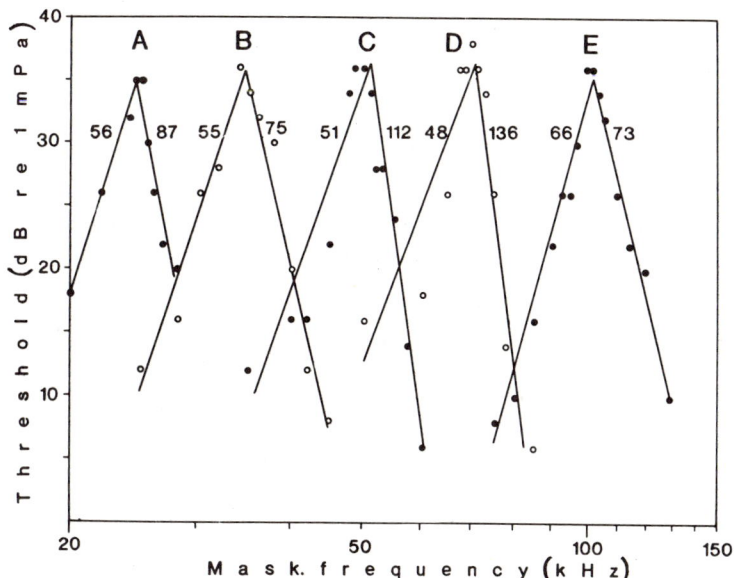

Fig.5. Approximation of masking curves with linear regression. A-E plots correspond to curves under the same indices in Fig.4. Near the regression lines their steepness is indicated in dB/octave.

The curves presented demonstrate frequency selectivity of tonal masking. The masking maximum (the highest thresholds) in all the curves was observed when the frequencies of the test and the masking tones coincided. The response thresholds in this case varied from 35 to 38 dB; that is from -25 to -22 dB relative to the masking level (60 dB). By increasing the difference between the masking frequency and the test frequencies, the masking was attenuated (thresholds decreased). When the difference was large enough, the thresholds approached those without a masker (these thresholds were different for different test frequencies as is shown in Fig.4).

The relative width of masking curves did not differ significantly for various test frequencies. This can be seen in Fig.4 with a logarithmic scale.

In all curves, the low-frequency branches (masker frequency is lower than the test one) were less steep than those of the high-frequency branches. The steepness of the curves could be estimated by approximating the lower and high-frequency branches with straight regression lines (Fig.5). The regression lines obtained had a slope from 48 to 66 db/octave (mean: 55 dB/oct) for low frequency branches and from 73 to 136 dB/octave (mean: 97 dB/oct) for high frequency branches.

Although the width of the masking curves varied to some extent, there was no significant systematic relationship between the test frequency and the width of the curve. The width of all the curves was from 0.06 to 0.08 at the -3 dB level from the top (see Fig.4).

Measurements of Frequency Resolving Power (FRP)

The experiments with stimulation by comb-filtered noise were based on the assumption that the reversal of a spectrum ripple phase can evoke ACR only when the auditory system is capable of recognizing the ripple structure of the spectrum (see Fig.1). Thus, the maximum ripple density recognizable could be determined by ACR recording. This maximum ripple density is a measure of the frequency resolving power (FRP) of hearing. The FRP depends on the acuity of the auditory frequency filters (Wilson and Evans, 1971; Popov and Supin, 1984; Supin and Popov, 1986).

ACR to the stimulation by the reversal of spectrum ripple phase are shown in Fig.6. The ACR amplitude was dependent on the ripple density. It was maximal when the spectral structure was rather coarse, i.e. the ripple density was low (relative density 2-3, absolute one 0.04-0.07/kHz); with increased ripple density the response amplitude fell, and when the ripple density was sufficiently high (relative density 20, absolute density 0.45/kHz) the response disappeared. Therefore, the FRP in this case was about 20 (relative FRP) or 0.45/kHz (absolute FRP).

Measurements in frequency bands from 8-16 kHz to 128-180 kHz were carried-out in a similar way. In Fig.7, the dependence of measured FRP on the lower boundary of the noise spectrum (see F_o in Fig.1) is shown in two ways: as absolute recognized ripple density $D=1/dF$ (A) and as relative density $R=F_o/dF$ (B). The curves contain two parts. At frequencies up to 64 kHz the absolute FRP was virtually constant and amounted to 0.4-0.5/kHz (Fig.7A). Accordingly, the relative FRP grew proportionally to frequency (Fig.7B). At frequencies from 64 kHz and higher the relative FRP was constant, being 25-30 (Fig.7B), and the absolute FRP fell with frequency.

DISCUSSION

According to the generally accepted assumptions, the tonal masking curves reflect the characteristics of the auditory frequency filters. Thus, the data presented here suggest that the relative pass band of the filters in the bottlenose dolphin's auditory system is 0.06-0.08 at the -3 dB level.

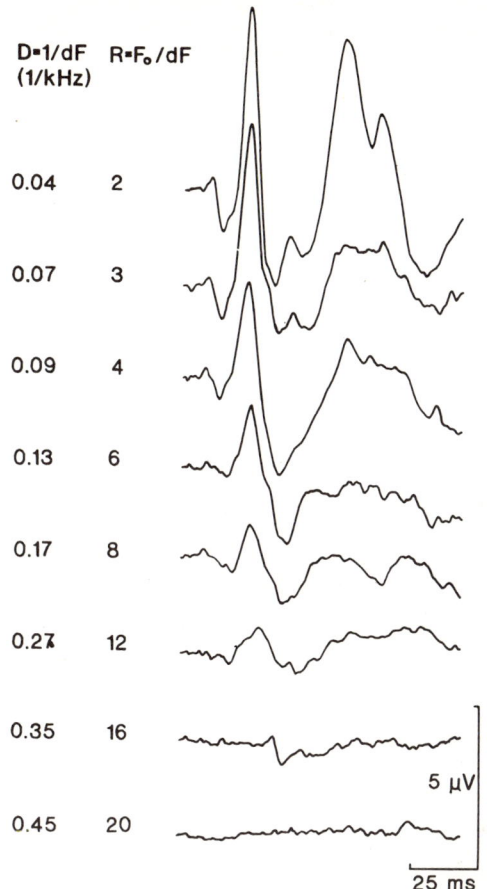

Fig.6. ACR to reversal of spectral ripples. Frequency band of comb-filtered noise is 45-90 kHz, intensity is 60 dB. Absolute ripple density (D=1/dF, 1/kHz) and relative density (R=F_o/dF, taking F_o=45 kHz) are indicated near the oscillograms.

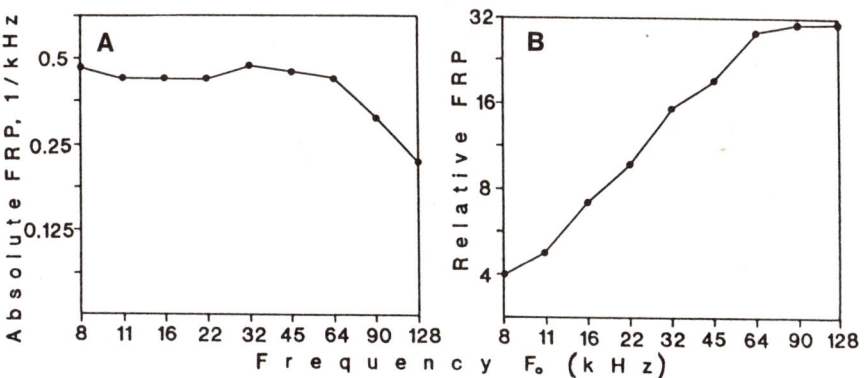

Fig.7. FRP of dolphin's hearing. F_o is the low boundary of the noise spectrum (see Fig.1). A - absolute FRP, B - relative FRP.

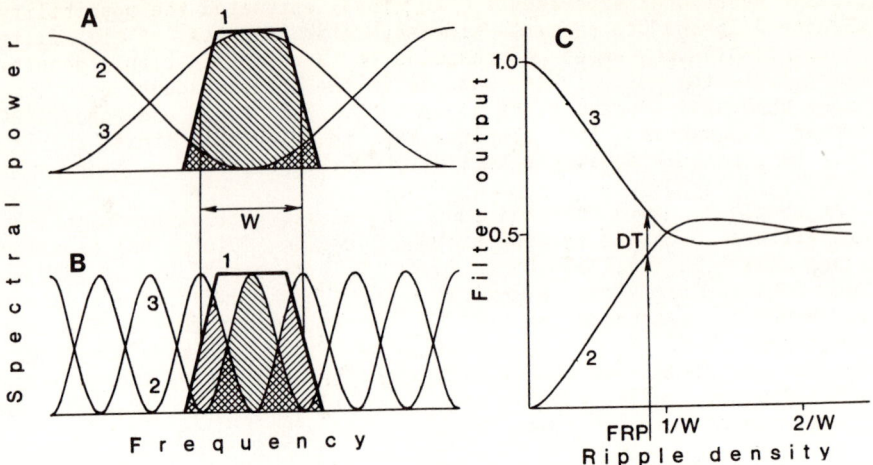

Fig.8. Relationship between pass band of frequency filter and frequency resolving power (FRP). A - characteristic of filter (1) and spectra with minimum (2) and maximum (3) of spectral power in the pass band of the filter. B - the same at higher ripple density. C - dependence of filter outputs on ripple density at the two input spectra (2 and 3). W - width of filter pass band, DT - differential threshold, FRP - frequency resolving power.

Determination of the frequency selectivity by FRP measurements requires some explanation. Let us assume that the filter has the characteristic 1 (Fig.8 A), and that the input signal has a rippled spectrum 2, which is then replaced by a signal with spectrum 3. The output signals will be proportional to the hatched areas under curves 2 and 3. In this case, they differ to a great extent, and changing the signals results in a large response of the filter. At greater ripple density (Fig.8B) the difference between output signals is smaller, i.e. the filter response is weaker.

Figure 8 C shows the dependence of the filter outputs on the ripple density in two cases: when the minimum (2) and the maximum (3) of the spectral power coincides with the center of the filter band. The difference between the two outputs is highest at zero ripple density, and the outputs are equal if the ripple density is $D=1/W$; where W is the pass band of the filter at the 0.5 (i.e. -3 dB) level. Consequently, the change of the signal spectrum from 2 to 3 becomes unrecognizable if the ripple density approaches $1/W$, and the filter response (the difference between curves 2 and 3 in Fig.8C) reaches a certain differential threshold (DT).

Hence, the minimum interval dF between the spectral ripples at which a response is possible is close to the width of a filter pass band, W. Then FRP values 25-30 (at frequencies of 64 kHz and higher) give a pass band of the bottlenose dolphin's auditory filter of 0.03-0.04.

At frequencies below 64 kHz, FRP values were lower. However, these values do not necessarily reflect the actual FRP. It has to be taken into account that the fine spectral structure can be recognized only by analyzing a sufficiently long signal; and for finer spectral structure analysis longer signals are needed. However, only the initial part of a stimulus (up to several ms) is effective for evoked potentials. This may restrict the absolute FRP at low frequencies to a level of 0.4-0.5/kHz, as seen in

Fig.7. In behavioral experiments, Au (1983) estimated the possibility of bottlenose dolphins to percept time-separation pitches of comb-filtered noise up to 0.5 ms, which corresponds to a spectral ripple density of 0.5/kHz, i.e. the value is the same as in the present study. However, using more prolonged EP gave a FRP of up to 1.5/kHz (Supin and Popov,1986b). At higher frequencies, the absolute FRP is below this limit, and it may reflect better the frequency selectivity of the dolphin's auditory system.

It should be noted that the pass band of the bottlenose dolphin's auditory filters obtained by the masking technique (0.06-0.08) is somewhat narrower than the critical bands (over 0.1) obtained by the behavioral method (Johnson, 1968; Moore and Au, 1983). This difference is not very large. However, measuring the FRP by comb-filtered noise yielded an even narrower pass band: 0.03-0.04, that is two times less than the value obtained in our study with the masking technique. So far we can not explain this discrepancy, but we suppose that it indicates the possible sharpening of the frequency selectivity to complex sound signals.

REFERENCE

Au W. W. L., 1988, The perception of time-separation pitch by dolphins. J. Acoust. Soc. Am., 83, Suppl.1, S51.
Johnson C. S., 1968, Masked tonal thresholds in the bottlenosed porpoise. J. Acoust. Soc. Amer., 44: 965-967.
Ladygina T. F. and Supin A. Ya, 1970,. Acoustic projection in the dolphin's cerebral cortex (in Russ.). Fiziol. Zh. SSSR, 56: 1554-1560.
Moore O. W. B. and Au W. W. L., 1983, Critical ratio and bandwidth of the Atlantic bottlenose dolphin (Tursiops truncatus). J. Acoust. Soc. Amer., 74: Suppl.1, S73.
Popov V. V., Ladygina T. F. and Supin A. Ya., 1986, Evoked potentials of the auditory cortex of the porpoise, Phocoena phocoena. J. comp. Physiol., 158: 705-711.
Popov V. V. and Supin A. Ya., 1976, Determination of the hearing characteristics in dolphin by evoked potentials method (in Russ.), Fiziol. Zh. SSSR, 62: 550-558.
Popov V. V. and Supin A. Ya., 1984, Measurements of frequency resolving power of human's hearing (in Russ.). Dokl. Acad. Nauk SSSR, 278: 1012-1016.
Popov V. V. and Supin A. Ya., 1985, Determination of characteristics of the dolphin's hearing with the brain stem evoked potentials (in Russ.). Dokl. Acad. Nauk SSSR, 283: 496-499.
Popov V. V. and Supin A. Ya., 1986, Evoked potentials of the dolphin's cerebral cortex recorded from the body surface (in Russ.). Dokl. Acad. Nauk SSSR, 288: 756-759.
Ridgway S. H., Bullock T. H., Carder D. A., Seeley R. L., Woods D. and Galambos R., 1981, Auditory brainstem response in dolphin. Proc. Nat. Acad. Sci. USA, 78: 1943-1947.
Supin A. Ya. and Popov V. V., 1985, Recovery cycles of the dolphin's brain stem responses to paired acoustic stimuli (in Russ.). Dokl. Acad. Nauk SSSR, 283: 740-743.
Supin A. Ya. and Popov V. V., 1986a, Tonal masking curves in the bottlenose dolphin (in Russ.). Dokl. Acad. Nauk SSSR, 289: 242-246.
Supin A. Ya. and Popov V. V., 1986b, Determination of frequency resolving power of dolphin's hearing by evoked potentials of the cerebral cortex, in: V. E. Sokolov (ed.), "Electrophysiology of Sensory System of Aquatic Mammals" (in Russ.), Nauka, Moscow, 106-129.
Wilson J. P. and Evans E. F., 1971, Grating acuity of the ear: psychophysical and neurophysiological measures of frequency resolving power, in: "7th Int. Congr. on Acoustics", Budapest, 397.

MASKED HEARING ABILITIES IN A FALSE KILLER WHALE
(PSEUDORCA CRASSIDENS)

Jeanette A. Thomas, Jeffrey L. Pawloski[1], and
Whitlow W. L. Au[2]

Western Illinois University, [1]Scientific
Applications International Corporation, [2]Naval
Ocean Systems Center, Kailua, Hawaii, USA

INTRODUCTION

Underwater audiograms are available for only a few odontocete species: Phocoena phocoena (Andersen, 1970), Inia geoffrensis (Jacobs and Hall, 1972), Tursiops truncatus (Johnson, 1967), Delphinapterus leucas (White et al., 1978), Orcinus orca (Hall and Johnson, 1971), and Pseudorca crassidens (Thomas et al., 1988). All odontocetes studied have the typical U-shaped mammalian hearing curve and hear over a wide range of frequencies (up to 120 kHz in belugas and up to 140 kHz in bottlenose dolphins). Low-frequency hearing among these species is comparable, but the high frequency cut-off is species-specific.

Little is known about the ability of odontocetes to hear in a noisy environment. Fletcher (1940) first developed the critical ratio as a measure of the width of the human auditory filter. The critical ratio is the ratio of signal power to noise spectrum level at masked threshold. A low critical ratio suggests a good ability to detect low amplitude signals in a noisy environment. Johnson (1968) reports masked hearing thresholds and critical ratios for 15 frequencies between 5 and 100 kHz in the bottlenose dolphin (Tursiops truncatus). Moore and Au (1982; 1983) also measured critical ratios for the bottlenose dolphin. These studies show that for comparable frequencies (5 to 30 kHz) the critical ratios of Tursiops are slightly less than in phocid seals (Moore and Schusterman, 1987), cats (Watson, 1963), chinchillas (Seaton and Trachiotis, 1975), rats (Gourevitch, 1965), and humans (Scharf, 1970). A recent study by Johnson et al. (1989) reports masked hearing thresholds and critical ratios for the beluga (Delphinapterus leucas). In that study, Johnson et al. (1989) report that belugas have a slightly smaller critical ratio than measured in bottlenose dolphins.

Thomas et al. (1988) measured the underwater hearing sensitivity of a false killer whale, Pseudorca crassidens in a quiet pool. The maximum sensitivity was from 16 to 64 kHz measured underwater masked hearing thresholds of a false killer whale (<u>Pseudorca crassidens</u>) with three levels of broadband white noise and compared the critical ratios with those from bottlenose dolphins and belugas.

METHODS

<u>Subject</u>

The subject was a female false killer whale (pc738f) who was about 5 years old. Because the whale was young and had not taken ototoxic medications, we believe that her hearing was normal. Tests were conducted in a 9 X 12 m area of a floating pen complex (Figure 1) in Kaneohe Bay, Hawaii. The whale trained for the task over a 5-month period and received a mixed diet of herring, mackerel, and smelt.

<u>Apparatus</u>

Test signals were generated using a Qua Tech, Inc. WSB-10 board installed in a Compaq 286 Portable III personal computer. A control box gated the signal with a 160-ms rise/fall time and controlled the duration of the signal and noise. Using this control box, the researcher controlled the signal level using a 1-dB step attenuator and initiated the noise/signal onset. The masking level was controlled from another box with noise-generating circuitry. Because we collected data over a broad frequency range, we used two different projectors; an F30 transducer with a linear response to 110 kHz and a TR-25C Massa transducer with a linear response to 25 kHz. Both transducers were equalized by a two-pole, low-pass filter to provide a relatively flat response over the frequencies of interest. The output of the noise and test signal were monitored on two channels of a Tektronix 2245 oscilloscope.

We measured the received level of the test signals and masking noise at the whale's station and found some fluctuation in projected signal levels, especially below 16 kHz. These fluctuations probably were caused by surface-reflections that caused the signal level to change in an unpredictable manner, depending on the smoothness of the water surface. As described by Thomas et al. (1988), we used an aluminum plate with neoprene suspended between the whale and the projector (Figure 1) to diminish surface reflections. This baffle reduced fluctuations in the received levels to less than 3 dB, regardless of the frequency. When fluctuations were present, the signal level fluctuated several times during a signal presentation so that the whale received a maximum projected amplitude several times during the projection.

We tested the whale's hearing threshold at three levels of broadband, white noise (65, 75, and 85 dB re 1 $\mu Pa/Hz^{1/2}$). All three masking noise levels were well above the noise from snapping shrimp in Kaneohe Bay (Figure 2). We obtained masked thresholds for pure-tone signals at 8, 16, 20, 24, 32, 64, 85, 105, 110, and 115 kHz. From these threshold values, we calculated critical ratios for each frequency at each of the three noise levels.

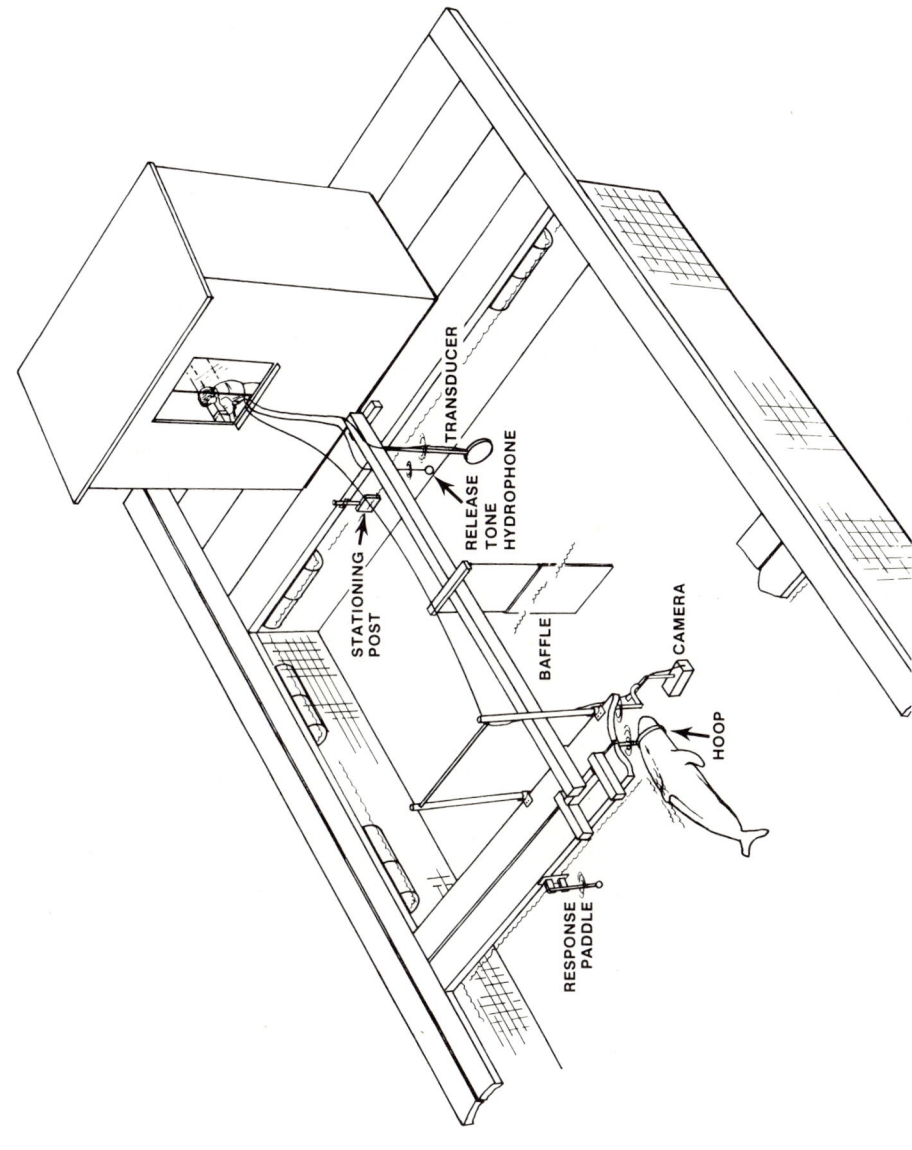

Figure 1. Floating pen complex and equipment in Kaneohe Bay, Hawaii.

Procedures

The whale reported the detection of signals using a go/no-go response paradigm. To initiate a trial, the whale rested her rostrum on a stationing post that faced the trainer (Figure 1). The trainer gave a 3 kHz "go-to station" tone and the whale swam to an underwater hoop station centered 1 m below the surface. An underwater video camera mounted to the right of the hoop allowed the trainer to watch the whale's position from a monitor inside the hut. When the whale was stationed properly in the hoop, the trainer started a 13-second projection of continuous, broadband noise. In half of the trials, only noise was projected. For a correct response, the whale remained at the station (no-go) until the trainer gave a 7 kHz "release" tone. During the other trials, a 2-second, pure-tone was projected two seconds into the noise cycle. If the whale heard the signal, the correct response was to back from the hoop (go) and touch the response paddle mounted to the left of the hoop.

The computer prompted the researcher for the proper equipment setup before a session and cataloged data during the session. The first ten trials of each session were designated as "warm-up" trials (five signal plus noise and five noise only trials). The signal level of the first warm-up was set about 20 dB above threshold and progressively made 2 dB quieter than the previous signal trial. Warm-ups were repeated if the whale responded incorrectly to more than 2 of 10 trials.

During a session, the computer selected the trial type (signal plus noise or noise only) based on a modified Gellerman random series table (Gellerman, 1933). Test signals were presented in an up/down staircase method (Robinson and Watson, 1973). The researcher attenuated the signal in 2-dB steps until the whale failed to respond to a signal plus noise trial (miss). After a miss, the researcher increased the signal in 1-dB steps for each signal plus noise trial until the whale again detected the signal (hit). The transition from miss to hit or hit to miss was called a reversal. Improper responses to noise only trials (false alarms) did not alter the attenuator settings. During the session, the computer displayed the trial number, attenuation level, signal type, whale's response, reversal number, threshold per reversal, average threshold, and number of false alarms. Eight reversals completed a session.

To verify the quality of data, ten "cool-off" trials (five signal plus noise and five noise only trials) were conducted at the end of each session. During a cool-off, the signal was made progressively 2 dB higher than the previous signal trial. A session was disregarded if the whale responded incorrectly to more than 2 of the 10 cool-off trials.

The computer averaged the threshold values at the eight reversals to calculate a mean threshold for the session. For each frequency, sessions were repeated until 3 consecutive sessions yielded mean thresholds which were $\pm$ 3 dB. The average of these three mean thresholds was used to calculate the critical ratio for a given frequency. Critical ratios were calculated for each of the 3 masking levels (65, 75, and

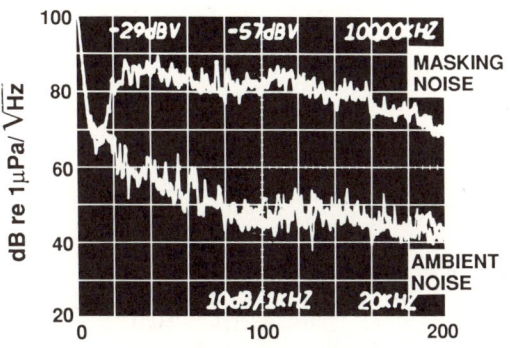

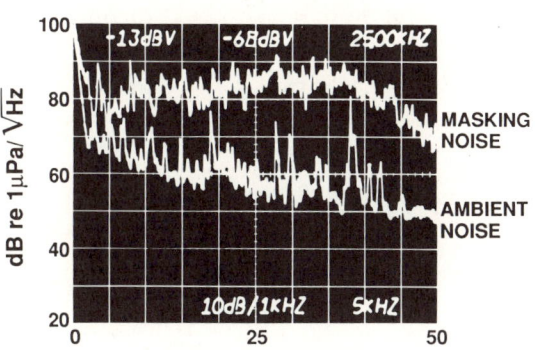

Figure 2. Spectrum of ambient noise in Kaneohe Bay compared with masking noise. Spectrum of test signals against masking noise. Measurements were taken with the F30 projector.

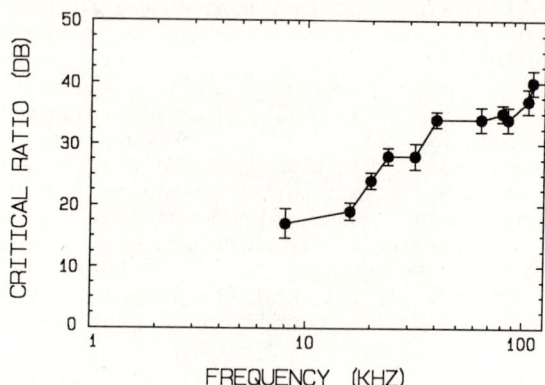

Figure 3. Mean critical ratios (± st. dev.) averaged over three noise levels for a given frequency.

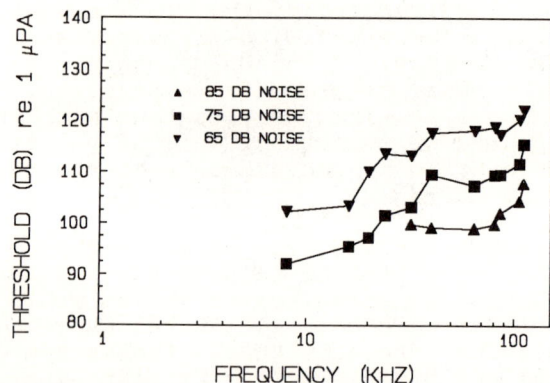

Figure 4. Average hearing thresholds based on the mean 24 reversals at each of the three different masking levels.

85 dB) before changing to another frequency. Measurements were taken with the F30 transducer first (32 kHz and above) and then with the TR-25C transducer (32kHz and below).

RESULTS AND DISCUSSION

Masked hearing thresholds for each noise level appear in Table 1 and critical ratios in Figure 3. The whale behaved reliably, repeating tests on successive days. Only one session was eliminated because the whale failed to meet the cool-off criteria. The false alarm rate increased slightly at lower frequencies (Table 1). Critical ratios ranged from 17 to 42 dB (Figure 3) and increased as a function of frequency (Table 1). Critical ratio measurements at 32 kHz ranged from 25 to 33 kHz. We do not know why there was some discrepancy between critical ratios measured with the two transducers. Because the signal-to-noise ratio is the important measurement, critical ratios should be relatively independent of transducer characteristics. The thresholds obtained at the three masking levels (Figure 4) produced parallel sensitivity curves and the critical ratios were in good agreement between masking levels (Table 1).

Using a second order polynomial to fit our data, we compared masked hearing abilities for the false killer whale, bottlenose dolphin (Johnson, 1968; Moore and Au, 1982; 1983), and beluga (Johnson et al., 1989). The species-specific differences (Figure 5) were slight. Tursiops consistently had the highest critical ratios, followed by Delphinapterus. Pseudorca invariably had the lowest critical ratios. We do not believe that Pseudorca had a lower performance than other species tested, but could not address this topic because our data were collected with a staircase method. The critical ratio for the Pseudorca at and above 32 kHz increased with frequency at a similar rate as the critical ratio for Tursiops. At lower frequencies the critical ratio for the false killer whale decreased with frequency at a much higher rate than for both Tursiops and Delphinapterus. The peculiar variation at the lower frequencies is puzzling. The whale's behavior during the low frequency testing with the TR25C transducer appeared to be normal, and the variations in the data were consistent with measurements at the higher frequencies (<32 kHz).

The high ambient noise from snapping shrimp in Kaneohe Bay prevented us from testing below 8 kHz. We do not know if the critical ratios in false killer whales would show a similar curvilinear decrease below 8 kHz as in belugas and bottlenose dolphins. Because these studies are each based on a single animal and had slightly different experimental designs, we do not know if these are significant species differences. With this low number of test subjects the differences could simply reflect individual variations.

False killer whales, belugas, bottlenose dolphins have slightly lower critical ratios than phocid seals (Moore and Schusterman, 1987), chinchillas (Seaton and Trachiotis, 1975), cats (Watson, 1963), rats (Gourevitch, 1965), and humans (Scharf, 1970). Because their critical ratios are smaller, Odontocetes are able to detect smaller amplitude signals in noise than most other mammals. It is tempting to

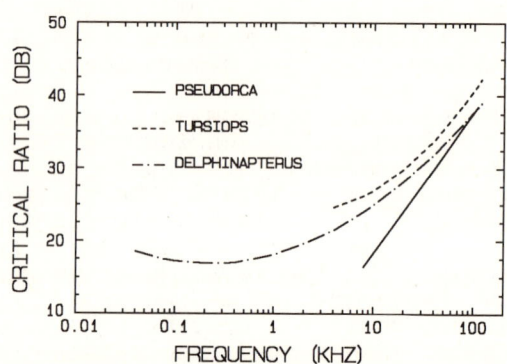

Figure 5. Second order polynomial fits of critical ratios by frequency for a false killer whale (Pseudocra crassidens), a bottlenose dolphin (Tursiops truncatus), and a beluga whale (Delphinatperus leucas). Data taken from Johnson (1967), Moore and Au (1982; 1983), and Johnson et al. (1989)

conclude that the slightly lower critical ratios of odontocetes may enhance acoustic perception of low level signals in aquatic environments where vision may be severely limited.

ACKNOWLEDGMENTS

We thank Rick Kahikina and his crew of NOSC employees for construction of the pen facility. Mark Stoermer developed the staircase program for the personal computer. We acknowledge Martha Guidry for her fine drawing of the experimental setup.

REFERENCES

Andersen, S. ,1970, Auditory sensitivity of the Harbour Porpoise, Phocoena, in: "Investigations on Cetacea", G. Pilleri, ed. University of Berne, Switzerland, Vol. 2, pp. 255-258.
Fletcher, H. , 1940, Auditory Patterns, Review of Modern Physics, 12: 47-65.
Gellermann, L. W. , 1933, Chance orders of alternating stimuli in visual discrimination experiments, Journal of Genetic Psychology 42: 206-208.
Gourevitch, G., 1965, Auditory masking in the rat, J. Acoust. Soc. Am. 37: 439-443.
Hall, J. and Johnson, C. S., 1971, Auditory thresholds of a killer whale, J. Acoust. Soc. Am. 51: 515-517.
Jacobs, D. and Hall, J., 1972, Auditory thresholds of a fresh water dolphin, Inia geofrensis, J. Acoust. Soc. Am. 51: 530-533.
Johnson,C. S., 1967, Sound detection thresholds in marine mammals, in:"Marine Bio-acoustics" Tavolga, ed. Pergamon Press, New York, Vol. 2, pp. 247-260.
Johnson, C. S. , 1968, Masked tonal thresholds in the bottlenosed porpoise, J. Acoust. Soc. Am. 44: 965-967.
Johnson, C. S., McManus, M. W. and Skaar, D. , 1989, Masked tonal threshold in the belukha whale, J. Acoust. Soc. Am.
Moore, P. W. B. and Au, W. W. L., 1982, Masked pure-tone thresholds of the bottlenosed dolphin (Tursiops truncatus) at extended frequencies, J. Acoust. Soc. Am. 72: Suppl. 1, S42.
Moore, P. W. B. and Au, W. W. L. , 1983, Critical ratio and bandwidth of the Atlantic bottlenose dolphin (Tursiops truncatus), J. Acoust. Soc. Am. 74: Suppl. 1, S73.
Moore, P. W. B. and Schusterman, R. J., 1987, Audiometric assessment of northern fur seals, Callorhinus ursinus, Marine Mammal Science. 3(1): 31-53.
Robinson, D. and Watson, C., 19973, Psychophysical methods in modern Psychoacoustics, in: "Foundations of Modern Auditory Theory" J. V. Tobias, ed. Academic, New York, Vol. 2, pp. 99-131.
Scharf, B. , 1970, Critical bands, in: "Foundations of Modern Auditory Theory", J. V. Tobias, ed. Academic Press, New York. Vol. 1, pp. 159-202.
Seaton, W. H., and Trahiotis, C. ,1975, Comparison of critical ratios and critical bands in the monaural chinchilla, J. Acoust. Soc. Am. 57: 193-199.

Thomas,J., Chun, N. Au, W., and Pugh, K. ,1988, Underwater audiogram of a false killer whale (Pseudorca crassidens). J. Acoust. Soc. Am. 84(3): 936-940.

Watson, C. S. ,1963, Masking of tones by noise for the cat," J. Acoust. Soc. Am. 35: 167-172.

White, M., Jr., Norris, J., Ljungblad, D., Barton, K. and di Sciara, G., 1978, Auditory thresholds of two belukha whales (Delphinapterus leucas), in HSWRI Techn. Rep. No. 78-109, Hubbs Marine Research Institute, 1700 S. Shores Road, San Diego, CA.

ELECTROPHYSIOLOGICAL STUDIES OF HEARING IN SOME CETACEANS AND A MANATEE

Vladimir Popov and Alexander Supin

Severtsov Institute of Evolutionary Morphology
and Ecology of Animals
USSR Academy of Sciences
Leninsky prosp, 33, 117071, Moscow, USSR

INTRODUCTION

Auditory brain stem response (ABR) is used extensively for studies of hearing. The possibility to record activity of auditory nerve centers without any surgery makes this evoked response useful for investigation of auditory physiology and pathology, particularly, for comparative studies of hearing in various groups of animals.

Investigation of ABR in Cetaceans originates from the work of Bullock et al. (1968); they recorded the evoked potentials by electrodes introduced into the dolphin's brain stem. Then similar evoked potentials were observed with electrodes placed intracranially, but not in the brain stem (Ladygina and Supin, 1970 ; Bullock and Ridgway, 1972); some features of these responses were described (Bullock and Ridgway, 1972; Supin et al., 1978). Ridgway et al. (1981) described the ABR recorded extracranially and studied in detail a number of its features. Finally, the possibility to record ABR from the body surface in dolphins was shown in the harbor porpoise (Bibikov et al., 1986) and in the bottlenosed dolphin (Popov and Supin, 1985), and some characteristics of the dolphin's hearing were measured in such a way (Supin, Popov, 1985; Popov, Supin, 1988, 1990).

Recording of ABR from the body surface enabled us to study hearing of some Cetacean species and to compare them. This was the subject of the present paper. Besides, some data on manatee (Sirenian) are presented for comparison.

MATERIALS AND METHODS

Experiments were carried out on four bottlenosed dolphins, Tursiops truncatus (Delphinidae), two Tucuxi dolphins, Sotalia fluviatilis (Delphinidae), four Amazon river dolphins, Inia geoffrensis (Platanistide), two beluga dolphins, Delphinapterus leucas (Monodontidae), and one manatee, Trichechus inunguis (Sirenia).

Experiments with bottlenosed dolphins were carried out at the Utrish Sea Station of the USSR Academy of Science, on the Black Sea Cost; the animals were well adapted to captivity. The experiments with beluga dolphins were run at the TINRO Biostation of the USSR Ministry of Fishery on the Japan Sea; the animals were caught shortly before. The Amazon dolphins and Tucuxi dolphins were caught in the Ucayaly river (Peru); the experiments were performed at the Soviet-Peruan Biostation, Pucallpa, Peru. The experiments on manatee were run at the Biostation of the Institute of Investigation of Peruvian Amazony (IIAP), Iquitos, Peru.

The experiments with ABR recording required neither anaesthesia nor curarization. It was sufficient to place a dolphin on a stretcher. The animal on the stretcher was placed in an experimental bath or pool in such a way that the main part of the body was in water and only the dorsal part of the head with the blowhole and the back were above water. In such a position an animal could remain rather quitely for 2-3 hours.

During the experiments with bottlenosed dolphins, Tucuxi dolphins, Amazon dolphins and manatee, the animals was placed in the bath 4 x 0.6 x 0.6 m or in the round pool filled with water. The experiments with beluga dolphin were carried out in an enclosure in a sea bay.

The electrodes for recording ABR from the body surface were thin needles 0.4-0.6 mm in diameter . The electrodes were inserted into the skin 3-5 mm deep. The active electrode was placed on the dorsal head surface 6-9 cm caudal from the blowhole, the reference electrode was placed on the back near the dorsal fin. A signal from the electrodes was fed to an amplifier and then to an averager of evoked potentials. The pass band of the channel was 5-5000 Hz.

The acoustic stimuli were clicks, noise and tone bursts emitted into the water by piezoceramic transducers. To produce a click, the transducer was activated by a 5 us long rectangular pulse; to produce noise it was activated by a pseudo-random binary signal with a pace duration 5 us. Thus clicks and noise had identical spectra. Parallel connection of spheric transducers of 20, 30 and 50 mm diameter produced noise and clicks with a spectral width from 10 to 100 kHz (at a -10 dB level) with spectrum irregularity up to 10 dB in this range. Frequencies of tone bursts ranged from 5 to 160 kHz. Noise and tone bursts were formed by an acoustic key. Noise bursts had abrupt rise and fall, tone bursts had linear rise and fall 0.25 ms long.

The transducers were placed 30 cm deep in water, 1 to 2 m from the animal's head. All the characteristics of acoustic signals were checked by a hydrophone with a pass band of 200 kHz located near the animal's head. Intensity of the stimuli was designated as dB relative 1 mPa of effective (for noise and tone) or peak (for clicks) sound pressure.

RESULTS

ABR Characteristics

Acoustic stimuli, such as clicks, noise, or tone bursts evoked ABR recorded from the body surface in all the dolphins under study. Fig.1 shows ABRs to sound clicks in four dolphin species. Both the form of the responses and their temporal characteristics were similar in some respects in all the dolphins . ABRs consisted of a sequence of fast waves lasting mainly less then 1 ms.

We designated the main waves of ABR by ordinal numbers from P1 to N5 (P - positive, N - negative waves). We consider this designation as preliminary and arbitrary, but acceptable since the origin of ABR waves in dolphins is still unclear.

Some difference in the ratio of ABR waves was observed between Cetaceans studied. The waves P3, P4 and N5 were most stable and similar in all the dolphins. The earlier wave P1-N2 was rather similar in the bottlenosed dolphin and Tucuxi dolphin, but in the Amazon dolphin it fell into two oscillations, and in the beluga dolphin this wave was virtually absent. The amplitude of ABR was usually measured between the peaks of the greatest waves P4-N5.

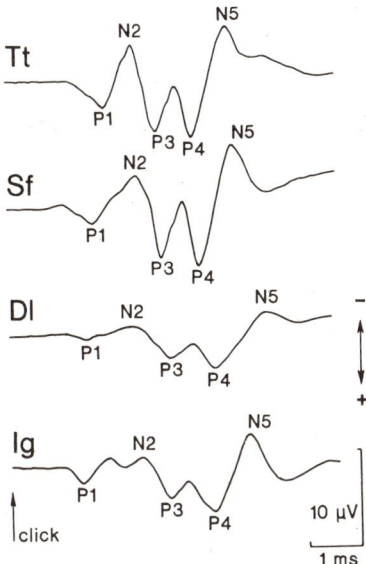

Fig.1. ABRs to sound click in four dolphin species. P1-N5: designation of the main waves (P - positive, N - negative). Tt - bottlenosed dolphin, Tursiops truncatus, Sf - Tucuxi dolphin, Sotalia fluviatilis, Dl - beluga dolphin, Delphinapterus leucas, Ig - Amazon river dolphin, Inia geoffrensis.

In all the dolphins the best site for recording ABR was located at the middorsal line, 5-10 cm caudally from the blowhole. Fig.2 shows ABRs recorded from some points on the dorsal head surface in a bottlenosed dolphin. The maximum amplitude was observed in the point P6 at the middorsal line. This position of the active electrode was used in subsequent recordings from all dolphins.

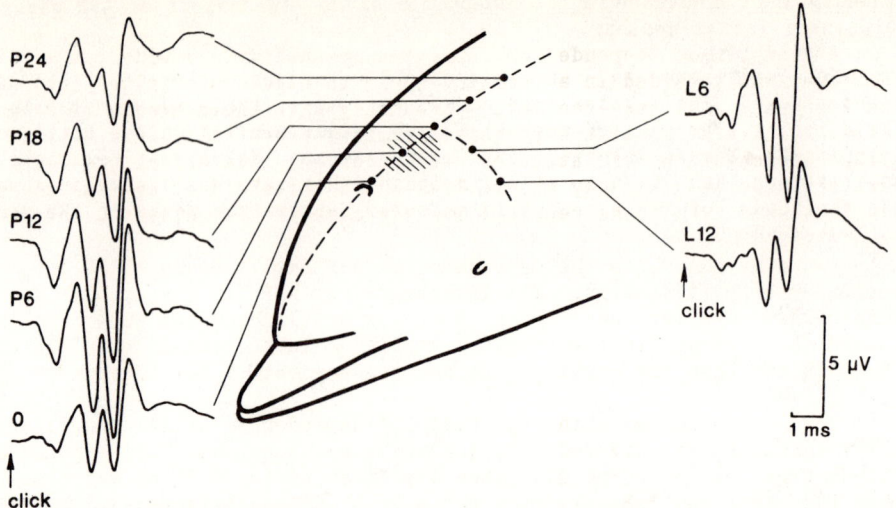

Fig.2. ABRs recorded from some sites on the body surface in a bottlenosed dolphin. 0-P24 - points 0 to 24 cm posteriorly the blowhole, L6-L12 - points 6 to 12 cm laterally from the middorsal line. The area of the greatest ABR is shaded.

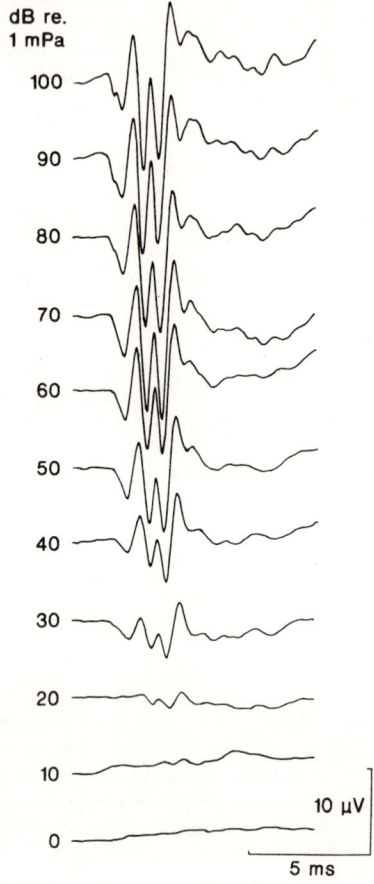

Fig.3. ABRs of a Tucuxi dolphin to clicks of various intensity. Stimulus intensity is indicated near the oscillograms.

Dependence of ABR on Stimulus Intensity

ABR amplitude depended on stimulus intensity in a wide range. Fig.3 presents ABRs recorded in a Tucuxi dolphin to clicks of various intensity. The response amplitude decreased gradually with the decrease of stimulus intensity. The form of the response remained practically invariable. Only slight prolongation of latencies was observed; this effect was described earlier in detail (Ridgway et al.,1981). The threshold of the ABR could be distinguished from these records; in this case it was between 0 and 10 dB.

Fig.4 demonstrates the dependence of ABR amplitude on click intensity in three dolphin species. The thresholds of ABR in these three species were rather similar: near 10 dB. Both the bottlenosed and Tucuxi dolphins showed an increase in the response amplitude in the intensity range up to 60-70 dB and then the amplitude became independent of the intensity.

In the Amazon dolphin the plot of dependence of ABR amplitude on click intensity was divided into four distinct sections. The response amplitude grew up to 50-60 dB; then a plateau up to 65-70 dB was observed; then the amplitude sharply rose again with intensity increasing up to 85-90 dB; finally, there is the second plateau at high intensities.

Dependence of ABR on Intensity and Frequency of Tone Bursts. Audiograms

We determined ABR thresholds using tone bursts of varying intensity and frequency as stimuli. A rather high amplitude of ABRs and their stability in dolphins made it possible to determine thresholds with an accuracy of 3-5 dB. So the audiograms (frequency-threshold functions) of a few dolphin species could be obtained in the similar conditions.

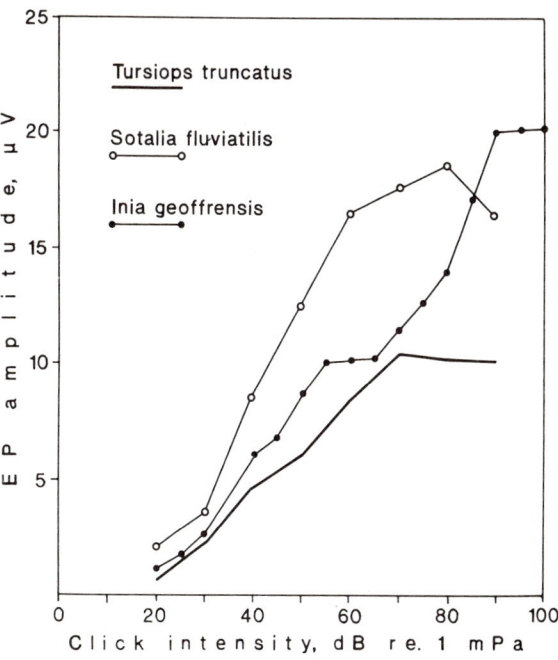

Fig.4. Dependence of the ABR amplitude on click intensity in the three dolphin species.

Such audiograms of 5 dolphin species are presented in Fig.5. All results are based on ABR recording; an exception is the audiogram of the harbor porpoise obtained by recording cortical evoked potentials with implanted electrodes (Popov et al., 1986). The audiogram of the manatee is also shown for comparison.

The audiograms of the bottlenosed, Tucuxi and beluga dolphins were mainly similar: they were U-shaped with a steeper high-frequency branch. The frequency bands of the audiograms were from 110 kHz (beluga dolphin) to 140 kHz (Tucuxi dolphin) at the level 60 dB above the minimum thresholds. The minimum thresholds were observed at frequencies from 60 to 80 kHz. In the bottlenosed and Tucuxi dolphins the minimum thresholds were below 1 mPa (below 0 dB), in the beluga dolphin they were slightly higher.

The audiograms of the harbor porpoise and Amazon river dolphin had essentially the same minimum thresholds levels and frequency range as those of the dolphins mentioned above. The widest frequency range was observed in the porpoise: up to 160 kHz. However, the audiograms of the harbor porpoise and Amazon dolphin was of an unusual W-shape. The specific feature of these audiograms was the presence of two sensitive frequency regions with low thresholds. In the harbor porpoise these regions were at frequencies near 30 and 125 kHz, in the Amazon dolphin, at frequencies 20-25 and 70-80 kHz respectively. This feature was especially pronounces in the Amazon dolphin: the two low-thresholds regions were separated by a deep high-threshold hiatus.

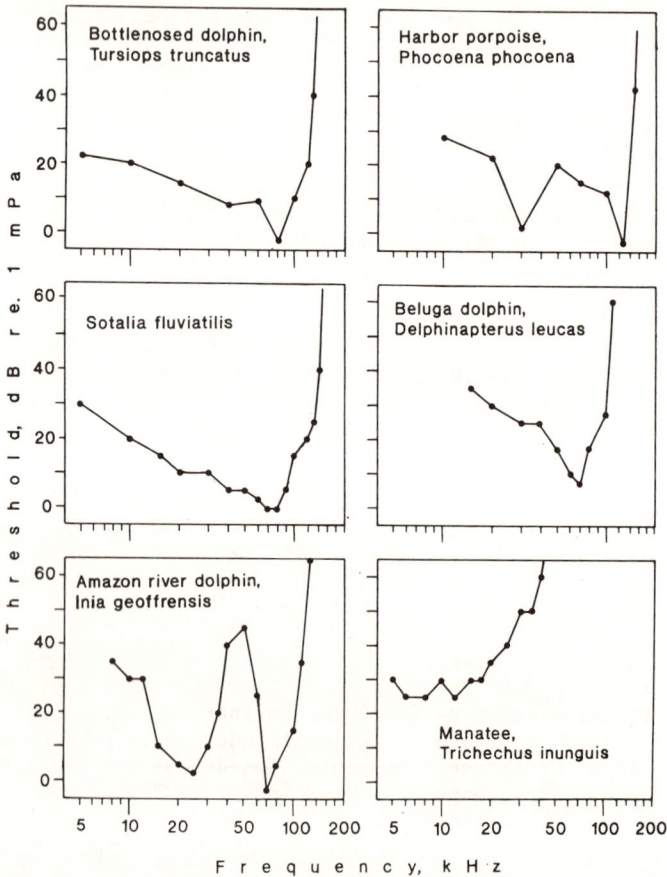

Fig.5. Audiograms of the five dolphin species and of the manatee.

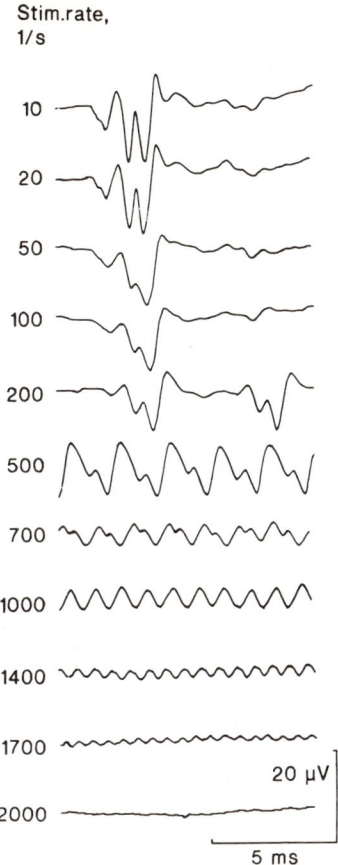

Fig.6. ABRs of a Tucuxi dolphin to rhythmic clicks of various rates. The click intensity 80 dB. Stimulus rate (clicks per s) is inducated near the oscillograms.

Contrary to dolphins, the audiogram of the manatee was characterized by rather high thresholds (no less then 25-30 dB re. 1 mPa) and a far more narrow frequency range (up to 40 kHz at the level 40 dB above the minimum thresholds).

Dependence of ABR on Rhythmic Stimulus Rate

The ability of responses to follow rhythmic stimulation appears to be an important parameter of the auditory system. For measuring this characteristic in dolphins we recorded ABR evoked by rhythmically repeated clicks of various rate.

Fig.6 presents an example of ABR of a Tucuxi dolphin to rhythmic clicks of varying rate. The response shape and amplitude were slightly changed at stimulus rates up to 200/s. Further increasing of the rate resulted in a gradual decrease of the amplitude and in smoothing of the response; individual response components fused together forming a sinusosidal response. Such response could follow the stimulus rate up to 1700/s in the case presented here.

Fig.7 presents plots of the amplitude of auditory responses in four dolphin species versus the rate of rhythmic stimulation. For comparison

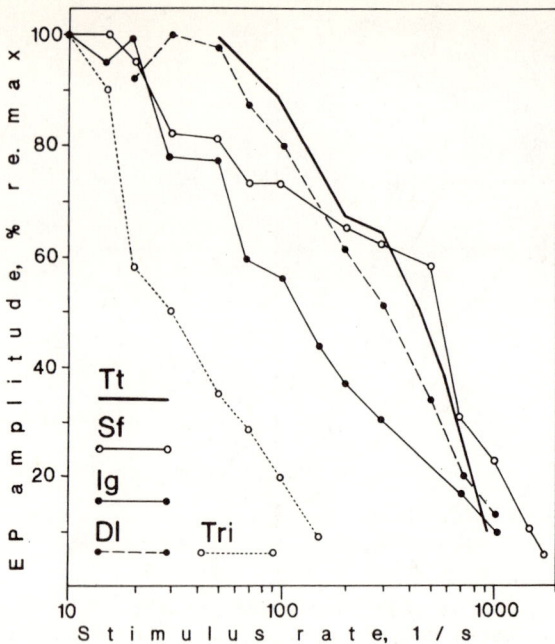

Fig.7. Dependence of the ABR amplitude on rhythmic click rate in the four dolphin species and in the manatee. Tt - bottlenosed dolphin, Tursiops truncatus, Sf - Tucuxi dolphin, Sotalia fluviatilis, Ig - Amazon dolphin, Inia geoffrensis, Tri - manatee, Trichechus inunguis.

the same characteristic of a manatee is given. The auditory responses of dolphins demonstrated a very high ability to follow rhythmic stimulation. In all the dolphins the limit of the response following was near or above 1000/s, particularly, in Tucuxi dolphin it was up to 1700/s. On the contrary, in manatee this ability was rather low: up to 150/s.

Directional Sensitivity of ABR

The directional sensitivity of hearing was studied by recording ABR in three dolphin species, the bottlenosed dolphin, Tucuxi dolphin, and Amazon river dolphin. The usual difficulty in studying the directional sensitivity of hearing is the necessity to create echoless field. The advantage of the ABR recording method is that the effect of echo on results of measurements can be eliminated. Indeed, the duration of ABR is very short, therefore it is only sufficient to make the distance from the animal to the pool walls at least 2-3 m. At such a distance the echo delay is longer than the ABR duration, so the response to a direct signal does not interfere with that to echo sounds.

Experiments on determination of hearing directionality were carried out in round pools 7 or 10 m in diameter, either 3 m deep, or with flat water layer 0.4 m deep. A sound source was transposed in a horizontal plane around the animal's head by 15° steps. The thresholds of ABR were measured for each of the source position.

Fig.8 presents data on ABR thresholds depending on the direction to the sound source in three dolphin species. The stimulus was a click with a wide-band spectrum. The plots demonstrate a considerable dependence of the auditory thresholds on the direction. The lowest thresholds were observed

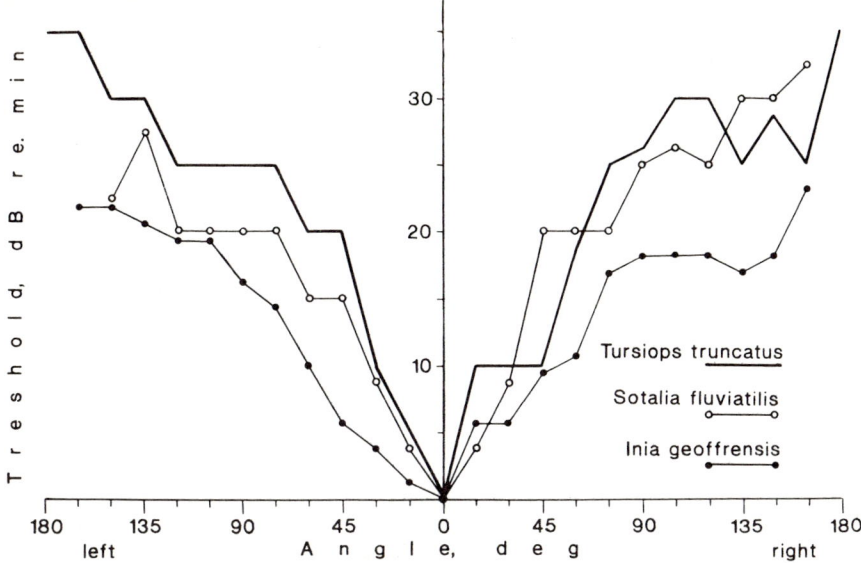

Fig.8. Dependence of the ABR thresholds on the direction to the sound source in the three dolphin species. 0 - the frontal direction, 180 - the caudal direction.

for sources located in front of the animal (0° angle), the highest thresholds were observed at a caudal position of a source (150-180°). The difference between these two directions was 30-35 dB. No significant difference between the three dolphins studied was observed.

DISCUSSION

It may be pointed out that the ABR recording is rather productive in studying the dolphin's hearing. This method is harmless for an animal if the electrodes are arranged at the body surface. The amplitude of ABR in dolphins is very large; probably it is due to the extraordinary development of the brain auditory nuclei. While in humans as a rule the ABR is less than 1 μV, in dolphins it can reach 10-20 μV. It provides for precise measurements and does not require much time for distinguishing the response by averaging. Recording ABR from the body surface makes it possible to carry out experiments with an animal that was caught shortly before and then to set it free. That is what we have done with some of the animals.

The designation that we suggested for the ABR waves can be regarded as preliminary, since we know little as yet about the generation of these waves. The similarity in the forms of the main ABR components in all the dolphin species studied suggests that these components reflect a similar sequence of events in the auditory system.

The results presented show that ABR can be used for testing the frequency range and thresholds of dolphin's hearing. At the same time it is necessary to take into account some specific features of this response. Indeed, the ABR is a very short response, and its temporal summation at high stimulus intensities is also short, less than 0.1 ms (Popov, Supin, in press). So the effective part of a tone stimulus that evokes the response is short, and, accordingly, the frequency spectrum of this effective part is wide. It is a disadvantage in studying the frequency-selective characteristic of the auditory system. However, at near-threshold intensities of a stimulus the temporal summation was more prolonged - up to 0.5 ms. Tone stimuli of this duration have the speectrum narrow enough for measuring a frequency-threshold dependence. Therefore it is possible to evaluate frequency characteristics of the auditory system by measuring the ABR thresholds. It could be supposed therefore that the audiograms presented above characterize thereal dolphin's hearing.

These audiograms are in good agreement with the data of behavioral studies (see reviews: Popper, 1980; Fobes and Smock, 1981). The relatively higher thresholds obtained by us for beluga dolphin may result from a higher noise level in the bay, where the measurements were carried out. All the audiograms of dolphins show a rather wide fréquency band - more than 100 kHz, up to 160 kHz in the harbor porpoise.

Of special interest is the audiogram of the Amazon dolphin that had two separated frequency regions of high sensitivity. At present we can not explain the significance of this fact. However such a feature is not a unique peculiarity of the Amazon dolphin; the audiogram of the harbor porpoise had also the two threshold minima, although not so prominent. Some peculiarities of the Amazon dolphin's hearing may be associated with the organization of the audiogram described. In particular, we suppose that an intermediate plateau in the dependency curve of the ABR amplitude on click intensity (see Fig.4) was due to the audiogram form in the Amazon dolphin: the responses evoked by the two separated frequency regions of high sensitivity may reach their maxima at different click intensities if nonequal spectral power of the stimulus falls into these regions.

The marked difference between audiograms of all the dolphins and the manatee (the latter had much higher thresholds and a narrower frequency band) shows the efficiency of the method employed for comparative studies of hearing of aquatic mammals.

Our data on repeated rhythmic stimulation point to the ability of the auditory nuclei in dolphins for a fast recovery of their responsiveness. This feature was described for some dolphin species (Supin et al., 1978; Ridgway et al., 1981). The data presented herein show that this property is characteristic of all the dolphin species studied, but not of all aquatic mammals: in the manatee ABRs follow the stimulus rate that was nearly 10-times less than in dolphins. It is possible that the ability of dolphin's ABRs to recover their responsiveness exclusively quickly is associated with echolocation.

Data on directional sensitivity of ABR showed non-uniform directional sensitivity of hearing in the dolphins studied. The variation of sensitivity due to the direction of the sound source was significant, up to 30 dB or more. Probably it is not sufficient for explaining all the ability of dolphins for spatial differentiation of sound sources (Renaud and Popper, 1975), however, it may be one of the existing mechanisms.

So the method of recording ABR offers some possibilities for comparative studies of various aspects of hearing in aquatic mammals.

REFERENCE

Bibikov N.G., Rimskaya-Korsakova L.K., Zanin A.V. and Dubrovsky N.A., 1986, Investigation and modelling of auditory brainstem evoked potentials in the harbor porpoise, in: V.E.Sokolov (ed.), "Electrophysiology of Sensory Systems in Marine Mammals" (in Russ.), Nauka, Moscow, 56-84.

Bullock T.H., Grinnell A.D., Ikezono E., Kameda K., Katsuki J., Nomota M., Sato O., Suga N. and Yanagisawa K., 1968, Electrophysiological studies of central auditory mechanisms in Cetaceans. Z. Vergl. Physiol., 59: 117-156.

Bullock T.H. and Ridgway S.H., 1972, Evoked potentials in the central auditory system of alert porpoises to their own and artificial sounds. J. Neurobiol., 3: 79-99.

Fobes J.L. and Smock C.C., 1981, Sensory capacities of marine mammals. Psychol. Bull., 89: 288-307.

Ladygina T.F. and Supin A.Ya., 1970, Acoustic projection in the dolphin's cerebral cortex (in Russ.). Fisiol. z. SSSR, 56: 1554-1560.

Popov V.V. and Supin A.Ya., 1985, Determination of characteristics of the dolphin's hearing with the brain stem evoked potentials (in Russ.). Dokl. Acad. Nauk SSSR, 283: 496-499.

Popov V.V., Ladygina T.F. and Supin A.Ya., 1986, Evoked potentials of the auditory cortex of the porpoise Phocoena phocoena. J. comp. Physiol., 158: 705-711.

Popov V.V. and Supin A.Ya., 1988, Hearing directional diagram of the dolphin Tursiops truncatus (in Russ.). Dokl. Acad. Nauk SSSR, 300: 756-760.

Popov V.V. and Supin A.Ya., 1990, Auditory brain stem responses in characterization of dolphin hearing. J. comp. Physiol., 166: 385-393.

Popper A.N., 1980, Behavioral measures of Odontocete hearing. in: R.G. Busnel and J.F.Fish (eds.),"Animal Sonar Systems", Plenum Press, N.Y., 496-481.

Renaud D.L. and Popper A.N., 1975, Sound localisation by the bottlenose porpoise, Tursiops truncatus. J. Exp. Biol., 63: 569-585.

Ridgway S.H., Bullock T.N., Carder D.A., Seeley R.L., Woods D. and Galambos R., 1981, Auditory brainstem response in dolphin. Proc. Nac. Acad. Sci. USA, 78: 1943-1947.

Supin A.Ya., Mukhametov L.M., Ladygina T.F., Popov V.V., Mass A.M. and Polyakova I.G., 1978, Electrophysiological studies of the dolphin's brain (in Russ.). Nauka, Moscow.

Supin A.Ya. and Popov V.V., 1985, Recovery cycles of the dolphin's brain stem responses to paired acoustic stimuli (in Russ.). Dokl. Acad. Nauk SSSR, 283: 740-743.

LOCALIZATION OF THE ACOUSTIC WINDOW AT THE DOLPHIN'S HEAD

Vladimir Popov and Alexander Supin

Severtsov Institute of Evolutionary Morphology
and Ecology of Animals,
USSR Academy of Sciences
Leninsky prosp. 33, 117071, Moscow, USSR

INTRODUCTION

The mechanism of sound transmission to the ear in aquatic mammals, particularly in dolphins, remains obscure. It has been suggested that sound reaches the inner ear directly via head tissues, or sound transmission involves the middle ear and the closed auditory meatus (Fraser and Purves, 1959; Fleischer, 1978). This assumption was supported by calculations indicating that the closed auditory meatus can serve as an effective acoustic transformer for the sounds to be transmitted to the middle ear (Lipatov, 1978).

On the other hand, there is the popular so-called "mandibular" hypothesis emphasizing the key role of the lower jaw with its peculiar fat body as a specific sound-conducting pathway (Norris, 1964, 1968). Sounds are believed to reach the fat body through a thin bone plate on the lateral mandibular surface and are transmitted further via the fat body to the bulla. Therefore, the hypothesis postulates the presence of an "acoustic window" on the lower jaw which allows for the perception of sounds.

The validity of the mandibular hypothesis was supported by data on lower thresholds of evoked potentials to the contact acoustic stimulation of the mandible as compared with those obtained after a similar stimulation of other body sites (Bullock and al., 1968). Similar data were obtained upon recording the microphonic potentials (McCormic et al., 1970, 1980). In other experiments, it has been shown that a sound-proof hood attached to the lower jaw impaired the dolphin's ability to recognize echolocation signals (Brill et al., 1988).

These data can, however, be interpreted in several ways. The contact acoustic stimulation in the air is not adequate for dolphins, and its effect depends to a large extent on the properties of the tissue with which a hydrophone is in contact. The sound-proof hood may affect the acoustic field near the head and influence auditory perception even if the acoustic window is not located on the lower jaw. Therefore, the acoustic window problem remains to be elucidated.

We have approach this problem in a different way, that is, we tried to determine the spatial position of an acoustic window by measuring acoustic delays of evoked auditory potentials at different positions of sound sources.

To employ this idea for the location of the acoustic window, the acoustic delays must be measured with accuracy within 20-30 µs which corresponds to a 3-5 cm of sound spreading in water. Such an accuracy of measurements can be ensured by recording short-latency evoked potentials, e.g. auditory brainstem response (ABR), from either the skull or the body surface of dolphins (Ridgway et al., 1981; Popov and Supin, 1985). This ABR contains very short waves (about 0.5 ms). The peak latency of these waves can be measured with an accuracy up to 20 µs.

METHODS

Principle of the method

The general idea of the method originates from the following. When an evoked potential is recorded, its latency is the sum of the proper response time (physiological latency) and the acoustic delay (the time of sound transmission from the source to the receiver). The acoustic delay depends on the mutual position of the sound source and the sound receiver, or the acoustic window, if it exists. So, measuring the acoustic delays at various source positions makes it possible to determine the position of a sound receiver or an acoustic window.

For measuring of the acoustic delays, the proper response time has to be known. However, <u>difference</u> of the delays can be measured if the own response time is unknown, but constant. It is possible when the auditory evoked responses are recorded.

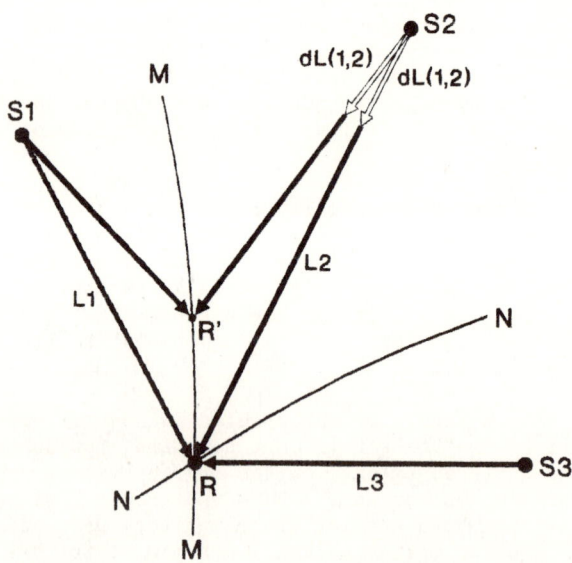

Fig.1. Calculation of the position of the receiver from the difference between acoustic delays. S1, S2, S3 - sound sources, R - receiver, dL(1,2) - difference between distances L(1) and L(2) from S1 and S2 to R, dL(2,3) - difference between distances from S2 and S3 to R

The principle of the method is illustrated in Fig.1. Two sources of sound signals S1 and S2 are placed in the known positions, the position of receiver R is to be determined. The latency of the response of the receiver R to signals S1 and S2 can be measured. This latency is the sum of the acoustic delay and the proper response time. Let us assume that the latter is a constant. It is thus possible to determine the difference dT(1,2) of the two acoustic delays for sources S1 and S2 from measurement of the two response latencies. The speed of sound transmission (C) being known, the difference between acoustic delays is a measure of the difference between the distances: dL(1,2)=C*dT(1,2). Receiver R would therefore be located on a MM line defined by the equation: L(2)-L(1)=C*dT(1,2). So we can't obtain the exact position of the receiver R; for example, for the point R' (and for all other points of the MM line) the delay dL(1,2) is the same. However, one coordinate of the receiver (the horizontal one in Fig.1) can be obtained in such a way with a good approximation. Moreover, the use of one more sound source (S3) allows to draw another line (NN) from the difference dL(2,3) which is also the site of the location of the receiver. The intersection point of the MM and NN lines gives the exact position of the receiver R, or the acoustic window through which sounds are specifically transmitted to the receiver.

Material

Experiments were carried out on four Amazon river dolphins, <u>Inia geoffrensis</u>, two bottlenosed dolphins, <u>Tursiops truncatus</u>, and two Tucuxi dolphins, <u>Sotalia fluviatilis</u>.

Experiments with bottlenosed dolphins were carried out at the Utrish Sea Station of the USSR Academy of Science, on the Black Sea Cost; the animals were adapted to captivity. Amazon and Tucuxi dolphins were caught in the Ucayaly river (Peru); this work was performed at the Soviet-Peruan Biostation, Pucallpa, Peru, in accordance with the Agreement between the USSR Academy of Science and the National San-Marcos University of Peru.

Evoked potentials recording

For measuring acoustic delays, the auditory brainstem responses (ABR) were recorded from the dolphin's head surface. The experiments with ABR recording involved no surgical procedure and required neither anaesthesia nor curarization. It was sufficient to place a dolphin on a stretcher. The animal on the stretcher was placed in an experimental pool in such a way that the main part of the body was in water and only the dorsal part of the head with the blowhole and the back were above water.

The electrodes for recording ABR from the body surface were thin needles 0.4-0.6 mm in diameter. The electrodes were inserted into the skin 3-5 mm deep. The active electrode was placed on the dorsal head surface 6-9 cm caudally from the blowhole, the reference electrode was placed on the back near the dorsal fin. A signal from the electrodes was fed to an amplifier and then to an averager of evoked potentials. The pass band of the channel was 50-5000 Hz. The peak latency of ABR was measured using an averaging processor at either 10 or 20 µs discretion.

Acoustic stimulation

The measurements were made in a round pool 7 or 10 m in diameter. The water column was only 40 cm high to minimize the echo from both the pool bottom and the water surface, that is, the sound transmission occurred as if in a plane layer. An animal was supported by a stretcher in such a way that its head was positioned in the center of the pool. The stretcher was made of a sound-transparent material (fine net).

Short clicks were used as acoustic stimuli. The frequency band of the click was up to 120 kHz, the maximum spectral power at 30-50 kHz. The stimuli were delivered through spherical piezoelectric transducers immersed in water. Sound intensity from 0.1 to 300 Pa of the peak sound pressure were used for stimulation, i.e. from 40 to 110 dB relative 1 mPa (relative level for sound intensity indication used in this paper was 1 mPa).

Position of the sound source (hydrophone) was determined relative to a reference point which coincided with the rostral tip of the melon (point O in Fig.3 and 5). The hydrophone was situated at a distance of 1 m from the reference point, at an angle of 0 to $\pm 165°$ relative to the longitudinal axis.

RESULTS

The Amazon river dolphin

ABRs elicited in an Amazon dolphin by sound clicks are shown in Fig.2. The characteristic feature of the ABR of the Amazon dolphin was presence of the clearly defined earliest positive wave with the peak latency (including an acoustic delay) 1.5-1.7 ms; this wave of the three responses is marked by cursors in Fig.2. The peak latency of this wave can be measured with a good accuracy: 10 to 20 µs. Specifically this wave was used to measure acoustic delays in the Amazon dolphin.

It is evident from Fig.2 that responses induced by stimuli from different sources S1, S2, and S3 (positions of these sources are shown in Fig.3) had somewhat different latencies due to different acoustic delays. The peak latency of the first positive wave of the three responses is indicated by cursors in the Fig.2. The differences dT(1,3) between the response latencies to the S1-S3 stimuli and dT(2,3) between the latencies to S2-S3 stimuli are also shown.

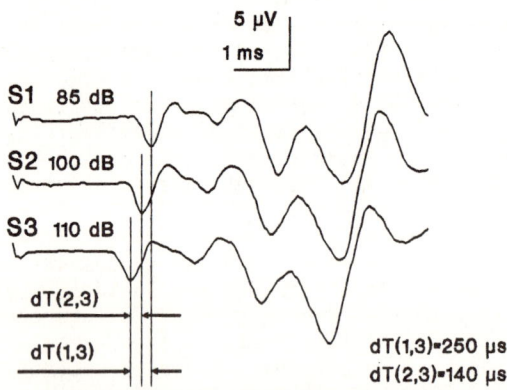

Fig.2. ABRs of an Amazon dolphin elicited by stimulation from three sources S1, S2, and S3; position of the sources is shown in Fig.3. Vertical lines are temporal cursors indicating the peak of the first wave; dT(1,3) - the difference between latencies of responses to stimuli S1 and S3; dT(2,3) - that to stimuli S2 and S3. Stimulus intensity is indicated near the responses.

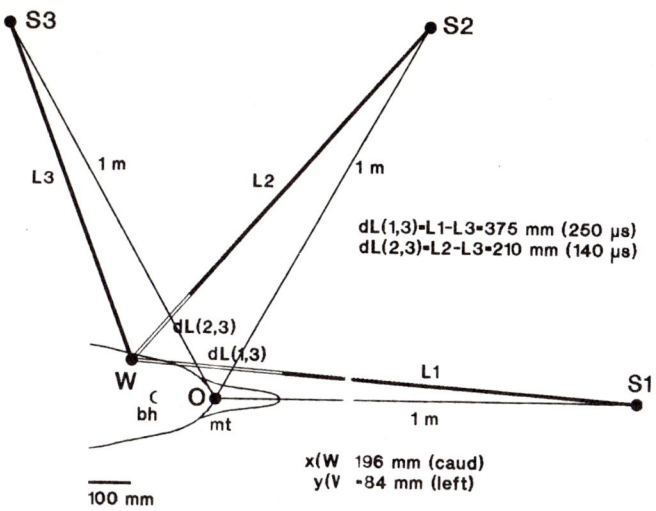

Fig.3. Positions of the sound sources relative to the head of an Amazon dolphin (dorsal view); O - reference point (the melon tip), S1-S3 - sound sources, L1-L3 - distances from the sources S1-S3 to an acoustic window W. The dark part of lines L1 and L2 is equal to L3; the light parts are difference dL(1,3) and dL(2,3). x(W) and y(W) - coordinates of the point W relative to the reference point O. bh - blowhole, mt - melon tip.

When recording ABRs, we took into account that hearing sensitivity of dolphins depends on the position of a sound source. The highest ABRs were obtained when the source was situated in front of the animal's head, and they diminished when the source was moved laterally and further caudally (Popov and Supin, 1988). That is, at an equal physical intensity of the stimulus the sound in front of the animal was the loudest. On the other hand, a change of loudness evidently affected not only the amplitude but also the latency of the ABR (Ridgway et al., 1981).

However, it was important for our purpose that the proper response time of the ABR remain unaltered. Therefore, the click intensity at each position of the sound source was chosen to elicit ABRs of the equal amplitude. Fig.2 shows click intensity in dB re. 1 mPa: the lowest was the intensity of the frontally located sound source S1 (85 dB), the highest - that of the caudally located source S3 (110 dB), so the amplitude of all the three response was virtually equal.

The difference between acoustic delays observed was used to calculate the location of an acoustic window. In an example shown in Fig.3, the sources S1, S2, and S3 were located 1 m far from the reference point (the melon tip), differing by 60°. The single point W with coordinates x=196 mm and y=84 mm (relative to the reference point O) is defined by the requirement that dT(1,3)=250 μs and dT(2,3)=140 μs, that is dL(1,3)=375 mm and dL(2,3)=210 mm.

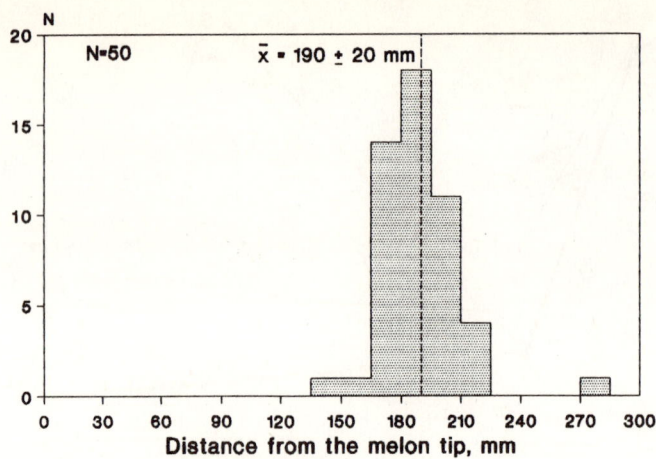

Fig.4. Distribution of the observed distances between the melon tip and the acoustic window in the Amazon river dolphin.

A total of 50 measurements was available for the calculation of the acoustic window coordinates in the Amazon dolphin. The angles between directions to sound sources were varied from 60 to 150 to exclude influence of any specific source position. Of special interest is the longitudinal X coordinate along the body axis because it is this parameter that is likely to indicate whether the acoustic window is located on the lower jaw or near the bulla and the auditory meatus.

Taken together, these data are presented in Fig. 4 showing the distribution of the observed distances between the reference point (melon tip) and the acoustic window. The mean distance and the standard deviation of the distribution were 191±20 mm, with more than 95% measurements (48 out of the 50 available) exceeding 165 mm.

The bottlenosed dolphin

ABRs elicited in an bottlenosed dolphin by sound clicks are shown in Fig.5. Contrary to ABRs of the Amazon dolphin, their first positive wave has no sharp peak, and the peak latency of this wave can not be measured with high accuracy. Due to this reason we used the other way of measuring the latency difference of the responses. The latency of six waves from 1 to 6 (Fig.5) of every response was measured, and for each of these waves the latency difference of the two response compared was calculated. The mean of the six differences was used as the measure of the acoustic delay difference of the two responses compared.

As for Amazon dolphin, the click intensity at each position of the sound source was chosen to elicit ABRs of the equal amplitude; the click intensities are shown near the responses in Fig.5.

The difference of acoustic delays obtained in such a way was used to calculate the location of an acoustic window. In an example shown in Fig. 5, the sources S1, S2 and S3 were located 1 m from the reference point (the melon tip) at the angles 30, 90 and 150 from the frontal direction (Fig.6). The latency differences between the responses are shown by

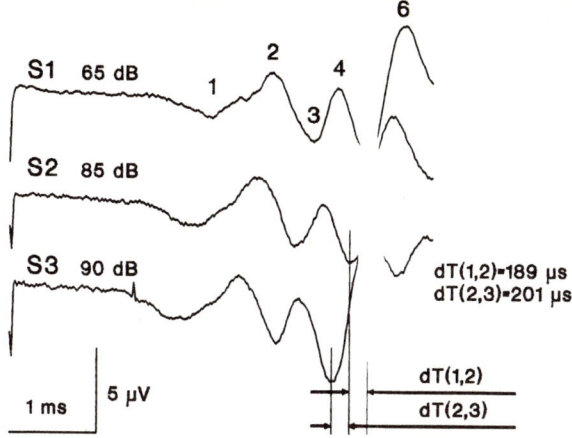

Fig.5. ABRs of a bottlenosed dolphin elicited by stimulation from three sources S1, S2, and S3; position of the sources is shown in Fig.6. Vertical lines are temporal cursors indicating the peak of the wave 5; dT(1,2) - the difference between latencies of responses to stimuli S1 and S2; dT(2,3) - that to stimuli S2 and S3. Stimulus intensity is indicated near the responses.

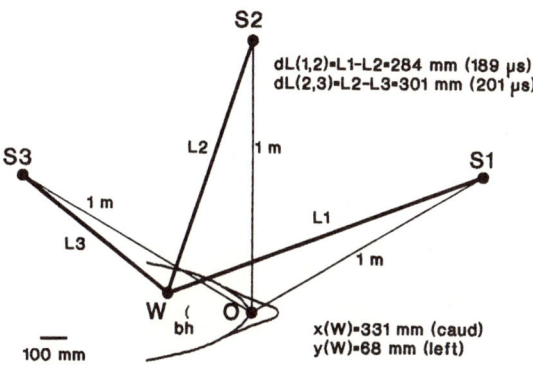

Fig.6. Positions of the sound sources relative to the head of a bottlenosed dolphin (dorsal view). Designations are the same as in Fig.3.

cursors only for wave 5, however the mean latency difference of the six waves (1 to 6) are indicated in Fig.5 as dT(1,2) and dT(2,3). As shown in Fig.6, a single point W with coordinates x=331 mm and y=68 mm (relative to the reference point O) is defined by the requirement that dT(1,2)=189 µs and dT(2,3)=201 µs, that is dL(1,2)=284 mm and dL(2,3)=301 mm.

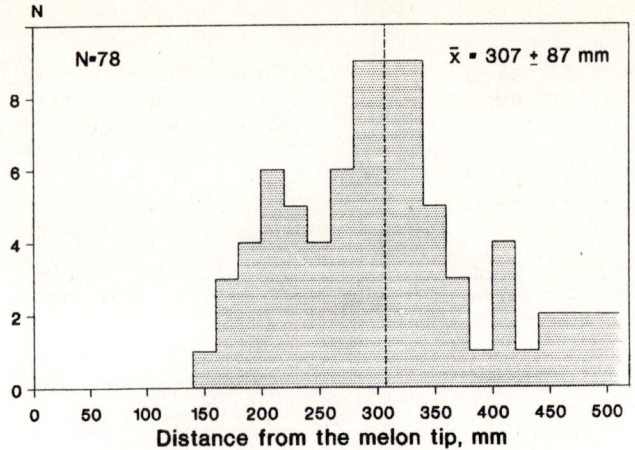

Fig.7. Distribution of the observed distances between the melon tip and the acoustic window in the bottlenosed dolphin.

A total of 78 measurements was available for the calculation of the acoustic window coordinates in the bottlenosed dolphin. The angles between directions to sound sources were varied in these measurements from 60 to 150°. The results are presented in **Fig.7** as the distribution of the obtained distances between the reference point (the melon tip) and the acoustic window. The mean distance and the standard deviation of the distribution were 307±87 mm, with 95% measurements (74 out of 78) exceeding 180 mm.

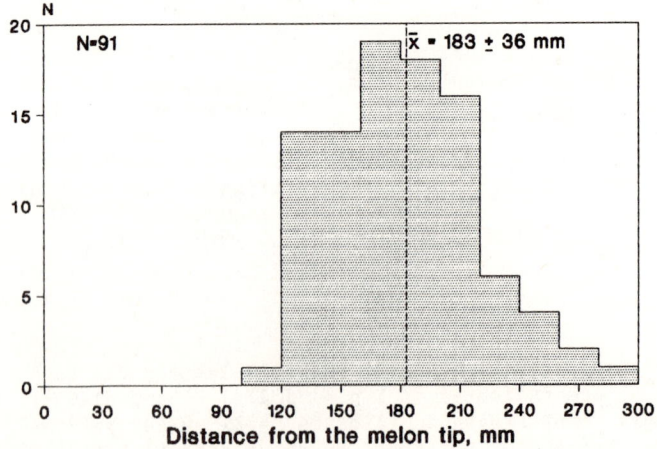

Fig.8. Distribution of the observed distances between the melon tip and the acoustic window in the Tucuxi dolphin.

The Tucuxi dolphin

The ABRs recorded in Tucuxi dolphins were quite similar to that observed in the bottlenosed dolphin, so they are not illustrated here. The method of calculation of the acoustic delays in the Tucuxi dolphin was the same as in the bottlenosed dolphin.

A total of 91 measurements was available in the Tucuxi dolphin. The results are presented in Fig.8. The mean distance from the melon tip to the acoustic window and the standard deviation were 183±36 mm with 99% measurements (90 out of 91) exceeding 120 mm.

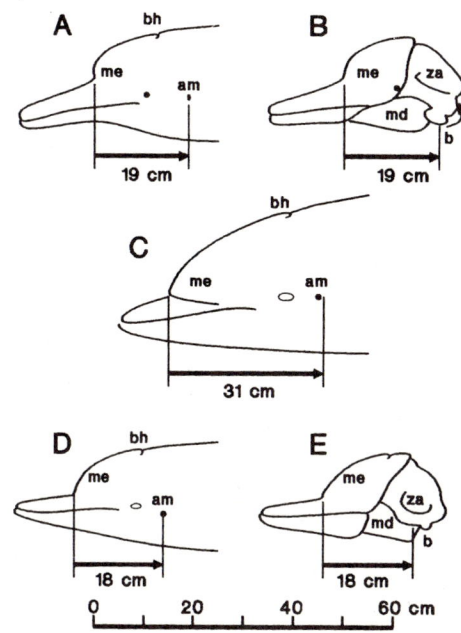

Fig.9. The side-view of the intact and the partially dissected dolphin's head; A & B - an Amazon dolphin, C - a bottlenosed dolphin, D & E - a Tucuxi dolphin. A, C and D - intact heads of the experimental animals, B and E - partially dissected head of dead animals. Arrows indicate the mean distance from the reference point (the melon tip) to the acoustic window. bh - blowhole, me - melon, am - auditory meatus, b - bulla, za - zygomatic arc, md - mandible.

Comparison of results with the head anatomy

To compare these results with the basic head and cranial measurements, Fig. 8 A, C and D present the side-view of the experimental dolphin's heads. Additionally Fig. 8 B and E show the side-view of the partially dissected head of the dead Amazon and Tucuxi dolphins with the intact melon and exposed lower jaw, basal and occipital parts of the skull.

The Figure shows that the sites indicated by our results - near 19 cm from the melon tip in the Amazon dolphin, near 31 cm in the bottlenosed dolphin and near 18 cm in the Tucuxi dolphin, lye near the auditory meatus and the bulla, and outside the lower jaw.

DISCUSSION

The results of all our measurements in three dolphin species show the presence of the acoustic window in the vicinity of the outer and middle ear (the auditory meatus and the bulla).

On our opinion, these results can not be explained if the unique acoustic window is located in the lower jaw, i.e. if sounds can reach the middle ear only through the lower jaw. In this case acoustic delays would be dependent only on distance from a sound source to the mandibular window (the transmission time from the window to the bulla would be equal for all sounds), and calculations by our method must indicate this mandibular window. Really this was not observed. Our results can be explained only if the direct sound transmission is possible to the outer and(or) middle ear.

On the other hand, we don't think that the problem of the acoustic window is solved completely by our results. We don't exclude the possibility of additional ways of specific sound transmission, especially from the frontal direction, where the auditory sensitivity is maximal.

REFERENCES

Brill R.L., Sevenich M.L., Sullivan T.J., Sustman J.D. and Witt R.E., 1988, Behavioral evidence for hearing through the lower jaw by an echolocating dolphin (Tursiops truncatus). Mar. Mammals Sci., 4: 223-230.
Bullock T.H., Grinnell A.D., Ikezono E., Kameda K., Katsuki J., Nomota M., Sato O., Suga N. and Yanagisawa K., 1968, Electrophysiological studies of central auditory mechanisms in Cetaceans. Z. Vergl. Physiol., 59, 117-156.
Fraser F.C. and Purves P.E., 1954, Hearing in cetaceans. Bull. Brit. Mus. (N.H.), 2: 103-116.
Fraser F.C. and Purves P.E., 1960, Hearing in cetaceans. Bull. Brit. Mus. (N.H.), 1: 1-140.
Fleischer G., 1978, Evolutionary principles of the mammalian middle ear, in: "Advances in Anatomy, Embryology and Cell Biology", Springer Verl., Berlin, 55.
Lipatov N.V., 1978, The functional role of the external auditory meatus under water (in Russ.), in: V.E.Sokolov (ed.), "Morskiye mlecopitayushchie", Nauka, Moscow, 112-124.
McCormic J.G., Wever E.G., Palin J. and Ridgway S.H., 1970, Sound conduction in the dolphin ear. J. Acoust. Soc. Amer., 48, 1418-1428.
Norris K.S., 1968, The evolution of acoustic mechanisms in odontocete cetaceans, in: Drake E.T. (ed.) "Evolution and environment", New Haven-Lond., Jail Univ. Press, 297-324.
Popov V.V. and Supin A.Ya., 1985, Determination of characteristics of the dolphin's hearing with the brain stem evoked potentials (in Russ.). Dokl. Acad. Nauk SSSR, 283, 496-499.
Popov V.V. and Supin A.Ya., 1988, Hearing directivity diagram of the dolphin Tursiops truncatus (in Russ.). Dokl. Acad. Nauk SSSR, 300, 756-760.
Ridgway S.H., Bullock T.H., Carder D.A., Seeley R.L., Woods D. and Galambos R., 1981, Auditory brainstem response in dolphins. Proc. Nat. Acad. Sci. USA, 78, 1943-1947.

CONCLUDING COMMENTS ON CETACEAN HEARING AND ECHOLOCATION

Frank T. Awbrey

Biology Department, San Diego State University and Sea World Research Institute,
San Diego, CA

INTRODUCTION

About forty years ago biologists began to learn that dolphins have an extraordinary ability to use sound to find food and get precise information about their environment. That line of research proved to be very fruitful. Hundreds of studies cover a range of topics from simple descriptions of sounds to complex theoretical models of cetacean echolocation and hearing. The result of all that effort is that we know quite a bit about some aspects of echolocation and hearing in a few odontocete species but almost nothing about it in most cetacean species, especially the mysticetes.

PRODUCTION OF ECHOLOCATION SIGNALS

Our charge, in this workshop, was to identify some of the main unanswered questions and controversies in cetacean echolocation and hearing and to think about directions and approaches for future research. Our group focused first on the production of echolocation signals. From the beginning, investigators divided odontocete sounds into those thought to serve for communication and the broadband clicks that serve for echolocation (Schevill and Lawrence, 1949, 1950; Wood, 1953). Determining the nature of those signals always has been a prominent part of studies because understanding how odontocetes echolocate requires knowing how acoustical energy is distributed over time and among frequencies. Most of the small cetaceans whose signals have been recorded (see Watkins and Wartzok, 1985, for a summary) produce trains of very brief broadband clicks with emphasized frequency bands. Porpoises, and dolphins in the genus <u>Cephalorhynchus</u>, differ in that most of the energy in their clicks is concentrated in a band of frequencies about 20 kHz wide, centered between 110 and 140 kHz (Møhl and Andersen, 1973; Kamminga and Wiersma, 1982; Dawson, 1988).

We have a good start on understanding the physics of cetacean echolocation. Still, it is just a start and we have a long way to go before we can say that we really understand how these animals use echolocation. The important questions now concern what we need to know and how best to go about learning it. In two reviews, Johnson (1980; 1986) suggested some important problems for future study. There has been some progress on the questions he posed. For example, in this volume, Au reports using synthesized phantom echoes in detection experiment and Thomas et al. (1990) extend noise masking experiments to another species. Nevertheless, many fundamental questions Johnson raised, plus many others, remain unanswered, including:

Exactly Where and How are the Clicks Produced?

The consensus now is that clicks, while dependent on gas pressure in the nasal passages, are produced in soft tissues (see Cranford, 1988 and Amundin, this volume, for a summary of the evidence). None of the experiments, so far, show how the structures implicated make clicks. In fact, the work of Amundin et al. (1988) implies that Cephalorhynchus commersonii and Phocoena phocoena make very similar narrowband clicks with substantially different apparatus. We do not know how the click-producing mechanisms of these species compare to those of species like Tursiops, which make broadband clicks. Definitive answers to these questions probably will require new or vastly improved, non-invasive instruments that can show detailed changes in structures while live animals make clicks.

How Much do Clicks Vary Among Species in Waveform, Frequency, Sound Pressure Level and Beam Patterns?

Good understanding of cetacean echolocation depends on the ability to generalize. For that, we need comparative studies, but lack necessary data. As yet, most species have not been recorded at all. Many species need to be re-recorded with broadband instruments and with known orientation toward the hydrophone.

How Much Control Over Their Clicks do Odontocetes Have?

Romanenko (this volume) and Moore (this volume) report progress on this issue, showing that Tursiops has some control over sound pressure level, pulse shape and peak frequency of its clicks. Belugas also change levels and frequencies in different conditions (Au, et al., 1985). Markov (this volume) has data supporting the hypothesis that dolphins (Tursiops) have more than one click generator, and that control is inversely related to the number of generators employed in generating a signal. Still unanswered are such questions as: How do species that make narrowband pulses cope with noisy environments? Can they also vary SPL or frequency? What do the larger species do? How much, if any, can odontocetes control their beam patterns? Can they steer the beam or vary its width? How does this ability, if it exists, compare among species? What is the mechanism for beam forming? Is the melon involved, as generally assumed, or can odontocetes use some nasal structure to shape and direct the beam in the manner proposed for leafnose bats by Pye (1988)? The phase shift patterns he shows resemble those of the narrowband odontocetes. If that happened to be the way Phocoena or Cephalorhynchus shaped their beams, would that imply the same mechanism for broadband clicks? Answers to these questions could direct research to help settle the question of whether odontocetes use powerful "bangs" to stun prey (Norris and Møhl, 1983).

PERCEPTION

A second area needing attention is perception because, obviously, echolocation systems need both a signal and a receiver. We assume that all odontocetes echolocate, but have verification for only 13 species. Even so, nearly all of the detailed information on odontocete echolocation capability comes from Tursiops, Delphinapterus and Phocoena. Again, we need comparative data on more species, especially the larger ones.

Audiometry

More comparative data on basic audiometry would let us compare the characteristics of odontocete sounds with what they can hear. Beyond that, critical bands and critical ratios, or, better, filter shapes are needed for inferences about how the animals cope with noisy environments. Dubrovski (personal communication) postulates a "panoramic" filter in the dolphin's central nervous system. Essentially, this system waits for any frequency that comes in, then builds a filter around it. He and other Soviet scientists are using neurophysiological methods, as well as behavioral tests, to look at filter characteristics (Supin and Popov, this volume) and other attributes of hearing (Supin, this volume). There is a need for better, more portable systems to collect neurophysiological data in ways that are harmless to the animals and with results directly comparable to those obtained in behavioral experiments with conditioned animals.

Frequency and Time Discrimination

In addition to more comparative data on their basic ability to distinguish between frequencies, we need to know more about how odontocetes use frequency-modulated and amplitude-modulated signals. When dolphins change the structure of their clicks, and when clicks reflect off targets, their spectral content changes. How small a change can odontocetes detect? What is the basis for that discrimination; time, frequency or phase change? What neural and cochlear mechanisms are involved and what are their relative importance? Are the changes in signals related to the nature of the target or to some environmental condition? Are they consistent?

Localization of Sound Sources

We know that dolphins are very good at discriminating and locating targets, but we still do not understand how they do it. The need to have dolphins work from a fixed position so we can measure their signals precisely hampers progress here. If dolphins were free to swim and approach targets as they would in nature, we could learn much more about the dynamics of the echolocation process. We need to know how the production and processing of pulses changes during detection and approach to targets. Do odontocetes consistently use certain kinds of signals in particular situations? Just how flexible is their sonar? What are the mechanisms of spatial hearing in odontocetes? No one yet has investigated binaural hearing in odontocetes. Is our knowledge about human binaural hearing applicable, or do odontocetes use a different mechanism for spatial localization? Studies involving synthesized phantom echoes or human listeners and electronically transformed echoes should prove useful in these investigations.

Signal Analysis

One great difficulty in understanding how odontocetes echolocate comes from the difficulty of understanding how they process and analyze echoes. Theorists propose various models, usually based on man-made sonar, and experimenters then are challenged to test those models. Investigators have done well here, especially considering the difficulties involved in training animals, and then measuring their signals and returning echoes. Dolphins are too valuable to use in the kinds of combined neurophysiological and neuro-anatomical approaches that have proved so fruitful with bats, so investigators have to use less direct approaches. Emerging techniques for medical imaging will be used more and more as investigators collaborate with colleagues in medical facilities. One challenge here is to use resources wisely. When animals die, either in captivity or in strandings, we should preserve the heads as quickly as possible and send them the anatomists who need them to map auditory pathways (Zook, this volume) and reconstruct details of the hearing apparatus (Ketten, this volume).

The Hearing Receptor

The question of the location of the odontocete hearing receptor still is not answered to everyone's satisfaction. In this volume, Romanenko and Supin interpret their data to mean that the auditory window is located near the auditory meatus or bulla, rather than on the lower jaw (Brill et al, 1988). Goodson and Klinowska argue for a role of tooth spacing and jaw asymmetry in receiving high-frequency pulses. Behrmann proposes that Phocoena has three organs for receiving sound. Resolution of this controversy will require critical experiments that do not depend on assumptions about the density of soft tissues in the head or that one or the other of these mechanisms is exclusive. The solution will come faster through collaboration with anatomists who use computed X-ray tomography (CT scans) to build detailed models showing precise anatomical relationships of all structures near the ear in live animals. These may give clues about how to prevent conduction to the ear by all pathways but the one of interest in the experiment.

Learning and Development

Learning and attentiveness play significant roles in signal detection. The phenomena are interesting in themselves as problems in cognition, but that also means that investigators have to control for them in experiments. Now that dolphins are breeding so successfully in captivity, there is an opportunity for detailed studies on the ontogeny of echolocation. It may be possible to couple these studies with non-invasive methods, such as CT scans, for following the morphological development of associated structures.

Echolocation vs. Listening

We often assume that odontocetes must use echolocation to get useful information about their surroundings when vision is limited. However, they probably get a great deal of information simply by listening (I suggest using "listening" rather than an oxymoronic term like "passive sonar" or the misleading "passive hearing". The animals are not passive when they are listening intently). How much of the time do dolphins move silently? When in schools, do all echolocate actively, or do only certain ones, such as the leaders echolocate? Do they use either or both of the two modes differently in pools than in the open ocean?

INSTRUMENTATION AND STANDARDS

Progress toward understanding odontocete echolocation continues, but some improvements and innovations in instrumentation and adherance to established standards could speed things along.

Evoked Potentials

Getting audiometric data usually entails long periods for operant conditioning and testing of captive animals. Often, however, an electrophysiological approach may be a better way to get the data. Evoked potentials, for example, give reliable data quickly (Seely, et al., 1976; Voronov and Susman, 1986; Supin, this volume). Sometimes, as with stranded, entangled, or temporarily restrained wild animals, electrophysiology may be the only way to get data. In fact, its procedures may become the method of choice for much of the work now done with conditioned animals that work from a fixed position. Light-weight, versatile systems based on laptop computers will let investigators measure hearing whenever and wherever animals are available.

Data Collection

Many of the questions about odontocete sonar need acoustical or electrophysiological data recorded from precise positions around unrestrained animals. Progress here depends on developing small packages that dolphins can wear. These packages would hold various combinations of amplifiers and signal conditioners for hydrophones or electrodes, a video camera and recorder for documenting the exact position of the animal relative to the target, and a recorder for acoustical and electrophysiolgical data. Digital memory is becoming a feasible alternative to analog recorders for storing data in these applications. If the package includes facilities for processing signals, digital storage becomes even more attractive. At first these methods would work only with captive animals, but eventually they may be applicable to animals in the wild. If so, we finally might start to learn how odontocetes use sound in coping with the demands of their environments.

Standards

We all spend much time reading scientific papers and reports keep up with what is happening at the forefronts of our fields. The task of understanding and comparing findings would be a little easier if everyone used the standards published by international standards committees. For example, computing sound pressure levels with a non-standard reference pressure forces readers to convert in order to compare results. The International Electrochemical Committee (IEC) standard for underwater reference pressure is 1 micropascal and, for the sake of clear communication, we should all use it rather than 20 micropascals, 1-millipascal, 1 dyne per cm^2, 1 pound per square foot, or anything else. The IEC has set another standard that few people know of, yet it would simplify comparing acoustical data at least as much as a common reference pressure would. Graphs prepared with a ratio of 2 mm per decibel on the ordinate to 50 mm per frequency decade on the abscissa would all have the same proportions, facilitating comparison of the shapes of hearing curves or sound spectra, even if the absolute sizes of the graphs were different. And think of the time we all would save if every journal and book used the same format for references.

Through all of this, keep in mind that odontocetes are animals and that echolocation is adaptive. We expect that species from different environments might have different systems, or might use the system in different ways. Consequently, generalizing from too limited a database can lead us astray. Proper understanding of odontocete echolocation requires comparative information from as many species in as many different habitats as possible and collaboration among acousticians, morphologists, behaviorists, ecologists and all others involved in this research.

ACKNOWLEDGMENT

Patrick Moore, in a moment of inspiration, produced such a nice outline that I used it to organize this summary.

REFERENCES

Amundin, M., Kallin, E. and Kallin, S., 1988, The study of the sound production apparatus in the harbour porpoise, Phocoena phoecena, and the jacobita, Cephalorhynchus commersoni by means of serial cryo-microtome sectioning and 3-D computer graphics, in: "Animal Sonar: Processes and Performance," P. E. Nachtigall and P. W. B. Moore eds., Plenum Press, New York and London.

Au, W. W. L., Carder, D. A., Penner, R. H., and Scronce, B. L., 1985, Demonstration of adaptation in beluga whale echolocation signals, J. Acoust. Soc. Am., 77:726.

Brill, R. L., Sevenich, M. L., Sullivan, T. J., Sustman, J. D., and Witt, R. E., 1988, Behavioral evidence for hearing through the lower jaw by an echolocating dolphin (Tursiops truncatus), Mar. Mamm. Sci., 4:223.

Cranford, T. W., 1988, The anatomy of acoustic structures in the spinner dolphin forehead as shown by X-ray computed tomography and computer graphics, in: "Animal Sonar: Processes and Performance," P. E. Nachtigall and P. W. B. Moore eds., Plenum Press, New York and London.

Dawson, S. M., 1988, The high frequency sounds of free-ranging Hector's dolphins, Cephalorhynchus hectori, in: "Reports of the International Whaling Commission (Special Issue 9). Biology of the Genus Cephalorhynchus," R. L. Brownell and G. P. Donovan, eds., International Whaling Commission, Cambridge.

Johnson, C. S., 1980, Important areas for future cetacean auditory study, in: "Animal Sonar Systems," R. G. Busnel and J. F. Fish, eds., Plenum Press, New York.

Johnson, C. S., 1986, Dolphin audition and echolocation capacities, in: "Dolphin Cognition and Behavior: a Comparative Approach," R. J. Schusterman, J. A. Thomas and F. G. Wood, eds., Lawrence Erlbaum Associates, Hillsdale, NJ.

Kamminga, C. and Wiersma, H., 1982, Investigations on cetacean sonar V. The true nature of the sonar sound of Cephalorhynchus commersonii, Aquat. Mamm., 9:95.

Møhl B. and Andersen, S. A., 1973, Echolocation; high frequency component in the click of the harbor porpoise, J. Acoust. Soc. Am., 54:1368.

Norris, K. S. and Møhl B., 1983, Can odontocetes debilitate prey with sound?, Am. Nat., 122:85.

Pye, J. D., 1988, Noseleaves and bat pulses, in: "Animal Sonar: Processes and Performance," Nachtigall, P. E. and P. W. B. Moore, eds. Plenum Press, New York and London.

Schevill, W. E., and Lawrence, B., 1953, Auditory response of a bottlenosed porpoise, Tursiops truncatus, to frequencies above 100 kc, J. Exp. Zool., 124:147-165.

Schevill, W. E., and Lawrence, B., 1956, Food-finding by a captive porpoise (Tursiops truncatus), Breviora. Mus. Comp. Zool., 53:1-15.

Seeley, R. L., Flanigan, W. F. Jr., and Ridgway, S. H., 1976, A technique for rapidly assessing the hearing of the bottlenosed porpoise, Tursiops truncatus, "NUC TP 522," Naval Undersea Center, San Diego, CA.

Thomas, J. A., Pawloski, J., and Au, W. W. L., 1990, Masked hearing abilities in a False Killer Whale (Pseudorca crassidens), J. Acoust. Soc. Am., in press.

Voronov, V. H. and Stosman, I. T., 1986, Electrical responses of the stem structures of the acoustic system of Phocoena phocoena to tonal stimuli, "The Electrophysiology of the Sensory Systems of Marine Mammals," V. E. Sokolov, ed., Nauk, Moscow (in Russian).

Watkins, W. A., 1980, Click sounds from animals at sea, in: "Animal Sonar Systems, Biology and Bionics," R. G. Busnel and J. F. Fish, eds., Plenum Publishing Corp, New York.

Wood, F. G., 1953, Underwater sound production and concurrent behavior of captive porpoises, Tursiops truncatus and Stenella plagiodon, Bull. Mar. Sci. Gulf and Caribbean, 3:120-133.

PRELIMINARY RESULTS FROM PSYCHOPHYSICAL STUDIES ON THE TACTILE SENSITIVITY

IN MARINE MAMMALS

Guido Dehnhardt*

University of Münster
Department of Zoology
Badestrasse 9
4400 Münster
Federal Republic of Germany

INTRODUCTION

Psychophysical studies with animals that investigated their capability to identify objects within the environment by tactile senses are rare. Nevertheless, there do exist studies on mammals, especially on species with prehensile extremities like hands and paws, as in primates and procyonidae (Rensch and Dücker, 1963; Davenport et al., 1973; Jarvis and Ettlinger, 1977). However, many other terrestrial mammals, for example pigs, do not use their extremities for tactile investigation of the environment, but the region around the snout (Fuchs and Dücker, 1988).

Marine mammals, whose extremities are reduced and adapted to the aquatic environment, use mainly parts of the snout or rostrum to touch objects. In most cetaceans there are a lot of mechanoreceptors within the hairless skin of the rostrum. These receptors have been identified as Pacinian corpuscles and free-nerve endings (Ling, 1974; Herman and Tavolga, 1980). In addition, river dolphins and mysticete whales have vibrissae along the upper lip area. These special hairs are thought to be used also for touching and may help the animals in detecting food (Layne and Caldwell, 1964; Ling, 1977).

Comparing all mammals, the vibrissae of pinnipeds are the most developed. In some behavioral studies, it already has been mentioned that these stiff hairs in the region around the snout of pinnipeds serve as sensitive tactile organs (Schusterman, 1968; Miller, 1975). In discussions of whether pinnipeds possess an active sonar system (Poulter, 1963, 1969; Hobson, 1966; Oliver, 1978) it has been speculated that the vibrissae may play a significant role. The vibrissae, together with a weakly-developed sonar system, may help the animals in detecting, identifying and catching prey (Schusterman, 1968; Oliver, 1978; Renouf, 1989). It has been presumed that vibrissae are mainly sensitive to vibrations, e.g. caused by fleeing prey (Stephens et al., 1973). However, neurophysiological results by Dykes (1972) suggest that vibrissae by their kind of innervation especially are fit for discrimination of texture and shape of touched

---

*Prof. Dr. Dr. h.c. Bernhard Rensch dedicated to his 90th birthday.

objects. A psychophysical study by Kastelein and van Gaalen (1988) confirms these results for the first time for the walrus (Odobenus rosmarus). Wearing eye caps the walrus was able to distinguish between a circle and a triangle using its (approximately 450) vibrissae, even when the surface area of the stimuli gradually decreased down to 0.4 cm².

The purpose of this study was to investigate whether a California sea lion (Zalophus californianus) is able to achieve a complex form-discrimination by the use of its vibrissae only. This research was based on former experiments in which a female California sea lion discriminated 6 three-dimensional geometric figures each representing a different reward (Dehnhardt, 1988). At discrimination tests with differentially rewarded stimuli the choice of the most rewarded stimulus passed as a correct response. The sea lion was able to discriminate all forms employed in these studies. Although the use of vision was the primary sensory modality in these former studies, it was obvious that the sea lion also touched the figures with her vibrissae. The purpose of the present study was to explore to what extent the female sea lion was able to achieve the same complex form-discrimination by the use of her vibrissae only.

METHODS

The study was accomplished like the former one (Dehnhardt, 1988) at the Dolphinarium Münster with a ten-years-old female Californian sea lion, named Fleur. To eliminate visual discrimination, the sea lion was trained to wear eye caps developed for human medicine. The caps were covered with water-proof plaster. A 4 mm-thick rubber band at the edge of the inner side fixed the caps on the head of the sea lion. When putting on the eye caps, the sea lion had to pass her head through elastic bands which were sewed on to them. These elastics were fixed behind the external ears. This training was difficult and took about two months time.

Stimuli

Five geometrical figures made of 20 mm-thick polyethylen served as stimuli. Each stimulus was rewarded with a different amount of fish (Fig. 1). A semicircle was unrewarded; a hexagon was rewarded with one whole sprat, a "sandglass" was rewarded with two half sprats, a rectangle with two whole sprats and a square was rewarded with four whole sprats. Consequently, the total reward for the two stimuli was the same when presenting hexagon versus sand-glass (1/1 sprat versus 2/2 sprats), just differing in the number of pieces. When presenting the stimuli sandglass versus rectangle (2/2 sprats versus 2/1 sprats), the number of pieces was the same, but the amount of fish reward was different.

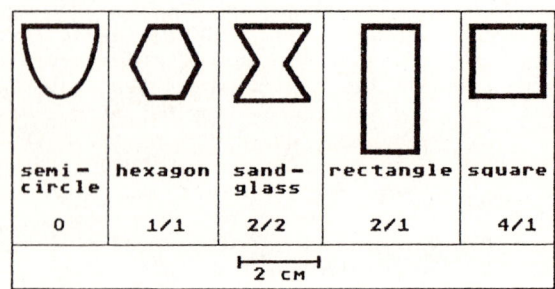

Fig. 1. Five stimuli used in the present study and the associated magnitude of the fish;
(1/1 = one whole sprat; 2/2 = two half sprats)

The weight of each stimulus (semicircle, hexagon, sand-glass and square) was equalized (about 25 g) by little lead balls through drill holes in the stimulus edge. The holes were closed with silicon. The difference in weight between the rectangle and the other stimuli could not be equalized (too many drill holes). In the middle of the upper edge of each stimulus there was a notch of 1 cm to be suspended in clamps at the apparatus. The 5 stimuli lead to 10 two-choice discrimination tasks.

Apparatus

The apparatus was constructed for two-choice discrimination tasks. It consisted of a particleboard covered with white plastic, 1.60 m (H) x 1.00 m (W). Two planks at the sides of the apparatus supported the slightly backwards inclined board (Fig. 2). At a height of 90 cm there was an open window (50 cm (H) x 90 cm (W)) in the board. Above the window two S-shaped, 34 cm long metal bars were fixed at a distance of 46 cm, with plastic clamps at the free ends. Stimuli could be fastened by these clamps and then suspended in front of the window of the apparatus at a distance of 24 cm. Below the window, in the middle of the board, a muzzle was fixed as a stationing point for the sea lion. Because the muzzle was 30 cm in front of the apparatus, the animal stationed at 6 cm distance to the stimuli. From the stationing point, the sea lion could reach the stimuli by just swinging her head from left to right.

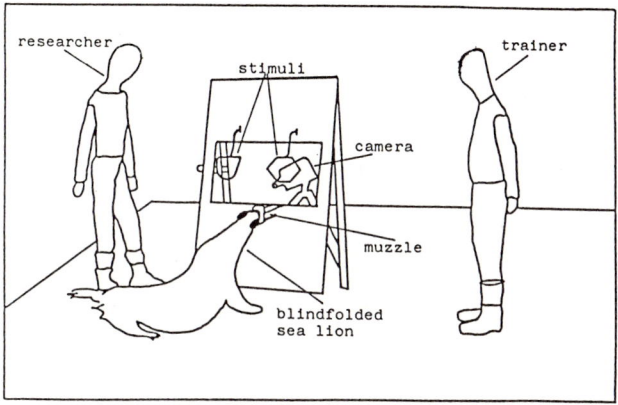

Fig. 2. Experimental situation

Two m behind the apparatus a video camera (SABA CVC 76 SL) was placed. It recorded through the window of the apparatus so that the stimuli, the stationing point, and the sea lion were in view. Therefore, each trial could be evaluated in detail afterwards. Because the camera had a digital clock (1/10 sec. accuracy), it was possible to measure the time the sea lion needed for one trial. With a video recorder (JVC-HR-D 725 EG) the recordings could be studied frame by frame. Therefore, it was possible to gain information about which parts of the stimuli were touched by the sea lion for discrimination and which vibrissae were used.

Experimental Procedures

Phase 1 : Visual Discrimination

After having learned to wear eye caps, the animal was trained for the procedure of the study. This training was carried out without eye caps so that the sea lion was able to use her vision. The complete procedure of the study was as follows:

First, the sea lion stationed in front of the apparatus and pushed her snout against the muzzle. While she was in that position, a trainer, standing at the right-hand side of the sea lion, suspended two stimuli into the clamps of the apparatus. After he had stepped back, the researcher, standing to the left of the sea lion, gave the start signal (a short whistle). When hearing the signal, the sea lion left the stationing point at once. She bit into the lower edge of the most rewarded stimulus of the presented combination and pulled it out of the clamp. With the stimulus in her mouth she turned left to the researcher who took the stimulus and gave the sea lion the appropriate amount of fish.

Because Fleur had not been exposed to the stimuli for over six month, it had to be checked if she was still able to discriminate by vision. Therefore, she was confronted with each of the 10 discrimination tasks. The animal was confronted with a new task not before it had reached criterion during two successive sessions. Every day one session of 16 trials was held. According to the binomial probability table by Bernoulli, 13 correct responses of 16 trials (81.5%) are significant (p = 0.008). As criterion for a successful discrimination at least 14 correct responses (87.5%) were required during a session.

Phase 2 : Tactile Discrimination

In order to perform the final tests, the animal had to wear eye caps. At the beginning of a session the eye caps were put right behind the eyes of the sea lion so that she could orientate by vision. Only when stationing, the eyes of the animal were covered (Fig. 3). When the sea lion was blindfolded, the trainer mounted the stimuli at the apparatus and the start signal was given. As soon as the sea lion handed over the chosen stimulus to the researcher, the trainer removed the stimuli. Then, the researcher took off the eye caps and gave her the appropriate reward. Because of this the animal never saw the 'chosen' or 'not chosen' stimuli.

At first, the sea lion was confronted with each of the 10 discrimination tasks. The number of trials per session and the successful discrimination criterion were the same as it is described for Phase 1. If the animal reached criterion during the first session, the same task was presented for a second session in order gain more data.

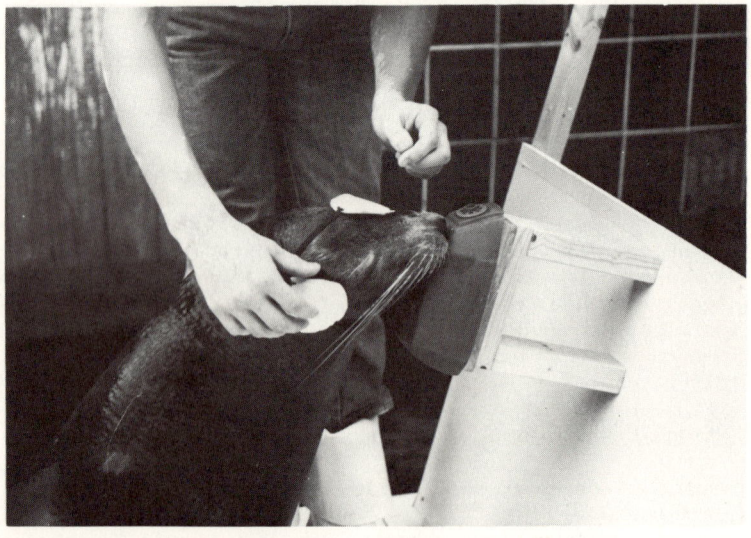

Fig. 3. Sea lion, stationing at the muzzle.

At the end of Phase 2, the sea lion was confronted with the 10 discrimination tasks in random order. One session consisted of 20 trials. This way, each task was presented twice per session. According to the binomial probability table by Bernoulli, 16 correct responses of 20 trials are significant ($p = 0.005$).

Each session was evaluated by analysing the video tape recordings concerning the following aspects:

1. Number of correct responses:
Since at the task hexagon versus sand-glass (1/1 fish versus 2/2 fish) the amount of reward was identical, the stimulus preferred by the sea lion passed as correct response.

2. The time needed per trial (selection time):
The time was measured from the moment the sea lion left the stationing point to that of pulling the chosen stimulus out of the clamp. This period contained both, the "touch time" and the time, needed for decision making (e.g. VTE- behavior).

3. The number of scanning movements before selecting a stimulus:
Playing the video tape recordings in slow motion, the scans were counted for each trial.

4. Which vibrissae were used for touching:
The mystacial vibrissae were devided into two areas: the short vibrissae at the top of the snout and the long vibrissae at the sides of the snout. During a frame by frame analysis it was evaluated which vibrissae came into contact with a stimulus first and which vibrissae were used for further investigation of the stimulus.

5. How the vibrissae were applied:
When playing the video tape in slow motion it was evaluated whether the sea lion moved her vibrissae, parts of the snout or the head.

6. Which part of a stimulus was touched by the sea lion:
The animal had to investigate two stimuli successively during a trial. During a frame by frame analysis it was noted for each trial which parts of the stimuli were touched by the sea lion.

RESULTS

Phase 1 : Visual Discrimination

Although the sea lion had not seen the 5 stimuli for over 6 months, her visual discrimination was faultless in every single task. After the start signal, she moved straight away to the most rewarded stimulus and took it out of the clamp. Only at combination 6 (hexagon versus sand-glass) she swang several times between the stimuli before choosing one. At this combination she always decided for the sand-glass, rewarded with 2/2 sprats. It was obvious that the sea lion did not touch the stimuli with her vibrissae, as she had done during the former study (Dehnhardt, 1988). The faultless visual discrimination was a requirement for the following Phase 2.

Phase 2 : Tactile Discrimination

The results of the sea lion's tactile discrimination of all 10 tasks are summarized in Table 1. During the first session of the first task (semicircle versus hexagon), she merely chose the stimulus suspended in the left position and, therefore, inevitably made mistakes. Having chosen the semicircle four times and, therefore, having received no reward, she gave up "position" preference. From then on, the sea lion always touched both stimuli, first the left and then the right sided stimulus. The sea lion displayed this behavior during the entire study.

From the beginning of the second session of the first task, the sea lion already differentiated between the semicircle and hexagon faultlessly. The sea lion differentiated the stimuli of tasks 2-10 faultlessly from the very beginning on. This result remained constant during the second session of each task.

There were clear differences between the tasks in time needed per trial (average of both sessions; Fig. 4). The sea lion required for tasks 1 and 6 twice as much time and for task 10 even more than twice as much time as for the other tasks. It is obvious that for all tasks including the square (2,5,7,8) the time for decision making was the shortest one.

Table 1. Tactile Discrimination: Percentage of Correct Responses for the 10 Consecutive Tasks.

TASK NO.	1	2	3	4	5	6	7	8	9	10
STIMULI COMBINATION	▽◯	□□	□◯	▽▽	▽□	◯▽	▽□	□◯	▽□	▽□
RELATION OF REWARD	$0:\frac{1}{1}$	$\frac{2}{1}:\frac{4}{1}$	$\frac{1}{1}:\frac{2}{1}$	$0:\frac{2}{2}$	$\frac{2}{2}:\frac{4}{1}$	$\frac{1}{1}:\frac{2}{2}$	$0:\frac{4}{1}$	$\frac{1}{1}:\frac{4}{1}$	$0:\frac{2}{1}$	$\frac{2}{2}:\frac{2}{1}$
CORRECT RESPONSES I. SESSION (%)	68.8	100	100	100	100	100	100	100	100	100
CORRECT RESPONSES II. SESSION (%)	100	100	100	100	100	100	100	100	100	100

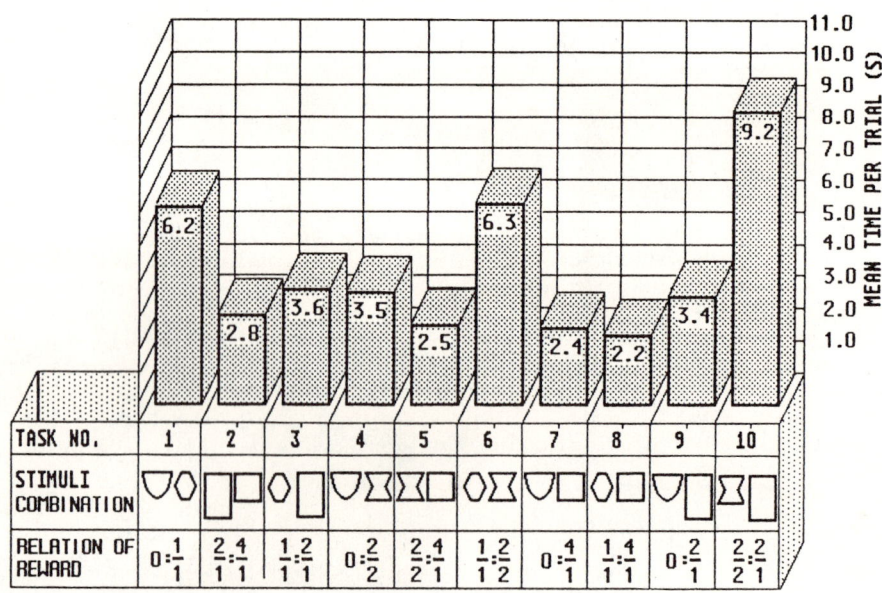

Fig. 4. Mean selection time per trial for the 10 tasks

Table 2. Mean Number of Scans per Trial for the 10 Tasks

task No.	1	2	3	4	5	6	7	8	9	10
$\overline{X}$ number of scans	5.5	2.8	2.9	3.3	2.2	7.5	1.4	2.0	3.2	10.3

The number of scans (average of both sessions = 32 trials) between both stimuli also differed between tasks. The highest figures were for task 1, 6 and 10, the lowest for tasks including squares (Table 2). During all tasks the animal swang her head several times between the stimuli. Only at tasks 2, 5, 7 and 8 it happened that she chose the square directly from the left position or after scanning once (left to right), from the right position.

In order to touch a stimulus, the sea lion used her vibrissae always in the same manner. While stationing, the mystacial vibrissae lie close at her snout (see: Fig. 3). As soon as she left the stationing point the vibrissae were erected so that the lateral parts of her snout vaulted round (Fig. 5). The first contact with the stimuli always happened with the long posterior vibrissae of the snout. Having contact with a stimulus in that way, the sea lion touched it only with the shorter vibrissae at the top of the snout. She laid the erected vibrissae slightly on the edge of the stimulus and made jerky vertical and horizontal head movements. The sea lion did not move the individual vibrissae while touching. The sea lion touched the stimuli only with the vibrissae, not with the snout. The touching was so lightly that the stimuli did not move.

The sea lion always touched the lower part of a stimulus only (Table 3). She glided with her short vibrissae along the lower edge until she reached one of the side edges. She touched the angle or round passage between the lower edge and side edge of the stimulus by head jerks. When doing so, the short vibrissae slid over the angle or round passage several times. Touching the rectangle was slightly different. As all other stimuli

Fig. 5. The blindfolded sea lion, touching the hexagon.

Table 3. Parts of Stimuli Touched and the
Corresponding Angles Between Lower
and Vertical Sides of Stimuli

	SEMI - CIRCLE	HEXA - GON	"SAND - GLASS"	RECT - ANGLE	SQUARE
	▽	⬡	⧖	▯	□
TOUCHED PART OF STIMULI	∨	ᴗ	⌣	L	⌴
ANGLE	∞	115°	60°	90°	90°

had nearly the same length (about 2 cm), the sea lion always searched at this height after the start signal. When in contact with the rectangle with the long lateral vibrissae, she could not find the lower edge of the pattern with the short vibrissae at the top of the snout. With a quick, jerky head movement she glided along a side edge of the rectangle until she reached the lower edge. There she mainly touched the lower right corner.

At tasks 6 and 10, it sometimes happened that the sea lion bit into the lower edge of a stimulus, pulled it slightly, but did not take it. Instead, she once again touched the other stimulus. This behavior could be repeated several times during a trial, before a stimulus was chosen.

When the 10 discrimination tasks were presented in random order within a session, the behavior of the sea lion was the same as described for the sessions where one task was presented only. During 10 sessions with 20 trials each, she made no mistakes.

DISCUSSION

The results of this study show that a California sea lion is able to distinguish 5, three-dimensional objects by means of its vibrissae. The sea lion's tactile discrimination did not differ much from the visual discrimination of the first part of the study. Yet, it is not possible to assess this discrimination as a cross-modal transfer (see: Milner and Ettlinger, 1970; Elliott, 1975) because the sea lion touched the stimuli with her vibrissae in the beginning of visual discrimination tasks too. Therefore, the results do not answer the question of whether sea lions use different cerebral centers for tactile and visual information (Ettlinger and Blakemore, 1966).

Only during the first task (semicircle versus hexagon) of tactile discrimination some kind of learning became visible because the sea lion made mistakes only then. These initial false responses were caused by her position preference. They may be explained by the adaptation from simultaneous visual discrimination in the first phase of the study to successive tactile discrimination in the second phase.

Although the sea lion mastered all other tasks faultlessly, she had different degrees of difficulty with them. This can be seen clearly in the different selection times. For tasks 1, 6 and 10 it took her more time to make a decision than for the other tasks. The novelty of the experimental situation connected with a lack of practice in tactile discrimination may explain the longer time per trial for task 1, but not for the subsequent tasks (6, 10) of the second phase of the study. Therefore, one can suppose that for these tasks the different selection times can be explained by differences in the sea lions ability to discriminate the stimuli. For the successful discrimination of the stimuli in the present study the following processes are necessary:

1) discrimination of form
2) discrimination of reward
3) association of reward to the correct stimulus
4) different valuation of a stimulus in different tasks

Apart from task 1, the longest selection times occurred in those tasks in which the amount of reward for the stimuli was difficult to distinguish (task 6: same amount of reward; task 10: same number of pieces of reward). Because the sea lion quickly identified the stimuli of these tasks during other tasks, it is likely that the longer selection times in tasks 6 and 10 did not result from difficulties in form recognition, but from the fact that the amount of reward was harder to distinguish.

The number of scans (Vicarious trial and error behavior) made by the sea lion per trial may be considered as a degree of difficulty in decision making (Klix, 1971). As it can be seen in Table 2, the highest degrees are in tasks 1, 6 and 10. Both, time per trial and scans per trial were the lowest in the tasks including the square. This suggests that identifying the most rewarded stimulus was the easiest task for the sea lion.

Like the walrus of the study by Kastelein and van Gaalen (1988), the sea lion used its vibrissae for touching actively. If the vibrissae were not used, they lay close to the snout. Miller (1975) described the erection of mystacial vibrissae for other pinnipeds. Beside the function of touching, bristling also is used in different social behavior. It is caused by contraction of fine muscles which are inserted at the follicles of the vibrissae (Andres, 1966).

Depending on the way in which the sea lion uses her vibrissae for touching, two different functional areas of the latter can be described. The long lateral vibrissae are used to find an object. The sea lion then uses the short vibrissae at the top of the snout for identifying it. Kastelein and van Gaalen (1988) also found this regional functional difference in the walrus. While touching, the sea lion did not move its vibrissae individually. While bristling, the vibrissae are visibly stiffened by the contracted follicle muscles; may be the animal is not able to move them in this situation. Touching a stimulus with her vibrissae (see: Fig. 5) the sea lion's jerky head movements led to a passive movement of the vibrissae within the follicles. This stretching and jolting may lead to an excitation of the cutaneous receptors of the snout (Andres, 1966).

The sea lion always touched the lower part of a stimulus, and especially the existing angles, or curve, whereas she probably identified the rectangle by its lower edge which was not hanging at the same height as those of the other stimuli. Whether the sea lion used the described characteristics of the stimuli for visual discrimination too, can only be presumed because of the nearly faultless tactile discrimination.

The sea lion's ability to distinguish between 5 patterns just by touching them, supports Dykes' (1965) assumption that vibrissae are used in the first place to identify shape and texture of an object. This result leads to the question, in which way a blind sea lion (or other pinnipeds) in the wild is able to discover fish (Poulter, 1963; Hewer, 1974), if the vibrissae just give information about objects close to the head. As long as we do not know exactly if pinnipeds have an active sonar (Schusterman, 1981), the answer to this question remains speculative. Provided that pinnipeds did not develop a sonar system, it is conceivable that the survival of a blind sea lion in the wild depends on a change of nourishment. Young harbour seals for example hunt only for shrimp and ground living fish just after weaning (Havinga, 1933). If this is similar for Zalophus, an adult blind animal may change to this youth nourishment. In this case no sonar is needed to find the ground living prey (see: Poulter, 1963) and the vibrisae also are not needed for receiving vibrations caused by fleeing prey (Stephens et al., 1973; Renouf, 1989). The ability to identify shape and texture of an object by means of the vibrissae could enable a blind sea lion to find sufficient food. This hypotheses also means that sea lions searching for food in murky water can find prey with the help of their vibrissae.

The results of this study show that the tested sea lion has a very good tactile form discrimination capability. Further studies will deal with the size discrimination threshold and with the relation between tactile sense and visual sense in discrimination learning.

As for pinnipeds, a lot of suppositions have been made about the well-developed tactile sense in cetaceans (Bryden and Molyneux, 1986). These are based on morphological findings about the innervation of the skin as described first by Palmer and Weddell (1964). It has been speculated that mechanoreceptors in the dermis of cetaceans may serve for the detection of low-frequency vibrations (Bryden and Molyneux, 1986), or may help the animal to detect prey on the river bed (Liu Renjun et al., 1986).

Since functions of a sensory system can only be postulated on the basis of structure, Ling demanded:

"Experimental verification of suggestions based on anatomical findings as to the possible sensory role of the cetacean integument is urgently required." (Ling, 1974, page 41)

It would be difficult to carry out the experiment described in the present paper with the bottlenose dolphin (Tursiops truncatus), because it is suited to the vibrissae of pinnipeds. For this reason the author designed a roughness-discrimination task, which requires a bottlenose dolphin to touch a rough or smooth tray with parts of its rostrum. Since cetacean cutaneous innervation is richer than that from comparable regions in man (Palmer and Weddell, 1964), it is expected that the tested dolphin will show a very fine perception of roughness.

REFERENCES

Andres, K. H., 1966, Über die Feinstruktur der Rezeptoren an Sinushaaren, Zeitschrift für Zellforschung, 75: 339-365.
Bryden, M. M. and Molyneux, G. S., 1986, Ultrastructure of encapsulated mechanoreceptor organs in the region of the nares, in: "Research on Dolphins", M. M. Bryden and R. J. Harrison, eds., Clarendon Press, Oxford, pp. 99-107.

Davenport, R. K., Rogers, C. M., and Russel, J. S., 1973, Cross-modal perception in apes, Neuropsychologia, 11: 21-28.
Dehnhardt, G., 1988, Investigations on the stimulus evaluation problem considering different amounts of reward in the California sea lion (Zalophus californianus), Sonderheft zum Band 53 der Zeitschrift für Säugetierkunde. p.10.
Dykes, R. L., 1972, What the seal's vibrissae tell the seal's brain, Proceedings of the 9th annual conference on biological sonar and diving mammals, Biological Sonar Lab., Fremont, Calif. pp. 123-136.
Elliott, R. C., 1977, Cross-modal recognition in three primates, Neuropsychologia, 15: 183-186.
Fuchs, A., and Dücker, G., 1988, Manipulative problem-solving in a group of collared-peccaries (Tayassu tajacu L.), Sonderheft zum Band 53 der Zeitschrift für Säugetierkunde. p.12.
Havinga, B., 1933, Der Seehund (Phoca vitulina L.) in holländischen Gewässern, Tijdschr. Nederl. Dierkund. Vereen, 3: 79-111.
Herman, L. M., and Tavolga, W. N., 1980, The communication systems of cetaceans, in: "Cetacean behavior", L. M. Herman, ed., Wiley & Sons, pp. 149-209.
Hewer, H. R., 1974, "British seals", Taplinger Publishing, New York. p.256.
Hobson, E. S., 1966, Visual orientation and feeding in seals and sea lions, Nature, 210: 326-327.
Jarvis, M. J., and Ettlinger, G., 1977, Cross-modal recognition in chimpanzees and monkeys, Neuropsychologia, 15: 499-506.
Kastelein, R. A., and v. Gaalen, M. A., 1988, The sensitivity of the vibrissae of a Pacific walrus (Odobenus rosmarus d.), Part 1, Aquatic Mammals, 14(3): 123-133.
Klix, F., 1974, "Information und Verhalten", Huber-Verlag, Bern, Switzerland
Layne, J. N., and Caldwell, D. K., 1964, Behavior of the Amazon dolphin, Inia geoffrensis B., in captivity, Zoologia, 49: 81-108.
Ling, J.K., 1974, The integument of marine mammals, in: "Functional anatomy of marine mammals", Vol. II., R. J. Harrison, ed., Academic Press, pp. 1-44.
Ling, J. K., 1977, Vibrissae of marine mammals, in: "Functional anatomy of marine mammals", Vol. III, R. J. Harrison, ed., Academic Press, pp. 387-415.
Liu Renjun, Harrison, R. J., Thurley, K. W., 1986, Characteristics of the skin of Neophocaena phocaenoides from the Changjiang (Yagtse River), China, in: "Research on Dolphins", M. M. Bryden and R. J. Harrison, eds., Clarendon Press, Oxford, pp. 23-31
Miller, E. H., 1975, A comparative study of facial expressions of two species of pinnipeds, Behaviour, 53: 168-184.
Milner, A. D., and Ettlinger, G., 1970, Cross-modal transfer of serial reversal learning in the monkey, Neuropsychologia, 8: 251-258.
Oliver, G. W., 1978, Navigation in mazes by a grey seal, Halichoerus grypus (Fabricius), Behaviour, 67(1-2): 97-114.
Palmer, E. and Weddell, G., 1964, The relationship between structure, innervation and function of the skin of the bottlenose dolphin (Tursiops truncatus), Proc. Zool. Soc. Lond., 143: 553-567
Poulter, T. C., 1963, Sonar signals of the sea lion, Science, 139: 753-755.
Poulter, T. C., 1969, Sonar of penguins and fur seals, Proc. Calif. Acad. Sci., 36(13): 363-380.
Renouf, D., 1989, Sensory function in the harbour seal, Scientific American, 260 (4): 62-67.
Rensch, B., and Dücker, G., 1963, Haptisches Lern- und Unterscheidungsvermögen bei einem Waschbären, Z.f. Tierpsychologie, 20: 608-615.
Schusterman, R. J., 1968, Experimental laboratory studies of Pinniped Behavior. in: "The behavior and physiology of Pinnipeds", Harrison, R. J., Hubbard, R. C., Peterson, R. S., Rice, C. E., Schusterman, R. J., eds., Appleton-Century-Crofts, New York. pp. 87-171.

Schusterman, R. J., 1981, Behavioral capabilities of seals and sea lions: a review of their hearing, visual learning and diving skills, Psychol. Rec., 31: 125-143.

Stephens, R. J., Beebe, I. J., and Poulter, T. C., 1973, Innervation of the vibrissae of a California sea lion, Zalophus californianus, Anatomical Record, 176(4): 421-442.

TASTE RECEPTION IN THE PACIFIC BOTTLENOSE DOLPHIN (TURSIOPS TRUNCATUS GILLI)

AND THE CALIFORNIA SEA LION (ZALOPHUS CALIFORNIANUS)

William A. Friedl, Paul E. Nachtigall, Patrick W. B. Moore,
Norman K. W. Chun, Jeffrey E. Haun, Richard W. Hall[1],
and James L. Richards[1]

Naval Ocean Systems Center, Kailua, HI 96734
[1]Naval Ocean Systems Center, San Diego, CA 92152

INTRODUCTION

The literature contains relatively scant information on marine mammals' abilities to taste. Kuznetsov (1974), Lowell and Flanigan (1980) and Nachtigall (1986) reviewed research on marine mammal taste reception. Kuznetsov (1978) reports that Black Sea bottlenose dolphins (Tursiops truncatus) can distinguish acidic (sour), bitter, and salty substances but the Tursiops were "sensitive only to bitter stimuli" (Kuznetsov, 1974, p. 177). Kuznetsov's work is discussed in Nachtigall's 1986 review.

Sea lion taste abilities have not been studied extensively, either. Kuznetsov (1982) used a conditioned reflex technique to investigate taste perception of the Steller sea lion (Eumetopas jubatus). He concluded that "...sea lions sense bitter and saline, have a good sensitivity to acid and are indifferent to sweet" (Kuznetsov, 1982, p. 191). California sea lion (Zalophus californianus) gustation has not been studied (Fobes and Smock, 1981).

Anatomical studies (Suchowskaja, 1972; Donaldson, 1977; Valiulina and Khomenko, 1976) report structures that resemble taste buds in small pits in the root region of dolphin tongues. Herman and Tavolga (1980) reviewed the anatomical evidence and concluded the reports were contradictory, but suggested a "gustatory capability for some species"; they also suggested that the taste sensitivity is best assessed behaviorally. Gilevich and Khomenko (1975) and later Agarkov and Gilevich (1979) concluded that odontocete chemoreception was highly specialized because the structure and number of lingual taste receptors was reduced and different from other mammals. Functionally, dolphins may use taste reception in a form of chemically-mediated communication (Yablokov, 1961; Sokolov and Kuznetsov, 1971).

We studied the abilities of a Pacific bottlenose dolphin (Tursiops truncatus gilli) and a California sea lion (Zalophus californianus) to detect what humans perceive as four major tastes: sour, bitter, salty, and sweet. Both animals were male, had been in captivity for several years prior to this experiment, and had been part of other research studies. The dolphin weighed 154 kg and the sea lion 136 kg. Some of the results from the work with the dolphin have been discussed previously (Nachtigall and

Hall, 1984; Nachtigall, 1986). This paper describes the details of those experiments and of the experiments with the California sea lion and discusses comparative aspects of the results.

METHODS AND MATERIALS

The test equipment was functionally similar for both animals, with structural differences to accommodate each species. The dolphin was tested in a floating enclosure; the sea lion on a platform next to his enclosed pool. A floating pen similar to those used in this experiment is illustrated in Figure 5, in Murchison (1980).

Each animal was stationed on a biteplate during testing. The biteplates were made from a nylon composite material (Polymide). Each biteplate had a single opening in the upper leading edge. Solutions flowed through the opening and irrigated the test animal's tongue. The biteplates were mounted on frames and positioned so the test animal's mouth was horizontal during irrigation. Solutions were delivered from polyethylene reservoirs through silicone tubing (Tygon) via peristaltic pumps. The electric pumps had a quick start and stop control and dispensed consistent volumes in a given interval. The tubing was 0.95 cm outer diameter, 0.64 cm inner diameter and nonreactive with our solutions. A four-position valve controlled flow to each biteplate. The stainless steel valve controlled delivery of: (1) distilled water standard solution, (2) test solution, (3) compressed air, and (4) distilled water rinse solution. Between trials, the mouthpiece was purged with air and then rinsed with distilled water. An operator controlled the pumps, coordinated the trial sequence, and recorded results from a small enclosure between the test areas while the experimenter worked the valve and conducted the trials with the animal.

The experimenter sat (for tests with the dolphin) or stood (for tests with the sea lion) behind a visually opaque screen and operated the valve out of the test animal's view. When stationed on the biteplate, the test animal could not see the operator or the experimenter. The enclosure was physically separate from the test areas, so pump vibration was not transmitted to the test animals. A radio was played during testing and we ran an air compressor pump whenever the standard or test solution pump was operated, to mask any audible cues. We also exchanged standard and test solution pumps periodically and changed the tubing connection positions on the four-position valves. Regular checks confirmed that the animals responded to stimulus solutions only.

The dolphin was tested on a fiberglass tray that elevated the animal from the water, but kept the dolphin's body horizontal. The tray was 1.2 m square. The dolphin's horizontal ("beached") position eliminated residual seawater from the dolphin's mouth and prevented seawater contamination of the tests. Between trials, the dolphin stationed opposite the tray to control activity and to minimize behavioral bias. After hearing a 5 kHz underwater tone activated by the experimenter, the dolphin swam from the station position to the tray, beached onto the tray, and placed his mouth on the biteplate. The experimenter could see the biteplate, the dolphin's mouth and rostrum in a strategically-placed mirror but the mirror's position prevented the dolphin from seeing the experimenter. The test sequence began after the experimenter told the operator that the animal was "set" in proper test position.

During testing, the sea lion was positioned on a stand with his forelimbs on a raised platform and his hindlimbs on the deck. Between trials, the sea lion stationed on the stand with his head erect. On a signal from the experimenter, the sea lion leaned forward and placed his mouth on the biteplate. The experimenter stepped behind the screen after telling the operator the animal was "set". Thus, at the start of a test sequence, the test animal was positioned on the biteplate with his head horizontal.

We used a two-irrigation comparison cycle as a test sequence. All test solutions were compared to distilled water. In the sequence, the animal received a first irrigation of distilled water and, after a 5-second pause received a second, comparison irrigation of either distilled water or a test solution. During the pause, the pump was briefly reversed to remove distilled water from the biteplate and the tubing between the valve and biteplate. Thus, second irrigations in test trials were undiluted by residual distilled water, but 10 ml less than the first irrigation.

After the second irrigation, we allowed an additional 5-second response period. The proper response to a test solution comparison irrigation was releasing the biteplate, and to a distilled water comparison irrigation was to remain on the biteplate. We sounded an acoustic signal to reinforce a correct response and rewarded a correct response with fish. The dolphin slid backward from the tray and received his fish in the water. The sea lion remained on the stand and raised his head for his reward.

We designated the distilled water/distilled water comparison a "standard" trial and the distilled water/test solution comparison a "test" trial. The sequence was the same for both animals, but irrigation duration, volumes and bridge tone frequencies differed. The dolphin's 8-second irrigations delivered 80 ml during the first irrigation and 70 ml during the comparison; the bridge tone was 9-kHz. The sea lion's 5-second irrigations delivered 40 ml and 30 ml, respectively; the tone was 2 kHz. The intertrial interval was 45 to 90 seconds for the dolphin and 30 to 45 seconds for the sea lion.

Between trials, we flushed the system with compressed air and distilled water. The tubing and biteplate were filled with distilled water before the first irrigation, but were empty before the second. For test trials, we defined the latency period as the interval from the onset of second irrigation flow from the biteplate until the biteplate was released as a response.

The test solutions were citric acid (sour), sodium chloride (salty), quinine sulfate or quinine monohydrochloride dihydrate (QMD) (bitter), and sucrose (sweet); the solvent was distilled water. Initially, we shaped the animals' response to the test solution by pairing test solution taste with a flow volume cue. We used 0.5-molar (9 ppt) citric acid as the test solution during the shaping, and, initially, 120 ml (dolphin) and 70 ml (sea lion) irrigation volumes. Citric acid solution at 0.5M (9 ppt) is distinctly sour to human taste, but not so acidic as to injure the subjects' oral tissues. After establishing baseline performance with "standard" trials, the "test" solution was presented; release response was immediate. With the behavior established, we decreased test volume flow gradually, over a series of sessions, to nominal irrigation volumes i.e., 70 ml for the dolphin and 30 ml for the sea lion. During the flow volume reduction series, the minimum criterion for correct detection was 80%. With the behaviors established, we then tested the animals' detection ability, using a random series of test solution presentations (Gellermann, 1933) and a distilled water standard solution. The Gellermann series equalizes the number of "test" and "standard" trials within a session and presents no more than 3 consecutive trials of the same condition.

We used the go (release biteplate) no go (remain on biteplate) procedure to test the animals' abilities to detect the four tastes. After taste detection was established, we bracketed the approximate taste thresholds. We conducted a series of "probe trials" to estimate a range of concentrations that would be above and below the threshold level. The probe series were 6 trials each, 3 test and 3 standard, randomly presented. In the threshold determination series, we used a modified method of constants. With the method of constants, we presented successive blocks of trials for each test concentration. Each block was 6 trials; 3 standard and 3 test solutions. Each session had 3 blocks (18 trials) and we ran 2 sessions per day. Daily block presentation order was randomized; but for each session, the highest concentration was tested in the first block. Except for the sucrose and quinine sulfate tests with the sea lion, we presented six concentrations of each test solution and monitored the animals' detection performance until at least 100 trials were run at each concentration. After collecting data for six concentrations, we defined taste thresholds by arbitrarily establishing 75% correct response as the detection criterion. The 75% threshold was selected because half the presentations were standards. If the test animal failed to respond (i.e. release the biteplate), a 50% performance level would result. Our experience indicated that the animals' correct detection performance was 80% or better when detection was distinct and 60% or less when the taste stimulus was detected no longer. The 75% level thus indicated a concentration where detection performance was somewhat below maximum, but not totally stimulus-independent. We graphed performance (percent correct detection) as a function of test solution concentration and, from the graph, estimated the threshold concentration.

We established behavior, tested detection ability, and estimated detection thresholds initially for citric acid solutions with both animals. We tested detection abilities and measured thresholds (as appropriate) for the other solutions successively. For each change in substance tested, we stabilized the test animals' behavior using a detectable concentration of previously detected substance and introduced the new test substance at a relatively high concentration. Before the transfer session, we checked the animal's performance with the previously studied substance at a level above threshold; criterion was 90% correct detection. In the transfer session, the animal received two, 10-trial blocks (5 test, 5 standard) at random. The previously tested substance was the "test solution" in the first block, the substance to be introduced was the test solution in the second. We introduced the new solution at a concentration that was distinctly detectable to humans.

RESULTS

Our results indicate this bottlenose dolphin detected the four primary tastes in the solutions tested. The California sea lion detected the salty and sour tastes and the bitter taste of quinine monohydrochloride dihydrate (QMD) solutions, but did not detect the bitter taste of quinine sulfate or the sweet taste of sucrose.

Table 1 summarizes our results for the dolphin and Table 2 summarizes results from the sea lion. Values for the correct response in Tables 1 and 2 are normalized means from the total correct responses from the test sessions with each solution. Except for the sucrose solution with the sea lion, each concentration of test solution was presented 102 times in 17 sessions. For the sea lion, sucrose was presented 51 times at 0.5 M, 130 times at 1.0 M, and 107 times at 2.0 M.

Table 1. Results from Taste Detection Test Series for the Pacific Bottlenose Dolphin.

Test Solution	Test Concentration (M)	Correct Responses (%)	Response Latency Mean and (st. dev.) in Seconds
Citric Acid	.026	94.7	3.25 (0.59)
	.024	99.0	3.41 (0.68)
	.022	96.0	3.54 (0.68)
	.020	96.0	4.03 (1.08)
	.018	77.5	4.10 (1.02)
	.016	69.7	4.47 (1.02)
Quinine Sulfate	2.86×10^{-5}	98.0	3.03 (0.64)
	2.45×10^{-5}	89.2	3.42 (0.89)
	2.04×10^{-5}	84.3	3.40 (0.78)
	1.63×10^{-5}	80.4	3.61 (0.81)
	1.22×10^{-5}	71.6	3.66 (1.24)
	0.81×10^{-5}	53.9	3.77 (1.17)
Sodium Chloride	.30	97.1	3.06 (0.34)
	.20	95.1	3.51 (0.56)
	.15	82.4	4.24 (0.89)
	.10	77.5	4.15 (0.89)
	.05	69.6	4.31 (0.80)
	.03	60.8	4.48 (0.88)
Sucrose	.20	92.2	3.12 (0.34)
	.15	96.1	3.57 (0.82)
	.08	71.6	5.55 (1.64)
	.06	81.4	5.50 (1.29)
	.04	75.5	5.77 (1.57)
	.02	64.7	5.65 (1.63)

Response latency values are mean intervals to response for all correct detections in test trials at each concentration. Sample size varied for each concentration because the number of correct responses was different for the different solutions. The irrigation interval varied slightly with solution and depended on the latency of the response. The maximum total response interval (irrigation time plus response period) was 12 seconds. Latencies ranged from 2 to 9 seconds for the dolphin and from 1 to 12 seconds for the sea lion.

The California sea lion's bitter taste detection ability was tested with QMD solutions rather than quinine sulfate, the bitter test solution which we used with the dolphin. The sea lion's behavior was balky and erratic throughout the attempted transition to quinine sulfate test solution and response rates were inconsistent. The sea lion's performance indicated that he was not detecting the quinine sulfate, so the attempt was abandoned and the sea lion's baseline behavior re-established. The sea lion maintained consistent behavior with QMD test solutions. The QMD solutions ranged in concentration from 9 parts per million (ppm) to 0.7 parts per thousand (ppt). The sea lion's 75% detection threshold was 0.396 ppt (0.022 molar).

Table 2. Results from Taste Detection Test Series for the California Sea Lion.

Test Solution	Test Concentration (M)	Correct Responses (%)	Response Latency Mean and (st. dev.) in Seconds
Citric Acid	.0200	91.0	3.91 (0.76)
	.0150	90.0	3.83 (1.07)
	.0100	88.0	4.25 (0.92)
	.0050	85.0	4.78 (1.83)
	.0010	67.0	4.14 (0.79)
	.0005	60.0	3.73 (0.64)
Quinine Monohydrochloride Dihydrate (QMD)	.0400	80.5	4.59 (1.96)
	.0300	82.4	4.71 (1.55)
	.0200	73.1	5.14 (2.19)
	.0100	64.8	5.06 (1.80)
	.0050	63.0	4.75 (1.73)
	.0005	46.3	4.43 (1.71)
Sodium Chloride	.60	88.2	4.43 (1.75)
	.40	87.2	4.14 (1.59)
	.20	75.5	3.76 (1.55)
	.10	63.7	4.37 (1.90)
	.05	60.8	3.61 (1.45)
	.01	53.9	3.86 (1.85)
Sucrose	0.5	57.1	6.12 (1.28)
	1.0	52.6	5.40 (1.20)
	2.0	40.0	7.20 (2.97)

DISCUSSION

The bottlenose dolphin's taste acuity is roughly an order of magnitude less than humans (Pfaffman, 1959). The California sea lion's threshold for sodium chloride is 20 times the human standard, but the threshold for citric acid is nearly the same. Seawater is isotonic with a 0.6 M sodium chloride solution (Nicol, 1960). Our results with sodium chloride solutions indicate that sea lions can detect salt solutions of even lower salinity. To check this result, we tested the sea lion with a seawater test solution against a distilled water standard. The seawater was from Kaneohe Bay, Hawaii, with approximately 35 ppt salinity. The sea lion discriminated seawater at a performance level above 80%. This result implies that sea lions have the ability to discriminate fresh and salt water.

Table 3 compares the calculated Pacific bottlenose dolphin and California sea lion threshold values with human thresholds. Kuznetsov (1978) reported the following thresholds for Black Sea bottlenose dolphins: citric acid, 0.99 ppt; quinine monohydrochloride dihydrate (QMD), 0.2 to 0.6 ppm; sodium chloride, 5.4 ppt; and no detection for sucrose. Thus, the reported taste reception abilities and thresholds for Pacific and Black Sea bottlenose dolphins differ, but the causes of the differences are not obvious from the available data.

Table 3. Detection Thresholds for Distilled Water Solutions; NT = not tested, ND = not detected.

Test Solution	Human (ppt)	Dolphin (ppt)	Sea Lion (ppt)
Citric Acid	0.04	0.3	0.05
Quinine Sulfate	1.4×10^{-4}	2.4×10^{-4}	ND
Quinine Monohydrochloride Dihydrate	5.4×10^{-4}	NT	0.40
Sodium Chloride	0.18	1.6	3.6
Sucrose	0.18	< 1.6	ND

The dolphin's latency periods increased as test substance concentration (and the correct response percentage) decreased for all solutions tested. The sea lion's latency periods were unrelated to test solution concentration. This observation may indicate that somewhat different sensing processes operate in sea lion and dolphin taste discrimination.

Our results confirm other studies which considered bottlenose dolphins' taste reception abilities from anatomical, physiological, or behavioral evidence. However, we demonstrated for the first time that bottlenose dolphins can detect sweet substances and that California sea lions have gustatory senses. Our results show that the California sea lion's ability to detect basic tastes resembles that reported for the Steller sea lion by Kuznetsov (1982).

This study, and those of Kuznetsov (1974, 1978, 1982) establish the primary taste abilities of bottlenose dolphins and California and Steller sea lions, but more work is required to identify how these abilities function as part of the animals' natural social and behavioral biology (see Watkins and Wartzok, 1985, p. 241). Ceruti et al. (1985) identified a number of chemoreceptively active compounds in secretions, excretions and tissues of bottlenose dolphins and California sea lions. Yablokov (1961) and Sokolov and Kuznetsov (1971) discuss possible functions of chemoreception ability in marine mammals, but definitive studies have not been conducted. The curious tests reported by Klinowska et al. (1987) indicate the difficulty associated with studies of dolphin chemoreception. So, the way marine mammals use their ability to taste remains unclear and, as Nachtigall (1986) observed, much work remains for patient researchers in the area of marine mammal chemoreception.

ACKNOWLEDGMENTS

We gratefully acknowledge the assistance of Ron Schusterman in the experimental design and Ignatius Lopez, Jr., Gail Peiterson, Dorian Dunbar and Ronald Yamada in training.

REFERENCES

Agarkov, G. B. and Gilevich, S. A., 1979, On the question of chemoreception by dolphins, Vestnik Zool. 1979:3-11.

Ceruti, M. G., Fennessey, P. V. and Tjoa, S. S., 1985, Chemoreceptively active compounds in secretions, excretions and tissue extracts of marine mammals, Comp. Biochem. Physiol., 82A (3):505-514.

Donaldson, B. J., 1977, The tongue of the bottlenosed dolphin (Tursiops truncatus), in: "Functional Anatomy of Marine Mammals, Vol. 3," R. J. Harrison, ed., Academic Press, New York.

Fobes, J. L., and Smock, C. C., 1981, Sensory capacities of marine mammals, Psychol. Bull., 89:288-307.

Gellermann, L. W., 1933, Chance orders of alternating stimuli in visual discrimination experiments, J. Gen. Psychol., 42:206-208.

Gilevich, S. A., and Khomenko, B. G., 1975, On the question of intraorganic innervation of the lips, oral and lingual mucosa of dolphins, in: "Morskiye Mlekopitayushchiye," G. B. Agarkov, ed., Naukova Dumka, Kiev.

Herman, L. M. and Tavolga, W. N., 1980, The communication systems of cetaceans, in: "Cetacean Behavior: Mechanisms and Functions," L. M. Herman, ed., Wiley, New York.

Klinowska, M., Lockyer, C., and Morris, R. J., 1987, Chemoreception in wild and captive dolphins, in: "Report of the European Cetacean Society," J. W. Broekma and C. Smeenk, eds., Hirtshals, Denmark.

Kuznetsov, V. B., 1974, A method of studying chemoreception in the Black Sea bottlenose dolphin (Tursiops truncatus), in: "Morfologiya, Fiziologiya I Akustika Morskikh Mlekopitayushchikh," V. Ye. Sokolov, ed., Nauka, Moscow.

Kuznetsov, V. B. 1978. Certain characteristics of the chemoreception of dolphins, in: "Morskiye Mlekopitayushchiye," G. B. Agarkov, ed., Central Scientific Research Institute for Information and Technical and Economic Studies of Fisheries, Moscow.

Kuznetsov, V.B., 1982, Taste perception of sea lions, in: "Izucheniye, Okhrana, I Ratstional'noye ispol'zovaniye Morskikh Mlekopitayushchikh, V. A. Zemskiy, ed., Ministry of Fisheries, USSR, Ichthyology Commission, VNIRO, and the Academy of Sciences, USSR, Astrakhan'.

Lowell, W. R. and Flanigan, W. F., Jr., 1980, Marine mammal chemoreception, Mam. Rev., 10:53-59.

Murchison, A. E., 1980, Detection range and range resolution of echolocating bottlenose porpoise (Tursiops truncatus), in: "Animal Sonar Systems," R. G. Busnel amd J. F. Fish, eds., Plenum Press, New York.

Nachtigall, P. E., 1986, Vision, audition, and chemoreception in dolphins and other marine mammals, in: "Dolphin Cognition and Behavior: A Comparative Approach," R. J. Schusterman, J. A. Thomas and F. G. Wood, eds., Lawrence Erlbaum Associates, Hillsdale, NJ.

Nachtigall, P. E. and Hall, R. W., 1984, Taste reception in the bottlenose dolphin, Acta Zool. Fennica, 172:147-148.

Nicol, J. A. C., 1960, "The biology of marine animals," Sir Isaac Pitman and Sons, Ltd., London.

Pfaffmann, C, 1959, The sense of taste, in: "Handbook of Physiology: Neurophysiology, Vol.I," J. Field, H. W. Macoun and V. E. Hall, eds., Williams and Wilkins, Baltimore.

Sokolov, V. Ye. and Kuznetsov, V. B., 1971, Chemoreception in the Black Sea dolphin Tursiops truncatus Mont. Dokl. Akad. Nauk SSSR, 201: 998-1000.

Suchowskaja, L. I., 1972, The morphology of the taste organs in dolphins, in: "Investigations on Cetacea, Vol. 4," G. Pilleri, ed., Der Bund, Berne.

Valiulina, F. G. and Khomenko, B. G., 1976, Micromorphology of the receptor system of the tongue in the bottlenose dolphin, Tursiops truncatus, Zool. Zh., 3:467-470.

Watkins, W.A. and Wartzok, D., 1985, Sensory biophysics of marine mammals, Mar. Mam. Sci., 1(3):219-260.

Yablokov, A. V., 1961, The "sense of smell" in marine mammals. Trudy Soveshch. Ikhtiol. Komm., Akad. Nauk SSSR, 12:87-93.

COGNITIVE PERFORMANCE OF DOLPHINS IN VISUALLY-GUIDED TASKS

Louis M. Herman

Kewalo Basin Marine Mammal Laboratory and Department of
Psychology, University of Hawaii
1129 Ala Moana Boulevard, Honolulu, Hawaii 96814 USA

INTRODUCTION

Sensory systems function to monitor events relevant to an animal's well-being and success. In some cases, a sensory system also may serve as a valuable interface between the real world and higher cognitive centers that deal with abstractions, knowledge, generalizations, and representations. It is important to distinguish between these two functions--the strict biological and the cognitive--as they separate the relatively rigid, constrained system from the more open, flexible system. There is a difference, for example, between seeing a fish and then beginning a capture strategy, and seeing a television scene of a fish being captured and recognizing it as a representation of a real-world event.

One common view in the animal cognition literature is that while all sensory systems are available for the biological function, not all, and typically only one--the "dominant" sensory system--contribute to the cognitive function. For example, monkeys (e.g., *Macaca* sp. or *Cebus* sp.) generally are regarded as visual specialists (Fobes and King, 1982). These animals appear to have a relatively limited ability to use sounds as information for solving complex cognitive tasks, but are quite adept at using visual information for this purpose (D'Amato, 1973; D'Amato and Salmon, 1982; Colombo and D'Amato, 1986).

Dolphins are typically classified as acoustic specialists, because of the remarkable development of their hearing and sound-production systems (Popper, 1980; Nachtigall and Moore, 1988; Ralston and Herman, 1989), and because the underwater world generally favors sound as an information or communication channel (Herman and Tavolga, 1980). Sound-production and hearing support vital biological functions, and are used in such basic tasks as prey and predator detection, environmental scanning, and social recognition of individuals in the herd through their unique sounds (Herman and Tavolga, 1980). Studies also have shown that the auditory system can support the cognitive function, since dolphins are able to carry-out a variety of complex cognitive tasks using only sonar emissions and returns as information (Roitblat et al., 1989) or sounds heard passively (Herman and Arbeit, 1973; Herman and Gordon, 1974; Thompson and Herman, 1977; Herman and Thompson, 1982; Ralston and Herman, 1989).

The visual system of the dolphin is also well-developed, although it would be an exaggeration to say that its development matches that of the auditory system in relative sensitivity, resolution, range, or other attributes of a strictly sensory nature. Yet, there is reasonably good resolution acuity, brightness and contrast sensitivity, distance perception, and likely, movement detection capabilities (Dral, 1972; Herman et al., 1975; Dawson, 1980; Madsen and Herman, 1980; Mobley, 1984). Vision serves many important biological functions for the dolphin: for example, prey detection and capture, conspecific and individual identification, and migration and orientation (Madsen and Herman, 1980).

The remaining question is whether the dolphin's visual system can support the cognitive function as well as the biological function. An early view of dolphins, which I held at one time, was that the visual system primarily was constrained to serving the biological function (Herman, 1980).

In the remainder of this chapter, I will try to convince the reader that this view is incorrect. I will review several areas of study from our laboratory, studies which grow progressively more complex in character, and which have convinced me of the power of the visual system as a source of information for complex cognitive behaviors and mental processes. The subjects of these studies were two adolescent female bottlenosed dolphins (_Tursiops truncatus_), named Phoenix and Akeakamai.

COGNITIVE TASKS

Matching-to-Sample

In the most common form of this test, the animal is required to view a "sample" stimulus, generally for only a brief time. The sample is then withdrawn and the animal is presented with two or more comparison stimuli, one of which matches the sample. The animal is rewarded if it chooses the matching comparison stimulus. The task seems elementary, but can present various degrees of visual difficulty depending on the types of samples and comparison stimuli. Cognitive issues come into play when new samples are introduced. The use of new samples allows the experimenter to test whether the animal has abstracted the general matching-rule from the given limited set of training problems. If it has, this will be evidenced by the immediate solution of new matching problems. Monkeys using visual materials (D'Amato, 1973) and dolphins using auditory materials (Herman and Gordon, 1974; Thompson and Herman, 1977; Herman and Thompson, 1982) immediately solved new matching problems and hence demonstrated that they had abstracted the matching-rule.

Our earliest work with visual materials suggested that the visual matching-to-sample test was difficult for the dolphin to learn (Herman, 1980; Forestell and Herman, 1988). Subsequent work, however, identified an important procedural problem that appeared to be limiting the performance levels reached in earlier tests. Basically, clear distinctions between figure and background are essential for the ready solution of visual matching problems. Three-dimensional, real-world objects presented against a broad diffuse background meet this criterion of distinctiveness. Such objects were used in our recent studies (Herman et al., 1989a). The objects, such as a child's toy lawn mower, a set of plastic fan blades, a metal tool box, and a Halloween goblin, were shown as samples to the dolphin subject (Phoenix) in air in her home pool. During the initial training and during the subsequent tests, the objects were held by the trainer so that the background for the dolphin, as it looked upward at the objects, was the open sky. When the sample object was removed, two comparison objects were placed in the water by two assistants, one located to each side of the trainer. Phoenix then was required to swim to the object of her choice. Under these conditions, Phoenix learned the matching task readily and performed matches reliably, as evidenced by typical performance levels of 80% correct responses or better during testing sessions. Phoenix also showed rapid transfer of the matching-rule to new objects shown her for the first time. Thus, Phoenix chose the correct comparison object on the first trial of 16 of 18 matching tests, each involving a novel sample object.

In other conditions studied, comparison objects were displayed in air rather than placed in the water, and sample objects were displayed inside a large box that was open at the top and on the side facing the dolphin's pool. These changes produced no adverse effects on visual matching performance. Also, during delayed matching tests in which delays of as long as 80 seconds were interposed between removal of the sample object and the display of the comparison stimuli, Phoenix continued to match successfully at performance levels well above chance. Successful matching after such long delays implies that the dolphin constructed and accessed a mental representation of the displayed sample object.

The same principle--an emphasis on the distinction between figure and background--was applied subsequently to two-dimensional figures consisting of white arbitrary shapes displayed on a large black board (Hunter, 1988). Phoenix was able to match a variety of these two-dimensional figures accurately even when the number of comparison stimuli was increased to six. In further tests matching performance was evaluated as a function of the area of the displayed figures. There was no performance decrement when the area was reduced from the initial training value of 645 cm^2 to 161 cm^2, 91 cm^2, and 40 cm^2, successively. Performance decreased significantly only when the area was finally decreased to 19 cm^2. The limitation on performance was thus strictly perceptual and not conceptual. Performance with these two-dimensional patterns remained significantly above chance, even after delays in presenting the comparison stimuli of 60 seconds. The overall results thus show that there was little difference in performance using either two- or three-dimensional visual materials. The good performance after delays indicates that mental representations of abstract, two-dimensional shapes may be formed as reliably as those for three-dimensional objects.

Behavioral Mimicry

In some of our earlier work, we studied the ability of a dolphin to imitate a variety of computer-generated sounds broadcast into her pool (Richards et al., 1984). A generalized capability for vocal imitation was shown in that novel sounds not heard previously by the dolphin usually were imitated on the first attempt. More recent work at our laboratory has focused on behavioral mimicry (Xitco, 1988). This work has demonstrated that a dolphin can be taught to attend to the behavior of a model--either another dolphin or a human--and on command imitate that behavior. To imitate successfully, visual observation of the behavior is necessary. If a delay is introduced before the imitation can be carried out, then a mental representation of the observed behavior must be maintained by the dolphin and accessed when commanded to imitate.

The dolphins Phoenix and Akeakamai served as subjects of these studies. They were able to exchange roles as models and mimics, and were accomplished in each role. In the simplest paradigm, the two dolphins stationed in the water side by side, each facing her trainer. One dolphin served as the model, the other as the imitator. A board projecting over the pool wall and above the water surface restricted the view of each dolphin to her own trainer. The dolphins were able to view each other beneath the board. The trainer controlling the model directed her dolphin to perform some behavior, using gestural signals that could not be observed by the imitator. The behaviors that were demonstrated included some that were completed at the training station, such as somersaulting, some that were completed away from the station, such as a porpoising leap, and some that involved an object, such as tossing a ball in the air. Either during the model's behavior, immediately thereafter, or after a longer delay, the trainer controlling the imitator directed her dolphin to imitate the model or to perform some other behavior. Imitation attempts only occurred if requested; otherwise, the alternative behavior directed by the imitator's trainer was performed. Accuracy in imitation was between 90% to 95% correct during simultaneous mimicry and showed almost no decline after delays as long as 20 seconds. For delays between 50 to 80 seconds, the longest delays tested, accuracy declined to about 75% correct imitations for Akeakamai and to approximately 59% for Phoenix, still well above chance considering that seven or more different behaviors might be modeled during a session. Both familiar and novel behaviors were imitated successfully, although imitations of familiar behaviors were more reliable than those of novel behaviors.

Most recently, we demonstrated that a dolphin can attend selectively to the behavior of either a human model or another dolphin, while both models are simultaneously demonstrating behaviors (Herman et al., 1989b). In this procedure, each model performs a different behavior. The imitating dolphin, Akeakamai, is then directed through gestures meaning "PERSON" or "PHOENIX" to imitate only that person or dolphin, or to do something entirely different, if so asked. A test of selective-mimicry ability was run consisting of 24 trials during which each of 12 different behaviors was demonstrated twice. Akeakamai chose the correct model--the one specified by the trainer on 19 of 24 trials (79.2%). Of these 19 occasions, Akeakamai performed an exact copy of the modeled behavior 12 times (63.2%). On an additional four trials, the imitation resembled the model's behavior in that the same object was manipulated or a similar action was taken to that of the model. Thus, on 16 of 24 trials (84.2%) the imitation matched or resembled the model's behavior.

A related question is: did the dolphin imitate the behavior of one of the models correctly, regardless of whether it was the designated model? On 17 of 24 trials (70.8%) the dolphin produced exact imitations of one or the other model, and on an additional four trials she produced a behavior resembling that of one of the models.

In general, this imitation ability seems an extension of the natural behavior of synchronous swimming and leaps of dolphins, but carried over into more abstract and arbitrary behaviors and situations. The overall findings from the set of behavioral mimicry studies--(1) imitation of both conspecifics and another species; (2) imitation of novel behaviors, (3) imitation after delays; and (4) selective imitation--indicate the presence of mental representations that guide the dolphins' interpretations of what it sees and that lead to the development of a rich concept of arbitrary, generalized mimicry.

Gestural Language Comprehension

Mastering a language-like system is certainly one of the most challenging cognitive tasks that can be presented to an animal. In work reviewed in several publications (Herman, 1980; Herman et al., 1984; Herman

1986, 1987), we demonstrated that artificial languages consisting of a vocabulary of words and of word-order rules governing the grammatical function of these words, were understood by dolphins. This work emphasized language comprehension rather than language production. The receptive competencies of animals (their ability to process and interpret information) appear to exceed their abilities to produce (generate) information (Herman, 1986; Herman and Morrel-Samuels, 1990). Tests of language comprehension may therefore be better indicators of an animal's ability to learn some of the fundamental properties of a language system than are tests of language production.

A gestural language was used with the dolphin Akeakamai, in which gestures of the trainer's arms and hands were used to refer to objects, actions, and the modifiers LEFT and RIGHT. The grammatical rules allowed for the construction of several-thousand unique sentences consisting of two to five sequential gestures. The sentences were imperatives instructing the dolphin to take named actions relative to named objects and their named modifiers. For many of the sentences, both the meaning of the gestures and their order of appearance in the sequence determined the correct interpretation of the sentence. Different sequences of gestures conveyed different meanings, as in the contrasting English sentences "Take the ball to the basket" and "Take the basket to the ball." These same instructions are expressed, respectively, as BASKET BALL FETCH and BALL BASKET FETCH in our artificial gestural language. Akeakamai's comprehension of these and other instructions was measured by the accuracy of her responses. Comprehension was good, as indicated by accurate responding to greater than 90% of the simpler two- and three-word sentences and to 60% or greater of the more complex and longer sentences (Herman et al., 1984, Table 7). Because of the many different response alternatives available to the dolphin, the probability of responding accurately to any sentence by chance alone is very low (about 4% or less).

Interrogative sentences also could be constructed within the artificial gestural language (Herman and Forestell, 1985). These sentences asked Akeakamai about the status of objects in her pool--that is, whether or not a particular named object was present in the pool. For example, a sequence of two gestures, such as BALL QUESTION, asks whether the named object is present in the pool. Akeakamai could respond "present" ("yes") or "absent" ("no") by pressing one of two paddles. Again, Akeakamai's accuracy was high, averaging about 80% correct reports of presence or absence with no significant difference in accuracy for negative and positive reports.

Thus, two factors are involved in the accurate performance of the dolphin in these language comprehension tasks: (1) a perceptual factor relying on the visual discrimination of multiple gestures given in rapid sequence and of the various objects to which some of these gestures referred, and (2) a cognitive factor relying on the understanding of the referents of the gestures, and of the combinatorial rules for sequences of gestures. Akeakamai's good level of comprehension shows that visual materials can be used as information for carrying out these highly complex cognitive tasks.

Processing of Visual Information Provided through Television

The dolphins Akeakamai and Phoenix were exposed to scenes on a television screen displayed through an underwater window in their pool. In the initial study, the scenes showed a trainer giving gestural signals comparable to those that the dolphins experienced in the live situation. Akeakamai was shown sequences of gestures from the artificial language she had mastered. For Phoenix, who was not versed in the gestural language, single gestural signs were used that asked her to carry out some simple behavior, such as a porpoising leap, a swim on her back, or a tail-walk. Both dolphins immediately understood the large majority of gestures or gestural sequences they viewed, as evidenced by their correct responses (Herman et al., 1990).

In subsequent tests, the images shown to the dolphins were degraded in several steps. These included, first, a display of only the trainer's arms and hands, then hands alone, and finally two spots of white light ("point-light" displays) tracing-out the movement paths of a gesture. In each case, the dolphins were able to interpret these images accurately. There virtually was no decrement in performance accuracy during the first two steps (arms and hands, and hands alone) relative to the accuracy obtained with the undegraded image. There was a significant performance decrement under the point-light display, but the level of performance was still well above chance. In fact, the performance of the dolphins was comparable with that achieved by our intermediate-level trainers tested with the same displays viewed by the dolphins.

In further work with television images, we displayed scenes of either a human or a dolphin performing some behavior, and studied the ability of the dolphins viewing the scenes to imitate those behaviors (Herman et al., 1989b). Both Akeakamai and Phoenix were able to imitate the behaviors seen on television, although with less reliability than those demonstrated by live models. Given the demands of the experiment and the viewing constraints imposed by underwater television images, it seems remarkable that the dolphins were able to use this type of visual information to carry out demanding cognitive tasks.

CONCLUSIONS

Our work suggests that the bottlenosed dolphin enjoys a visually-rich world as well as the acoustically-rich world traditionally attributed to it. The visual system is not limited to the detection and monitoring of biologically relevant events but, as our research shows, can serve as a valuable interface to higher cognitive centers that deal with abstractions, generalizations, and representations. This suggests that the dolphin has not relinquished the power of the visual sense, which was undoubtedly of fundamental importance in its earlier evolution.

These results may force us to change some of our views of dolphins. For example, should we continue to regard the hypertrophy of the brain of the dolphin as mainly a response to a need for acoustic processing power since visual processing power is also well developed? It may be better to consider the development of the brain as a response to pressures of a more general cognitive character, such as the advantages accruing from the ability to construct representations of the real world, to form abstractions, and to generalize from old situations to new. Elsewhere (Herman, 1980, 1986), I have suggested that the requirements for social learning, social adaptations, and intra-specific communication that occur in a socially-living species, such as the dolphin, favor a general increase in information-processing power mediated by any available sensory apparatus.

Finally, there are many research questions that remain to be explored about visual skills and cognitive processes. One important and interesting area of research would be to explore the interface between visual and auditory processing: For example, can ensonified targets be recognized visually, or observed objects be recognized through echolocation? Such studies, which deal with the degree to which representations achieved by one sense can be integrated with those achieved by another sense, can lead to an increased appreciation of the full role of sensory systems in the control of dolphin cognitive behavior.

ACKNOWLEDGMENTS

Preparation of this paper was supported by Contract N00014-85-K-0210 from the Office of Naval Research and from a grant from the Center for Field Research (Earthwatch). Release time provided by an appointment to the Social Science Research Institute of the University of Hawaii provided the opportunity to prepare this paper.

REFERENCES

Colombo, M. and D'Amato, M. R., 1986, A comparison of visual and auditory short-term memory in monkeys (Cebus apella). Quarterly Journal of Experimental Psychology, 38B: 425-448.

D'Amato, M. R., 1973, Delayed matching and short-term memory in monkeys. in: "The psychology of learning and motivation: Advances in research and theory" (Vol. 7) G. H. Bower (ed.). New York: Academic Press.

D'Amato, M. R. and Salmon, D. P., 1982, Tune discrimination in monkeys (Cebus apella) and in rats. Animal Learning and Behavior, 10: 126-134.

Dawson, W. W., 1980, The cetacean eye. in: "Cetacean Behavior: Mechanisms and Functions" L. M. Herman (Ed.), (pp. 53-100). New York: Wiley Interscience.

Dral, A. D. G., 1972, Aquatic and aerial vision in the bottle-nosed dolphin. Netherlands Journal of Sea Research, 5: 510-513.

Fobes, J. L. and King, J. E., 1982, Vision: The dominant primate modality. in: "Primate Behavior" J. L. Fobes and J. E. King (eds.) (pp. 219-243). New York: Academic Press.

Forestell, P. H. and Herman, L. M., 1988, Delayed matching of visual materials by a bottlenosed dolphin aided by auditory symbols. Animal Learning and Behavior, 16: 137-147.

Herman, L. M., 1980, Cognitive characteristics of dolphins. in: "Cetacean Behavior: Mechanisms and Functions" L. M. Herman (ed.) (pp. 363-429). New York: Wiley Interscience.

Herman, L. M., 1986, Cognition and language competencies of bottlenosed dolphins. in: "Dolphin Cognition and Behavior: A Comparative Approach" R. J. Schusterman, J. A. Thomas, and F. G. Wood (eds.) (pp. 221-252). Hillsdale, New Jersey: Erlbaum.

Herman, L. M., 1987, Receptive competencies of language-trained animals. in: "Advances in the Study of Behavior" (Vol. 17) J. S. Rosenblatt, C. Beer, M. C. Busnel, and P. J. B. Slater (Eds.) (pp. 1-60). New York: Academic Press.

Herman, L. M. and Arbeit, W. R., 1973, Stimulus control and auditory discrimination learning sets in the bottlenosed dolphin. Journal of the Experimental Analysis of Behavior, 19: 379-394.

Herman, L. M. and Forestell, P. H., 1985, Reporting presence or absence of named objects by a language-trained dolphin. Neuroscience and Biobehavioral Reviews, 9: 667-681.

Herman, L. M. and Gordon, J. A., 1974, Auditory delayed matching in the bottlenosed dolphin. Journal of the Experimental Analysis of Behavior, 21: 19-26.

Herman, L. M., Hovancik, J. R., Gory, J. D. and Bradshaw, G. L., 1989a, Generalization of visual matching by a bottlenosed dolphin (Tursiops truncatus): Evidence for invariance of cognitive performance with visual or auditory materials. Journal of Experimental Psychology: Animal Behavior Processes, 15: 124-136.

Herman, L. M. and Morrel-Samuels, P., 1990, Knowledge acquisition and asymmetry between language comprehension and production: Dolphins and apes as a general model for animals. in: "Interpretation and Explanation in the Study of Behavior: Comparative Perspectives" M. Bekoff and D. Jamieson (eds.). Boulder, Colorado: Westview Press, in press.

Herman, L. M., Morrel-Samuels, P. and Brown, L. A., 1989b, Dolphins recall and imitate action sequences. Paper presented at the Psychonomics Society Meeting, Atlanta, Georgia.

Herman, L. M., Morrel-Samuels, P. and Pack, A. A., 1990. Bottlenosed dolphin and human recognition of veridical and degraded video displays of an artificial gestural language. Journal of Experimental Psychology: General, in press.

Herman, L. M., Peacock, M. F., Yunker, M. P., and Madsen, C., 1975, Bottlenosed dolphin: Double-slit pupil yields equivalent aerial and underwater diurnal acuity. Science, 139: 650-652.

Herman, L. M., Richards, D. G., and Wolz, J. P., 1984, Comprehension of sentences by bottlenosed dolphins. Cognition, 16: 129-219.

Herman, L. M. and Tavolga, W. N., 1980, The communication systems of cetaceans. in: "Cetacean Behavior: Mechanisms and Functions" L.M. Herman (ed.) (pp. 149-209). New York: Wiley Interscience.

Herman, L. M. and Thompson, R. K. R., 1982, Symbolic, identity, and probe delayed matching of sounds by the bottlenosed dolphin. Animal Learning and Behavior, 10: 22-34.

Hunter, G. A., 1988, Visual delayed matching of two-dimensional forms by a bottlenosed dolphin. Unpublished Master's thesis, University of Hawaii.

Madsen, C. J. and Herman, L. M., 1980, Social and ecological correlates of cetacean vision and visual appearance. in: "Cetacean Behavior: Mechanisms and Functions" L. M. Herman (ed.) (pp. 101-147). New York: Wiley Interscience.

Mobley, J. R., 1984, Visual discrimination of relative distance by a bottlenosed dolphin (Tursiops truncatus)--Evidence for relational learning. Unpublished Ph.D. thesis, University of Hawaii.

Nachtigall, P. E. and Moore, P. W. B., 1988 (eds.), "Animal Sonar Systems: Processes and Performance." New York: Plenum.

Popper, A. N., 1980, Sound emission and detection by delphinids. in: "Cetacean Behavior: Mechanisms and Functions" L.M. Herman (ed.) (pp. 1-52). New York: Wiley Interscience.

Ralston, J. V. and Herman, L. M., 1989, Dolphin auditory perception. in: "The comparative psychology of audition: Perceiving complex sounds" J. R. Dooling and S. H. Hulse (eds.) (pp. 295-328). Hillsdale, New Jersey: Erlbaum.

Richards, D. G., Wolz, J. P. and Herman, L. M., 1984, Mimicry of computer-generated sounds and vocal labeling of objects by a bottlenosed dolphin. Journal of Comparative Psychology, 98: 10-28.

Roitblat, H. L., Penner, R. H. and Nachtigall, P. E., 1989, Matching-to-sample by an echolocating dolphin. Journal of Experimental Psychology: Animal Behavior Processes, (in press).

Thompson, R. K. R. and Herman, L. M., 1977, Memory for lists of sounds by the bottlenosed dolphin: Convergence of memory processes with humans? Science, 195: 501-503.

Xitco, M. J., Jr., 1988, Mimicry of modeled behaviors by bottlenose dolphins. Unpublished Masters thesis, University of Hawaii.

ANATOMICAL AND HISTOLOGICAL CHARACTERISTICS OF THE EYES OF A

MONTH-OLD AND AN ADULT HARBOR PORPOISE (PHOCOENA PHOCOENA)

Ronald A. Kastelein, Ruud C.V.J. Zweypfenning[1]
and Henk Spekreijse[1]

Zeedierenpark Harderwijk (Harderwijk Marine Mammal Park),
Strandboulevard-Oost 1, 3841 AB Harderwijk, The Netherlands
[1] The Netherlands Ophthalmic Research Institute, P.O. Box 12141,
Amsterdam Zuidoost, The Netherlands

SUMMARY

Several anatomical characteristics of the cornea, lens, iris and retina of the eyes of a month-old and an adult harbor porpoise (Phocoena phocoena) were investigated by scanning, transmission and light microscopy.

Also, the refractive power and chemical composition of the eye mucous were determined. The results are compared to those of other cetacean and human eyes, and the ecological significance of the differences is discussed.

INTRODUCTION

Cetaceans have developed special sensory systems to survive in an aquatic environment. Vision is one of these senses and is well-adapted to perform both in air and water. This requires special adaptations because:

(1) The refractive power of the eye is stronger in air than in water, since the eye media have almost the same optical density as water.
(2) Water conducts heat better than air, so in water eyes cool-off faster.
(3) Retinal illumination is always lower in water than in air; furthermore, the short wavelength part of the visible spectrum is underrepresented in water.
(4) The pressure on the eye ball is greater in water than in air.

Knowledge of cetacean senses may not only widen our insight into general sensory processes but also can be useful for management purposes. Knowledge on audition of cetaceans can be used in management of background noise levels in pools. Knowledge of color vision and visual acuity may be useful for training purposes. Also for management of species in the field, information on the senses can be used to reduce the impact between human and animal activities. For example, information on the gustatory and auditory capabilities possibly could help in developing enticements or repellents to keep the animals away from fishing nets and oil slicks. Sounds already are used to keep whales from entering certain areas. Knowledge on the quality of their echolocation system could be useful in manufacturing fishing nets that are detectable by dolphins. Data on senses also can be used in the development of technical equipment, such as sonar systems and underwater cameras.

Nachtigall (1986) and Dawson (1988) recently have summarized all available data on vision of cetaceans. The eyes of only few species have been investigated. Apart from similarities between the species so far investigated, also differences were seen. These differences may point to special ecological adaptations and hence research on vision of other cetacean species should be continued. For that reason, the eyes of a young and an adult harbor porpoise were preserved for anatomical and histological examination to add new data to the information already accumulated by Hulke (1868) and Pütter (1903). Furthermore, the refractive index and chemical composition of the eye mucous was determined from two live harbor porpoises that were rehabilitating at the Harderwijk Marine Mammal Park. Detailed knowledge of the composition of mucous can be useful for selecting ointments to treat eye infections in dolphins.

MATERIAL AND METHODS

A young, male harbor porpoise, Phocoena phocoena, (PpSH015) was stranded on the Island Sylt, West Germany, on 24 June 1988, and was flown to Harderwijk Park the same day. The animal was estimated to be one month-old and must still have been suckling. The animal unfortunately died from pneumonia three days after arrival. Examination revealed that one of the eyes had a severe corneal infection.

An adult stranded harbor porpoise (PpSH014) arrived at the park on March 28, 1988 and was treated successfully (Kastelein et al., 1990). This animal died on October 5, 1988 from diamond disease (Erysepelothrix rhusiopathiae), a bacterial infection.
The eyes were enucleated within an hour after death, and preserved in Peters' fixative (1% paraform-aldehyde and 1.5% glutare-aldehyde in cacodylate buffer; pH 7.6). Several characteristics of the eyes were examined with light microscopy (LM), scanning electron microscopy (SEM) and transmission electron microscopy (TEM).

Furthermore, mucous samples of both eyes of a live female and male harbor porpoise (PpSH012 and PpSH013) were collected when they were taken-out of the water for ten minutes (Fig. 1). The refractive indices of the mucous samples were measured with a microrefractometer of Jelley within two minutes after collection. This refractometer was chosen because it allows determination of the refractive index of small samples of fluid. The chemical composition was analyzed later on in the Biochemical Laboratory of the Netherlands Ophthalmic Research Institute.

RESULTS

Outer Dimensions Eye

A schematic overview of a cross-section of a harbor porpoise eye is given in Figure 2. The optic axis of the eye is relatively short. The dimensions of the various compartments are listed in Table 1.

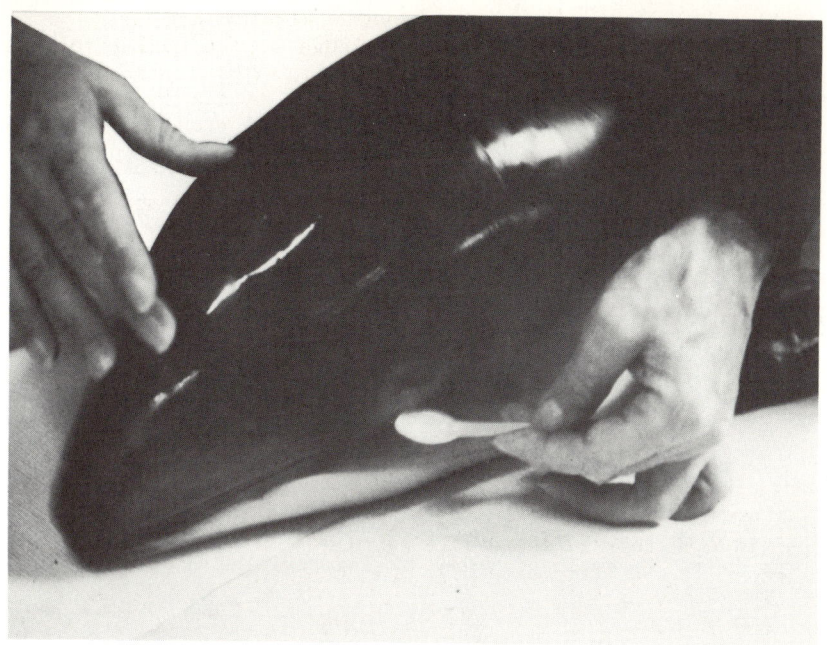

Fig. 1. The collection of the eye mucous from harbor porpoise 013.

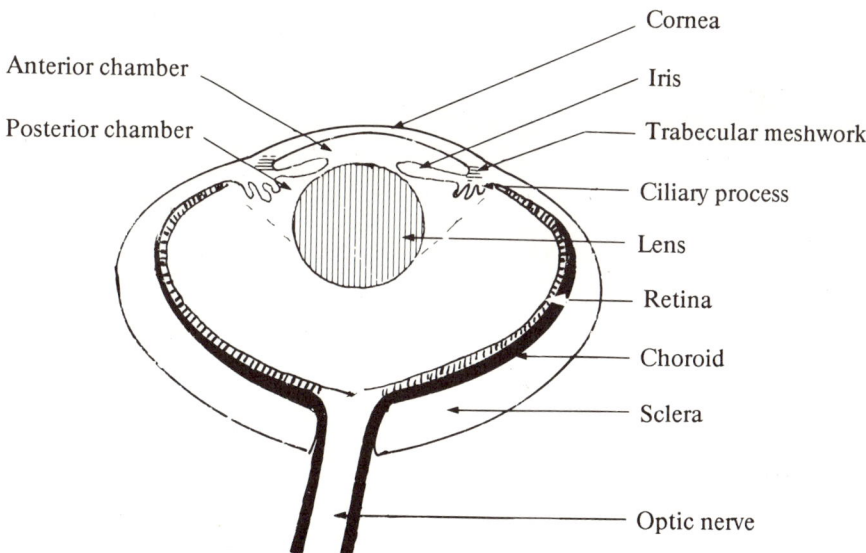

Fig. 2. A cross section of a harbor porpoise eye.

Table 1. Dimensions of Bodies and Eyes of Two Harbor Porpoises.

Sex	Male (015)	Male (014)
Age	Juvenile	Adult
Body weight (kg)	7.5	34
Body length (cm)	78	139
Eye length dorso-ventral direction (mm)	21	23.5
Eye length rostro-caudal direction (mm)	25	28
Eye length sagittal (axial length) (mm)	18	19.7
Diameter optic nerve (mm)	2.1	2.3
Scleral thickness near optic nerve (mm)	-	2.3
Scleral thickness near iris (mm)	-	1.2

The eyes and optic nerve are embedded in a thick layer of connective tissue with glands. The number of axons was estimated at 136.000. The optic nerve is enclosed fully by intertwining bloodvessels (Fig. 3).

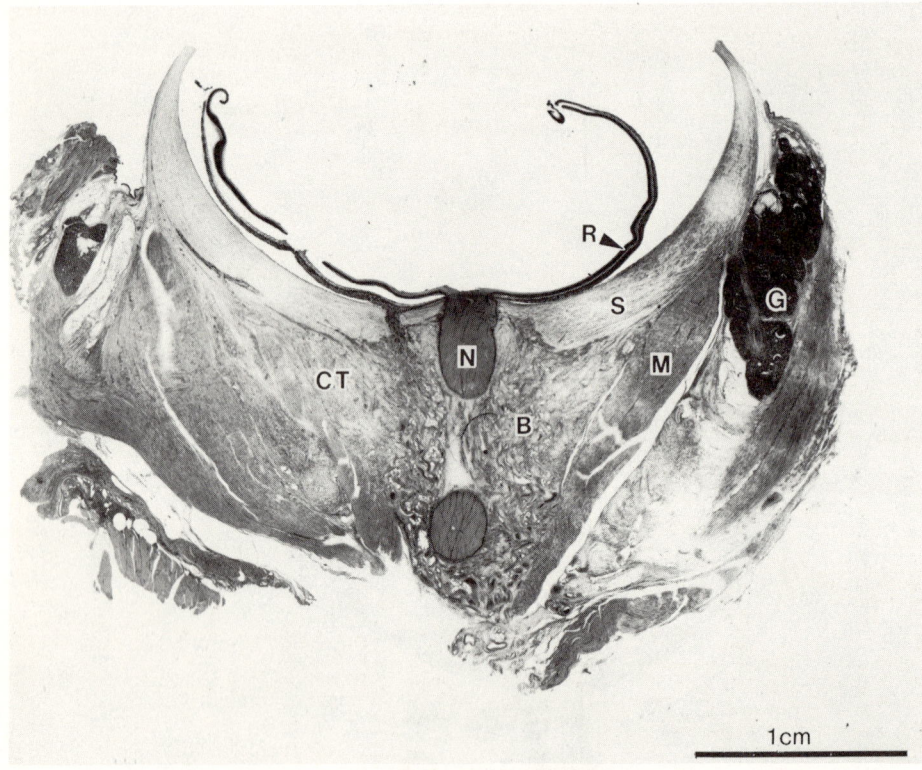

Fig. 3. Anterior part of the eye with surrounding tissue. Retina (R), sclera (S), optic nerve (N), bloodvessels (B), eye muscles (M), connective tissue (CT), and eye gland (G). (LM, scale = 1 cm).

Cornea

The cornea consists of a transparent, viscoelastic tissue, whose thickness increases gradually from 0.35 mm in the center to 1.2 mm at the chamber angle in the adult. Structurally, it consists of an epithelial layer, followed by the stroma with, on its anterior side, the modified zone of Bowman and, on the posterior side, the Descemet's membrane and an endothelial layer.
The outside of the cornea is covered with 9 to 10 layers of squamous epithelium (50 to 55 µm). A single row of basal cells covers the inside (Fig.4). The constant mitotic activity of the basal cells generates the middle and upper zone of cells. The superficial zone of one to two cell layers is formed by flattened polygonal cells (Fig.5).

The basal cells are attached firmly to the underlying Bowman membrane, which consists of a homogeneous, acellular, collagenous feltwork of 10 to 12 µm. More posterior, the stroma becomes organized in layers of parallel collagen bundles. The ground substance of the stroma consists of collagen and glycoproteins which are secreted by the keratocytes. The keratocytes have a thin irregular outline and are oriented parallel to the corneal surface (Fig. 4).

The cornea is bordered at the posterior side by the Descemet's membrane, a thin homogeneous layer of collagen of 1 to 2 µm. This membrane is a secretion product of the endothelial cells (Fig. 6).

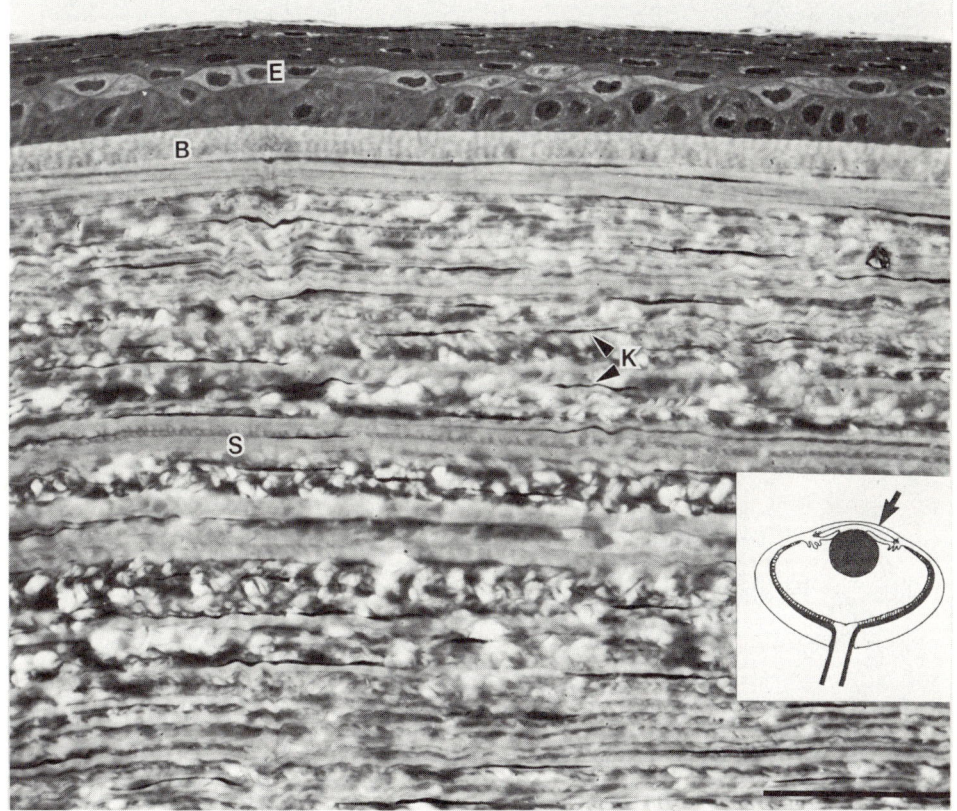

Fig. 4. Anterior side of the cornea with epithelium (E), Bowman membrane (B), stroma (S), and keratocyte (K). (LM, scale = 100 µm).

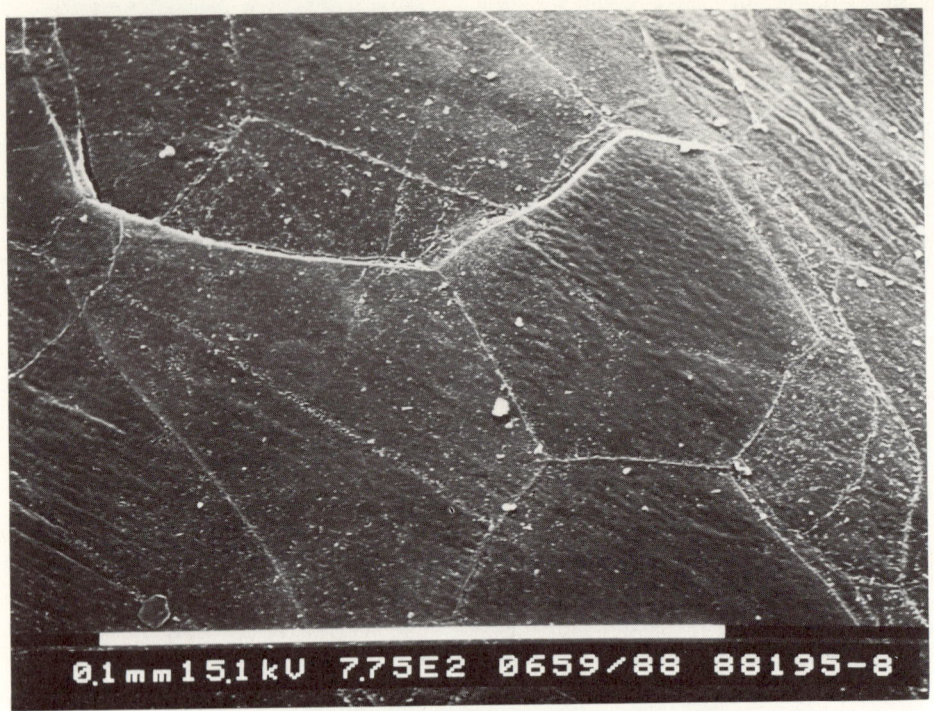

Fig. 5. Flat epithelium. (TEM, scale = 100 μm).

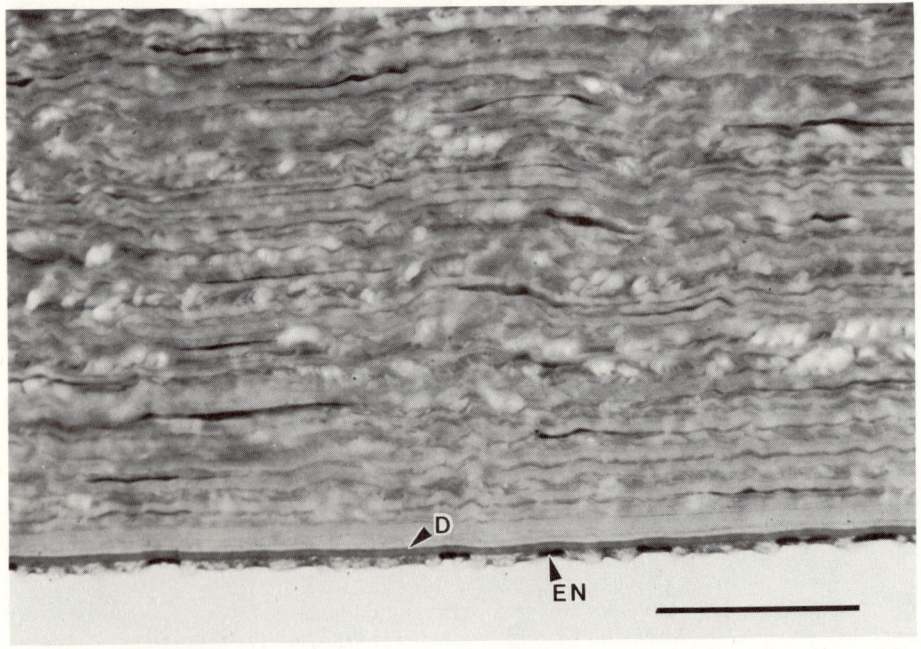

Fig. 6. Posterior side of the cornea with the Descemet membrane (D) and endothelium (EN). (LM, scale = 100 μm).

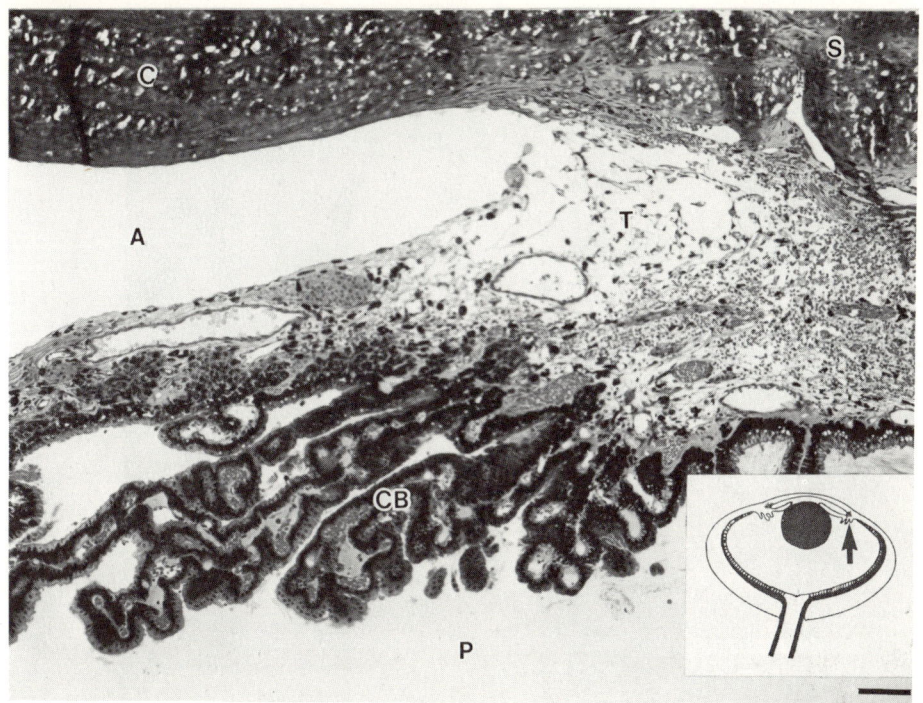

Fig. 7. Corneascleral limbus with the trabecular meshwork (T), cornea (C), sclera (S). The cornea and the iris (I) border the anterior eye chamber (A). The posterior eye chamber (P) and ciliary body (CB) are situated at the other side of the iris. (LM, scale = 100 μm).

Chamber Angle

The trabecular meshwork is situated at the corneascleral limbus, the transition zone of the cornea and the sclera (Fig. 7). This meshwork serves as drainage for the aqueous humor. The iris also is attached in this corner. The iris and the cornea enclose the anterior eye chamber. The posterior eye chamber is located at the posterior side of the iris.

The aqueous humor is secreted by the ciliary body which is situated at the attachment of the iris in the posterior eye chamber. The ciliary body consists of strongly folded processes. Each process contains numerous thin-walled capillaries covered by two layers of epithelium. At the capillaries the epithelium is pigmented. This layer is covered with non-pigmented ciliary epithelium (Fig. 8).

Iris

The following layers can be distinguished when going from the anterior to posterior side of the iris: first the stroma, with rather loose connective tissue with many chromatophores and thick-walled blood vessels, then the sphincter followed by the dilator and finally, on the posterior surface, a pigmented epithelium (Fig. 9).

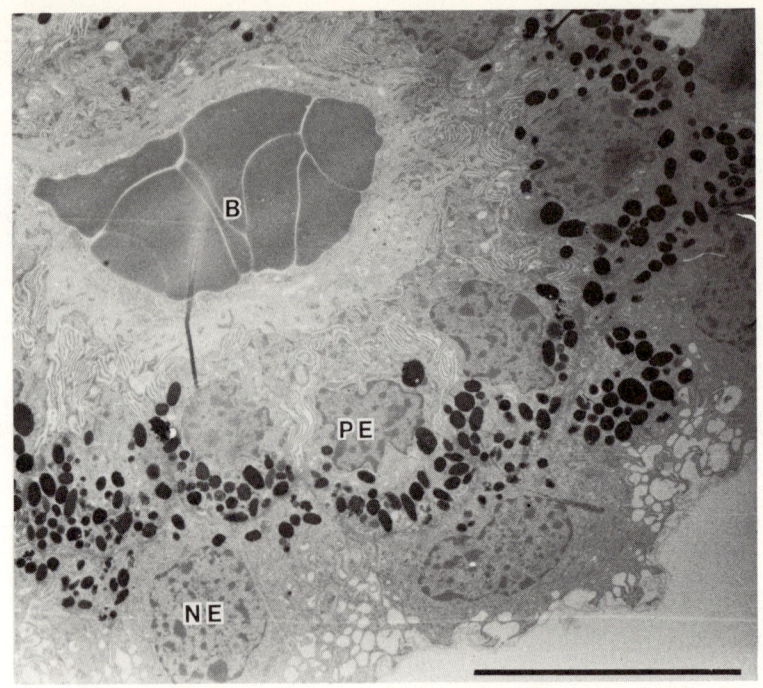

Fig. 8. Ciliary body with capillaries (B), pigmented epithelium (PE), and non-pigmented epithelium (NE). (SEM, scale = 10 μm).

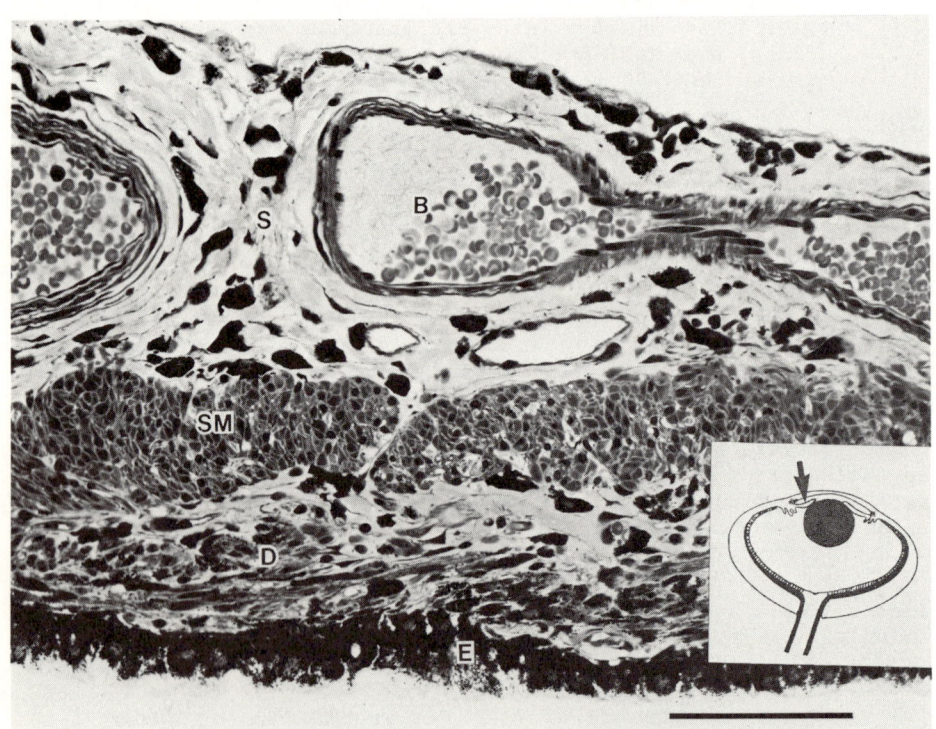

Fig. 9. The iris. Stroma (S) with bloodvessels (B), the sphincter muscle (SM), dilator muscle (D) and pigmented epithelium (E). (LM, scale = 100 μm).

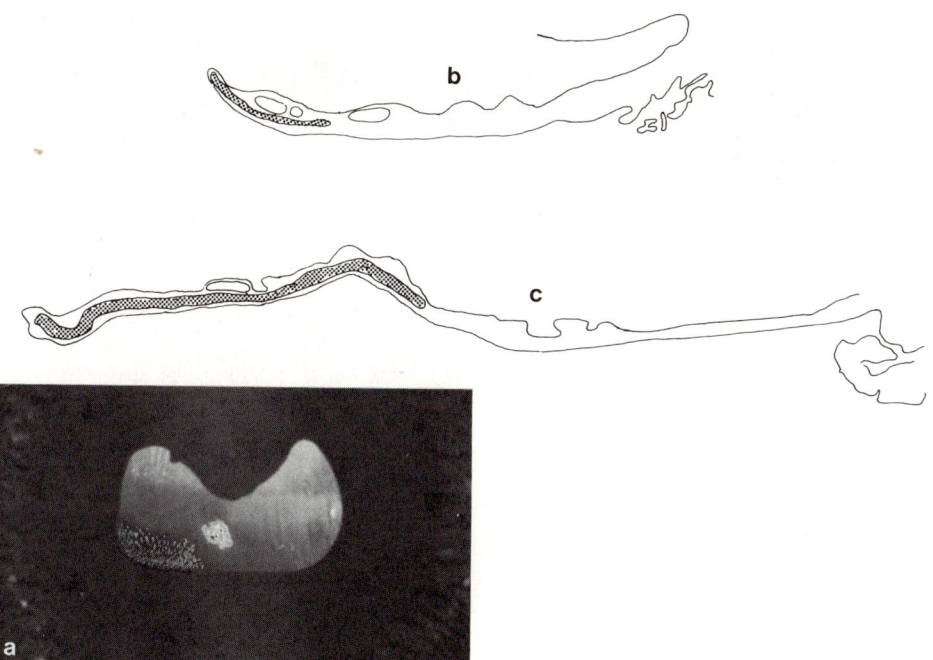

Fig. 10. a) U-shaped iris. b) Rostro-caudal section through the iris. c) Dorso-ventral section. The dotted area represents the sphincter muscle. (Scale = 1 mm).

The iris functions as a diaphragm, operated by sphincter and dilator muscles. In contrast to most mammals the pupil is not circular, but U-shaped (Fig. 10a). A lobe or operculum is present at the dorsal side. At high light intensities this operculum narrows the pupil to a double slit (Dawson, 1988). In the operculum the sphincter muscle is broader (3 mm) than in the remainder of the iris (1 mm) (Fig. 10b and c). The sphincter muscle consists of bundles of smooth muscle fibers, while the dilator consists of myoepithelium. The myoepithelial cells often are pigmented.

At the anterior chamber side of the iris, the veins of the juvenile are conspicuously thick (Fig. 11). This may be partly a juvenile characteristic, but in the adult animal the veins are also more prominent than in most other mammals.

Lens

Like in fish, the lens of the harbor porpoise is colorless, almost round (7.5 by 7 mm in the adult) and attached to the eye wall with ligaments (Fig. 12). No ciliary muscles were observed. In comparison to the human lens, the lens of the porpoise is very rigid. Scanning photographs were taken at three different levels of the lens. The peripheral fibers are almost smooth, with only a few balls and sockets (Fig. 13a). When the fibers grow older (deeper layers), the sides of the fibers develop protrusions, which interlock and create a bondage between the fibers. The number of balls and sockets decreases in these layers (Fig. 13b). In the center the balls and sockets have disappeared completely (Fig. 13c).

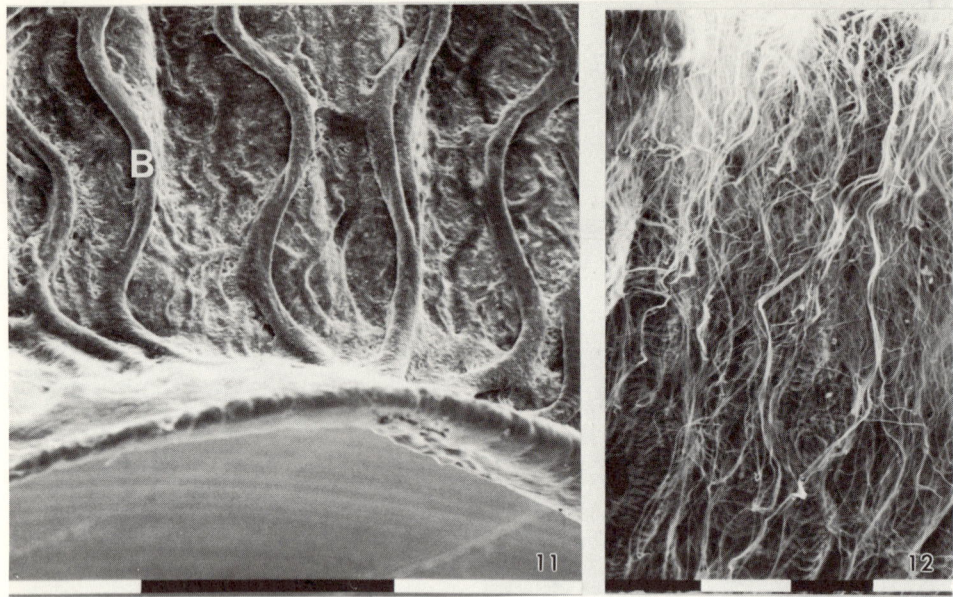

Fig. 11. The iris of the juvenile animal with thick veins (B). (TEM, scale = 1 mm).

Fig. 12. Zonula fibres. (TEM, scale = 100 µm).

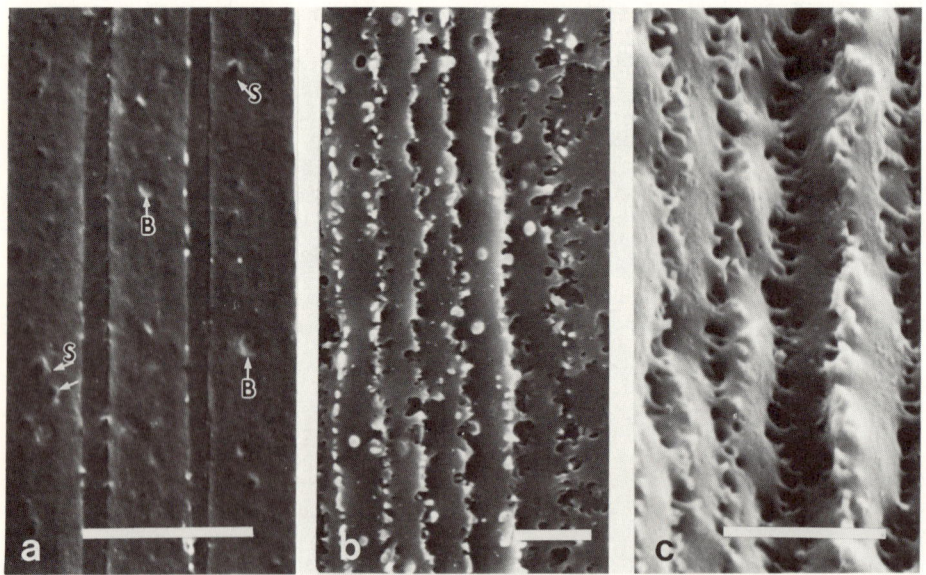

Fig. 13. Lens at three different levels. a) Peripheral fibers (F) with balls (B) and sockets (S). b) Deeper layers with protrusions. c) Central fibers. (TEM, scale = 10 µm).

Retina

The blood vessels of the fundus are rather thick in comparison to those of terrestrial mammals. Also many small blood vessels are found in the retina. Behind the retina a tapetum lucidum is found.
Although the fixation of the retina was not optimal, outer segments of cones could be seen in the entire retina (Fig. 14) while no fovea was observed. In the outer segment of a cone, the light-receptive membranes, are folds of the surrounding membrane, while in a rod outer segment the light-receptive membranes consist of isolated disks surrounded at all sides by a membrane (Fig. 15). Other light microscopic preparations showed that part of the most outer nuclei of the outer nuclear layer are slightly larger and probably belong to cones. The ratio of rods and cones could not be estimated. In one of the retinas at the nasal and caudal side of the optic nerve, the retina was slightly different in color, which suggests a higher number of cones.

The outer nuclear layer is very thick (12 cell layers, Fig. 16) compared to the human outer nuclear layer, which consists of only 6 cell layers. The bulk of the nuclei belong to rod cells.

In the ganglion layer most cells have a diameter of 20 to 30 μm, while some cells have a diameter of 40 to 50 μm. Also a displaced ganglion cell with a diameter of 40 μm was observed in the inner plexiform layer. The ganglion cells of 40 to 50 μm are confined to areas with low ganglion cell densities.

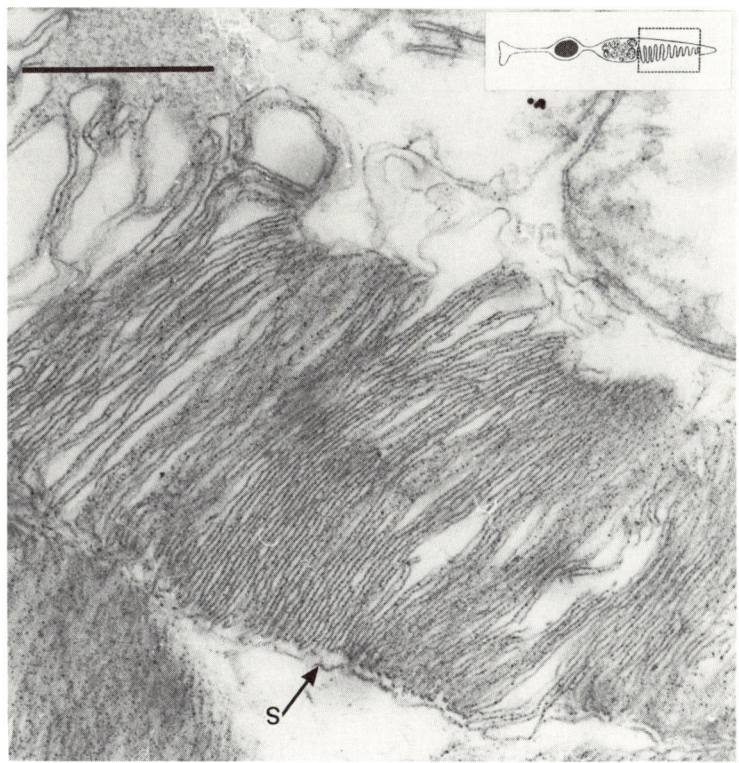

Fig. 14. Cone outer segment with surrounding membrane (S), folded light-receptive membrane. (SEM, scale = 0.5 μm).

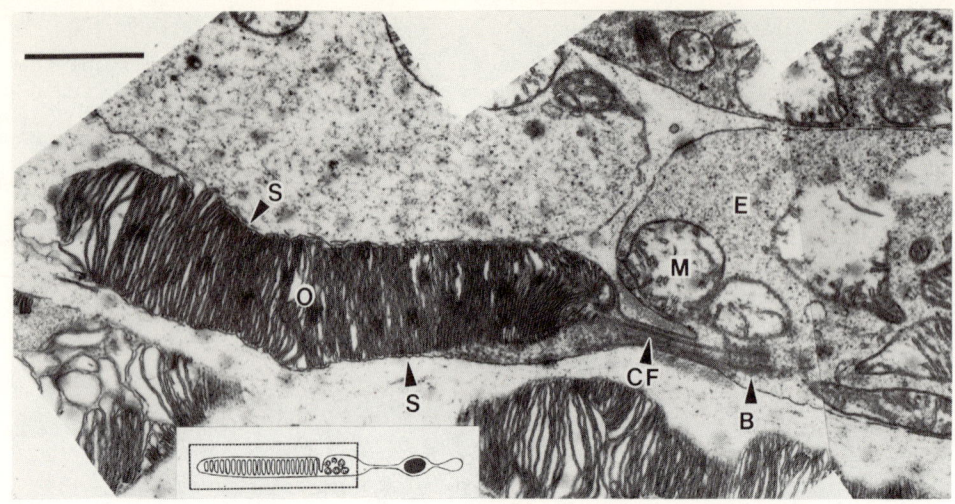

Fig. 15. Rod receptor. Ellipsoid (E) with swollen mitochondria (MI), outer segment (O) with disks (D), and surrounding membrane (M). At the connection of the outer segment with the ellipsoid the ciliary filaments (CF) with the basal body (BD) can be seen. (SEM, scale = 0.5 µm).

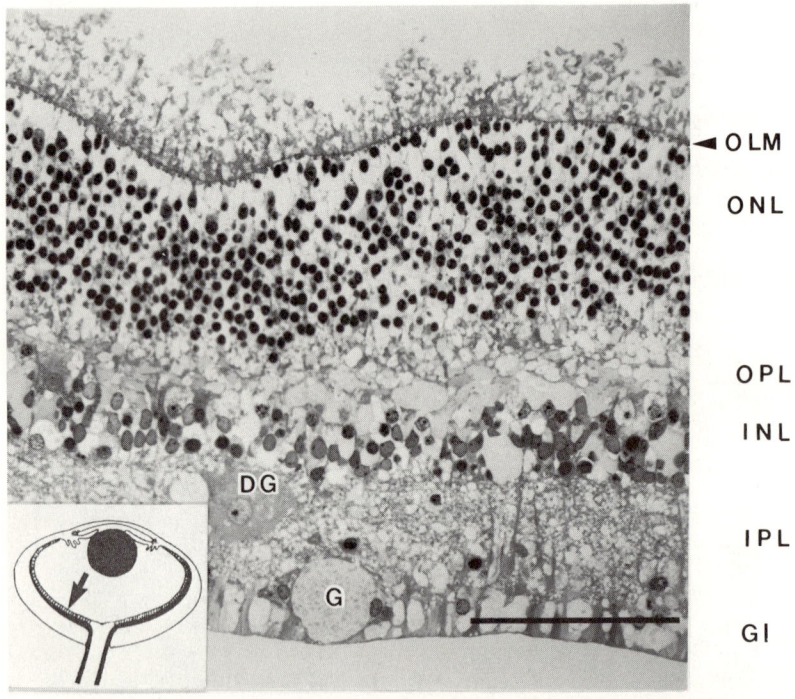

Fig. 16. Retina. Outer limiting membrane (OLM), outer nuclear layer (ONL), outer plexiform layer (OPL) inner nuclear layer (INL), inner plexiform layer (IPL), ganglion layer (GL) with ganglion cell (G) and a displaced ganglion cell (DG). (LM, scale = 100 µm).

Refractive Index of Eye Mucous

In cetaceans, the production of eye mucous occurs in the Hardian glands (Waller and Harrison, 1978). Freshly collected eye mucous of the two harbor porpoises had a transparant and highly viscous nature, and had the same refractive index as the artificial saltwater (2.25%) of the pool in which the animals were kept.

Chemical Composition of Eye Mucous

The similarity of the refractive indices of mucous and the salt water of the pool also is supported by chemical analysis (Table 2).

Table 2. Chemical Composition of mucous of the left (OS) and right (OD) eyes of two harbor porpoises and some terrestrial mammals.

		Porpoise 013	Porpoise 012	Human	Rat	Rabbit
Protein (g/l)	OD	0.28	1.10	5	35	7
	OS	0.30	0.66			
Osmolarity (mOsm/kg)	OD	641	669	280	280	280
	OS	648	492			
Enzymes:						
Lysozyme (mg HEL/ml)		0	0	3	0	0.015
N-acetylglucosaminidase (U/l)		0	0	80	2200	80
Peroxidase (U/ml)		0	0	5	500	0
Plasmine (mg/l)		0	0	0-5	-	-

The mucous contains about 10% of the protein content of terrestrial mammals, including humans. No enzymes could be found. The salt content, as is evident from the osmolarity measure, is high. For comparison, the osmolarity of seawater near the coast of the North Sea is 850 mOsm/kg.

DISCUSSION AND CONCLUSIONS

Outer Dimensions

Measurements from the present study and those reported by Pütter (1903) show that, during growth, the eye size of cetaceans remains rather constant simular to other mammals. When including the summarized data in Dawson (1988) the following rule of thumb is evident: the larger the animal, the smaller the axial length of the eyes compared to body length.

The outside of the eye is a tough protective coat (sclera) which is modified in front to form the transparent cornea. The relative thick sclera in cetaceans probably is not to cope with hydrostatic pressure, but to withstand hydrodynamic pressure during fast swimming. The sclera also may insulate the eye against the water which conducts heat 25 times better than air. The bloodvessels around the optic nerve also were noted by Hulke (1868) and probably serve a thermoregulatory function. The optic nerve axon density of the animals in the present study was 36.000 per mm^2. This is comparable to the findings in the harbor porpoise (Pütter, 1903) and in bottlenose dolphins (Dawson et al., 1983). This relative low density is not due to thick axons, but to glial elaboration and increased extraneuronal space. It is suggested that they serve as energy reserves for long periods of apnea during dives.

Cornea

Because the refractive index of the cornea is similar to that of sea water, the cornea can not function as an optical device, but serves only to keep water out (Dawson, 1988). The teardrop shaping would tend to minimize distortion of the eye due to differential water pressure at the rostral edge of the cornea (Jamieson, 1971).

Chamber Angle

Like in all mammalian eyes there is a constant flow of aqueous humor from the ciliary body in the posterior eye chamber to the trabecular meshwork in the anterior eye chamber (Cole, 1984). The eye pressure could not be determined non-invasively in a live harbor porpoise because the cornea is covered with a thick irregular layer of mucous. The ciliary body and the trabecular meshwork play an important role in the maintenance of the eye pressure. The ciliary body consists of strongly folded processes. Each process contains numerous thin-walled capillaries covered by two layers of epithelium. These cells are tightly packed and serve as an active pump (Cole, 1984). Dawson (1988) reports that the eye pressure in living bottlenose dolphins (Tursiops truncatus) is at least 50 mm Hg above the normal human level. Since this pressure declines rapidly after death, he suggests that the intraocular pressure in cetaceans is maintained by a more powerful active process than in terrestrial mammals.

Iris

The amount of light entering the eye through the pupil is controlled by the iris, which is pigmented heavily and covers part of the lens. The large blood vessels which also are described by Pütter (1903) in the iris probably serve a thermoregulatory function when the animals dive. The U-shaped pupil found in the harbor porpoises also has been seen in other cetaceans, but the function of this shape is not fully understood (Dawson, 1988). Ravimonte (1976) proposes a theory in which the operculum obscures the core of the lens, while the remaining slit leaves a passway for the light mainly through the margin of the lens. Because natural lenses consist of layers, the refractive index at the periphery can differ from that in the center of the lens.

Lens

The round lens is composed of transparent material and serves to refract the incoming light rays to form a clear image on the retina. In cetaceans the anterior chamber is smaller than in terrestrial mammals (Dawson, 1988). The eyes of the animals in the present study had no accommodative muscles. This also was seen in other cetaceans by Dral (1975) and by Yablokov et al. (1972). Because of the rigidity of the lens, accommodative muscles would be useless. However, visual acuity of the marine cetaceans so far investigated was good both under water as in air (Kellogg and Rice, 1963; Sprong and White, 1971; White et al., 1971; Pepper et al., 1972; Noordenbos and Boogh, 1974; Dral and Dudock van Heel, 1974). A stenopaic pupil (in which cetaceans reduce their pupils to a pinhole) has been suggested as a partial aid to form a clear image on the retina in air (Walls, 1963; Dawson et al., 1972; Herman et al., 1975).

Retina

The extremely thick layer of rod nuclei is probably an adaptation to the poor light conditions in deep and/or murky water. Also the lack of a fovea points in this direction. Moreover, the pressure of the presence of the tapetum lucidum provides additional useful light to the receptors (Dawson, 1988). Cones also are found in most other cetaceans that so far have been investigated (Perez et al., 1972; van Esch and de Wolf, 1979; Waller, 1982; Dawson, 1988). In dolphins a duplex retina, consisting of nocturnal (high-sensitivity) and diurnal (high-resolution) receptors, seems useful. Whether cetaceans can discriminate colors, has not yet been demonstrated in carefully designed experiments (McFarland, 1971; Simons and Huigen, 1977; Nachtigall, 1986).

Hulke (1868) observed giant ganglion cells in the retina of harbor porpoises as was confirmed by the present study. Based on counts of large and small ganglion cells, Dral (1974) identified two zones with concentrations over 500 cells/mm^2 in the bottlenose dolphin. These zones seem to be the areas of best vision. One area was found dorsal to the geometrical median and close to the temporal equator, and the other was located ventrally of the median and at the rostral side of the fundus. Dral (1983) observed also two areas with higher densities of ganglion cells in the temporal and rostral side of the eyes of a common dolphin (Delphinus delphis). Mass et al. (1986) found two zones with high density ganglion cells in the harbor porpoise. They found these zones in the temporal and rostral areas of the eyes. Mass and Supin (1989) found a high density of ganglion cells in the ventral part of the eyes of the Amazon river dolphin (Inia geoffrensis). Dawson and Perez (1973) suggest that the giant ganglion cells in the retina of the bottlenose dolphin facilitate rapid communication between the retina and the brain.

Because the large ganglion cells are confined to areas with low cell densities, it is likely that they support a large dendritic tree. So the large ganglion cells integrate the signals of the photoreceptors over a largr area than the smaller ganglion cells. I t is suggested that one of the functions of the ophthalmic rete is to keep the retina warm under water (Dawson, 1988).

Binocular Vision

Harbor porpoises often swim up-side-down, and sometimes rotate around their axis when looking at people at the edge of a pool (Kastelein et al., 1990). When underwater, bottlenose dolphins, killer whales and false killer whales at the Harderwijk Park also try to look at people who stand at the edge of a pool with only one eye at the time. This seems to indicate that the animals have no binocular vision in the rostral-dorsal direction. However, when false killer whales approach a person at an underwater viewing window they often try to approach head on, and just from above. The human observer can then see both eyes of the animal at the same time. This seems to suggest that at least in some cetaceans binocular vision is possible in the rostral-ventral direction. Dral (1972) suggests that the bottlenose dolphin has binocular vision in the rostral-ventral direction, and therefore, looks at people who stand at the edge of a pool with both eyes when its head is out of the water, and with only one eye when its head is in the water.

Refractive Index of Eye Mucous

The refractive index of the eye mucous was the same as that of the pool water in which the animals were kept. So, the irregularity of the outer surface of the mucous film on the cornea does not interfere with the optical imaging on the retina when the animals are underwater. This suggests that this thick gelatine film mainly has a protective and insulating function. Yablokov et al. (1972) report that the eye mucous has the same refractive index as the surrounding sea water. Because the salt concentration in the pool in the present study was only 2.25%, which is less than seawater (3.5%, Flemming, 1977), our finding suggests that the harbor porpoise is able to adapt the osmolarity of the mucous to the osmolarity of the surrounding water, so that the irregular mucous layer does not interfere with image focussing.

Chemical Composition of Eye Mucous

Yablokov et al. (1972) report that the mucous contains a high level of protein. However, the chemical analysis in the present study showed that the protein level is very low compared to terrestrial mammals. Also, Waller and Harrison (1978) report that the eye mucous of a bottlenose dolphin contained 97% water by weight, and that sodium chloride was detected in the residue.

When out of the water, the harbor porpoise produces more mucous per time unit than bottlenose dolphins, killer whales (Orcinus orca) and false killer whales (Pseudorca crassidens) that also are kept at the Harderwijk Park. The animals in the present study produced around 1 ml per 5 minutes per eye when out of the water, which is more than the 2 to 3 ml per hour collected from bottlenose dolphins (Waller and Harrison, 1978). Maybe harbor porpoises encounter more particles in their natural environment than the other species mentioned, and so they need a thicker protective mucous layer. That corneas of cetaceans can be damaged was shown by Dawson et al. (1987) who reported corneas of bottlenose dolphins with pronounced, random, local curvature changes typical of old keratitis and scarring. Furthermore, a thick mucous layer serves also as a thermal insulator. This is particularly important for small cetaceans, like the porpoise, which have a very unfavorable surface area to body weight ratio (Kanwisher and Sundnes, 1965). This is probably also the reason why the dolphins metabolism is about three times higher than measured in terrestrial mammals of same body weight.

ACKNOWLEDGMENTS

We thank the employees of the Seal Rescue Renter of Pieterburen for the transport of the young harbor porpoise to the Harderwijk Park. We thank Ben Willikens for the scanning micrographs and N. J. van Haeringen for the chemistry on the mucous samples.

REFERENCES

Cole, D. F., 1984, Ocular fluids, chapter 2, in: "The Eye", Volume IA Vegatative Physilogy and Biochemistry, Ed. H. Dawson.
Dawson, W. W., Birndorf, L. A.and Perez, J. M., 1972, Gross anatomy and optics of the dolphin eye (Tursiops truncatus), Cetology, 10: 1-12.

Dawson, W. W. and Perez, J. M., 1973, Unusual retinal cells in the dolphin eye, Science, 181: 747-749.

Dawson, W. W., Hope, G. M., Ulschafer, R. J., Hawthorne, M. N. and Jenkins, R. L., 1983, Contents of the optic nerve of a small cetacean. Aquatic Mammals, 10(2): 45-56.

Dawson, W. W., Schroeder, J. P. and Sharpe, S. N., 1987, Corneal properties of two marine mammal species, Marine Mammal Science, 3(2): 186-197.

Dawson, W. W., 1988, The cetacean eye. In: Cetacean behavior: mechanisms and functions (Ed. L.M. Herman). Robert E. Krieger Publishing Company, Malabar, Florida.

Dral, A. D. G., 1972, Aquatic and aerial vision in the bottle-nosed dolphin, Neth. J. of Sea Res., 5(4): 510-513.

Dral, A. D. G., 1974, Some quantative aspects of the retina of Tursiops truncatus. Aquatic Mammals, 2(2): 28-31.

Dral, A. D. G., 1975, The atrophic eye of Platanista gangetica. Aquatic Mammals, 3(1): 1-4.

Dral, A. D. G., 1983, The retinal ganglion cells of Delphinus delphis and their distribution, Aquatic Mammals, 10(2): 57-68.

Dral, A. D. G. and Dudok van Heel, W. H., 1974, Problems in image-focusing and astigmatism in cetacea- a state if affairs. Aquatic Mammals, 2(1): 22-28.

Esch, A. van, and de Wolf, J., 1979, Evidence for cone function in the dolphin retina - a preliminary report, Aquatic Mammals, 7(2): 35-36.

Flemming, N. C., 1977, "The undersea." McMillan Publish. Co. New York.

Herman, L. M., Peacock, M. F., Yunker, M. P. and Madsen, C. J., 1975, Bottlenosed Dolphin: Double-slit pupil yields equivalent aerial and underwater diurnal acuity, Science, 189: 650-652.

Hulke, J., 1868, Notes on the anatomy of the retina of the common porpoise (Phocoena communis), J. Anat. Physiol., 2: 19-25.

Jamieson, G., 1971, The functional significance of corneal distortion in marine mammals, Can. J. Zool., 49: 421-423.

Kanwisher, J. and Sundness, G., 1965, Physiology of a small cetacean, Hvalradets Skrift., 48: 45-53.

Kastelein, R. A., Bakker, M. A. and Dokter, T., 1990, The rehabilitation of 3 stranded harbor porpoises (Phocoena phocoena), Aquatic Mammals, 15(4).

Kellogg, W. N. and Rice, C. E., 1963, Visual discrimination in a bottlenose porpoise, Psychological Record, 13: 483-498.

Mass, A. M., Supin, A. Ya., and Severtsov, A. N., 1986, Topographic distribution of sizes and density of ganglion cells in the retina of a porpoise, Phocoena phocoena, Aquatic Mammals, 12(3): 95-102.

Mass, A. M. and Supin, A. Ya., 1989, Distribution of ganglion cells in the retina of an amazon river dolphin, Inia geofrensis. Aquatic Mammals, 15(2): 49-56.

McFarland, W. N., 1971, Cetacean visual pigments, Vision Res., 11: 1065-1076.

Nachtigall, P. E., 1986, Vision, audition and chemoreception in dolphins and other marine mammals, in: "Dolphin cognition and behavior: a comparative approach." (eds. R. J. Schusterman, J. A. Thomas and F. G. Wood) Lawrence Erlbaum Associates Publishers, London.

Noordenbos, J. W. and Boogh, C. J., 1974, Underwater visual acuity in bottle-nosed dolphin (Tursiops truncatus Mont.), Aquatic Mammals, 2(2): 15-24.

Pepper, R. L., Simmons, J. V. Jr, Beach, F. A. III and Nachtigall, P. E., 1972, In-air visual acuity of the bottlenose dolphin, Proc. of the 9th Ann. Conf. on Biol. Sonar and Diving Mamm. Fremont, California.

Perez, J. M., Dawson, W. W. and Landau, D., 1972, Retinal anatomy of the bottlenose dolphin (Tursiops truncatus), Cetology, 11: 1-11.

Pütter, A., 1903, Die Augen der Wassersaugetiere, in: "Zoologische Jahrbücher." (ed. J.W. Spengel) Jena, Verlag von Gustav Fisher, 17: 232-247.

Ravimonte, L. A., 1976, Eye model to account for comparable aerial and underwater acuities of the bottlenose dolphin, Neth. J. Sea Res., 10(4): 491-498.

Simons, D. and Huigen, M., 1977, Analysis of an experiment on colour vision in dolphins. Aquatic Mammals, 5(2): 27-33.

Sprong, P. and White, D., 1971, Visual acuity and discrimination learning in the dolphin (Lagenorhynchus obliquidens), Experimental Neurology, 31: 431-436.

Waller, G. N. H., 1982, Retinal ultrastructure of the amazon river dolphin (Inia geoffrensis), Aquatic Mammals, 9(1): 17-28.

Waller, G. H. and Harrison, R. J., 1978, The significance of eyelid glands in delphinids, Aquatic Mammals, 6(1): 1-9.

Walls, G. L., 1963, The vertebrate eye. Hafner Publishing Co. New York, chapter 11: 368-461.

White, D., Cameron, N., Sprong, P., and Bradford. J., 1971, Visual acuity of the Killer whale (Orcinus orca), Experimental Neurology, 32: 230-236.

Yablokov, A. V., Bel'kovich, V. M. and Borisov, V. I., 1972, "Whales and dolphins." Jerusalem: Israel program for scientific translations.

CHEMICAL SENSE OF DOLPHINS: QUASI-OLFACTION

Vitaly B. Kuznetzov

Department of Vertebrate Zoology
Faculty of Biology
Moscow State University
119899 Moscow, USSR

INTRODUCTION

Vertebrates possess two types of chemoreception-olfaction and taste. The distinction between these senses is often believed to be determined by different media of perception. While the olfaction perceives chemicals in the gaseous state, the taste perception takes place when the receptor contacts with the dissolved chemicals. But it is the seeming difference. Fish perceive the chemical solutions both by olfaction and taste. The sensitivity of both senses is higher in fish than in terrestrial vertebrates (Kleerekoper, 1969; Atema, 1980).

There is little difference between the olfaction in water and air media. In terrestrial animals the vapor of chemicals in order to be perceived should be dissolved in mucus, which covers the receptive surface. The same process takes place in aquatic animals hence here structure of olfactory epithelium is similar. Moreover several newt species live in different stages in water and terrestrial habitats. The same olfactory organ is functioning in both media (Matthes, 1927). Thus, taste and olfaction differ in the types of molecular receptors, in the mechanisms of receptor-ligand interaction, in brain structures which process the information, as well as in their functional role in animals' life.

Aquatic animals have certain advantages over terrestrial forms in using the olfaction. In the water medium the processes of diffusion and intermixing are slower than in the air due to higher density of the former. On this basis odorous traces stay in water for a longer time than in air. Numerous local streams exist in seas and oceans transferring chemicals for long distances.

The chemical communication is well-developed in fish (Liley, 1982), but there exist some special features of chemical communication patterns (solid substrate scent making) in the terrestrial animals, absent in aquatic animals.

The organ of olfaction, as well as the vomeronasal organ,

are not found in toothed whales. The information collected by these organs undergoes the first stage of processing in the olfactory bulbs. Toothed whales have no olfactory bulbs (Langworthy, 1932; Filimonov, 1949, 1965; Breathnach, 1953, 1960, and Jacobs et al., 1971) and thus, they have no olfaction in its typical form.

However, male toothed whales have openings of perianal glands. Toothed whales also display a high frequency of urination. These facts lead Yablokov (1961) to suggest that, as in other mammals, "gland secretions and urine are used in toothed whales for chemical communication and that these signals are perceived by means of olfaction". The author suggested that the "olfaction" organ is localized in small pits in the tongue root. Further study of tongue root pits in beluga (Delphinapterus leucas) by the author of this hypothesis demonstrated that the epithelium of these pits have few layers and is different from that of other tongue parts. At the same time neither taste nor olfactory receptors in the pits were described (Kleinenberg et al., 1964).

It is more difficult to determine the presence or absence of taste sensitivity in toothed whales. The presence nerves and brain structures connected with taste information processing could not serve alone as a basis for the statement that toothed whales possess taste sensitivity.

The structure of taste buds in all vertebrates is quite similar. They are oval or circular voluminous bodies 30-100μm in length, 17-70μm width. Mammal taste buds are placed mainly in taste papillae on the tongue. Cetacea, unlike other mammals, have no taste papillae on the tongue (Sonntag, 1922). This fact leads to suggestion that these animals lack taste sensitivity. Although more sound conclusion could be drawn only on the basis of histological investigation of the tongue.

The tongues of 4 out of 5 families of the Odontoceti suborder were investigated. There are 4 river dolphin species family Platanistidae. No taste buds were found in Platanista gangetica (Arvy and Pilleri, 1970) in and Pontoporia blainville (Yamasaki et al., 1976). In Lipotes vexillifer taste buds were present in the calf which was 860 mm long and there were not so many taste buds in a young female of 2060 mm. They were absent in an old female with a body length of 2410 mm (Li Yemin, 1983). There are only two species in the Monodontidae family. A histological investigation of tongue was performed only in one of them, the beluga. But no taste buds were found (Kleinenberg et al., 1964). In the Physeteridae family 3 species are known. The histological investigation of tongue was performed in the sperm whale (Physeter catodon). No taste buds were revealed (Berzin, 1971).

The picture is more complicated in the Delphinidae family, consisting of 46 species (Lowell and Flanigan, 1980). The histological examination of tongues was performed in 5 species. Several authors revealed taste buds while other found no taste buds in the very same species. No taste buds were found in the striped dolphin (Stenella coeruleoalbus) from the Pacific as well as in the Black Sea species of bottlenose dolphin (Tursiops truncatus), common dolphin (Delphinus), and harbor porpoise (Phocoena) (Sokolov and

Volkova, 1971; Sokolov and Kuznetsov, 1971). Taste buds were found in the Black Sea dolphin species-bottlenose dolphin and common dolphin (Sukhovskaya, 1972).

Nerve elements and receptor-like structures in 19 bottlenosed dolphin tongues were investigated by means of a silver impregnation method and no chemoreceptors were discovered (Valiulina and Chomenko, 1976). Donaldson (1977) revealed taste buds in the bottlenose dolphin. The taste buds were found in bottlenose fetuses and neonates (Gilevitch, 1978), while "structures, resembling taste buds" in adults animals, are reported to be rare (Agarkov and Gilevitch, 1979).

Certain elucidation of the problem occurred after the (Yamasaki et at. 1978) investigation of striped dolphin tongues of different ages. They demonstrated that taste buds which are present in young dolphins suffer complete atrophy in adults. This reduction proceeds most intensively at the age of 1.5 to 2.0 years. This clarifies why taste buds were difficult to discover inspite of their big size and specific pattern. The authors suggest that Sukhovskaya (1972) and Donaldson (1977) succeeded in discovering taste buds because the histological material investigated belonged to young animals. It seems probable that taste buds are absent in the tongues of adult toothed whales.

As olfactory bulbs are reduced in toothed whales long before the moment of birth, (Sinclair, 1966; Oelschläger and Buhl, 1985) they possess no olfaction. Taste buds are found in neonates and young animals, but then they are reduced (Yamasaki et al., 1978; Li Yemin, 1983) so that the adults seem to possess no gustatory perception. Morphological data made it possible to conclude that toothed whales possess no typical vertebrate senses - olfaction and taste.

Questions arisen - what is the cause of this phenomenon and what is the possible compensation for the absence of chemical senses? The solution of the chemosensitivity problem is of primary importance for our understanding of the toothed whales, biology, behavior and evolution - because these animals had lost an important channel of information.

The unique situation - the absence of chemosensory perception in adult toothed whales is the reason why the experimental investigation of chemical perception in dolphins is a very important issue for the solution of a number of general chemoreception problems in vertebrates. It is also important for the solution of such general neurological problems as structure-function relationships in the vertebrate brain. In particular, it is not clear what the reason of all olfactory brain structures present in dolphins (with the exception of anterior olfactory nuclei). At the same time olfactory bulbs and olfactory tracts and hence the olfaction are absent in dolphins (Filimonov, 1949 1965; Breathnach, 1953, 1960; and Jacobs et al., 1971). This fact was taken as the basis for wide speculations. Filimonov (1965) considered a complete change in the function of dolphin olfactory brain structures the possible explanation for such phenomenon. This fact gave birth to the opinion that, in general, in mammals the olfactory function is actually much less represented in the olfactory brain than it is usually thought, on the basis of experimental data.

All this signifies that the question of chemoreceptivity in these animals is a problem integrating the interests of different biological fields. This study represents the main results of the author's 20 year investigation of dolphin chemoreceptivity as well as some theoretical concepts proposed by the author on the basis of data collected.

DOLPHIN SENSITIVITY TO CHEMICAL STIMULI

Conditioning Experiments

Conditioning experiments were performed in Outrish Station on The Black Sea with adult bottlenosed dolphins only. The following procedure of chemical cues presentation was used. The experimental animal was kept isolated in a special enclosure. After the animals had been caught in the wild they needed some time for taming. The animal was trained to approach the experimenter's platform after a sound signal and had to stay there in a vertical posture with the head above the water and mouth opened. All test solution (40 ml chemical stimuli in sea or fresh water) were compared to 40 ml sea or fresh water. These liquids were poured into the oral cavity of the animal (Fig. 1). The animals were trained to perform different instrumental reactions. For example, to toss up the ball after receiving the chemical solution, or to swim away in response to water presentation. The order of chemical stimuli, or water presentation was random (Gellermann, 1933). We rewarded a correct response with fish. The percentage of correct responses in the sessions of each concentration of chemicals was calculated. This number was compared to a 50% correct response score (which would mean that the animal could not distinguish chemical solution from water) by means of coefficient test (Glass and Stanley, 1970).

Our first experiments demonstrated that dolphins are capable of distinguishing sea water and sea water solutions of trimethylamine, indole, camphor and valeric acid in concentrations of millimoles (Sokolov and Kuznetzov, 1971; Kuznetsov, 1974). All these chemicals are the odorants. At the same time these substances also are capable to invoke taste perceptions. The questions is how to separate these two types of perception from each other and whether dolphins are capable to detect the odorous substances.

There exist a class of chemicals which are capable to invoke the primary taste sensation - the sour taste - and at the same time they can be "smelled". They are carbonic acids -butyric, valeric, capronic acids and others. The taste sensitivity to these acids in several mammal species is lower than that to hydrochloric and citric acids, while their olfactory sensitivity to these chemicals is several orders higher than the gustatory threshold.

Based on these data, we determined sensitivity threshold in dolphins to primary taste stimuli to hydrochloric, citric and carbonic acids. The objective was to determine whether dolphin sensitivity to carbonic acids was higher than that to hydrochloric and citric acids. This would signify that

Fig. 1. Presentation of chemical stimulus to bottlenose dolphin.

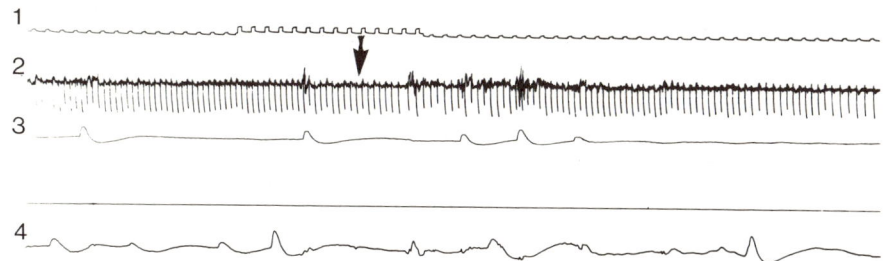

Fig. 2 Polygraph records of autonomous response: electrocardiogram (2), respiration (3) and galvanic skin response (4) in bottlenose dolphin to 10 ml valeric acid in concentration of 1.10^{-4} M. Time marking = 1 sec. The increase of time marking height shows the duration stimulus injection in the steady flow of sea water. A pointer indicates stimulus outflow on dolphin tongue.

Table 1. The Sensitivity of 2 Black Sea Bottlenose Dolphins to Sea Water Solutions of Chemicals

Chemical Stimuli	Concentration (M) Elza		Philip		Level of Significance
Sucrose	0.3*	0.6*	0.3*	0.6*	
Picric acid	---------		4.3×10^{-5}		<0.05
Citric acid	0.2		--------		<0.05
Citric acid	0.05*		0.5*		
Capronic acid	1×10^{-4}		1×10^{-4}		<0.05

* = The proportion of correct choices is about 50% level.

animals were able to perceive these acids not by taste, but by means of some other type of sensitivity.

Experimental data on dolphin sensitivity to different sea water solutions are presented in Table 1. Only the sensitivity to bitter substances was similar to that of most other animals. Dolphins were insensitive to sucrose. They discriminated fresh water from sea water while they were unable to distinguish sea water from twice diluted sea water. Dolphins were insensitive to citric acid in fifty millimoles concentration which is perceived as very sour by humans and its pH is 2.5. At the same time, dolphins were capable of sensing capronic acid solutions in concentrations which were 200 times lower. The pH was equal to the pH of sea water.

Dolphin sensitivity to fresh water solutions of standard taste chemicals is presented in Table 2. The animals showed low sensitivity to sodium chloride and acids. At the same time their sensitivity to odorous substances - to carbonic acids and to indole - is several hundred times higher than that of most other mammals. We do not know any terrestrial mammals possessing taste sensitivity of that order.

It is possible to suggest that these abilities of taste reception are inherent to sea mammals. To investigate this hypothesis conditioning experiments with pinnipeds were performed. Northern fur seal (Callorhinus ursinus) sensitivity to citric and valeric acids was tested. The taste sensitivity of this animal to valeric acid (0.01 M) was 5 times lower than that to citric acid (0.002 M). In Steller sea lions (Eumetopias jubatus) and Caspian seals (Phoca caspica) the threshold concentration of citric acid was 0.002 M. It means that peculiarities of dolphin chemical sense is significantly different from taste sensitivity of other sea mammals (Kuznetzov, 1982; 1984).

Autonomous Reaction Experiments

The chemical sensitivity in other dolphin species was also evaluated. The autonomous reactions is response to odorous chemical presentations were recorded in harbor porpoises and common dolphins. The bottlenose dolphins (Fig. 2) were tested also to the solutions used in the conditioning experiments. This series served as a control series.

Table 2. The Sensitivity of 2 Black Sea Bottlenose Dolphins to Fresh Water Solution of Chemicals

Chemical Stimuli	Concentration (M) Bogdan		Daryal		Level of Significance
Sucrose	0.6*	0.3*	0.6*	0.3*	
Glucose	1.0*	0.5*	1.0*	0.5*	
NaCl	0.3		---------		<0.01
Quinine HCl	3×10^{-5}		1×10^{-5}		<0.01
Picric acid	5×10^{-5}		2×10^{-5}		<0.01
HCl	0.1		---------		<0.01
Citric acid	0.05		0.05		<0.001
Oxalic acid	0.03		---------		<0.01
Capronic acid	1×10^{-5}		5×10^{-5}		<0.01
Valeric acid	1×10^{-5}		1×10^{-5}		<0.001
Indole	1×10^{-6}		1×10^{-6}		<0.001

* = The proportion of correct choices is about 50% level.

The dolphin was placed into the basin, the tube was inserted into its mouth cavity. A steady flow of sea water was provided to wash the mouth cavity. The chemical solution was introduced into the tube with variable intervals. No chemical was given as a control stimulus. The reactions to pure sea water and the reactions to chemicals were compared.

The results are shown in Table 3. The table shows that dolphins can discriminate odorant solutions in concentration ranging from 1.10^{-4} to $8.5.10^{-7}$ M. Common dolphins detected the faeces extract solution diluted in proportion 1 on 10 millions. Bottlenose dolphins were able to discriminate the presence of urine components in the water when urine was diluted one to one hundred. The table shows that the order of chemical sensitivity in all three species investigated is similar and rather high. We conclude that this type of chemical sense resembles the sensitivity of olfaction (Kuznetsov, 1979).

Nachtigall and Hall (1982) reported the results of similar investigations. The interesting technique they introduced permitted them to establish threshold concentrations which Pacific bottlenose dolphins could perceive. It was 0.17 M for citric acid and $1.36.10^{-5}$ M for quinine-sulfate. These values coincided closely with those obtained in our work, which were obtained by an other technique. These authors concluded that dolphins possess well developed sense of taste.

This conclusion seems to be drawn on the basis of the high sensitivity of dolphin to quinine solutions. As the same time, it is established now that the sensitivity to bitter substances is not necessarily connected with taste reception, and the bitter taste can be perceived by olfaction. The investigations of Kurihava (1973) showed that in contrast to salts and acids, bitter substances are capable to bind with hydrophobic regions of receptor cell membrane as well as to olfactory receptors (Koyama and Kurihara, 1972).

Table 3. The Sensitivity of Black Sea Dolphins: Bottlenose Dolphins, Common Dolphins and Harbor Porpoise to Solutions of Chemicals. Autonomous Reaction Data

Dolphin species	Number of subjects	Chemical Stimuli	Concentration	Level of significance
Bottlenose dolphin	3	Valeric acid	1×10^{-4} M	<0.001
	3	Quinine HCl	1×10^{-4} M	<0.005
	2	HCl	0.15	<0.01
	1	Urine	1×10^{-2}	<0.01
Common dolphin	2	Faeces suspension	1×10^{-7}	<0.001
Harbor porpoise	4	Scatole	1.7×10^{-6} M	<0.01
	4	Camphor	3.0×10^{-6} M	<0.01
	4	Trimethylamine	8.5×10^{-6} M	<0.001
	2	Trimethylamine	8.5×10^{-7} M	<0.05

CHEMORECEPTION AND BEHAVIOR IN DOLPHINS

We may conclude that dolphin chemical sense has no analogy in other vertebrates. It is not taste or olfaction in their typical form. The question is whether in dolphins this type of sensitivity influences their behavior or that this is a rudimentary type of perception. Observations on dolphin behavior in the wild and in captivity gave no answer. The next series of our experiments was aimed at elucidating this point.

It is well known that due to the gradual complication of experimental paradigm, an animal can solve the more complicated tasks than they solve in the wild. In case of a positive discrimination task, the situation is reversed. The animal cannot solve such problem unless it is provided by natural capacities to do so. Without such capacities the animal is unable to acquire two conditioned reflexes and perform accurate reactions during multiple trials. According to what has been mentioned above, we based our experimental techniques on a positive discrimination learning paradigm, aiming to investigate dolphin chemoreception. We considered reasonable to present the animal with a very difficult task, keeping in mind that if the dolphin has no capacity to form the adequate behavior on the basis of chemical information, it would not be able to make multiple correct choices in the experiment.

Seven dolphins were conditioned to perform rather difficult instrumental actions - positive discrimination task. The animal had to perform one type of action in response to one chemical cue and to perform a second type of action in response to another chemical. For instance, in 10 sec. after the introduction of chemical stimulus (sea water or chemical solution in sea water) the special frame with two objects of different shape were lowered into the water (Table 4, first experiment). The animal was trained to touch one of them after receiving sea water or another one after chemical stimuli. The left and right positions of the objects were changed randomly. Correct choices were reinforced by fish.

Table 4. Instrumental Discrimination Conditioning of Chemical Stimuli in Dolphins

Exp. No.	Task Paradigm	Number of Subjects	Level Stimuli Presentation	Correct Solution	Significance
1.	Chemical stimulus visual identification of different targets	3	320	205	<0.001
2.	Chemical stimulus echolocation targets	2	140	88	<0.001
3.	Chemical stimulus definite instrumental movement pattern	1	87	57	<0.01
4.	Chemical stimulus instrumental movement to definite place	6	721	483	<0.001

The learning criterion was nine correct responses out of ten. All animals reached this criterion after a different amount of presentations (30-100 trials).

The main types of paradigms which were used in the experiment are presented in the Table 4. Dolphins discriminated chemical stimuli successfully. Thus, they proved to be capable to associate chemical and visual as well as echolocation information. One may suggest that they can perceive the chemical trace of the object as the object's signal. The capacity of a dolphin to associate a chemical stimulus with a definite spatial location, makes it possible to suggest that dolphins are capable to use chemoreception for orientation. Experiment number three demonstrates that dolphins are capable to modify their behavior according to chemical cues perceived.

Dolphins proved to be capable to solve the very difficult tasks on the basis of chemical stimuli which included compounds of six chemical classes (Table 5). Not all dogs could solve these positive discrimination tasks (Lawika, 1969). These facts signify that dolphins possess a well-developed system of chemical sensation which is functionally connected with those brain systems which are relevant to behavioral manifestations.

Atema (1980) demonstrated that fishes solve complex problems using exclusively the olfaction sense, although their taste and olfactory capabilities are equally well-developed. Fishes can use taste sensation when they look for dead prey or when they are presented with the most simple learning paradigm of attraction-aversion. The chemical communication signals and chemical signals of living prey are perceived by means of olfaction.

Table 5. Chemicals Presented to Dolphins

Chemical Stimuli	Concentration (M)	Class of Substances
Valeric Acid	5×10^{-4}	Carbonic acids
Trimethylamine	3.4×10^{-2}	Amines
Phenylethanol	5×10^{-4}	Aromatic Compounds
Camphor	5×10^{-3}	Terpends
Indole	9×10^{-3}	5-atom-heterocyclic compounds
Pyridine	1×10^{-4}	6-atom-heterocyclic compounds

Our experiments demonstrated that dolphins are sensitive to chemical communication signals (Kuznetsov, 1978a; 1979). These data proved our suggestion that dolphin chemical sense and olfaction have many features in common.

COMPARISON OF TONGUES IN DOLPHINS, PINNIPEDS AND UNGULATES

The next problem was the investigation of the histological structure of dolphin tongues, as well as those of several other mammal species. The tongues of bottlenose dolphins (Tursipos truncatus) of different ages, harbor porpoises (Phocoena), Amazon river dolphins (Inia geoffrensis) and beluga (Delphinapterus leucas). Also the tongues of pinnipeds: the Caspian seal (Phoca caspica) and fur seal (Callorhinus ursinus), as well as a cow and pig. (The latter animals tongues also were taken because as dolphins they are supposed to have evolved from primitive ungulates).

In contrast to the tongues of other animals, the adult dolphin tongue is devoid of papillae. At the same time bottlenose dolphin fetuses and neonates possess fungiform and circumvallate papillae in the region of tongue root (Fig. 3A, B). These structures resemble those seen on the cow tongue.

Histological investigation of the papillae demonstrates typical taste buds with taste pores and innervation (Fig. 3C). The taste pores are numerous (Fig. 3D). There are about 1600 taste pores in the bottlenose dolphin neonate's tongue. Thus, taste reception is well-developed in neonatal dolphin and perhaps provides the normal feeding during suckling.

In 2 to 3 year old dolphin the circumvallate papillae already are transformed into pits: the specialized structures of the dolphin tongue which is filled with mucous secreted by special glands (Fig. 4A). At the bottom of such pit one may see the irregularly formed papillae. The taste buds are reduced. No taste buds are present in the tongue of the adult animal.

In the tongue of ungulates and pinnipeds which were investigated, the circumvallate papillae are covered by cornified epithelial cells (Fig. 5A). The surface structure of pit papillae in the dolphin tongue root is drastically different. The surface is covered by microvilli (Fig. 3B, C).

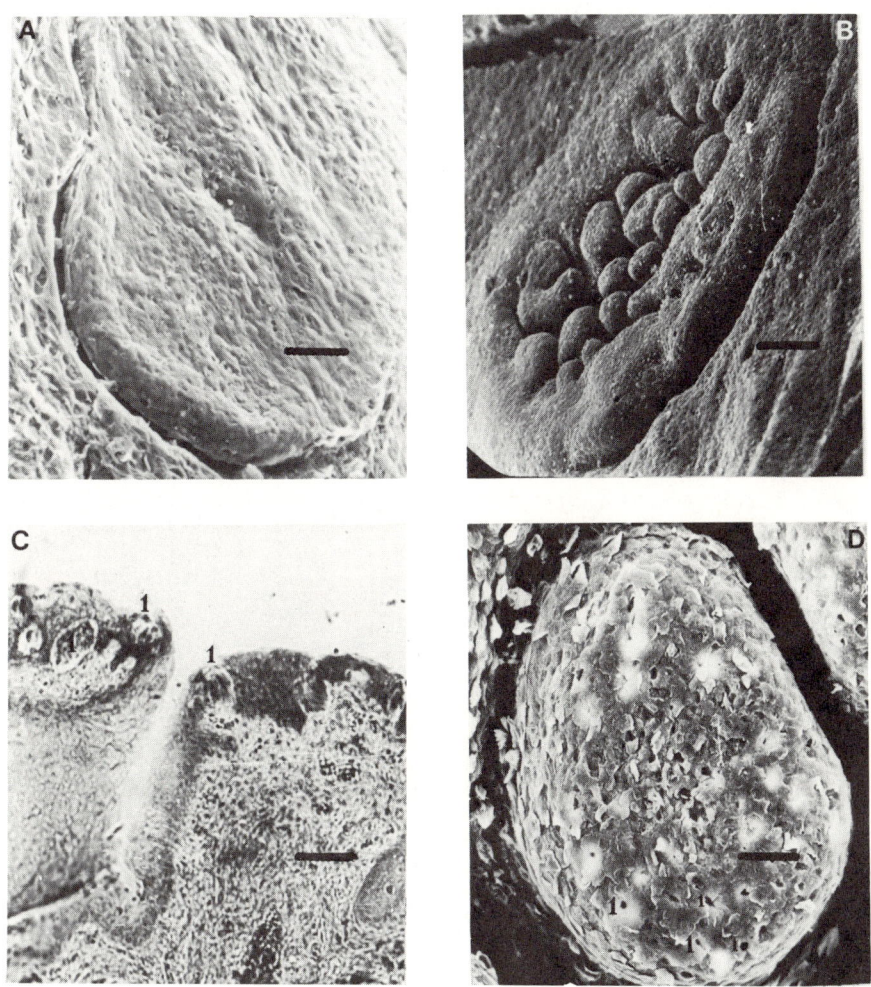

Fig. 3. The circumvallate papillae of bottlenose dolphin.
A: Scanning electron micrograph (SEM) of papillae of a fetus with a body length of 900 mm. Bar = 100 25μm. B: SEMs of the papillae of the neonate of 1100 mm body length. Bar = 330 μm.
C: Photomicrograph of the epithelial structure of the circumvallate papillae of the neonate, I = taste bud. Bar = 50 μm. D: SEMs of the taste pores in the papillae, I = taste pore. Bar = 50 μm.

491

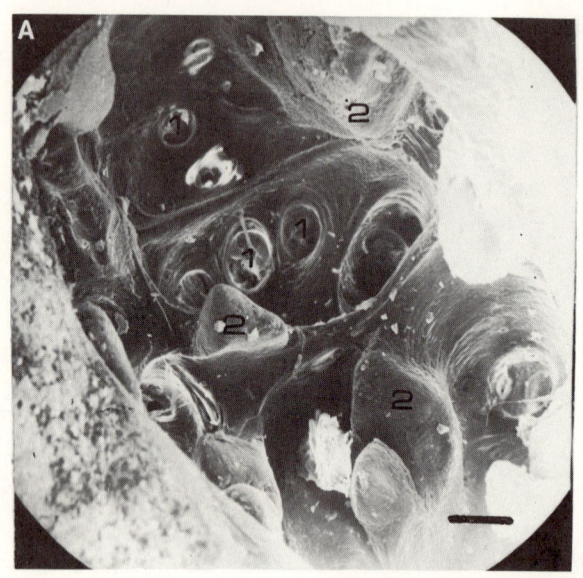
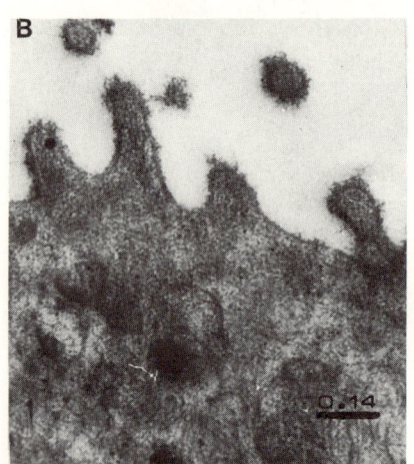
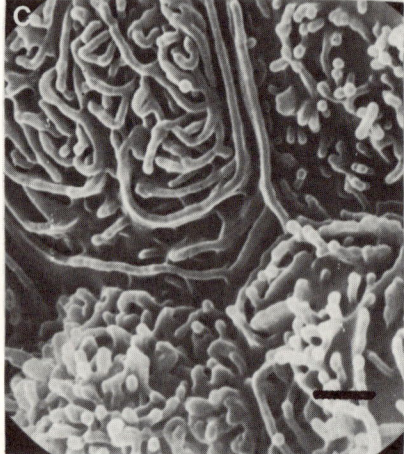

Fig. 4. The pit of an adult bottlenose dolphin tongue. A: SEMs of the pit, I = the opening of the gland duct, 2 = the irregular form of papillae. Bar = 165 μm. B: High voltage electron micrograph (HVEM). The microvilli of the epithelial cell of the irregular form papillae. Bar = 0.14 μm. C: SEMS of the microvillar cells of the papillae. Bar = 0.67 μm.

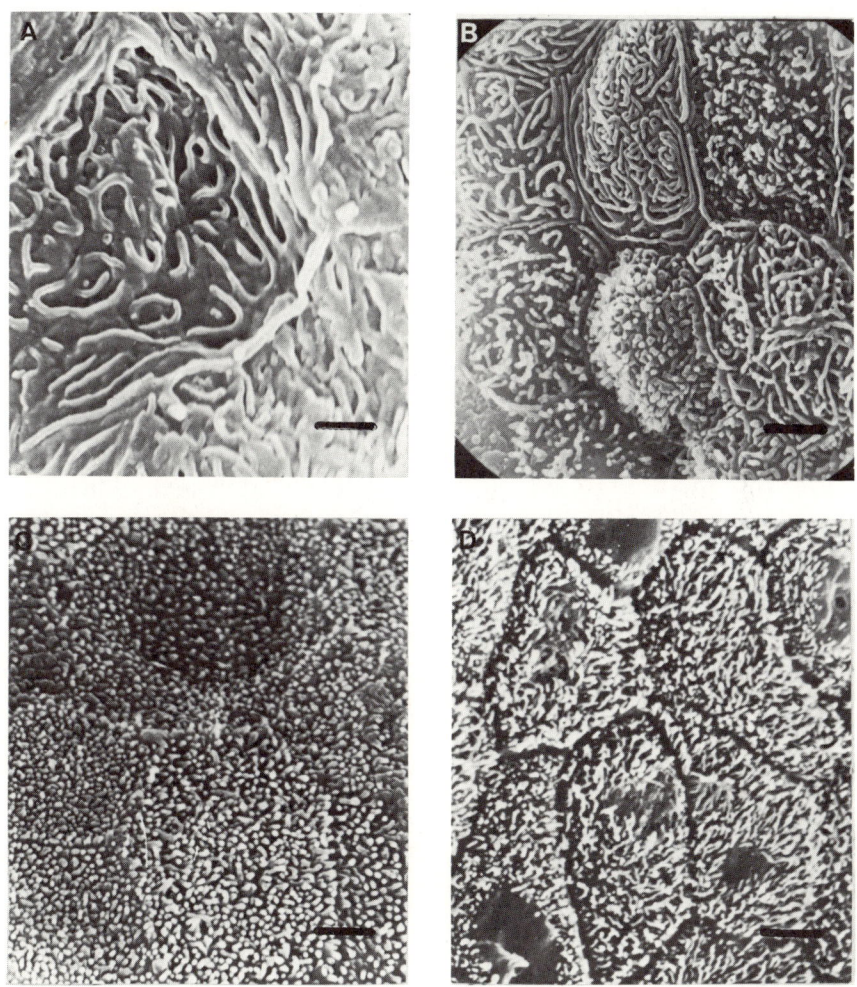

Fig. 5. A: SEMs of the circumvallate papillae of the Caspian seal. Bar = 1 μm. B, C, D: SEMs of the microvillar cells of the pit's papillae of different dolphin species. Bottlenose dolphin. Bar = 2 μm (B). Harbor porpoise. Bar = 1.67 μm (C). Amazon river dolphin. Bar = 3.75 μm (D).

In some regions the microvillar structure is variable (Fig. 4B, 5B). Nine types of microvilli cells could be detected in the proximity of the pit papillae. Their diameter was 0.1 μm, their length varies from 0.2 to 2.0 μm.

The papillae cells in the harbor porpoise tongue also are covered by microvilli (Fig. 5C), as well as those of the Amazon river dolphin (Fig. 5D). Preliminary data suggest that in the beluga tongue a similar microvillar system exists. It is possible that such peculiarities are characteristic for all toothed whales. No structures comparable to these were revealed in tongues of other mammals in our investigation or in the data of other authors.

CHEMICAL SENSE OF DOLPHINS - QUASI-OLFACTION

We conclude that adult dolphins have no taste buds and that their taste sensitivity is of low level. At the same time dolphins demonstrate the sensitivity to odorous substances and the morphology of receptors is radically different from that of taste buds.

Vertebrates possess two types of chemoreceptor-olfaction and taste. Which one do dolphins possess?

The main function of taste is the homeostatic one - to provide the stability of internal medium of the organism. Taste is crucial for acceptance of rejection of different food items. Animals can be easily conditioned to a taste aversion reaction while other types of reflexes are much more difficult to establish.

The main function of olfaction is the communicative one. It is used during reproductive, territorial and social interactions. Olfaction serves for individual recognition of chemicals.

The dolphin chemical sense possesses a relatively high sensitivity to different odorous substances. Those invoked in communication are among them. Using their chemical sense, dolphins probably direct behavioral actions. All this makes us conclude that dolphin chemical sense substitutes the role of olfaction. At the same time it substitutes the role of olfaction. This is why we call it "quasi-olfaction".

The microvillar cells of tongue root pits are the most probable candidates for the role of quasi-olfaction receptors. Innervation of these receptors are provided by fibers of nervus glossopharyngeus and perhaps by the trigeminus. The information enters the nucleus tractus solitarius. This nucleus is known to have fiber connections with the structures of olfactory brain (Giachetti and MacLeod, 1977). It is probable that in dolphins the chemical information is processed in the olfactory brain structures considering that their olfactory tuberculum being surprisingly well-developed (Filimonov, 1965).

We have no direct experimental data showing that the olfactory brain of dolphins processes chemical information. However, the presence of the islands of Calleja in dolphin olfactory tubercles (Jacobs et al., 1971) could be a serious foundation for such a conclusion. According to the most current ideas, the islands of Calleja influence emotional behavior actions such as a reproductive on the basis of chemical (odorous) information in other mammals (Zvegintseva, 1986; Millhouse, 1987; Talbot et al., 1988).

PROBLEM OF OLFACTORY REDUCTION IN TOOTHED WHALES

Why did the reduction of olfactory perception occurred in toothed whales in course of evolution?

The most widely accepted explanation is that in dolphins the respiratory and gastrointestinal systems are separated from one another (Lawerence and Schevill, 1965; Yablokov et al., 1972) (see Fig. 7). On the basis of this hypothesis a new hypothesis found a wide exception that, in the open sea, the clear air contained no smell molecules and thus bore no information crucial for an animal. Possibly this led a complete reduction of olfaction (Moulton, 1967).

We carried out direct experiments in bottlenose dolphins in order to investigate the completeness of separation of oral and nasal cavities (Kuznetzov, 1978b; 1988). For this purpose we used special plastic tube 120mm length, 10mm diameter. A stiff plate was placed into this tube and straightened it; as a spring (Fig. 6). The ends of the tube were closed and small holes (1 mm) were made in it.

If such tube was lowered under water the air in the tube was pressed and some amount of water entered the tube replacing the air. When the tube was lifted, water remained inside.

In a bended position this device was inserted into nasopharyngeal cavity of dolphin where it straightened up and was kept tight by walls. Capron filament fixed this tube to the sucker on the head of the dolphin. The dolphins were swimming for 15 min. with this device in enclosure (depth 4 metre). The behavior was natural: they were searching for fish using echolocation and then they were eating this fish.

After the tube was withdrawn from the dolphin's blowhole, sea water was in it. It seems that the nasopharyngeal sphincter does not retard the filtration of sea water to nasopharyngeal cavity from oral cavity. The pressure in this cavity is a hydrostatic one because sea water enters nasopharyngeal cavity. It is possible that the fountains of toothed whales consist of this water. The water can not enter the trachea, because there is no large pressure difference between nasopharyngeal cavity and trachea, and special preventing mechanisms exist in the larynx (Grachova, 1971). Trachea, bronchia, bronchioles and alveolus are filled with air.

Thus, the experiment showed that in dolphins as well as other mammals the respiratory and gastrointestinal systems are not separated, so both hypotheses are not right.

First, let us consider the function of bone nares in modern toothed whales - that is the place where the olfactory epithelium resided in toothed whales ancestors (Fig. 7, 2). Direct measurements demonstrated that in the cavities of the bony nares (Fig. 7, 2) the pressure increases during clicks

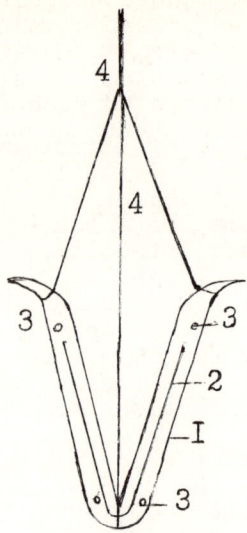

Fig. 6. The device used in experiment for investigation of the completeness of separation of oral and nasal cavities in bottlenose dolphin. 1 = plastic tube. 2 = stiff plate. 3 = holes. 4 = capron filaments.

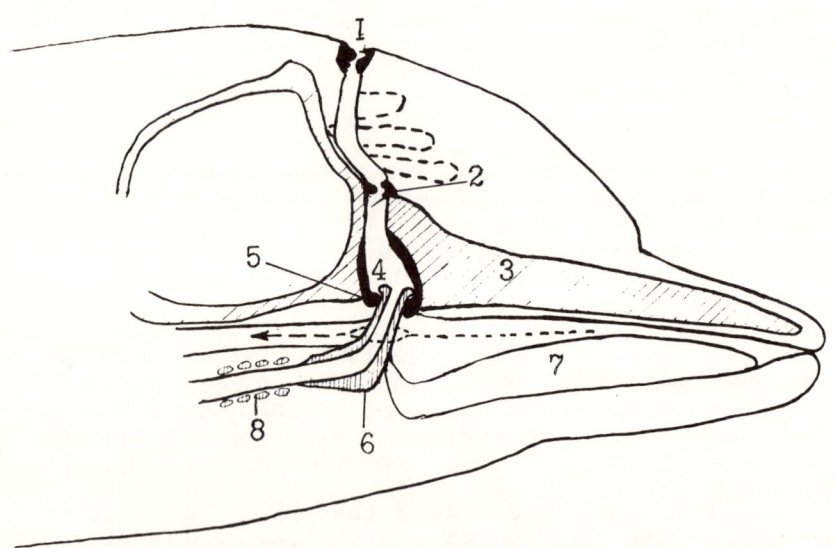

Fig. 7. Schematic drawings of the upper parts of the respiratory and gastrointestinal tract of the bottlenose dolphin. 1 = blowhole, 2 = bone nares, 3 = skull, 4 = Nasopharyngeal cavity, 5 = sphincter, 6 = larynx, 7 = tongue, 8 = trachea.

generation from the zero level up to 54 kPa in harbor porpoise and to 81 kPa in the bottlenose dolphin respectively (Amundin and Anderson, 1983). The pressure values mentioned above developed during ordinary click generation. They increase during distant location scanning (Au et al., 1974). Because olfactory cells are sensitive to mechanical stimuli (less than 1 gram per cm^2), the normal functioning of olfactory epithelium seems impossible in bony nares of dolphins.

Apparently, the main function of the nasopharyngeal sphincter is preventing air out from nasopharyngeal in oral cavity. Hence, apart from breath the constructions of all structures from blowhole to larynx is determined by effectuation of echolocation function.

Probably the ancestor of toothed whales possessed olfaction as well as low power echolocation. Fossil data demonstrate that the increase of trigeminal and auditory nerve systems in the whales took place parallel with the decrease in the size of olfactory nerve openings and the olfactory bulbs (Dart, 1923; Edinger, 1955).

Mchedlidze (1964) suggested that the increase of locomotion in ancestor whales, which permitted them to colonize the open ocean, was followed by the reduction of olfaction. These animals probably began to chase very motile prey and the success of such hunting was provided by echolocation which provides more possibilities to discover and to catch the prey in cases of low visibility. At the same time, the increase of echolocation effectively was accompanied by an increase of pulse power.

The increase of pulse power was in conflict with the safety of the olfactory epithelium while the need of olfaction prevented the improvement of echolocation. The final reduction of olfactory nerve openings took place about 25-12 million years ago. It means that the typical olfaction was lost totally (Edinger, 1955). So this competition was solved in favor of echolocation.

Thus, during adaptation of toothed whales proceeded on the way of echolocation sophistication which led to the reduction of the typical olfaction. At the same time in vertebrates specific chemical stimuli are the impact of reproduction. That is the why the loss of olfaction in toothed whales could injure the reproductive process. Quasi-olfaction emerged in these animals.

The process of evolution resulted in total spatial separation of these two functions - pulse generation and chemical sensitivity. The tongue receptors of special chemical sense developed which began to fulfill the olfactory and possible gustatorial functions.

The receptors are placed in pits and protected from mechanical contacts and have a stable chemical medium due to gland secretion. Microvillar receptor cells in dolphins differ from taste receptors of mammals. We suggest that they are formed of epithelial cells by developing the potential capacity of cells to sense chemicals. The presence of such capacity is shown in mouse neuroblastoma cells (N 18 clone) which are not chemoreceptors. These cells reacted to odorous and bitter substances in concentration 10^{-5}-10^{-6} M (Kashiwaganagi, Kurihara, 1984; and Kumazawa et al., 1985).

The root of the dolphin tongue is innervated by nerves of the glossopharyngeus and perhaps n. trigeminus. The nuclei of these nerves in mammals have developed connections with the nucleus tractus solitarius which in turn has structural connections with olfactory brain (Giachetti and MacLeod, 1977). It seems probable that in the process of reduction of olfactory bulbs the central structure of the olfactory brain more and more switched to the analysis of chemical information received from tongue receptors. Thus, maybe quasi-olfaction emerged in dolphins with the sense receptors on the tongue and in the center the olfaction brain structures.

CONCLUSIONS

This article represents a review of the literature as well as the main results and the theoretical approaches of the author. Chemical sense of dolphins was investigated by 3 general experimental methods: conditioning experiments, autonomous reaction experiments and communication experiment (the latter see Kuznetsov, 1978a), in which 10 different experimental techniques were used. Some histological methods were used as well. At the same time the author recognizes that this is only the beginning of investigation of dolphin chemical sense and not all his conclusions are undoubtedly correct.

The basic conclusions concerning the chemical sense of dolphins are discussed in proper parts of this article. In this section we want to draw attention to: 1. some biological problems, and 2. some possibilities to understand the biology of dolphins and probably toothed whales in general.

1. The development of quasi-olfaction in dolphins and morphofunctional peculiarities of this sense permit some new viewpoint for various fields of biology.

General Zoology. Toothed whales were considered the only vertebrates unable to perceive smells, i.e. anosmatics (Turner, 1891). The presence of quasi-olfaction in dolphins testifies that there are no anosmatics among vertebrates.

General Neurology. The development of quasi-olfaction in dolphins makes the current opinion doubtful that the olfactory brain structure is not involved in the treatment of chemical information (smells). So, different suppositions and hypotheses of olfactory brain function in dolphins and mammals in general based on this opinion may prove to be wrong, in particular the hypothesis of a complete change of olfactory brain functions (Filimonov, 1965).

Sensory Physiology, Chemoreception. Trigeminal nerve chemoreception is accomplished by fibers which innervate olfactory epithelium and oral cavity. It is believed that this type of chemoreception is not a specific one and carries out accessory functions. Until now according to Parcer (1922) trigeminal nerve chemoreception is related to general chemical sense in vertebrates. However, it was known long ago that chemical perception is sensitive to odorous substances in mean concentration (Silver and Maruniak, 1980). In fish the sensitivity of trigeminal chemoreception is 10^{-5}-10^{-6} M for different odorous substances (Belousova et al., 1983). Among vertebrates only the toothed whales that have no olfaction. In dolphins this type of chemoreception is the single specific chemical sense (quasi-olfaction) which fulfills the olfactory functions. This suggests that in other vertebrates trigeminal chemoreception also may be specific chemical sense relevant to quasi-olfaction.

The unique development of ontogenesis in dolphin's chemoreception organs is confirmation of the fact that functional possibilities of taste and olfaction are different in principle. As the taste reception can not perform the functions of olfaction, the quasi-olfaction emerges in dolphins instead of taste reception.

The question of whether specific chemoreception can be performed by nerve fibers or that special structures are necessary for that, has been discussed for a long time. In case of dolphins, we can see that special structures (microvillar cells) are necessary for specific chemical sense. Thus, similar cells in other vertebrates could perform these functions as well.

General Histology. Microvillar cells similar to dolphin cells are found in brain ventricles in different parts of reproductive system and other structures. We suggest that these cells can perform chemoreceptive functions and are interchemoreceptors.

2. As for functional possibilities, the chemical sense of dolphins coincides with olfaction and can be used by them for solving the same tasks where the vertebrates use olfaction. At the same time, the sensitivity of quasi-olfaction is lower than that of olfaction.

Feeding Behavior. Many species of fish unite in big schools of thousands of animals. After the migration of such a school an odorous trace like an odorous road remains. The chemical sensitivity of dolphins is strong enough to perceive that trace and find the school. Thus, the quasi-olfaction of dolphins may play an important role in their feeding behavior.

Orientation. There are many currents in the sea with a particular odour such as of animals, kelps, gas seeps, etc. It is possible that quasi-olfaction helps the dolphins to orient themselves in those currents.

Social Behavior of Dolphins. Dolphin schools can be numerous. The bladder of cetacea is small. Urination occurs often in bottlenose dolphins: once every 7-10 minutes (Yablokov, 1961). Thus, the migrating school of dolphins chemical trace remains which makes it possible for an animal or a small group to find the dolphin school by using their sense of quasi-olfaction.

Reproduction. The most important factor here is the influence of chemical stimuli of animals in close contact. Chemical stimuli are known to play a great part in invoking the physiological and psychological processes involved with preparing for reproduction. Thus, the capacity of dolphins to find a school and physiological influence of chemical stimuli may play an important role in reproduction.

Diseases, Infections, Invasions. The presence of sea water in the nasopharyngeal cavity seems to take place. This route may be of significance in developing infections and invasions in dolphins. Especially at present due to the pollution of seas with industrial flow and sewage.

Bacteria presenting in unpurified sewage and in dolphinarium water may be the cause of respiratory diseases of wild and captive dolphins in coastal waters.

REFERENCES

Agarkov, G. B., and Gilevich, S. A., 1979, On the question of chemoreception by dolphins, Vestnik Zoologii, 3:3 (Russian).
Amundin, M., and Andersen, S. H., 1983, Bony nares air pressure and nasal plug muscle activity during click production in the harbor porpoise, Phocoena and the bottlenose dolphin, Tursiops truncatus, J. Exp. Biol., 105:275.
Arvy, L., and Pilleri, G., 1970, The tongue of Platanista gangetica and remarks on the cetacea tongue, in: "Investigations on Cetacea", Vol. 2, G. Pilleri, ed., Berne, Switzerland.
Atema, J., 1980, Chemical senses, chemical signals, and feeding behavior in fishes. in: "Fish behavior and its use in the capture and culture of fishes", J. E. Bardach, J. J. Magnuson, R. C. May, J. M. Reinhart, eds., Manila.
Au, W. W. L., Floyd, R. G., Penner, R. H., and Murchison, A. E., 1974, Measurement of echolocation signals of the Atlantic bottlenose dolphin, Tursiops truncatus, Montagu, in open waters, J. Acoust. Soc. Amer., 56:1280.
Belousova, T. A., Devitsina, G. V., and Malyukina, G. A., 1983, Functional peculiarities of fish trigeminal system, Chemical Senses, 8 N 2:121.
Berzin, A. A., 1971, "Sperm whale", Pisch. promishlennost, Moscow (Russian).
Breathnach, A. S., 1953, The olfactory tubercule, prepyriform cortex and precommissural region of the porpoise (Phocoena phocoena), J. Anat., 87:96.
Breathnach, A. S., 1960, The cetacean central nervous system, Biol. Rev., 35:187.
Dart, R., 1923, The brain of the Zeuglodontidae, Proc. Zool. Soc., London, p. 615.
Donaldson, B. I., 1977, The tongue of the bottlenose dolphin (Tursiops truncatus), in: "Functional anatomy of marine mammals", R. Harrison, ed., Academic Press, London.
Edinger, T., 1955, Hearing and smell in cetacean history, Monatsch. Psychiat. Neurol., 129:37.
Filimonov, I. N., 1949, "Comparative anatomy of the cerebral cortex of mammals. Paleocortex, Archicortex, and Intermediate Cortex", Publ. Acad. Med. Sci., Moscow (Russian).
Filimonov, I. N., 1965, On the so-called Rhinencephalon in the dolphin, J. Hirnforsch., 8:1.
Gellerman, L. W., 1933, Chance ordr of alternating stimuli in visual discrimination experiments, J. Genet. Psychol., 42:207.

Giachetti, J. and MacLeod, P., 1977, Olfactory input to the thalamus: evidence for ventroposteromedial projection, Brain Res., 125:166.

Gilevitch, S. A., 1978, Histostructure of papillae of fetuses and neonates bottlenose dolphins, in: "Abstracts of Reports of the Seventh All-Union conference, Simferopol", (Russian).

Glass, G. V., and Stanley, J. C., 1970, "Statistical Methods in Education and Psychology", Prentice-Hall, Inc., Englewood Cliffs, New Jersey.

Grachova, M. S., 1971, Some peculiarities of the structure of the bottlenose dolphin, Zoologicheskii Zhurnal, 50:1539 (Russian).

Jacobs, M. S., Morgane, J., and McFarland, W., 1971, The anatomy of the brain of the bottlenose dolphin (Tursiops truncatus), Rhinic Lobe (Rhinencephalon) 1. The paleocortex, J. Comp. Neurol., 141:205.

Kashiwayanagi, M., and Kurihara, K., 1984, Neuroblastoma cell as model for olfactory cell: mechanism of depolarization in response to various odorants, Brain Res., 293:251.

Kleereloper, H., 1969, "Olfaction in Fish", Pergamon Press, UK.

Kleynenberg, S. E., Yablokov, A. V., Bel'kovich, V. M., and Tarasevich, M. N., 1964, "Beluga", Nauka, Moscow (Russian).

Koyama, N. and Kurihara, K., 1972, Effect of odorants on lipid monolayers from olfactory epithelium, Nature, 236:402.

Kumazawa, T., Kashiwayanagi, M., and Kurihara, K., 1985, Neuroblastoma cell as a model for a taste cell: mechanism of depolarization in response to various bitter substances, Brain Res., 333:27.

Kurihara, Y., 1973, Effect of taste stimuli on the extraction of lipids from bovine taste papillae, Biochem.Biophys. Acta, 306: 478.

Kuznetsov, V. B., 1974, A method of studying chemoreception in the Black Sea bottlenose dolphin (Tursiops truncatus), in: "Morphology, Physiology and Acoustic of Marine Mammals", Nauka, Moscow (Russian).

Kuznetsov, V. B., 1978a, Chemical communication and ability of bottlenose dolphins to send information about chemical stimulus, in: "Marine Mammals. Results and Methods of Investigation", Nauka, Moscow (Russian).

Kuznetsov, V. B., 1978b, The peculiarities of functioning of the upper respiratory tract under water, in: "Abstracts of Reports of the Seventh All-Union Conference, Simferopol", (Russian).

Kuznetsov, V. B., 1979, Chemoreception in dolphins of the Black Sea: bottlenose (Tursiops truncatus), common dolphins (Delphinus delphis) and common porpoise (Phocoena), Dokl. Acad. Nauk. SSSR, 49:1498 (Russian).

Kuznetsov, V. B., 1982, Taste perception of sea lions, in: "Abstracts of Reports of the Eighth, All-Union Conference, Astrakhan". (Russian).

Kuznetsov, V. B., 1984, Chemoreception of dolphins, "Avtoredferat dissertatsii kand. biol. nauk", Moscow. (Russian).

Kuznetsov, V. B., 1986, Chemical sense in dolphins, in: "Chemical Communication of Animals. Theory and Experience", Nauka, Moscow. (Russian)

Langworthy, O. R., 1932, A discription of the central nervous system of the porpoise (Tursiops truncatus), J. Comp.Neur., 54:437.
Lawika, W. D., 1969, Differing effectiveness of auditory quality and location cues in the form of differentiation learning, Acta biologiae experimentale, 29: N 1.
Lawrence, B., and Schevill, W. E., 1965, Gular musculature in delphininds, Bull. Mus. Comp. Zool., 133:1.
Liley, N. R., 1982, Chemical communication in fish, Can. J. Fish Aquat. Sci., 39:22.
Lowell, W. R., and Flanigan Ir. W. F., 1980, Marine Mammal chemoreception, Mammal Rev.., 10:53.
Li Yuemin, 1983, The tongue of Baiji, Acta zool. sin., 29:35.
Matthes, E., 1927, Der Finfluss des medium-wechsels auf das Geruchsvermögen von Triton, Z. Vergl. Phsiol., 5, N 1:83.
Mchedlidze, G. A., 1964, "Cetacean Paleospecies of the Caucasus" Metsniereba, Tbilisi.
Millhouse, O. E., 1987, Granule cells of the olfactory tubercle and the question of the island so Calleja, J.Comp. Neurol., 265:1.
Moulton, D. G., 1976, Olfaction in mammals, Am. Zool., 7:421.
Nachtigall, P. E., and Hall, R. W., 1982, Taste reception in bottlenose dolphin. in: "Third international theriological congress. Helsinki, Abstracts of papers", Helsinki.
Oelschläger, H., and Buhl, E. H., 1985, Development and rudimentation of the habor porpoise, Phocoena (Mammalia: Cetacea), J. Morphol. 184:360.
Parker, G. H., 1922, "Smell, Taste and Allied Senses in Vertebrates", Lippincott, Philadelphia.
Silver, W. L., and Marunial, J. A., 1981, Trigeminal chemoreception in the nasal and oral cavities, Chemical Senses, 6, N 4:121.
Sinclair, J. G., 1966, The olfactory complex of dolphin embryos, Texas Rep. Biol. Med., 24:426.
Sokolov, V. E., and Kuznetsov, V. B., 1971, Chemoreception in the Black Sea dolphin (Tursiops truncatus Mont.), Dokl. Akad. Nauk SSSR, 204:998 (Russian).
Sokolov, V. E., and Volkova, O. V., 1971, The structure of the tongue in dolphins, in: "Morphology and Ecology of Marine Mammals", Nauka, Moscow (Russian).
Sonntag, C. F., 1922, The comparative anatomy of the tongues of the Mammalia, VII, Cetacea, Sirenia, and Ungulata, Proc. Zool. Soc., London, p. 639.
Suchowskaja, L. I., 1972, The morphology of the taste organs in dolphins, in: "Invest. Cetacea. Vol. 4", Berne.
Talbot, K., Woolf, N. J. and Butcher, L. L., 1988, Feline islands of Calleja complex: II. cholinergic and cholinesterasic features, J. Com. Neurol., 275:580.
Turner, W., 1891, The convolutions of the brain: a study in comparative anatomy, J. Anat. Physiol., 25:105.
Valiulina, F. G., and Khomenko, B. G., 1976, Micromorphology of the tongue receptor system in Tursiops truncatus, Zoologicheskii Zhurnal, 55:467.
Yamasaki, F., Satomi, H., and Kamiya, T., 1976, The tongue of Franciscana (La Plata dolphin) Pontoporia blainvillei, Okajimas Folia Anatomica Japonica, 53:77.
Yamasaki, F., Komatsu, S., and Kamiya, T., 1978, Taste buds in the pits at the posterior dorsum of the tongue of Stenella coeruleoalba. in: "Sci. Rep. Whales Res. Int. Tokyo", Tokyo.

Yablokov, A. V., 1961, About "the olfction" in marine mammals, in: Trudy Soveshchaniy Ikhtiologicheskoj Komissii A. N. SSSR, 12:87. Publ. AN SSSR, Moscow (Russian).

Zvegintseva, E. G., 1986, Structural organization and connections of cellular ensembles of the cat cerebral olfactory tubercle, Archiv anatomy histology i embriology, 40, N 5:5 (Russian).

BEST VISION ZONES IN THE RETINAE OF SOME CETACEANS

Alla Mass and Alexander Supin

Severtsov Institute of Evolutionary Morphology and
Ecology of Animals
USSR Academy of Sciences
Leninsky prosp.33, 117071, Moscow, USSR

INTRODUCTION

Organization of the visual system in Cetacea still remains unclear in many respects. Of special interest are data on ganglion cell distribution in the retina. It is related closely with visual acuity (Van Buren, 1963; Frisen and Frisen, 1976).

Of particular interest are the retinal zones with high ganglion cell density. They are known in terrestrial mammals as the area centralis or the visual streak. These retinal areas project to visual centers of the brain with the highest magnification factor and are characterized by a higher visual acuity than other retinal sites (Van Burgen, 1963; Rolls and Cowey, 1970; Perry, Cowey, 1985). The location of these zones in the retinae of terrestrial mammals, and the shape, size, and density of ganglion cells in them are associated with the ecology of the species (Hughes, 1977; Shkolnik-Jarros and Kalinina, 1986).

Dral (1975; 1977; 1983), using total retinal preparations, obtained data on the topography of ganglion cells in two species of dolphins, Tursiops truncatus and Delphinus delphis. Somewhat later such investigations were carried-out on Phocoena phocoena (Mass and Supin, 1986), Neophocaena phocaenoides and Lipotes vexillifer (Gao and Zhou, 1987). These investigations revealed two zones of high ganglion cell density in Cetacea retinae. Such retinal organization was unusual for mammals. In the northern fur seal, Callorhinus ursinus, the same method showed one zone of high ganglion cell density, similar to the area centralis in terrestrial mammals (Mass, 1987).

To understand the organization of the best vision zones of Cetacea, it is necessary to have analogous data on other representatives of this order. The objective of this study was to investigate the spatial distribution of ganglion cell density in three Cetaceans, differing in their taxonomy and ecology: the harbor porpoise, Phocoena phocoena, the grey whale, Eschrichtius gibbosus, and the Amazon river dolphin, Inia geoffrensis.

MATERIALS AND METHODS

Four retinae of harbor porpoises, 3 retinae of grey whales, and 6 retinae of Amazon river dolphins were used in this study. Studies were carried-out on total retinal preparation. The material was fixed in 10% formalin. Methods of Stone (1965) and its modifications for Cetaceans (Mass and Supin, 1986) were used. The preparations were stained by Pishinger's technique with 0.06 % methylene blue solution and mounted in Apathy's gumsyrup without dehydration. The method allowed total staining of the ganglion cells and mapping of the topography of the cells in the retina.

The ganglion cells were identified according to standard criteria (Stone, 1965; Hughes, 1975; Provis, 1979; Perry et al., 1984; Wong and Hughes, 1987; Wassle et al., 1987). The polygonal neurons, located in the retinal surface layer and having a broad rim of the cytoplasm with abundance of well-stained Nissl substance and a nucleus with a prominent nucleolus, were regarded as ganglion.

In the harbor porpoise and the grey whale preparations, the ganglion cells were counted across the whole retinal surface at 1 mm intervals in 0.15 mm² squares. In the Amazon river dolphin, the cells were counted at 0.5 mm intervals in 0.084 mm² squares. The counts were converted to cell/mm². Data obtained were used for mapping the distribution of ganglion cell density in the retina, as well as for calculating the total number of ganglion cells. The cell size was measured along two mutually perpendicular diameters and the mean of the two values was considered.

If necessary, the smoothing of the maps was carried-out by averaging the number of cells in 3 x 3 squares. Maps were transformed by computer in such a way as to project them onto the hemisphere approximating the entire retina. In these spherical maps, the cell density was specified in spherical coordinates, cell/deg². The averaging of the maps of several preparations was possible due to a common coordinate system of the spherical maps.

RESULTS

Harbor Porpoise

The harbor porpoise's retina contained mainly large ganglion cells; their size ranged from 8 to 50 μm. Figure 1 shows fragments of total retinal preparation in the regions of high and low cell density. Most cells were of a typical polygonal shape with several (3-6) clearly visible processes. The cell bodies had a large amount of cytoplasm with well-stained Nissl substance and a pale-stained nucleus with a prominent nucleolus. In the high density region, the ganglion cells were regularly arranged in radially-directed rows, with cell bodies stretched in the same direction. In the low density regions, no such regular arrangement of cells was observed. In the harbor porpoise retina, we could not distinguish the two ganglion cell types described in the <u>Tursiops truncatus</u> and <u>Delphinus delphis</u> (Dral, 1975; 1983).

With a retinal area of 675 to 747 mm², the total number of ganglion cells amounted to 96,500-133,000. Hence the mean density for the preparations was from 140 to 170 cell/mm². All preparations showed that there are different cell densities throughout the surface. Figure 2 illustrates an example of a ganglion cell density map in one preparation. As the map shows, the porpoise retina had two regions of high ganglion cell density; one in the temporal and one in the nasal areas. The highest cell density was observed in the temporal area, where it amounted to 700 cells per mm².

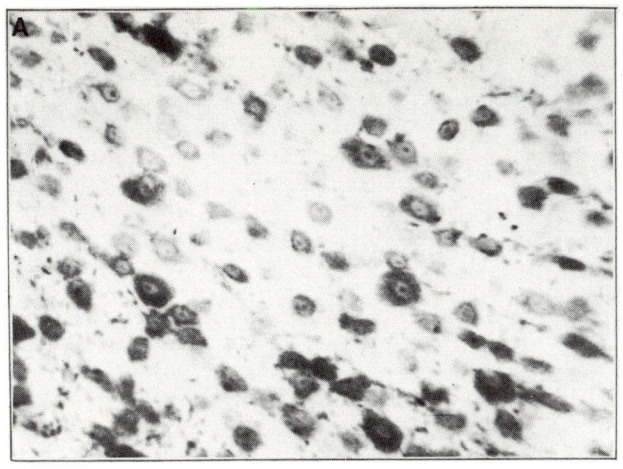

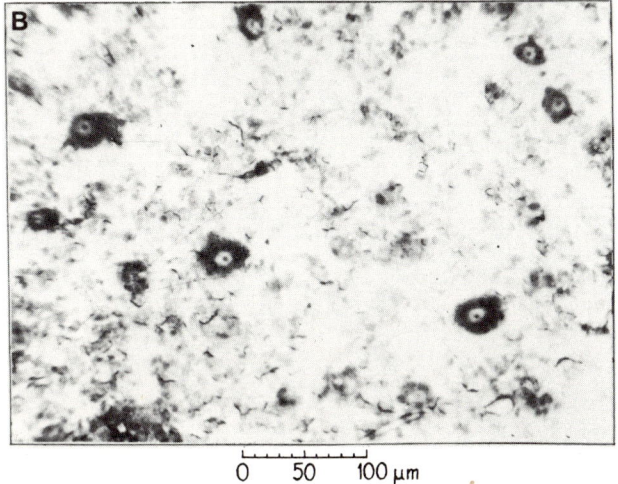

Fig. 1. Ganglion cells in the retina of a harbor porpoise. A: in the temporal high density zone; B: in the central low density zone.

In the nasal area, the cell density was somewhat lower; up to 500 cell/mm². With increasing distance from these zones, the cell density decreased. In the central part of the retina, near the optic disc, the cell density was about zero, only individual cells were observed.

According to our measurements, in the harbor porpoise the radius of the hemisphere approximating the retina, was 11.5 mm, with 1° corresponding to 0.2 mm (i.e., 1 deg² corresponding to 0.04 mm²). Continuous spherical retinal maps were constructed taking into account this value. As seen from the map (Fig.3), the zones with high cell density were located in the temporal and nasal retinal areas near the horizontal diameter, at a distance 50-70° from the optic disc. The highest cell density in the temporal zone was about 24 /deg². In the nasal zone, the maximum density was about 18/deg². Around the optic disc, in the dorsal and ventral areas, the cell density ranged from 0 to 6/deg². As appears from Fig.3, the high density zones might be connected by a "visual streak" under the optic disc with a cell density 6-12/deg². Such maps constructed for preparations of

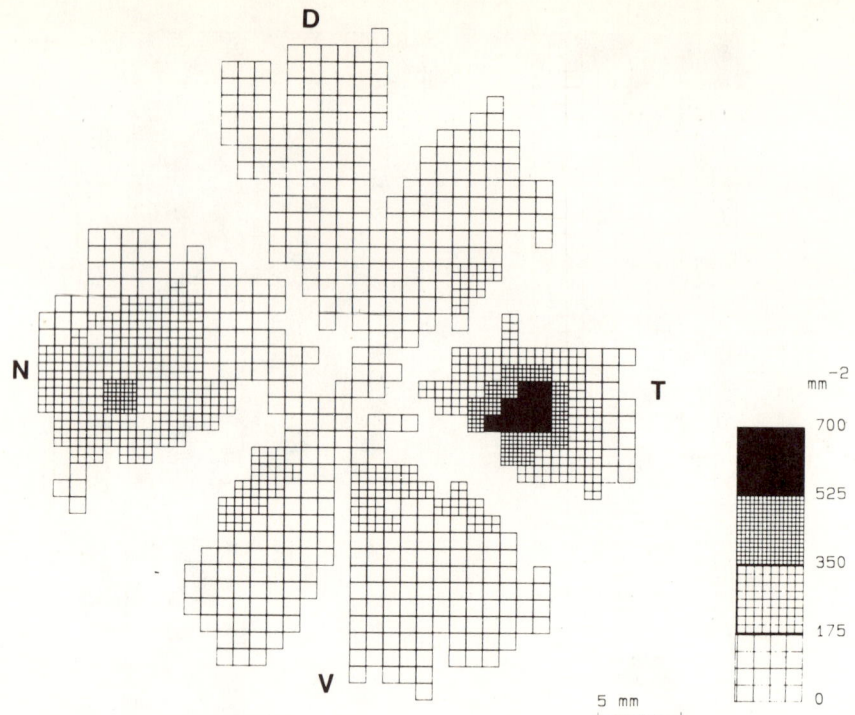

Fig. 2. Map of ganglion cell density in the total retinal preparation of a harbor porpoise, Phocoena phocoena. The map was smoothed. Cell density is designated according to scale on the right. N, T, D, V - nasal, temporal, dorsal and ventral poles of the retina.

other eyes of the same species showed that the position of high density zones of ganglion cells was very similar in all the preparations.

Grey Whale

A conspicuous feature of the retina of the grey whale was a large size of ganglion cells, ranging from 15 to 74 µm. The cells show a broad rim of the cytoplasm with a large amount of intensely stained Nissl substance, a pale-stained nucleus with a clear-cut nucleolus. The cells varied in shape; most having several processes. In many cases, the ganglion cells were distributed in groups. The averaged retinal area in three preparations was 2520 mm². The total cell number in the three preparations ranged from 165,000 to 184,000, with a mean of 174,000. The mean ganglion cell density over the entire retina was 70 cells per mm².

Figure 4 presents a characteristic map of ganglion cell density in the retina of the grey whale. The map shows two zones of high cell density: one zone was in the temporal area of the retina and the other in the nasal area. The highest cell density amounting to 200/mm² was observed in the temporal zone. The averaged highest cell density for the three preparations in the temporal zone was 183/mm².

In the nasal high density zone, this value was lower than in the temporal one; it was not higher than 142/mm². The averaged highest cell density for the three preparations in the nasal zone was 130/mm². In the dorsal, ventral and central parts of the retina the cell density was low, particularly in the region of the optic disc.

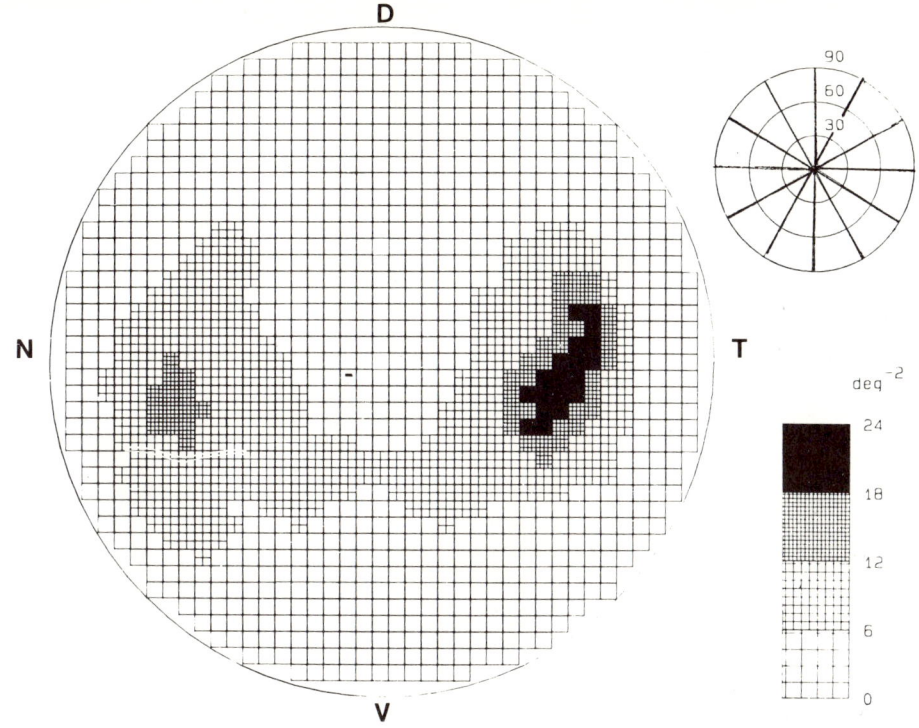

Fig. 3. The continuous spherical map of ganglion cell density in the harbor porpoise, Phocoena phocoena. The preparation shown in Fig.2 was used for transformation. The spheric coordinates are shown in the right top corner. N, T, D, V - nasal, temporal, dorsal and ventral poles of the retina.

Upon constructing spherical maps of the retina, we assumed that the mean radius of the eyecup hemisphere was 23 mm. Hence, the distance of 1° corresponds to 0.4 mm and 1 deg² to 0.16 mm². Figure 5 presents a spherical map obtained by the transformation of the preparation shown in Fig.4. High cell density zones were located in the temporal and nasal sectors near the horizontal diameter of the retina, 50-70° from the geometric center. The cell density in the temporal and nasal sectors is up to 32/deg² and 21/deg², respectively.

Such maps were made for three preparations. Figure 6 illustrates an averaged spherical map for these preparations. The map shows the high density zones in the temporal and the nasal retinal areas, near the horizontal diameter, 50-70° from the center. It was evident that the position of high density zones was similar in all the preparation. The maximum density in the temporal zone averaged for three preparations was 28/deg², in the nasal zone it was 21/deg². In the periphery of the retinae and around the optic disc, the cell density was rather low - up to 7/deg².

Amazon River Dolphin

The retina of the Amazon river dolphin (Inia geoffrensis), like that of the harbor porpoise and the grey whale, was characterized by large ganglion cells. The cell size ranged from 10 to 42 µm. The cells showed abundance of cytoplasm with well-stained Nissl substance, and a pale nucleus with a clear-cut nucleolus. The nucleus might be located both in the center of a cell and eccentrically. Two groups of ganglion cells could be distinguished in the retina of the Amazon dolphin (Fig. 7). The main

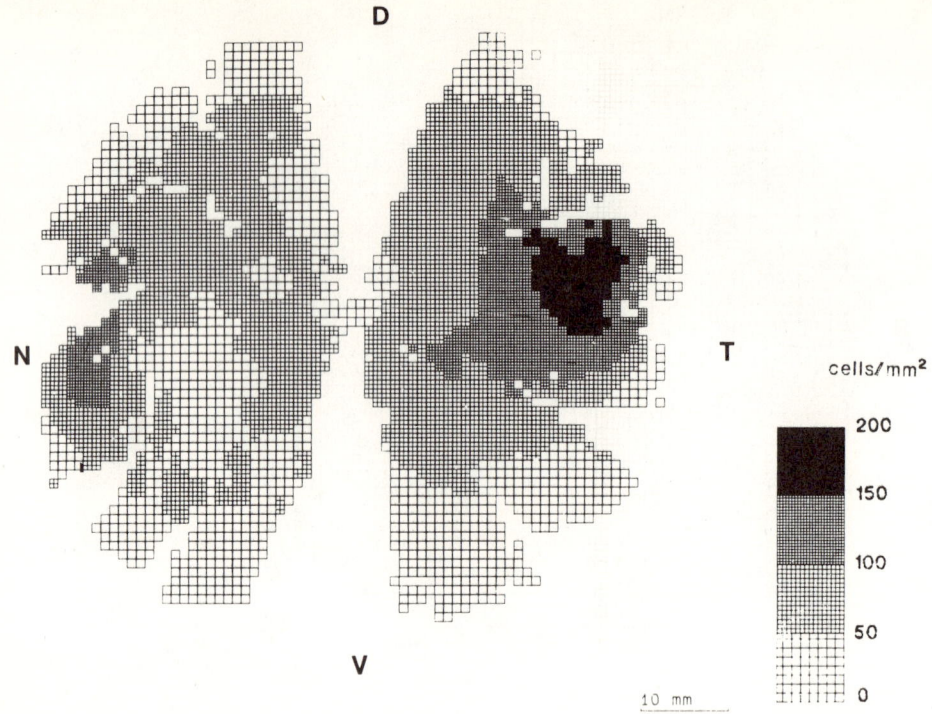

Fig. 4. Map of ganglion cell density in the retina of the grey whale, Es-chrichtius gibbosus. The designations are as in Fig.2.

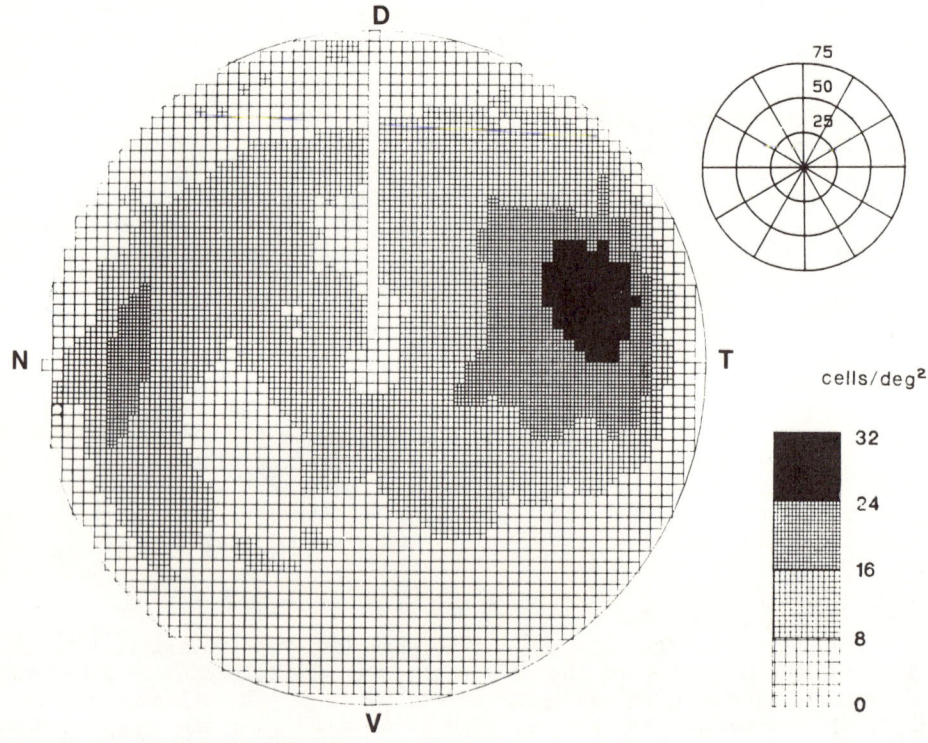

Fig. 5. The continuous spherical map of ganglion cell density in the grey whale, Escrichtius gibbosus. The preparation shown in Fig.4 was used for transformation. The designations are as in Fig.3.

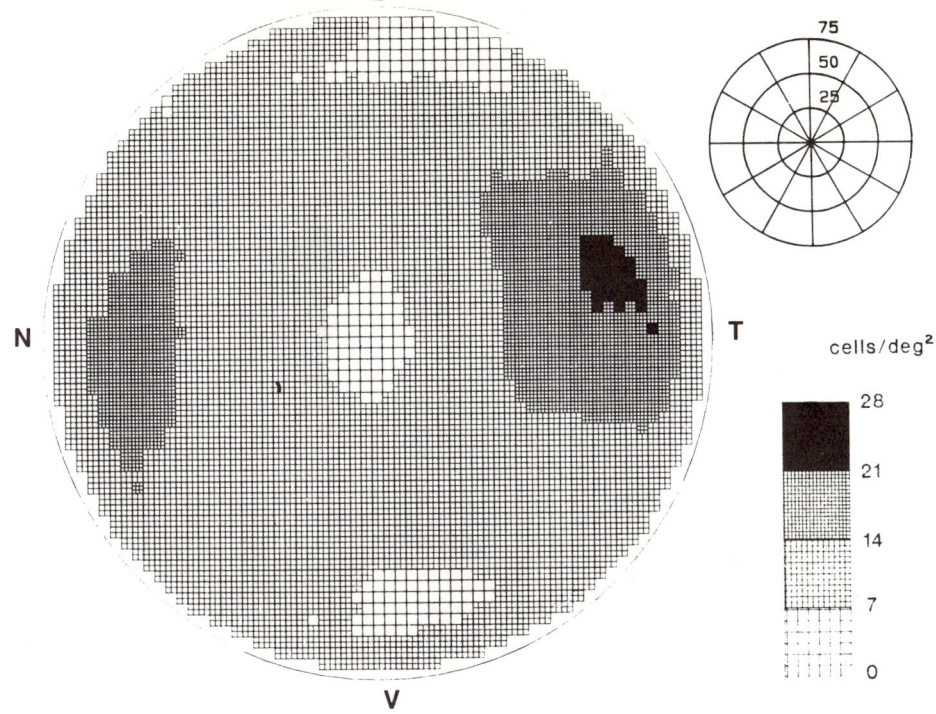

Fig. 6. Averaged map of ganglion cell density of 3 retinal preparations of the grey whale, Eschrichtius gibbosus. Designations are as in Fig.3.

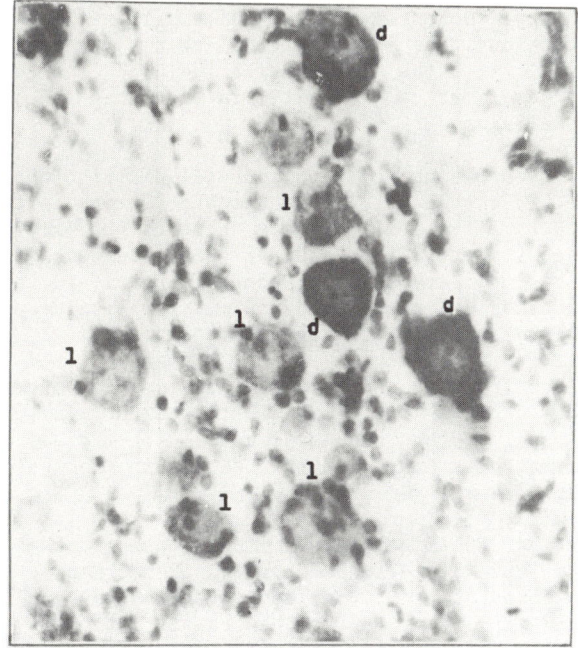

Fig. 7. Two types of cells in the retina of Amazon river dolphin, Inia geoffrensis. l - light cells, d - dark cells.

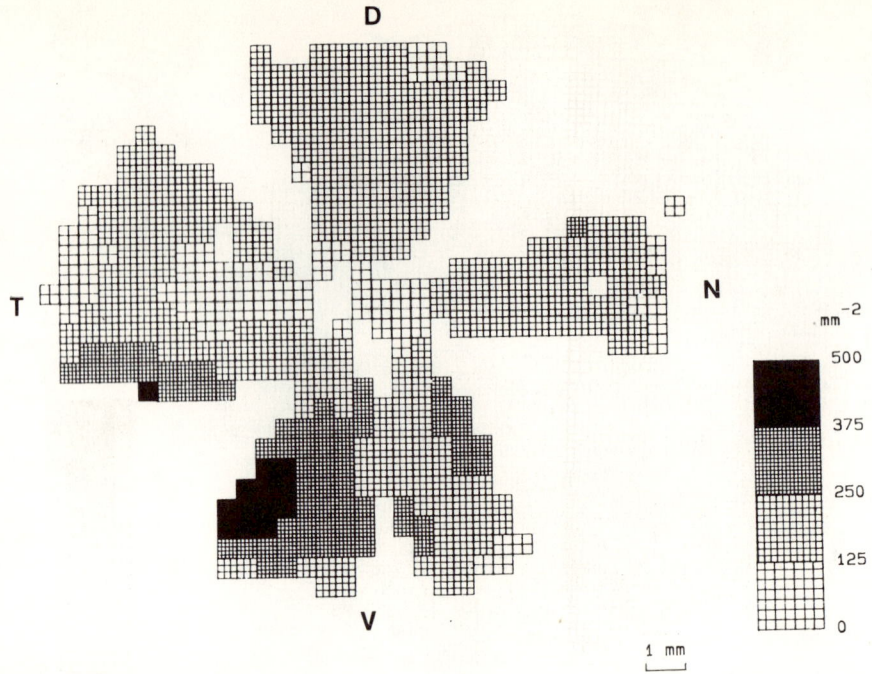

Fig. 8. Map of ganglion cell density in the total retinal preparation of Amazon river dolphin, *Inia geoffrensis*. The designations are as in Fig.2.

group (light cells) included cells from 10 to 40 μm with light-stained cytoplasm. The other less numerous (16 % of the population) group of dark cells included large (more than 20 μm) cells with well-stained Nissl substance in the cytoplasm. The total retinal area averaged for 6 preparations was 102 mm², with the total number of ganglion cells at 14,000. Hence, the mean ganglion cell density in the retina of the Amazon river dolphin was 180 cells/mm².

Figure 8 presents a map of ganglion cell distribution in one retinal preparation. The highest density of ganglion cells was in the ventral area of the retina. Here, the cell density exceeded 500/mm². The rather high ganglion cell density of 375 to 500/mm² was observed here as a horizontally stretched region. In most of the dorsal, nasal and temporal areas of the retina the density did not exceed 250/mm². This value particularly was low near the optic disc and in the retinal periphery.

The other preparations showed a similar location of the high density zone. The maximum density was from 220 to 380/mm². In some preparations the dorsal and nasal sectors had small areas with increased cell density. However, these areas were small and the density in them was far lower than in the ventral zone.

In constructing spherical maps, it was assumed that the radius of eyecup hemisphere was from 4 to 5 mm, mean 4.5 mm. Hence, the transformation factor was found to be 0.08 mm/deg or 0.0064 mm²/deg². Figure 9 presents a spherical map obtained by the transformation of the preparation shown in Fig.8. The maximum cell density was observed ventrally 50° from the retinal center. The cell density there was from 2.4 to 3.2/deg². The

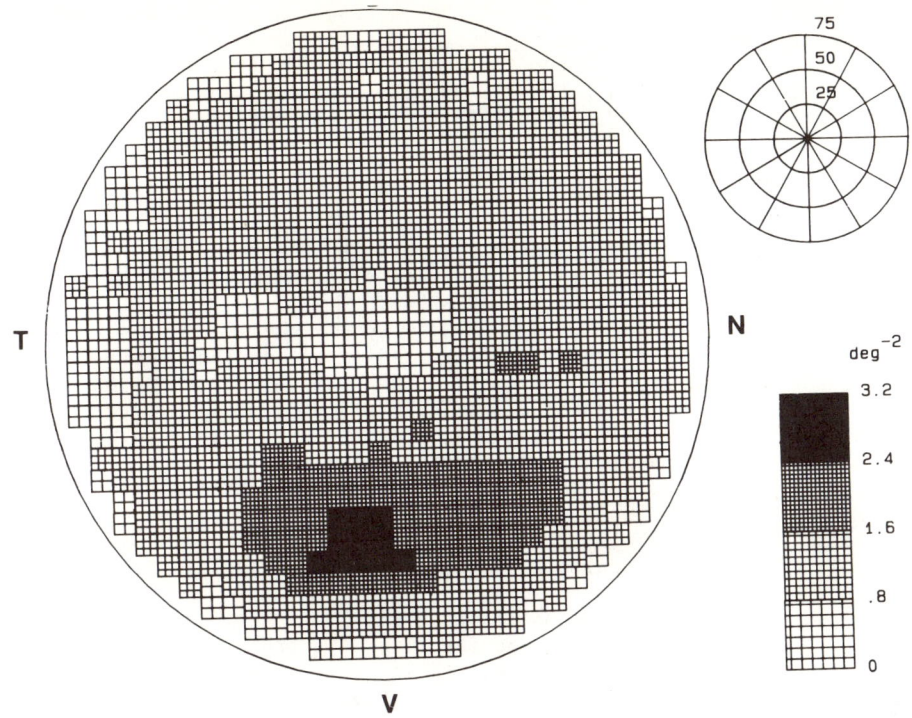

Fig. 9. Continuous spherical map of ganglion cell density in Amazon river dolphin, <u>Inia geoffrensis</u>. The preparation shown in Fig.8 was used for transformation. The designations are as in Fig.3.

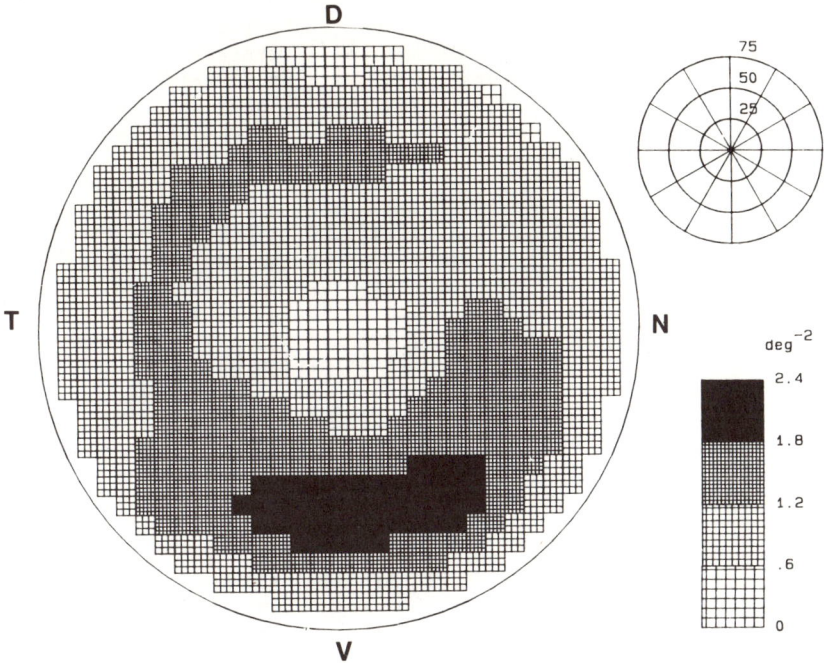

Fig. 10. Averaged map of ganglion cell density of 5 retinal preparations of Amazon river dolphin, <u>Inia geoffrensis</u>. Designations are as in Fig.3.

cell density was from 2.4 to 3.2/deg². The averaged spherical map obtained for 5 preparations is shown in Fig.10. Here the maximum density zone also was located in the ventral part of the retina, 50° from the center. The highest density there was 2.4/deg². In the periphery and near the optic disc, the cell density ranged from 0 to 6 cells per deg².

DISCUSSION

Data obtained in the present study revealed the presence of zones of high ganglion cell density in the retinae of the Cetaceans investigated. These zones seem to be the best vision areas.

Cetacean retinae have no avascularized areas in high cell density sites (Dawson, 1987). High cell density zones were unknown until recently. Attempts to localize these zones in marine mammals by studying cross-retinal cuts yielded no results (Landau and Dawson, 1970). This gave rise to the hypothesis that the retina of marine mammals has no area centralis. However, Peers (1971) observed some irregularity of ganglion cell distribution in the retina of Tursiops truncatus. Only studies on the topography of ganglion cells in total retinal preparations made it possible to discover zones of high ganglion cell density in marine mammals.

The existence of two zones of high ganglion cell density was described earlier for two dolphin species: Tursiops truncatus and Delphinus delphis (Dral, 1977; 1983). However, these data suggested a higher cell density in the nasal zone, while we found (Mass and Supin, 1986, and this paper) the highest cell density in the temporal zone of the retina from the harbor porpoise and grey whale. The reasons of this discrepancy are not yet clear; maybe it is dependent on the precision of counting.

A higher ganglion cell density in the temporal zone of the retina in the harbor porpoise and grey whale suggests a functional role of this zone and the possibility of its use for binocular vision. This possibility was doubted both in Odontocetes (Langworthy, 1931; Kellogg and Rice, 1963) and Mysticetes (Walls, 1963). The nasal zone of high ganglion cell density may be supposed to be responsible for panoramic vision. The position of high cell density zones is similar in all the cetaceans studied, except in the Amazon river dolphin. We suppose that such retinal organization is common in most marine cetaceans.

This retinal organization is correlated with features of the eye optics in cetaceans. A quantitative study of the eye optics in Mysticetes showed that the greatest amount of light falls on the retinal periphery through the horizontally-stretched pupil (Matthiessen, 1893); the high ganglion cell density zones also are located far from the retinal center. The pupil of Odontocetes under high illumination has two slits located on both sides from the eye center (Dral, 1972; 1974; Herman et al., 1975), like the two high ganglion cell density zones in their retina.

The retina of the Amazon river dolphin demonstrates an unusual example of organization. Unlike the marine Cetaceans studied, it has only one zone of high cell density located in the ventral part of the retina. A variety of features, such as microphthalmia, the shape of the eyecup, round pupil without operculum, absence of tapetum (Pilleri, 1977; Dawson, 1980; Dral, 1981) and the unusual position of the high ganglion cell density area shown here and earlier (Mass and Supin, 1989), point to the specific organization of the vision in the Amazon river dolphin. These features possibly reflect an adaptation to the inhabiting turbid water (Waller, 1982).

Data on the retinal topography of other fresh-water dolphins are conflicting. Some authors failed to find areas with a high density of ganglion cells in the retina of Lipotes vexillifer (Li, 1983) while others pointed to the existence of two high density zones in the retina in this dolphin (Gao and Zhou, 1987).

The spherical retinal maps presented in the present study are given in coordinates of the hemisphere approximating the retina, not in the coordinates of the visual field. However, it is expected that the difference between these two coordinate systems is not significant: the optical center of the eye in cetaceans is located near the center of the retinal hemisphere (Rivamonte, 1976). Therefore, as an approximation, the spherical retinal coordinates, rotated at 180°, can be considered to be the visual field coordinates.

Hence, zones of high ganglion cell density correspond to the areas in the visual field at a distance 50-70° from the visual axis. The cell density in the harbor porpoise and the grey whale reached 24 and 28/deg², respectively in the front of the visual field, 18 and 21/deg² in the temporal part.

The mean distance between ganglion cells may be assumed to be $1/\sqrt{D}$, where D is the cell density. So in the harbor porpoise this distance amounted to 11-12' in the frontal part of the visual field and 13-14' in the temporal part. In the grey whale, these values were 10-11' and 12-13', respectively. Since visual acuity depends on the distance between the adjacent ganglion cells, the values presented characterize the visual acuity of the Harbor porpoise and the Grey whale in the best vision zones.

These values are similar to those of some Odontocetes obtained in the same way: 9.5-10' for Tursiops truncatus, and 8-9.5' for Delphinus delphis (Dral, 1975, 1983). Data on the visual resolution of dolphins obtained in behavioral experiments are conflicting and vary from 5' to 18' (White et al., 1971; Spong and White, 1971; Pepper and Simmons, 1973; Herman et al., 1975). As for the Mysticetes, no data on their visual acuity are available. Our data showed the visual acuity of the grey whale to be virtually the same as that of the harbor porpoise. These data disagree with Walls' (1963) hypothesis of regression of the visual function in Mysticetes due to their "trawling" method of feeding (Walls, 1963) and agree better with optimistic estimations of the visual ability of Mysticetes (Hosokawa, 1951; Slyper, 1962; Jacobs and Jensen, 1964; Jansen and Jansen, 1969; Kinne, 1975; Madsen and Herman, 1980; Waller, 1984).

In the Amazon river dolphin, the visual acuity, calculated in the above mentioned way, is about 0.7° (40-45'). Although this acuity is low, the existence of a zone with higher ganglion cells density points to the possibility for the river dolphin use it for perception of visual images. This also is supported by behavioral observations in captivity (Caldwell et al, 1966; Pilleri et al., 1980) and in particular by the ability to differentiate gratings (Phillips and McCain, 1964). The low visual acuity may be sufficient for vision in turbid water, where only large objects at short distance can be distinguished.

Thus, our data suggest that the specific zones of high ganglion cell density exist in the retina of the investigated cetaceans. These zones may be a structural basis for the perception of clear visual images.

REFERENCES

Caldwell M. C., Caldwell D. K. and Evans W. E., 1966, Sound and behavior of captive Amazon fresh water dolphins, Inia geoffrensis. Los Ang. Mus.Nat.Hist. Contrib.Sci. 108: 1-24.
Dawson W. W., 1980, The Cetacean eye. In: "Cetacean behavior: Mechanisms and functions", L. M. Herman (ed.), John Wiley & Sons, N.Y. 53-100.
Dawson W. W., 1987, The ocular fundus of two Cetaceans, Marine Mammal Science, 3: 1-13.
Dral A. D. G., 1972, Aquatic and aerial vision in the bottlenosed dolphin, Neth.J.Sea Res. 5: 510-513.
Dral A. D. G., 1974, Problems of image-focusing and astigmatism in Cetacea - a state of affairs, J.Anat.Mamm. 2: 22-28.
Dral A. D. G., 1975, Some quantitative aspects of the retina of Tursiops truncatus, Aquatic Mammals 2: 28-31.
Dral A. D. G., 1977, On the retinal anatomy of Cetacea (mainly Tursiops truncatus), in: "Functional Anatomy of Marine Mammals", R. J. Harrison (ed.), Acad. Press, Lond. III, 81-134.
Dral A. D. G., 1981, Ophthalmoscopical observations on the Amazon dolphin, Inia geoffrensis, Aquatic mammals 8: 40.
Dral A. D. G., 1983, The retinal ganglion cells of Delphinus delphis and their distribution, Aquatic mammals 10: 57-68.
Frisen L. and Frisen M.,1976, A simple relationship between the probability distribution of visual acuity and density of retinal output channels, Acta ophtalm. 54: 437-443.
Gao A. and Zhou K., 1987, On the retinal ganglion cells of Neophocaena and Lipotes, Acta Zool.Sin., 33: 316-322.
Herman L. M., Peacock M. F., Yunker M. P and Madsen C. J., 1975, Bottlenosed dolphin: double slit pupil yields equivalent aerial and underwater diurnal acuity, Science 189: 650-652.
Hosokawa H., 1951, On the extrinsic eye muscles of the whale with special remarks upon the innervation and function of the musculus retractor bulbi. Sci.Rep.Whales Res.Inst., Tokyo 6: 1-31.
Hughes A., 1975, A quantitative analysis of the cat retinal ganglion cell topography. J.Comp.Neurol. 163: 107-128.
Hughes A.,1977, The topography of vision in mammals of contrasting life style: Comparative optics and retinal organization. In: "Handbook of Sensory Physiology: The Visual System in Vertebrates", F. Crescitelli (ed.), 613-756.
Jacobs M. S. and Jensen A. V., 1964, Gross aspects of the brain and a fiber analysis of central nerves in the great whale, J.comp.Neurol. 123: 55-65.
Jansen J. and Jansen J. K. S., 1969, The nervous system of Cetacea. In: "The Biology of Marine Mammals", H. T. Andersen (ed.), Acad.Press, N.Y., 175-252.
Kellogg W. N. and Rice C. E., 1963, Visual discrimination in a bottlenosed porpoise, Psychol.Rec., 13: 483-498.
Kinne O., 1975, Orientation in space: Animals: mammals. In: "Marine Ecology", O. Kinne (ed.), 2, 702-852.
Landau D. and Dawson W. W., 1970, The histology of retinae from the pinnipedia, Vision Res. 10: 691-702.
Langworthy O. R., 1931, Factors determining the differentiation of the cerebral cortex in sea-living mammals (the Cetacea). A study of the brain of the porpoise, Tursiops truncatus. Brain 54: 225-236.
Li J. F., 1983, Retina of river dolphin Lipotes vexillifer, Sci.Sin. 26: 145-150.
Madsen C. J. and Herman L. M., 1980, Social and ecological correlates of cetacean vision and visual appearance. In: "Cetacean Behavior: Mechanisms and Function", L.M.Herman (ed.), Wiley Interscience, N.Y. 101-147.

Mass A. M., 1987, Topographic organization of the ganglion layer in the retina of a northern fur seal, Callorhinus ursinus (in Russ.), Dokl.Akad.Nauk SSSR, 294: 1469-1472.
Mass A. M. and Supin A. Ya., 1986, Topographic distribution of sizes and density of ganglion cells in the retina of a porpoise, Phocoena phocoena, Aquatic Mammals 12: 95-102.
Mass A. M. and Supin A. Ya., 1989, Distribution of ganglion cells in the retina of an Amazon river dolphin, Inia geoffrensis, Aquatic mammals, 15: 49-56.
Matthissen L., 1983, Uber den physicalisch-optischen Bauder Augen von Knoelwal (Megaptera boops, Fabr.) und Finnwal (Balaenoptera musculus, comp). Z.vergl.Augenheilk, 7: 77-101.
Peers B. A., 1971, Retinal histology of the Atlantic Bottlenosed dolphin Tursiops truncatus (Montagu, 1821). Thesis Univ. of Guelph.
Pepper R. L. and Simmons J. V., 1973, In air visual acuity of the bottlenose dolphin, Exp.Neurol. 41: 371-276.
Perry V. H. and Cowey A., 1985, The ganglion cell and cone distribution in the monkey's retina: implications for central magnification factor, Vision Res. 25: 1795-1810.
Perry V. H., Oehler R. and Cowey A., 1984, Retinal ganglion cells that project to the dorsal lateral geniculate nucleus in the macaque monkey, Neuroscience, 12: 1101-1123.
Phillips J. and McCain G., 1964, Black-white visual discrimination in the Amazon porpoise Inia geoffrensis, Amer.Psychologist, 19: 503.
Pilleri G., 1977, The eye of Pontoporia blainvillei and Inia boliviensis and some remarks on the problem of regressive evolution of eye in Platanistoidea. Invest.on Cetacea, 8: 149-159.
Pilleri G., Gihr M. and Kraus C., 1980, Play behaviour in the Indus and Orinoco dolphin (Platanista indi and Inia geoffrensis). Invest.on Cetacea, 9: 58-108.
Provis J. M., 1979, The distribution and size of ganglion cells in the retina of the pigmented rabbit: a quantitative analysis, J.comp. Neurol., 185: 121-137.
Rivamonte A., 1976, Eye model to account for comparable aerial and underwater acuities of the bottlenose dolphin. Neth.J.Sea Res., 10: 491-498.
Rolls E. T. and Cowey A., 1970, Topography of the retina and striate cortex and its relationship to visual activity in rhesus monkeys and squirrel monkeys, Exp.Brain Res., 10: 298-310.
Shkolnik-Jarros E. G. and Kalinina A. V., 1986, "Neurons of a retina" (in Russ.), Nauka, Moscow.
Slyper E. J., 1962, "Whales", Hutchinson, London.
Spong P. and White D., 1971, Visual acuity and discrimination learning in the dolphin (Lagenorhynchus obliquidens). Exp.Neurol. 32: 431-436.
Stone J., 1965, A quantitative analysis of the distribution of ganglion cells in the cat's retina. J.comp.Neurol., 124: 337-352.
Van Buren K. M., 1963, "The retinal ganglion cell layer", Charles Thomas, Springfield (Ill).
Waller G. N. H., 1982, Retinal ultrastructure of the Amazon river dolphin (Inia geoffrensis), Aquatic mammals, 9: 17-28.
Waller G. N. H., 1984, The ocular anatomy of cetacea: an historical perspective, Invest.of Cetacea, 16: 138-148.
Walls G. L., 1963, "The vertebrate eye", Hafner, N.Y.
Wassle H. Chun Myung Hoon and Muller F., 1987, Amacrine cells in the ganglion cell layer of the cat retina, J.comp.Neurol., 265: 391-408.
White D., Cameron N., Spong P. and Bradford J., 1971, Visual acuity in the Killer whale (Orcinus orca), Exp.Neurol., 32: 230-236.
Wong R. O. L. and Hughes A., 1987, The morphology, number, and distribution of a large population of confirmed displaced amacrine cells in the adult cat retina, J.comp.Neurol., 255: 159-177.

VISUAL ECOLOGY AND COGNITION IN CETACEANS

Joseph R. Mobley, Jr. and David A. Helweg[1]

Dept. Social Sciences, University of Hawaii-West Oahu, 96-043 Ala Ike, Pearl City, HI 96782 and [1]Kewalo Basin Marine Mammal Laboratory, 1129 Ala Moana Blvd, Honolulu, HI 96814

INTRODUCTION

Observations of captive and wild cetaceans provide abundant evidence that vision is an important sensory modality that functions in proximal social and predator/prey interactions, despite the limitations imposed on optic signals in water (Mitchell, 1970, 1972; Otte, 1974; Herman and Tavolga, 1980; Madsen and Herman, 1980). New views of the importance of vision in the life of the dolphin and of the ability to use visual information for carrying out complex cognitive tasks have been advanced by Herman and his colleagues (Madsen & Herman, 1980; Mobley & Herman, 1985; Herman, 1986, 1987; Herman et al., 1989a, 1989b; also see Herman, this volume for an extensive review and analysis of the cognitive data). These new views form the basis for much of the discussion in this chapter.

Early studies of cetacean cognition appeared to produce evidence to confirm auditory dominance in cognitive tasks (for a review, see Herman, 1980). The current weight of data no longer allows the simplifying assumption of a "unimodal" dolphin (e.g., Herman et al., 1989a). This paper reviews the role of vision in cetacean ecology and the relationship of these observations to cognitive abilities as revealed by delphinids performing visual tasks.

We discuss the mechanisms and functions of visual adaptation in cetaceans using a comparative approach. First, we present a review of the physical specializations observed in cetacean eyes and then present a brief review of the functional role of vision in cetacean life. We conclude by discussing current results of cognitive testing in delphinids that have used visual tasks in light of the importance of vision to free-ranging cetaceans.

CETACEAN EYE MORPHOLOGY

Ancestors of modern-day cetaceans possessed sense organs that evolved in a terrestrial environment. The terrestrial mammalian eye evolved in an environment with a wide range of light intensities and a predominance of yellow-green wavelength light. Specialization to the marine environment required morphological adaptations, and the sensory systems evolved to meet the demands of aquatic life. Eyes of cetaceans necessarily must collect information from a world characterized by low light levels and temperature and pressure conditions that fluctuate widely between the surface and deeper waters. Cetacean eyes have a thickened sclera for protection against increased pressure (Yablokov et al., 1972), and prevent excessive heat loss in cold water by the opthalmic rete (a highly efficient vascular radiator) and fatty insulation in the eyelids (Dawson, 1980; Dawson et al., 1987a).

The environment immediately surrounding the eye is subject to hostile biochemical and immunological conditions from dissolved ions, particulate matter and microorganisms, as well as to hydrodynamic deformation due to swimming (Dawson, 1980; Fobes and Smock, 1981; Watkins and Wartzok, 1985; Nachtigall, 1986). Secretions from the Harderian glands continuously bathe the eyes with a layer of fluid which probably reduces hydrodynamic distortion of the cornea by shedding easily and in doing so flushes away any irritants which may have settled onto the eye (Yablokov et al., 1972; Dawson, 1980; Dawson et al., 1987a; Dawson et al., 1987b).

Unlike the oceanic delphinid species, the river dolphins (Platanistidae) live in murky rivers, where vision is restricted primarily to the detection of light and dark for orientation to the water surface. As a result, the eyes of platanistids have no lens and have vascularized corneas, trading reduced light transmission for the immunological support to the cornea necessary for a healthy life in microbe-rich waters (Dawson, 1980). It is interesting to note that, in contrast to delphinids, several Platanistid species (e.g., _Platanista minor_ and _Platanista gangetica_) have been described as echolocating continuously throughout a 24-hour cycle, suggesting heavy reliance on echolocation for orientation and navigation (DeFran and Pryor, 1980).

Except near the surface, the underwater world is nearly monochromatic, with bluish-green light penetrating to about 100 m. Cetacean retinae provide vision under dim, monochromatic conditions with the following adaptations: (1) the retina which essentially is all rods interspersed with modified "conelike" structures, (2) a tapetum fibrosum that blankets the entire retinal surface (Dawson, 1980, for _Tursiops_, _Stenella_, _Inia_, and _Kogia_). The absorption spectra of cetacean photopigment is shifted towards blue wavelengths, with maximal absorption at a median value of approximately 490 nm, as compared to about 498 nm for "terrestrial" rhodopsin (McFarland, 1971, reporting on two mysticete and seven odontocete species). These photopigments are sensitive to lower illuminance levels, providing vision in both scotopic and photopic conditions (Madsen and Herman, 1980).

The tapetum lucidem is a reflective structure that maximizes the ability of the retina to function under low light conditions by reflecting light onto the photoreceptors with minimal scattering, thereby functionally increasing light levels available for vision with minimal loss of resolution. This type of structure is useful for cetaceans that live in dim or aphotic light conditions. The cetacean retina is completely tapetalized, notable because the tapeta in terrestrial species normally lie only on the upper visual field of the retina (Young et al., 1988). A completely tapetalized retina presumably is more effective at capturing available light because of the horizonless environment of cetaceans.

Histological examination of the tapeta from a pygmy sperm whale (_Kogia breviceps_) and a bottlenose dolphin (_Tursiops truncatus_) indicate that the tapeta provided extremely pure reflectance of bluish light near the center of the optic fundus and more greenish wavelengths at the superiotemporal aspect of the fundus (Young et al., 1988). Examination of the ocular fundus of two species of cetaceans provided evidence of uniform tapetal reflectance, with the pelagic Grampus griseus transmitting more bluish wavelengths than the neritic Tursiops (Dawson et al., 1987a). Parenthetically, Young et al. (1988) made the observation that cetaceans are the only carnivores with a tapetum fibrosum, which is found in terrestrial ungulates. Carnivore eyes typically are lined with a tapetum cellulosum. This observation increases the evidence for a common phylogenetic ancestry among cetaceans and ungulates.

ANATOMY AND PSYCHOPHYSICS

Anatomical and Physiological Features Relevant to Visual Acuity

Terrestrial mammals' eyes utilize two focusing structures to project a clear image upon the photosensitive retina. The cornea provides gross focus as a product of refraction not under neuromuscular control. The lens provides fine focus which is under control from a system of ciliary muscles, contraction of which flex the lens, producing change in focus. In contrast to the general terrestrial model, the cetacean cornea functionally is absent in water because it has nearly the same index of refraction as water. Examination of corneas of Tursiops suggest that the corneas have irregular surfaces with uneven radii of curvature (Dral, 1974; Dawson et al., 1987b). The uneven quality of the cetacean cornea is pertinent only to discussions of in-air vision because it functionally is nonexistent underwater.

Lenses in terrestrial eyes tend to be ellipsoid but may be nearly spherical in the eyes of cetaceans (Yablokov et al., 1972). The possibility of dynamic accommodation (i.e., change in focal length through muscular deformation of the lens) has been discounted, at least in Tursiops, due to the absence of ciliary muscles surrounding the lens (Yablokov et al., 1972; Dral, 1975) and the lack of any observed accommodative movements during opthalmoscopic observations (Dawson, 1980). As a result, many authors have been quick to conclude that this characteristic, in combination with the irregularly-shaped cornea, necessarily produce an animal with both severe in-air myopia and astigmatism (Walls, 1942; Fobes and Smock, 1981). However, these observations contrast with behavioral evidence for delphinid species that suggest relatively good visual acuity for those species tested.

Observations of the ocular fundus of some cetacean species have failed to reveal clear evidence of an area centralis--the retinal area found in many terrestrial mammals that appears to be specialized for high resolution of detail. Dral (1977, 1983), examining Tursiops truncatus and Delphinus delphis specimens, outlined two areas possessing relatively high ganglion cell densities--an area temporalis and area centralis--that showed maximum ganglion cell densities around $500/mm^2$. Mass et al. (1986) similarly found two such areas in retinal preparations obtained from Phocoena phocoena--one in the nasal region ($500/mm^2$) and an area of greatest concentration in the temporal region ($700/mm^2$). However, the estimates of both authors are relatively low as compared to estimates for domestic cats ($4000/mm^2$, Stone, 1965) and domestic rabbits ($10,000/mm^2$; Rodieck, 1973). Additionally, recent examinations of the ocular fundus of Tursiops truncatus and Grampus griseus failed to show the reduced vascularization in these areas normally characteristic of an area centralis, thus suggesting that the areas proposed by Dral and Mass et al. may not function as areas of high resolution (Dawson et al., 1987a).

Psychophysical Evidence Concerning Visual Acuity

Behavioral evidence points to more refined capabilities of the delphinid eye than anatomical evidence would suggest. Spong and White (1971) obtained in-water estimates of visual acuity for Lagenorhynchus obliquidens using the method of constant stimuli for a Minimum Angle of Resolution (MAR) of 6 min in bright light--the latter referring to the mimimum visual angle that the animal can discriminate 75% of the time. Using a modified version of the same procedure, White et al. (1971) derived a similar MAR of 5.5 min for an Orca subject, also using underwater viewing. These acuity estimates are comparable to those obtained for pinnipeds (e.g., Jamieson and Fisher, 1972) and for cats (Walls, 1942). For humans, the clinical criterion for normal visual acuity (20/20 vision) is an MAR of 1.0 min at 20 feet (Spoehr and Lehmkuhle, 1982).

Despite the previously cited claims of severe in-air myopia for at least Tursiops, observations of cetacean behavior in the wild and in oceanariums provide contradictory evidence regarding in-air visual acuity (Yablokov et al., 1972; Chun, 1978; Dawson, 1980; Fobes and Smock, 1981; Watkins and Wartzok, 1985). Herman et al. (1975), using a bottlenose dolphin subject, found "best acuity" estimates of 12.6 and 9.0 min of arc for in-air and in-water viewing, respectively. These estimates are considerably better than those obtained in former studies and, when corrected for differential size in the two mediums, indicated nearly identical acuity of in-air and in-water vision under optimal conditions for each. Further, the behavioral acuity estimates of Herman et al. (1975) correspond to those of Dral (1975, 1977) who predicted 10 and 9.5 min of arc for temporal and nasal areas, respectively, based upon analysis of ganglion cell densities in the Tursiops retina. A somewhat divergent estimate was reported by Nachtigall (1989) who derived an in-air visual acuity estimate of 18.7 minutes at 1 m (left eye only) using a Risso's dolphin (Grampus griseus).

One possible explanation for the inconsistency between anatomical- and behavioral-based indicators of in-air acuity was provided by Dral (1974) who noted a small emmetropic (clearly-focused) field of view in the rostroventral direction during opthalmoscopic examination of a live Tursiops eye. This is consistent with the often-observed tendency for dolphins to view objects in this regard during in-air viewing, as compared to the tendency for lateral viewing in-water. Dral cited the irregular curvature of the delphinid cornea to explain this finding. Herman et al. (1975) pointed to the possibility of stenopaic vision created by extreme contraction of the pupil under bright light conditions. Stenopaic vision functions in an analogous fashion to pinhole cameras, where a very small aperture provides fairly good depth-of-field. Herman et al. (1975) conjectured that the double-slit pupil of Tursiops could provide two focal planes, selected by the dolphin via head movements. The authors proposed that the combination of more delicate oculomuscular control and pupillary specialization produce good in-air visual acuity (as well as in-water).

Anatomical and Physiological Features Relevant to Motion Detection

Histological examination of retinae from Tursiops (Dral, 1977; Dawson, 1980; Nachtigall, 1986) and Phocoena phocoena (Mass et al., 1986) revealed unusually large "giant" ganglion cells with large receptive fields. The soma of the giant ganglion cells are as big as the entire dendritic fields of analogous ganglion cells in the retinae of terrestrial species. Speed of conduction of a neural impulse is proportional to the diameter of the neuron and degree of myelination, thus the gigantic ganglion cells probably are characterized by extremely rapid impulse propagation (Levinthal, 1983). Each of the giant cells was reported to generate a large axon that continued to form one fiber of the optic nerve tract. Dawson (1980) inferred that these unusually large axons served as high-speed neural pathways for the communication of motion cues in the peripheral field, allowing for rapid detection of moving targets.

Further anatomical support for the importance of motion cues derives from several physical features of the delphinid eye. The shape of the eye of <u>Tursiops</u> deviates from the spherical shape typical of terrestrial vertebrates, describing an ellipse with the shortest axis from pupil to retina (Yablokov et al., 1972; Dawson, 1980; Madsen and Herman, 1980). Walls (1942) describes such horizontally-elliptical eyes as serving to enhance movement detection capabilities--a phenomenon he referred to as the "barrel distortion effect." Objects moving in the peripheral visual field of the horizontally ellipsoid eye sweep across a broader area of the retina than in the case of the spherical eye, resulting in an optical enhancement of the size and speed of the object's projection. Together, these features suggest an enhanced motion detection system that arguably would improve the detection of elusive prey, avoidance of predation, and coordination of rapid social group movements (Madsen and Herman, 1980; Norris and Dohl, 1980).

Depth Perception

The visual scaling of distance finds importance in a variety of survival concerns to any species--the pursuit and capture of elusive prey, evasion of predators, and for terrestrial species, estimation of jumping distances. There are many possible depth cues available (see review in Hochberg, 1971 and Fox, 1978). Monocular cues include accomodation cues, overlap, texture gradients, linear perspective, and relative size cues. Binocular cues are available to some species, such as convergence and retinal disparity. Several species have been shown to be capable of using retinal disparity cues and thus are said to demonstrate "stereopsis" (e.g., Sarmiento, 1975 in rhesus monkeys; Fox et al., 1977 in falcons; Mitchell et al., 1979 in cats).

A number of previous reviewers have pointed to the possibility of binocular vision in dolphins based on their often-observed tendency to view objects in the rostro-ventral (forward and downward) direction (e.g., Slijper, 1962; Yablokov et al., 1972; Dawson, 1980). Based on such informal observations, Caldwell and Caldwell (1972) proposed the possibility of binocular vision in <u>Tursiops truncatus</u>, <u>Stenella plagiodon</u>, <u>Lagenorhynchus obliquidens</u>, <u>Grampus griseus</u>, and <u>Globicephala scammoni</u>. Considerations of the meaning of "binocularity" range from simple fusion of the images from both eyes to true stereopsis where utilization of disparity information can be shown to enhance depth acuity beyond that possible with each eye separately (see Fox, 1978). For purposes of clarity here, we will differentiate between the terms "binocularity" and "stereopsis."

Anatomical and Physiological Features Relevant to Depth Perception

The more traditional view of the requisite features for stereopsis included coordinated movements of the eyes and pupils (consensual response) as well as a significant degree of overlap of the right and left visual fields (Walls, 1942; Duke-Elder, 1958; Fox, 1978). Dawson (1980), however, noted that the pupils of <u>Tursiops</u> are only "partially consensual." Additionally, Dawson (1980) noted no evidence of correlated eye-movements based on observations drawn from two thousand sampling periods.

As concerns the degree of overlap of the visual fields, Dawson (1980) commented on the unusually large orbit of the Tursiops eye, as well as its heavily developed external musculature. Yablokov et al. (1972) ascribed a protraction-retraction function to these muscles, based on observations that the eyes protrude substantially when viewing forward. Yablokov et al. (1972) further calculated that, at maximum protrusion, there may be as much as 20-30 degrees of visual field overlap, which is sufficient to allow for at least part of their visual field to be binocular, as was proposed earlier by Slijper (1962) and Dral (1972).

One final hallmark of stereopsis traditionally has been the presence of "partial decussation" of the optic nerve fibers (Walls, 1942; Duke-Elder, 1958). As noted by Walls (1942, p.32): "The great majority of physiological opticists have seen in partial decussation the essential basis of fusion and stereopsis." There is, however, no evidence for partial decussation in Tursiops. Jacobs et al. (1975) found total decussation (total crossover) of the optic fibers in a Tursiops specimen using retrograde staining techniques. Based on past assumptions of anatomical requirements, then, there is little evidence that might support stereopsis in dolphins, as represented by Tursiops.

Recent evidence has caused some investigators of animal vision to revise their assumptions regarding the anatomical requirements of stereoptic vision. As noted by Collett and Harkness (1982): "Until recently it was often thought that frontal eyes, partial decussation at the optic chiasm, and convergent eye movements were all essential for effective stereopsis. But the demonstration of stereopsis in birds and amphibians has shown that none of these properties are really necessary (pp. 149-150)." Experimental manipulations involving mammalian species similarly have yielded unexpected results. For example, rabbits, with their laterally-placed eyes, independent eye-movements and low proportion of same-side optic fibers (10%), are capable of binocular fusion of images presented separately to each eye (Van Hof and Russell, 1977) (Note: presumptions of stereoptic capabilities, as noted above, are not necesarily implied by these results). These behavioral tests further were supported by the finding of binocularly-driven cortical cells in the rabbit's occipital cortex (Van Sluyters and Stewart, 1974). Together, these results suggest that binocularity and/or stereopsis are derived from a number of convergent evolutionary pathways involving divergent mechanisms. The final proof of stereoptic capabilities in delphinids, therefore, must rely upon behavioral demonstrations.

Behavioral Evidence Concerning Depth Acuity

Observations of trained odontocetes (e.g., Tursiops spp., Orcinus orca, and Pseudorca crassidens) have shown remarkable abilities to gauge precisely distances in air, as when jumping over ropes or snatching fish from a trainer's hand up-to twenty feet above the water's surface (Kellogg and Rice, 1966; personal observations). Because echolocation in-air is improbable given the impedance mismatch between the delphinid auditory system and air, such informal evidence strongly is suggestive of the ability to estimate distance visually.

Experimental evidence has confirmed a readiness to discriminate relative distance of visual targets (Mobley, 1984; Mobley and Herman,

1985). A bottlenose dolphin ("Keola") was trained to go to the nearer of two visual targets. In contrast to the difficulty encountered in attempting to train a discrimination between simple geometric shapes, the discrimination of relative distance (closer/farther) was trained in only a few trials. Average performance remained above 90% correct across hundreds of trials. Using the Method of Constant Stimuli, a depth acuity estimate of 16.2 min of visual arc was derived--comparable to those derived from monocular presentations to terrestrial mammals shown to have stereopsis (e.g., Fuji and Kojima, 1981, using Japanese macaques, derived estimates of 16.17 min and 4.93 min for monocular and binocular viewing conditions, respectively).

There are a number of possible paradigms for demonstrating stereoptic capabilities (see review in Collett and Harkness, 1982). In order of increasing experimental control, they include: (a) systematically reducing or eliminating other possible cues and observing the resultant effects on depth acuity; (b) showing binocular depth acuity to be significantly superior to monocular depth acuity; or (c) the use of "stereograms" (Julesz, 1971), where the target stimulus is only visible through the use of disparity cues.

Experimental efforts by Mobley (1984; Mobley and Herman, 1985) involved the first technique--that of cue reduction. Comparisons of performance on stationary and free-viewing of size-equated targets (where other possible cues either had been ruled out theoretically, or controlled for), indicated a general superiority of the free-viewing performance. The latter was taken to suggest a reliance on motion parallax cues, because the animal was free to move relative to the visual targets during the free-viewing condition, whereas in the stationary condition he was not. This interpretation is consistent with the anatomical features of the delphinid eye, reviewed earlier, suggesting heightened sensitivity to motion cues.

However, since it is difficult to control for all the possible cues for depth, a more definitive test would be to perform either a test of the relative effectiveness of monocular versus binocular performance, or presentation of stereogram images. Thus, when ruling-on the question of stereopsis in dolphins, we must conclude that the answer awaits further experimental research.

ECOLOGICAL FUNCTION OF VISION

In this section, we discuss the function of vision in cetacean foraging strategies and social interaction. Madsen and Herman (1980) and Herman and Tavolga (1980) previously have provided extensive reviews of the functions of vision and visual appearance in cetaceans, including the role of vision in foraging and in social relationships. As we have discussed in the earlier sections, behavioral evidence suggests relatively good visual acuity for odontocete species tested thus far (Tursiops truncatus, Orcinus orca, Lagenorhynchus obliquidens, Grampus griseus), as well as comparable in-air and in-water visual acuity for Tursiops. Cetacean visual acuity in-water almost certainly is superior to that of most terrestrial mammals. Lateral placement of the eyes insures a wide visual field adaptive for surveillance. The ability to judge precisely relative distance of nearby visual stimuli and good visual acuity potentially are useful for prey capture and proximal social interaction, where visual signals must be detected rapidly and acted upon (Madsen and Herman, 1980).

Functions of Vision in Foraging

Two evolutionary pressures related to foraging/swimming act on cetacean vision: (1) individuals must locate and capture prey and (2) individuals must avoid being eaten. Many cetaceans forage using acoustic cues and echolocation (Popper, 1980; Fobes and Smock, 1981). Visual appearance or visual surveillance also may be important in locating food (Mitchell, 1970, 1972; Madsen and Herman, 1980; Wursig and Wursig, 1980; Watkins and Wartzok, 1985).

Foraging Using Vision In-Air. Visual communication may play an important role in species that utilize cooperative or group feeding strategies (Otte, 1974; Dawson, 1980; Madsen and Herman, 1980). Wursig and Wursig (1980) observed that an increase in leaping behavior in Dusky dolphins (<u>Lagenorhynchus obscurus</u>) was related to the onset of foraging. They proposed that leaping increased when prey were encountered and served to communicate the location of prey to other pod members. Similarly, Yablokov et al. (1972) observed bottlenose dolphins herding grey mullet and jumping-out of water to catch them in mid-air. <u>Orca</u> have been observed to lunge-out of the water onto beaches in pursuit of prey, as have bottlenose dolphins foraging in salt marshes (Harrison and Bryden, 1988). Madsen and Herman (1980) note that the presence of flocks of birds above schools of fish in coastal regions are a visual cue to coastal-dwelling cetaceans, although no evidence exists that support this suggestion.

Foraging Using Vision In-Water. During foraging, vision may be used in-water for: (1) navigation and location and (2) prey capture. Norris and Dohl (1980) propose that cetaceans may recognize optimal feeding areas using visual cues present around seamounts. Once prey are located, tracking is probably coordinated via echolocation. However, Norris et al. (1961) found the majority of high-frequency acoustic energy (approximately 100 kHz) was directed in a narrow beam forward of the head of an echolocating bottlenose dolphin. When visual search by the dolphin was occluded through the use of eyecups, fish fragments drifting below the level of the mandible were not detected, suggesting that sound emission or detection was limited in the animal's vertical plane. Au and Moore (1984) extended these earlier findings in psychophysical studies examining sound localization capabilities of a bottlenose dolphin using 30, 60, and 120 kHz sounds presented in the vertical- and horizontal-planes. They found the animal localized best when sounds were presented 5 to 10 degrees above the midline of its mandible in the vertical plane. Sound localization ability dropped-off rapidly below this level.

A capability to gauge distances accurately using vision while pursuing elusive prey would especially be useful in complementing the directional limitations of the dolphin's biosonar. The sensitivity of the visual system to motion maximizes the probability of prey detection and capture at close range (Dawson, 1980; Watkins and Wartzok, 1985). Visual cues probably are important when prey are close, picking-up prey items where the forward-focused biosonar field leaves-off. The psychophysical evidence reviewed earlier for visual depth acuity suggests that <u>Tursiops</u>, is capable of relying on motion parallax or other cues, for judging depth.

Aerial Surveillance. Many cetaceans perform behaviors above the surface which may function as aerial surveillance (Herman and Tavolga, 1980; Madsen and Herman, 1980). These behaviors either are stationary, as in headrises or "spyhopping", or kinetic, which encompass the full range of leaping behaviors from porpoising to breaching.

Spyhopping has been observed in mysticetes and odontocetes. In spyhopping, an individual rises-out of the water up-to about the pectoral flippers, often bringing the eye above the surface. The animal may slip back under or may scull and rotate the body, and appear to investigate activity at the surface (Madsen and Herman, 1980). This behavior was observed in Orca which were driven-up against a rock wall ("cornered"), with all animals orienting towards a harassing vessel and assuming the in-air visual posture associated with rostro-ventral viewing (MacAskie, 1966). Among baleen whales, a gray whale (Eschrichtius robustus) that had been exposed to acoustic playback of Orca "screams" was observed to spyhop after fleeing into a kelp bed (Cummings and Thompson, 1971), suggesting the whale was searching for signs of Orca from the protection of the kelp bed.

Many cetacean species perform aerial leaping behaviors. Porpoising is observed in many odontocete species, involving a series of shallow arcing leaps that often are coordinated among pod members. Breaching involves leaps which are more vertically-oriented than porpoising, which carry the animal higher into the air and may involve twisting or spinning motions.

Some species of dolphins repeatedly leap when travelling from one location to another. This may, among other functions, allow individuals to assay the group size and extent (Herman, personal communication). Wursig and Wursig (1980) proposed that breaching by dusky dolphins facilitated coalescence of the feeding herd near schools of prey. Spinner dolphins (Stenella longirostris) were named for the robust spinning leaps performed by this species. Spinning may combine visual and acoustic components in communication designed to maintain herd cohesion (Norris and Dohl, 1980). Several authors note that breaching humpback whales (Megaptera novaeangliae) tend to be oriented towards nearby vessels, implying that breaching may be used for in-air visual search (Madsen and Herman, 1980; Forestell et al., 1985). However, breaching requires a tremendous expenditure of energy, particularly in mysticetes, and thus its primary function is not likely aerial surveillance, when a less energetic behavior such as spyhopping may suffice.

Visual Signals and Social Interaction

Visual signals may be used by cetaceans during social interactions for many different functions (Otte, 1974). For example, orientational signals help individuals coordinate activity in a group, using aposematically colored "blazes" that appear only when an individual reorients itself (Norris and Dohl, 1980). Visual signals also may be used to convey information about the state of the sender, such as sexual receptivity, through changes in coloration around the anogenital region. Coloration patterns, sexual dimorphism, and other visual markings, such as scars, may be used by animals for species-specific or individual recognition. In this section, we discuss the role of: (1) coloration and (2) posture or movement in cetacean social interactions.

Yablokov et al. (1972) suggests that cetacean coloration is primarily cryptic and thus was shaped by foraging pressures. This hypothesis was carried to a logical-end in discussion of adaptive coloration in cetaceans by Mitchell (1970, 1972) and by Madsen and Herman (1980). We will not discuss the role of coloration in crypsis, but rather will concentrate on detection of signals.

Mitchell (1970, 1972) suggests that patchy coloration of some delphinid species is a case of Mullerian mimicry to reduce detection.

For example, Mitchell (1970) suggests that Heaviside's dolphin (<u>Cephalorhynchus heavisidei</u>) has evolved coloration that mimicks <u>Orca</u>, thereby reducing the probability of <u>Orcas</u> identifying and eating the Heaviside's. Mitchell had difficulties rationalizing this seeming convergence of coloration, because <u>Orcas</u> are not sympatric with Heaviside's dolphins. We speculate that the patchy coloration of <u>Cephalorhynchids</u> and <u>Lagenorhynchids</u> may have evolved as species-specific differences among sympatric species, i.e., where Mitchell proposed adaptive convergence, we propose enough divergence to allow discrimination of sympatric species.

<u>Orcas</u> show substantial variation in the extent and shape of the white/black blaze patterns, as well as in the size and shape of their dorsal fins. These variations have been used to track individuals and have been used to describe the membership of <u>Orca</u> pods over substantial periods of time (e.g., Balcomb et al., 1980). It is possible that <u>Orcas</u> use individual variations in coloration not only as species-recognition signals, but as a means of individual recognition (Perry et al., 1988).

Mixed schooling of spotted (<u>Stenella attenuata</u>) and spinner dolphins (<u>Stenella longirostris</u>) in the Eastern North Pacific provides further examples of morphology and coloration which may have evolved for visual recognition. The spinners have black-tipped rostrums whereas the spotted dolphins have white tips (Norris and Dohl, 1980). Additionally, the males of spinner populations that mix with spotted dolphins develop postanal keels and forward-canted dorsal fins (Mitchell, 1970; Madsen and Herman, 1980). These characteristics may assist in species recognition and reproductive isolation.

Blazes and patches of pigmentation may serve as orientation signals. Many cetacean species have areas of contrasting color around the pectoral flipper. Appearance of this blaze could serve as a signal that the animal is changing direction, and may help to coordinate schooling (Norris and Dohl, 1980). Similarly, the ventral surface of the long pectoral flippers of humpback whales is white and can signal changes in direction when sculled or reoriented (Herman and Tavolga, 1980; Madsen and Herman, 1980; personal observations). The distinctive patches of contrasting pigmentation on or near the genital region of many cetaceans may serve as signals for reproductive and nursing behavior (Mitchell, 1970; Norris and Dohl, 1980).

Visual Displays

Postures, or "body language" are important components of social interactions in most terrestrial mammals. Most postures are performed with stereotypical topography and may convey information through intensity of performance (Otte, 1974). Relatively few ritualized visual signals have been proposed for cetaceans compared to terrestrial mammals (Herman and Tavolga, 1980). An "S-shaped posture" has been observed in several delphinids (Herman and Tavolga, 1980) and in humpback whales (Bauer, 1986; Helweg, personal observation). This posture may increase apparent size by exaggeration of the dorsal fin (by arching the back) and spreading of the pectoral flippers, and has been reported in contexts related to reproduction and/or agonism (Madsen and Herman, 1980; Bauer, 1986).

Baker and Herman (1984) propose at least two visual displays associated with agonism in humpback whales. These authors note that agonistic encounters typically unfold in a series of escalations, beginning with simple interception of one whale's path by another in the form of a "broadside display." The most common agonistic behavior that follows is the "headlunge," which may involve engorgement of the ventral pleats with water or air. Baker and Herman (1984) suggest this is analogous to piloerection in terrestrial mammals, serving to increase the apparent size of the displaying whale to conspecific competitors.

COGNITIVE TASKS WITH VISUAL DEMANDS

Early Results (1980 and Before)

Visual acuity and discrimination abilities were the first targets of visual cognitive studies in captive odontocetes (Kellogg and Rice, 1966; Spong and White, 1971; White et al., 1971; Chun, 1978; Herman, 1980). The authors of the early cognitive literature made common observations that the subjects found visual tasks extremely difficult, especially if the task was performed in-air (Kellogg and Rice, 1966; Herman, 1980; Nachtigall, 1986). The review by Herman (1980) revealed in-air visual discrimination of shape and relative brightness to be characterized by poor performance levels, relative to those shown for auditory tasks, and a lack of any evidence of positive transfer across successive problems. This poor performance seemed unlikely attributable to poor visual acuity in light of the psychophysical results suggesting "good" visual acuity for the species in question (Spong and White, 1971; White et al., 1971; Herman et al., 1975), as well as near equal in-air and in-water acuity for Tursiops (Herman et al., 1975).

Kellogg and Rice (1966) generally are credited with the first demonstration of visual discrimination of shape by a bottlenose dolphin subject ("Paddy"). They reported that Paddy performed "surprisingly well" when the brass stencil stimuli were presented underwater, but rarely exceeded chance performance levels when the same stimuli were presented in-air. Part of the discrepancy in performance may have been due to the accessibility of the underwater stimuli to the use of echolocation, as suggested by Herman (1980).

Spong and White (1971) report difficulty in training their Pacific white-sided dolphin (Lagenorhynchus obliquidens) to discriminate between a blank white card and one containing two black lines. Performance remained at chance levels until a "simpler" task was substituted involving discriminating a solid white versus a solid black card. The initial discrimination task was successfully reintroduced following this simple task. Forestell and Herman (1988) report similar difficulties in training the dolphin subject "Puka" to discriminate a white versus black circle until auditory "names" were associated with the stimuli. These authors claim that the use of auditory cues helped to "release the constraint" of the visual task.

One early report of success using in-air visual discrimination of shape was that of Chun (1978), who trained a bottlenose dolphin to perform a simultaneous matching-to-sample (MTS) task. The task required the animal to view a linear array of three two-dimensional shapes, where the middle shape was designated as the "sample" stimulus to be matched against one of the "comparison" stimuli on either side. Overall performance across six sets of matching problems was high, with an average greater than 80% correct. Chun (1978) may have had better success because the stimuli were presented embedded in a large display apparatus, which probably increased the dolphins' ability to discriminate figure from background, as suggested by later results of Hunter (1988). Chun did note, however, a high degree of variability in performance across successive problems, thus evidencing a lack of transfer of the MTS rule, consistent with other reports (e.g., Herman, 1980).

Chun's (1978) analysis further suggested that increasing differences in perimeter length correlated with increasing discrimination performance, as would be predicted by information theory (Attneave, 1954). This finding supports the possible function of black-and-white pigmentation patterns that we proposed earlier. One parsimonious strategy permitting individuals to perceive and categorize animals with disruptive or patchy coloration would involve attending to the shape of the color boundaries.

Recent Results (1981 to Present)

Reexamination of earlier results indicate that the subjects were not attending to the task stimuli (Spong and White, 1971) and that, as suggested by Herman (1987, 1989a), discrimination was hampered by inadequate distinctions between the figure and the background material. Subsequent studies (Mobley, 1984; Mobley and Herman, 1985; Hunter, 1988; Herman et al., 1989a) suggest that cognitive representation of two- and three-dimensional stimuli can be rich if the task is designed carefully.

Mobley (1984) and Mobley and Herman (1985) reported performance levels exceeding 90% correct for a bottlenose dolphin discriminating relative distance between visual targets presented in-air. On each trial, the dolphin's task was to observe the targets from a stationary viewing bar and choose the nearer of the two by moving to the display board and deflecting one of two response switches. Moreover, evidence of transfer of the "go to the nearest target" rule was found across successive problems involving a large assortment of visual targets. Such rule-governed transfer had not been noted in prior studies. Further evidence of rule-governed learning in the form of conditional discrimination was noted using a rotating array of targets. The dolphin quickly learned the conditional rule, "if near target on top, go right; if near target on bottom, go left." Performance remained above 80% correct throughout all rotational trials.

The miniature artificial language (MAL) studies underway at the Kewalo Basin Marine Mammal Laboratory require _Tursiops_ to respond to imperative and interrogative sentences (Herman, 1980; Herman et al., 1984; Herman et al., 1989a). The earlier report of Herman et al. (1984) involved two bottlenose dolphins--one ("Phoenix") trained to use an underwater acoustic language, and the other ("Akeakamai") a dynamic gestural language as performed by human trainers. Comprehension of sentence content by the dolphins is indicated by accurate performance on various cognitive tasks. Contrary to initial expectations, performance on both the visual and auditory MALs proceded in comparable fashion with only minor differences in performance levels. This fact, coupled with earlier problems with static visual targets, led Herman et al. (1984) to suggest that dynamic presentation was important for visual discrimination.

Herman et al. (1988) demonstrated high levels of matching-to-sample (MTS) performance using three-dimensional objects presented at delays ranging from 0 to 80 sec. Objects were held by human trainers and presented in both static and dynamic presentation modes. Results indicated high overall levels of performance (i.e., typically greater than 80% correct) and analysis of first-trial performance across successive problems indicated successful transfer of the MTS rule, in contrast to the lack of MTS transfer reported by Chun (1978). Unexpectedly, overall performance on dynamic presentation trials was not significantly different from that observed on static trials. However, the dynamic mode did significantly aid matching performance during the early stages of practice, relative to the static mode. By the end of the third and final block of tests sessions, performance was equivalent for the two modes of presentation. As noted above, earlier language-training results had suggested that moving visual signals could be more readily perceived (Herman, 1980). Based on these results, Herman et al. (1988, p. 134) concluded that "...the modality through which information arrives, acoustic or visual, is not necessarily a limitation on cognitive performance in this species." The findings from Herman (1989a), together with other visual work of Herman and colleagues (Herman, 1986, 1987; Herman et al., 1984, 1989b) forced a major departure from earlier views of the bottlenosed dolphin as limited in its ability to carry out complex cognitive tasks using visual information (Herman, 1980).

Hunter (1988) similarly was able to demonstrate successful MTS performance using two- and three-dimensional stimuli at delays ranging from 0 to 60 seconds. Hunter's results also evidenced transfer of the MTS rule across successive problems involving novel stimuli. In carrying out his study, Hunter (1988) followed the suggestion of Herman that enhanced figure-ground (stimulus-background) relationships should maximize attention of the dolphin to the relevant stimulus material. To achieve this goal, Hunter displayed large white stimuli (100 cm^2) against a large black, homogeneous background (i.e., 1.22 x 7.31 m black plywood wall). This resulted in the rapid success of the dolphin in discriminating among the white stimuli and in solving the various matching problems it was given.

The high performance levels found by Mobley (1984) and Herman et al. (1989a) similarly could be explained by appeal to the enhanced figure-ground relationships posed in the stimulus arrays of these studies. Herman, et al. (1989a) used trainers to present three-dimensional stimuli held extended in front of their bodies, presenting an overall stimulus field displaced in depth, with the relevant stimuli foremost. In the series of experiments by Mobley, the targets were presented in different planes of depth. In light of the sensitivity shown by the bottlenose dolphin to visually discriminate relative distance, depth is apparently a very salient feature of the visual world. Thus, the difference between earlier difficulties with visual discrimination performance and more recent success is the presentation of visual stimuli in a manner which clearly distinguishes the relevant from irrelevant (figure-ground) aspects of the stimulus array. This apparently may be done by: (a) embedding the stimuli in a large homogenous contrasting field (as per Hunter, 1988; Herman et al., 1989a); (b) moving the stimuli across a static field (as per Herman et al., 1989a); or (c) displacing the visual targets in depth (as per Mobley, 1984).

One recent methodological breakthrough involves the controlled presentation of visual stimuli through the use of two-dimensional video projections. Recent data indicate that a dolphin trained to use a gestural MAL can utilize gestures presented on a video screen (Herman, 1987; Herman et al., 1989b; Morrell-Samuels et al., 1989). The trainer signs to the dolphin from a remote location. The dolphin watches the trainer on a video monitor through an underwater viewing port in the side of the tank. Morrell-Samuels et al. (1989) found evidence suggestive of asymmetrical functioning of the <u>Tursiops</u> brain, based on analysis of the dolphin's response to gestural commands. His conclusions were based on the assumption that the dolphin's laterally-placed eyes and total decussation of the optic fibers insured that stimuli presented by video to one eye initially will be processed entirely in the contralateral cerebral hemisphere. Assymetries were detected as differences in reaction time to comprehend and begin execution of a signed request presented only to one eye, with a right eye (left hemisphere) preference.

A second series of experiments using video displays, Herman (1987; Herman, Morrel-Samuels and Brown, 1989b) demonstrated a bottlenose dolphin's ability to comprehend degraded versions of its gestural MAL. In the most extreme example, a signer is dressed entirely in black and stands behind a black curtain while transmitting signs to the dolphin subject by video. The trainer holds a white sphere on a stick in each hand, which appear as moving white dots in a homogenous black background that correspond to the hand and arm movements in the video image. The original MAL uses space extensively (Herman et al., 1984; Shyan and Herman, 1987), which is compressed into lateral and vertical movements of the dots on the two-dimensional video screen. Performance on a comprehension task was almost perfect when using a full video image of the trainer signing and remained well above chance when guided by the extremely degraded "moving dots" condition.

SUMMARY AND CONCLUSIONS

The eyes of most cetaceans do not limit their visual experience to blurry indications of movement and brightness. The cetacean eye appears to be designed to enhance any visual information derived from a dark environment. The specialized photopigments and tapetum insure that any ambient light will be used. Visual information is quickly relayed to the CNS by extremely large ganglion cells. Psychophysical data for some odontocete species suggest that visual acuity is relatively good in air and in water.

Disruptive coloration or blaze markings on many species suggests that pattern recognition is important in cetacean social interactions. Other than crypsis, the areas of highly-contrasting coloration in some species may serve for purposes of species and/or individual identification.

The ability of cetaceans to recognize and categorize two- and three-dimensional forms is important to their survival. Earlier difficulties in training discrimination tasks with visual demands appear to have stemmed more from limitations of the manner of presentation rather than from limitations of the animals' visual processing capabilities. Review of the more recent successes with visual discrimination using _Tursiops_ subjects suggest that performance benefits from maximizing "figure-ground" discriminability. This may be done by: (a) presenting the visual targets on a large homogenous display board; (b) moving the targets against a stationary background; or (c) displacing the targets in depth.

The apparent readiness of the bottlenose dolphin to discriminate visual depth suggests a possible adaptive benefit of this capability--perhaps in complementing the directional limitations of its forward-directing biosonar. It was proposed that this would prove particularly advantageous during the final stages of closure when pursuing elusive prey.

One recent contribution to the experimental investigation of delphinid vision has been the discovery that dolphins can readily discriminate images presented on a video screen. This permits precise experimental control of the qualities of the visual stimulus and manner of presentation. The ability of dolphins to recognize degraded versions of familiar visual patterns would serve to partially overcome the optical limitations of seawater. Such capabilities presumably find application in the utilization of coloration patterns as part of species and/or individual recognition, as well as recognition of prey species, among other possibilities.

REFERENCES

Attneave, F., 1954, Some informational aspects of visual perception. Psychological Review, 61:183-193.
Au, W. W. L. and Moore, P. W. B., 1984, Receiving beam patterns and directivity indices of the Atlantic bottlenosed dolphin, Tursiops truncatus. Journal of the Acoustical Society of America, 75:255-262.
Baker, C. S. and Herman, L. M., 1984, Aggressive behavior between humpback whales (Megaptera novaeangliae) wintering in Hawaiian waters. Canadian Journal of Zoology, 62:1922-37.
Balcomb, K. C., Boran, J. R., Osbourne, R. W. and Haenel, N. J., 1980, Observations of killer whales (Orcinus orca) in Greater Puget Sound, State of Washington, Final Report to the Marine Mammal Commission, Washington, D.C.: Marine Mammal Commission.
Bauer, G. B., 1986, The behavior of humpback whales in Hawaii and modifications of behavior induced by human interventions, Unpublished Ph.D. Dissertation, University of Hawaii at Manoa, Honolulu, Hawaii.
Caldwell, D. K. and Caldwell, M. C., 1972, Senses and communication, in: "Mammals of the Sea," S.H. Ridgway, ed. (pp. 466-502), Springfield, IL: Thomas.
Chun, N. K. W., 1978, Aerial visual shape discrimination and matching-to-sample problem solving ability of an Atlantic bottlenosed dolphin, San Diego, CA: NOSC Technical Report 236.
Collett, T. S. and Harkness, L., 1982, Depth vision in animals, in: "Analysis of Visual Behavior," D. J. Ingle, M. A. Goodale and R. J. W. Mansfield, eds. (pp. 111-176), Cambridge, MA: MIT Press.
Cumming, W. C. and Thompson, P. O., 1971, Grey whales, Eschrictius robustus, avoid the underwater sounds of killer whales, Orcinus orca, U.S. Fisheries Bulletin, 69:525-530.
Dawson, W. W., 1980, The cetacean eye, in: "Cetacean Behavior: Mechanisms and Functions," L. M. Herman, ed. (pp. 53-100), New York: Wiley Interscience.
Dawson, W. W., Schroeder, J. P., and Dawson, J. F., 1987a, The ocular fundus of two cetaceans, Marine Mammal Science, 3(1):1-13.
Dawson, W. W., Schroeder, J. P., and Sharpe, S. N., 1987b, Corneal surface properties of two marine mammal species, Marine Mammal Science, 3(2): 186-197.
DeFran, R. H. and Pryor, K., 1980, The behavior and training of cetaceans in captivity, in: "Cetacean Behavior: Mechanisms and Functions," L. M. Herman, ed. (pp. 319-362), New York: Wiley Interscience.
Dral, A. D. G., 1972, Aquatic and aerial vision in the bottlenosed dolphin, Netherlands Journal of Sea Research, 5:510-513.
Dral, A. D. G., 1974, Problems in image-focusing and astigmatism in cetacea--a state of affairs, Aquatic Mammals, 2:22-28.
Dral, A. D. G., 1975, Vision in cetacea, Journal of Zoo Animal Medicine, 6:17-21.
Dral, A. D. G., 1977, On the retinal anatomy of cetacea (mainly Tursiops truncatus), in: "Functional Anatomy of Marine Mammals," Vol. 3, R. J. Harrison, ed. (pp. 81-134), New York: Academic.
Dral, A. D. G., 1983, The retinal ganglion cells of Delphinus delphis and their distribution. Aquatic Mammals, 10:57-68.
Duke-Elder, S., 1958, "System of Opthamology," Vol. 1, St. Louis, MO: Mosby.
Fobes, J. L. and Smock, C. C., 1981, Sensory capacities of marine mammals, Psychological Bulletin, 89:288-307.
Forestell, P. H. and Herman, L. M., 1988, Delayed matching of visual materials by a bottlenosed dolphin aided by auditory symbols, Animal Learning and Behavior, 16:137-147.

Forestell, P. H., Veghte, M. B., and Herman, L. M., 1985, Evidence for visual search by breaching humpbacks, Paper presented at the Sixth Biennial Conference on the Biology of Marine Mammals, Vancouver, BC.

Fox, R., 1978, Binocularity and stereopsis in the evolution of vertebrate vision, in: "Frontiers in Visual Science," S. J. Cool and E. L. Smith, III, eds. (pp. 316-327), New York: Springer-Verlag.

Fox, R., Lemkuhle, S. W. and Bush, R. C., 1977, Stereopsis in the falcon, Science, 197:79-80.

Fuji, K. and Kojima, S., 1981, Acquisition of depth discrimination in a Japanese macaque: A preliminary study, Perceptual and Motor Skills, 52:827-830.

Harrison, Sir R. and Bryden, M. M., eds., 1988, "Whales, Dolphins and Porpoises: An Illustrated Encyclopedic Survey by International Experts," New York: Facts on File Publications.

Herman, L. M., 1980, Cognitive characteristics of dolphins, in: "Cetacean Behavior: Mechanisms and Functions," L. M. Herman, ed. (pp. 363-430), New York: Wiley Interscience.

Herman, L. M., 1987, The visual dolphin, Abstracts, Seventh Biennial Conference on the Biology of Marine Mammals, Miami, Florida.

Herman, L. M. and Forestell, P. H., 1985, Short-term memory in pigeons: modality-specific or code-specific effects? Animal Learning and Behavior, 13:463-465.

Herman, L. M., Hovancik, J. R., Gory, J. D., and Bradshaw, G. L., 1989a, Generalization of visual matching by a bottlenosed dolphin (Tursiops truncatus): Evidence for invariance of cognitive performance with visual and auditory materials, Journal of Experimental Pscyhology-- Animal Behavior Processes, 15:124-136.

Herman, L. M., Morrel-Samuels, P. and Brown, L. A., 1989b, Recognition and imitation of television scenes by bottlenosed dolphins, Abstracts, Eighth Biennial Conference on the Biology of Marine Mammals, Monterey, CA.

Herman, L. M., Peacock, M. F., Yunker, M. P. and Madsen, C. J., 1975, Bottlenosed dolphin: Double-slit pupil yields equivalent aerial and underwater acuity, Science, 139:650-652.

Herman, L. M., Richards, D. G., and Wolz, J. P., 1984, Comprehension of sentences by bottlenosed dolphins, Cognition, 16:129-219.

Herman, L. M. and Tavolga, W. N., 1980, The communication systems of cetaceans, in: "Cetacean Behavior: Mechanisms and Functions," L. M. Herman, ed. (pp. 149-210), New York: Wiley Interscience.

Hochberg, J., 1971, Perception. II. Space and movement, in: "Experimental Psychology" (3rd ed.), J. W. Kling and L. A. Riggs, eds., (pp. 475-531), New York: Holt, Rinehart and Winston.

Hunter, G., 1988, Visual delayed matching of two-dimensional forms by a bottlenosed dolphin, Unpublished M.A. thesis, University of Hawaii Manoa.

Jacobs, M., Morgane, P. and McFarland, W., 1975, Degeneration of visual pathways in the bottlenosed dolphin, Brain Research, 88:346-352.

Jamieson, F. S. and Fisher, H. D., 1972, The pinniped eye: A review, in: "Functional anatomy of marine mammals," Vol. 1, R. J. Harrison, ed. (pp. 245-261), New York: Academic.

Julesz, B., 1971, "The Foundations of Cyclopean Perception," Chicago, IL: University of Chicago Press.

Kellogg, W. N. and Rice, C. F., 1966, Visual discrimination and problem solving in a bottlenosed dolphin, in: "Whales, Dolphins, and Porpoises," K. S. Norris, ed. (pp. 731-754), Berkeley: University of California Press.

Levinthal, C. F., 1983, "Introduction to Physiological Psychology," 2nd ed., Englewood Cliffs, NJ: Prentice-Hall.

MacAskie, I. V., 1966, Unusual example of group behavior by killer whales (Orcinus rectipinna), Murrelet, 47:38.

MacFarland, W. N., 1971, Cetacean visual pigments, Vision Research, 11:1065-1076.
Madsen, C. J. and Herman, L. M., 1980, Social and ecological correlates of cetacean vision and visual appearance, in: "Cetacean Behavior: Mechanisms and Functions," L. M. Herman, ed. (pp. 101-148), New York: Wiley Interscience.
Mass, A. M., Supin, A. Ya., Severtsov, A. N., 1986, topographic distribution of sizes and density of ganglion cells in the retina of a porpoise, Phocoena phocoena, Aquatic Mammals, 12:95-102.
Mitchell, D. E., Kaye, M. and Timney, B., 1979, Assessment of depth perception in cats, Perception, 8:389-396.
Mitchell, E., 1970, Pigmentation pattern evolution in delphinid cetaceans: An essay in adaptive coloration, Canadian Journal of Zoology, 48:717-740.
Mitchell, E., 1972, Whale pigmentation and behavior, American Zoologist, 12:60.
Mobley, J. R., 1984, Visual discrimination of relative distance by a bottlenosed dolphin (Tursiops truncatus)--evidence for relational learning, Unpublished Ph.D. Dissertation, University of Hawaii at Manoa, Honolulu, Hawaii.
Mobley, J. R. and Herman, L. M., 1985, Visual discrimination of relative distance by a bottlenosed dolphin (Tursiops truncatus), Abstracts, Sixth Biennial Conference on the Biology of Marine Mammals, Vancouver, B.C.
Morrel-Samuels, P., Herman, L. M., Bever, T. G. and Rettig, E. J., 1989, Cerebral assymetries for gesture recognition in the dolphin, Abstract, Eighth Biennial Conference on the Biology of Marine Mammals, Monterey, CA.
Nachtigall, P. E., 1986, Vision, audition, and chemoreception in dolphins and other marine mammals, in: "Dolphin Cognition and Behavior: a Comparative Approach," R. J. Schusterman, J. A. Thomas, and F. G. Wood, eds. (pp. 79-113), Hillsdale: Lawrence Erlbaum Associates.
Nachtigall, P. E., 1989, Risso's dolphin (Grampus griseus) vision. Abstracts, Eighth Biennial Conference on the Biology of Marine Mammals, Monterey, CA.
Norris, K. S. and Dohl, T. P., 1980, The structure and function of cetacean schools, in: "Cetacean Behavior: Mechanisms and Functions," L. M. Herman, ed. (pp. 211-262), New York: Wiley Interscience.
Norris, K. S., Prescott, J. H., Asa-Dorian, P. V. and Perkins, P., 1961, An experimental demonstration of echolocation behavior in the porpoise Tursiops truncatus (Montagu), Biology Bulletin, 120:163-176.
Otte, D., 1974, Effects and functions in the evolution of signaling systems, Annual Review of Ecology and Systematics, 5:385-417.
Perry, A., Mobley, J. R., Baker, C. S. and Herman, L. M., 1988, "Humpback Whales of the Central and Eastern North Pacific: A Catalog of Individual Identification Photographs," Honolulu, HI: Sea Grant College Program Publications.
Popper, A. N., 1980, Sound emission and detection by delphinids, in: "Cetacean Behavior: Mechanisms and Functions," L. M. Herman, ed. (pp. 1-52), New York: Wiley Interscience.
Rodieck, R. W., 1973, "The Vertebrate Retina," San Francisco: Freeman.
Sarmiento, R. F., 1975, The stereoacuity of the macaque monkey, Vision Research, 15:493-498.
Shyan, M. R. and Herman, L. M., 1987, Determinants of recognition of gestural signs in an artificial language by Atlantic bottlenosed dolphins (Tursiops truncatus) and humans (Homo sapiens), Journal of Comparative Psychology, 101:112-125.
Slijper, E. J., 1962, "Whales," London: Hutchinson.

Spoehr, K. T. and Lemkuhle, S. W., 1982, "Visual Information Processing," San Francisco, CA: W.H. Freeman.

Spong, P. and White, D., 1971, Visual acuity and discrimination learning in the dolphin (Lagenorhynchus obliquidens), Experimental Neurology, 31:431-436.

Stone, J., 1965, A quantitative analysis of the distribution of ganglion cells in the cat's retina, Journal of Comparative Neurology, 124:337-352.

Townsend, J. T., Hu, G. G., and Kadlec, H., 1987, Feature sensitivity, bias, and interdependencies as a function of intensity and payoffs, Technical Report 87-4, West Lafayette, Indiana: Purdue University.

Van Hof, M. W. and Russell, I. S., 1977, Binocular vision in the rabbit, Physiology and Behavior, 19:121-128.

Van Sluyters, R. C. and Stewart, D. L., 1974, Binocular neurons of the rabbit's visual cortex: Receptive field characteristics, Explorations in Brain Research, 19:166-195.

Walls, G. L., 1942, "The Vertebrate Eye and its Adaptive Radiation," Bloomfield Hills, MI: Cranbrook Press.

Watkins, W. A. and Wartzok, D., 1985, Sensory biophysics of marine mammals, Marine Mammal Science, 1:219-260.

White, D., Cameron, N., Spong, P. and Bradford, J., 1971, Visual acuity in the killer whale (Orcinus orca), Experimental Neurology, 32:230-236.

Wursig, B. and Wursig, M., 1980, Behavior and ecology of dusky dolphins, Lagenorhynchus obscurus, in the south Atlantic, Fisheries Bulletin, 77:871-890.

Yablokov, A. V., Bel'kovich, V. M. and Borisov, V. I., 1972, "Whales and Dolphins," Jerusalem: Israel Program for Scientific Translations.

Young, N. M., Hope, G. M., Dawson, W. W., and Jenkins, R. L., 1988, The tapetum fibrosum in the eyes of two small whales, Marine Mammal Science, 4:281-290.

NON-ACOUSTIC COMMUNICATION IN SMALL CETACEANS: GLANCE, TOUCH, POSITION, GESTURE, AND BUBBLES

Karen W. Pryor

44811 S.E. 166th St., North Bend, WA 98045, U S.A.

INTRODUCTION

Behavior enables an animal to interact with and survive in its environment. In cetaceans, as in all other animals, sensory systems exist to serve behavior. Perhaps more than most animals, cetaceans may be said to live in two worlds: their physical universe of air and water, and the social universe of the other dolphins around them. Their sensory systems serve them in both. In the physical universe, sensory systems are used in locomotion, foraging, maintaining physical and physiological equilibrium, and so on. In the social universe, sensory systems are used in communication In fact, it might be said that all social behavior constitutes communication.

For many years, researchers interested in small cetaceans have concentrated on acoustic communication, partly because it is a conspicuous feature of dolphin behavior, and partly because of the interesting specialization of the dolphin acoustic system for echolocation. Perhaps because of this research emphasis, a common supposition has arisen that the acoustic system is the primary or even the only mode of communication in cetaceans. The assumption has been that life in the water precludes visual and gestural communication (facial expression, for example) and that the acoustic output of dolphins is elaborated, at least partly, due to the necessity of cramming into a single mode the social information that terrestrial mammals convey in many ways.

In fact, in cetaceans as in other mammals all the sensory systems are used in social communication. Chemoreception, the sense of taste and smell in other mammals, has been assumed to be non-existent in dolphins, because dolphins have no olfactory lobe in their brain. Recent work has shown that even chemoreception not only exists, but may play an important role in social communication (Kastelein and Spekreyse, 1990; Kuznetsov, 1989, 1990; Pryor, 1990). And the communication functions of the remaining sensory systems, visual and tactile, are of enormous importance to small cetaceans.

Visual systems

Dolphins were once assumed to have poor eyesight; for example, some early experimenters thought they could only see moving objects. In fact, most species appear to see very well indeed, both above and under water (Herman, 1990; Mass, 1990; Nachtagall, 1986).

Cetaceans see monocularly laterally, but have binocular vision downward (and in some species to the rear; Pryor, 1973). In either mode, they readily make eye contact with humans and with each other. This tendency may be the cause of some popular misconceptions about dolphins. In the dolphin, as in many primates, the focus of the eye is obvious; when a dolphin "looks you in the eye" you know it. In cetaceans, however, eye contact is usually brief; the eye does not stare, but quickly moves on. Also, when two cetaceans make eye contact, the dominant animal normally looks away first. In primates, however, like ourselves, dominant animals stare, and submissive individuals break eye contact or look away. Humans, therefore, who are apt to look at dolphins longer than the dolphins look back, tend to feel that the dolphins are "friendly." Their intense and yet seemingly nonthreatening eye contact may be a major reason why people develop the feeling that dolphins are in some way human-like or even magical.

Since dolphins literally swim eye-to-eye, and often leap and breathe in unison, they have ample opportunity to communicate with each other by eye contact, both above and under water. Cetaceans convey many intentions and internal states via the eye and surrounding tissues; for example in the killer whale, Orcinus orca, the sclerae turn red in aggression. Experienced dolphin trainers can predict behavioral events from eye expression: one can learn to tell when an animal is feeling ill, or when it is about to make mischief, by the look in the eye (Defran and Pryor, 1980). If humans can learn to "read" the dolphin's eye, presumably the dolphins themselves are adept at interpreting each others' visually-signalled information.

Many species of small cetaceans have striking or elaborate color patterns. While such patterns may serve in fish-catching or in predator avoidance they undoubedly also serve in a social context. Color patterns enable animals to recognize conspecifics. Dramatic or "flashy" color patterns, like the black and white of Dall's porpoise (Phocoenoides dalli) may help animals locate each other visually in the open sea or in murky water (Würsig and Kieckhefer, 1990). Individual variations in color patterns, as seen in killer whales (Orcinus orca), enable humans to identify individuals, and might serve the same function for the whales themselves (Felleman et al., 1990). The Pacific spotted dolphin (Stenella attenuata) passes through five different color phases from birth to sexual maturity (Perrin, 1969). In this species, patterns give visual cues to age, status, and social role, as well as to species (Pryor and Shallenberger, 1990). For example, dominant males are marked by conspicuous white jaw tips, which in tropical waters can be seen at 50 m or more: these "white noses" are as much a visual indication of age and status as is the mane of a male lion (Figure 1).

In addition, dolphins can give each other information through visually-perceived gestures and postural displays. An S-shaped body posture is a threat display in the bottlenose dolphin, Tursiops truncatus (Tavolga, 1966). In the genus Stenella threat displays include halting, spreading the pectoral fins, an S-shaped posture, and nodding or shaking the head with open jaws. Gaping the jaws even slightly (with or without accompanying sound production) is also a common threat display, easily perceived visually at close range (Fig. 2). Gestures and postural displays also may be affiliative behavior; in the genus Stenella a greeting gesture consists of tilting sidewise to flash the white or light belly at another animal (Norris et al., 1985).

Figure 1. Conspicuous white jaw tips identify this group of Pacific spotted dolphins (Stenella attenuata) as dominant males. Another group of four adult males passes behind them. (Photographs by Karen Pryor).

Leaps and aerial actions of some species of dolphins also may serve as visual displays with a communicative function. Different species leap in different ways; for example spinner dolphins (Stenella longirostris) popularly are named for their exotic and species-specific rotating leap. Specific leaps also can be related to specific internal states: in many species breaches and headslaps can sometimes be correlated with excitement and aggression (Defran and Pryor, 1980); and spotted and spinner dolphins, pursued in the process of purse-seining for tuna, exhibit an unusual vertical leap when released from the net or upon evading pursuit, that indicates successful escape (Norris et al., 1985; Pryor and Kang, 1980.)

It has been argued that such displays are not visible to other dolphins, since they are likely to be under water most of the time and unable to see through the water surface except directly overhead; and therefore, any social information in leaps or other surface activity is probably contained in the splashing or slapping noises accompanying the activity. The accompanying noises certainly play a part in the communicative value of some aerial behavior. Dolphins however often leap simultaneously or in groups, and can see above water whenever they surface to breathe. It seems likely that the panoply of highly stereotyped aerial displays exhibited by many species do serve as visual signals which are perceived in fleeting glimpses while above the water. Such signals could communicate, perhaps at considerable distance, what species is present and what, to some extent, is going on.

Figure 2. Photographed in the net of a tuna purse-seining vessel, a male Pacific spotted dolphin, rising toward the surface, gapes in threat at a group of males passing by overhead.

Bubbling

Stereotyped patterns of bubble production constitute a mode of communication in dolphins which is not available to terrestrial animals (Pryor, 1973; Pryor, 1986; Pryor and Kang, 1982). Bubble formations, barely noticeable above the surface, are conspicuous under water both visually and to echolocation. Perhaps the commonest type of bubble display in small cetaceans is the whistle-trail, in which a stream of bubbles is emitted from the blowhole in synchrony with an audible whistle. Dolphins do not need to emit air when they whistle; the bubbles seem to be "added for emphasis". If the whistle is interrupted, the bubble stream will be interrupted too (Fig. 3).

Whistle-trails often are emitted during social interactions and may serve to demarcate the signalling individual. Whistle-trails are produced most abundantly in species which occur in aggregations of hundreds of animals, such as the spinner dolphin (Stenella longirostris).

A dolphin which is surprised by an unfamiliar sight or sound may release a sphere of bubbles under water, a "query balloon." Spheres may also be released during play. Released bubbles may themselves be used as toys. During underwater observations within the purse seine net of a tuna vessel, Pacific spotted dolphins (S. attenuata) were seen releasing torus formations during dominance disputes (Pryor and Kang, 1980). A young adult male engaged in an aggressive interaction with an older male was observed to leap in the air and fall back, creating a large cloud of bubbles on re-entry, and then to escape his opponent behind this "smokescreen." Future underwater observations may reveal other types and functions of bubble displays.

Figure 3. A juvenile Pacific spotted dolphin emits a trail of bubbles in two sections, accompanying an interrupted whistle sound.

Some aerial displays create almost no bubbles or splash above water, while others produce a lot of spray. Against the dark surface of the ocean, spray produced by surface activity heightens the visibility of many sorts of aerial behavior, and may increase the utility of these actions for visual communication.

Tactile Senses

The skin of small cetaceans is well innervated and extremely sensitive. Most species seem to enjoy being touched; acclimated captive bottlenose dolphins (Tursiops truncatus) can be trained using touch alone as a reinforcement (Defran and Pryor, 1980). Dolphins frequently touch each other. They may stroke or pat each other with pectoral fins, flukes, or rostrum, rub bodies, or swim with fins or bodies in physical contact. Touching is common between females and young, among groups of juveniles, and between males and females.

Tactile contact in small cetaceans also may be aggressive. Tooth-raking occurs in many species: one animal draws the open jaws across another animal's body or extremities, often leaving parallel lines or even drawing blood. In aggressive encounters males may strike each other with the fins or flukes or ram an adversary head-on.

Positioning

The tactile sense is undoubtedly important to cetaceans in judging water flow and movement, and maintaining position in the water. Dolphins are remarkably stable in the water: a captive dolphin can, if it wishes, render itself virtually immobile in the water, so resistant to a human's push or pull that it seems like a rock in concrete. The combination of mechanoreception and fine-tuned muscular control enables dolphins to coast in the pressure-wave of the bows of travelling ships, and to travel in storm waves.

Dolphins also may use mechanoreception to coordinate movements with each other. Unison swimming and synchronicity of movement is a conspicuous feature of cetacean behavior. Closely associated individuals often exhibit unison behavior, such as simultaneous respiration and matched leaps or breaching. Unison behavior, in fact, is in itself a statement of relationship. Mechanoreception, particularly pressure sensitivity, may enable dolphins both to perceive and give signals relating to movement, facilitating their ability to synchronize even complex patterns of activity (Fig. 4).

Dominance hierarchies are a major feature of cetacean behavior both in captivity and in the wild. Relative dominance can be signalled by position; dominant animals or groups may be slightly ahead of those nearby, or above other groups, or separated from others by wide inter-animal distances. Dominant animals displace others as they pass among them. Small inter-animal distances (a body-width or less) can indicate close or long-term association, and are typical of females and young, juvenile bands, and sometimes adult male pairs and bands. Mechanoreception of water pressure and movement may be vital in the maintenance of inter-animal distance, especially at night, in murky water, or at great depths, in a society where the relative position between animals has important social consequences and functions.

Figure 4. In the Eastern tropical Pacific a group of spinner dolphins (Stenella longirostris) leap in unison during rapid travel.

It is not unusual for some species of dolphins to travel in silence; killer whales (Orcinus orca) in Puget Sound, for example, neither echolocate nor whistle when hunting seals (Felleman et al., 1990), and bottlenose dolphins are often quiet while travelling. If vision, also, is limited during periods of silence, tactile senses then may help to fill the information gap, providing the sensory input necessary to coordinate behavior, signal intentions, and maintain group structure, in the demanding physical universe of a totally aquatic life.

REFERENCES

Defran, R. H., and Pryor, K., 1980, The behavior and training of cetaceans in captivity, in: "Cetacean Behavior: Mechanisms and Functions," L. Herman, ed., Wiley-Interscience, N.Y.

Felleman, F., Heimlich-Boran, J. and Osborne, R. 1990, The feeding ecology of killer whales (Orcinus orca) in the Pacific Northwest, in: "Dolphin Societies: Discoveries and Puzzles," K. Pryor and K. Norris, eds., University of California Press, Berkeley.

Herman, L., 1990, Visual performance in the bottlenosed dolphin (Tursiops truncatus), in: "Sensory Methods of Cetaceans: Laboratory and Field Evidence," J. A. Thomas and R. A. Kastelein, eds., Plenum Press, N.Y.

Kuznetsov, V.B., 1990, Chemical sense in dolphins: quasiolfaction, in: "Sensory Methods of Cetaceans: Laboratory and Field Evidence," J. Thomas and R. Kastelein, eds., Plenum Press, N.Y.

Kuznetsov, V.B., 1989, Chemoreception and communication in dolphins, in: Abstracts, Fifth International Theriological Congress, 22-29 August 1989. Rome, Italy.

Kastelein, R., 1990, Marginal papillae on the tongue of harbor porpoises and bottlenose dolphins, in: "Sensory Methods of Cetaceans: Laboratory and Field Evidence," J. A. Thomas and R. A. Kastelein, eds., Plenum Press, N.Y.

Kastelein, R. A., and Dubbeldam, J. L., 1990, Marginal papillae on the tongues of the harbor porpoise (Phocoena phocoena), bottlenose dolphin (Tursips truncatus) and Commerson's dolphins (Cephalorhynchus commersonii). Aquatic Mammals 15:4.

Kastelein, R., and Spekreyse H., 1990, The anatomical characteristics of the eyes of a young and an adult harbour porpoise, in: "Sensory Methods of Cetaceans: Laboratory and Field Evidence," J. A. Thomas and R. A. Kastelein, eds., Plenum Press, N.Y.

Mass, A. M., 1990, The best vision areas in the retina of some cetaceans, in: "Sensory Methods of Cetaceans: Laboratory and Field Evidence," J. A. Thomas and R. A. Kastelein, eds., Plenum Press, N.Y.

Nachtagall, P. E., 1986, Vision, audition and chemoreception in dolphins and other marine mammals, in: "Dolphin Cognition and Behavior: A Comparative Approach," R. J. Schusterman, J. A. Thomas, and F. C. Wood, ed., Lawrence Erlbaum Associates, Hillsdale, N.J.

Norris, K. S., Würsig, B., Wells, R., Würsig, M., Brownlee, S., Johnson, C., and Solow, J., 1985, "The Behavior of the Hawaiian Spinner Dolphin, Stenella longirostris," National Marine Fisheries Service Southwest Fisheries Center Administrative Bulletin LJ-85-06C, La Jolla, CA.

Perrin, W., 1969, Color pattern of the eastern Pacific spotted porpoise, Stenella graffmani Lonnberg, Zoologica, 54:12.

Pryor, K., 1973, Behavior and learning in porpoises and whales, Naturwissenschaften 60:412.

Pryor, K., 1986, Non-acoustic communicative behavior of the great whales: origins, comparisons, and implications for management, in: "Report of the International Whaling Commission, Special Issue 8," 89-96.

Pryor, K., 1990, Report of the working group on non-acoustic sensory systems, in: "Sensory Methods of Cetaceans: Laboratory and Field Evidence," J. A. Thomas and R. A. Kastelein, eds., Plenum Press, N.Y.

Pryor, K., and Kang, I., 1980, School Structure and Social Behavior in Pelagic Porpoises (Stenella attenuata and Stenella longirostris) during Purse Seining for Tuna," NMFS SW. Fish. Cent. Rept. LJ-80-11C, La Jolla, CA.

Pryor, K., and Shallenberger, I. K., 1990, in: "Dolphin Societies, Discoveries and Puzzles," K. Pryor and K. S. Norris, eds., University of California Press, Berkeley.

Tavolga, M., 1966, Behavior of the bottlenose dolphin (Tursiops truncatus): social integration in a captive colony, in: "Whales, Dolphins, and Porpoises," K. Norris, ed., University of California Press, Berkeley.

Würsig, B., and Kieckhefer, T., 1990. Visual displays for communication in cetaceans, in: "Sensory Methods for Cetaceans: Laboratory and Field Evidence," J. A. Thomas and R. A. Kastelein, eds., Plenum Press, N.Y.

VISUAL DISPLAYS FOR COMMUNICATION IN CETACEANS

Bernd Würsig, Thomas R. Kieckhefer[1], and Thomas A. Jefferson

Marine Mammal Research Program, Texas A&M University at
 Galveston, Department of Marine Biology, P.O. Box 1675,
 Galveston, TX 77553, USA
[1]Moss Landing Marine Laboratories, P.O. Box 450, Moss Landing,
 CA 95039, USA

INTRODUCTION

Social mammals use facial signals and body postures, often highlighted by coloration, for well-developed visual communication. African forest monkeys (Cercopithecus sp.), for example, display an impressive repertoire of stereotyped head movements and facial expressions for courtship, aggression, fear, and appeasement (Kingdon, 1980). Thomson's gazelles (Gazella thomsoni) rapidly alert conspecifics to danger by a tense, upright stance, directed gaze, and often a twitching of their edge-receptive flank, which is marked with an eye-catching black longitudinal stripe (Estes, 1967; Walther, 1969). This sequence of subtle body movements produces a "Morse code" of visual information about the potential threat of an approaching predator. To the untrained observer only obvious alarm signals and warnings, such as running and stotting (a stiff-legged, bounding gait), are appreciable (Caro, 1988). It recently has been found that some mammalian species display different warning signals for different predators (Seyfarth et al., 1980; Sherman, 1985), and this extra sophistication beyond mere communication of danger is probably widespread among animals.

IMPORTANCE OF COMMUNICATION

Cetaceans have relatively nonexpressive faces, frozen rigid during evolution towards streamlining in water. Also, most species lack great flexibility of the neck and forelimbs, body parts that are used expressively in many terrestrial vertebrates. Yet visual communication is important to cetaceans, and this is likely true especially for social cetaceans in clear surface waters, where highly complex coloration patterns can transmit refined information. Cetaceans have remarkable abilities to communicate and to scan their environment by using sound. However, it is potentially disadvantageous to make sounds that prey or predators might hear (Myrberg, 1981); and we have found that Hawaiian spinner dolphins (Stenella longirostris), dusky dolphins (Lagenorhynchus obscurus), and killer whales (Orcinus orca) all can be remarkably silent during daytime. This is especially true when dolphins are resting and when killer whales are travelling in apparent search of marine mammal prey. In British Columbia and Washington, for example, resident killer whales that feed on salmon are quite noisy, while transient killer whales that feed largely on pinnipeds tend to be quiet (Ford and Fisher, 1983). We assume that their social communication at these times is largely by sight.

Visual communication is effected by the interactive traits of morphology, coloration, and postures. For visual communication to be useful, water clarity, light level, and visual acuity must be sufficient for efficient transfer of messages (Lythgoe, 1979). All cetaceans have a functional sense of sight. It is reduced in the platanistoids, but even the Amazon river dolphin, or boutu (Inia geoffrensis), can see flash patterns of contrasting white and black; and its yellow-eye lens (Dawson, 1980) is believed to filter out the glare and dazzle of back-scattered light in murky water (Lythgoe, 1979; K. S. Norris, pers. comm., 1983; Walls, 1967). The Ganges and Indus river dolphins (Platanista spp.) often are referred to as blind because they have no lens at all (Dawson, 1980); nevertheless, their eyes are thought to be capable of light detection and possibly forming crude images, since the eye opening is extremely small and may operate like a pinhole camera (Purves and Pilleri, 1973). In some environments sight probably is useful only above the surface and at extremely close range. The Chinese river dolphin, or baiji (Lipotes vexillifer), for example, lives in water that is essentially opaque. It often brings its head and eyes out of water when surfacing to breathe, possibly scanning the shoreline, river vessel traffic, and conspecifics while doing so (B. Würsig, pers. observ.). However, we assume that cetaceans that live in extremely murky waters use visual communication in a less refined way than those occupying clearer habitats. Cetaceans lack some of the underwater sensing mechanisms used by many other vertebrates. For example, sophisticated chemical and electrical sensing systems and a very sensitive hydrodynamic sense are common to many fishes and amphibians, and some reptiles are extremely thermosensitive. We would expect that without the availability of a wide range of sensory capabilities, cetacean communication by sight will have evolved to the highest possible degree, given the constraints of the environment.

Visual communication has not been explored in detail for cetaceans. One of the major reasons is that visual communication and sound communication have been difficult to separate because underwater acoustic data have not been recorded with consistency during behavioral observations. Furthermore, from an observer's standpoint, subtle posturing and movements that reflect flash patterns of communicative light are often obscured by distance and various properties of the aqueous environment (i.e., refraction, turbidity, wave-induced surface scatter); and even when animals are within close range for behavioral observations, hydrophone recordings often are difficult to synchronize. Also, it is not always possible to separate the visual from the tactile modes often used by cetaceans. Yet, it is interesting to attempt to separate these sensory modes, for each has its advantages and disadvantages; and the relative use of each mode may tell us much about the behavioral and perhaps emotional states of these interacting animals. However, very few successful behavioral observations have been carried-out under water, where cetaceans spend most of their time. Our observations are usually carried-out from our own environment, and we describe those fleeting times and interactions when whales and dolphins are at the surface. What we need are more long-term behavioral observations under water, such as those on Hawaiian spinner dolphins (Norris and Dohl, 1980a; Norris et al., 1985) and those now being done by several research groups working with Atlantic spotted dolphins (Stenella frontalis) in the Bahamas. Such observations are restricted to clear waters and to animals that consistently allow approach by humans under water. Although fleeting glimpses of underwater visual communication have been gathered for humpback whales (Megaptera novaeangliae; Baker and Herman, 1984; Bauer, 1986; Madsen and Herman, 1980; Tyack, 1981; Tyack and Whitehead, 1983), very little is known about visual communication of any of the large whales.

ELEMENTS OF VISUAL COMMUNICATION

Morphology

Sexual dimorphism is subtle in most species of cetaceans, with the majority of odontocete males being slightly larger than females, while the reverse is true for mysticetes. In odontocetes, this size disparity hints at a polygynous mating system, size and other secondary sexual characteristics being a likely display of condition in males. In most species, polygyny does not appear to be harem-like, where one male has exclusive access to several females. Instead, it appears that much mate sharing is taking place and that sexual partners may shift over a matter of hours, days, and longer times, thus combining elements of promiscuity or a multi-male/multi-female social system (Baker, 1985; Darling et al., 1983; Mobley and Herman, 1985; Norris et al., 1985; Tyack and Whitehead, 1983). Males in rut have very large testes and presumably are able to produce large quantities of sperm per ejaculation (Ridgway and Green, 1967; Kenagy and Trombulak, 1986). In these species, competition may therefore take place at the level of the sperm rather than in the direct competitive sense of a stronger male displacing a weaker one or gaining easier access to a female due to his strength or size (Kenagy and Trombulak, 1986; Brownell and Ralls, 1986). Nevertheless, size disparity may be important for communicating reproductive stamina from male to female or for sorting out access of males to females. Although male toothed whales in captivity are often aggressive, the establishment of "dominance hierarchies" has not been linked solidly to size (Tavolga, 1966). Sperm whale (Physeter macrocephalus) and killer whale males are very much larger than females of the same age. Sperm whale males are thought to battle for access to female groups (Caldwell et al., 1966; Shaler, 1873) or possibly to employ a "searching" strategy like that of African elephants (Loxodonta africana; Whitehead and Arnbom, 1987).

The reverse sexual dimorphism of the baleen whales may have to do less with social signalling than with the need for females to build up particularly large fat reserves while developing their young during long seasonal fasts.

Several cetacean males also have secondary sexual characteristics that clearly mark them as males. Male spinner dolphins have a postanal keel of connective tissue and a triangular dorsal fin (Perrin, 1972, 1975). The extent of development of these characteristics is different for different populations and reaches its extreme in the eastern race of spinner dolphin in the Pacific; older males have developed a large postanal keel and a dorsal fin which is so erect and pointed forwards that it appears to be stuck on backwards (Figure 16 of Perrin, 1972). A similar, though less extreme, situation exists in Dall's porpoise (Phocoenoides dalli), and adult males of this species can be readily distinguished (Jefferson, 1990). It has been suggested that the male's postanal keel (which exists in many species of small cetaceans) may help provide leverage to thrust the penis into the female, thus acting like a "penile anchor," or that it acts as a means of stimulating the female during courtship by bumping the keel against the female's genitals with a quick "snapping action" (T. R. Kieckhefer, pers. observ.). Many beaked whale (e.g., species of the genus Mesoplodon) males develop particularly large and modified paired teeth that are displayed outside the mouth and are thought to be used in male/male competition (Heyning, 1984; McCann, 1974; Mead et al., 1982), and Heyning (1984) has suggested that white areas on the lower jaws of male Mesoplodon carlhubbsi may function to exaggerate apparent tooth size. The male narwhal (Monodon monoceros) apparently makes use of its spiralled tusk for ritualized sparring (Best, 1981).

Less obvious characteristics consist of the right whale (Eubalaena australis) callosity pattern differences between males and females (Payne and Dorsey, 1983); the hemispherical lobe of female humpback whales (Glockner, 1983); curved flippers of old male belugas (Delphinapterus leucas; Vladykov, 1943); serrated leading flipper edges, mainly in male Commerson's dolphins (Cephalorhynchus commersonii; Goodall et al., 1988); extensive white scarring of adult Risso's dolphin (Grampus griseus; Leatherwood et al., 1982); and a host of other morphologic distinctions between adult males and females and between adults and young. We assume that these developmental and secondary sexual characteristics are involved in social signalling and in helping to regulate mating systems; however, presently we have little data to support this assumption.

Coloration

Much has been written about cetacean body coloration, which consists mainly of shades of white and black (e.g., Evans and Yablokov, 1983; Mitchell, 1970; Perrin, 1972; Yablokov, 1963), but little of this attempts to address visual signalling of body color as communication. Coloration exists in three basic forms (Yablokov, 1963): uniform, countershaded with dark dorsum and light ventrum, and disruptive. Uniform and countershaded coloration types are probably most useful for hiding. Largely uniform coloration, such as in blue whales (Balaenoptera musculus), pilot whales (Globicephala spp.), and white whales, may be related to feeding on small invertebrate prey or in deep or murky water where countershading is not necessary. Countershading would seem especially useful for "sneaking up" on larger fish and squid prey which have high visual acuity and rapid escape capabilities, as well as for avoiding predators. Bottlenose dolphins (Tursiops truncatus), harbor porpoises (Phocoena phocoena), and some of the rorqual whales (e.g., fin whales, Balaenoptera physalus, and minke whales, Balaenoptera acutorostrata) may be appropriate examples.

Strong countershading does not appear to be needed in murky inshore waters, as demonstrated by the muted contrasts of the harbor porpoise, bottlenose dolphin, Burmeister's porpoise (Phocoena spinipinnis), and the river dolphins. Munz and McFarland (1977) found a similar trend of reduced countershading of fishes in silt-laden or productive waters; they attributed the decrease to the more uniform lighting in these waters. Less heavily countershaded animals have lighter backs; therefore, more light reflects off the back, resulting in a better match of their surroundings, which have brighter horizontal light. In brief, countershading appears to hide the animals from prey and possibly predators, and is probably of little communicative value.

Disruptive coloration, as in patches of white on a dark background, may also serve to hide an animal behind an illusion of visual after-images each time the white patches alight (Marler and Hamilton, 1966). On the other hand, spots and longitudinal stripes may serve to break-up the outlines of the body as a whole. For example, at close range, the conspicuous black and white bars or stripes on terrestrial animals, such as zebras (Equus grevyi), do not suggest protective value. However, it has been resolved that at some critical distance or light level, the stripes blend into the background, thus camouflaging the animal (Colt, 1957; Marler and Hamilton, 1966; see Fig. 1). We believe that conspicuous disruptive patterns of numerous delphinids may have a similar function (for example, the Pacific white-sided dolphin, Lagenorhynchus obliquidens; striped dolphin, Stenella coeruleoalba; common dolphin, Delphinus delphis; and Fraser's dolphin, Lagenodelphis hosei), and at close range their distinctive markings may facilitate intraspecific signalling and schooling. This form of crypsis is most valuable in clear waters and complex habitats (such as coral reefs) or close to the surface where the flickering of down-dwelling sunlight entering

Fig. 1. An illustration of the principle of disruptive margins. As one backs away from the photo, the zebra (as opposed to the "pseudozebra") and the figures with disruptive margins disappear first. (After Cott, 1957)

the water can cause disruptively colored animals to be difficult to see (Munz and McFarland, 1977). McFarland and Loew (1983) point-out that the stripe patterns of some fishes, particularly those that live just below the water's surface, also may serve to camouflage the fish against the natural light-flickering action at shallow depths. Numerous dolphins have recurrent types of body marks (Mitchell, 1970). Common are spots, saddle patches, capes, and longitudinal striping, which are associated with those animals living in clear tropical and temperate pelagic waters where surface flickering may be of special importance. However, much of cetacean disruptive coloration probably has evolved for social signalling and perhaps to communicate false information to prey.

While many pelagic dolphins of the tropics have color patterns of varying shades of gray, those dolphins and porpoises that live in offshore areas in higher latitudes tend to have very bold patterns of distinct black and white areas (Heyning, 1988). There also is a tendency for those species that live in the largest schools and that most often engage in interspecies associations (e.g., species of Stenella, Delphinus, and Lagenorhynchus) to have the most complex color patterns of the small cetaceans. This complexity may be important for individual recognition in schools of many individuals, and for efficient species recognition in multi-species aggregations.

Social signalling by disruptive coloration is particularly likely for those species which have patches of white on a dark background that can be rapidly exposed or "flashed" at a conspecific. The belly patches of killer whales and right whales, for example, can be exposed or hidden by the tilt of the body; and the small axillary white patches of the Chilean dolphin (Cephalorhynchus eutropia) can be exposed or hidden by small movements of the flippers (Norris and Dohl, 1980b).

Commerson's dolphin has body coloration that probably obscures its true shape in poor light levels and murky waters; and the same situation is certainly true for Dall's porpoises, southern right whale dolphins (Lissodelphis peronii), hourglass dolphins (Lagenorhynchus cruciger), killer whales, and Hector's dolphin (Cephalorhynchus hectori). We suspect that these striking markings (e.g., a large white spot behind the eye) may serve to confuse predators and prey, perhaps by obscuring the position of the animal's mouth. But such blatant markings may also serve to coordinate activities of dolphins swimming side-by-side (Norris and Schilt, 1988) and may even serve as individual identification cues, for there are slight differences in angles and extents of coloration between most individuals (Mitchell, 1970; Yablokov, 1963).

Disruptive coloration is likely useful for interspecies recognition. It has been pointed-out (Norris and Dohl, 1980b) that the white snout tips of adult pan-tropical spotted dolphins (Stenella attenuata) contrast remarkably well with the black snout tips of spinner dolphins, with whom they often travel in multispecies aggregations. Indeed, one of us (B. Würsig) has seen both species together at a distance > 100 m under water, and all that could be seen upon first approach were the "flashlight noses" of the spotted dolphins. Upon closer approach, it was easy to make-out the species, despite a generally similar body morphology. It also is possible that the particularly enlarged postanal keel and high dorsal fin of adult male spinner dolphins, which travel with other species in the tropical Pacific, help to distinguish those animals not only as males, but emphatically as spinner dolphin males in multispecies aggregations (Perrin, 1972).

Many cetaceans are shaded differently when they are young, consistent with the general trend for birds and mammals. Coloration tends to be muted, and young animals usually blend into the surrounding habitat. Appropriate examples are Dall's porpoises (Jefferson, 1990) and northern right whale dolphins (Lissodelphis borealis; Leatherwood and Walker, 1979). A striking young-adult difference is shown by the spotted dolphin (Perrin, 1969). The young are a uniform or lightly countershaded gray, while the adult coloring consists of a profusion of black and white mottling. The difference in color probably helps to rapidly identify a calf or subadult as a nonreproductive member of the social unit and may thereby alleviate aggression and, perhaps, identify it as one of lower social status. The crisscross "hourglass pattern" found on common dolphins, and to lesser degree on some other delphinids (Mitchell, 1970), may serve to countershade and probably provide position cues to conspecifics of a school (Norris, et al., 1985). But it also may tend to hide the very young calves, for they ride above or below the midline of the body (usually above it, just behind the dorsal fin) at precisely the location where the hourglass pattern widens and presents a large, uniformly shaded patch similar in size to a newborn (Yablokov, 1963; Mitchell, 1970). In many other species as well, uniform coloration of the young may provide camouflage against the background color in a particular position along the mother's body.

Quite a few darkly shaded cetaceans that feed, at least partly, on squid in deep water have white noses, heads, or lips. The sperm whale (which also has a white inner mouth), many of the beaked whales (e.g., Cuvier's beaked whales, Ziphius cavirostris; melon headed whales Peponocephala electra; and pygmy killer whales, Feresa attenuata) are particularly good examples. Gaskin (1967) suggested that sperm whale lips might attract bioluminescent squid. Indeed, bioluminescence (probably transferred from squid) has been noticed on sperm whale lips; and the luminescent effect was intensified due to the whiteness of the lips (Gaskin, 1967). Many squid have light-sensitive vesicles that monitor light from themselves and others, and this adaptation may further serve to attract prey (Young, 1978). It is

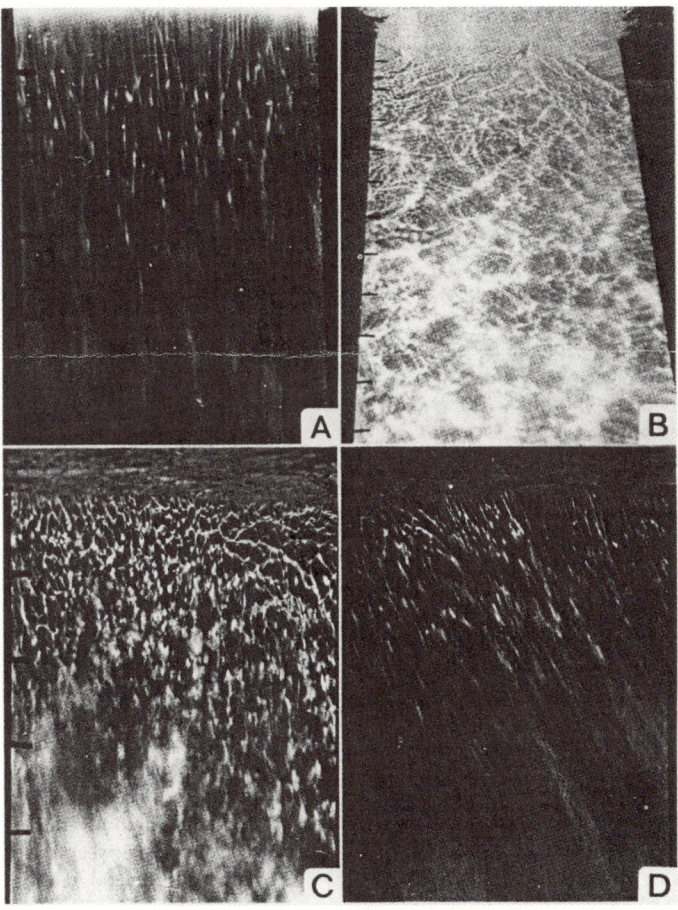

Fig. 2. Wave-induced flickering patterns produced below the surface. A. Vertical orientation of a white reflector with sun directly overhead. B. Reflector 30° from vertical, sun overhead. C. Reflector vertical, sun 45°. D. Reflector 45°, matched with sun altitude of 45°. Two basic patterns occur: vertical linear grating and reticulated mosaic. (After McFarland and Loew, 1983)

presently unknown whether this head coloration is predator-related in any of the squid feeders, including the sperm whale, but experiments with squid in controlled environments could provide better information.

A device for preventing the escape of prey may exist in the long, often partially white, flippers of humpback whales and in the white chin patches of bowhead whales (Balaena mysticetus). It has been suggested (Brodie, 1977) that humpback whales throw their flippers forwards to effectively corral prey during feeding lunges, and the flashing white of the flippers would tend to keep small schooling fishes and euphausiid prey concentrated in front of the lunging mouth. In addition, during a laboratory experiment in Southeast Alaska, one of the authors (T. R. Kieckhefer) found that krill (Euphausia pacifica) were attracted to the reflection of their own bioluminescence off a white sheet of paper placed at either end of a tank, which may provide a possible hint to an alluring nature behind the brilliant white undersides of humpbacks' flippers. The fin whale's asymmetrically colored lower jaw, with a white right lower lip and palate, may serve to keep prey concentrated in a feeding lunge. It also is possible that white marks near the mouth, including the white flipper bands of northern hemisphere minke whales, are used to coordinate actions in a group. Humpback whales and fin whales, which often lunge in groups (Baker, 1985; D'Vincent et al., 1985; Tershy, 1990) may coordinate lunges by these easily seen points of reference.

Furthermore, the disruptive white flank patches on numerous delphinid species, such as the dusky dolphin, may serve as a mechanism to attract and/or herd prey. Würsig and Würsig (1980) found that dusky dolphins in the South Atlantic fed primarily on large schools of anchovy (Engraulis anchoita), driving them up to the surface waters from below around mid-day, in contrast to the daily twilight or crepuscular periods that are critical for many predator-prey interactions (e.g., Major, 1977). A possible explanation may be that changes in orientation of the reflective flank patch of the dolphin to the sun and water surface will produce two primary patterns--vertical linear grating and reticulated mosaic (McFarland and Loew, 1983; Fig. 2). Vertical linear grating occurs when the sun is overhead (at mid-day) and the dolphin is oriented vertically or swimming on its side (reflecting its white underside) in the water column. However, with sun angled (at dawn and dusk) or when the reflective surface of the dolphin is angled upwards towards the water's surface, a reticulated mosaic scattering of light is created. We suspect that this flickering light off the white flank patches of numerous delphinid species not only may increase the efficiency of their visual communication system by enhancing the near-field contrast between schooling members, but also may be used in cooperative herding of fish and squid. This would operate by attraction of prey when using vertical linear grating and repulsion when reticulated mosaic light is employed, controlled by shifting and turning of the body axis (Fig. 3).

In a laboratory study, Koike (1985) found a remarkable difference in fish-school behavior of horse-mackerel (Trachuras japonicus) in response to reflected intermittent or flickering light compared to that in constant light. Flickering light provoked an erratic response, with fish observed chasing after the disappearing light, very similar to the earlier findings of Shaw and Tucker (1965) wherein schooling carangid fishes followed vertically moving light and dark lines, termed the "optomotor reaction."

The ecological importance of white flank patches is that they may serve as a valuable predatory tool in mimicking the attracting stimulus of schooling prey, since one of the primary sensory mechanisms that attract fish to other companions is vision (Breder, 1959; Morrow, 1948; Radakov, 1972; Shaw, 1961). There is, in fact, some evidence to indicate that conspicuous black and white patterns in piscivores can cause depolarization in schools of prey

Fig. 3. An example of a vertical grating along a dusky dolphin's flank.

fish (Wilson et al., 1987), thus enhancing their predatory efficiency. However, Pitcher et al. (1976) found that temporarily blinded fish could maintain a polarized state within a school despite their clumsiness and slowness of reaction. Detailed field work is necessary, possibly with the use of an underwater stereo-video (or "video-grammetry") system to measure distances between predator and prey, to determine if reflected flickering light off the white patches of a specific cetacean plays a role in coordinating and facilitating the capture of elusive prey.

Postures

When we think of communication, we generally assume a purposeful action to obtain some kind of result (Myrberg, 1981). Morphology and coloration are not behavioral communication tools unless used posturally by the animals. Dolphins use their shape and color patterns for communication by changing positions relative to conspecifics and, possibly, to predators and prey. We assume that large whales do this as well, but we have less information for them.

Changes of posture may signify turning or reorganizing of a school, with animals paying attention to the sides of nearby conspecifics. This use of patterns to synchronize social action is well known for schooling fishes (Partridge, 1981, 1982) and flocking birds (Major and Dill, 1978). It is even likely that fast response time to one individual's turning is mediated by a type of sensory integration system called the "chorus line effect"

(Potts, 1984), whereby animals pay attention not just to nearest neighbors but to conspecifics some distance away. Hence, they can anticipate the timing and extent of position changes necessary to remain synchronized members of the group (Norris and Dohl, 1980b; Norris and Schilt, 1988). As mentioned previously, the slight movement of a flipper, especially if enhanced by white on the flipper or a white patch behind it, may signify a turn or other change in position.

Dolphins and whales are almost always on the move and as a consequence are usually oriented in the same direction. It appears that in dolphins a direct facing towards another animal is usually an intentional act, charged with aggressiveness; it may be accompanied by opening of the mouth, exposing the teeth, and arching of the back. This arch presumably makes the approaching dolphin look somewhat bigger. Aggressive posturing reaches its extreme in the shark-like "S" posture of, for example, spinner dolphins, which "hang" in front of the object of threat (such as a human swimmer) with arched back, rostrum pointed towards the object, rapid head movements, and often jaw clapping. Because of the similarity to a threat position of several sharks, Norris et al. (1985) suggested that dolphins mimic sharks in order to make themselves appear more fierce. We think it doubtful that aggressive dolphins have evolved to behave like sharks per se, but we can visualize the aggressive stance as being related to the humpbacked stance of any threatening (and often fearful) terrestrial mammal.

Fig. 4. The uro-genital color pattern of a female Dall's porpoise (the head is to the left), showing black "fingers" pointing to the genital opening and to the mammary slits.

Submissive posturing is indicated by closed mouth, facing side or belly towards a conspecific, and a straight back. Norris and Møhl (1983) propose that if dolphins are able to stun prey by projecting intense sounds beamed at them, it may be a threatening act to point the snout towards another dolphin simply because of the potential of inflicting a painful sound; and that, therefore, normal interactions as well as submissive gestures should never be carried out head-on. In fact, such "echolocation manners" have been observed in spinner dolphins (Norris et al., 1985). It is difficult, as in many other mammals, to separate some submissive gestures from those of sexual intent or solicitation, where a belly might be submitted for view or touch or a side nuzzled with head or body. Sexual solicitation by displaying the genitals--in dolphins, often by rapidly swimming ahead of another animal, then turning or curving the body so that the genital area is exposed (Caldwell and Caldwell, 1977)--certainly is enhanced by the very disruptive marks many species have near their genitals. In fact, several species have patterns in the uro-genital area that appear almost as "pointing fingers," leading towards the genitals or mammaries (for example, killer whales, Bigg et al., 1987; Dall's porpoises, Morejohn et al., 1973; Fig. 4).

Visual signalling beyond coordinating movements and bouts of aggression or submission tends to grade into the tactile sense for dolphins. This is because dolphins are usually quite close together and are very tactile creatures with a sensitive skin (Ridgway, 1986). For example, the act of approaching a neighbor can rapidly turn into a bout of rubbing, with one animal placing a flipper along the side of another or--as often happens in spinner dolphins during social-sexual bouts--inserting the rostrum or tip of the dorsal fin into the genital slit of a partner (Norris and Dohl, 1980b).

It is likely that vision and tactile sense are especially closely linked in baleen whales, for their sheer size makes it important that they be close together in order to see a portion of each other during social interactions. It is very difficult for observers to differentiate between vision and touch when observing a group of up to one dozen rolling, twisting, turning right whales (Payne, in press) or bowhead whales (Everitt and Krogman, 1979) in mating aggregations. We suspect that the light-reflective callosity patterns of right whales, the white chin patches of bowhead whales, and the highly variable white marks of both species around the genitals function for recognition of position and of individuals.

OVERVIEW

We know lamentably little about visual communication in cetaceans, although we suspect that body shape, coloration, and body movements are important for recognition of species, sex, age, individual identity, behavioral intent, and emotional state. Our interpretation of visual communication patterns is clouded not only by the fact that little has been done in this field for odontocetes, and even less for mysticetes, but also by the fact that the capabilities of expression of cetaceans are so different from those of most four-legged terrestrial mammals with which we are familiar. Even with terrestrial mammals, much observational time needs to be expended before we can hope to describe some of the nuances of expressions and actions which make-up the visual communication modality. With cetaceans, we are still in a very primitive stage of describing blatantly obvious events, comparable to, in human terms, the frantic waving of a hand. When we can learn to recognize a subtle wink, blush, or eye movement with consistency of expression and intent, then we can truly say that inroads are being made into understanding the visual communication of these most fascinating creatures.

ACKNOWLEDGMENTS

We thank Myrna Benson for typing various drafts, Sheila Baldridge for help in obtaining references, and Carl Schilt, Jeanette Thomas, Ronald Kastelein, and Peter Rudolph for reviewing the manuscript. Discussions with numerous colleagues were pivotal in developing some of the ideas presented here, and we thank them for their input. This represents Contribution No. 3 of the Marine Mammal Research Program, Texas A&M University at Galveston.

REFERENCES

Baker, C. S. 1985. The population structure and social organization of humpback whales (Megaptera novaeangliae) in the Central and Eastern North Pacific. Ph.D. dissertation, Univ. of Hawaii, Honolulu, 306 pp.

Baker, C. S., and L. M. Herman. 1984. Aggressive behavior between humpback whales (Megaptera novaeangliae) wintering in Hawaii waters. Can. J. Zool. 62:1922-1936.

Bauer, G. B. 1986. The behavior of humpback whales in Hawaii and modifications of behavior induced by human interventions. Ph. D. dissertation, Univ. of Hawaii, Honolulu, 314 pp.

Best, R. C. 1981. The tusk of the narwhal (Monodon monoceros L.): interpretation of its function (Mammalia: Cetacea). Can. J. Zool. 59:2386-2393.

Bigg, M. A., G. M. Ellis, J. K. B. Ford, and K. C. Balcomb. 1987. Killer Whales: A Study of Their Identification, Genealogy, and Natural History in British Columbia and Washington. Phantom Press, 79 pp.

Breder, C. M., Jr. 1959. Studies on the social grouping in fishes. Bull. Am. Mus. Nat. Hist. 117:393-482.

Brodie, P. F. 1977. Form, function and energetics of Cetacea: A discussion. In R. J. Harrison (ed.), Functional Anatomy of Marine Mammals. Acad. Press, New York, Vol. 3, pp. 45-58.

Brownell, R. L., Jr., and K. R. Ralls. 1986. Potential for sperm competition in baleen whales. Rep. Int. Whal. Comm. (Spec. Iss. 8):97-112.

Caldwell, D. K., and M. C. Caldwell. 1977. Cetaceans. In T. A. Sebeok (ed.), How Animals Communicate. Indiana Univ. Press, pp. 794-808.

Caldwell, D. K., M. C. Caldwell, and D. W. Rice. 1966. Behavior of sperm whale, Physeter catodon L. In K. S. Norris (ed.), Whales, Dolphins, and Porpoises. Univ. of California Press, Berkeley, pp. 677-717.

Caro, T. 1988. Why do Tommies stott? Nat. Hist. 97:26-30.

Cott, H. B. 1957. Adaptive Coloration in Animals. Methaen and Co., London, pp. 68-113.

Darling, J. D., K. M. Gibson, and G. K. Silber. 1983. Observations on the abundance and behavior of humpback whales (Megaptera novaeangliae) off West Maui, Hawaii, 1977-1979. In R. Payne (ed.), Communication and Behavior of Whales. Westview Press, Inc., Boulder, CO, pp. 201-222.

Dawson, W. W. 1980. The cetacean eye. In L. M. Herman (ed.), Cetacean Behavior: Mechanisms and Functions. Wiley, New York, pp. 53-100.

D'Vincent, C. G., R. M. Nilson, and R. E. Hanna. 1985. Vocalization and coordinated feeding behavior of the humpback whale in southeast Alaska. Sci. Rep. Whales Res. Inst., Tokyo 36:41-47.

Estes, R. D. 1967. The comparative behavior of Grant's and Thomson's gazelles. J. Mamm. 48:189-209.

Evans, W. E., and A. V. Yablokov. 1983. Variability Cetacean Colour Pattern: A New Approach to Mammalian Colouration Study. Nauka, Moscow, U.S.S.R. (in Russian).

Everitt, R. D., and B. D. Krogman. 1979. Sexual behavior of bowhead whales observed off the north coast of Alaska. Arctic 32:277-280.

Ford, J. K. B., and H. D. Fisher. 1983. Group-specific dialects of killer whales (Orcinus orca) in British Columbia. In R. Payne (ed.), Communication and Behavior of Whales. Westview Press, Inc., Boulder, CO, pp. 129-161.

Gaskin, D. E. 1967. Luminescence in a squid and a possible feeding mechanism in the sperm whale. Tuatara 15:86-88.

Glockner, D. A. 1983. Determining the sex of humpback whales (Megaptera novaeangliae) in their natural environment. In R. Payne (ed.), Communication and Behavior of Whales. Westview Press, Inc., Boulder, CO, pp. 223-258.

Goodall, R. N. P., A. R. Galeazzi, and A. P. Sobral. 1988. Flipper serration in Cephalorhynchus commersonii. Rep. Int. Whal. Comm. (Spec. Iss. 9):161-171.

Heyning, J. E. 1984. Functional morphology involved in intraspecific fighting of the beaked whale, Mesoplodon carlhubbsi. Can. J. Ecol. 62:1645-1654.

Heyning, J. E. 1988. Whales past and present. In S. Taylor (ed.), The World's Whales: A Closer Look. American Cetacean Society, San Pedro, CA, pp. 51-58.

Jefferson, T. A. 1990. Sexual dimorphism and development of external features in Dall's porpoise (Phocoenoides dalli). Fish. Bull, U.S. (in press).

Kenagy, G. J., and S. C. Trombulak. 1986. Size and function of mammalian testes in relation to body size. J. Mamm. 67:1-22.

Kingdon, J. S. 1980. The role of visual signals and face patterns in African forest monkeys (guenons) of the genus Cercopithecus. Trans. Zool. Soc. Lond. 35:425-475.

Koike, T. 1985. A study on fish schools' reaction in response to intermittent light. Bull. Jap. Soc. Sci. Fish. 51:1097-1102 (translated by Chizuko Inoue Hallows).

Leatherwood, S., R. R. Reeves, W. F. Perrin, and W. E. Evans. 1982. Whales, dolphins and porpoises of the eastern North Pacific and adjacent Arctic waters. NOAA Tech. Rep. NMFS Circ. 444, 245 pp.

Leatherwood, S., and W. A. Walker. 1979. The northern right whale dolphin, Lissodelphis borealis Peale in the eastern North Pacific. In H. E. Winn and B. L. Olla (eds.), Behavior of marine animals, Volume 3: Cetaceans. Plenum Press, New York, pp. 85-141.

Lythgoe, J. N. 1979. Vision through scattering media. In J. N. Lithgoe (ed.), The Ecology of Vision. Clarendon Press, Oxford, pp. 112-127.

Madsen, C. J., and L. M. Herman. 1980. Social and ecological correlates of cetacean vision and visual appearance. In L. M. Herman (ed.), Cetacean Behavior: Mechanisms and Functions. Wiley, New York, pp. 101-147.

Major, P. F. 1977. Predator-prey interactions in schooling fishes during periods of twilight: A study of the silverside Pronesus insularum in Hawaii. Fish. Bull., U.S. 75:415-426.

Major, P. F., and L. M. Dill. 1978. The three-dimensional structure of airborne bird flocks. Behav. Ecol. Sociobiol. 4:111-122.

Marler, P., and W. J. Hamilton. 1966. Behavior functions of vision. In P. Marler and W. J. Hamilton (eds.), Mechanisms of Animal Behavior. Wiley, New York, pp. 357-396.

McCann, C. 1974. Body scarring on Cetacea-Odontocetes. Sci. Rep. Whales Res. Inst., Tokyo 26:145-155.

McFarland, W. N., and E. R. Loew. 1983. Wave produced changes in underwater light and their relations to vision. Environ. Biol. Fishes 8:173-184.

Mead, J. G., W. A. Walker, and W. J. Houck. 1982. Biological observations on Mesoplodon carlhubbsi (Cetacea: Ziphiidae). Smithson. Contrib. Zool. 344, 25 pp.

Mitchell, E. 1970. Pigmentation pattern evolution in delphinid cetaceans: An essay in adaptive coloration. Can. J. Zool. 48:717-740.

Mobley, J. R., and L. M. Herman. 1985. Transience of social affiliations among humpback whales (Megaptera novaeangliae) on the Hawaiian wintering grounds. Can. J. Zool. 63:762-772.

Morejohn, G. V., V. Loeb, and D. M. Baltz. 1973. Coloration and sexual dimorphism in the Dall porpoise. J. Mammal. 54:977-982.

Morrow, J. E., Jr. 1948. Schooling behavior in fishes. Quart. Rev. Biol. 23:27-38.

Munz, F. W., and W. N. McFarland. 1977. Evolutionary adaptations of fishes to the photic environment. In F. Crescitelli (ed.), Handbook of Sensory Physiology. Springer-Verlag, Heidelberg, pp. 193-274.

Myrberg, A. A., Jr. 1981. Sound communication and interception in fishes. In W. N. Tavolga, A. N. Popper, and R. R. Fay (eds.), Hearing and Sound Communication in Fishes. Springer-Verlag, New York, pp. 365-426.

Norris, K. S., and T. P. Dohl. 1980a. Behavior of the Hawaiian spinner dolphin, Stenella langirostris. Fish. Bull. 77:821-849.

Norris, K. S., and T. P. Dohl. 1980b. The structure and functions of cetacean schools. In L. M. Herman (ed.), Cetacean Behavior: Mechanisms and Functions. Wiley, New York, pp. 211-261.

Norris, K. S., and B. Møhl. 1983. Can odontocetes debilitate prey with sound? Am. Nat. 122:85-104.

Norris, K. S., and C. R. Schilt. 1988. Cooperative societies in three-dimensional space: On the origins of aggregations, flocks, and schools, with special reference to dolphins and fish. Ethol. Sociobiol. 9:149-179.

Norris, K. S., B. Würsig, R. S. Wells, M. Würsig, S. Brownlee, C. Johnson, and J. Solow. 1985. The behavior of the Hawaiian Spinner Dolphin, Stenella longirostris. Southwest Fish. Cent. Admin. Rep. LJ-85-06C, 213 pp.

Partridge, B. L. 1981. Lateral line function and the internal dynamics of fish schools. In W. N. Tavolga, A. N. Popper, and R. R. Fay (eds.), Hearing and Sound Communications in Fishes. Springer-Verlag, New York, pp. 515-522.

Partridge, B. L. 1982. The structure and function of fish schools. Sci. Am. 245:114-123.

Payne, R. In press. Behavior of Southern Right Whales. Univ. of Chicago Press, Chicago, Ill.

Payne, R., and E. M. Dorsey. 1983. Sexual dimorphism and aggressive use of callosities in right whales (Eubalaena australis). In R. Payne (ed.), Communication and Behavior of Whales. Westview Press, Inc., Boulder, CO, pp. 295-329.

Perrin, W. F. 1969. Color pattern of the eastern Pacific spotted porpoise, Stenella graffmani, Lonnberg (Cetacea: Delphinidae). Zoologica 54:135-142.

Perrin, W. F. 1972. Color patterns of spinner porpoises (Stenella cf. S. longirostris) of the eastern Pacific and Hawaii, with comments on delphinid pigmentation. Fish. Bull, U.S. 70:983-1003.

Perrin, W. F. 1975. Variation of spotted and spinner porpoise (genus Stenella) in the eastern tropical Pacific and Hawaii. Bull. Scripps Inst. Oceanogr. 21, 206 pp.

Pitcher, J. J., B. L. Partridge, and C. S. Wardle. 1976. A blind fish can school. Science 194:963-965.

Potts, W. K. 1984. The chorus-line hypothesis of manoeuvre coordination in avian flocks. Nature 309:344-345.

Purves, P. E., and G. Pilleri. 1973. Observations on the ear, nose, throat and eye of Platanista indi. Invest. Cetacea 5:13-57.

Radakov, D. U. 1972. Schooling in the Ecology of Fish (trans. from Russian by Israel Program Sci. Transl. Publ.). Wiley. New York, 173 pp.

Ridgway, S. H. 1986. Physiological observations on dolphin brains. In R. J. Schusterman, J. A. Thomas, and F. G. Wood (eds.), Dolphin Cognition and Behavior: A Comparative Approach. Lawrence Erlbaum Assoc., Publ., Hillsdale, New Jersey, pp. 31-59.

Ridgway, S. H., and R. Green. 1967. Evidence for a sexual rhythm in male porpoises, Lagenorhynchus obliquidens and Delphinus delphis bairdi. Norsk Hval.-Tid. 57:1-8.

Seyfarth, R. M., D. L. Cheney, and P. Marler. 1980. Monkey responses to three different alarm calls: Evidence of predator classification and semantic communication. Science 210:801-803.

Shaler, N. S. 1873. Notes on the right and sperm whales. Am. Nat. 7:1-4.

Shaw, E. 1961. Minimal light intensity and dispersal of schooling fish. Bull. Oceanogr. Monaco 1213:1-18.

Shaw, E., and A. Tucker. 1965. The optomotor reactions of schooling carangid fishes. Anim. Behav. 13:330-336.

Sherman, P. W. 1985. Alarm calls of Belding's ground squirrels to aerial predators: Nepotism or self-preservation? Behav. Ecol. Sociobiol. 17:313-323.

Tavolga, M. C. 1966. Behavior of the bottlenosed dolphin (Tursiops truncatus): Social interactions in a captive colony. In K. S. Norris (ed.), Whales, Dolphins and Porpoises. Univ. of California Press, Berkeley, pp. 718-730.

Tershy, B. R. 1990. Body size, diet, habitat use, and social behavior of four sympatric Balaenoptera whales. M. Sc. Thesis, Moss Landing Marine Laboratories, California.

Tyack, P. 1981. Interactions between singing Hawaiian humpback whales and conspecifics nearby. Behav. Ecol. Sociobiol. 8:105-116.

Tyack, P., and H. Whitehead. 1983. Male competition in large groups of wintering humpback whales. Behaviour 83:132-154.

Vladykov, V. D. 1943. A modification of the pectoral fins in the beluga from the St. Lawrence River. Naturaliste Canadien 70:23-40.

Walls, G. L. 1967. Adaptations to diurnal activity. In G. C. Walls (ed.), The Vertebrate Eye and its Adaptive Radiation. Hafner Publ. Co., New York, pp. 169-205.

Walther, F. R. 1969. Flight behaviour and avoidance of predators in Thomson's gazelle (Gazella thomsoni Guenther 1884). Behaviour 34:184-221.

Whitehead, H., and T. Arnbom. 1987. Social organization of sperm whales off the Galapagos Islands, February-April 1985. Can. J. Zool. 65:913-919.

Wilson, R. P., P. G. Ryan, A. James, and M. T. Wilson. 1987. Conspicuous coloration may enhance prey capture in some piscivores. Anim. Behav. 35:1558-1560.

Würsig, B., and M. Würsig. 1980. Behavior and ecology of the dusky dolphin, Lagenorhynchus obscurus, in the south Atlantic. Fish. Bull., U.S. 77:871-890.

Yablokov, A. V. 1963. Types of colour of the Cetacea. Bull. Moscow Soc. Nat. Biolog. 68:27-41 (translated by Fish. Res. Bd. Can. Transl. Series 1239).

Young, R. E. 1981. Color of bioluminescence in pelagic organisms. In K. H. Nealson, (ed.), Bioluminescence, Current Perspectives. Burgess Publ. Co., Minneapolis, Minn., pp. 72-81.

CONCLUDING COMMENTS ON VISION, TACTITION, AND CHEMORECEPTION

Karen Pryor

44811 S.E. 166 Street

North Bend, Washington, U.S.A. 98045

INTRODUCTION

Throughout history, people have marvelled at the physical adaptations that enable cetaceans, though they are mammals, not fish, to live an entirely aquatic life. Many of these adaptations are obvious externally; the hairless skin, the insulating blubber, the streamlined shape; loss of limbs and evolution of flukes, a new propulsion system; and the replacement of nostrils with blowholes, and related restructuring of the respiratory system.

Other, more subtle adaptations were not recognized until we began to keep cetaceans in captivity. The first astonishing discovery was the existence, at least in some small cetaceans, of a modified acoustic perception system for echolocation. The development of echolocation capability was accompanied by the development of new physical structures -- air sacs modified for sound production, the oil-filled "lens" of the melon, and skull and brain changes -- not seen in any terrestrial animal.

Perhaps because so much cetacean behavioral and sensory research has focussed on acoustics and related areas, it has become apparent only recently that other sensory systems in cetaceans are also highly modified. These systems may prove to be equally bizarre when compared to terrestrial equivalents, and also may include structural innovations with significant implications in both the applied and basic sciences.

CHEMORECEPTION

Current Knowledge

Conventional wisdom has long held that dolphins differ from terrestrial mammals in having no sense of taste or smell. This assumption has been based on some obvious anatomical facts: (1) As any neuroanatomist presented with a dolphin brain can see, the olfactory bulb is missing completely. (2) Although the newborn dolphin has tastebuds on the tongue, like other mammals, these structures disappear after the first few months of life; the adult tongue lacks tastebuds altogether. (3) Since the dolphin breathes air, it presumably does not sense chemicals in the water through the respiratory system, as a fish does. (4) Since external air passes directly into the pharynx without traversing the nasal passages and sinuses, the dolphin has no opportunity to smell even airborne chemical stimuli. In fact, the elaboration of acoustic sensing and communication in dolphins has been assumed to result from the absence of the chemical sensory channel which is so important to most other animals.

Nevertheless, from a pragmatic standpoint dolphins appear to be able to taste. Captive dolphins have definite food preferences, and are well-known to reject bad-tasting food, such as a fish in which an inserted vitamin capsule has broken open. Work in the Soviet Union has demonstrated that bottlenose dolphins (Tursiops truncatus) do have taste sensitivity (Kuznetsov, 1974; Sokolov and Kuznetsov, 1971). In a series of subsequent experiments with trained animals at the Naval Ocean Systems Center in Hawaii, Nachtigall and colleagues demonstrated that the bottlenose dolphin can indeed detect the four basic taste stimuli -- sweet, sour, bitter, and salt -- although they may not react to them as terrestrial mammals do (Nachtigall, 1986; Nachtigall and Hall, 1984).

The work of Kuznetsov, reported at this meeting, has extended greatly our understanding of chemoreception in dolphins (See Kuznetsov, this volume; Kuznetsov, 1988; Kuznetsov, 1989a). Drawing on published and unpublished findings, our Working Group summarized our present state of knowledge of chemoreception in dolphins.

Quasiolfaction

Far from being impervious to chemical stimuli, the bottlenose dolphin in fact possesses a highly effective secondary chemoreception system which Kuznetsov has labeled quasiolfaction. Chemical stimuli are perceived in pits on the back and roots of the tongue, through a system of closely-packed microscopic villi. A widened buccal cavity above the tongue roots surrounds the pharyngeal tube leading to the blowhole. Sea water is taken into this cavity and can be expelled, along with respired air, through the blowhole on exhalation. (This explains the occasional presence of real water, not just vapor, in the "spout" or exhalation of normal healthy dolphins. This phenomenon has long puzzled those who maintain dolphins in captivity, since the spray seems to be too ample in quantity to be simply water collected on the animal's dorsal surface, and yet certainly could not be coming from the lungs).

Possible roles of quasiolfaction

Our Working Group proposed four possible functions for adaptations for chemoreception in dolphins:

Food-finding. Since a large school of fish leaves a distinct chemical trail, quasiolfaction might be used in locating food; and chemoreception has the advantage of being long-term, in that the stimulus may be present for hours, rather than brief, as are acoustic and visual stimuli.

Locating other dolphins. The Soviet work indicates that dolphins are especially sensitive to biological substances, such as those found in mammalian urine and feces. Each dolphin typically urinates every ten minutes; dolphin schools therefore also leave a chemical trail or "road" in the water which could persist for many hours. Quasiolfaction might thus enable dolphins to locate, follow, or avoid other dolphins.

Reproduction. Quasiolfaction also almost certainly serves a reproductive role. Captive females in estrus become extremely attractive to males, without any observed external or behavioral changes, a circumstance parsimoniously explained by the release of chemical stimuli. Kuznetsov's studies of the perianal gland in male bottlenose dolphins suggest that reproductive communication is an important role of chemoreception in both sexes of dolphins, as in other mammals (Kuznetsov, 1989b). This gland, incomplete in juvenile males, is supplied with striated muscle fibers in adult males, and becomes enlarged during the breeding season. It is presumably under voluntary muscular control. The existence of this organ strongly suggests the use of chemical stimuli in reproductive behavior, perhaps in the same ways that male terrestrial mammals use scent to attract females, to warn or challenge other males, and to mark territory.

Sensing of Stress. Since bodily byproducts, especially urine, may contain chemical indicators of physiological stress, quasiolfaction might serve to alert dolphins to the physical state of other animals; similarly, Yablakov has reported that belugas evince alarm at the presence of blood in the water, apparently sensing it through chemoreception (Yablakov, 1961).

The substances involved in chemical signalling in dolphins are not yet known. Almost certainly the signalling substances are complex, and perhaps consist of combinations of several compounds. Ceruti, Fennessy, and Tjoa (1985) have identified a long list of compounds present in dolphine urine and feces. Probably the compounds actively used by dolphins are similar to those used for social communication in terrestrial mammals.

The evolution and neuroanatomy of the quasiolfaction system is not well understood. The system is an old one, in the evolutionary sense, and appears to have arisen early in the history of the Odontoceti. Theoretically, the development of the echolocation system displaced the ancestral olfaction system. The slower turbulent flow of water, compared to air, may have contributed to the development of a structure capable of capturing and holding chemical-bearing seawater. Current anatomical studies suggest that with the loss of the olfactory lobe, other areas of the brain have been usurped to function with the new chemoreceptors, and that the terminalis and trigaminal nerves may provide pathways between the quasiolfaction receptor system and the brain (Oelschlager, 1989).

Chemoreception: Directions for Research

Three principle areas of research were suggested: (1) basic biological research; (2) field and ethological studies, and (3) applied research. From a basic research standpoint, the discovery of a completely new sensory system, with previously unrecognized receptors and neuroanatomy, is of profound interest. Physiologists and anatomists in many areas of work might be interested in fundamental questions indicated by this discovery; anatomical material should be made available for research outside the field of cetacean studies. The nature of the chemical compounds involved in cetacean chemo-signalling also awaits basic research.

In behavioral research, the usual procedure consists of observing living animals and formulating hypotheses that are tested in the laboratory. In the case of dolphin chemoreception, this approach may be reversed; it would appear to be fruitful to test the experimental and anatomical laboratory findings, and the resultant hypotheses, with behavioral studies in captivity and in the field. Such studies might provide information about the existence and use of chemical "roads" in the sea by dolphins or migrating great whales; the transmittal of information such as stress levels by quasiolfaction; and the role of quasiolfaction in cetacean reproduction.

From an applied standpoint, the newly-recognized chemical sensing abilities of dolphins might be used, as we presently use the acoustic skills of trained cetaceans, in the open ocean. Dolphins might be able to sense gas and oil leaks in the water, and to locate the source of pollutants. If stress- or reproduction-related compounds prove to constitute alarm signals, such compounds might be used as continuous warning signals, to direct animals away from drift nets and similar hazards. Chemoreception studies also may be of value in further understanding of the tuna-porpoise bond in the eastern tropical Pacific purse-seine tuna fishery.

VISION

Current Knowledge

As with chemoreception, the conventional view of vision in dolphins has been that their eyesight is not good; that vision is not especially important to dolphins underwater; that they do not see well in air, and that perhaps they can see only moving objects. In fact, with the possible exception of some of the freshwater dolphins which live in muddy rivers, small cetaceans appear to have excellent vision, both in water and in air (Dawson, 1980; Herman, this volume; Nachtigall, 1986). Many have binocular vision over at least part of the visual field. Psychophysical studies have demonstrated that the bottlenose dolphin is well able to make fine discriminations of both moving and still patterns, in air and in the water. Underwater vision is more acute at short distances (1 m); at longer distances (2.5 m) vision is more acute in the air. While dolphins have few cones in their eyes, and are insensitive to primary colors, their spectral sensitivity is highest at the blue end of the spectrum (Dawson, Schroeder, and Dawson, 1987; Madsen and Herman, 1980).

A remarkable aspect of dolphin vision is the apparently imperfect consensuality of the two eyes. It may be possible for one eye to accommodate to bright light while the other accommodates partially to dim light, as a dolphin may need to do while swimming on its side, with one eye looking at the bright surface or into the air, and the other at the darkness below (Dawson, 1980). Some evidence suggests that the eyes may be able to move independently, and possibly even to focus independently.

Soviet studies have demonstrated that dolphins exhibit unihemispherical sleep. While one side of the brain is in deep sleep, the other side remains in light sleep or awake (Mukametov, 1989). During periods of sleep, one eye remains open and responsive to visual stimuli (it need not be the eye associated with the wakened side of the brain).

Anatomically, the dolphin eye differs in many respects from the eye of terrestrial mammals (Dawson et al., 1972). Intraocular pressure is extremely high. The optic nerve is very large, perhaps providing an ability to perceive visual stimuli extremely rapidly, or perhaps simply as an accommodation to the necessity of operating in a low temperature environment. The eye is coated with a thick mucus, a gel constructed of long-chain polysaccharides, which protects the cornea both from salt water and from air; it may also act as a lens for accommodation to in-air vision.

The dolphin eye appears to have two axes of gaze; in bright light the pupil narrows to a double slit, or double pinhole system, facilitating in-air visual acuity (Dawson, 1980; Herman et al., 1975). Studies of the retinas of several species of cetaceans show that the extremely large ganglia are concentrated in two areas in each eye, and that the locations of these areas differ from species to species (see Mass, this volume). The large ganglia, and interneural connections between them, are constituents of a visual system able to compensate for low light.

The positioning of the eye, and resulting field of vision, varies from species to species. Observations of captive animals show that bottlenose dolphins, for example, have approximately $180°$ of vision, forward, to the side, and back, but no visual range upward; however the false killer whale (Pseudorca crassidens) can look upward behind the head with both eyes, and the pigmy killer whale (Feresa attenuata) has binocular vision directly backward. Presumably these anatomical differences, like the species-specific areas of ganglion concentration in the retina, are related to the ecology of each species, but in ways that presently are not known.

The utility of these excellent visual capabilities is evident in water, where vision is used in orientation, in social communication, and in feeding (Madsen and Herman, 1980; Pryor, this volume; Würsig et al., 1989). Observation of animals trained to wear eye-cups for echolocation studies indicate that while targets may be located by echolocation, homing in on objects in the near field, i.e. the last meter or so, is accomplished visually.

The usefulness of excellent vision in air is perhaps more perplexing; for animals which spend most of their time submerged, it would seem that marginal vision would suffice. However, field observations of travelling dolphin schools indicate that animals often look above the surface briefly during respiration. These repeated glimpses allow them to exchange glances with each other, and to survey the environment for orientation and feeding clues. Coastal animals, for example, may be able to identify coastal features visually at some distance. Circling seabirds and aerial behavior of other dolphins may indicate the presence of prey; Würsig has reported dusky dolphins (Lagenorhynchus obscurus) turning from the coast and heading out to sea, toward the visible activity of distant dolphins, when no acoustic stimuli were received on hydrophones (Würsig and Würsig 1980). Many species of dolphins give highly specific aerial displays, such as the lateral revolutions of the spinner dolphin (Stenella attenuata) which could be perceived visually in air by conspecifics. Some aerial displays have acoustic components (perceivable over a short range, only) while others do not. Grampus griseus, for example, while capable of noisy breaches or leaps, occasionally hangs motionless with the tail in the air, in a silent but visually conspicuous display.

Vision: Directions for Research

Compared to our knowledge of vision in terrestrial animals, our fundamental knowledge of the cetacean's highly evolved and very different eye is "pathetic," in the words of one researcher. We do not know if the visual fields indeed correspond to the concentrations of ganglia in the retina; we do not know exactly how the eye accommodates to focus in both air and water, and we know almost nothing about the extent of and mechanisms involved in independent functioning of the two eyes. Future research in cetacean vision should include comparative measures of vision in species other than the bottlenose dolphin. We need to combine psychophysical and anatomical studies and put them into an ecological context. Key species should be chosen from broad ecological divisions: mysticetes vs. odontocetes, pelagic vs. coastal animals, and riverine vs. oceanic. New technologies may allow more sophisticated investigations both on the anatomical and psychophysical levels.

Understanding cetacean vision may lead to direct practical benefits, such as an explanation for how the dolphin tolerates intraocular pressures which would cause glaucoma in humans. The processing of visual information in dolphins may differ significantly from image-processing in the human eye, given a system in which even one eye can apparently look at two objects at once. Vision in cetaceans may provide a new kind of biological model for the design of mechanical vision enhancement devices (in low light, for example) and for the further development of computer vision.

TACTILE SENSES

Current Knowledge

Experience with captive animals suggests that dolphins have highly sensitive skin (although perhaps no more so than other mammals). Mechanoreception enables dolphins to sense turbulence and water flow, and to position themselves with great accuracy in the water (see Pryor, this volume). Ridgway reports that the surface of the dolphin's body, while apparently smooth, is in some areas surfaced with extremely fine ridges, like our fingertips (Ridgway and Carder, 1989). The extent and patterning of these ridges varies from species to species; whether they serve a tactile function, as do our fingertip ridges, or are merely of mechanical advantage, is not known.

Probably dolphins are sensitive to changes in their own boundary layer; this sense would contribute to the facility with which dolphins can position themselves relative to another animal or to a hard surface such as the bow of a moving ship. Many species of cetaceans have been observed swimming in echelon formation, one benefit of which presumably is a hydrodynamic increase in swimming efficiency. Dolphins can perform remarkable feats through mechanoreception: just in play, a dolphin can carry quite a large, unwieldy object, such as a wooden stick, on rostrum or flipper for hours, balancing it against the sensed water flow. Dolphin calves are able to position themselves against their mother's flank in such a way as to be able to coast within the pressure wave caused by the mother's forward motion; neonate calves in this location need not move their tails at all, except when surfacing to breathe, even when the mother is swimming rapidly. It has been suggested that the small vibrissae present in calves may help in sensing the right position. The vibrissae however may be completely involuted by birth time, in bottlenosed dolphins, while persisting through the first month of life in harbor porpoise (Phocoena phocoena).

Tactile Senses: Directions for Research

Research on tactile sensitivity or mechanoreception is hampered by our lack of knowledge as to the location and nature of receptors (which might be small and widely scattered, like heat receptors in human skin) and by an absence of knowledge of the neural pathways and areas of the brain involved in processing this information. Consequently, marine mammal researchers are studying mechanoreception in pinnipeds, a group equipped with stout vibrissae which are used extensively as tactile organs (Kastelein, this volume). These stout structures act as a multi-organ palp with which seals and walruses can identify shapes and locate small objects and prey items. An understanding of the function and neural pathways of these accessible mechanoreceptors may lay the groundwork for research in the perhaps more subtle and extensive mechanoreception systems in cetaceans.

CONCLUSIONS AND RECOMMENDATIONS

Our Working Group on non-acoustic sensory systems included researchers from the United States, the European community, and the Soviet Union. While their published studies have fueled each others' imagination for decades, many were meeting in person for the first time. The exchange of current research results was immeasurably enriched by this opportunity for informal discussion, speculation, and exploration of ideas.

Cetacean research has proceeded to a level at which specific problems are no longer the province of one researcher, one laboratory, or even one nation. Remarkable new findings in chemoreception and vision, with extensive biological implications, have developed through many simultaneous programs. Now what is needed are ways of fusing our efforts. Attention needs to be given to the economics of various research approaches, as sensory-system research requires ingenuity if it is not to be time-consuming and expensive.

International cooperation between laboratories is extremely important. Researchers on different continents have access to different species of animals. Organizations have different areas of expertise, funding, and equipment, and differing constraints on research. For example, studies of brain function in the Soviet Union are made possible by the use of internally-monitored animals not available in the United States. Techniques for behavioral and psychophysical studies have been most highly developed in the United States, but are transferable. And open-ocean research is feasible in some locations and not others.

The Working Group made specific recommendations regarding chemoreception research. Since quasiolfaction appears to be a unique sensory capability, the mechanisms of which are poorly understood, the questions are broad and varied. The Working Group proposed that the next step should be a small and select international meeting of scientists working on marine mammal chemoreception, in conjunction with scientists actively studying chemoreception and related topics in ecologically or morphologically related species. Such a meeting might greatly increase cost-effectiveness and speed of results, through the development of joint research programs among several laboratories and nations.

The Working Group also recommended four areas of increased communication within the international scientific community;

<u>Informal Exhange and Contact</u>. The meeting reported on herein was viewed as highly valuable by all participants because of the opportunity to talk, question, discuss, and exchange information in an informal setting. A desireable next step would be the organization of exchange visits between investigators from marine mammal laboratories in different countries, with the goals of collective enhancement of knowledge, and of development of new and mutually productive research programs.

<u>Replication Experiments</u>. Cooperation would be desirable in setting up more extensive programs for replicating other laboratories' research, to confirm results, particularly in the areas of vision and chemoreception.

<u>Exchange Visits</u>. Development of visiting investigator and joint research programs should be encouraged. The Working Group pointed out, however, that the welcoming of a visiting investigator to any facility should be preceded by ample planning and communication; much time is lost if necessary equipment is not available or if the visitor's expectations and needs were not understood or cannot be met.

Future Conferences. In addition to a small and select meeting on chemoreception and quasiolfaction, planning should begin immediately for future multi-disciplinary cetacean conferences. As far as possible these conferences should include representation from the whole international marine mammalogy community. The planners should also strive to include, as speakers and as panelists, influential participants from related disciplines outside the marine mammal field, such as physiology, neurology, systematics, ecology, energetics, and so on. Such participation will help to disseminate knowledge and insights achieved by marine mammalogists; and will also broaden the resources available to us from other fields.

REFERENCES

Ceruti, M. G., Fennessey, P. V., and Tjoa, S., 1985, Chemoreceptively active compounds in secretions, excretions and tissue extracts of marine mammals. Comp. Biochem. Physiol., 82A(3): 505-514.

Dawson, W. W., 1980, The cetacean eye, in: "Cetacean Behavior: Mechanisms and Functions," L. M. Herman, ed., John Wiley and Sons, New York.

Dawson, W. W., Birndorf, I. A., and Perez, J. M., 1972, Gross anatomy and optics of the dolphin's eye (Tursiops truncatus), Cetology, 10: 1-11.

Dawson, W. W., Schroeder, J. P., and Dawson, J. F., 1987, The ocular fundus of two cetaceans, Marine Mammal Science, 1987, 3(1): 1-13.

Herman, L. M., Peacock, M. F., Yunker, M. P., and Madsen, C. J., 1975, Bottlenosed dolphin: Double-slit pupil yields equivalent aerial and underwater diurnal acuity, Science, 189: 650-652.

Kuznetsov, V. B., 1974, A method of studying chemoreception in the Black Sea Dolphin (Tursiops truncatus), in: "Morphologiya, Fiziologiya, I Akustika Morskikh Mlekopitayushchikh," V. Ye. Sokolov, ed., Nauka, Moscow.

Kuznetsov, V. B., 1988, Problem of olfaction reduction in odontoceti toothed whales, Zh. Obsh. Biol., 49(1): 128-135.

Kuznetsov, V. B., 1989a, Chemical senses of dolphins - quasiolfaction, Abstract, Fifth International Theriological Congress, Rome.

Kuznetsov, V. B., 1989b, Chemoreception and communication in dolphins, Abstract, Fifth International Theriological Congress, Rome.

Madsen, C. J., and Herman, L. M., 1986, Social and ecological correlates of cetacean vision and visual appearance, in: "Dolphin Cognition and Behavior: A Comparative Approach," R. J. Schusterman, J. A. Thomas, and F. G. Wood, eds., Lawrence Erlbaum Associates, Hillsdale, NJ.

Mukhametov, L. M., 1989, Sensory contact with the environment during sleep in dolphins, Abstract, Fifth International Theriological Congress, Rome.

Nachtigall, P. E., 1986, Vision, audition and chemoreception in dolphins and other marine mammals, in: "Dolphin Cognition and Behavior: A Comparative Approach," R. J. Schusterman, J. A. Thomas, and F. G. Wood, eds., Lawrence Erlbaum Associates, Hillsdale, NJ.

Nachtigall, P. E., and Hall R. W., 1984, Taste reception in the bottlenosed dolphin, Acta Zoologica Fennica, 172: 147-148.

Oelschlager, H. A., 1989, Evolutionary morphology and acoustics in the dolphin skull, Abstract, Fifth International Theriological Congress, Rome.

Ridgway, S. H., and Carder, D. A., 1989, Tactile sensitivity, somatosensory responses, skin vibrations, and the skin surface ridges of the bottlenosed dolphin, Tursiops truncatus, Abstract, Fifth International Theriological Congress, Rome.

Sokolov, V. Ye, and Kuznetsov, V. B., 1971, Chemoreception in the Black Sea dolphin (Tursiops truncatus), Dokl. Acad. Nauk. SSSR, 201: 990-1000.

Würsig, B., Jefferson, T., and Kieckhefer, T., 1989, Visual displays for communication in cetaceans, Abstract, Fifth International Theriological Congress, Rome.

Würsig, B., and Würsig, M., 1980, Behavior and ecology of dusky porpoises, Lagenorhynchus obscurus, in the south Atlantic, Fish. Bull., U.S., 77: 871-890.

Yablokov, A. V., 1961, the "sense of smell" in marine mammals, Trudy Sovesch. Ikhtiol. Komm. Akad. Nauk SSSR, 12: 87-93.

ACOUSTIC BEHAVIOR OF MYSTICETE WHALES

Christopher W. Clark

Cornell University
Laboratory of Ornithology
Bioacoustics Research Program
Sapsucker Woods Rd.
Ithaca, New York 14850 USA

INTRODUCTION

There are eleven species of mysticetes (baleen whales), and sounds have been recorded from all but the pygmy right whale (Caperea marginata). The greatest amount of acoustic information has been gathered for the bowhead (Balaena mysticetus), gray (Eschrichtius robustus), humpback (Megaptera novaeangliae), and right (Eubalaena australis and E.glacialis) whales, because they are coastal, relatively vocal, and more social among the mysticetes. The more pelagic species, which include the blue (Balaenoptera musculus), fin (Balaenoptera physalus), Bryde's (Balaenoptera edeni), sei (Balaenoptera borealis), and minke (Balaenoptera acutorostrata) whales, are more difficult to observe and are less vocal than the coastal species. The most recent review of mysticete sounds was presented by Thompson et al. (1979; but see Winn and Perkins, 1976). Herman and Tavolga (1980) discuss mysticete sounds from the perspective of communication. Ridgway and Harrison (1985) provide some further descriptions of sound production for some of the mysticetes, while Payne (1983) presents a number of chapters on specific aspects of acoustic behavior in southern right whales and humpback whales.

Overall, knowledge of sound repertoires and acoustic functions in the different species is uneven. This is a result of differences in our access to the whales, in the amount and sophistication of the equipment required to study their acoustic behavior, and in the backgrounds and motivations of the different investigators.

This paper is not intended to be an in-depth review of existing descriptions of mysticete sounds, since that mostly has been covered by earlier reviews. Instead, I will begin with a synthesis of mysticete sounds and acoustic repertoires, and present some new material on acoustic behavior, particularly as it relates to communication.

TYPES OF MYSTICETE SOUNDS

Mysticete sounds can be divided into two major classes; non-vocal and vocal sounds. Non-vocal sounds include blow, slap, and miscellaneous sounds. Vocal sounds include calls and songs.

Blow sounds are produced during expiration and inspiration of air through the nares. Blows typically are noisy, broadband (up to 2 kHz), low intensity (not quantified) signals lasting several seconds generated in air at one or both nostrils during normal respiration. However, the acoustic characteristics of blow sounds can vary a great deal depending upon the social context (e.g. sexual activity, agonistic encounter, harassment) and whether they are produced underwater, or close to the air-water interface. There is evidence that some blow sounds are communicative (Watkins, 1967; Clark, 1983; Tyack and Whitehead, 1983; Watkins and Wartzok, 1985; Silber, 1986).

Slap sounds are produced at the surface of the water as a result of aerial displays (breaching, lob-tailing, and pectoral-slapping) or below the water as a result of two surfaces being struck against each other. Slap sounds typically are broadband (up to 4 kHz), intense (not quantified), short duration (< 0.2 s) signals with sharp acoustic onset (< 1-2 ms risetime). Some slap sounds are assumed to be communicative (Clark, 1983; Tyack and Whitehead, 1983; Silber, 1986)

Miscellaneous sounds are adventitious sounds (e.g. baleen rattle, flatulence, or rubbing against an object) and are considered to have no intentional communicative function.

Vocal sounds do not imply that sounds are produced by vocal cords, but rather by some internal mechanism under voluntary control involving the mouth cavity and respiratory system, especially the larynx. Under situations when the whale producing the call has been under close observation, no air is seen escaping from the nares or mouth, implying that air is recycled during sound production (Clark, 1982). There is evidence that calls and songs are communicative (Clark and Clark, 1980; Tyack, 1981, 1983; Clark, 1983; Clark et al., 1986; Mobley et al., 1988), and some circumstantial evidence that calls aid in orientation and navigation (Ellison et al., 1987; George et al., 1989).

Calls often can be subdivided into call types based upon the acoustic features most salient to the human ear and eye. Calls considered as belonging to the same call type do not necessarily share similar biological functions, especially in cases where the calls are complex and the discriminating features subjective. There are three general types of mysticetes calls: 1) simple calls; 2) complex calls; and 3) clicks, pulses, knocks, and grunts.

Simple calls are usually low-frequency, frequency-modulated (FM) signals with narrow instantaneous bandwidth, and often are referred to as moans. In some cases simple calls contain a number of harmonics or some amount of amplitude modulation (AM), but the band of principle energy is below 1000 Hz.

Complex calls are broadband, pulsive signals which consist

of variable mixtures of amplitude modulation of noise and/or a frequency-modulated fundamental. Typical bandwidths for complex pulsive signals are in the 500-5000 Hz range. Complex calls often are described as sounding like screams, roars, and growls.

Clicks, pulses, knocks and grunts are short duration (< 0.1 s) signals with little to no frequency modulation. Narrowband, broadband, low-frequency and high-frequency signals have been reported. Clicks and pulses usually refer to very short (< 2 ms) signals in the 3-31 kHz range, while grunts and knocks refer to longer (50-100 ms) signals in the 100-1000 Hz range. Clicks and pulses are by far the most controversial signals described for mysticetes since recordings similar to the original ones have never been obtained. These types of sounds can also be recorded as an artifact of equipment responses, so extreme caution is needed when assigning them to whales (Watkins and Wartzok, 1985).

A song refers to sequences of notes occurring in a regular sequence and patterned in time (see Payne and McVay, 1971). In most cases song notes are recognizably different from calls, both aurally and visually (spectrograms).

DESCRIPTIVE OVERVIEW

All mysticetes that have been recorded, with the exception of the sei whale, produce at least some form of simple call. Often these signals are very intense, with estimated levels above 180 dB (re 1 µPa at 1 m). Some of the most remarkable of these low FM calls are those from the blue and fin whales, whose signals essentially are infrasonic (blue, 14-20 Hz; fin, 18-23 Hz) and extremely well-adapted for long-range transmission (Payne and Webb, 1971). Simple calls for the other species cover a broader bandwidth; bowhead (50-400 Hz), Bryde's (75-245 Hz), gray (20-200 Hz), humpback (50-800 Hz), minke (115-160 Hz), and right whales (50-400 Hz). For the bowhead and right whales, simple calls have been divided into different types depending on the call's FM contour.

Complex calls have been recorded from bowhead, humpback and right whales. There is a remarkable similarity in the complex calls of the three species, with greatest similarity between the calls of the more closely related bowhead and right whales.

Descriptions of call repertoires for bowhead, fin, gray, humpback, and right whales probably are complete. The one caveat to this optimistic statement is that there are no published reports on sounds from bowhead whales during their breeding season (January - March) in the Bering Sea or from sexually active fin whales. The call repertoire in the bowhead, humpback and right whales is believed to represent a continuum of signals ranging from simple, FM-moans to complex mixtures of AM and FM sound (Clark, 1982; Silber, 1986; Würsig and Clark, 1990). The repertoire of the gray whale consists of simple calls and broadband knocks and pulses (Thompson et al., 1979; Dahlheim, 1984) while the repertoire of the fin whale consists of a small variety of simple calls and broadband low-frequency pulses (Watkins, 1981; Edds, 1988). There are no records of complex calls as defined here from either gray or fin whales.

The call repertoires for blue, Bryde's, sei, and minke whales probably are not documented fully due to limited opportunities for good recordings and observations.

In some cases, there is not total agreement as to whether the sounds attributed to a species are identified correctly. This is especially true for high-frequency clicks and pulses recorded in the presence of blue, Bryde's, fin, minke and sei whales (Beamish and Mitchell, 1971, 1973; Thompson et al., 1979).

Songs have now been reported for blue (Alling and Payne, 1990), bowhead (Ljungblad et al., 1982), fin (Watkins et al., 1987), and humpback whales (Payne and McVay, 1971). Other balaenopterids might sing relatively simple songs like the fin and blue whales, but there are too few recordings from these species to say definitively whether they sing. Both gray and right whales have been recorded extensively during their mating seasons but no sounds resembling song have been recorded; it is assumed these species do not sing.

BEHAVIOR AND COMMUNICATIVE FUNCTION

It is assumed generally that almost all sounds produced by whales serve some communicative function, yet only within the last decade have attempts been made to study specifically acoustic communication in mysticetes. Initial recording efforts were aimed at documenting mysticete sounds, since they were largely unknown, and not designed as long-term studies on communication.

Research on acoustic communication has concentrated on coastal species, because they are more accessible, more social and provide a greater spectrum of activities to observe with richer data, and the cost is much less than for ship-based studies. Approaching the topic of communication in large whales is difficult. To demonstrate communication one must show that sounds from one whale modify the behavior of another whale in a predictable manner, and this requires following the behaviors of individuals and correctly identifying the sounds produced by those individuals. It is not by chance that the first two species on which long-term research efforts have been successful, right and humpback whales, are the two in which individuals are recognized easily from photographs (Katona et al., 1979; Payne et al., 1983).

Four species have received considerable attention relative to acoustic communication in the last decade. These have been the bowhead (Clark and Johnson, 1984; Cummings and Holliday, 1985; Clark et al., 1986; Clark and Ellison, 1988), fin (Watkins, 1981; Watkins et al., 1987; Edds, 1988), humpback (Tyack 1981, 1983; Chabot, 1984; Silber, 1986; Mobley et al., 1988), and southern right (Clark and Clark, 1980; Clark, 1982, 1983) whales. Some effort also has been made to study the blue (Alling and Payne, 1990; Edds, 1982), gray (Dahlheim, 1984), and minke (Edds, 1980) whales.

The following are brief summaries of research efforts as they relate to acoustic communication:

Blue Whale

Work on blue whales in the Indian Ocean has revealed that they sing (Alling and Payne, 1990). Their songs consist of four distinct notes, which together last about two minutes. The first, second, and fourth notes are pulsive and the third note is a pure tone. One singer was recorded in each of two different years and the song was basically the same for both years. Long, low-frequency calls, similar to those from the Pacific Ocean, were reported by Edds (1982) from the St. Lawrence River, Canada.

Fin Whale

Studies of the fin whale are a noteworthy exception to the paucity of behavioral research accomplished on pelagic species. Watkins and co-workers have done an extraordinary study on fin whale acoustic behavior by analyzing a massive data set of 20-Hz sounds from the North Atlantic, (Watkins, 1981; Watkins et al., 1987). These sounds are produced in stereotypic series with regular sequences of repetition and periodic rests between sequences. The signals sometimes are observed to stop when the calling whale is approached by other fin whales, a reaction similar to that observed in humpbacks (Tyack, 1981) and southern right whales (Clark, 1983). From the association of the occurrence of the 20-Hz signals with the finback reproductive season, Watkins et al. (1987) tentatively conclude that calling whales are males, and that the signals function as a form of song. They speculate that differences in the repetition patterns and intervals between the 20-Hz signals from different geographic areas could reflect separate populations, while variations in signal detail from the same local area could indicate that either individual whales produce different signals or whales change signal characteristics over time.

For fin whales in the Gulf of St. Lawrence, Edds (1988) showed that characteristics for downswept signals could be different depending upon the number of whales in a group, but discriminant analysis could not distinguish between individual whales based on specific signal features of their calls. There is some evidence associating sounds other than the 20-Hz downsweep with particular contexts (Watkins, 1981; Edds, 1988). Higher frequency calls are heard mostly when two or more whales are in close proximity, a context for the fin whale which could be labelled as social. Low-frequency rumbles occur when two fin whales approach each other, during close approaches to boats and during social interactions; contexts which both authors suggested were agonistic.

Humpback Whale

Studies on the acoustic behavior of humpback whales have concentrated on their long, complex songs (Payne and McVay, 1971; Winn and Winn, 1978; Winn et al., 1981; Payne, 1983; Payne and Payne, 1985; Matilla et al., 1987; McSweeney et al., 1989). The results are well-known and show that: (1) songs from the Pacific and North Atlantic populations are different, (2) whales within a population sing the same basic song in any one year, (3) song changes occur during the season when females are calving and presumably mating, (4) singers are almost always males, and (5) some singing occurs during the summer and fall. The function of song has been deduced by observing the whales and their reactions to playback of song and other conspecific sounds (Tyack, 1981, 1983; Mobley et al., 1988). Interestingly, song does not attract females or males. In fact, the most attractive acoustic signals are the calls recorded from groups of humpbacks feeding in southeast Alaska (Mobley et al., 1988). Darling (1983) and Tyack (1982) agree that song serves a reproductive function, but it is not clear whether males sing to advertize their sexual status and location to females, to dominate and displace other males, or both.

Relatively little work has been carried-out on other types of humpback sounds (in Newfoundland by Chabot, 1984; in Hawaii by Silber, 1986), and these studies have concentrated on active groups. Social sounds in Hawaii are recorded almost exclusively from large aggregations that engage in bouts of surface activities, and never from mother/calve pairs or single adults. Silber (1986) viewed the repertoire as a continuum from simple to complex calls similar to that of right whales. There was significant correlation between group size and call rate, and there was a tendency for call rate to increase after a new whale joined the group (Silber, 1986). Unlike right whales, it was common to hear several whales vocalizing simultaneously. This would suggest that males, not females, were responsible for many of the sounds from active groups.

Humpbacks often are seen striking each other with flukes and pectorals, and underwater exhalations of long bubble streams also are observed in agonistic situations (Tyack and Whitehead, 1983; Baker and Herman, 1984). Humpback aerial displays could produce sounds that might serve as threats (Whitehead, 1985). But underwater bubble displays, although they do produce some sound, predominantly seem to be visual. Humpbacks do make long whistle-like blow sounds (Watkins, 1967) that have been associated with disturbance (Baker and Herman, 1984) and could be a form of threat.

Southern Right Whale

The first descriptions of the southern right whale repertoire came in the early 1970's (Payne and Payne, 1971; Cummings et al., 1972) from studies on a population of 500-600 whales which frequented the bays of Peninsula Valdes, Argentina. During this period Payne developed two critically important methods for studying whales. He adapted the surveyor's theodolite as a tool for visually tracking the movements of whales, and he demonstrated conclusively that whales could be re-identified over many years from photographs. These two techniques now are routinely used in studies on cetaceans.

In the mid 1970's, I began a two year project specifically designed to study acoustic communication in the southern right whales of Golfo San Jose, Peninsula Valdes. By using the theodolite and photo-identification techniques of Payne, and a phased hydrophone array mounted in the gulf (Clark, 1980), sounds were associated with whales in real time. The final results of this acoustic study provided a detailed analysis of the southern right whale's acoustic repertoire (Clark, 1982), and some understanding of their acoustic communication. The repertoire is a continuum consisting of two functional subdivisions; a set of simple FM, discrete calls associated with resting and transitting whales, and a set of highly variable, intergraded signals associated with groups of active whales (Clark, 1983). The most common form of discrete call is an FM upsweep, or "up-call". All whales produce this call, including newborn calves. Whales counter-call with up-calls over many miles, and join other whales who call back. Calling between the two ceases once they are together. From this I concluded that the up-call was a contact call which served to keep animals in acoustic range and bring them together. Subtle features of this contact call might encode information such as the caller's identity, but I have been unsuccessful in finding systematic differences in the shapes of up-call frequency contours that were associated with different individuals (unpublished data).

Animals in close, active (often sexually active) groups produce a much richer set of acoustic signals, and there is good agreement between the complexity of the social context (group size, number of males, sexual activity) and the complexity of signals. These results agree with acoustic grading strategies as discussed by Morton (1977). Active group sounds typically are produced in a rapid series with only one whale in the active group vocalizing at any one time.

Blow sounds also play a role in southern right whale communication (Clark, 1983). There are distinct varieties of blow sounds, and these often clearly are associated with social contexts. Resting or swimming whales sometimes make loud, harsh blow sounds when they are disturbed by dolphins (Lagenorhynchus obscurus, Tursiops truncatus) or sea lions (Otaria flavescens). Females in sexually active groups or mothers whose calves are kidnapped temporarily by another whale make this same type of harsh blow. These sounds are so distinctive and loud that we often used them to identify the presence of an active group that was not immediately visible from our observation hut.

Slap sounds resulting from aerial displays are remarkably variable. The loudness of slap sounds from successive breaches, lob-tails or pectoral-slaps is not predictable, but appears to vary depending upon the orientation of the whale as it strikes the surface. Variability in the level of successive breaches also has been noted for gray whales by Dahlheim (1984), for fin whales by Watkins (1981), and for bowhead whales by Würsig et al. (1984). Most aerial displays are performed by single, mildly active whales. The context in which these displays are performed are numerous, and whether they serve a communicative function remains uncertain (Whitehead, 1985). The most intriguing slap sound is referred to as a "gunshot" slap (Cummings et al., 1974; Clark, 1983). This sound is very intense but, is not produced by a whale striking the surface of the water with its body. Instead this slap sound is produced under water. Gunshot slaps have been associated with two whales slapping their bellies together, a lone adult female thrashing at bottlenose dolphins with her rostrum well-out of the water or fully active (including sexually active) groups (Clark, 1983). Because of these contexts and the auditory pain caused by the sound when heard by a human, I believe that gunshot slaps are threats. Their similarity to the loud explosive pulses of certain odontocetes (Norris and Møhl, 1983) is intriguing.

Behavioral responses of right whales to playback experiments (Clark and Clark, 1980) further support the conclusion that calls are communicative; whales responded to playbacks of conspecific sounds by orienting and moving toward the loudspeaker and calling as they approached.

Northern Right Whale

Acoustic studies on southern right whales are no longer in progress, but some acoustic work is being carried-out on northern right whales under the direction of Scott Kraus and the New England Aquarium. There is a remarkable similarity between the two species in terms of their repertoires and the contexts in which the various sound types are found (S. Kraus, pers. comm.). Rich assortments of complex pulsive sounds are heard in rapid series from large, sexually active groups, and only one whale in the group appears to be calling at any one time. Kraus believes that the whale responsible for the screams is a female who is advertizing her reproductive condition to attract males. Kraus also has heard gunshot slaps from these active groups. Perhaps females are using sounds to attract males, and competing males produce gunshot slaps to threaten and intimidate other males.

Bowhead Whale

The bowhead has a repertoire similar to that of the right whales and many of their calls sound identical. There is some evidence based on acoustic studies (Würsig and Clark, 1990) to suggest that different sound types are associated with different contexts, but properly associating sounds from a single hydrophone with whale activities has been problematic. The few observed associations between sounds and activities are similar to those for right whales; a single whale in an active group often produces rapid sequences of complex pulsive calls, while less active whales produce mostly simple calls. Some of the clearest observations synchronizing bowhead activities with sounds have come from Isabella Bay where bowheads have been observed and recorded from a kayak (Richardson and Finley, 1989).

Insights into bowhead acoustic behavior and in particular, communication, recently have emerged from the acoustic studies off Pt. Barrow, Alaska during the spring migration (Clark et al., 1986; Clark and Ellison, 1988). By acoustically tracking whales using linear hydrophone arrays, it has been shown that bowheads often adopt signature calls (Clark, 1989). These calls typically are simple FM signals and an individual whale repeatedly produces calls with identical characteristics for up to several hours. Bowheads sometimes countercall antiphonally such that if one whale calls, an answer is given within 5-30 seconds. Often these countercalling episodes will involve as many 3-5 whales. Whales also imitate the calls of other members of the countercalling group, or sometimes the entire group produces the same call.

Acoustic observations of bowheads migrating in the ice off Point Barrow, Alaska suggest that the whales use the surface reverberation (echo) of their sounds to navigate (Ellison et al., 1987; George et al., 1989). This is the first observation of low-frequency navigation as proposed by Norris (1967). The evidence for this behavior comes from observations made between 28 April to 1 May, 1985 from the shorefast ice off Point Barrow, Alaska. During these observations a large, 10 km^2 multi-year ice floe with 5-10 m surface ridges was fixed approximately 2 km offshore of the array in 20-30 m of water. Singers swam inshore of this floe and came to within about 500 m to its southern edge (Clark, 1989, Fig. 1). Calling whales avoided the floe and in some cases there was a change in calling behavior associated with this avoidance. As the lead whales in the herd approached the floe, they increased their call rates until 250-500 m from its southern edge, at which point they swam either inshore or offshore of the floe. Whales behind these lead whales by 1-2 km did not increase their calling rates, but simply initiated their detour around the floe at a greater distance from its edge. Ellison et al. (1987) calculated that the echo of a typical bowhead FM call is as much as 20 dB greater off ice with pronounced underwater keels (a common characteristic of multi-year ice) than the echo off uniform new ice. This suggests that bowheads actively use the echoes of their sounds off the ice to navigate around large masses of ice.

Bowheads also sing during their migration off Pt. Barrow in the spring. Their songs consist of one or two phrases which together last about one minute. Two harmonically unrelated notes often are sung simultaneously suggesting either that the singer is using two different sound sources or that two whales are duetting. Whales sing in bouts which can last for several minutes or many hours (10 h maximum to date). Songs are different from year to year, and all whales in the same year sing the same general song. Most song is recorded in the early part of the migration (mid April through the first week in May). There are no visual observations accompanying these songs so we do not know the size, sex or behavior of the singers, except that they were swimming in the migration direction.

SUMMARY

Despite the difficulties inherent in studying mysticete acoustics, a great deal has been learned in the last ten years concerning their acoustic behavior. We now have some clear examples of acoustic communication in the bowhead, fin, humpback and right whales. Low-frequency, simple calls are used for long distance contact, while the more complex signals often are associated with more complex social situations. There is now evidence that three species, other than the humpback, sing. The finback song is composed from a single note sung repeatedly in a long, patterned sequence. The singing behavior of the bowhead shows some remarkable similarities to the song of the humpback. The bowhead song is composed from repeated phrases, all the whales within a given year sing the same basic song, and the song changes annually. There is some evidence showing that certain types of blow sounds are communicative, while the idea that sounds resulting from aerial displays are communicative still needs verification. Together, these studies indicate that mysticetes possess a relatively rich lexicon of discrete and graded acoustic signals which are associated with social context in much the same way as in terrestrial animals.

More work still is needed to fully describe the repertoires of the balaenopterids. Earlier reports describing high-frequency clicks and pulses in the presence of balaenopterids should not be cast aside simply because time has past. Further efforts must be made to record these animals with high-frequency systems. Research on mechanisms of sound production is long overdue. Greater progress toward understanding the biological significance of mysticete sounds will come through long-term studies on well-known populations in which the behaviors (including acoustic behavior) of individuals can be monitored for long periods. Acoustic communication is an important component in the life of a whale but must be placed in the broader contexts of evolutionary biology, behavior, and ecology to be fully appreciated. For this reason, it is critical that students be encouraged to incorporate acoustic studies into their research goals. The techniques of sound playback, acoustic tracking and acoustic telemetry will be critical tools in future studies of mysticete acoustic behavior and should help to open-up many new opportunities for all of us.

REFERENCES

Alling, A. and Payne, R., 1990, Song of the Indian Ocean blue whale, Balaenoptera musculus, in: "Special Issue on the Indian Ocean Sanctuary," S. Leatherwood, ed., Int. Whal. Commn., Cambridge.

Baker, C. S., and Herman, L. M., 1984, Aggressive behavior between humpback whale, Megaptera novaeangliae, wintering in Hawaiian waters, Can. J. Zool., 62:1922-1937.

Beamish, P., and Mitchell, E., 1971, Ultrasonic sounds recorded in the presence of a blue whale, Balaenoptera musculus, Deep-sea Research, 18:803-809.

Beamish, P., and Mitchell, E., 1973, Short pulse length audio frequency sounds recorded in the presence of a minke whale, Balaenoptera acutorostrata, Deep-sea Research, 20:375-386.

Chabot, D., 1984, Sound production of the humpback whale, Megaptera novaeangliae Borowski, in Newfoundland waters, M.Sc. thesis, Memorial University, St. John's, Newfoundland, Canada.

Clark, C. W., 1980, A real-time direction finding device for determining the bearing to the underwater sounds of Southern Right Whales, Eubalaena australis, J. Acoust. Soc. Am., 68:508-511.

Clark, C. W., 1982, The acoustic repertoire of the southern right whale: a quantitative analysis, Anim. Behav., 30: 1060-1071.

Clark, C. W., 1983, Acoustic communication and behavior of the southern right whale, in: "Behavior and Communication of Whales," R. S. Payne, ed., Westview Press, Boulder.

Clark, C. W., 1989, Call tracks of bowhead whales based on call characteristics as an independent means of determining tracking parameters, Rep. int. Whal. Commn., 39:111-112.

Clark, C. W., and Clark, J. M., 1980, Sound playback experiments with southern right whales (Eubalaena australis), Science, 207:663-665.

Clark, C. W., and Ellison, W. T., 1988, Numbers and distributions of bowhead whales, Balaena mysticetus, based on the 1985 acoustic study off Pt. Barrow, Alaska, Rep. int. Whal. Commn., 38:312-320.

Clark, C. W., and Johnson, J. H., 1984, The sounds of the bowhead whale, *Balaena mysticetus*, during the spring migrations o 1979 and 1980, *Can. J. Zool.*, 62:1436-1441.

Clark, C. W., Ellison, W. T., and Beeman, K., 1986, Acoustic tracking of migrating bowhead whales, *Oceans 86*, IEEE Oceanic Eng. Soc., 341-346.

Cummings, W. C., and Holliday, D. V., 1985, Passive acoustic location of bowhead whales in a population census off Point Barrow, Alaska, *J. Acoust. Soc. Am.*, 78:1163-1169.

Cummings, W. C., Fish, J. F., and Thompson, P. O., 1972, Sound production and other behavior of southern right whales, *Eubalaena glacialis*, *Trans. San Diego Soc. Nat. Hist.*, 17:1-14.

Cummings, W. C., Thompson, P. O., and Fish, J. F., 1974, Behavior of southern right whales: R/V Hero cruise 72-3. *Antarct. J. U.S.*, 9:33-38.

Dahlheim, M. E., Fisher, H. D., and Schempp, J. D., 1984, Sound production by the gray whale and ambient noise levels in Laguna San Ignacio, Baja California Sur, Mexico. in: "The Gray Whale *Eschrichtius robustus*," M. L. Jones, S. L. Swartz, and S. Leatherwood, eds., Academic Press, Inc., New York.

Darling, J. D., 1983, Migration, abundance and behavior of Hawaiian humpback whales, *Megaptera novaeangliae*, Ph.D. thesis, University of California, Santa Cruz.

Edds, P. L., 1980, Variations in the vocalizations of fin whales, *Balaenoptera physalus*, in the St. Lawrence River, M.S. thesis, Univ. Maryland, College Park, Maryland.

Edds, P. L., 1982, Vocalizations of the blue whale, *Balaenoptera musculus*, in the St. Lawrence River, *J. Mamm.*, 63:345-347.

Edds, P. L., 1988, Characteristics of finback, *Balaenoptera physalus*, vocalizations in the St. Lawrence estuary, *Bioacoustics*, 1:131-149.

Ellison, W. T., Clark, C. W. and Bishop, G. C., 1987, Potential use of surface reverberation by bowhead whales, *Balaena mysticetus*, in under-ice navigation: preliminary considerations, *Rep. int. Whal. Commn.*, 37:329-332.

George, J. C., Clark C., Carroll, G. M., and Ellison, W. T., 1989, Observations on the ice-breaking and ice navigation behavior of migrating bowhead whales (*Balaena mysticetus*) near Point Barrow, Alaska, spring 1985, *Arctic*, 42:24-30.

Herman, L. M. and Tavolga, W. N., 1980, The communication systems of cetaceans, in: "*Cetacean Behavior; Mechanisms and Function*," L. M. Herman, ed., John Wiley & Sons, New York.

Katona, S., Baxter, B., Brazier, O., Kraus, S., Perkins, J., and Whitehead, H., 1979, Identification of humpback whales by fluke photographs. in: "Behavior of Marine Mammals - current perspectives in research. Vol. 3, Cetaceans," H. E. Winn and B. L. Olla, eds., Plenum Press, New York.

Ljungblad, D. K., Thompson, P. O., and Moore, S. E., 1982, Underwater sounds recorded from migrating bowhead whales, *Balaena mysticetus*, in 1979, *J. Acoust. Soc. Am.*, 71:477-482.

Mattila, D., Guinee, L. N., and Mayo, C. A., 1987, Humpback whale songs on a North Atlantic feeding ground, *J. Mammal.*, 68:880-883.

McSweeney, D. J., Chu, K. C., Dolphin, W. F., and Guinee, L. N., 1989, North Pacific humpback whale songs: a comparison of Southeast Alaskan feeding ground songs with Hawaiian

wintering ground songs, Mar. Mammal.Sci., 5:139-148.
Mobley Jr., J. R., Herman, L. M., and Frankel, A. S., 1988, Responses of wintering humpback whales, Megaptera novaeangliae, to playback of recordings of winter and summer vocalizations and of synthetic sound, Behav. Ecol. and Sociobio., 23:211-223
Morton, E. S., 1977, On the occurrence and significance of motivational-structural rules in some bird and mammal sounds, Am. Nat., 111:855-869.
Norris, K. S., 1967, Some observations on the migration and orientation of marine mammals, in: "Animal orientation and navigation. Proceedings of the twenty-seventh annual biology colloquium," R. M. Storm, ed., Oregon State University Press, Corvallis.
Norris, K. S., and Møhl, B., 1983, Can odontocetes debilitate prey with sound?, The American Naturalist, 122:85-104.
Payne, R., 1983, "Communication and Behavior of Whales," R. Payne, ed., Westview Press, Inc., Boulder.
Payne, R., and McVay, S., 1971, Songs of humpback whales, Science, 173:583-597.
Payne, R. S., and Payne, K., 1971, Underwater sounds of southern right whales, Zoological, 58:159-165.
Payne, K., and Payne, R., 1985, Large scale changes over 19 years in songs of humpback whales in Bermuda, Z. Tierpsychol., 68:89-114.
Payne, R., and Webb, D., 1971, Orientation by means of long range acoustic signalling in baleen whales, Ann. N.Y. Acad. Sci., 188:110-141.
Payne, R., Brazier, O., Dorsey, E. M., Perkins, J. S., Rowntree, V. J., and Titus, A., 1983, External features in southern right whales, Eubalaena australis, and their use in identifying individuals, in: "Communication and Behavior of Whales," R. Payne, ed., Westview Press, Inc., Boulder.
Richardson, W. John, and Finley, K. J., 1989, Comparison of behavior of bowhead whales of the Davis Straight and Bering/Beaufort stocks, Report prepared by LGL Limited, 22 Fisher St., P.O.B. 280 King City, Ontario, Canada for the Minerals Management Service, 1110 Herndon Parkway, Herndon, VA 22070.
Ridgway, S. H., and Harrison, R., 1985, "Handbook of Marine Mammals, Vol. 3: The Sirenians and Baleen Whales," Academic Press, Orlando.
Silber, G. K., 1986, The relationship of social vocalizations to surface behavior and aggression in the Hawaiian humpback whale, Megaptera novaeangliae, Can. J. Zool., 64:2075-2080.
Thompson, T. J., Winn, H., and Perkins, P.J., 1979, Mysticete sounds, in: "Behavior of Marine Animals - current perspectives in research. Vol. 3: Cetaceans," H. E. Winn and B. L. Olla, eds., Plenum Press, New York.
Tyack, P., 1981, Interactions between singing Hawaiian humpback whales and conspecifics nearby, Behav. Ecol. Sociobiol., 8:105-116.
Tyack, P., 1982, Humpback whales respond to the sounds of their neighbors, Ph.D. thesis, The Rockefeller University, New York.
Tyack, P., 1983, Differential response of humpback whales, Megaptera novaeangliae, to playback of song or social sounds, Behav. Ecol. Sociobiol., 13:49-55.
Tyack, P., and Whitehead, H., 1983, Male competition in large groups of wintering humpback whales, Behavior, 83:132-154.
Watkins, W. A., 1967, Air-borne sounds of the humpback whale,

Megaptera novaeangliae, J. Mammal., 48:573-578.
Watkins, W. A., 1981, The activities and underwater sounds of fin whales, Sci. Rep. Whales Res. Inst., 33:83-117.
Watkins, W. A., Tyack, P., Moore, K. E., and Bird, J. E., 1987, The 20-Hz signals of finback whales (Balaenoptera physalus), J. Acoust. Soc. Am., 82:1901-1912.
Watkins, W. A., and Wartzok, D., 1985, Sensory biophysics of marine mammals, Mar. Mammal. Sci., 1:219-260.
Whitehead, H., 1985, Why whales leap, Scientific American, 252:84-93.
Winn, H. E., and Perkins, P. J., 1976, Distributions and sounds of the minke whale, with a review of mysticete sounds, Cetology, 19:1-12.
Winn, H. E., and Winn, L. K., 1978, The song of the humpback whale, Megaptera novaeangliae, in the West Indies, Mar. Biol., 47:97-114.
Winn, H. E., Thompson, T. J., Cummings, W. C., Hain, J., Hudnall, J., Hays, H., and Steiner, W. W., 1981, Song of the humpback whale - population comparisons, Behav. Ecol. Sociobiol., 8:41-46.
Würsig, B., and Clark, C. W., 1990, Behavior of bowhead whales in the Western Arctic, in: "The Bowhead Whale Book," J. Burns and J. Montague, eds., Allen Press, Lawrence.
Würsig, B., Dorsey, E. M., Richardson, W. John, Clark, C. W., Payne, R., and Wells, R. S., 1984, Normal behavior of Bowheads, 1983, in: "Behavior, disturbance responses and distribution of bowhead whales Balaena mysticetus in the eastern Beaufort Sea, 1983," W. John Richardson, ed., LGL Ecol. Res. Assoc., Inc., Bryan, Texas.

ACOUSTIC BEHAVIOR IN A LOCAL POPULATION OF BOTTLENOSE DOLPHINS

Manuel E. dos Santos, Giorgio Caporin[1],
H. Onofre Moreira[1], António J. Ferreira[1]
and J. L. Bento Coelho[1]

Instituto Superior de Psicologia Aplicada
Rua Jardim do Tabaco, 44 - 1100 Lisboa
[1] Centro de Análise e Processamento de Sinais
I.N.I.C. - I.S.T, Av. Rovisco Pais
1096 Lisboa Codex, Portugal

INTRODUCTION

Groups of bottlenose dolphins are a common sight in the Sado estuary, near Lisbon, Portugal. They make up a small resident population of about 40 animals, usually divided in groups of about 15, that move in and out of the estuary. They exploit the faunal riches upstream and also engage in frequent excursions to sea, covering distances yet unknown but following, as far as we can tell, routes close to shore.

This population has been the subject of an opportunistic study which started in 1984. It has a long-term perspective, and has been developing in terms of research methods. Photographs of dorsal fins allow the recognition of most individuals, and confirmed the resident character of the population (dos Santos and Lacerda, 1987). Movement patterns are being analysed and, more recently, acoustical monitoring and recording of underwater calls has been added to the other methods of observation and recording of behavior. In this paper, we present the preliminary results of this acoustical monitoring program.

Some difficulties should be mentioned. These dolphins rarely approach boats or hydrophones. Compared to other geographical forms, they are very robust (we have found a stranded female that was nearly 380 cm long) and it does not seem feasible or desirable to corral them. This means we cannot determine their sexes and their age can only be estimated afterwards if we find them dead.

The literature on wild odontocetes usually presents the activities of the animals neatly divided into a few functional categories of behavior, typically traveling, feeding, social interactions and resting (see Shane et al., 1986, for a review in the case of *Tursiops*).

We found this functional categorization inappropriate to our observations, and for that reason we have tried to define formal categories for which hypothetical interpretations are proposed. The whole set of our categories will be presented elsewhere (Harzen et al., in prep.). Here we will only discuss those activity patterns (and their tentative interpretations) which were observed during our acoustical sessions.

The objectives of this preliminary study were:

(1) to collect as many different sounds as possible, to describe and classify them (it should be noted that non-English speakers often are confused by the terms used in the literature to name dolphin vocalizations). It is of great interest to compare representations of sounds produced by free-ranging dolphins of different populations, especially if behavioral information is also available. We thus decided to present our results so far, although our sample size is still too small and allows no more than a rough qualitative discussion.

(2) to analyse variations in the acoustic behavior of the animals.

(3) to try to relate it to general activity patterns, as observed at the surface.

METHODS

The Sado estuary is located on the Atlantic coast of Portugal. Its mouth (at approximately 38° 29 N, 08° 55 W) is about 2 km wide, but the estuary is a vast body of water, more than 20 km long and up to 5 km wide. Its inner part is very shallow, with extensive mud flats exposed at low tide. More details appear in dos Santos and Lacerda (1987).

Dolphins were monitored during 30 boat-surveys (corresponding to 30 dolphin sightings) in the Sado estuary and adjacent coastal waters. These boat-surveys were opportunistic and occurred between July 1987 and June 1989, covering all months of the year except November and December.

We used a fiberglass boat propelled by two 25-HP outboard engines. Boat-surveys involved at least three people, one driving the boat and taping behavior observations on a dictating machine, one photographing the animals and the third operating the sound recorder inside the cabin. Both the driver and the recorder operator monitored underwater sound through headphones. Our Brüel & Kjær 8104 hydrophone was towed by a 30 m-long cable, mounted inside a plastic hose to avoid traction on the cable and to ensure flotation. The last 1.5 m before the hydrophone was hanging free. The cable was connected to a Brüel & Kjær 2635 charge amplifier which fed channel 1 of a Nagra IV-SJ recorder. Comments were taped on channel 2.

As soon as a group of dolphins was detected and approached, the hydrophone would be lowered and one engine turned off. Thus we followed the dolphins as close as possible, just beyond their flushing distance (usually about 50 m), monitoring underwater sound while observing and photographing the animals using a 300 mm lens.

Comments on underwater sounds were added to the description of the animals' behavior at the surface. The loudest sources of ambient noise were the engines of our boat and also of the many boats that often criss-cross the estuary (small leisure-boats, fishing boats, ferry-boats and large ships such as tankers). Whenever the animals were calling, and signal-to-noise ratio seemed acceptable, we would turn off the engine and record. As we were only interested in the audible range, recordings were made at a tape speed of 19 cm/s (frequency response up to 20 kHz).

Various methods were used to analyse the recordings. A Kay analogical Sona-Graph produced spectrograms of signals selected from the tapes, using a 300-Hz bandwidth filter. By visually inspecting the spectrograms, whistle sounds were classified according to different contour types and their variation was analysed by measuring the following parameters: duration, initial and final frequencies, minimum and maximum frequencies. We also produced time-expanded spectrograms of some pulsed sounds (playing them at 1/5 of their recorded speed), which was useful to determine pulse repetition rates, bearing in mind the problems noted by Thomas et al. (1986).

We also did several digital analyses of the signals. These were processed by a Data Translation 12-bit A/D converter board feeding a Compaq 386/20 PC. Waveforms, power spectra and spectrograms were produced using the Hypersignal Workstation processing package.

RESULTS AND DISCUSSION

During these 30 sightings we were able to monitor the underwater sound near dolphin groups for about 76 hours (average session, 153 min). As to recordings, 342 min of tape were made.

Activity patterns

The concurrent behavior of the dolphins during these monitoring sessions may be summarized in the following general activity patterns:

Fast Directional Moving - the animals surface close together (although often in more than one unit), showing fast movements and covering considerable distances in the same direction. There may be some long dives (i.e., over 2 min), or some front leaps, either occasionally or in sequences ("porpoising"), but no other behaviors at the surface occur. This pattern is considered equivalent to "traveling".

Slow Directional Moving - the animals surface close together (although often in more than one unit), showing slow movements and covering relatively short distances. They move in the same general direction, but there might be a zigzag pattern or some interruptions in the linear displacement.

There are long dives, occasionally some leaps (frontal, lateral or high leaps), and some other behaviors might occur at the surface, such as tail slaps. This seems to be a combination of traveling and searching for prey.

Erratic Group Movements - similar in form to the previous pattern, except that either there isn`t a total displacement of the group or else it occurs in a variable and unpredictable fashion. Although prey is usually not visible, the animals seem to be searching for bottom prey.

Spread Erratic Movements - the animals surface in a disperse fashion, mostly alone but also in dyads or triads, spreading over a wide area. Different animals move at the surface with different speeds and levels of arousal, and dives are usually short. The group may stay in the same area for a long time or it may drift in an erratic manner. Groups may gradually split in the course of this type of activity. Front, lateral or high leaps are occasional, and it is possible to see animals with prey in their mouths (fish or cuttlefish, Sepia officinalis). Even when prey is not visible, such groups seem to be feeding or searching for prey.

Localized Surface Feeding - the animals surface close together, but in different directions, showing fast movements. All dives appear to be very short, and there is not any directional displacement of the group except sometimes a slow drift. Many leaps of different types are visible, as well as a variety of behaviors at the surface, including some which involve more than one animal. Commonly, fish are seen leaping between the dolphins and gulls frequently circle above the group. This seems to be a collective attack on a fish school, and usually we can recognize the fish as mullets.

Social Interactions at the Surface, unrelated to feeding - similar in form to the previous pattern, except that no prey can be detected and there are more contacts between animals, at the surface or even in the air. These are considered episodes of social interactions and play.

Types of calls

Based on visual inspection of the spectrograms and on aural impressions, various types of signals could be distinguished: whistles, pulsed sounds of variable repetition rate and a few other more peculiar sounds, all easily recognizable on audition, and which are presented below.

The observation protocols stated the occurrence or the absence of calls. If the animals were close to the boat, noise was relatively low and nothing could be heard, it was considered that the animals were silent.

Whistles. These are continuous, tonal sounds, occuring in a variety of frequency-modulated contours, sometimes repeated in series. We were able to recognize 17 different contour types from the spectrograms, and they are presented on Table 1, together with the ranges of various parameters.

TABLE 1. Whistle characteristics.

Type	Contour	Frequency (kHz)				Duration (ms)
		Minimum	Maximum	Initial	Final	
AZ		5 - 8.5	16 - >16	5 - 9	8.5 - 16	650 - 1500
US		4 - 7.5	10 - >16	4 - 7.5	10 - >16	600 - 1100
SU		5 - 6	9 - 13	5 - 6	9 - 13	450 - 550
AR		7 - 9	>16	7 - 9.5	7.5 - 9.5	225 - 325
ME		4 - 5.5	>16	4 - 5.5	5.5 - 14	950 - 1900
SM		4 - 5.5	16	4 - 5.5	16	370 - 650
AP		3.5 - 7	14 - 16	3.5 - 11	5 - 13	600 - 2400
SP		6.5 - 7.5	11.5 - 13.5	6.5 - 8.5	7 - 7.5	275 - 700
CA		5.5 - 8	15 - 16	5.5 - 15	6 - 14.5	450 - 700
CM		7.5 - 8	15 - >16	7.5 - 13	9 - >16	350 - 600
SC		4.5 - 7.5	10.5 - 13.5	5 - 13.5	4.5 - 7.5	475 - 800
DD		5.5 - 8.5	9.5 - 10.5	9.5 - 10.5	4.5 - 8.5	825 - 1000
SD		3.5 - 5.5	12 - 16	3.5 - 10.5	12 - 16	550 - 900
AA		6 - 8	13 - >16	6 - 9.5	13 - >16	550 - 1725
CS		5.5 - 6	14 - 14.5	14	5.5 - 6	775 - 800
PU		7.5 - 9.5	12.5 - 15.5	11.5 - 14	9 - 15.5	650 - 700
KE		3 - 4.5	8.5 - 9	3 - 4.5	4.5 - 9	550 - 825

TABLE 2. Occurrence of recognized whistles.

Date	US	AP	SU	CM	SP	ME	KE	CA	DD	CS	AZ	SD	SC	AA	AR	PV	SM
25.07.87	X	X	X														
26.07.87		X		X													
01.08.88		X															
24.09.88				X													
23.10.88					X												
30.10.88		X				X	X										
29.01.89		X															
16.02.89			X														
17.02.89	X								X								
25.04.89	X	X								X							
07.05.89				X		X	X										
14.05.89				X							X						
15.05.89	X											X	X				
19.05.89				X									X	X	X		
28.06.89	X	X				X							X	X			X
29.06.89		X												X			
30.06.89	X	X	X		X		X	X			X	X	X			X	X

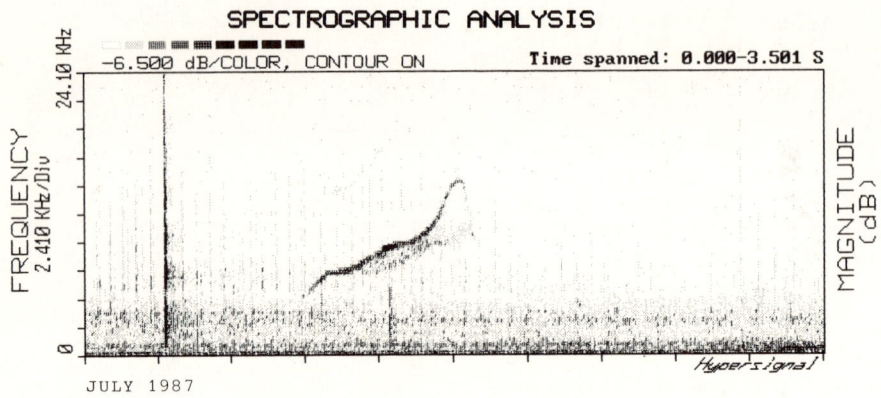
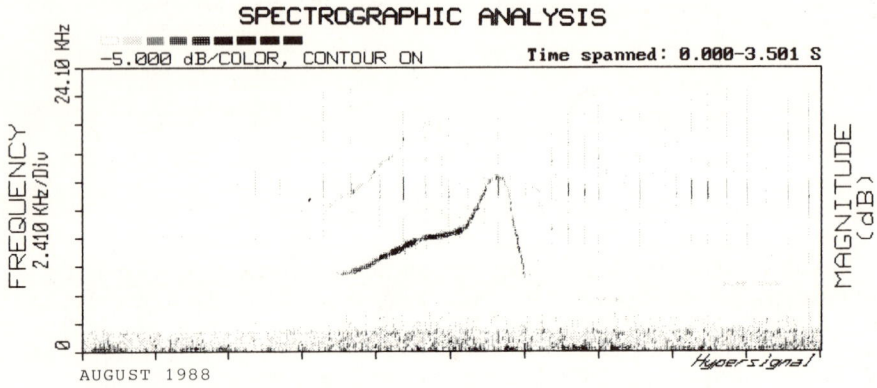
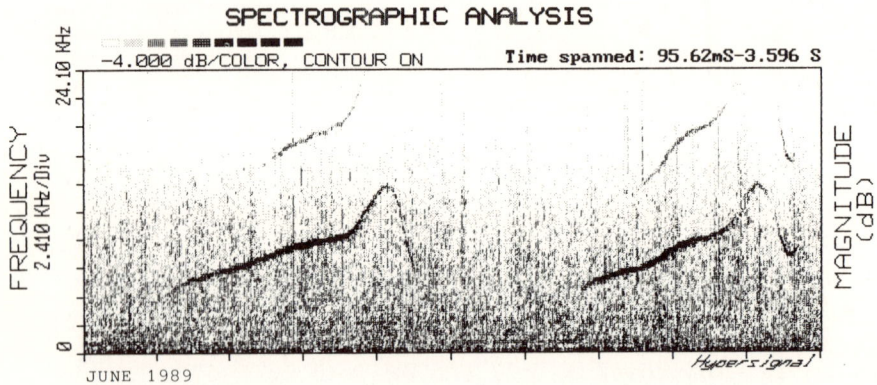

Fig. 1. Whistle contours from different years.

Also, we present the occurrences of these recognized contours throughout this study period (Table 2). The literature suggests that whistle contours are associated with individual dolphins and that they are stable in time (see also Fig. 1). Whistles are possibly used as signature signals and may be mimicked by other animals (Caldwell and Caldwell, 1965; Herman and Tavolga, 1980; Tyack, 1986, 1987; Sayigh and Tyack, 1987; Ralston et al., 1987).

The possiblity that signature whistles are mimicked by other animals may be irrelevant in the context of field recordings. Presumably, the animals are mimicking the whistles of other individuals in the same group. If this assumption is correct, the repeated recognition of contours, in a more extended scale could perhaps be compared with group composition data based on dorsal fin identification.

Whistles occurred in all activity patterns but were especially abundant in Localized Surface Feeding and Social Interactions at the Surface. These two patterns, although related to quite different motivational states, involve a high level of arousal. In fact, several other calls are heard together with whistles during such activities (see below). In more calm activities, whistles are often accompanied just by click trains.

Pulsed sounds. These include the broadband *clicks* generally assumed to be associated with echolocation, and which come in trains of variable length and repetition rate. As the pulse rate increases, these trains sound like *creaks* (such as those of a rusty hinge), *low creaks* and *moans*.

Creaks, which usually are broadband, have a pulse rate extending over 40 p.p.s. (pulses per second). An example, with a rate varying between 40 and 140 pps, is shown in Fig. 2-a. Low creaks have most of their energy below 2.5 kHz (see example in Fig. 2-b). They appear to be similar to sounds presented by Sjare and Smith (1986, their Fig. 3-j) for the beluga. Popper (1980) refers the possibility that these narrow-band, low-frequency pulse trains are a particular type of echolocation emission. Regarding behavioral circumstances, our records show creaks and low creaks evenly distributed by the various activity patterns.

At higher repetition rates, it becomes harder to discern granularity and the sounds assume a moaning appearance, while several harmonics appear on the spectrogram (see Fig. 2-c). Although scarce, our sample of moans suggests that this sound type may be related to more aroused motivational states.

Besides naming these different sounds according to our aural impressions, we quantified their pulse repetition rate. Fig. 3 shows an attempt to summarize the gradual variation in the repetition rate of the pulsed sounds, and also the ranges of each pulsed call.

Bangs. Some instances of a relatively loud broadband pulse (duration, about 20 ms) were recorded (Fig. 4-a). Although on a larger time-scale, the waveforms of these bangs resemble a typical click waveform (Fig. 4-b).

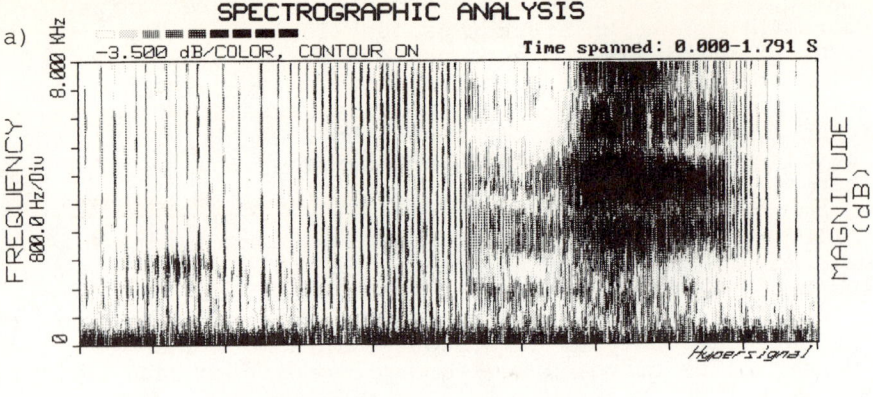

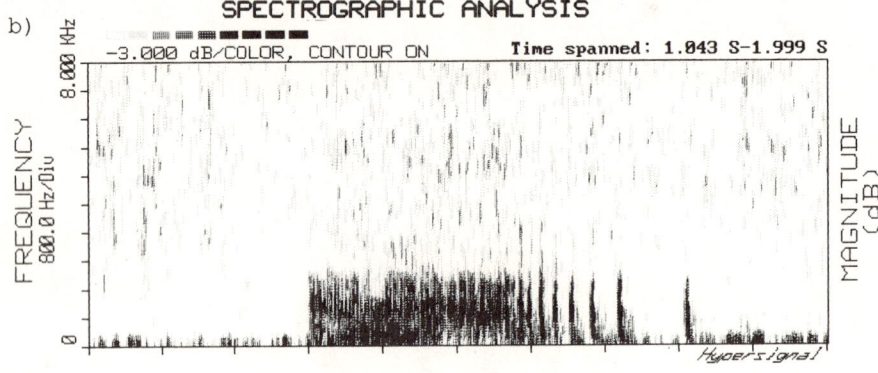

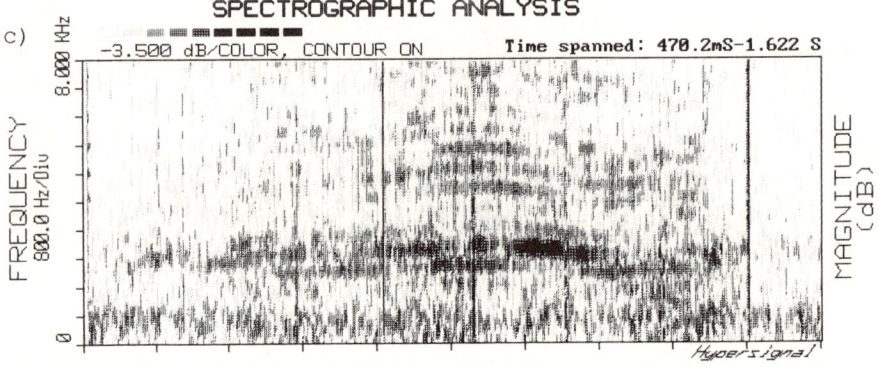

Fig. 2. Pulsed sounds with different repetition rates.
a) Creak.   b) Low creak.   c) Moan.

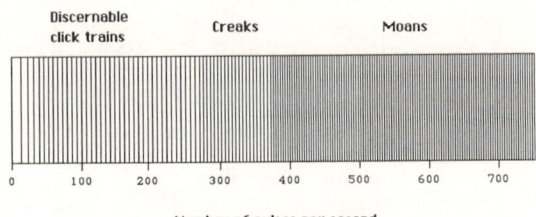

Fig. 3. Variation in the pulsed sounds repetition rate

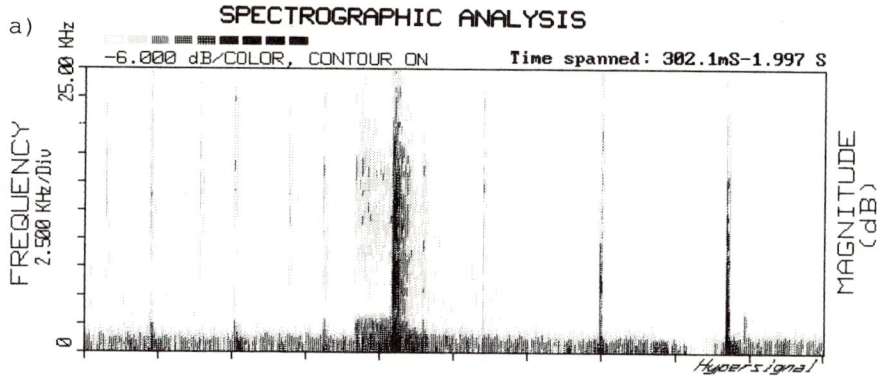
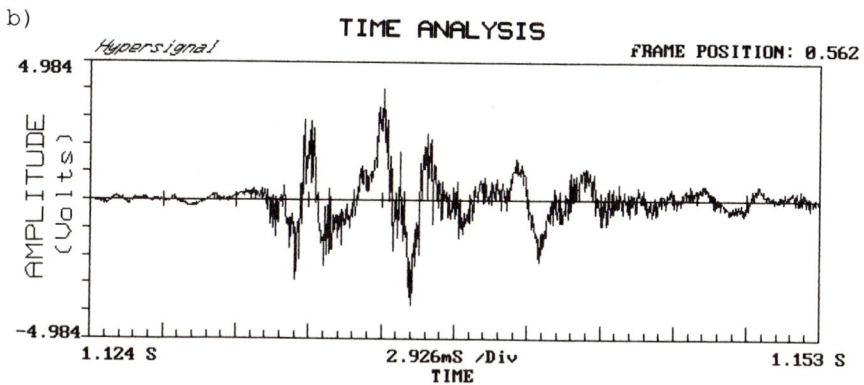
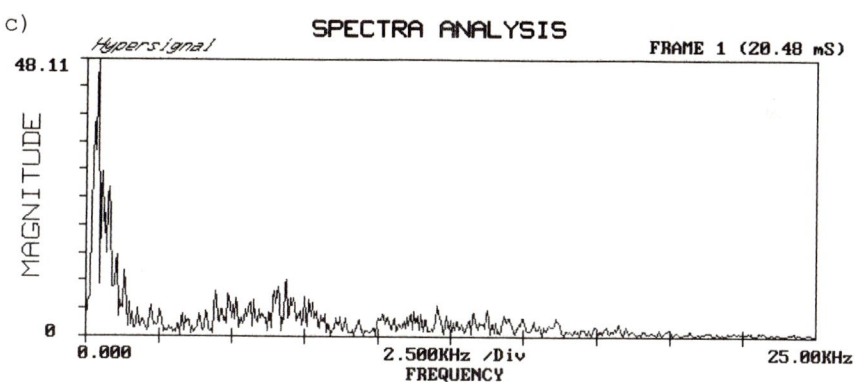

Fig. 4 a) Bang spectrogram. b) Bang waveform.
c) Bang power spectrum.

It actually looks like some of the loud sounds produced by several delphinids, predatory or social, discussed by Marten et al. (1988), in particular the "jaw clap" (their Fig. 3-b), which is used in social contexts. The spectrograms of the bangs in our tapes resemble those of some sounds reported by Caldwell and Caldwell (1967, their Figs. 30 and 31), which they referred to as "cracks" and "pops".

Bangs were recorded during episodes of Localized Surface Feeding and Spread Erratic Movements. The fact that these seem to be feeding activities does not eliminate the possibility of a social, agonistic role. It would be interesting to find out whether mullets (the main prey of these dolphins) are at all disturbed by such sounds.

<u>Brays</u>. This sound, which has the alternating appearance of a donkey`s bray, is frequently heard. It consists of sequences of squeak-like sounds followed by grunts (see Fig. 5-a). The interval between the squeak part and the grunt lasts an average of 390 ms. Sometimes the pairs are preceded by a creak or a noisy sound (Fig. 5-b). Brays are very conspicuous even at distance, but we have not been able to find any description of such a call.

Bray sequences tend to occur concomitantly with aroused activities such as Localized Surface Feeding and Social Interactions at the Surface. We have never been able to record or monitor one such episode during Fast Directional Moving. This pattern of occurrence suggests that the signal has a social function. When used in feeding contexts, often it is mixed with sounds such as whistles, creaks and bangs. It is uncertain whether the bray series have an agonistic character (as the lower-frequency grunt suggested) or whether they elicit aggregation and perhaps even cooperation.

<u>"Buzz Effect"</u>. Some bursts of variable duration seem to be modifications of other sounds, making them noisy, with the appearance of a buzzing wasp. Modified signals such as "buzzed creaks" or "buzzed moans" could be distinguished in the spectrograms (Fig. 6-a). Even if the sound was analysed at 3.8 cm/s, it was not possible to find a click structure in the sound, as opposed to what occurs in the case of normal moans. The "buzz effect" also appears to be associated with activities when the animals are close together and aroused.

<u>Blasts</u>. These noisy sounds appear in sequences of 3 bursts each lasting about 150 ms, the whole set lasting about 700 ms (Fig. 6-b).

A somewhat unexpected result of our monitoring efforts was a high incidence of underwater silence close to dolphin groups. Apart from the episodes of Localized Surface Feeding or Social Interactions at the Surface, which involved various types of calls and were always quite noisy, dolphin groups often were silent. This is noticeable especially when they travel (Fast Directional Moving), when passing through sensitive areas like the mouth of the river or when they had to pass close to our stationary boat. We also monitored absence of calls in contexts involving movement presumably related to prey-searching, like Erratic Group Movements or Slow Directional Moving.

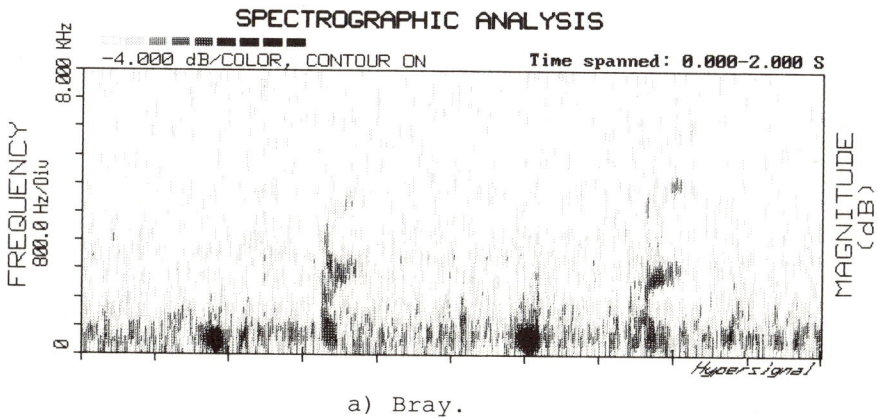

a) Bray.

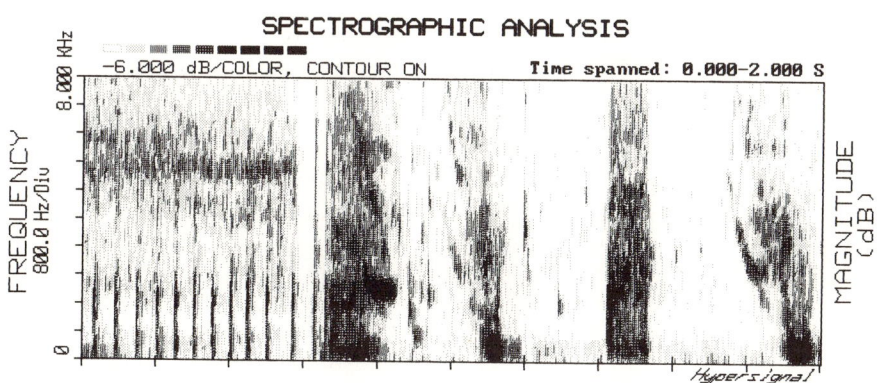

b) Bray preceeded by a creak and a "buzzed" sound.

Fig. 5.

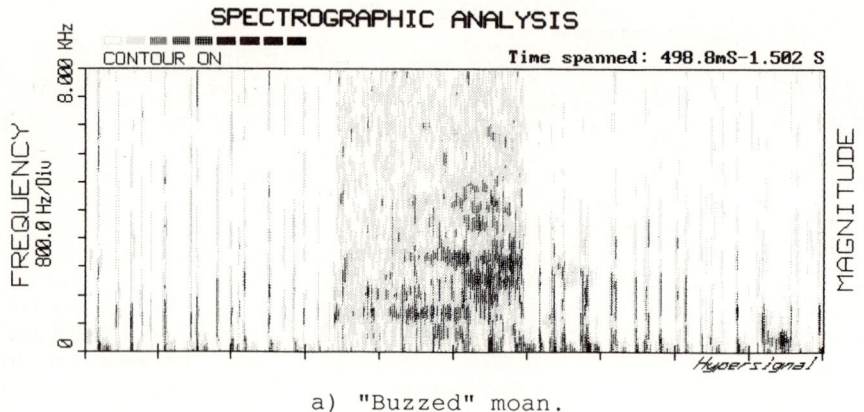

a) "Buzzed" moan.

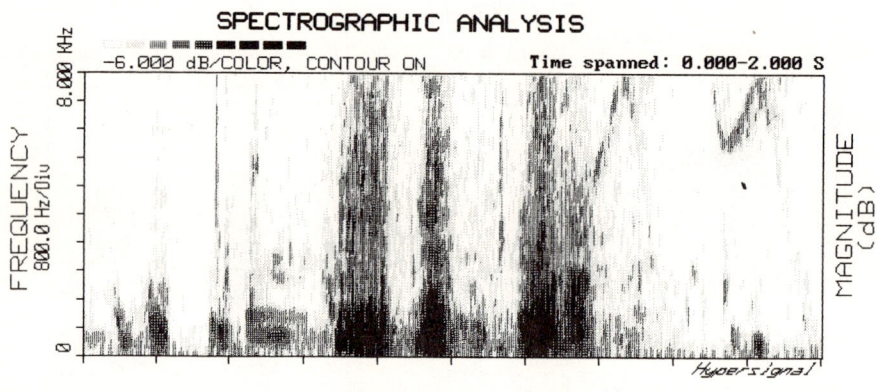

b) Series of blasts.

Fig. 6.

Underwater visibility in the Sado estuary usually is low, and it is unknown to what extent they rely on passive audition for prey detection. It seems clear, anyway, that listening is an important component of their displacement activities.

The study of the acoustic behavior of the bottlenose dolphins in the Sado has been enriching to the behavioral observations of these animals. It is now necessary to obtain significant quantitative correlations among sounds, behavior and ecological circumstances.

ACKNOWLEDGMENTS

This research was supported by the Portuguese Board for Scientific and Technological Research (JNICT) (Project 87/446). The following entities have provided logistic support: TORRALTA, Club Naval de Setúbal and Clube Náutico de Tróia. We thank the Navy Yard of Alfeite for having lent us some equipment and the Port Authority of Setúbal for permitting the use of nautical facilities. The Linguistics Center of the University of Lisbon kindly allowed us to use the Kay Sona-Graph. The following people participated in the field work: Stefan Harzen (who also allowed us to use here some of our common unpublished material), Rui Porteiro, Paulo Xavier, Alexandra Lázaro and Mimi Mendonça. Luís Azevedo, Jeanette Thomas and an anonymous reviewer made valuable comments on the manuscript. Our participation in this Symposium was funded by the Portuguese Office of NATO Scientific Affairs (INVOTAN).

REFERENCES

Caldwell, M. C. and Caldwell, D. K., 1965, Individualized whistle contours in bottlenose dolphins, Tursiops truncatus, Nature, 207:434.
Caldwell, M. C. and Caldwell, D. K., 1967, Intraspecific transfer of information via the pulsed sound in captive odontocete cetaceans, in: "Animal Sonar Systems. Vol. II", R. G. Busnel, ed., Laboratoire de Physiologie Acoustique, Jouy-en-Josas, France.
dos Santos, M. E. and Lacerda, M., 1987, Preliminary observations of the bottlenose dolphin (Tursiops truncatus) in the Sado estuary (Portugal), Aquatic Mammals, 13:65.
Herman, L. M. and Tavolga, W. N., 1980, The Communication Systems of Cetaceans, in: "Cetacean Behavior: Mechanisms and Functions", L. M. Herman, ed., Wyley-Interscience, New York.
Marten, K., Norris, K. S., Moore, P W. B. and Englund, K. A., 1988, Loud impulse sounds in odontocete predation and social behavior, in: "Animal Sonar: Processes and Performance", P. Nachtigall, ed., Plenum, New York.
Popper, A. N., 1980, Sound Emission and Detection by Delphinids, in: "Cetacean Behavior: Mechanisms and Functions", L. M. Herman, ed., Wiley-Interscience, New York.
Ralston, J. V., Williams, H. N. and Herman, L. M. 1987, Vocalizations of stressed and unstressed Atlantic bottlenose dolphins (Tursiops truncatus), Abstract, Seventh Biennial Conference on the Biology of Marine Mammals, December 5-7, Miami.
Sayigh, L. S. and Tyack, P. L., 1987, Development of signature whistles in wild bottlenose dolphins, Abstract, Seventh

Biennial Conference on the Biology of Marine Mammals, 5-7 December, Miami.

Shane, S. H., Wells, R. S. and Würsig, B., 1986, Ecology, behavior and social organization of the bottlenose dolphin: a review, Mar. Mamm. Sci., 2(1):34.

Sjare, B. L. and Smith, T. G., 1986, The vocal repertoire of white whales, Delphinapterus leucas, summering in Cunningham Inlet, Northwest Territories, Can J. Zool, 64:407.

Thomas, J. A., Fisher, S. R. and Awbrey, F. A., 1986, Use of acoustic techniques in studying whale behavior, Rep Int. Whal. Comm., Special Issue #8:121.

Tyack, P. L., 1986, Whistle repertoires of two bottlenose dolphins, Tursiops truncatus, mimicry of signature whistles?, Behav. Ecol. Sociobiol., 18:251.

Tyack, P. L., 1987, Do untrained dolphins imitate signature whistles to call each other?, Abstract, Seventh Biennial Conference on the Biology of Marine Mammals, 5-7 December, Miami.

ORGANIZATION OF COMMUNICATION SYSTEM IN TURSIOPS TRUNCATUS MONTAGU

Vladimir I. Markov and Vera M. Ostrovskaya

A. N. Severtsov Institute of Evolutionary Morphology and
Ecology of Animals, USSR Academy of Sciences, 33 Leninsky
Prospect, Moscow, 117071, USSR

INTRODUCTION

The problem of the degree of complexity and semantic capabilities of the acoustic communicative system in bottlenose dolphins has been under discussion for over a quarter of a century, ever since John Lilly published his book "Man and Dolphin" (Lilly, 1962). Nowadays, there is an abundance of literature, with different viewpoints, but no consensus has been reached so far among researchers. The problem proved to be very complicated, both methodologically and experimentally, while the methods used turned-out to be labor-consuming and, on the whole, inefficient; all kinds of straightforward attacks failed. Meanwhile, one can try and assess potential capabilities of communicative system by analysing dolphins' mechanisms ensuring its productivity, i.e. the creation of signals and messages in amounts nessesary for communication. These mechanisms ensure the encoding of information and, in accordance with the theory, their functioning, in this way or other, affects the structure of signals and their sequences, in other words, it affects the organization of communicative system.

The increase of communicative system productivity in the course of evolution was associated with the expansion of the species' semantic field which, in its turn, depended on the increasing level of higher nervous activity and more complex intraspecific social relations. Therefore, there exists a close relationship between semantic capabilities of communicative systems and their organization. One can hardly imagine the existence of a developed communicacative system in a species which does not need such a system or can not use it properly. The functioning of the system is ensured by adequately developed functions of the brain which must be capable of solving both organizational problems of communicative system control and semantic problems. Therefore, if one knows how the communicative svstem of a species is organized, what mechanisms ensure its productivity and how fully it uses capabilities provided by the above mechanisms, one can make an overall assassment of its semantic capabilities.

The study of communicative system organization is inhibited strongly by their heterogeneity caused by a simultaneous use of several sensory chanals for communication. Consequently, signals with different physical natures, multimodal signals and motor behavioral acts used as independent signals, can be used within one communicative system. One can overcome this difficulty if one bears in mind that communicative systems transfer information, hence, natural "sign" characteristics. This allows one to apply a semiotic approach to analysis, to ignore modality in the physical essence of signs and to study only the p r i n c i p l e s of the formation of vocabulary units and text* used in the communicative system under investigation. This is done with the help of assessment tools developed within general theory of sign systems and information theory. The analysis of the physical structure of signals will be necessary only when singling out meaningful structural units. The logic of such an approach and the analysis procedure were described in a separate paper (Markov et al., 1983).

The comparative analysis of communicative systems in various animal species has shown that there are only several types of communicative system organizations that range from the simplest monosignal systems intended to ensure the encounter of males and females during the reproduction cycle, to complex lexical systems, such as human speech. The productivity of the above types does not depend on characteristics of communication channels, or the physical nature of signals or signal generation systems. Each type can be described as having a certain set of vocabulary-formation mechanisms, a certain level of polysemy development, a certain ratio between congenital signalization components and those acquired in the course of training, the nature of relationship between the structural complexity of vocabulary units and semantics and the presence or absence of syntax organization in signal sequences. Different taxonomic groups - classes in the vertebrates and families in the invertebrates - have the same types of communicative systems organization. Within each group they regularly succeed each other and can be regarded as stages in the development of communication inside a given group. More productive types are encountered in species with a higher level of mental activity (Markov, 1976; 1984).

However, one can not reveal one common way development of communicative systems in the course of evolution. The problem of communication arises and is solved in each group independently, with the available morphological and physiological basis. Therefore, groups with a high phylogenetic status can lose least productive types of communicative systems and develop most productive systems. But the uniformity of possible solutions is amazing. It must be due to the fact that, information requirements are the key factor determining the direction of communicative systems development, and communicative systems develop in conformity with the logics of information systems development, to extend their capabilities and to enhance reliability and flexibility of information exchange.

What kind of communicative system organization can one expect to find in bottlenose dolphins? This species has a large, well-developed brain, with its cephalization index, absolute neocortex volume, relative area of nonprojetion fields and other "intelligence" indices close to those of the human brain (Ladygina and Supin, 1974; Morgane, 1978; Yablokov, 1983). Bottlenose dolphins can solve various, sometimes complicated, logical problems, and their cognitive abilities reach a high level (Herman, 1980, 1986, 1987; Herman et al., 1984; Krushinskaya, 1983). They have complex behavioral patterns in social groups which are typical for mammals with a rather highly-developed psychics; various patterns of cooperative behavior are widely developed (Bel'kovitch et al., 1978a, 1978b). Observations and experiments show that the coordination of individuals' actions in cooperative behavior is achieved through exchange of acoustic signals (Bel'kovitch et al., 1978a; Morosov, 1970; Zanin et al., 1990). All this indicates that the communicative system of bottlenose dolphins should posses great capabilities, which suggests a great complexity of organization. Therefore, bottlenose dolphins are most likely

---

*The terms "vocabulary", "syntax", "semantics", "text", etc., should be regarded semiotically.

to have a syntactic or hierarchic communicative system (Markov, 1976).

METHODS AND MATERIALS

Bottlenose dolphins use several sensor modalities for communication: visual, tactile, chemical and acoustic. Because of specific conditions of the marine environment, the first three modes can be used in close proximity, while the acoustic mode is actively used both in close proximity and at long distances. We made a series of observations making it possible to assume that the acoustic mode is sufficient for communication. We believe that acoustic signalization has all features typical of the communication system of bottlenose dolphins on the whole.

The study of organization of any communicative system has to address the following questions: how is the structure of animals' signals organized and are there organized sequences of signals generated by animals in the course of communication? One has to make an in-depth structural analysis of signals, to reveal the set of constituent structural elements and combination rules during the formation of the final structure of signals, to develop structural classification of signals, and to establish the relationships between signals in the sequence.

By the time we began our work, it had been already established that two sound generators can participate in the generation of dolphin signals (Lilly and Miller, 1961). That is why we recorded signals in such a way as to make it possible, at a later stage, to attribute each signal to a definite dolphin and to make sure that the recorded structure was produced by one animal. Signals were recorded on magnetic tape in situations allowing for the above conclusions: in free swimming isolated animals and during communication of isolated animals by the electro-acoustic link. We recorded the signalization of animals in sea enclosures and in pools. In all cases we got reverberation characteristics of the pools and sea enclosures for the working position of hydrophones. We used calibrated specially devised spherical ceramic hydrophones. The signals were recorded on uni-channel or multi-channel tape recorders with the frequency response from 0.5 to 120 kHz. We also used tape recorders with working bandwidth from 0.05 to 20 kHz for the registration of low-frequency signals. In all cases recording paths were tuned-up in accordance with individual characteristics of hydrophones, so that irregularity of frequency characteristic of the whole recording path in working bandwidth was not more than 3 dB. Dolphin sounds underwent dynamic spectral, contour-graphic and oscillographic analysis with the help of serial and specially designed instruments, as well as auditory analysis. A symbolic and graphic language capable of describing fine details of structure was developed to describe the structure of signals. Its application helped to obtain signals structural formulas which were further analysed and classified. If signals formulas turned-out to be similar, we performed a direct comparison of analytical material to make sure the description was correct. Twenty adult dolphins (7 males and 13 females) and three calves were used in the experiments; the total number of examined signals exceeds 300 000.

ORGANIZATION OF SIGNAL STRUCTURE

Bottlenose dolphins can produce signal structure with the help of 1 to 4 sound generators, each of them capable of working in a tonal or pulse regime (Markov, 1977, 1983; Markov and Ostrovskaya, 1983; Markov and Tarchevskaya, 1978). When working in the tonal regime, the generator produces narrow-band frequency-modulated signals (whistles). By varying the direction and rate of frequency variation, a dolphin can produce diverse and sometimes "queer" acoustic structures (Fig.1). When analysing them, one notices that they are

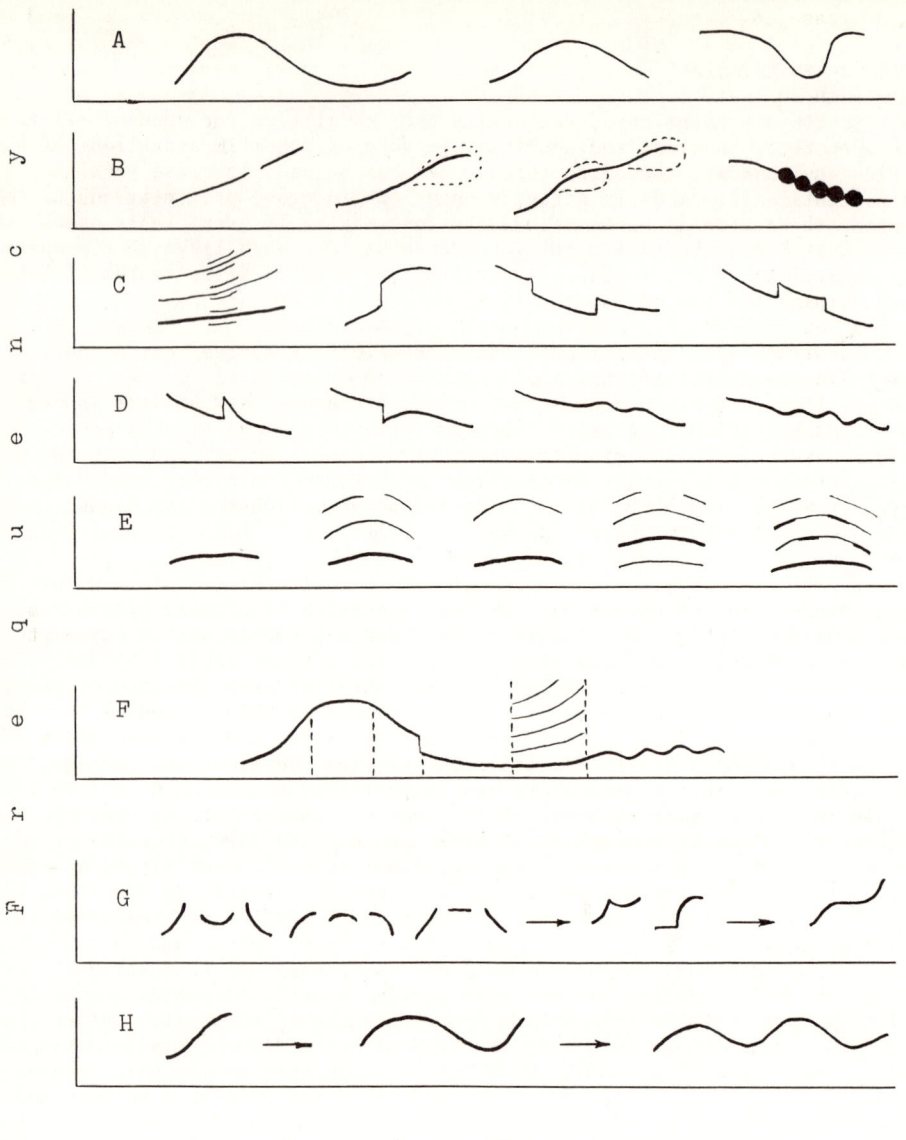

Fig. 1. Operation of one sound generator in the whistle regime. Graphic presentation of the sonagrams. (A) Generation of contours with various shapes. (B-E) Ways of modifying the structure: (B) pauses in the contour, local spectral expansions, rhythmical pulsations of the contour; (C) local amplitude modulation of the contour by constant frequency, single and double "steps"; (D) surges and waves in contours; (E) control of the number of overtones, generation of complete and incomplete harmonic rows, strengthening of the second harmonic, control of energy in harmonics. (F) Discrete structure of the whistle obtained after successive application of sound generation and modification methods. (G) A set of simplest structural elements and generation of two- and three-element signals out of them. (H) Successive stages in the generation of a complex rhythmical signal.

produced by arbitrarily alternating sections with a rather rapid increase or decrease of frequency and sections with a constant or slowly chanding frequency. As a result, signal structure becomes a chain of acoustically different elements and develops a contrast necessary for information encoding. Information capacities of such a system can be enhanced by changing the steepness of countour sections, by changing the limits frequency range, the register (the position of contour on the frequency axe) and duration, as well as by increasing the total number of elements in signal structure (its length). Long, gently sloping sections of contours which are less informative, according to the information theory, can be diversified by local contour modification. Bottlenose dolphins posses a wide range of modification methods such as (Figs. 1-B,C,D,E) pauses (short-duration cessation of sound generation), local expansions and periodic pulsations of the spectrum, "steps" and surges of frequency in the contour, frequency and amplitude vibrato. It is important for the discretization of monotonous contour sections that bottlenose dolphins can change the structure of the overtone part of the signal (change the number of overtones, create a complete or incomplete harmonic row and overtones which are not multiples of the principal tone, to strengthen or weaken certain overtones, to "transfer" energy from one harmonic to the other). As a result of successive one-shot or repeated applications of various contour generating means, the structure of whistles becomes discrete and consists of several blocks with different phonations (Fig. 1-F). On the average, one can identify 5-7 blocks in the signal, though their number can reach 12.

Observations show that the structure of blocks is formed by the combination of simpler elements, i.e. there is a gradual complication of the structure. Following the analysis of very short signals, we managed to identify 9 types of contours which can be regarded as analogs of initial structural elements which form the structure of signals (Fig. 1-G). They represent whistles ranging from 17 to 80 ms. Actually, a whistle of any type amounts to an entire class of signals, since, while having the same shape of the contour, they can differ in duration, frequency range, register and the rate of frequency modulation. Successive combinations of several elements of the same type, but differing in register, makes it possible to generate signals or structural elements (blocks) of a second level which maintain the same operating regime of the sound generator throughout their duration (homotypical combination). It is such blocks which we examined as discrete signal structure elements.

The boundaries between initial elements in homotypical signals and second-level blocks can be noticed only in very short structures consisting of 2-3 elements, if the sound generator is operating to its utmost capacity. Each element has a surge of amplitude. With the increase of overall duration of second-level blocks and, consequently, with the increasing number in their structures, the frequency range of elements and the rate of frequency alteration the signal are decreasing. So, the boundaries between elements are gradually smoothed and become less pronounced (Markov and Ostrovskaya, 1975).

Third-level structural blocks always are formed by a combination of second-level blocks with different characteristics (heterotypical combination); the attachment between them is always smooth, with the existence of the structure transformation zone. However, the boundaries between blocks are pronounced. At this stage, the signal's amplitude envelope becomes extremely complicated, because different second-level blocks have different amplitudes and amplitude peaks can be associated both with blocks and boundaries between them. In general, heterotypical combining is characterized by clear-cut boundaries between structural elements, irrespective of their level; they also are fairly-well pronounced in very short second-level signals if those are formed by combining initial elements of different types (Fig 1-G). Third-level blocks are used for creating blocks of the next level or final signal structure.

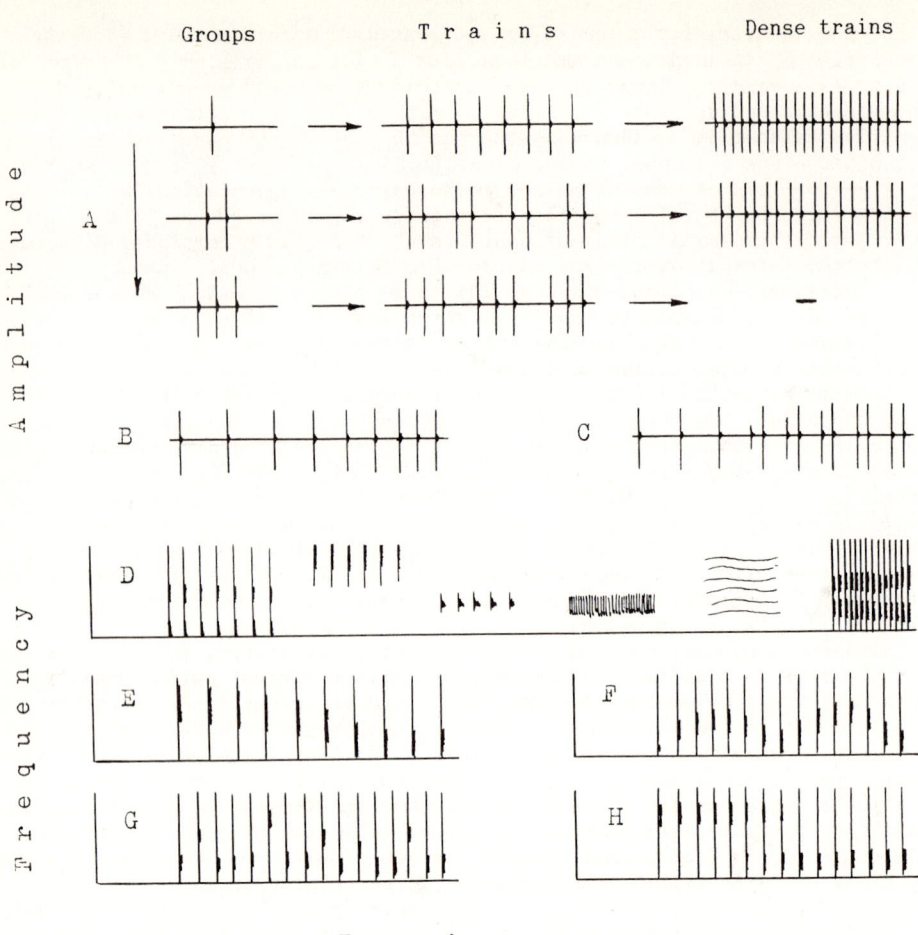

Fig. 2. Operation of one sound generator in the pulse regime. (A-C) - Graphic presentation of the oscillograms,(D-H) - sonagrams. (A) Temporal organization of the trains of pulses (primary grouping); (B) alteration in repetition rate of pulses; (C) alteration in the grouping of pulses; (D) examples of pulsed signals: three fragments of trains with different spectra, narrow-band noise signal, spectrum of a dense modulated train, dense train of pulses with a formant structure; (E) shift in the location of the spectral peak; (F) "spectral wave"; (G) energy "bursts"; (H) "pumping" of energy from one spectral maximum to the other.

When operating in the pulse regime, the sound generator can produce batches of pulses, whose spectral and temporal characteristics in most cases depend on the preceding tuning of the generation system. One can single-out three basic classes of pulsed sounds (below - "pulses"): clicks, clear blows and prolonged blows. Clicks represent an aggregate class of 0.1-3 ms long pulses with various spectra, mostly differing in the distribution of energy in the spectrum and in duration: there exist wide-band and narrow-band clicks, clicks with several maxima in the spectrum. Clear blows are also an aggregate group. They are high-frequency pulses, 4-9 ms long; they are characterized by a triangular shape of the amplitude envelope. Prolonged blows are

Table 1. Transformations of Pulsed Signals During the Operation of One Sound Generator.

Transformed Pulse Characteristic	N	%
Density of pulses in train	229	25,9
Grouping of pulses	250	28.3
Spectrum of pulses	251	28.4
Density in a train and grouping of pulses	102	11.5
Density in a train and spectrum of pulses	16	1.8
Grouping and spectrum of pulses	31	3.5
Density, grouping and spectrum of pulses	5	o.6
T o t a l :	884	100.0

dense packages of strong low-frequency pulses whose duration can reach 60 ms. By means of "primary grouping" (Fig. 2-A), pulses of the same type and with identical spectra can be organized in time in a certain way: single pulses are combined in groups containing 2-10 pulses; these groups, as well as single pulses, can be used to form train-sequences with different pulses density. Though individual pulses and groups can be used occasionally as independent signals, it is trains which play the major role in pulsed signalization of bottlenose dolphins. During sound generation, bottlenose dolphins can, either successively or simultaneously, change the density in a train, the grouping of constituent pulses or their spectrum, and such changes can be very rapid. As a result, the ultimate structure of the signal will consist of different fragments connected by transformation zones, which result from successive combining of blocks with stable temporal and spectral characteristics.

The analysis of events in transformation zones shows that when only one characteristic in a train has been changed (Table 1) various types of transformations occur with nearly the same frequency. Simultaneous transformation of two or three characteristics is observed far less frequently which suggests limited combinatory capabilities of the sound generation system. The data in the table allow us to make the conclusion that bottlenose dolphins have two different mechanisms of signal structure control, one of which controls temporal organization of the train, while the other controls the spectrum of pulses. The complexity of simultaneous control of their operation must be the cause of the above-mentioned limitations.

In addition bottlenose dolphins have several specific types of spectral transformations. They always are observed in prolonged simple (non-grouped) signal fragments with stabilized repetition rate of pulses and should be regarded as a way of modifying the train. They include (Fig. 2-E,F,G,H) the shift of one of spectral peaks along the scale of frequencies, generation of a "spectral wave" in a train, "bursts" of energy, "pumping" of energy from one spectral peak to the other. It might be possible that parts with such transformations should be treated as independent structural blocks of a signal.

It is interesting to note the existence of short signals in bottlenose dolphins signals, which are very dense trains (more than 200 pulses/s) wherein the area of spectral transformations covers the entire structure. One can identify 4 major classes of such signals (Fig. 2-D): (1) narrow-band noise signals which, depending on the width of the spectral band, can sound like hoots, hoarse whistles or frequency-coloured noises. They can have a certain of the second class can become trains of pulses (and back). Signals from the third and fourth classes never make part of complex signals if their structure is formed by changing the sound generator's regime (they might become part of complex signals only if they are formed by two successively working sound generators).

The structure formation process of most pulsed signals is very similar to that of tonal signals, though, due to discreteness of initial elements, there are certain differences. The primary grouping, which creates stable structural elements of the second (groups and non-grouped trains) or the third (grouped trains) level, can be regarded as an analogue for homotypical combining of initial tonal elements and the combination of those elements into higher-level blocks - as an analogue of heterotypical combining. One can discover blocks of still higher levels in signalization.

If two sound generators participate in the formation of signal structures they can operate jointly in two ways - one after the other or simultaneously. If two generators operate in succession, signal structure is formed by fragments alternately produced by different generators. These fragments can be very diverse and are represented by trains of pulses differing in their grouping, spectral or density characterictics or the array thereof; tonal signals differing in register, contour shape, number of harmonics; noise signals, modulated trains. In the formation of the ultimate structure of a signal, they are used as constructive elements. Since there are no apparent restraints as to the compatibility of different types of elements and the nature of final construction is controlled only by requirements to information encoding, successive combining frequently results in generation of signals with peculiar structures (Fig. 3). They are characterized by great internal contrasts on the boundary between fragments resulting from the rapid switch of generators.

In trains of pulses, the generators' switch-over time is extremely short. That is why, when different trains are joined together, generators' switching zone is typically unpronounced on signal structure; it sometimes can be detected through a certain change in the amplitude of pulses. However, the junction of fragments must be difficult if it is necessary to connect two fragments and to maintain the trend of the preceding fragment's contour (Fig. 3-B). In such cases, dolphins often make a 10-60 ms pause in signal, to give the second generator time for a finer tuning, or switch the second generator a little earlier, so that there exists a little zone of overlapping fragments in the signal. The last-mentioned way is used to generate signals whose structure emphasizes the discreteness of its elements (Fig. 3-B,C,D). One should bear in mind that each successively operating generators preserves all of its earlier-described capabilities, therefore one can also observe all earlier-described trasformations and structural modifications within fragments.

In parallel operation, both generators work simultaneously for at least part of a signal's duration. New types of structural blocks are created in the zone of their joint operation, with combined levels higher than that of their components. Since each generator can work in the tonal or pulse regime, such blocks can consist not only of either tonal or pulse component, but can include both, thus emerges a possibility to generate combined signals.

When generators are operating in parallel, they produce signals wich are superimposed. This enabled us (Markov et al., 1974) to give the name "superimposition" to this method of generating complex structures. We observed no cases of both generators starting simultaneously: one of them always started earlier than the other, thus forming the basis of the signal which was later overlaid by the output from the second generator ("superstructure"). Irrespective of the type of signals (pulsed or tonal) produced by a generator, the number of relationships between them might be limited: (1) the superstructure

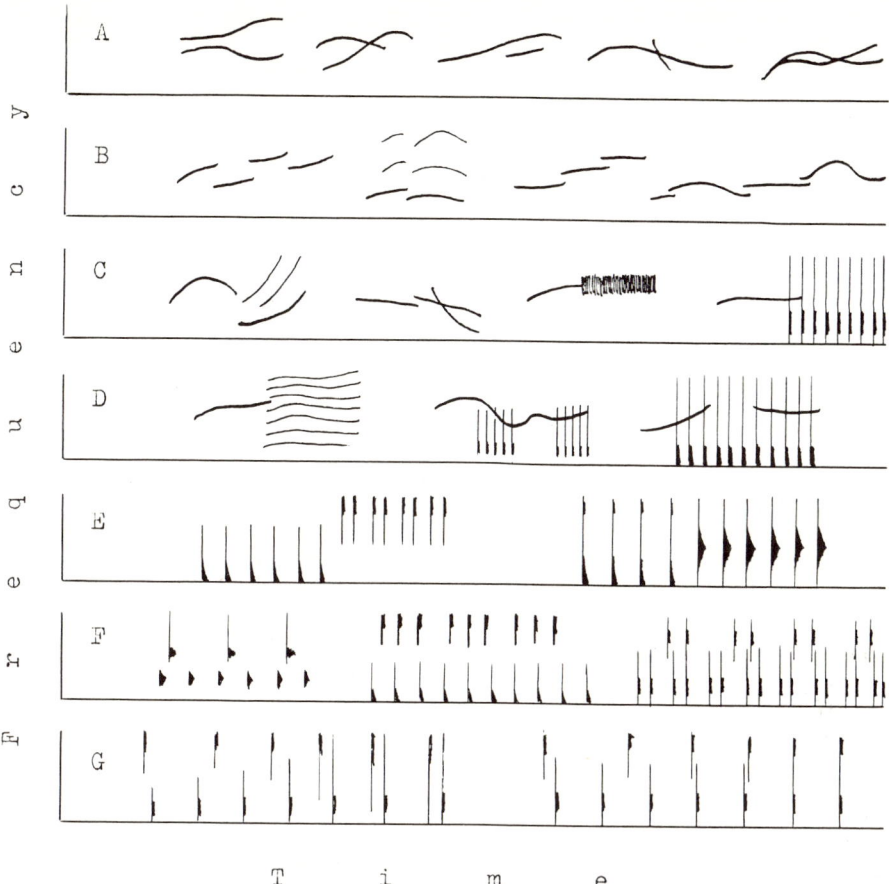

Fig. 3. Operation of two sound generators. Graphic presentation of the sonagrams. (A) Simultaneously (parallel) operation of two generators in tonal regime; (B) successive operation; (C) junction of different elements: various tonal components, tonal component and train of pulses; (D) junction of a tonal component and a modulated train of pulses, examples of signals formed during simultaneous operatition of two generators, one working in the tonal regime, the other - in the pulse regime; (E) junction of two trains of pulses with different groupings and pulse spectra; (F) parallel operation of both generators in the pulse regime; (G) conjugation and synchronization.

alteration of the average frequency of pulse spectra; (2) dense wide-band trains pronounced formant structure. The ration between formant frequencies can be multiple or not, and frequencies of formants can either remain constant or change in the course of signal duration; (3) modulated trains of pulses which have a discrete spectrum and sound like bleating; (4) dense trains of narrow-band pulses wherein the width of spectral peaks and location thereof on the frequency axis change simultaneously during signal generation. Depending on the way these characteristics change, signals of this class sound like growling, howling, barking, mewing. Signals from the first two classes can be used occasionally as second-level structural blocks in the formation of complex signals. In this, first class signals can turn into whistles, while those

Table 2. Transformations of Pulsed Signals During Simultaneous Operation of Two Sound Generators.

Transformed Pulse Characteristics of the first generator	Transformed Pulse Characteristics of the second generator	N	%
No transformations	– Density in a train	241	35.3
	Spectrum of pulses	77	11.3
	Density in a train and spectrum of pulses	157	23.0
Density in a train	– Density in a train	38	5.6
	Spectrum of pulses	142	20.8
	Density in a train and spectrum of pulses	9	1.3
Spectrum of pulses	– Spectrum of pulses	–	–
	Density in a train and spectrum of pulses	7	1.0
Density in a train and spectrum of pulses	– Density in a train and spectrum of pulses	11	1.6
	T o t a l :	682	100.0

is superimposed on the basis in such a way that it stops sounding simultaneously with the basis or earlier; (2) the superstructure is superimposed onto the end of the basis so that it stops sounding after the basis in which case inversion is possible (That is, later on the superstructure can be used as the basis); (3) a pause in the superstructure in which case the sounding of the superstructure is interrupted for some time and is resumed, the signal having the same or absolutely different parameters after being resumed. These relationships make it possible to create complex signal structures, especially if they are used repeatedly.

Observations and study of signal structure make it possible to conclude that dolphins use these relationships between generators in a meaningful way. Thus, in combined signals groups of pulses are confined to a certain contour area, and there are many instances when the same combinations of structural blocks with a high level of complexity are used in different signals. However, the most illustrative example is the use of synchronization and conjugation (Markov and Tarchevckaya, 1978; Markov and Osrovskaya, 1983) - two methods of mutual adjustment of generators aimed at creating rigid acoustic constructions (Fig. 3-G). This is possible only when both generators are working in the pulse regime and producing trains with different spectra of pulses and their frequency repetition rate. During synchronization, the frequency repetition rate pulses in one train changes to become equal to their frequency repetition rate in a second train; at the same time, pulses produced by different generators are precisely matched to form pulses with an integral spectrum Synchronization of trains consisting of single pulses (or of trains with the same grouping results in the formation of trains with a more complex spectrum than in initial trains. But if synchronization involves a train of single pulses and a grouped train, then pulses from the first train always are timed with the first pulse of the second train to form a grouped train where the first pulses in a group are spectrally different from other pulses; the total number of pulses in groups of the resulting train is equal to their number in groups of the initial grouped train. The dolphins use synchronization abun-

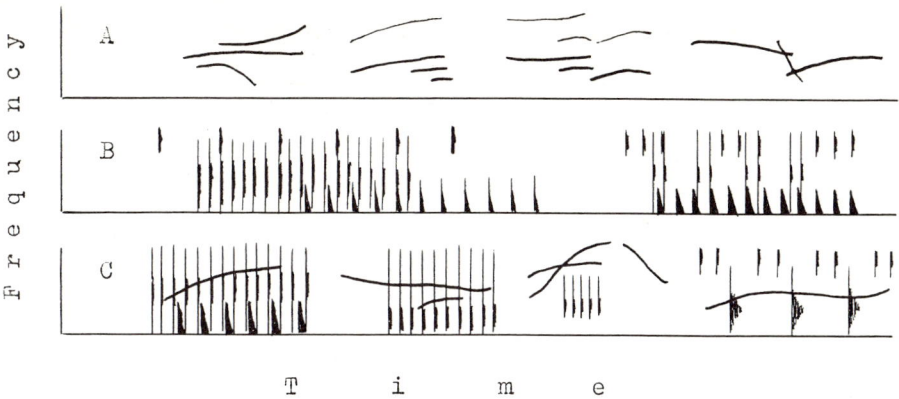

Fig. 4. Operation of three sound generators. Graphic presentation of the sonagrams. (A) All generators working in the tonal regime; (B) all generators working in the pulse regime; (C) different versions of combined signals.

dantly: about 22% of pulsed signals with superimposition involve synchronization. The mechanism of conjugation is very similar to that of synchronization, but in conjugation, pulses are not matched absolutely and the process consists in the formation of a train out of groups which are composed of pulses produced by different generators. Two trains can be conjugated only if one of them is simple (that is, composed of single pulses). If both rows are simple, conjugation results in the formation of a train composed of pairs of spectrally different pulses. But if one of them is a grouped train, groups in final sequence will contain one additional pulse, and pulses from the simple train always will be first and will be spectrally different from others. Sometimes conjugation involves trains consisting of spectrally similar pulses.

Simultaneous use of two sound generators considerably enhances dolphins' combination capabilities. However, this also brings about problems of sound generation control which inevitably results in certain losses. Even a cursory analysis shows that if both generators are operating in the tonal regime, each of them loses the possibility to dynamically control the number and energy of harmonics. If they are operating in the pulse regime, they lose the posibility to change the grouping of pulses in the process of signal generation. The latter partially is made-up for by the possibility to pre-tune generators to any group regime. In combined signals, if complex whistling is used as the basis, superimposition typically involves simple trains and if the basis is made-up of a train of pulses, superimposition involves tonal components with uncomplicated contour shapes (usually second-level blocks). A more careful analysis reveals further constraints which are most seen clearly in train transformation zones when both generators are operating in the pulse regime. Data in Table 2 show that it is so difficult to control the work of two generators that, in order to make any transformation in the structure of a train produced by one generator, dolphins stabilize the work of the second generator in 70% of cases. Simultaneous alternation of two characteristics in one train or of one characteristic in each train (if these characteristics are associated with the work of different mechanisms) occur with the same frequency. But if these transformations are associated with the operation of the same mechanisms, the frequency repetition rate of corresponding combinations goes down. Up to 4 characteristics (for two in each train instead of three during the work of a single generator) can be altered simultaneously in the signal which more than makes-up for the losses, since it creates 9 versions of signal transformation.

If three generators participate in the formation of signal's structure, then, like in the case of two generators, they can work in succession or in parallel. Unfortunately, the structure of signals formed by successive switching of generators does not allow one to determine how many generators are working. This can be reliably determined only in some cases (if all generators produce tonal signals and are switched on with a certain lead-time to create an overlapping zone at the juncture of fragments - Fig. 4-A). When working in parallel, generators can produce tonal or pulsed signals and, consequently, create structural blocks formed by superimposition of three tonal, three pulsed, two tonal and one pulsed, one tonal and two pulsed components ("double superimposition", Fig. 4-B,C). Naturally, such blocks are more complex than their components and blocks formed by ordinary superimposition. The analysis of generators' capabilities shows that in case of double superimposition of tonal signals, each generator retains its ability to change signal frequency, though the range and the rate of alteration of frequency in most cases are reduced. In case of double superimposition of pulse trains, only one characteristic can change at a time - density in a train or spectrum of pulses, since only one generator can change its work, while other generators work steadily in the pre-tuning regime. Such superimposition rules out the possibility to conjugate trains of pulses, but creates the possibility of successive double synchronization (i.e. synchronization of a simple train of pulses with a synchronized train). Blocks by combining tonal and pulsed components are most diverse (Fig. 4-C), but their pulsed components have very stable characteristics. Synchronized trains are extremely widely used as part of combined blocks (up to 40% out of the number of blocks consisting of one tonal and two pulsed components).

Signals whose structure is formed with the participation of four sound generators are observed rarely in bottlenose dolphins. While the share of signals formed by one generator accounts for about 55%, by two generators - 40%, by three generators - somewhat less than 5%, the share of signals formed by four generators constitute less than 0.1%. This is in good correlation with observed losses in structure transformation capabilities which the growing number of simultaneously working generators, and undoubtedly is associated with increasing difficulties in the sound generation system control. Evidently, these difficulties become maximum during simultaneous operation of four generators. Therefore, one should not expect generators to have great capabilities in such case. Unfortunately, we actually could not assess them, since all identified signals were a combination of tonal components with trains formed by double synchronization.

One can conclude that bottlenose dolphins have a wide array of tools to ensure operational control of the sound generation system. Dolphins can create signals with a very complex structure by changing the working regime of certain elements of the system in the course of sound generation. As it has been shown, this structure is discrete and consists of blocks with different levels of organization which were constructed by successive combining of simple "initial" elements from some basic set. The complication of block structure can be traced reliably in the analysed material only as far as the first stages of heterotypical combining. The identification of higher-level blocks (except when they are formed by ordinary and double superimposition) and estimation of their stability and complexity (i.e. the number of combination levels) proves difficult because of a vast volume of analytical material and also because it is impossible to make direct comparisons of data to identify blocks. That is why we made of graphic-symbolic language mentioned earlier in Chapter "Methods and Materials" which makes it possible to make a non-ambiguous description of signals and to construct their structural formulas within the limits of our assumptions.

Ways of solving the above problems can be illustrated without a detailed description of principles used for constructing the language and for developing exact structural formulas of signals. Therefore, here we will describe the simplest version. However, let us agree that whenever we say that certain blocks are similar, the reader is to understand that we have made the required direct comparison of analytical material and proved the identity of structures.

In order to describe the structure of signals, one has to introduce markers for participating elements and symbols for describing lincages between them. Here we only will introduce symbols which are necessary for reading the examples given below (for instance, we will omit the description of whistle contours and of spectral characteristics of pulses). Let us agree that: (W) - stands for tonal components of signals, irrespective of the complexity of their structure; (C) - pulses from the class of clicks; (B) - pulses from the class of "clear blows"; (P) - pulses from the class of "prolonged blows". Stages of primary grouping and the structure of trains of pulses will be described by general formula I_n^m, where (I) - is the class of pulses, (n) - the number of pulses in a group, (m) - the train density index. Let us agree that m=0, if we are descibing single pulses or groups (in which case the index is not mentioned); m=1, if train density does not exceed 40 pulses/s; m=2, if train density ranges from 41 to 100 pulses/s. Trains with still higher density (from 101 to 1200 pulses/s) will be denoted by (Cr). Thus, a single group of 4 clicks will be described as C_4, a dense train consisting of pairs of prolonged blows - P_2^2, etc. Symbol ($\rightarrow$) will denote a smooth changes in characteristics of trains of pulses (density, grouping, spectrum of pulses or a simultaneous change of two or three of them); (+) - junction of stable fragments of pulse trains produced by differnt sound generators, or junction of tonal and pulse components; ($\frac{K2}{K1}$) - superimposition (K1 and K2 - are components produced by different sound generators, ⟨K1⟩ - being the basis, ⟨K2⟩ - the superstructure); ($\frac{K2}{K1} \lrcorner \frac{K3}{K2}$) - inversion (i.e. the use of the superstructure as the basis of the next structural block; the signal always has a section with only K2 sounding); (:) - pause in the superstructure (that is, a short break in the work of its generator). Using these and other symbols, we described the structure of great number of signals produced by several adult animals.

With the help of analysis of formulas, we established that signals produced by bottlenose dolphins contain stable blocks of fourth and higher levels of complexity which are used as an entity in different signals. Sometimes such blocks are used as independent signals. For instance, block $\boxed{C_1^2 \rightarrow C_2^{1\,\overline{W}}}$ can be used both as an independent signal and as a constructive element in various parts of complex signals:

(1) $C_3^1 \rightarrow \boxed{C_1^2 \rightarrow C_2^{1\,\overline{W}}} \rightarrow C_2^2 \rightarrow C_1^1$;

(2) $C_2^{2\,\overline{P_1^2}} \rightarrow \boxed{C_1^2 \rightarrow C_2^{1\,\overline{W}}} \lrcorner W$;

(3) $\boxed{C_1^2 \rightarrow C_2^{1\,\overline{W}}} \rightarrow \dfrac{P_1^1 : W + C_2^2 + C_2^2 \rightarrow C_1^2 \rightarrow Cr}{C_4^2}$;

(4) $\dfrac{C_2^1}{W} \mathrel{\llcorner} \boxed{C_2^1 \to C_1^z \to C_2^1} \to \dfrac{W:W:W:W:P_2^1}{C_4^z} \mathrel{\llcorner} P_2^1 \to C_1^1$;

(5) $C_4^z \to \dfrac{\dfrac{P_1^1}{W}}{C_2^1} \to \boxed{\dfrac{\dfrac{P_1^1}{W}}{C_1^z} \to \dfrac{W+B_z:B_z^z}{C_2^1}}$;

(6) $\boxed{C_1^z \to \dfrac{\dfrac{C_1^z}{W}}{C_2^1}} \mathrel{\llcorner} C_1^z \to \dfrac{W:W}{Cr}$.

Stable blocks with a more complex structure are also rather frequently observed, for instance, $\boxed{C_2^1 \to \dfrac{W}{Cr} \to \dfrac{C_1^z}{C_1^z}}$, as a structural unit makes-up part of $C_2^1 \to \dfrac{Cr}{C_1^1} \mathrel{\llcorner} Cr \to \boxed{\dfrac{W}{C_2^1} \to Cr \to \dfrac{C_1^z}{C_1^z}}$. Analysis shows that when producing a sequence of purely tonal signals, dolphins tend to combine two or three signals into one construction (uniting blocks of the 4th and 5th levels), the same type of combination also has been observed in pulses and combined signals. It is quite possible that most signals are a combination of several large blocks.

The last-mentioned fact made us assume that hierarchical combining of structural elements observed at earlier stages, is a common principle used for creating signal structures and that any structural block can be used both as an independent part of a signal, and as a structural element for the formation of blocks with a higher degree of complexity. The above examples support this idea, nevertheless, we conducted additional analysis of data to construct the entire hierarchical staircase for large blocks, starting from initial elements. One example is the formation of block $\boxed{\dfrac{Cr}{B_1^z}}$ from Cr and B_1^z, which also can be observed as independent signals. This block is used both as an independent signal and as a structural unit in the formation of complex $\boxed{\dfrac{\dfrac{Cr}{B_1^z}}{C_2^1}}$ which, as a stable block, forms part of complex signals:

(1) $C_3^z \to C_1^1 \to \boxed{\dfrac{\dfrac{Cr}{B_1^z}}{C_2^1}}$; (2) $Cr + C_2^1 + Cr + \boxed{\dfrac{\dfrac{Cr}{B_1^z}}{C_2^1}}$, etc. We identified structural blocks formed with the participation of seven combination levels, but in most cases dolphins use simpler blocks (with three to six combination levels). The structure of real signals is a combination of blocks with various degrees of complexity.

Multi-level combination is a tool for creating a variety of acoustic constructions with different qualities. If we know the set of initial elements, the number of combination levels, the indices describing the participation of blocks from different levels in the formation of signal structure, as well as the indices describing the increase in the number of combinations attainable at each combination level, we can estimate the potentially attainable vocabulary, but at present we do not have enough data for such estimations. That is why we will apply a different method.

Since signal structure is generated in time, a signal is a chain of differntly sounding blocks (irrespective of their complexity). The number of such blocks in a signal varies from 1 to 24, averaging to 5-7. The number of structural types of blocks has not been established definitely, but it is well over one hundred. Using this data and standard formulas from Games Theory, one can calculate easily that 10^{12} signals could be produced by means of free combining. True, it is a potential estimate, but even if we assume that, because of code-associated or physical block combination bans, only one ten-millionth of them actually is used, still there remain 10^5 signals available for communication, which certainly is more than required for actual communication. All this makes it possible to think that the communicative system of bottlenose dolphins is "open" in terms of vocabulary formation. This conclusion is indirectly supported by the fact that dolphins use hundreds of structural types of signals for communication (see, for instance, Table 3).

ORGANIZATION OF SIGNAL SEQUENCES

It is rare that bottlenose dolphins produce single signals. As a rule, this is typical of very young or isolated adult animals. In normal communication, the intensity of signalization is very high, reaching sometimes 50 signals per minute. In free dialogue (for instance, during communication of isolated animals through elecro-acoustic communication link), signals with different structures are combined into groups, the way human words are combined to construct phrases. Grouping is well-pronounced in normal conditions of communication between calm animals, but it drastically changes or disappears in stressed situations, when the frequency range of communication is severely restricted or communication between individuals is broken.

In a number of situations, when dolphins mostly are using tonal signals, one can indentify sets of tonal signals with a common structure component. The analyses of variability of signals from the set, has shown that their middle sections are most stable, while edge sections (especially those at the end) are extremely variable. Variability behaves differently in groups composed of different signals. This allows one to assume that the order in which signals follow each other in groups, is meaningful for the animals and that the described variability depends on the interaction of signals and, consequently, on the existence of organization in a sequence of signals. This assumption is supported indirectly by dolphins producing groups with identical composition, sometimes consisting of signals with a very complicated structure.

This assumption is rather non-trivial and actually recognizes the ability of bottlenose dolphins to generate organized messages (texts). Certainly, it still has to be proved. Unfortunately, semantic control of messages is impossible for our case, so such proof has to be obtained with the help of a different method. We will apply the rank distribution method which is used widely in systems analysis.

If we classify a set according to some feature or the sum of features, establish the frequency of occurence of identified taxa, and rank the taxa according to their decline of frequency of occurence, we can obtain a distribution in which a taxon's frequency of occurence is associated with its place (rank) in the ranked row. While there is a great variety of such distributions, there is one type which attracted our attention. It has long been known that many events in different sciences can be described by the same type of rank distribution, described by equation

$$f_r = Cr^{-\gamma}$$

where C and γ are constant, and r is a taxon's rank. Numerous studies have used this equation (which bears the name of Pareto's law in economics, Zipf's law - in linguistics, Lotka's law - in sciencemetry, Bradfort' law - in informatics). This function has a broad nature and can describe various events, such as urban growth, distribution of annual incomes, ranking of scientists according to the number of published articles and contacts with colleagues, structure of musical pieces, and lexical structure of texts. In biology, this distribution was studied in connection with the mathematical Theory of Evolution; it governs the composition of biocenoses, the number of endemic genera in families, the number of species in genera, distribution of soil fauna species according their numerical strength.

Hiperbola, as a rank distribution type, first was established purely empirically, and it was quite recently (Kozachkov, 1973, 1978; Sreyder and Sharov, 1982; Yablonsky, 1976) that it was substantiated theoretically and meaningfully interpreted. It became evident that a hiperbolic rank distribution encompasses a multi-level organizational structure of a complex integral system. It also was shown (Shreyder and Sharov, 1982) that hiperbolic distribution is a consequence of the minimum-of-symmetry principle applied in the entire system, and has a criterion-setting status, that is, it provides an operational method for integrity diagnosis, a method for empirical verification of the existence of system organization. Therefore, the task of establishing the presence of organization in signal sequences produced by bottlenose dolphins is reduced to obtaining sequences containing, in as much as possible, integral texts, to compiling frequency of occurence dictionaries for them and verifing their correspondence to theoretical distributions. Since complex systems have semiotic (that is, really language-oriented) informational relationships and information exchange in them takes place at the semantic level, the proximity of empirical and theoretical distributions also will testify to ability of bottlenose dolphins to produce semantically organized messages.

The array of analytical and contents-oriented assessments most comprehensively developed in liguistics, where hyperbolic frequence of occurence distribution of words in a text has the name of Zipf's disribution. That is why we also are going to use this name. However, it would be appropriate to make several comments before going-over to the analyses of the results obtained.

Various human-produced texts follow Zipf's law in practically the same manner (Zipf, 1949), because rank distribution reflects properties of a text as an integral system, rather than the nature and subject orientation of the text. Therefore, deformation of rank distribution should be considered and interpreted in terms of changes in the structure of the text and breach of its integrity. Words within a rank disribution are qualitatively different. Those which rarely are encountered in the text (1-2 times) make-up the bulk of the vocabulary (but a small part of the text) and impart originality to the text; words with medium frequency of occurence occupying middle ranks are "responsible" for the description of a theme, while words occupying first places in the rank are mostly link-words. The very existence of the distribution points to the fact that there is a certain equilibrium between those groups of words in the text, therefore the analysis of departure in the distribution from theoretical values allows one to make a judgement about qualitative characteristics of the text. The quality of the text can be assessed integrally through index γ, which usually is close to 1 in normal texts.

All constituent elements of a system are interrelated by structural links and there also are links between system levels. That is why, in contrast to random sampling, a mere increase in the volume of the text does not improve the shape of rank distribution. On the contrary, distribution becomes deformed if a certain critical volume is exceeded, since an infinitely large increase of volume makes the disribution curve tend to the curve described by a circle equation in bilogarithmic coordinates. This is associated with the breach of balance between the volume of the text and the volume of the vocabulary and can be explained by the weakening of structural links in the text. As a result, a bulge appears in the middle part of the rank distribution curve. The same is observed in texts with a "good" volume if they are stylistically non-uniform (poorly organised) or if their integrity is upset in this or that way (the text contains parts from another text, or some parts are missing, etc.). Therefore, the bulge appearing in the middle of rank distribution can be interpreted as an increase of informational noise level in the system.

In our case, there were two obstacles which hindered the use of simple methods and analytical tools of Zipf's law. First, it is usually insufficient volumes of texts, since it is difficult to obtain sufficient volumes (15-30 thousand signals). As for smaller texts, it was shown by Arapov and co-authors (1975), that their lexical composition tends to have a greater diversity of vocabulary, with γ index decreasing, and the hyperbola being spread along the rank axis. Secondly, empirical disributions always are integer ones, while the above-cited Zipf's equation is continuous. Whether should the theoretical curve lie with respect to empirical distribution and what corrections should be used to make-up for the whole-numbered feature of empirical distribution, is of paramount importance and resulted in a whole array of papers, but a cardinal solution seems to lie in the development of a computational method for a theoretical numerical curve for every empirical distribution being studied.

Such a method, making it possible to work with texts of unspecified volumes, and the associated assessment tools have been developed in our laboratory by M. A. Ostrovsky on the basis of the minimum-of-symmetry principle. Corresponding algorithms and computer programs have been developed, allowing one to estimate a corresponding theoretical rank distribution for any text and theoretical volume of vocabulary, and to assess the similarity between the shapes of empirical and theoretical distributions (parameter D), as well as the general inclination of the rank distribution curve in the bilogarithmic coordinate system (parameter γ). They have been tested on linguistic material (monologues and dialogues), following which they were used for the analysis of dolphin signals. Here we will mention only some of the results which are important for the subject of the paper (Table 3, Figs. 5 and 6).

In order to establish the presence of internal organization in signal sequences, signals produced by dolphins were recorded in the course of special experiments involving the communication of isolated animals via a wideband electro-acoustic communication link. We obtained a series of recorded signals produced by animals in different psychological states. Sets containing dialogues were considered as an integral text and were checked against "pseudo-dialogues" representing a sum of simultaneously recorded signals produced by animals when communication between them was broken. According to the results of structural analysis, the signals were classified; frequency of occurence dictionaries and ranked rows were constructed for each version of the experiment. The latter were fed into a computer and processed with the help of the above programs.

Empirical rank distribution of signals produced by calm animals, well-

Table 3. Major Parameters of Rank Distributions

C*	Vol. of text	Volume of vocabulary			Sum of frequency 1, 2 and 3 ranks			Sum of signals with freq.1,2,3			γ	Dx100
		E	T	K	E	T	K	E	T	K**		
A	810	198	207	95.7	198	216	91.7	223	232	96.1	0.89	0.114
B	6930	822	1057	77.8	1305	1686	77.4	797	1154	69.1	0.98	0.366
C	488	44	86	51.2	157	210	74.8	42	93	45.2	1.08	2.283
D	498	63	88	71.6	261	274	95.3	71	62	114.5	1.24	1.587

*Conditions: (A) calm animals, integral text; (B) calm animals, aggregate text; (C) animals in a stressed state; (D) pseudo-dialogue during interrupted communication.
**(E) empirical and (T) theoretical distributions, (K) is the ratio between data for empirical and theoretical distributions, %.

adapted to conditions of experiments, have a good fit with theoretical distributions (Table 3, Fig. 5A), if the analysis involved continuous sets of dialogues. As was expected, the middle part of rank distribution obtained for an aggregate set (consisting of several signalization fragments of the same animals, recorded in the course of one day) has a bulge caused by a breach of the texts integrity (Fig. 5B). The value of parameter D in this set is higher than in homogeneus text (i.e. the shape of its rank distribution is more different from that of theoretical distribution); it has a certain lack of signals from first ranks and a decreased diversity of vocabulary (assessed through a sum of signals observed 1,2 and 3 times in the set) - Table 3. In both cases the value of γ parameter virtually does not differ from that of normal human speech.

In order to make judgements about the behavior and similarity between empirical and theoretical distributions, Arapov and his co-authors (1975) proposed to use Z-estimate and estimate of the degree of vocabulary coverage by the text, i.e. the calculation of the sum of frequences accumulated during the movement along the rank axis. However, this method can not be applied in its "pure" form, because of different volumes of vocabularies in theoretical and empirical distributions. That is why we used rated indices, and Z-difference for detecting most diverging zones, i.e. we calculated the difference between relative accumulated frequences of empirical and theoretical disributions. Rating required the use of interpolation procedure which made computations somewhat more difficult, but it provided us with a convenient and illustrative presentation of divergencies. Figure 6 shows Z-difference curves for both of the above-described texts. One clearly sees that the heterogenic and theoretical sets are most different in the zone of small and medium ranks, but on the whole, both empirical distributions are very different from theoretical.

Interruption of communication between animals isolates them from one another and breaks the integrity of the text, which results in an abrupt deformation of empirical distribution obtained for "pseudo-dialogue" (Table 3, Figs. 5-C, 6). A considerable excess of initial rank frequencies signifies that the animals switched-over to stereotype signalization, while lack of medium rank frequencies points-to the destruction of principal links in the message (at the same time, signalization is more diverse than predicted by the theory

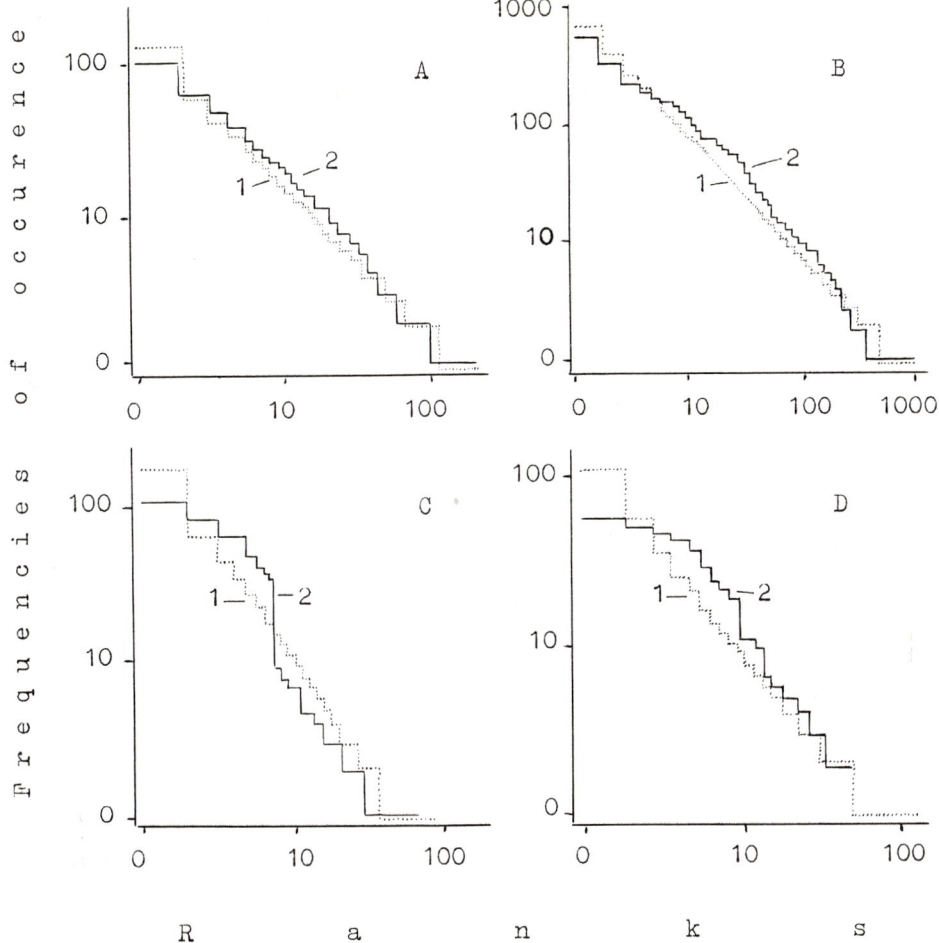

Fig. 5. Zipf's distribution plots for texts obtained during dolphins communication via elecro-acoustic communication link. For all the plots, ⟨1⟩ - theoretical, ⟨2⟩ - empirical disribution. (A) normal communication of calm animals, integral text; (B) normal communication of calm animals, aggregate text; (C) during interrupted communication and (D) animals in a stressed state.

which might be caused by the animals' attempts to restore interrupted communication). The values of D and γ parameters are very high, but not beyond the limits characterizing a completely destroyed system, therefore one might suggest the existence of organized fragments in individual signalization of the animals.

Stress is always accompanied by the use of stereotype signalization and abrupt decrease of dolphin vocabulary (Sidorova et al., 1990), which should have a strong deforming effect on empirical rank distributions. One can expect that the zone of initial ranks will have a group of signals with frequencies exceeding those predicted by the theory. For verification, we recorded two dialogues between very excited animals, and the analysis of those yielded similar results (Table 3, Figs. 5-D and 6 show results for only one dialogue).

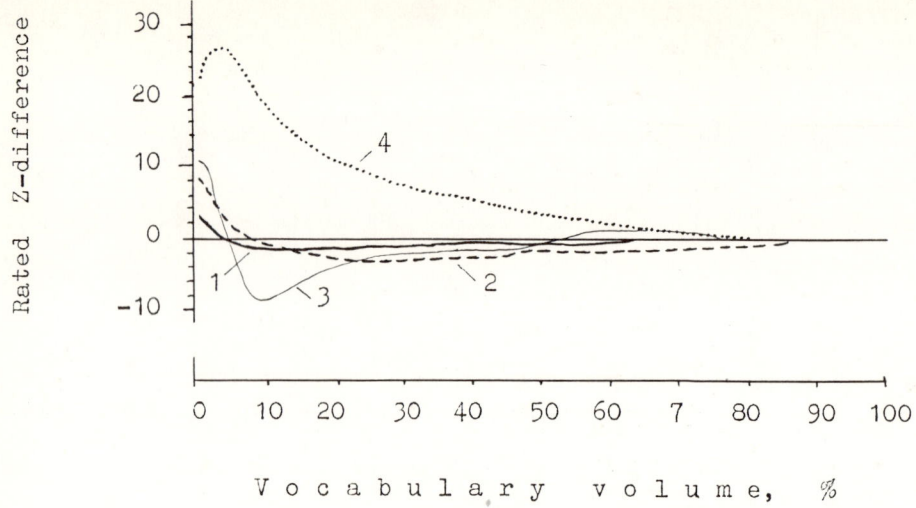

Fig. 6. Rated Z-difference for different texts. <1> normal communication of calm animals, integral text; <2> normal communication of calm animals, aggregate text; <3> during interrupted communication; <4> animals in stressed state.

Both cases displayed an increased frequency of first ranks and a strong decline of vocabulary (especially with respect to its "contens-oriented" part). Because of this, Z-difference plots show a strong departure of empirical curves from the theoretical axis of the plot, thus pointing to a great unbalance in the vocabulary-text system, nearly amounting to breach of communication. The value of D parameter also points to substantial differences in distributions.

Thus, signal sequences of bottlenose dolphins have a multi-level hierarchical organization, typical of complex systems. It is well-pronounced by calm adapted animals and is upset in isolated or very excited individuals. Since communicative system is an informational system, the existence of such organization signifies the ability of bottlenose dolphins to exchange organized messages or, at least, construct texts in the course of a dialogue. But certain laws typical of texts, should be observed in signalization. If we managed to prove the existence of even one such law, using criteria and metrics different from those used in Zipf's law, we could verify our conclusion.

As was already shown, any text contains various elements, equally important and necessary for message organization, but having different functions: rarely observerd words determining the novelty and originality of the topic of the message, more frequently observed words responsible for the description of the topic, and very frequently observed words performing link-functions and ensuring the arrangement of the text. It is even intuitively clear that, to ensure normal functioning of the communication system, it is preferable that link elements have smaller energy values, and, consequently, less complicated structure than, for instance, elements describing the novelty of the topic. In mathematical linguistics and semiotics, the complexity of words often is expressed through their length measured by the number of letters and syllables (Piotrovsky et. al., 1977). The length of a word used in the text

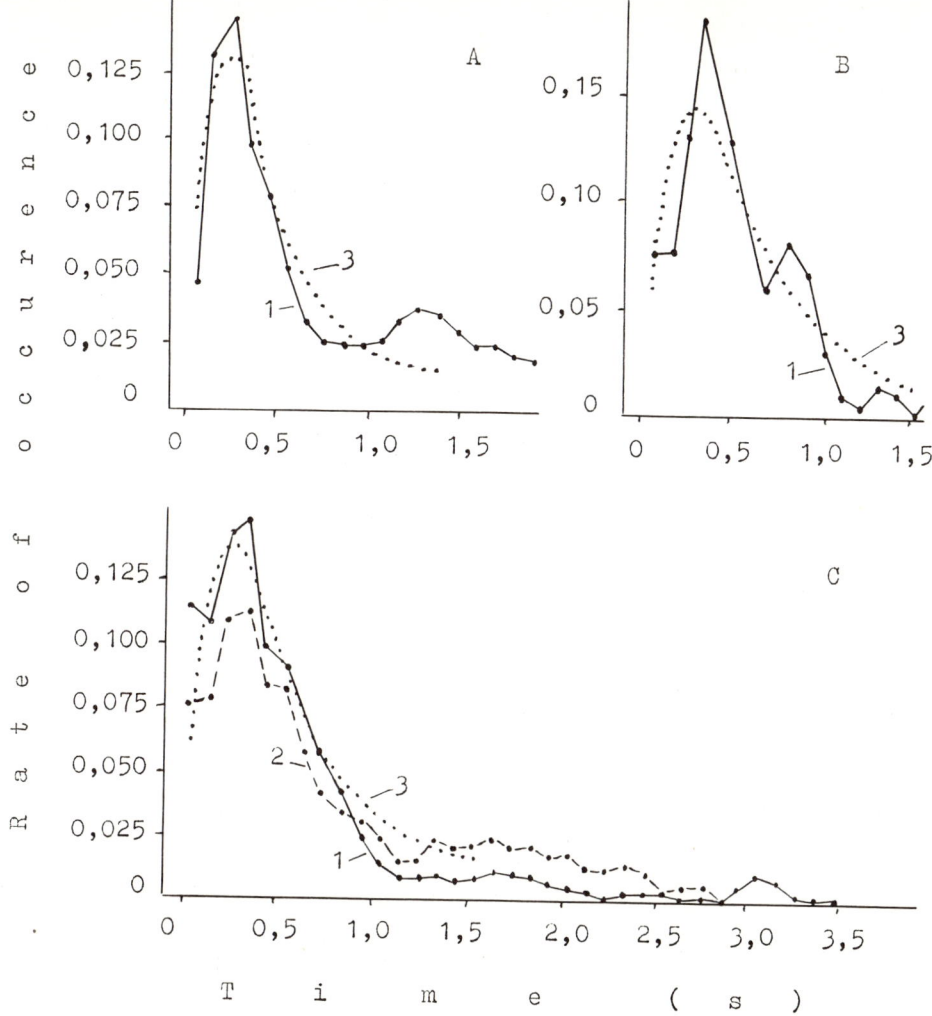

Fig. 7. Distribution of signal lengths in texts produced during normal communication of calm dolphins. (A) An adult female, (B) and (C) different adult males. ⟨1⟩ and ⟨2⟩ - empirical curves; ⟨3⟩ - theoretical curves (in part C the theoretical curve is for empirical curve N1). Number of samples: A1 = 426 signals, B1 = 102, C1 = 438, C2 = 664.

is not random and depends on its context and the length of the preceeding word. That is why, in contrast to a random set of unconnected words, those analysis yields a normal length distribution, a set containing a connection, organised message yields a lognormal disrtibution. Herdan (1964) believes that this distribution reflects the principle of optimal information encoding typical of natural languages.

Our verification procedure consisted in measuring lengths of signals in sets used for the analysis of rank and some other distributions, and in analyzing the obtained empirical distributions in terms of their fit with the lognormal law. The length of signals was measured in the number of 100-ms

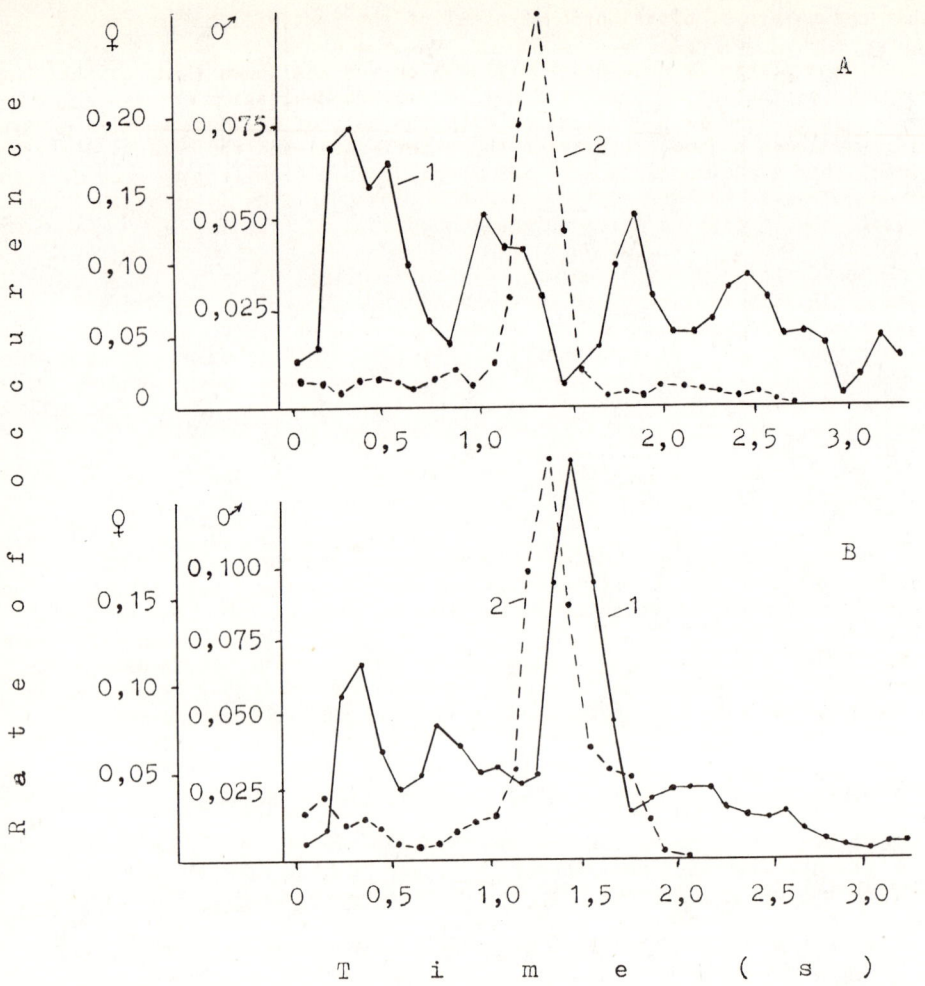

Fig. 8. Distribution of lengths of signals in texts produced by dolphins during interruption of communication (A) and in a stressed state (B). ⟨1⟩ An adult male, ⟨2⟩ an adult female. Number of samples: A1 = 152 signals, A2 = 272, B1 = 280, B2 = 258.

time intervals.

The analysis of aggregate data with lengths of signals produced by several animals, has shown that distribution curves always have two peaks. Later it was established that the origin of those peaks is different; the first of them being associated with informational interaction of animals, while the second is due to a stressed state. Therefore, we singled-out two groups of situations - situations associated with signalization in a stressed state and those associated with normal bilateral communication between animals. In each situation, the analysis was performed separately for every dolphin, because preliminary observations established that the rate of signalling and the average length of signals are individual features (in our case, it is admissible to single-out signals of specific individuals out of dialogues, because, with a sufficiently large amount of samples, the law of large numbers guarantees that the nature of distribution typical of the text will be preserved).

The analysis of obtained individual curves has shown (Fig. 7) that lognormal distribution is preserved well during communication between calm, well-adapted animals. Verification with the help of Kolmogorov-Smirnov criterion has shown a good fit between the experimental and theoretical distributions. This supports the conclusion about dolphin's ability to produce organized messages, besides, following Herdan, one can draw an additional conclusion about dolphins using optimal encoding.

As it was to be expected, the fit to lognormal distribution is upset in case of animals isolation and in stressed situations (and, as was shown by additional experiments, when the frequency range of the communication is severely restricted). In such cases, signal length distribution curves develop a peak caused by stereotype signalization (Fig.8) whose shape is described by a normal distribution curve. As the animals calm down, signalization changes and the lognormal signal length distribution is restored.

CONCLUSIONS

Summing-up data provided in the paper, one concludes that bottlenose dolphins have an open-type communicative system. Using a very flexible sound generation system, they can use multi-level combining for constructing a virtually unlimited vocabulary - a set of acoustic signals - and for forming a multitude of organized text messages out of its units. As to degree of complexity, the communicative system of bottlenose dolphins is unique, and nothing of the kind has so far been discovered in other animal species.

REFERENCES

Arapov, M. V., Yefimova, Ye. O., and Shreyder, Yu. A., 1975, On the essence of rank distributions, VINITI, NTI, ser. 2, I : 7.
Bel'kovitch, V. M., Ivanova, Ye. Ye., Yefremenkova, O. V., Kozarovitsky, L. B., and Kharitonov, S. P., 1978a, Features of searching and hunting behaviour in dolphins, in: "Povedeniye i bioakustika delfinov", V. M. Bel'kovitch, ed., Inst. Okeanol. AN SSSR, Moscow.
Bel'kovitch, V. M., Ivanova, Ye. Ye., Yefremenkova, O. V., Kozarovitsky, L.B., and Kharitonov, S. P., 1978b, Play behavior of dolphins, in:"Povedeniye i bioakustika delfinov", V. M. Bel'kovitch, ed., Inst. Okeanol. AN SSSR, Moscow.
Herdan, G., 1964, "Quantitative linguistics", Butter Worths, London.
Herman, L. M., 1980, Cognitive characteristics of dolphins, in: "Cetacean behavior. Mechanisms and functions", L. M. Herman, ed., Willey-Intersci. Publ., NY.
Herman L. M., 1986, Cognition and language competencies of bottlenose dolphins, in: "Dolphin cognition and behavior. A comparative approach", R. J. Schusterman, J. A. Thomas, F. G. Wood, eds., Hillsdale, NY.
Herman, L. M., 1987, Receptive competencies of language-trained animals, in: "Advances in the study of behavior", J. S. Rosenblatt, C . Beer, M-C. Busnel, P. J. B. Slater, eds., 17, Acad. Press, Petaluma, CA.
Herman, L. M., Richard, D. G., and Wolz, J. P., 1984, Comprehension of sentences by bottlenose dolphins, Cognition, 16:129.
Kozachkov, L. S., 1973, Certain integral properties of hierarchic information system, Kibernetika, I : 139.
Kozachkov, L. S., 1978, Information systems with a hierarchic ("ranked") structure, VINITI, NTI, ser. 2, 8 :15.
Krushinskaya, N. L., 1983, Cetacean behavior, in: N. L. Krushinskaya, T. Yu. Lisitsina, "Povedeniye morskikh mlekopitajuqikh", Nauka, Moscow.

Ladygina, T. F., and Supin, A. Ya., 1974, Evolution of cortex zones in terrestrial and aquatic mammals. in: "Morfologiya, fisiologiya i akustika morskikh mlekopitajuqikh", V. Ye. Sokolov, ed., Nauka, Moscow.
Lilly, J. C., 1962, "Man and dolphin", Victor Gollancz, London.
Lilly, J. C., Miller, A. M., 1961, Sound emitted by bottlenosed dolphins, Science, 3465 : 1689.
Markov, V. I., 1976, Organizational principles and development of communicative systems in animals, in; "Gruppovoje povedenije zhivotnykh", B. P. Manteyfel, ed., Nauka, Moscow.
Markov, V. I., 1977, Signals of bottlenose dolphins produced by simultaneous operation of three sound generators, Doklady IX Vsesojusnoj Akust. Konf., Ser. TS, Inst. Akust., Moscow, 41.
Markov, V. I., 1983, Peculiar features in the transfortmation of the structure of pulsed communicative signals during simultaneous operation of two sound generators of bottle-nosed dolphins, Doklady X Vsesojusnoj Akust. Konf., ser. CH, Inst. Akust., Moscow, 12.
Markov, V. I., 1984. Evolution of animals' communicative systems, in: "Makroevolutsija", A. L. Yanshin, ed., Nauka, Moscow.
Markov, V. I., and Ostrovskaya, V. M., 1975, On combinatory nature of whistles emitted by bottlenose dolphins, in:" Morskije mlekopitajuqije", G. B. Agarkov, ed., I, Kiev.
Markov, V. I., and Ostrovskaya, B. M., 1983, Synchronization of sound generators in the formation of pulsed sound spectra by bottlenose dolphins, Doklady X Vsesojusnoj Akust. Konf., ser. CH, Inst. Akust., Moscow, 16.
Markov, V. I., Ostrovskaya, V. M., and Ostrovsky, M. A., 1983, Structural system approach to the analysis of animals' communicative systems, in: "Povedenije zhivotnykh v soobqestvakh", A. A. Zakharov, ed., v. 2, Nauka, Moscow.
Markov, V. I., and Tarchevskaya, V. A., Estimating the capabilities of sound generation system in bottlenose dolphins (structure of signals formed by joint work of two sound generators), in: "Morskije mlekopitajuqije. Resultaty i metody issledovanij", V. Ye. Sokolov, ed., Nauka, Moscow.
Markov, V. I., Tarchevskaya, V. A., and Ostrovskaya, V. M., 1974, Organization of acoustic signals in Black Sea bottlenose dolphins, in: "Morfologija, fisiologija i akustika morskikh mlekopirtajuqikh", V. Ye. Sokolov, ed., Nauka, Moscow.
Morgane, P. J., 1978, Whale brain and their meaning for intelligence, in: "Whales and whailing", S. Frost, ed., Austral. Gov. Pull. Serv., v.2, Canberra.
Morosov, D. A., 1970, Dolphins are hunting, Rybnoje khozyajstvo, 46:378.
Piotrovsky, P. G., Bektayev, K. V., and Piotrovskaya, A. A., 1977, "Matematicheskaya lingvistika", Vysshaya shkola, Moscow.
Sidorova, I. E., Markov, V. I., and Ostrovskaya, V. M., 1990, Changed signalling of bottlenose dolphins during adaptation to captivity (in press).
Shreyder, Yu. A., and Sharov, A. A., 1982, "Systems and models", Radio i svyazj, Moscow.
Yablonsky, A. I., 1976, Stochastic models of research activities, Sistemnyje issledovanija, Ezhegodnik 1975, Nauka, Moscow, 5.
Yablokov, A. V., 1983, Features of brain structure and development of cognitive abilities in Cetaceans, in: N. A. Krushinskaya, T. Yu. Lisitsina "Povedenije morskikh mlekopitajuqikh", Nauka, Moscow.
Zanin, A. V., Markov, V. I., and Sidorova, I. E., 1990, Ability of bottlenose dolphins, Tursiops truncatus, to report arbitrary information (this book)
Zipf, K., 1949, "Human behaviour and the principle of least effort", Cambridge.

SIGNALIZATION OF THE BOTTLENOSE DOLPHIN DURING THE ADAPTATION TO DIFFERENT STRESSORS

Irina E. Sidorova, Vladimir I. Markov, and Vera M. Ostrovskaya

A. N. Severtsov Institute of Evolutionary Morphology and Ecology of Animals, USSR Academy of Sciences, Leninskiy Prospect 33, Moscow 117071, USSR

INTRODUCTION

The results of our investigations (Sidorova et al., 1986) and the analysis of literature (Caldwell and Caldwell, 1965; Caldwell and Caldwell, 1977; Herman and Tavolga, 1980) indicate that the acoustic signalization of the bottlenose dolphin becomes stereotyped in different stressful and exciting situations. In fact, in these cases each dolphin begins to reiterate the only one, "individual" (Caldwell and Caldwell, 1965) whistle. In normal conditions, signalization is much more divergent. This leads to a supposition that the diversity of dolphin signalization depends upon the conditions around the animals, and, evidently, is connected with their physiological state. The aim of our investigation was to study the possibility of using signalization parameters for the assessment of the animals' state. We were to study the four main points: (1) whether or not the stereotyped signalization appears only in stressful situations; (2) commonalities in the conditions under which it is evoked; (3) signalization changes during the adaptation of the animals to different stressors; and (4) whether the peculiarities of the situation and the general level of adaptation of a particular animal to the life in captivity influence the parameters of stress signalization. In order to answer these questions, we studied the behavior and signalization of two bottlenose dolphins in the period of their adaptation to the life in an oceanarium and to experimental work, and also in the repeated model stress situations.

MATERIALS AND METHODS

This work was conducted in the Karadag Department of the Institute of Biology of Southern Seas of the Ukrainian Academy of Sciences from July 1982 to March 1983. Our subjects were two adult females of the Black sea bottlenose dolphin (*Tursiops truncatus ponticus*, Barabash), named Jenny and Kora. These animals were captured on May 1982 and for five weeks held in the sea enclosure with the other members of the same school on the Utrish station of the Institute of Evolutionary Morphology and Ecology of Animals of the USSR Academy of Sciences. Afterwards, they were transported to the Karadag station and for three weeks were lodged in a pool with the other two dolphins. Then we moved our animals to individual pools, and began our observations.

Table 1. List of Situations and Volume of Material.

Situations	Dolphin	Number of signals
Separation of the dolphins into individual pools	Jenny	364
	Kora	129
Training procedure	Jenny	1801
	Kora	1082
Strange man in the pool instead of the trainer	Jenny	173
	Kora	264
Introducing into smaller pool	Jenny	592
	Kora	285
Dividing nets in the pools	Jenny	250
	Kora	210

Jenny and Kora were located in adjacent pools separated by a concrete wall, and connected with a window in a top corner of the wall. The depth of water was about 4.5 m in both pools. The surface areas were 34.1 m^2 in Jenny' pool and 58.8 m^2 in Kora's one.

We watched the behavior and signalization of the dolphins for eight month in the process of their adaptation to the life in captivity, training, and the complication of experimental tasks. The influence of these factors on the animals was assessed by Jenny's and Kora's behavior. We determined situations as stressful if they evoked a strong negative emotional exitement in the animals which was expressed in different neurotic reactions. The most typical of these reactions was the specific character of moving. The dolphins were stationary in a vertical position with their rostrums to the bottom, and were flexing their tails intensively. Both animals could stay in this position rising to the surface only for breathing. The "normal" swimming pattern could disappear completely as well as the reaction to fish. The rate of neurotic reactions in the total activity of the dolphins, the degree of expressiveness of neurotic reactions, and the food motivation were our criteria for the level of the animal emotional excitement.

In order to investigate the peculiarities of the stereotyped signalization, we chose the most typical cases (Table 1). The dynamics of signalization in the process of adaptation of the dolphins to stress factors was studied in the two repeated model situations: (1) the taming, which was the initial stage of training, and (2) the temporary location of the animals in a small, shallow pool (depth 3 m and a surface area of 13.5 m^3). The level of adaptation of Jenny and Kora was assessed by the complex of factors: the rate of neurotic reactions in their behavior, their attitude to the trainers (avoidance or active contact), and their food consumption. Changes in signalization had been observed up to the complete calming of the dolphins. On the initial stage of work trainers spent from 3.5 to 6 hours a day in the water (4-7 sessions) with each dolphin.

The receiver system included a spherical hydrophone, 20 mm in diameter, an amplifier, a tape-recorder and an oscillograph for the controlling of the passing of a signal through the system. A working bandwidth of the tape-recorder was from 0.1 to 25.0 kHz. Our previous studies had shown that this was quiet enough for the registration of the stereotyped signalization. In all cases, the taping of signals was accompanied by the registration of the corresponding behavior in a journal.

The signals were analyzed on the dynamic spectrograph "Spectr-1", which gave the synchronous pictures of dynamic spectrogram and oscillogram. We used the sharp-tuning of filters. In order to coordinate the bandwidth of recording and analyzing technique, we reduced the recording speed twice while introducing the signals into the spectrograph. In necessary cases, the signals were analyzed on the Kay Electric Sonograph of the 7029 ADC model. The data of the spectrograph were registered from the screen of a multichannel oscillograph on the special camera devised in our laboratory. Relative to the real recording speed the speed of the camera was 25 mm per s. In order to control the speed of the film, we used the light time-marks which were given on the screen of the multichannel oscillograph from the spectrograph and the source of marks independent from the system. The duration of signals was measured on the film with 20 mm accuracy. To classify the pure tone signals to different types, we compared their contours. Overall, we analyzed 3180 signals from Jenny and 1970 signals from Kora.

RESULTS

The stereotyped signalization appeared in the dolphins only in different novel, unfavorable and potentially dangerous situations. We registrated this type of signalization, when the animals were introduced into the new pools, to the strange people and some kinds of novel objects (dividing nets, for example), when they were separated, stranded and under the other exciting circumstances. The stereotyped signalization always was accompanied by the changes in the dolphins' behavior (appearance of neurotic reactions, extinction of reaction on fish, and different acoustic and visual signals from the trainers), which expressed clearly the high level of emotional excitement of the animals. This gives us evidences to regard the stereotyped signalization as an indication on the stressed state of bottlenose dolphins.

In all enumerated cases, signalization of the animals was, in fact, a reiteration by each dolphin of whistles which included one, or several organized into a rythmical structure, basic elements, with a typical contour for a particular individual (Fig. 1). In any stressful situation, each animal always used the same basic element. The signals built from such elements correspond to the individual signals described by Melba and David Caldwell (1965), so below they will be called individual signals (IS). Ninety-nine percent of Jenny's IS included only one basic element, thus, her IS and basic element coincided. Kora's signal could consist of a changing number of basic elements (1-8), and, consequently, she had a whole class of IS with a different number of basic elements. The percentage of IS in signalization of strongly excited animals could exceed 90%.

We shall consider typical parameters of the stereotyped signalization of dolphins just after their separation to the individual pools. The degree of stereotypy (IS rate), IS repetition rate and the total acoustic activity were very high in both animals (Table 2). The majority of intervals between

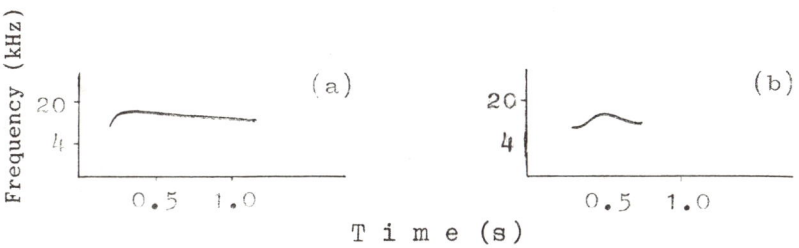

Fig. 1. Contours of Jenny's (a) and Kora's (b) basic elements of individual signals.

Table 2. General Characteristics of Jenny's and Kora's Signalization in Novel Stressful Situations.

Situation	Dolphin	Data	Total number of signals	Degree of stereotypy (%)	Average number of signals per min.	Average number of IS per min.
Separation	Jenny	20.07	364	99.7	19.7	19.7
	Kora	20.07	129	98.5	9.4	9.2
Training	Jenny	22.07	524	86.3	23.1	20.3
	Kora	21.07	443	91.7	14.7	13.4
Moving to smaller pool	Jenny	3.08	360	93.6	16.4	15.3
	Kora	29.07	285	84.9	12.4	10.5
Dividing nets	Jenny	20.09	250	80.0	13.7	11.0
	Kora	20.09	210	73.4	9.1	6.7

IS did not exceed 2 s (Fig. 2 a,c). In most cases, the duration of Jenny's intersignal intervals was 0.5-0.75 s, and of Kora's - 0.75-1.0 s (Fig. 2 b,d) IS divided by short intervals with less than 2 s duration formed series. The number of IS in series changed from 5 to 30 in Jenny and from 3 to 11 in Kora. Most intervals between series were not longer than 30 s (Fig. 2 a,c). The number of IS per minute varied from 5 to 34 in Jenny's signalization and from 0 to 19 in Kora's one (Fig. 3). The differences were connected with the changes in the number of signals in series and in the duration of interseries intervals. So there were rises and falls in the acoustic activity of the animals, but the average IS repetition rate was high.

The described type of signalization appeared in the dolphins immediately after the occurrence of any novel situation which caused excitement and passive-defensive behavior. Parameters of such signalization were very close in all such situations (Table 2). Only after the placement of dividing nets into the pools was the degree of stereotypy and IS repetition rate different. In our opinion, these differences were due to the animals being completely tamed rather, than due to the peculiarities of this situation.

The stereotyped signalization was destroying while the dolphins were getting used to the stressful at first situations. It found expression in the decrease of the total acoustic activity and IS rate, increase of the percentage and the total number of the other types of signals (Tables 3 and 4). The structure of the signal string also was changing: the number of IS in series was decreasing, the duration of interseries intervals was increasing and they were filling-up with the other types of signals.

The process of destruction of the stereotyped signalization was studied in two model situations: taming and the temporary location of the dolphin in a smaller pool. In the both cases, changes in signalization were connected closely with behavioral changes. On the whole, the process of destruction of the stereotyped signalization could be divided into 4 main phases, and each of them corresponded to the definite form of the dolphin behavior. We shall demonstrate the peculiarities of phases and of the corresponding forms of behavior in the example of training situation. In this case, the attitude of the dolphins to their trainers, i.e. the character of their contact with people, was a reliable criterion for the degree of adaptation of the animals to the situation.

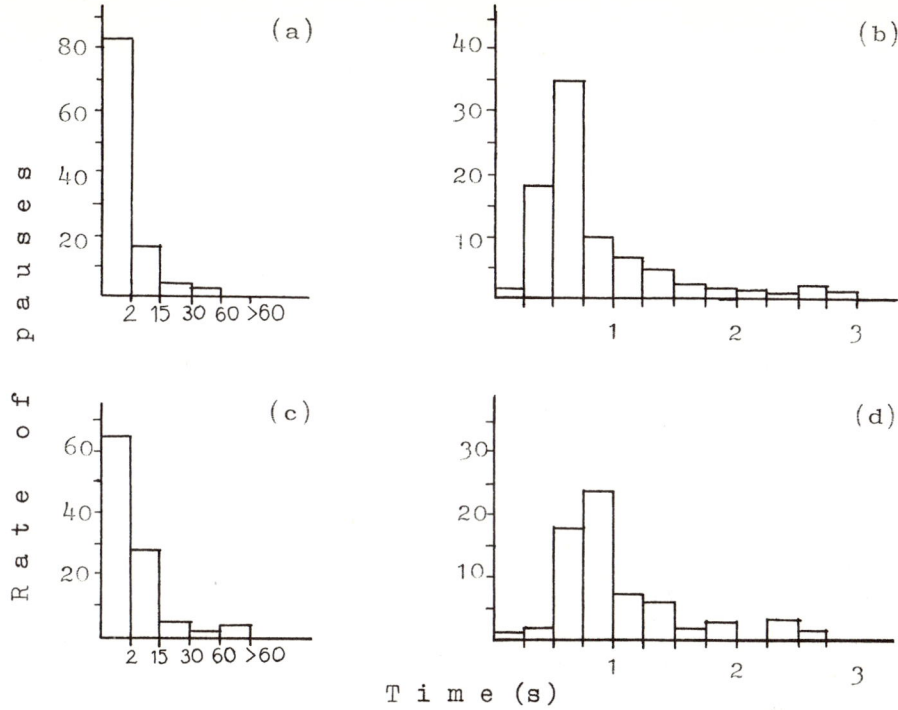

Fig. 2. Distribution of pauses between IS on duration after the separation of the dolphins; (a,b) Jenny, (c,d) Kora; (a,c) total, (b,d) short pauses.

Phase I. Acoustic activity is high, IS rate is about 90%. IS form rythmically organized series with the intersignal intervals not longer, than 2s. The average duration of interseries intervals is several seconds, in rare cases they last to 0.5 - 1.5 minutes. We observed such signalization in both females on the first two days of training simultaneously with the very intensive neurotic reactions and firm avoidance of the trainers. Jenny's and Kora's behavior and signalization were not changing during the session. Their food activity was very low, usually they did not take fish.

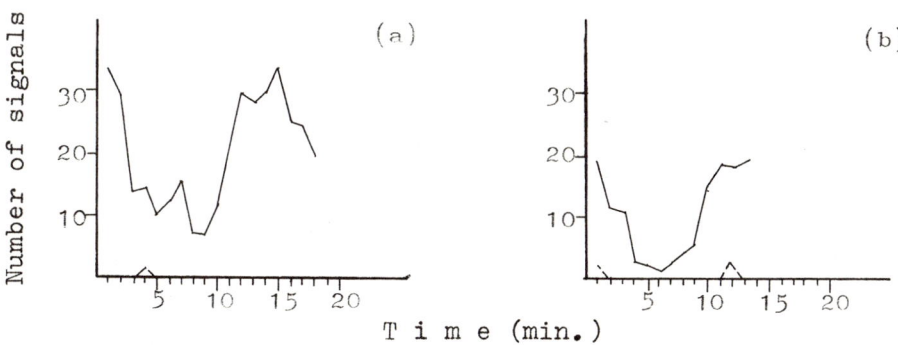

Fig. 3. Dynamics of Jenny's (a) and Kora's (b) signalization after their separation; ─── individual signals, ─ ─ ─ ─ other signals.

Table 3. Dynamics of Dolphin Signalization in the Process of Taming

Dolphin	Phase	Day of taming, session	Total number of signals	Degree of stereotypy (%)	Average number of signals per.min.	Average number of IS per.min.
Jenny	I	2 (1)	524	86.3	23.1	20.3
		2 (4)	288	91.0	26.1	23.6
	II	3 (1)	479	77.2	22.6	17.6
	IV	4 (1)	no signals			
	II	4 (2)*	173	68.2	17.3	11.8
	IV	4 (3)	no signals			
Kora	I	1 (2)	443	91.7	14.7	13.4
		2 (1)	266	91.0	12.6	11.4
	II	4 (1)	149	45.0	6.7	3.0
		4 (2)	192	55.2	9.3	5.1
		5 (2)*	264	61.0	9.6	7,6
	III	6 (1)	32	46.9	1.6	0.7
	IV	6 (2)	no signals			

*A strange man in a pool together with (Jenny) or instead of (Kora) the constant trainer

Phase II. Gradual decrease of the animals' acoustic activity, stereotypy of signalization, IS repetition rate and the number of IS in series (in our case the number of IS in series reduced from 3-30 to 2-5). This phase was observed on the third (Jenny) and the fourth (Kora) days of our work with the dolphins. Behavior and signalization of the animals were changing during the one training session. In the very beginning of the session, Jenny and Kora produced the pure tone, but not stereotyped signals, and the neurotic reactions in their behavior were not very intensive. The occurrence of stereotyped signalization and the simultaneous increase of the rate of neurotic reactions on the third - fifth minutes of the session were caused not by the appearance of the trainer in the tank (as on the first days), but by the enforcing of the contact by the trainer. After the firm avoidance of the trainers in the first 10 - 15 minutes of the session the dolphins began to swim around the trainers and avoided only direct tactual contact with them. From this time, the degree of stereotypy of signalization and the intensity of neurotic reactions began to decrease gradually, and the animals, who ignored fish initially, began to take it from the trainer's hands.

Phase III. Acoustic activity is very low. The signals of different types appear as well as solitary IS. About 60 % of intervals are longer, than 30 s. Such transitional type of signalization was observed in Kora on the sixth day of training. The dolphin did not avoid the trainer, but also did not search for a contact with him. Neurotic reactions were rare. Kora took fish from the trainer's hands, but her food activity was low.

Phase IV. "Silence". Dolphins practically do not produce sounds. The entire absence of the stereotyped signalization (on the fourth day of training in Jenny, and on the sixth day - in Kora) coincided with the transition of the dolphins to the active contact with people, and so with the accomplishing of taming. Both dolphins came-up to the trainers on their command signals, took fish, and initiated play with people. We did not observe neurotic reactions and the animal food motivation increased significantly.

Thus, different phases of signalization reflected different levels of

Table 4. General Characteristics of Jenny's Signalization After her Locations in the Small Tank.

Location	Phase	Total number of signals	Degree of stereotypy (%)	Average number of signals per min.	Average number of IS per min.
First (3.08)	I	360	93.61	16.36	15.32
Second (7.08)	total	232	36.21	10.55	3.82
	I (1-2 min.)	67	95.52	33.50	32.00
	II (3-9 min.)	124	13.71	17.71	2.43
	III (10-16 min.)	41	7.32	5.86	0.43
	IV (17-22 min.)	no signals			

the dolphins' emotional excitement and the degree of their adaptation to the situation. The silence period lasted for about a week, and only after it the divergent signals of different types began to appear. The normal divergent signalization of the animals returned only in two month of our work with them, when they were well-adapted to the keeping conditions and different experimental situations. The diversity of the dolphin behavior was increasing permanently during the whole time of our work.

In the beginning of our work, the appearance of strange people in the dolphins' pools instead of their constant trainers induced occurrence or increase in rate of IS and neurotic reactions, though they did not achieve top meanings. This shows that from the first days of taming the dolphins recognised their constant trainers. The case of Jenny was very demonstrative. On the fourth day of work the initial stage of her training was accomplished. There were no IS and neurotic reactions in her signalization and behavior either before and during the morning training session, even though the trainer for the first time worked with an aqua-lung. Jenny responded on her trainer's signals readily, took fish from his hands and played with him actively. When the strange man entered the pool, the dolphin went away from the trainer and since this moment began to avoid all contacts with both people. The appearance of the stranger immediately evoked stereotyped signalization and neurotic reactions in the animal. During the next sessions with the constant trainer Jenny did not use them again.

On the whole during the taming procedure, trainers transformed from a stressful factor to reducing or preventing stress. Up to the beginning of our work, stress of the capture and transport had still remained in both animals. In control recordings, made before the beginning of our work and between the training sessions, IS rate was about 40-50%. There were many neurotic reactions in the behavior of Jenny and Kora and there food activity was very low. After taming, no IS and neurotic reactions appeared during and between training sessions and the dolphin food motivation began to recover.

During the process of taming, we recorded all four phases of stress signalization only for Kora. The process off destruction of the stereotyped signalization in Jenny had accomplished very quickly, in three days, and we did not manage to registry the third phase. We observed all four phases on Jenny in the other situation, after her removal to the small pool which occured already after the finishing of taming.

When Jenny was moved to the small tank for the first time she continued to produce stereotyped signalization for two hours, until she was again taken to her usual tank. During the second location her stereotyped signalization disappeared completely in 16 minutes and afterwards the phase of "silence" began (Fig.4). Intensive stereotyped signalization (I phase) was observed during the first two minutes. On the third minute signalization changed significantly: the percentage of IS decreased from 95.5 to 20.8%, and their repetition rate reduced from 32 (average number for first two minutes) to 5 signals per minute. The rate and total number of other types of signals increased significantly (II phase). Such signalization continued up-to the ninth minute, but the level of total acoustic activity was gradually decreasing. From the tenth minute and up-to the sixteenth Jenny produced only solitary IS and very few signals of the other types (III phase). To the 17th minute signalization ceased completely (IV phase).

Thus, during the repeated location we watched in Jenny the same phases that during taming, and in this case we managed to registry also the third phase (Table 4). As well as during taming signalization was changing in accordance with behavioral changes. At first Jenny did not react at all on the appearance of the trainer near her pool. On the 8-9th minutes she began to come-up-to the trainer and on the 15th minute to take fish from his hands. Neurotic reactions in the character of moving were not observed as the level of water was only about 2 m and the animal could not take vertical position. During the next locations in the same small pool Jenny calmed quickly and signalization of a stress type ceased in 5-10 minutes or did not appear. The duration of "silence" phase depended upon the duration of the other phases and it could last from several minutes (as after the repeated removal to the small tank) to several days (during taming). Thus, total duration of all four phases is determined by the general stressfulness of the situation and the degree of adaptation of the animals to the situation. Nevertheless, in all situations we observed the same phases of signalization which corresponded to the definite levels of excitement and, consequently, to the definite stages of the animal adaptation and to their definite physiological state.

Now we shall examine if there was any regularity in the appearance of the other types of signals, which would confirm the existence of phases.

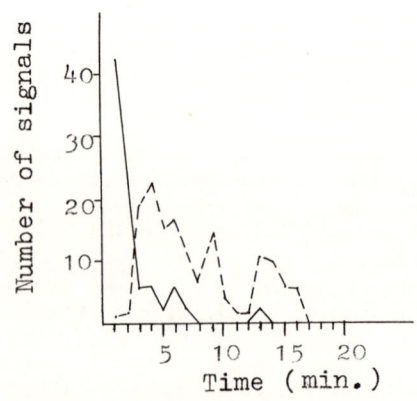

Fig. 4. Dynamics of Jenny's signalization during the second location; ——— individual signals, ----- other signals.

Preliminary analysis showed that whistle sounds, other than IS (OW), could be divided conditionally into three groups based on duration (naturally, we understood, that each group could include sounds of different structural types): up to 80 ms (we called them microwhistles, MW,); from 80 to 200 ms; and over 500 ms. MW were the most numerous. Their percentage achieved 82,4% in Jenny and 77,3% in Kora. All pulsed sounds were refered to one group, because they were very insignificant (less than 1%), and there was no connection between their use and changes in the animals' state.

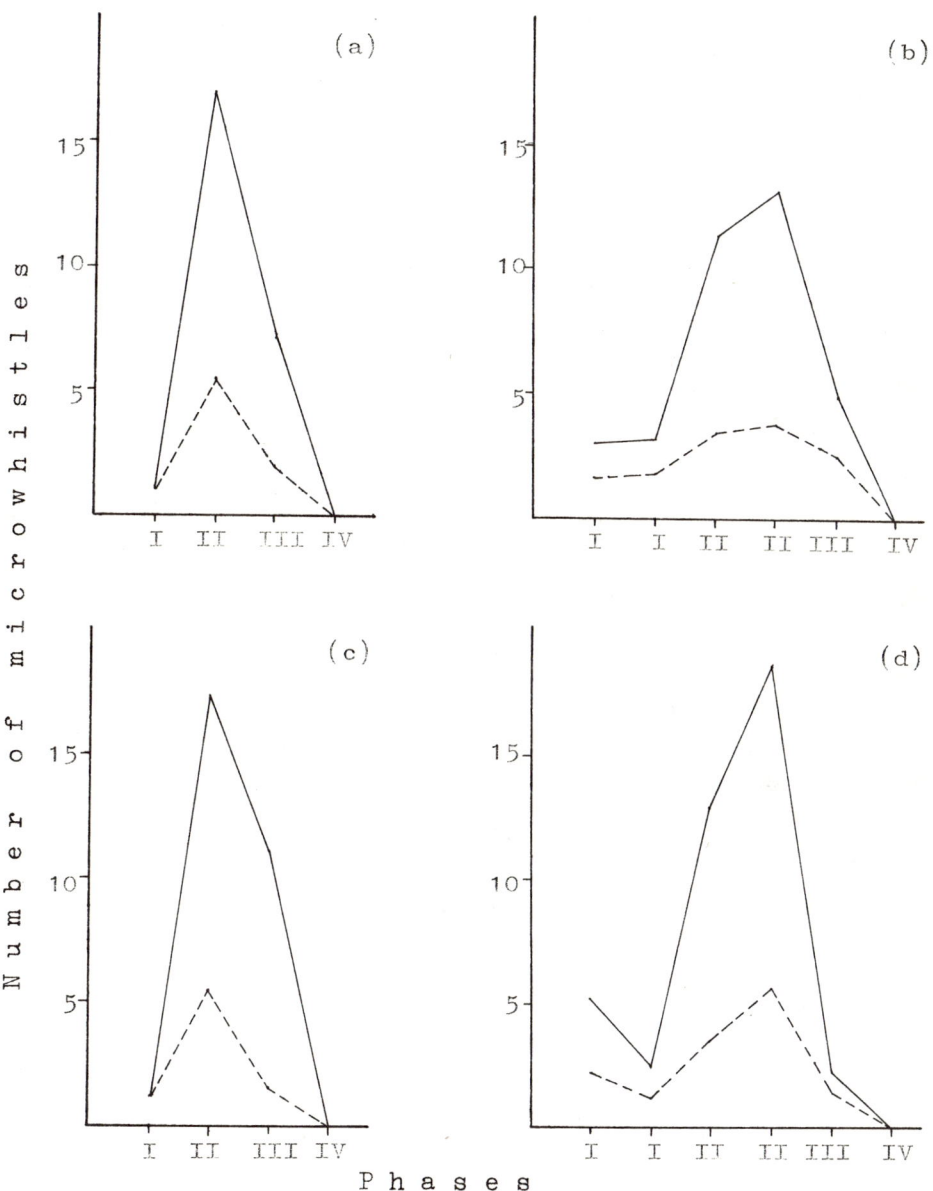

Fig. 5. Peculiarities of use of MW on different phases of stress signalization; (a,c) Jenny, repeated location to a small tank; (b,d) Kora, taming; (a,b) number of signals in chain; (c,d) number of MW of one contour type; ——— maximum number, ----- average number.

Table 5. Rate of Signals of Different Types in Signalization of the Dolphins.

Dolphin, situation	Phases	Total number of signals	IS	MW	K*	Whistles 80-200 ms	Whistles > 200 ms	Pulsed Sounds
Jenny, taming	I	524	86.3	8.2	10.5	0.9	2.1	2.5
		288	91.0	7.0	13.1	1.4	0.3	0.3
	II	479	77.2	20.6	3.7	0.9	1.3	-
	IV		n o	s i g n a l s				
location in small tank:								
first	I	360	93.6	5.0	18.7	-	0.6	0.8
second	total	232	36.2	53.0	0.7	8.0	2.2	0.4
	I	67	95.5	3.0	32.0	-	-	1.5
	II	124	13.7	76.6	0.2	5.7	4.0	-
	III	41	7.3	63.4	0.1	29.3	-	-
	IV		n o	s i g n a l s				
Kora, taming	I	443	91.7	3.8	23.9	2.0	1.6	0.9
		266	91.0	2.3	40.3	4.1	2.6	-
	II	149	45.0	40.3	1.1	7.4	1.3	6.0
		192	55.2	31.8	1.7	11.5	0.5	1.0
	III	32	46.9	31.3	1.5	10.9	-	10.9
	IV		n o	s i g n a l s				

* $K = \dfrac{\text{number of IS}}{\text{number of MW}}$

Data presented in the Table 5 show that in stressful situations the dolphins produced the signals of the two main groups: IS and MW, and that phases also differed in the IS/MW relation. This index could be similar on the second and the third phases. In such cases, additional information on the animal's acoustic activity should be used to determine the phase (Tables 3 and 4).

Increase of the MW rate in the second phase is very characteristic. On the first phase, which corresponded to the very high level of animal emotional excitement, only solitary MW appeared. On the second phase, when the animals were not so excited, the number of MW and their rate increased significantly, and the dolphins began to group them in chains analogous to the series of IS on the first phase. The number of MW in chains increased gradually to the middle of the second phase, and decreased again only in the third phase due to the reduction of the total acoustic activity (Fig. 5 a,b).

On the whole MW were rather divergent. Forty-four types of MW contours appeared in Jenny's signalization, and 35 - in Kora's one. On the first phase, solitary MW produced by the animals usually belonged to different contour types, and the stereotypy of signalization was achieved by the reiteration of IS. On the second phase, while the chains of MW were lengthening, the number of MW of the one contour type also was increasing (Fig. 5 c,d). Thus, the signalization was stereotyped again, but now due to the repeating of the limited number of MW contours, as well as IS. On the third phase, the number of MW of the same type reduced again. Only after the fourth phase ("silence") signalization of Jenny and Kora became really divergent and included diffe-

Table 6. General Characteristic of Phases of Stress Signalization of Dolphins and of the Corresponding Behavior.

Phases	Acoustic activity	Individual signals	Whistles < 80 ms	Other signals	Behavior
1	High level of acoustic activity	IS rate about 90%, IS form series with stable intervals about 2 s, interseries intervals usually < 15 s	There are solitary pure tone sounds of different types		Intensive neurotic reactions, animals avoid trainers and do not react on their signals and fish
2	Gradual decrease of the level of acoustic activity	Gradual decrease of IS rate and number of IS in series, interseries intervals are up to several minutes	MW rate is 20% and more, MW form chains, stereotypy of MW increases	No significant changes in the use of other signals	Decrease of intensity and rate of neurotic reactions, animals show interest to trainers, but avoid them, no reaction on signals, food motivation is very low
3	Acoustic activity is very low		Solitary signals of different types		Solitary neurotic reactions, animals do not avoid trainers and some times react on signals, food activity is low
4		"Silence", in fact, dolphins do not produce signals			No neurotic reactions, but the behavior is rather stereotyped, active contact with trainers, reaction on signals, food activity begins to recover

rent types of pulsed and pure tone signals usual for the dolphins in the normal state.

CONCLUSIONS

In the process of adaptation of the bottlenose dolphins to any novel, potentially dangerous, exciteful situation similar changes in signalization were observed. Definite parameters of signalization of the animals corresponded to the definite stages of their adaptation (Table 6). The time of the complete disappearence of the stereotyped signalization depended upon the specificity of the stressor and the general level of adaptation of the animal to the work with people. The duration of the "silence" phase depended upon the duration of the other three phases and it could last from several minutes to several days.

The work with the dolphins who produced the signalization of the first phase type was impossible. Animals did not react on the signals of the trainers and to fish. Signalization of the second and the third phase accompanied very unstable behavior of the animals, and the work with them was very difficult. However, attempts to conduct experiments could favor the calming of the dolphins. During the fourth phase the experimental work with the animals was possible, but after the recovery of normal divergent signalization reliability and effectiveness of the dolpfin work increased significantly. This was favored by the increase of diversity of the animal behavior and by the complete rehabilitation of their food activity.

Thus, the parameters of the bottlenose dolphin signalization can serve as a criterion for the assessment of the animal physiological state, and they can be used for prognosis of the possibility and to some extent the results of experimental work with the dolphins.

REFERENCES

Caldwell, D. K., and Caldwell M. C., 1977, Cetaceans, in: "How animals communicate," T. A. Sebeok, ed., Indiana Univ. Press, Bloomington.
Caldwell, M. C., and Caldwell D. K., 1965, Individualized whistle contours in bottlenosed dolphins, Tursiops truncatus, Nature, 207:434.
Herman, L. M., and Tavolga, W. N., 1980, The communication systems of cetaceans, in: "Cetacean behavior. Mechanisms and functions," L. M. Herman, ed., Wiley & Sons,Inc., New York.
Sidorova, I. E., Markov, V. I., and Ostrovskaya, V. M., 1986, Dynamics of signalization of the bottlenosed dolphin in the process of adaptation to stress factors, in: "Izuchenije, okhrana i ratsionalnoje ispolzovanije morskikh mlekopitajuqikh," V. A. Zemskiy, ed., Archangelsk.

CONCLUDING COMMENTS ON ACOUSTIC COMMUNICATION

Christopher W. Clark

Cornell Laboratory of Ornithology
Bioacoustics Research Program
159 Sapsucker Woods Road
Ithaca, New York 14850 USA

OVERVIEW

When consulting the recent history of marine mammal bioacoustics (for example, Tavolga 1964; Watkins and Wartzok 1985) one is struck by the remarkable similarity between the biological questions, experimental difficulties, scientific objectives, and technical requirements that existed in the 1960's and those that remain today. It is not so much that there has been no effort or nothing has been accomplished in cetacean acoustics over the past thirty years. It is just that relative to other research areas, there have been few accomplishments which provide insights into cetacean acoustic communication. We have gotten better at describing the acoustic repertoires for some species and done some intelligent speculation on the functions of certain signal classes, but we still are grappling with the basic mechanics of collecting high quality field recordings, calibrating recording and analysis systems, and accurately coordinating visual observations with acoustic records.

The reasons for the slow rate of progress certainly include the difficulties of working with marine mammals in their natural environment, as well as a basic dichotomy in backgrounds for the researchers involved. Biologists and psychologists, for the most part, are not trained in physical acoustics, mathematics, or engineering; disciplines which provide advantages when dealing with issues of underwater sound transmission, reflection, absorption, and refraction. Physical acousticians and engineers, in turn, are not trained to understand how living organisms function, survive and propagate under the constraints of genetic variation and natural selection. The result is that our collective emphasis is not integrated. We need greater cross-disciplinary dialogue between the more physical sciences and the life sciences. The tools of engineering and the rigors of physical acoustics must be focused constructively on describing the details of sound production, propagation and perception in cetaceans if we are going to more understand fully their acoustic communication systems.

For example, essentially nothing has been done to describe how large whales produce sounds. Yet, we have fairly good descriptions of the sound repertoires and sound-production anatomy (including laryngeal) systems of several toothed whales and large baleen whales. Since Roger Payne and Doug Webb's (1971) classic paper on long-range transmission of 20 Hz-fin whale sounds, no further details have emerged concerning long-range transmission of low-frequency baleen calls. Yet, we know much more about the characteristics of these biological signals and the properties of the ocean medium through which they propagate than we did 20 years ago.

We still have a great deal to learn from cetaceans regarding their solutions to underwater acoustic communication. It will take a concerted effort on the part of a spectrum of researchers to appreciate better the intricacies of cetacean sensory abilities and sound-production mechanisms. And, it will require human foresight and intelligence to modify our behaviors so that our impact on their acoustic environment is not detrimental to their continued survival.

DISCUSSION

It is appropriate that the composition of the bioacoustics workshop group was so varied; no two of us had similar backgrounds [1]. There were certainly participants in other sections of the workshop with some interest in acoustics, but acoustic communication is not their main research focus. Because of the somewhat mixed composition of the group, we were able to identify the common areas of concern, but could come to no definite conclusions as to solutions. Our deliberation of cetacean communication included discussions on the biological significance of sounds, basic descriptions of sound repertoires, limitations of equipment, techniques with the greatest potential, and effects of human activity. This report will summarize our one-day workshop discussions and synthesize a list of recommendations.

Biological Significance

As biologists, there is a basic interest in understanding the biological significance (function, meaning) of acoustic signals. This is done by combining visual and acoustic observations. Typically significance is deduced from lengthy observations or inferred indirectly using sound playback experiments. Field observations are affected seriously by environmental conditions, which reduces the number of reliable observations. Good underwater sound playback equipment is bulky and expensive, and has limited the number of playback experiments conducted with cetaceans. There has been inconsistency in recording methods, which has led to inconsistencies in descriptions of sound repertoires, especially as these relate to bandwidth and received levels.

Some people use broadband, calibrated equipment, but this is expensive and difficult to work with in small boats. Other researchers do not give acoustics a high priority in their research, only to learn later on that their recordings are of limited scientific value.

Acoustic Repertoires

It is clear that we have not adequately documented the acoustic repertoires of most cetaceans. In some cases there are preconceptions about what bandwidths are of interest, or recordings have been obtained during a limited set of social contexts (e.g. feeding, migrating). Access to

---

[1] Members of the working group included: Mats Amundin, Giorgio Caporin, Christopher Clark, Rune Dietz, Ron Mitson, H. Onofre Moreira, Manuel dos Santos and Don Woodward.

different species and the appropriate recording equipment are very different. The result has been that the recording efforts have been inconsistent and incomplete.

Techniques

There are a number of techniques, including acoustic playback, radio telemetry, electronic backpacks, and passive acoustic arrays which are well-suited for describing the acoustic behavior of individual animals, a step which is necessary for properly documenting communication. Sound playback is the most powerful method for conducting experiments on acoustic perception and sound function with free-ranging cetaceans. Satellite and UHF radio telemetry systems have improved considerably and should allow us to simultaneously follow several individuals for many months while continuously monitoring their discrete sound production activities. A harness system, consisting of a customized electronic package attached to an animal (similar to packages used with pinnipeds; Hill 1986), compresses and stores acoustic information in a memory device, but requires one to recover and replace the package at regular intervals. This type of system should work with semi-captive populations or habituated individuals. For passive acoustic arrays, digital signal processing capabilities have increased dramatically to the point where it is now feasible to build field-portable systems for locating and tracking sounds from cetaceans in real-time.

Human Activities

We have not described adequately the affects of noise, introduced into the water by human activities, on marine organisms, and this is the most pressing problem facing us today in cetacean bioacoustics. Since sound is transmitted so efficiently in water and cetaceans are particularly dependent on the sound modality for communication, there is considerable concern about the potential adverse affects that sounds from human activities have on cetaceans. To fully appreciate and anticipate what affects the introduction of noise might have on cetaceans, we must understand better the communicative function(s) of their acoustic signals, and their abilities to produce and perceive sounds.

Recommendations

The following recommendations would have significant impacts on improving our understanding of cetacean acoustic communication:

(1) A major comparative review of all existing acoustic material in conjunction with the establishment of a national marine mammal sound archive.

(2) Increased effort to obtain more high quality, broadband recordings from all cetaceans. This includes the mysticetes, for which there have been reports of ultrasonic signal production, as well as the odontocetes for which we continue to find discrepancies in descriptions of signal frequency content.

(3) Greater support for collaborations between physical acousticians, engineers and biologists to describe more rigorously mechanisms of sound production and the adaptive value of signal characteristics (for example for long-range transmission, surface reverberation, or bi-static acoustic imaging), and to further the application of realtime signal processing for signal classification, acoustic location and tracking, radio telemetry, and data acquisition/storage harness packs.

(4) A long-term program for documenting and monitoring ambient noise levels in the oceans and noises associated with human activities, with particular emphasis on those locations known to be of critical importance for breeding, migrating and feeding to cetaceans.

(5) A workshop on bioacoustic techniques which would bring together bioacousticians, psychologists, physical acousticians, oceanographers, and ocean engineers. The aim of the workshop would be to expose areas of critical interest and to identify areas of common research activity where collaboration could lead to timely, cost-effective solutions.

REFERENCES

Hill, R.D. 1986. Microcomputer monitor and blood sampler for freely ranging diving seals. J. Appl. Physiol. 61:1570-1576.

Payne, R. and D. Webb. 1971. Orientation by means of long range acoustic signalling in baleen whales. Ann. N.Y. Acad. Sci. 188:110-141.

Tavolga, W.N. 1964. Marine Bio-Acoustics. Pergamon Press, New York. 413 pp.

Watkins, W.A. and D. Wartzok. 1985. Sensory biophysics of marine mammals. Mar. Mammal Sci. 1:219-260.

GEOMAGNETIC SENSITIVITY IN CETACEANS: AN UPDATE WITH LIVE STRANDING RECORDS IN THE UNITED STATES

Joseph L. Kirschvink

Division of Geological and Planetary Sciences
The California Institute of Technology
Pasadena, California 91125, USA.

SUMMARY

Cetacean stranding sites have been linked to the presence of local magnetic anomalies in several widely-separated geographic areas, including the eastern coast of North America and the British Islands. Previous studies of this sort have been hampered largely by inadequate survey data for the magnetic field, as well as by incomplete records of cetacean stranding events. A major improvement in the geomagnetic anomaly data available for these studies has been the 1988 publication of the geomagnetic anomaly map of North America compiled by the Geological Society of America, and its subsequent public release in digital form. Compared with the records of cetacean live stranding events compiled by the Smithsonian Institution in Washington, D.C., these new magnetic anomaly data more than double the number of live stranding events in the United States which fall within the boundaries of geomagnetic surveys. These new data add further support to the hypothesis that cetaceans possess a geomagnetic sensory system comparable to that in other migratory and homing animals, and are consistent with previous suggestions that features of the geomagnetic field, in particular the marine magnetic lineations, play an important role in the long-distance navigation of marine mammals.

INTRODUCTION

Although migratory animals often display an uncanny ability to find their way over long distances of featureless terrain, how they navigate or pilot during these journeys remains a mystery. From behavioral and neurological studies conducted during the past 40 years, a wide range of different sensory modalities have been implicated or suggested as guidance mechanisms. These include the use of a sun compass (Kramer, 1952), a star compass (Sauer, 1957), skylight polarization (Kreithen and Keeton, 1974a), odor (Papi et al., 1972), infra-sound (Kreithen and Quine, 1979), UV-light (Kreithen and Eisner, 1978), electric fields (Kalmijn, 1974) and magnetism (Keeton, 1972; Walcott and Green, 1974). Few of these cues are available to aquatic animals, yet they too make accurate journeys across apparently featureless seas.

The question of whether or not geomagnetic stimuli were involved in the ability of organisms to find their way, or indeed, whether or not such sensitivity exists at all in animals, has been one of the most controversial topics in the field of animal behavior. Initial objections to the suggestion of geomagnetic sensitivity in animals were based on the apparent lack of this sense in humans, and on the lack of any known biophysical mechanism capable of transducing

Sensory Abilities of Cetaceans
Edited by J. Thomas and R. Kastelein
Plenum Press, New York, 1990

the weak geomagnetic field to an animals' nervous system (Griffin, 1944). Furthermore, many of the early behavioral results suggesting geomagnetic sensitivity were rather weak and difficult to reproduce (e.g., Kreithen and Keeton, 1974b; Griffin, 1982). However, the discovery of the magnetotactic bacteria (Blakemore, 1975) and of their 'biological bar magnets' (linear chains of membrane-bound, single-domain crystals of magnetite [Fe_3O_4], Frankel et al., 1979; Balkwill et al., 1980) provide clear examples of both geomagnetic sensitivity in a living organism, as well as a simple and elegant biophysical mechanism for transducing the geomagnetic field to the nervous system. In theory at least, the magnetite from a single bacterial magnetosome chain could provide a whale with an extraordinarily good geomagnetic compass receptor, although several million such organelles would be necessary to provide them with sensitivity adequate for the detection of geomagnetic anomalies at sea (Kirschvink and Gould, 1981).

Research in several separate fields during the past 5 years has provided a greater understanding of the geomagnetic sensitivity in living organisms. First, recent work has demonstrated that the magnetite formed by pelagic fish (e.g., Walker et al., 1984) is present in chains of single-domain magnetosomes, indistinguishable in many respects from those present in the magnetotactic bacteria (Kirschvink et al., 1985; Mann et al., 1988; Walker et al., 1988). The magnetite appears to be localized within the dermethmoid complex in a diverse variety of vertebrates (e.g., papers in Kirschvink et al., 1985). In the yellowfin tuna (Thunnus albacares) the tissue containing the single-domain magnetite also contains abundant axons and cells containing primary cilia which appear suitable for a role in magnetoreception (Walker and Kirschvink, unpbl.). Second, the paleontological record of the magnetotactic bacteria can be traced by the presence of the fossilized magnetite crystals in sedimentary rock (Kirschvink and Chang, 1984; Stolz et al., 1986; Vali and Kirschvink, 1989) these 'magnetofossils' have been recovered from sediments nearly 2×10^9 years old, which predates the origin of the eukaryotic cell (Chang and Kirschvink, 1989). This evolutionary history coupled with the wide phyletic diversity of magnetite biomineralizing organisms (3 of the 5 Kingdoms of living organisms, Lowenstam and Weiner, 1989) led Chang and Kirschvink (1989) to suggest that the magnetotactic bacteria were involved in the endosymbiotic origin of the eukaryotic cell. In turn, this suggests that magnetoreception may have been one of the most ancient sensory systems to evolve, and that organisms which apparently lack this sense may be relatively rare species (e.g., Homo sapiens) which have, for some reason, lost the ability.

The third and perhaps the most important advance in the understanding of how geomagnetic sensitivity works in animals has come from the development of robust psychological conditioning techniques to train animals to respond to weak magnetic fields in laboratory environments. Walker (1984) initially discovered that yellowfin tuna, Thunnus albacares, can be trained to discriminate the presence or absence of weak magnetic field gradients in large saltwater tanks using a classic reward/punishment scheme, but this training required several months per fish. In a series of recent papers, Walker and Bitterman (1985; 1989a,b,c) and Walker et al. (1989) report that similar techniques work extraordinarily well with honey bees, Apis mellifera, and that individual bees can be trained to respond to small magnetic anomalies during the course of only a few hours. Once established, the magnetic discrimination behavior of individual bees can be maintained for several days, which is ample time for a variety of psychophysical experiments to be performed. These training techniques are simple to perform, and several have been replicated independently in the author's laboratory (Kirschvink and Kirschvink, 1991). Two major results from these experiments include the measurement of the threshold sensitivity of the honeybee magnetoreceptor system and the rough localization of the receptors by mounted magnets. Threshold sensitivities were measured by progressively reducing the strength of an anomaly and finding the point at which the bees would loose the ability to find it. The median threshold value reported by Walker and Bitterman (1989b) is an anomaly of 0.6% of the ambient background (c.a., 250 nanotesla [nT] in the Hawaiian background field of about 42,000 nT). Several bees, however, maintained their ability to distinguish the presence of the anomalies at fields as low as 0.06% of the background, implying physiological threshold sensitivities of at least 25 nT! It is important to note that these values are within the range needed for migrating and homing animals to use magnetic anomalies or regional gradients in the geomagnetic field for navigation or piloting (e.g., Kirschvink and Walker, 1985). This extraordinary sensitivity also is consistent with a variety of correlational studies on honey bees, pigeons, and cetaceans (Keeton et al., 1974; Lindauer, 1977; Walcott, 1978; Kirschvink et al., 1986) which imply a similar sensory ability. Using conditioning experiments combined with small magnetized wires glued at various places on the bees, Walker and Bitterman (1989a,b) were able to demonstrate that the magnetoreceptors are located somewhere in the vicinity of the anteriordorsal abdomen, as this was the only location they found where the wires interfered with the animal's ability to detect magnetic stimuli. This location is far from the visual receptors in the head which have proposed as a possible site for an optical-pumping magnetoreceptor (Leask, 1977), but is very close to the location of the biogenic magnetite discovered by Gould et al. (1978).

Although these are dramatic advances in our understanding of how magnetoreception works in animals, they do not tell us why geomagnetic sensitivity is important, particularly for homing and migratory organisms. This question is best approached by comparing observations of an animal's behavior with the spatial or temporal variations in the geomagnetic field, as has been done in previous analyses of cetacean stranding locations (e.g., Klinowska, 1985a,b; Kirschvink et al., 1986) or their sighting observations at sea (Walker et al., 1986 and in review). This type of analysis, however, depends entirely upon the availability of dense geophysical survey data, as well as detailed observations of animal sighting or tracking observations within the same geographical area. Until fairly recently, the largest, publicly available body of digital magnetic anomaly data was from the United States Geological Survey's U.S. Atlantic Continental Margin study (Grimm et al., 1982), with a gridded pixel (picture element) spacing of 0.036°, or a square of about 4 km per side aligned along the latitude and longitude grid. A previous comparison of cetacean live stranding locations within the area of these data demonstrated that many species tend to strand at coastal locations with pronounced negative geomagnetic anomalies (Kirschvink et al., 1986), confirming the initial observation of Klinowska (1985a) of a similar tendency in the live stranding records from the United Kingdom. Walker et al. (1986) also combined these data with the extensive sighting observations of fin whales (*Balaenoptera physalus*) from the Cetacean and Turtle Assessment Program (CETAP) data base from the University of Rhode Island, and discovered similar tendencies for migrating whales to seek local geomagnetic minima, suggesting that they were using these features as part of their navigational map.

Unfortunately, the USGS aeromagnetic data set only covers the continental shelf and shoreline interval from Cape Canaveral, Florida through Cape Cod, Massachusetts, and there are many gaps in the data and places where the shoreline wanders in and out of the mapped area (Kirschvink et al., 1986). For the analysis of strandings, fewer than half of the well-documented live stranding locations along the U.S. coastline fall within the boundaries of this survey. Realizing the interdisciplinary scientific utility of large geophysical data bases, however, the Geological Society of America created, as part of its DNAG (Decade of North American Geology) project, a special working group to assemble all magnetic survey data now available for the entire North American continent, with the goal of compiling a coherent magnetic anomaly map of North America. As discussed below, this new data base covers most of the continental and offshore area of the U.S. with a pixel spacing of roughly 2×2 km. In this present study, I report results from a reanalysis of the U.S. stranding record data used in our previous study (Kirschvink et al., 1986) with this new DNAG data base. With over twice as many live stranding events included in the analysis, results of this study confirm previous results that many cetacean strandings occur at localities associated with negative geomagnetic anomalies.

METHODS AND DATA

All analyses used in the present work are patterned after those used by Kirschvink et al. (1986), except for improvements in the geophysical and stranding bases. These differences are discussed separately with regard to the magnetic, stranding, and coastline data in the following sections.

Magnetic Data base

As mentioned above, the DNAG project has compiled an exhaustive map of the magnetic anomalies measured in various portions of the North American Continent. The maps were compiled from the raw data obtained from a variety of public and private aeromagnetic surveys, as well as sea-surface magnetometer records obtained by marine research vessels. As these surveys differ greatly in terms of their track spacing, measurement densities, elevation, and date, it was necessary for the DNAG group to massage the data slightly to minimize boundary misfits between adjacent or overlapping survey areas. All of the original survey data were corrected with the new Definitive Geomagnetic Reference Field model (DGRF), also produced by the U.S. Geophysical Data Center, and scaled linearly to minimize the mismatch for adjacent areas and to fit the surveys to a common elevation surface. Visual examination of the new data base shows that the survey boundary mismatches are relatively small (usually less than 50 nT). Use of the new DGRF models has eliminated most of the problems associated with large-scale regional anomalies, which were of concern in our previous study along the U.S. east coast (Kirschvink et al., 1986), and was the reason for our extensive use of Monte-Carlo simulations to check the accuracy of the t-tests described below. Corrected residual magnetic anomaly data for all surveys were mapped onto a gridded surface using a Universal Transverse Mercator (UTM) projection centered on the 100° W longitude meridian, with approximately a 2 × 2 km pixel spacing. This UTM coordinate system minimizes geographic distortion of the mapping projection at high latitudes, which was a problem in our earlier study (Kirschvink et al., 1986).

For the present study, these maps and the associated gridded digital data on magnetic tape were purchased from the U.S. Geophysical Data Center in Colorado. For use in the stranding studies reported here, we removed a 1400 × 1400 (2800 × 2800 km) square image of these data, which cover the entire U.S. coastline from Western Texas, through Southern Florida and offshore islands, and up along the Eastern seaboard through Maine. As in our previous study, data were shifted linearly and converted to positive integer values that fit within two bytes of memory, thereby permitting the entire data frame to fit within the memory of our computing system (a microVAX II with an image driver). This linear shift does not alter the statistical analyses presented below, as all comparisons are done using the relative field changes along the coast adjacent to the stranding sites. A few areas within the image along the southern coastlines do not have survey data (presumably, the data exist, but have not been released for commercial or military reasons). Two smaller, 512 × 512 sub-frames, covering the California and Oregon/Washington coastlines also were removed to examine the West-Coast stranding records, but stranding data were not dense enough to warrant inclusion of these geographic areas in this analysis.

Stranding Data

Dr. James Mead of the Marine Mammal Program of the Smithsonian Institution in Washington D.C. kindly provided an extensive update of his stranding data base, which we had used in our earlier study (Kirschvink et al., 1986). These new data were merged with those used in our earlier study and cross-checked to eliminate duplicate records. Table 1 is a list of those species for which there are adequate records of live stranding events within the boundaries of the Texas to Maine data frame described earlier. Only stranding events which fell within 2 km (1 pixel) of the coastline were used; stranding locations which were far-out at sea or inland were assumed to be errors in the latitude or longitude values of the data base.

Coastline Data

As in our previous study, it is necessary to have a high-resolution digital representation of the coastline for use in comparing relative magnetic field values up and down the coastline surrounding each stranding event. For this, it was necessary to edit our previous high-resolution coastal data set (obtained from the program SUPMAP, distributed by the National Center for Atmospheric Research, NCAR) and delete all political boundaries and rivers. This step was not necessary in our previous study, because the USGS aeromagnetic set began at the coastline and did not extend significantly inland; hence, the political and river pixels were not included in the stranding analysis. In most areas, however, the DNAG data base extends continuously across the coastline, making it necessary to remove the political and river outlines before conducting statistical tests. Because this high-resolution outline does not extend into Canada or Mexico, no strandings are included from those areas. A total of 27,637 shoreline pixels were found within the area of our 1400 × 1400 magnetic data image. Coordinates for each of these points, as well as an estimate of the length of coastline within each pixel, were calculated for the large image and stored in separate files to eliminate the time-consuming process of finding them repeatedly from scratch.

Statistical Analysis

The problem of how to test the hypothesis that a group of stranding events is non-random with respect to the geomagnetic (or any other geopotential) field is discussed extensively by Kirschvink et al. (1986). In that study, the residual magnetic anomaly value at each stranding site was compared with similar field values from adjacent stretches of the coast on either side of the stranding event. For test purposes, a measure of the relative location of each stranding event with respect to the nearby high and low values of the magnetic field was devised as the parameter:

$$x_{i,r} = (B_{i,r,max} + B_{i,r,min})/2 - B_{ith\ stranding}, \qquad (1)$$

where

$B_{i,r,max}$ = Maximum field value within r km of stranding i, and

$B_{i,r,min}$ = Minimum field value within r km of stranding i.

For a random distribution of stranding sites on the coast, the expected mean of this parameter should be close to zero. On the other hand, strandings which occur near magnetic minima should have positive x values, and those near maxima should have negative values. If $<x_r>$ is the average of these values for a group of N whale strandings for a radius r, and s^2 is the associated variance, then the statistic

$$t = \frac{<x_r> \sqrt{N}}{s} \qquad (2)$$

will follow Student's t-distribution with (N-1) degrees of freedom (e.g., Sokal and Rohlf, 1981). As defined here and by Kirschvink et al. (1986), large magnitudes of t imply rejection of the null hypothesis, positive values imply that the strandings preferentially happen near local magnetic minima, and negative values imply strandings near local magnetic highs.

For all estimates of statistical significance in these analyses, it is appropriate to use two-tailed t-tests rather than the more usual one-tailed. There are two reasons for this, including the fact that it is a more conservative approach, and departures either towards higher or lower field values might have importance with regard to a geomagnetic navigation strategy. Thus, p-values listed in Table 1 have been doubled (e.g., decreasing the reported levels of significance) from those listed in standard tables of Student's t-distribution.

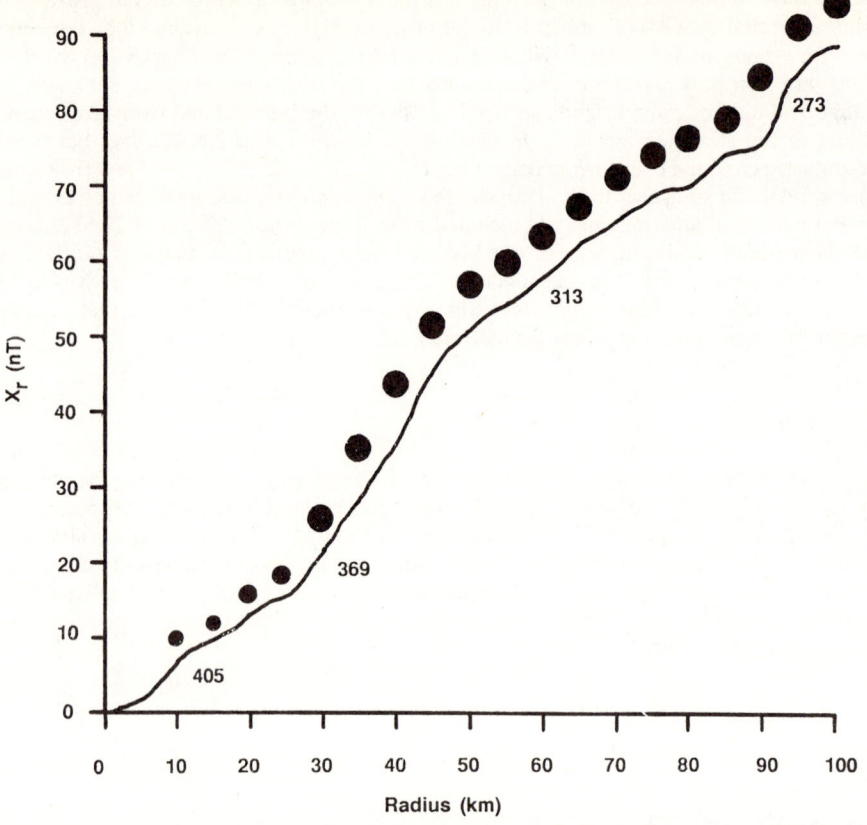

Fig. 1. Average magnetic field deviation parameters, X_r (in nT), calculated according to eq. (1) as a function of distance in 5 km intervals for the group of all 421 live cetacean stranding events on the coast from Texas through Maine. Small, medium, and large dots are positioned above points which show t-values from eq. (2) which are significant at the $p < 0.05$, $p < 0.01$, and $p < 0.001$ levels, respectively, on 2-tailed t-tests.

RESULTS

Table 1 shows representative results of these neighborhood analyses, grouped for all stranded whales in the data base, as well as for individual species, in a format similar to that used by Kirschvink et al. (1986). As in this previous study, the analysis which includes all of the 421 live stranding events for all species yields a very highly significant tendency ($p < .001$ using the t-statistic) for cetaceans to strand near coastal locations with slightly weaker total intensity of the geomagnetic field (e.g., negative magnetic anomalies). Figure 1 shows a plot of the average field deviation parameters calculated by eq. (1) as a function of distance from stranding events, showing that significant tendencies to strand at negative magnetic anomalies are achieved within a 10-km radius of the stranding sites, and are very highly significant ($p << .001$) beyond 30 km. When these data are analyzed on a species by species level, results similar to those observed by Kirschvink et al. (1986) emerge. As before, negative geomagnetic anomalies are associated with strandings of G. melaena, G. macrorhynchus, S. coeruleoalba, S. plagiodon, L. acutus, and B. physalus. In addition, the new analysis reveals similar associations with D. delphis, G. griseus, P. phocoena, and in the family Physeteridae, with P. macrocephalus and K. breviceps, (neither of which displayed significant tendencies previously). These differences probably arise from the larger number of events included in the new analysis, as well

Table 1. Neighborhood Analyses at Representative Radii for Various Cetacean Species. For the P values, the *, **, and *** symbols respectively indicate significance at the P < .05, P < .01, and P < .001 levels, respectively (on two-tailed t-tests), and correspond to the standard definitions for significant, highly significant, and very highly significant departures from the null hypothesis, respectively.

Species and common name	Number of Strandings Tot./@Rad.	Radius (km)	Average Local field Deviation (nT)	Standard Deviation (s, nT)	Student's t (N-1 d.f.)	P
All Species combined	421/334	45	44.6	126.9	6.42	<.001***
Globicephala melas (long-finned pilot Whale)	26/19	40	65.6	92.2	3.10	<.01**
Globicephala macrorhynchus (short-finned pilot whale)	22/18	50	104.9	150.3	2.96	<.01**
Stenella coeruleoalba (striped dolphin)	25/25	15	36.9	75.4	2.45	<.05*
Stenella frontalis (Atlantic spotted dolphin)	14/10	10	30.5	36.8	2.62	<.05*
Lagenorhynchus acutus (Atlantic white-sided dolphin)	34/29	30	84.3	89.6	5.10	<.001***
Tursiops truncatus (bottlenose dolphin)	50/32	75	20.6	103.0	1.13	>.20
Grampus griseus (Risso's Dolphin)	13/11	35	7.5	81.7	0.32	>.20
Delphinnus delphis (common dolphin)	14/14	5	28.2	31.3	3.49	<.01**
Physeter macrocephalus (sperm whale)	18/13	65	45.8	75.1	2.20	<.05*
Kogia breviceps (pygmy sperm whale)	98/70	100	70.5	164.5	3.59	<.001***
Kogia simus (dwarf sperm whale)	12/12	25	34.6	79.3	1.51	>.10
Phocoena phocoena (harbour porpoise)	21/19	55	161.4	231.8	3.04	<.01**
All family Ziphidae	6/4	40	-3.9	33.7	-0.26	>.2
Balaenoptera physalus (fin whale)	13/11	40	68.9	37.8	6.02	<.001***

as from better resolution of the magnetic data. The only surprise is found with T. truncatus, which in the earlier analysis of 17 live stranding events was associated significantly with negative anomalies, but shows no association with the 50 events in the new data base.

DISCUSSION

The comparison of live cetacean stranding events along the U.S. coastline with the new DNAG geomagnetic anomaly data base confirms the previous results of Kirschvink et al. (1986) that cetaceans tend to strand at coastal locations near negative magnetic anomalies. This new study, however, includes stranding events from Texas through Maine, whereas the previous U.S. analysis covered only the coastline from Cape Canaveral, Florida through Cape Cod, Massachusetts. These results are in general agreement with the initial results and interpretation of the U.K. stranding records made by Klinowska (1985a), and are consistent with her previous conclusions that cetaceans possess a highly developed magnetic sensory system.

It is important to emphasize here that studies of this sort can only tell something about the places where live stranding events happen; nothing can be inferred from these data concerning their cause (e.g., it addresses where, not why). The simplest and most conservative hypothesis which stems from these observations is that cetaceans may normally follow features in the geomagnetic field (like the marine magnetic lineations produced by the process of sea-floor spreading) for long-distance navigation or piloting. Walker et al. (1986 and in review) found that live fin whales (B. physalus) sighted over the continental shelf of off the eastern U.S. coast by the Cetacean and Turtle Assessment Program (CETAP) were found preferentially in places with both low total magnetic field intensities and low gradients, as would be expected at negative magnetic anomalies. Removal of sighting observations for whales which were observed to be engaged in feeding activities enhanced the geomagnetic associations; implying that the remaining behaviors, such as directed migration, were more strongly influenced by the geomagnetic field. As live stranding events are more frequent in species which rarely approach the coastline, the doomed whales involved in a stranding event may be trying to navigate in unfamiliar, shallow waters using a normal strategy of following paths of local geomagnetic minima. Failure to recognize the danger of shallow water and the barrier of the coastline then could lead to the observed association between magnetic anomalies and stranding locations, without the magnetic field playing a direct role in the process. Furthermore, the geographic association of a strong, linear magnetic anomaly with a large coastal embayment could act as a funnel to focus stranding events repeatedly to the same stretch of coastline. An example of this geometry is Cape Cod, Massachusetts, where numerous stranding events of L. acutus have occurred near the town of Eastham, where a linear negative anomaly runs perpendicular to the shore from the inner to the outer edge, midway through the blade of the Cape Cod 'sickle'.

In conclusion, it is clear that further tests of the geomagnetic hypothesis of cetacean navigation need to be conducted on whales which are actively migrating at sea. For this, it would be best to examine detailed tracking records over an area of true oceanic sea floor where the strong zebra-striped patterns of the marine magnetic lineations are present. The DNAG data base has superb coverage of the Juan de Fuca and Gorda ridges which lie offshore from Oregon, Washington, and Vancouver, B.C., and hence this portion of the Northeast Pacific ocean would be an ideal area to conduct such detailed tracking studies.

ACKNOWLEDGMENTS

I thank Dr. James Mead of the Smithsonian Marine Mammal Program for cheerfully providing his data files of cetacean stranding events, and the Geological Society of America for producing the magnetic anomaly map of North America. Contribution No. 4784 from the

Division of Geological and Planetary Sciences, the California Institute of Technology, was partially supported through NSF grant EAR83-51370.

REFERENCES

Balkwill D. L., Maratea D., and Blakemore, R. P., 1980, Ultrastructure of a magnetotactic spirillium. J. Bacteriol. 141:1399−1408.
Blakemore, R. P., 1975, Magnetotactic bacteria. Science 190:377−379.
Chang, S. R., and Kirschvink, J. L., 1989, Magnetofossils, the magnetization of sediments, and the evolution of magnetite biomineralization. Ann. Rev. Earth Planet. Sci. 17:69−95.
Frankel, R. B., Blakemore R. P., and Wolfe R. S., 1979, Magnetite in freshwater magnetotactic bacteria. Science 203:1355−1356.
Frankel, R. B., Papaefthymiou, G. C., and Blakemore, R. P., 1985, Mössbauer spectroscopy of iron biomineralization products in magnetotactic bacteria, in: "Magnetite Biomineralization and Magnetoreception in Organisms: A New Biomagnetism", Kirschvink J. L., D. S. Jones and B. J. MacFadden, eds., Plenum Press, New York, 269−287.
Gould J. L., Kirschvink J. L., and Deffeyes, K. S., 1978, Bees have magnetic remanence. Science 202:1026−1028.
Griffin, D. R., 1944, The sensory basis of bird navigation. Q. Rev. Biol. 19:15−31.
Griffin, D. R., 1982, Ecology of migration: Is magnetic orientation a reality? Q. Rev. Biol. 7:293−295.
Grim, M. S., Behrendt, J. C., and Klitgord, K. M., 1982, Description of digital aeromagnetic data, U.S. Atlantic Continental Margin, Survey of 1974−76. U.S. Geological Survey open-file report 82−189: pp. 1−11.
Kalmijn, A. J., 1974, The detection of electric fields from inanimate and animate sources other than electric organs, in: "Handbook of Sensory Physiology, v.9", A. Fessard, ed., 147−200.
Keeton, W. T., 1972, Effects of magnets on pigeon homing, in: "Animal Orientation and Navigation", S. E. Galler, ed., NASA SP-262, 579−594.
Keeton, W. T., Larkin, T. S., and Windsor, D. M., 1974, Normal fluctuations in the earth's magnetic field influence pigeon orientation. J. Comp. Physiol. 95:95−103.
Kirschvink, J. L., and Chang, S. R., 1984, Ultrafine-grained magnetite in deep-sea sediments: Possible bacterial magnetofossils. Geology 12:559−562.
Kirschvink, J. L., and Gould, J. L., 1981, Biogenic magnetite as a basis for magnetoreception in animals. Biosystems 13:181−201.
Kirschvink, J. L., Dizon, A. E., and Westphal, J. A., 1986, Evidence from strandings for geomagnetic sensitivity in cetaceans. J. Exp. Biol. 120:1−24.
Kirschvink, J. L., Walker, M. M., Chang, S-B. R., Dizon, A. E., and Peterson, K. A., 1985, Chains of Single-Domain Magnetite Particles in Chinook Salmon, Oncorhynchus tshawytscha. J. Comp. Phys. A., 157, 375−381.
Kirschvink, J. L. and Walker, M. M., 1985, Particle-size considerations for magnetite-based magnetoreceptors, in: "Magnetite Biomineralization and Magnetoreception in Animals: A New Biomagnetism", J. L. Kirschvink, D. S. Jones, and B. J. MacFadden, eds., Plenum Press, New York, 243−254.
Kirschvink, J. L. and Kobayashi-Kirschvink, A., 1991, Is Geomagnetic Sensitivity Real? Replication of the Walker-Bitterman Conditioning Experiment in Honey bees. American Zoologist, V. 31, January 1991 (in press).
Kirschvink J. L., Jones, D. S., and MacFadden, B. J., eds., 1985, "Magnetite Biomineralization and Magnetoreception in Organisms: A New Biomagnetism", Plenum Press, New York, 682 pp.
Klinowska, M., 1985a, Cetacean live stranding sites relate to geomagnetic topography. Aquatic Mammals 11(1):27−32.
Klinowska, M., 1985b, Cetacean live stranding dates relate to geomagnetic disturbances, Aquatic Mammals 11(3):109−119.

Kramer G., 1952, Experiments on bird orientation. Ibis 94:265–285.

Kreithen M. L. and Eisner T., 1978, Detection of ultraviolet light by the homing pigeon. Nature 272:347–348.

Kreithen, M. L. and Keeton, W. T., 1974a, Detection of polarized light by the homing pigeon, Columbia livia. J. Comp. Physiol. 89:83–92.

Kreithen, M. L. and Keeton, W. T., 1974b, Attempts to condition homing pigeons to magnetic stimuli. J. Comp. Physiol. 91:355–362.

Kreithen, M. L. and Quine, D., 1979, Infrasound detection by the homing pigeon: a behavioral audiogram. J. Comp. Physiol. 12:1–4.

Kuterbach, D., Walcott B., Reeder R. J., and Frankel R. B., 1982, Iron-containing cells in the honey bee (Apis mellifera), Science 218:695–697.

Leask, M. J. M., 1977, A physiochemical mechanism for magnetic field detection by migratory birds and homing pigeons. Nature 267:144.

Lindauer, M., 1977, Recent advances in the orientation and learning of honeybees. Proc. XV Int. Congr. Entomol., 450–460.

Lowenstam H. A. and Weiner S., 1989, "On Biomineralization", New York, Oxford: Oxford University Press, 324 pp.

Mann S., Sparks N. H. C., Walker M. M., and Kirschvink J. L., 1988, Ultrastructure, morphology and organization of biogenic magnetite from sockeye salmon, Oncorhynchus nerka: Implications for magnetoreception. J. Exp. Biol. 140:35–49.

Papi F., Fiore L., Fiaschi V., and Benvenuti S., 1972, Olfaction and homing in pigeons. Monit. Zool. Ital. (N.S.) 6:85–95.

Sauer, E. G. F., 1957, Die Sternenorientierung nachtlich ziehender Grasmucken (Sylvia atricapilla, borin und curruca). Z. Tierpsychol. 14:29–70.

Sokal, R. R. and Rohlf, F. J., 1981, "Biometry", New York: W. H. Freeman and Co., 859 pp.

Stolz, J. F., Chang, S. R., and Kirschvink, J. L., 1986, Magnetotactic bacteria and single-domain magnetite in hemipelagic sediments. Nature 321:849–851.

Vali H. and Kirschvink J. L., 1989, Magnetofossil dissolution in a Paleomagnetically unstable Deep-Sea Sediment. Nature 339:203–206.

Walcott, C., 1978, Anomalies in the earth's magnetic field increase the scatter of pigeon's vanishing bearings, in: "Animal Migration, Navigation, and Homing", K. Schmidt-Koenig and W.T. Keeton, eds., Springer-Verlag, Berlin, pp. 143–151.

Walcott, C. and Green, R. P., 1974, Orientation of homing pigeons altered by a change in the direction of an applied magnetic field. Science 184:180.

Walker, M. M., 1984, Learned Magnetic Field Discrimination in Yellowfin Tuna, Thunnus albacares. J. Comp. Phys. A. 155:673–679.

Walker, M. M. and Bitterman, M. E., 1985, Conditioned responding to magnetic fields by honeybees. J. Comp. Phys. A. 157:67–71.

Walker, M. M. and Bitterman, M. E., 1989a, Attached magnets impair magnetic field discrimination by honeybees. J. Exp. Biol. 141:447–451.

Walker, M. M. and Bitterman, M. E., 1989b, Honeybees can be trained to respond to very small changes in geomagnetic field intensity. J. Exp. Biol. 145:489–494.

Walker, M. M. and Bitterman, M. E., 1989c, Conditioning analysis of magnetoreception in honeybees. Bioelectromagnetics 10:261–276.

Walker, M. M., Baird, D. L., and Bitterman, M. E., 1989, Failure of stationary but not of flying honeybees to respond to magnetic field stimuli. J. comp. Psychol. 103:62–69.

Walker, M. M., Bitterman, M. E., and Kirschvink J. L., 1986, Experimental and correlational studies of responses to magnetic field stimuli by different species, in: "Biophysical Effects of Steady Magnetic Fields", Maret, G., Boccara, N., and Kiepenheuer, J., eds., Springer-Verlag, New York, 194–205.

Walker, M. M., Kirschvink, J. L., Chang, S-B. R., and Dizon, A. E., 1984, A candidate magnetic sense organ in the yellowfin tuna, Thunnus albacares. Science 224:751–753.

Walker, M. M., Kirschvink, J. L., Perry, A. S., and Dizon, A. E., 1985, Methods and techniques for the detection, extraction, and characterization of biogenic magnetite, in: "Magnetite Biomineralization and Magnetoreception in Organisms: A New Biomagnet ism", J. L. Kirschvink, D. S. Jones, and B. J. MacFadden, eds., Plenum Press, New York, 154–166.

Walker, M. M., Quinn, T. P., Kirschvink J. L., and Groot, T., 1988, "Production of single-domain magnetite throughout life by sockeye salmon, Oncorhynchus nerka." J. Exp. Biol. 140:51–63.

Walker, M. M., Kirschvink, J. L., Dizon, A. E., and Ahmed, G., "Evidence that fin whales respond to the geomagnetic field during migration", submitted to J. Exp. Biol.

GEOMAGNETIC ORIENTATION IN CETACEANS: BEHAVIOURAL EVIDENCE

Margaret Klinowska

Research Group in Mammalian Ecology and Reproduction
Physiological Laboratory, University of Cambridge
Downing Street, Cambridge CB2 3EG, UK.

ABSTRACT

Cetaceans appear to use the flux density of the earth's magnetic field (total field) in two ways as an aid to travel. The topography of the local field is used as a map, with the animals generally moving parallel to the contours. A timer, based on the regular fluctuations in this field, allows the animals to monitor position and progress on this map. Problems can arise leading to live strandings when the geomagnetic contour routes cross land or when the pattern of time information is disrupted by irregular field fluctuations. The animals do not use directional geomagnetic information, for example as we do with our magnetic compasses.

Statistical evidence for this travel strategy has come from the positions and timing of live strandings around the UK coast, and from similar studies elsewhere. Other sources of information on cetacean behaviour are being explored in order to gain further insights into this travel strategy. Among these are the records of the drive fisheries for pilot whales (<u>Globicephala</u> <u>melas</u>) in the North Atlantic. Traditional driving beaches generally have the same characteristic geomagnetic topography as live stranding sites, and it therefore appears that the whalers are exploiting the normal orientation strategy of the animals to facilitate their work. These whales will flee without reference to geomagnetic contour routes when alarmed - perhaps another means whereby the animals can take the "wrong turnings" which may lead to live strandings. Since they can be driven by experienced crews without reference to the geomagnetic contour route, this method could be used for averting live strandings or for rescue purposes. However, it is unlikely to be practical today, because of the lack of experienced driving crews.

INTRODUCTION

Cetaceans spend their entire lives in water. Typically, only about 5% of their time is spent at or near the surface, and then usually only the blow hole at the top of the head is exposed for respiration, the eyes remaining below water. They therefore appear to lack convenient access to position information from the sun and night sky and, unless the water is particularly clear or shallow, no information about the underwater surroundings can be gained by vision. Many species (but not all) have developed echolocation systems, which are used in the pursuit of prey and for monitoring the immediate surroundings, but do not appear adequate to provide reliable information for major journeys and for orientation in deep waters. However, some species, even those apparently without echolocation, travel up to thousands of kilometres every year, between summer feeding grounds in polar waters and wintering grounds in tropical waters, indicating that some efficient travel strategy has been evolved.

It is very difficult to observe cetaceans in their natural habitat, because so much time is spent under water. Also, it is not feasible to perform any of the classical release and re-capture experiments, which have told us so much about the travel strategies of species such as homing pigeons, with such large animals. However, in the final analysis, these experiments depend on inducing the animals to make mistakes, and from the mistakes information about their travel strategies can be obtained. For an aquatic mammal it seems possible that a major orientation mistake might result in the animal, or the whole social group, running into the land, instead of remaining safely at sea. This cetacean phenomenon, known as "live stranding," has intrigued observers at least from the time of Aristotle, and a great variety of explanations has been proposed, none of which could account for all observed cases.

Analysis of the geomagnetic topography of UK live stranding sites revealed that these events all occurred where geomagnetic contours were perpendicular to the coast, from the sea. The sites had no other physical features in common. Carcasses were washed up on shore wherever tides and currents took them; these sites merely reflecting the general distribution of areas with geomagnetic contours perpendicular and parallel to the coasts. The passive strandings reports also reflected the distribution of observers (H. M. Coastguard establishments), but the live strandings distribution did not. The simplest explanation is that the animals are using the geomagnetic topography as a map, and that their basic strategy is to travel parallel to the geomagnetic contours. Such a strategy would serve well at sea, but gives no indication of where land and sea meet, and hence could mislead unwary or sick animals or those unfamiliar with the area, producing the observed live stranding pattern.

Further investigations revealed that the dates of the live strandings were associated with dates when the normal patterns of mean daily geomagnetic fluctuations (as measured by the Sq index) were disrupted by irregular fluctuations (as measured by the three-hourly K index). The dates when carcasses were washed up showed no such relationship, nor did a sample of random dates which were taken as an additional control. The size of the irregular fluctuations was not important, only their pattern. The disruption associated with the live strandings dates was very precise; only loss of the morning fall in magnetic flux density (total field) appeared to be important, not disruptions at any other time of day. This would indicate that some timing mechanism is involved, rather than general difficulties through disturbed conditions, since transitions in natural phenomena are the features most readily available for re-setting biological clocks (for example, when the ratio of light to dark is the cue, dawn and dusk are the key times or Zeitgebers; midday or midnight are far more difficult to perceive without artificial aids). Regional differences in the relative importance of geomagnetic events on the live stranding and pre-stranding days, and inspection of local geomagnetic topography maps for areas where "wrong turnings" might occur, indicate that the primary mistakes leading to live strandings may happen some distance from land,

and that such animals appear to continue on the incorrect course for some days before blundering into the beach.

Thus, for general travel purposes, cetaceans appear to exploit the flux density of the earth's magnetic field in two ways: the topography of the local field is used as a map, with the animals generally moving parallel to the contours; and a timer, based on the regular fluctuations in this field, allows the animals to monitor position and progress on the map. The animals are not using directional geomagnetic information, for example as we do with our magnetic compasses (Klinowska, 1983; 1985a; 1985b; 1985c; 1986; 1988). Others (e.g. Kirschvink et al., 1984; Kirschvink et al., 1986; Walker et al., 1986) have analysed live stranding and sightings information to confirm that cetaceans are using geomagnetic information as a travel cue.

Despite early hopes for a system based on magnetite (Zoeger et al., 1981), the question of a suitable receptor in cetaceans for such geomagnetic information has yet to be answered (Bauer et al., 1985). However, Behrmann's (1988 and 1990) reports of structures in the tongue and integument of the lower jaw in harbour porpoises (Phocoena phocoena), similar to the electroreceptors in some fish (Szabo, 1974), which might be able to detect electric potential differences induced by swimming in the earth's magnetic field, could open a fruitful new avenue of research. Rosenblum et al. (1985) showed on theoretical grounds that, while a magnetic induction-based reception system is unlikely to be feasible for animals living in air, it is perfectly reasonable for those living in water. Further, as the authors themselves remark, an induction-based system would explain the failure of the behavioural experiments performed by Bauer et al. (1985) in their attempt to demonstrate magnetic sensitivity in captive bottlenose dolphins (Tursiops truncatus).

More sources of information on cetacean behaviour need to be explored to gain further insights into this travel strategy. Besides further controlled experiments, remote sensing techniques are one obvious answer, but the technology for the cost-effective, long-term monitoring via satellite, which would be required, is not yet sufficiently well-developed. Published tracking data of sufficient quality so far come either from geologically young areas (such as off California) where it is difficult to distinguish between bottom topography and geomagnetic topography, or from areas (such as off Iceland) where detailed geomagnetic survey data are not yet available (Klinowska, 1986; 1988).

However, it is possible that other records of cetaceans might provide further information on this travel strategy. Those of the long-finned pilot whale fisheries (Globicephala melas) in the North Atlantic (Hoydal, 1986; Joensen and Zachariassen, 1986; Klinowska, 1987; Sergeant, 1962 and 1982) are probably the longest and most detailed. It is likely that from the earliest times local people throughout the world have taken advantage of stranded cetaceans, in order to obtain food, oil and other products. It seems a small step from the use of animals already on the beach to using various methods to "encourage" live strandings - the drive fisheries. But how far do drive fisheries in fact resemble live strandings?

NORTH ATLANTIC LONG-FINNED PILOT WHALES

Pilot whales are distributed widely in the North Atlantic. They live in herds of about 50 animals, forming a close social unit which appears to facilitate drive fishing. It is thought that their travels are made, generally, in search of squid, the main food species. In both Canada (Sergeant, 1962) and the Faroe Islands (Hoydal, 1986) pilot whale and squid catches show a close relationship. Pilot whales and squid tend to frequent inshore waters in summer, but pilot whales can be found there at any time of the year, perhaps indicating other motives for travel than squid-hunting alone.

The oldest recorded pilot whale drive fishery, which still continues, is that of the Faroe Islands (Hoydal, 1986; Joensen and Zachariassen, 1986). Scattered written records go back to 1584, but the fishery is undoubtedly very much older. The records come mainly from tax accounts, relating to the elaborate traditions by which the catch was shared between participants, landowners and the church. The most detailed catch series runs from 1709 to the present, and contains the date, place and size of each school taken (although for various reasons records from before World War II are taken to underestimate the number of whales caught by at least 10%).

The pilot whale fishery in the northern British and Irish Islands, which died out about the turn of this century, may be equally old, but is less well documented (Klinowska, 1987). At least in the Orkney and Shetland Islands there was a tradition of catch sharing somewhat similar to that in the Faroe Islands. This was probably because the islands only came under Scottish jurisdiction in 1468, as security for the dowry of Princess Margaret, daughter of King Christian I of Denmark, on her marriage to King James III of Scotland. The cash value of the dowry was never paid, and the islands were administered under the old Norse laws until 1611. Although these laws were replaced by the laws of Scotland, various matters, including the division of whale catches, remained in dispute until almost the end of the 19th century. Elsewhere, except perhaps in the Hebrides where at least the church laid claim to shares, catch division was more informal.

In Shetland, the pilot whale was known as the "caa-ing whale" or "driving whale" (from a similar local word used to describe the driving of sheep). The name "caa-ing whale" has been misinterpreted frequently in the literature as a contraction of "calling whale" and written as "ca'aing". The earliest written records are those in the Court Book of Shetland (1602-1604) and relate to the catch shares demanded by Earl Patrick Stewart. They do not include any large pilot whale drives, but are likely to be accurate because the Earl was infamous for extorting every possible source of revenue from his lands. He was arrested and brought to trial in 1609 for his depredations in Orkney and Shetland. Later information on catches comes from scattered references, the investigations of local naturalists, and a few eye-witness accounts. However, even this limited data indicate a minimum annual take of a similar order of magnitude to that in the Faroes (Klinowska, 1987).

The main objective of the Faroese catch always has been food for human consumption, and this was also the case in the Hebrides, although here pilot whaling seems to have been less important. However, in Orkney and Shetland, oil for lighting was the only objective, and the meat, which was not considered fit for human consumption even in times of famine, was left to rot on shore.

An intensive commercial fishery for pilot whales existed in Newfoundland, Canada from 1947 to 1971 (Sergeant, 1962; 1982). The main technique was driving, but some harpooning of large animals at sea also occurred. This fishery was for oil and for meat as animal feed. The local population of pilot whales seems to have been over-exploited grossly, with about 54,000 animals taken, the majority of these in the early years. Although far fewer whales entered Newfoundland waters in the later years, there was no decline in the squid populations on which they fed. Catches in the Faroe Islands were reasonably steady over this period, indicating that a different population of pilot whales may have been involved.

The general method in these fisheries is similar (Hoydal, 1986; Klinowska, 1987; Sergeant, 1962). A look-out for the whales is kept, from the shore, from boats in adjacent waters, or both. Once a herd is sighted, a few boats slowly guide them shorewards. The phrase "driven like sheep" is frequently used to describe this stage, and it is possible to drive more than one herd together to form a larger group. However, herds of several thousand animals cannot be driven together and have to be split into smaller groups which can be moved towards the shore. The drive can be stopped, for example through the night, and it is even said to be difficult to keep the herd "awake" and moving during drives in summer twilight. As the whales near land they are met by many small boats, which surround the animals in a crescent and complete the drive to shore.

The long Faroese records would be the best for analysis, but unfortunately no detailed geomagnetic survey data are yet available for this area, nor have the basic catch data yet been published; only summaries and analyses. However, several Orkney, Shetland and Irish driving beaches are identified in the literature, there are eye witness accounts of drives, and detailed geomagnetic topographical maps. Similar, but less detailed information is available for Canada. It is therefore possible to make some preliminary investigations, with a view to testing the findings when the Faroese data are available.

ORKNEY AND SHETLAND

Materials and Methods

The 30 named driving beaches in Orkney and Shetland (Klinowska, 1987) provide the opportunity to explore their geomagnetic characteristics in a simple quantitative manner. All except three of these beaches are shown on the Ordnance Survey maps to have a sandy area, and all have some kind of road access and a settlement in the vicinity. As a control, the entire coastline of islands with at least one named driving beach was examined, and all other beaches with sandy areas, road access and a settlement in the vicinity noted. There were 90 such beaches in total. Another person was provided with the list of beaches (with no indication of which were driving beaches or control beaches), a magnetic map, and instructions to score each beach as to whether there was access from the sea if animals moved parallel to geomagnetic contours.

Results

Of the 27 named sandy driving beaches, 22 (81%) were scored as having access via parallel contours and 5 (19%) without. Of the 90 control beaches, 63 (70%) were scored as having such access and 27 (30%) did not. Chi-squared (d.f. 1) was 5.52 ($P < 0.03$), showing that the driving beaches were significantly more likely to have access via parallel geomagnetic contours.

Discussion

Two detailed eye-witness accounts of whale drives further illustrate the situation.

Gorrie (1868) describes his experiences when visiting the island of Stronsay, Orkney in the early 1860s. He was staying with the minister, and their breakfast was interrupted one morning by the excited servant, who panted out the news that the bay was full of whales. The minister and his guest ran to the garden, to see a crescent of boats about a mile from the shores of Mill Bay below, with a school of pilot whales in front (Figure 1).

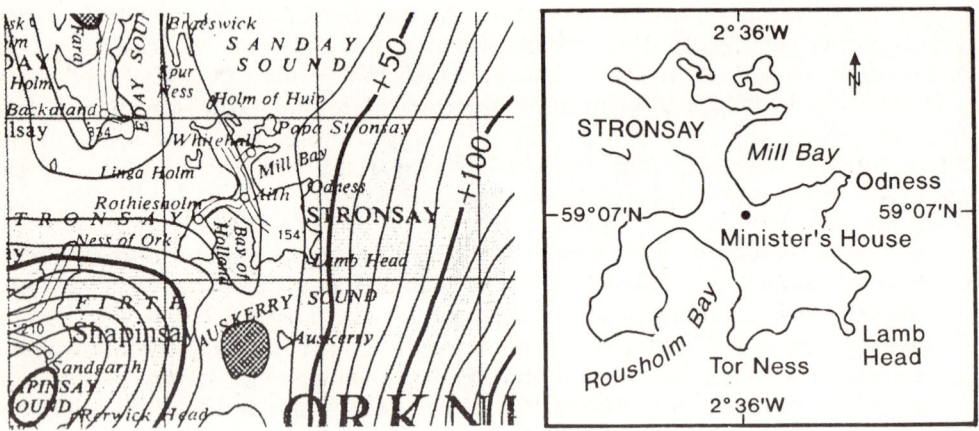

Fig. 1. Island of Stronsay, Orkney. Places mentioned in the text are shown in the sketch map on the right. The left hand map is a section from the British Geological Survey Aeromagnetic map. Geomagnetic contours are at 10 nanoTesla (nT) intervals and grid squares at 10km intervals.

More boats were arriving from all directions, and the local people were flocking to the shore. Unfortunately, when the whales were only a quarter of a mile away, some lads dashed out in a small rowing boat in front of the animals. At this, the school turned, rushed past the line of boats and away around the Odness headland. The boats followed, and having reached a new vantage point, Gorrie saw the whales calmly swimming in front of the boats towards Lamb Head. The minister said that they were "almost certain to take a snooze in Rousholm Bay, which is the best whale trap I know in Orkney". The gentlemen then repaired to a nearby farmhouse, where they "quenched their thirst with liberal draughts of home-brewed beer" and borrowed ponies to ride to Rousholm Bay (Bay of Holland on modern maps). On arrival, they saw that some 150 whales were being brought around Tor Ness Point, while the rest had disappeared west into Stronsay Firth. At the entrance to Rousholm Bay the whales again appeared ready to flee, but instead of turning and rushing to the open sea, they dashed rapidly forwards on to the beach where the killing took place.

From Figure 1 it can be seen that there is access to Mill Bay from the northwest following parallel geomagnetic contours, which seems to lead towards the southern corner of this bay, just below the minister's house. The flight path out to Odness crosses the contours, but once clear of land, the contours could be followed north or south. Perhaps the boats to the north encouraged the whales to move south on this occasion. Unfortunately, the observers were having their beer when the turn at Lamb Head was reached, and we have no information as to whether any difficulty was experienced in persuading the animals to move west rather than continue south, as would be expected from the set of the geomagnetic contours. One might also expect there to have been some difficulty in keeping the whales away from the sandy beaches between Lamb Head and Tor Ness. However, the loss of part of the school to the west after this area was passed fits in with the run of the contours, as does the slight difficulty described in persuading the animals to take the contour route into the bay. It is interesting to note that 90-100 pilot whales were live stranded in this bay on 22 April 1950 (Klinowska, 1987). A wider look at the general area shows that Stronsay Island appears to be right in the path of animals travelling north or south along geomagnetic contours. A slight deviation would bring southbound animals into Mill Bay, while another slight deviation would bring northbound animals towards Rousholm Bay (Bay of Holland). Here they could either move west or take the path into the bay.

Hibbert (1822) had just landed at Burra Voe on the island of Yell, Shetland in 1820 or 1821, when a fishing boat arrived with the news that a school of pilot whales had entered Yell Sound. The usual excitement followed, and the whales were soon seen at the entrance to the Sound, swimming quietly before a semi-circle of boats which followed them at a distance of about 50 yards. A second group of boats waited to intercept the whales, should they change their course. "The sable herd appeared to follow certain leaders; who, it was soon feared, were inclined to take any other route than that which led to the shallows (at Hamnavoe) on which they might ground. Immediately, the detached crews rowed with all their might, in order to drive back the fugitives, and, by means of loud cries and large stones thrown into the water, at last succeeded in causing them to resume their previous course." This happened again before the animals could be compelled to enter Hamnavoe harbour. Despite shouts and stone throwing, the whales turned several times and had to be driven back before some could be beached, and the rest followed (Figure 2).

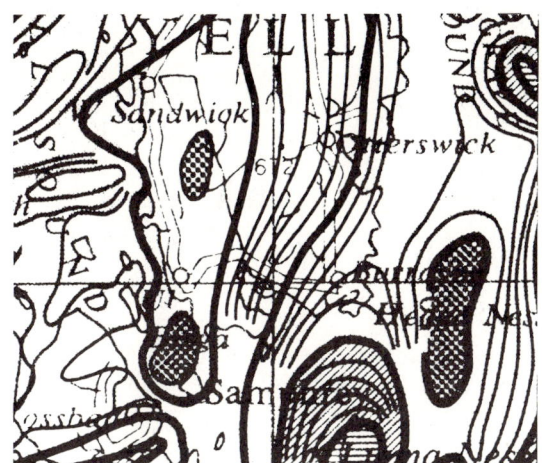 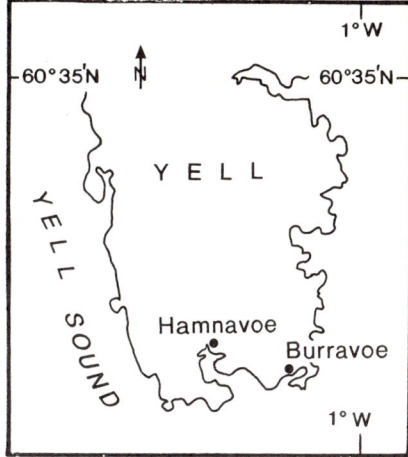

Fig. 2. Island of Yell, Shetland. Places mentioned in the text are shown on the sketch map to the right. The left hand map is a section of the British Geological Survey Map (see caption to Figure 1) magnified to demonstrate the situation more clearly.

It is not clear from the description whether the whales entered the part of Yell Sound in the vicinity of Hamnavoe from the south or from the north. However, the difficulty in driving the whales into Hamnavoe can be understood, because the geomagnetic contours do not lead in this direction (Figure 2). Indeed, the entire route from the entrance to the beach crosses contours. However, this is the only beach in the southern part of the island, which probably explains why the local people went to so much trouble to drive the whales in here. The fact that, almost uniquely in descriptions of pilot whale drives in the North Atlantic, two groups of boats were used, indicates that the people were familiar with the difficulties, and had learned to overcome them. No subsequent live strandings of pilot whales are recorded in Hamnavoe, but one animal stranded alive at Midyell Voe on the east coast on 31 October 1931, and a group of four at Basta Voe, also on the east coast, on 10 December 1982 (Klinowska, 1987).

A final example (Figure 3) is taken from an article in the "Shetland News" of 30 July 1898. This reports that on Saturday 23 July, between 4 and 5 a.m., about 200

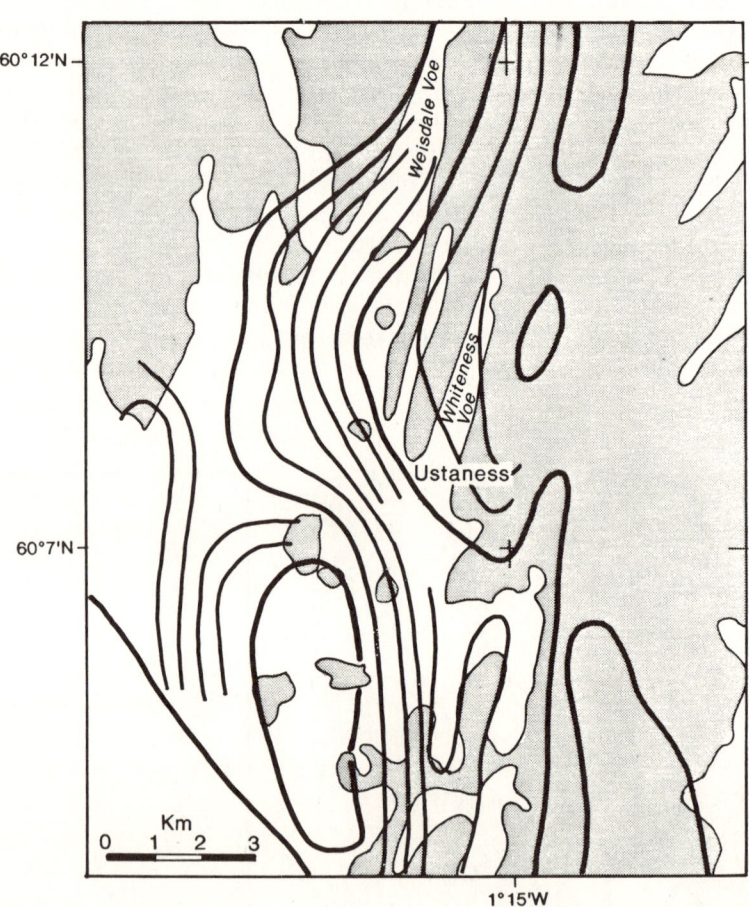

Fig. 3. Geomagnetic topography of the southwestern area of Mainland, Shetland, re-drawn and simplified from the British Geological Survey map. Land areas are shaded.

potheads (i.e. pilot whales) were seen off Ustaness, Whiteness Voe. They were chased, but would not strand in this Voe, unlike Weisdale Voe where they have been frequently captured. Two suckers (i.e. very young animals) were captured, and another animal struck with a harpoon but lost. Further on in the same article it is noted that some 500 animals went up Weisdale Voe later that day, but that there were not enough men or boats to follow them.

It is quite clear from this article that the unsuitability of Whiteness Voe for whale driving was well known, although this is attributed to the deepness of the water in comparison with the shallower (and "better") Weisdale Voe. It is also quite clear that the crews involved were inexperienced, ill equiped and too few for an effective drive. Indeed from the tone of the account, the incident seems more like an adventure than the serious (but exciting) socio-economic activity described in the earlier reports above. Since this took place at a time when whale driving was dying out, presumably because alternative fuels for lighting were more readily available, it is perhaps not surprising that the attempted drive was not taken very seriously.

From Figure 3 it can be seen that the geomagnetic topography can provide another explanation of the reluctance of pilot whales to enter Whiteness Voe, as well a good reason for them to enter Weisdale Voe readily. Besides general references, there are three other dated reports of whale drives in Weisdale Voe, including one on 7 February 1903, which is said to have been the last occasion when pilot whales were driven in Shetland. There are no other references to attempts to drive whales into Whiteness Voe. A school of about 200 was also sighted in Weisdale Voe on 8 June 1946 (Klinowska, 1987).

Conclusions

A simple quantitative test has shown that the geomagnetic topography of named pilot whale driving beaches is likely to resemble that of live stranding sites, in that animals travelling parallel to geomagnetic contours would have access. From the detailed descriptions of drives it can be seen that, although in general driving was easier when the animals were moved parallel to the geomagnetic contours (and thus the preference for driving beaches with this characteristic), it was possible for experienced crews to drive the whales more or less as required.

REPUBLIC OF IRELAND

From Figure 4, it can be seen that the beaches at Wexford harbour and Fethard can be approached by animals moving parallel to geomagnetic contours, as can the beaches at Brandon Bay and Bantry Bay. The only other named driving beach in the Republic of Ireland, described as "Dunfanaghy Co. Donegal", has not been located so far (Klinowska, 1987).

CANADA

The major Canadian fishery took place in Trinity Bay where whales encountered at the head of the bay were driven down to the small southern beaches at Chapel Arm, New Harbour, Old Shop and Bellevue (Fig. 5). Final choice of the beach to be used in a particular drive was said to depend on the behaviour of the whales, as well as on wind and tide. Catches were made also in the neighbouring Bonavista, Conception and Notre Dame Bays (the latter outside the area shown) (Sergeant, 1962). Even from the less detailed Canadian geomagnetic topography map, it is still reasonably clear that the drive routes generally run with the geomagnetic contours.

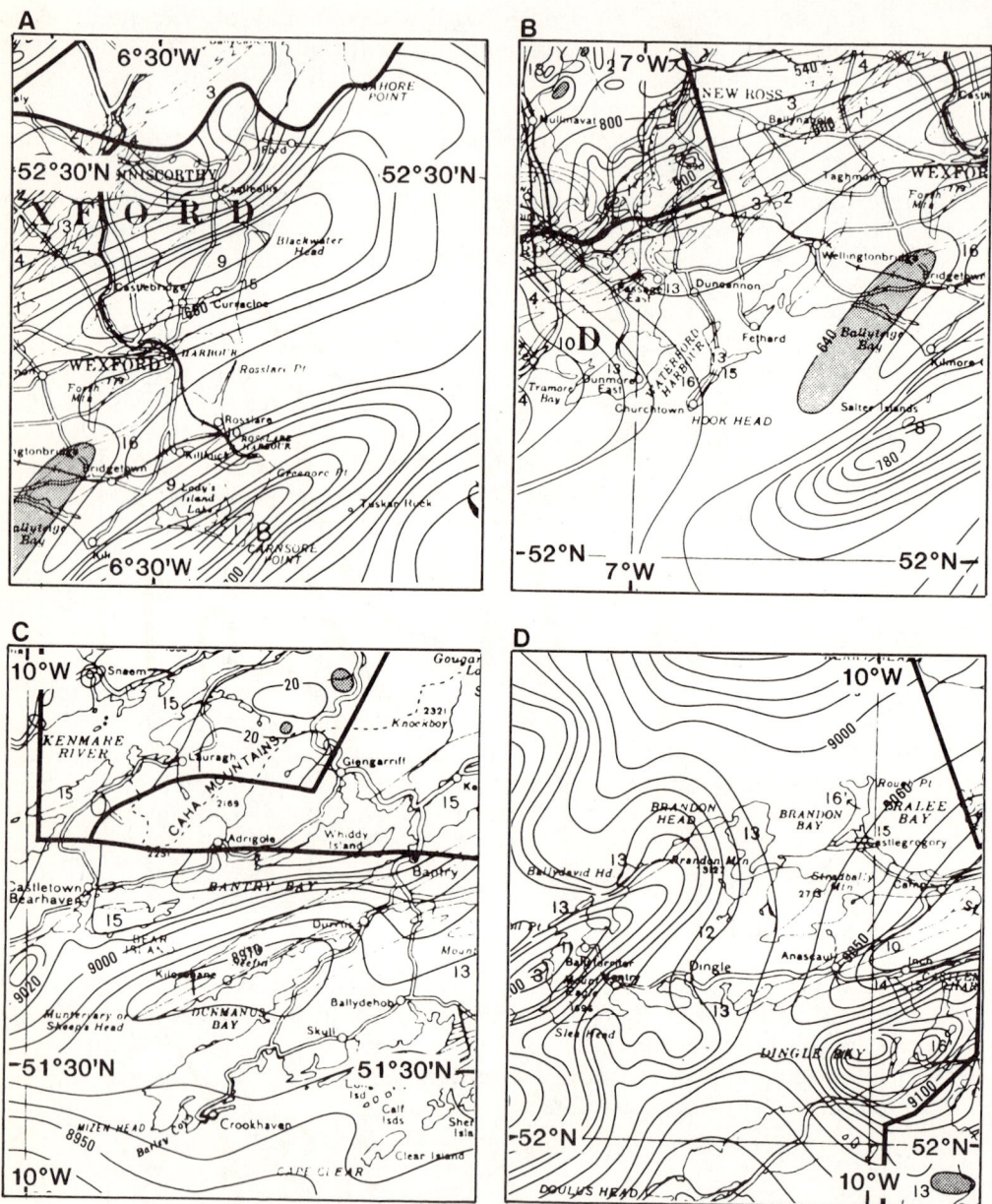

Fig. 4. Named pilot whale driving beaches in the Republic of Ireland. Sections of the Geological Survey of Ireland Compilation Geomagnetic Map, scale 1:750,000. A: Wexford harbour. B: Fethard. Geomagnetic contours at 20nT intervals. C: Bantry Bay, contours at 10nT intervals. D: Brandon Bay, contours at 50nT intervals.

A few live strandings of groups of pilot whales are noted in this area after the drive fishery ceased. As far as can be seen from the geomagnetic map, the sites all appear consistent with the idea that the animals are travelling parallel to the geomagnetic contours. Sergeant (1982) takes these events as evidence for an increase in the pilot whale population. However, while it is clear from the UK records (Klinowska, 1987) that reports of group strandings and sightings increase as catching ceases, this is hardly surprising.

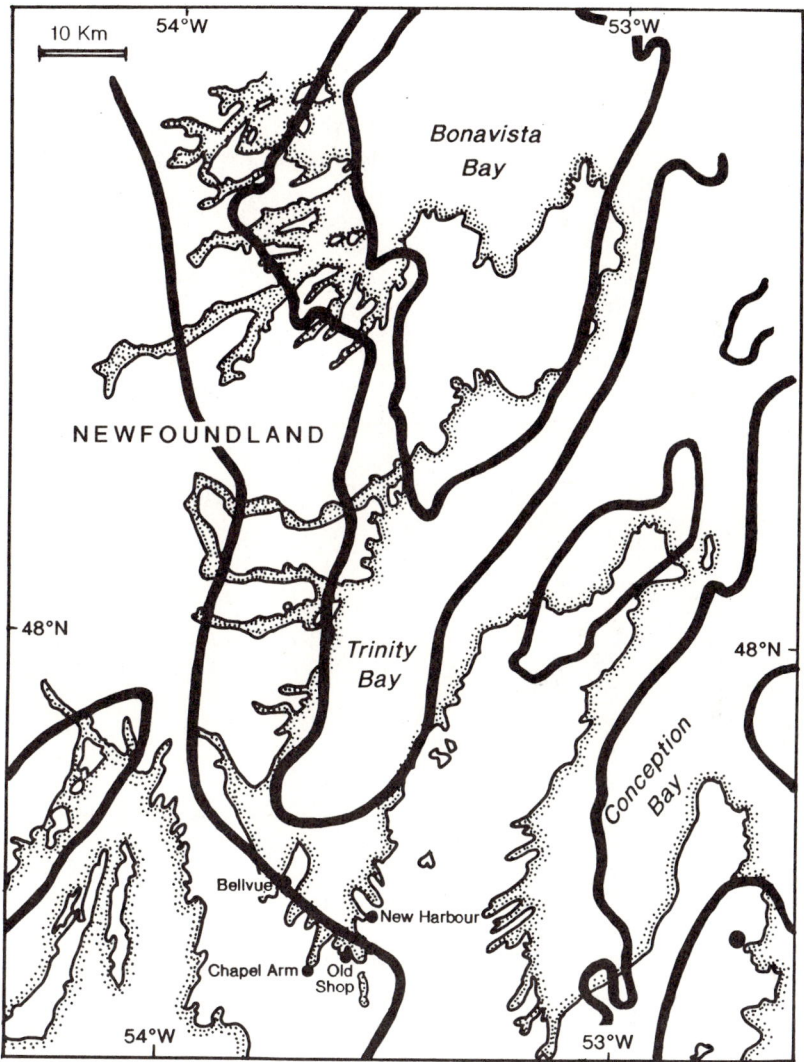

Fig. 5. Main pilot whale driving district in Newfoundland, Canada. Re-drawn area of the Memorial University of Newfoundland Magnetic Anomaly Map of the Appalachian Orogen. Geomagnetic contours at 200nT intervals. Original scale 1:1,000,000.

In former times, any school seen near the coast would have been pursued (although not usually on Sundays in Scotland), thus providing a "catch" record. When catching gradually ceased, observations are noted as "sightings" or "strandings" simply because nobody happened to pursue the animals. Even comparing the total number of schools reported before and after catching ceases is unlikely to give a good representation of the status of the local populations, because without the incentive of a potential drive, interest in reporting such events will be far lower.

DISCUSSION

The accounts of pilot whales submitting to driving and apparently approaching beaches willingly can be supported by reference to the local geomagnetic topography. Most driving beaches are like live stranding sites, in that they can be approached by animals travelling parallel to geomagnetic contours. Some live strandings are also recorded on old driving beaches, but not on such beaches without access via the geomagnetic contour route. Accounts of drives are generally consistent the idea that the drive fisheries exploit the weaknesses in the normal cetacean travel strategy.

Since the whales will flee without reference to the geomagnetic contour routes when alarmed, such disturbance could be another way in which animals come to take the "wrong turnings" which may end in live strandings. Experienced crews can drive the whales without reference to the geomagnetic contour route, indicating that this method might be useful for averting live strandings or for rescue purposes. However, since inexperienced crews (and even experienced crews on some occasions) fail to drive successfully in such circumstances, the method is unlikely to be practical today because there are few experienced whale drivers available.

ACKNOWLEDGMENTS

The Director of the British Geological Survey kindly gave permission for use of their geomagnetic map, copyright of which is reserved to the Natural Environment Research Council. Mrs D. Hughes and Mr P. Starling prepared the illustrations.

REFERENCES

Bauer, G. B., Fuller, M., Perry, A., Dunn, J. R. and Zoeger, J., 1985, Magnetoreception and biomineralization of magnetite in cetaceans, p. 489-508, in: "Magnetic Biomineralization and Magnetoreception in Organisms," J. L. Kirschvink, D. S. Jones and B. J. MacFadden, eds., Plenum Press, New York.

Behrmann, G., 1988, The peripheral nerve ends in the tongue of the harbour porpoise Phocoena phocoena (Linne, 1758), Aquatic Mammals 14: 107-112.

Behrmann, G., 1990, The tuberous organs of the harbour porpoise Phocoena phocoena (Linne, 1758), Aquatic Mammals 16: 33-35.

Gorrie, D., 1868, "Summers and Winters in the Orkneys," Hodder and Stoughton, London.

Hibbert, S., 1822, "A Description of the Shetland Islands," Constable and Co., Edinburgh.

Hoydal, K., 1986, Data on the long-finned pilot whale (Globicephala melaena Traill), in Faroe waters and an attempt to use the 274 years of time series of catches to assess the state of the stock, Document No. IWC/SC/38/SM7, submitted to the Scientific Committee of the International Whaling Commission.

Joensen, J. S. and Zachariassen, P., 1986, Statistics for pilot whale catches in the Faroe Islands 1584-1640 and 1709-1978, Document No. IWC/SC/38/SM20, submitted to the Scientific Committee of the International Whaling Commission. (Translation of: Joensen, J. S. and Zachariassen, P., 1982 Frodskaparrit (Annal. societ. scient. Faeroensis) 30: 71-102.)

Kirschvink, J. L., Dizon, A. E. and Westphal, J. A., 1986, Evidence from strandings for geomagnetic sensitivity in cetaceans, J. Exp. Biol. 120: 1-24.

Kirschvink, J. L., Westphal, J. A. and Dizon, A .E., 1984, Cetacean strandings along the North American Atlantic coast: evidence for a geomagnetic influence on pelagic navigation, EOS, Transactions of the American Geophysical Union, San Francisco, 65: 865.

Klinowska, M., 1983, Is the cetacean map geomagnetic ? Evidence from strandings, Aquatic Mammals, 10: 15.

Klinowska, M., 1985a, How do dolphins tell the time ? Aquatic Mammals, 11: 5.

Klinowska, M., 1985b, Cetacean live stranding sites relate to geomagnetic topography, Aquatic Mammals, 11: 27-32.

Klinowska, M., 1985c, Cetacean live stranding dates relate to geomagnetic disturbances, Aquatic Mammals, 11: 109-119.

Klinowska, M., 1986, The cetacean magnetic sense - evidence from strandings, p. 401-432, in: "Research on Dolphins," M. M. Bryden and R. J. Harrison, eds., Oxford University Press, Oxford.

Klinowska, M., 1987, Preliminary list of catches, live strandings and sightings of the pilot whale (Globicephala melaena) in the British and Irish islands, Document No. IWC/SC/39/SM2, submitted to the Scientific Committee of the International Whaling Commission.

Klinowska, M., 1988, Cetacean 'navigation' and the geomagnetic field, J. Navigation 41(1): 52-71.

Rosenblum, B., Jungerman, R. L. and Longfellow, L., 1985, Limits to induction-based magnetoreception, p. 223-232, in: "Magnetic Biomineralization and Magnetoreception in Organisms," J. L. Kirschvink, D. S. Jones and B. J. MacFadden, eds., Plenum Press, New York.

Sergeant, D. E., 1962, The biology of the pilot or pothead whale Globicephala melaena (Traill) in Newfoundland waters, Bull. Fish. Res. Board Canada 132. 84pp.

Sergeant, D. E., 1982, Mass strandings of toothed whales (Odontoceti) as a population phenomenon, Sci. Rep. Whales Res. Inst., Tokyo, 34: 1-47.

Szabo. T., 1974, Anatomy of the specialized lateral line organs of electroreception, p. 13-58, in: "Electroreceptors and Other Specialized Receptors in Lower Vertebrates," A. Fessard, ed., Springer-Verlag, Berlin.

Walker, M. M., Bitterman, M. E. and Kirschvink, J. L., 1986, Experimental and correlational studies of responses to magnetic field stimuli by different species, p. 194-205, in: "Biophysical Effects of Steady Magnetic Fields," Maret, G., Boccara, N. and Kiepenheuer, J., eds., Springer-Verlag, Berlin.

Zoeger, J., Dunn, J. R. and Fuller, M, 1981, Magnetic material in the head of a common Pacific dolphin, Science, 213: 892-894.

ATTENTION AND DECISION-MAKING IN ECHOLOCATION MATCHING-TO-
SAMPLE BY A BOTTLENOSE DOLPHIN (TURSIOPS TRUNCATUS): THE
MICROSTRUCTURE OF DECISION-MAKING

Herbert L. Roitblat[1], Ralph A. Penner[2] and Paul E. Nachtigall[2]

[1]University of Hawaii at Manoa
Department of Psychology, 2430 Campus Road
Honolulu, HI 96822 USA, [2]Naval Ocean Systems Center
Hawaii Laboratory, P. O. Box 997, Kailua, HI 96734
USA

INTRODUCTION

In delayed matching-to-sample (DMTS) the subject is presented with a sample at one time and must then pick a matching comparison stimulus from a set of alternatives. The choice the subject makes in discriminating among the comparison stimuli, is contingent on the identity of the sample that preceded them in the trial.

Despite a great deal of effort devoted to the investigation of the memory processes animals employ in DMTS (see Roitblat, 1982; 1987), very little information is available concerning the decision strategies they use when selecting a correct match. This relative lack of information is partly to the difficulty of collecting relevant data. The animal's decisions must depend on information it obtains from the sample and comparison stimuli, but measuring when and how much information an animal receives is extremely difficult. The echolocating dolphin provides an interesting exception to this generalization because each echo provides a discrete packet of information to the animal that can be measured relatively easily. As a result, we determine quite precisely when an animal is scanning a stimulus (for example, see Penner, 1989) and can examine the properties of the click and echo. In particular, we have been able to identify where the target is located and, on the basis of the echo, can identify the target that the animal is scanning.

MATERIALS AND METHODS

The subject of our ongoing investigation of echolocation DMTS performance is a male bottlenose dolphin, named Rake, housed in a floating pen in Kaneohe Bay at the Hawaii Laboratory of the Naval Ocean Systems Center. His eyes are covered during the daily tests with soft removeable eyecups that completely occlude his vision. He is highly experienced in DMTS and related tasks (e.g., Nachtigall et al., 1985).

In all of the experiments described here, the dolphin stationed 100 cm under water in the center of an observing aperture. The sample stimuli were located 5.3 m directly in front of the observing aperture. The comparison stimuli were suspended in three groups (arrays) from a bar located 4.3 m from the observing aperture. The left and right arrays were located approximately 1.6 m (22°) to the left and right respectively of the center comparison array. The sample arrays and each of the three comparison arrays contained one example of each stimulus. These stimuli were presented to the animal by lowering them into the water to a depth of approximately 100 cm. At other times they were maintained above the water's surface. The present studies employed three stimuli: (a) a PVC plastic tube open at both ends (15 cm long, 7.5 cm diameter, 30 mm wall thickness), (b) a water-filled stainless steel sphere (5 cm diameter), and (c) a solid aluminum cone (10 cm diameter base, 10 cm height).

A neoprene-covered aluminum panel located 25 cm in front of the observing aperture served as an acoustic shutter preventing the animal from echolocating when the screen was raised into position. Echolocation clicks were detected by three hydrophones located 2 m from the observing aperture in the direct path between the aperture and the stimuli. Peak amplitude and the time of each click were recorded by a computer. Occasional sessions were recorded using a RACAL store-4 tape recorder, with flat response up to 300 kHz, from which clicks and echoes were digitized at 1 MHz.

Each trial began with the dolphin stationed in the observing aperture with the acoustic screen closed. One of the sample stimuli was then lowered into the water, the screen was lowered, and the dolphin was allowed to echolocate. The acoustic screen was then raised back into position, the sample stimulus was removed from the water, and three comparison stimuli were presented in the same manner as the sample. All three comparison stimuli were presented on each trial, but their positions (left, right, or center) varied randomly between trials. The screen was then again lowered and the dolphin was allowed to echolocate on the comparison stimuli. The animal indicated his choice by touching a response wand approximately in contact with the water surface, and directly in front of each comparison stimulus. Contacting the wand in front of the comparison stimulus that matched the sample was designated the correct response and was reinforced with 3 smelt.

BEHAVIORAL RESULTS

The dolphin's DMTS choice accuracy experiment (Roitblat et al., 1990) averaged nearly 95% correct. So few errors were made, that the probability of an error is not likely to be informative concerning the processes dolphins use to select the correct match. The number of clicks emitted to each stimulus appears to provide more information about these processes. On average, the dolphin used 37.2 clicks to identify the sample, but the number of clicks depended reliably on the stimulus being scanned. Furthermore, the dolphin tended to scan multiple stimuli in selecting the matching comparison. An average of 4.2 scans were used per trial. A scan is a train of clicks to a single stimulus ended either by the initiation of a scan to another stimulus or by a cessation of echolocation clicking and the performance of a choice response. He tended to scan at least some of the targets more than once in a trial, and therefore, did not perform a strictly self-terminating search of the alternatives. To summarize the results of this experiment, the number and scan pattern of echolocation clicks were found to depend on the stimulus being scanned and on the identity of the sample stimulus.

A SEQUENTIAL-SAMPLING DECISION MODEL

These scanning patterns were modeled using sequential sampling theory (see also Roitblat, 1984). Sequential sampling theory applies when a subject uses multiple observations of a stimulus (looks) to identify it. The theory is especially applicable to echolocation because each click provides a discrete "look" at the item being scanned. Sequential sampling theory assumes that: (a) each look has some cost or effort, (b) each look provides stochastic information about the identity of the stimulus, (c) information from successive looks is combined to identify the stimulus, (d) the observer attempts to minimize the number of looks subject to meeting a confidence criterion for identifying the stimulus (i.e., the observer's confidence in the identification grows monotonically with the numbers of looks, but at the expense of the effort entailed in making those looks).

The information in an echo is a sample from a population of possible echoes that the object may return. (see Fig. 1. Because of noise in the environment and variability in click production, each echo from a particular object will be slightly different from the other echoes it returns. Echoes can be expected, therefore, to be samples from a random distribution about some prototypical value (all other things being equal). Sequential sampling theory assumes that the dolphin continues to send echolocation clicks until a sufficiently confident identification can be made. With successive observations, the accumulated information will be more similar to the prototype of one of the possible samples and less similar to others. Because there are three alternatives and one of these is guaranteed, by the rules of the task, to be the correct match, information obtained by echolocating on one of the comparison stimuli can be used to reduce uncertainty about the remaining comparison stimuli. Identification of one item reduces the possible set of stimuli that could be present in either of the other two comparison positions.

On average, the echo from the first click to a target provides considerably more information about the identity of the target (i.e., whether or not it matches the sample) than succeeding echoes. Each succeeding echo provides diminishing amounts of additional information about the identity of the

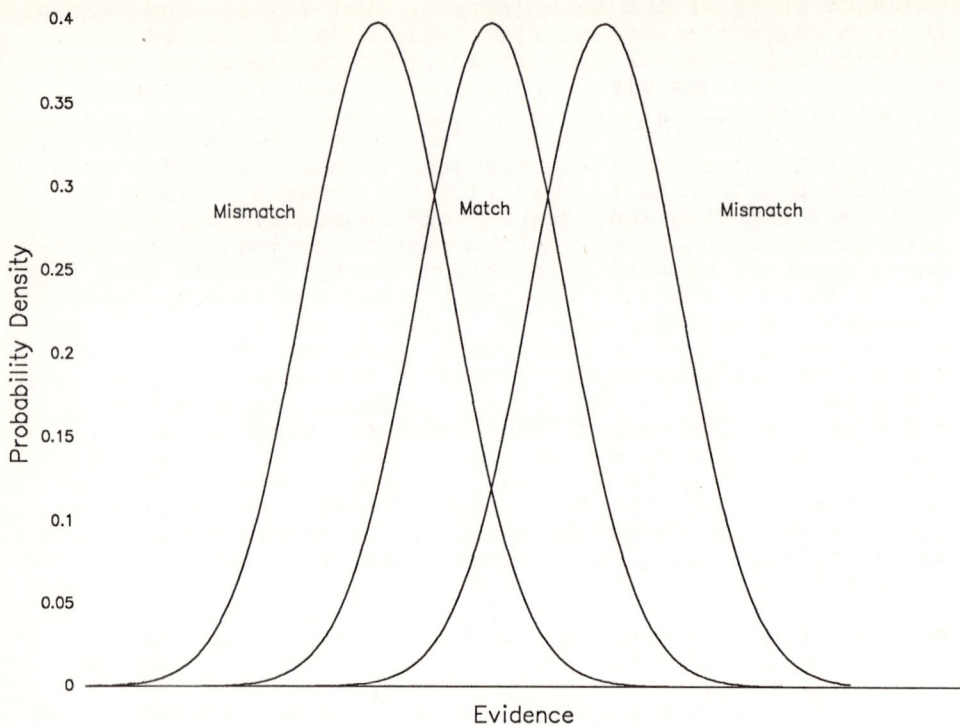

Fig. 1. The hypothetical distributions of echo-evidence returning from a matching or two mismatching comparison stimuli. Each echo provides evidence concerning the target that returned it, for example, in the form of a correlation between the received echo and a representation of a prototypical echo from each known object.

target (Dye and Hafter, 1984; Lindsay et al., 1968; Swets and Green, 1964).

Our model attempts to describe three related components of the decision task through the use of sequential sampling and Bayesian decision rules (as in signal detection theory). First, it describes how the dolphin identifies the correct match. Second, it describes the factors that control the number of scans the animal emits in identifying the correct choice. Third, it describes the number of echolocation clicks the animal emits during those scans.

Recognizing the target

All other things being equal, each click reflecting from a target gives rise to an echo from one of three overlapping distributions. (see Fig. 1) One distribution corresponds to the matching comparison, the other two distributions flank the match distribution and correspond to the two mismatching comparison stimuli. We have generally assumed that successive echoes from a particular stimulus are combined by averaging (Green and Swets, 1974 Roitblat, 1982, 1984). Additional echoes from the same object will tend toward the mean of one of these distributions, approximating it according to the standard error of the observations. Put simply, the mean of the echoes from a single object will move toward the prototypical echo for that object, but the speed of the approach decreases with the square root of the number of echoes received.

A likelihood ratio is computed from the averaged echo. The animal uses this likelihood ratio to make a Bayesian decision regarding the identity of the scanned stimulus. Bayesian decisions use the information obtained from the environment along with information about the prior probabilities of events (a priori probabilities) to infer the posterior probability that the scanned target is the correct match. Recall that the information obtained from scanning one comparison stimulus not only reduces the uncertainty regarding the scanned stimulus, but it also reduces the uncertainty regarding the identity of the other comparison stimuli. The Bayesian decision rule takes these changes into account. For example, if the first stimulus scanned is determined from several echoes as the likely correct match (but not yet so likely that a confident decision can be made), then the remaining two stimuli are correspondingly unlikely to be the correct match.

Scan Duration and Distribution

According to the model, the animal selects a target to scan and continues to scan it until one of four criteria is reached. Scans terminate when the probability that the target is the correct match exceeds the acceptance criterion, when the probability goes below a rejection criterion, when a maximum number of clicks is reached, or when the change in the posterior probability from successive clicks (the marginal rate of return from an echo) falls below a criterion. The animal then uses the information obtained from this and any previous scans to select another stimulus, and repeats the process until the posterior probability for one of the comparison stimuli exceeds the acceptance criterion. The animal then selects that stimulus.

Simulations based on this model provide a reasonably good approximation of the dolphin's performance. The simulation, however, tends to be somewhat less variable than the live dolphin and tends to have fewer long click trains. As a result, we have started to investigate the details of the dolphin's click production and of the echoes received.

A NEURAL NETWORK FOR STIMULUS RECOGNITION

We (Roitblat et al., 1989) recorded (on a Racal Store 4-DS high-speed, 60 ips tape recorder and then digitized on a Data Precision Data 6000 digital analyzer at 1 MHz) a sample of echoes during one of Rake's echolocation sessions. The center portions of the spectra from these echoes (from 40 to 138 kHz) were then divided into 20 frequency bins and range-normalized. We tested our ability to recognize the stimulus returning this echo by training a counterpropagation artificial neural network (Hecht-Nielsen, 1987, 1988) to

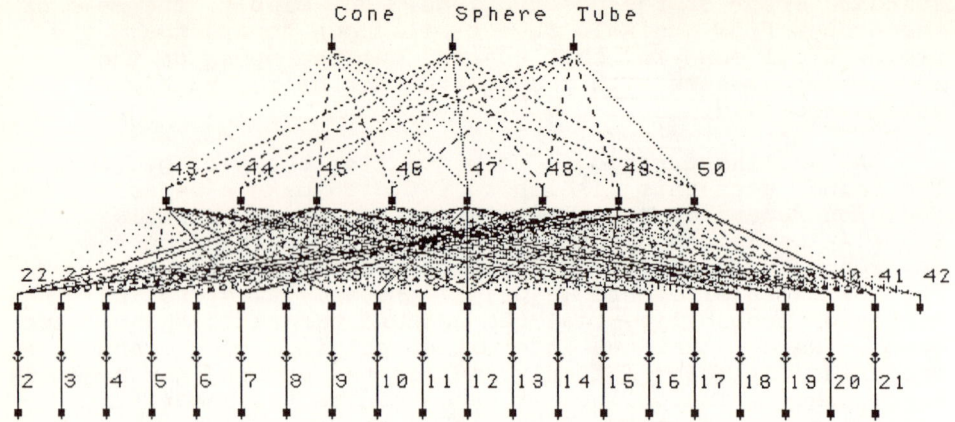

Fig. 2. Schematic of the counterpropagation artificial neural network used to classify targets on the basis of their echoes. Spectral stimulus patterns were presented to the bottom layer and the top layer was trained to classify the pattern. The network was then tested with novel patterns from the same stimuli, which it also classified.

classify the echoes into categories corresponding to each of the stimuli.

An artificial neural network is a parallel computational system metaphorically based on the kind of processing apparently performed by the brain (McClelland and Rumelhart, 1986; McClelland et al., 1986; Smolensky, 1988). Unlike standard computer systems that consist of a single central processor operating serially, the brain consists of many simple processing units or neurons that are organized into a highly interconnected network. Each neuron receives information from many (even thousands) of other neurons and transmits information, in turn, to many other neurons. The output of each neuron depends in a complex way on the pattern of inputs it receives.

Artificial neural networks are structured similarly, although still on a relatively tiny scale, when compared to biological brains. A network consists of many simple processing units, each of which receives inputs from other units or from the environment and produces an output value according to some usually nonlinear transfer function. When organized into networks, these collections of simple processing elements are capable of substantial and complex processing.

Artificial neural networks have proved quite powerful for a number of pattern-recognition problems (e.g., Gorman and Sejnowski, 1988). We used a so-called counterpropagation type of artificial neural network (Hecht-Nielsen, 1987). The counterpropagation network contains four layers of processing elements. See Fig. 2. The bottom layer consists of 20 input

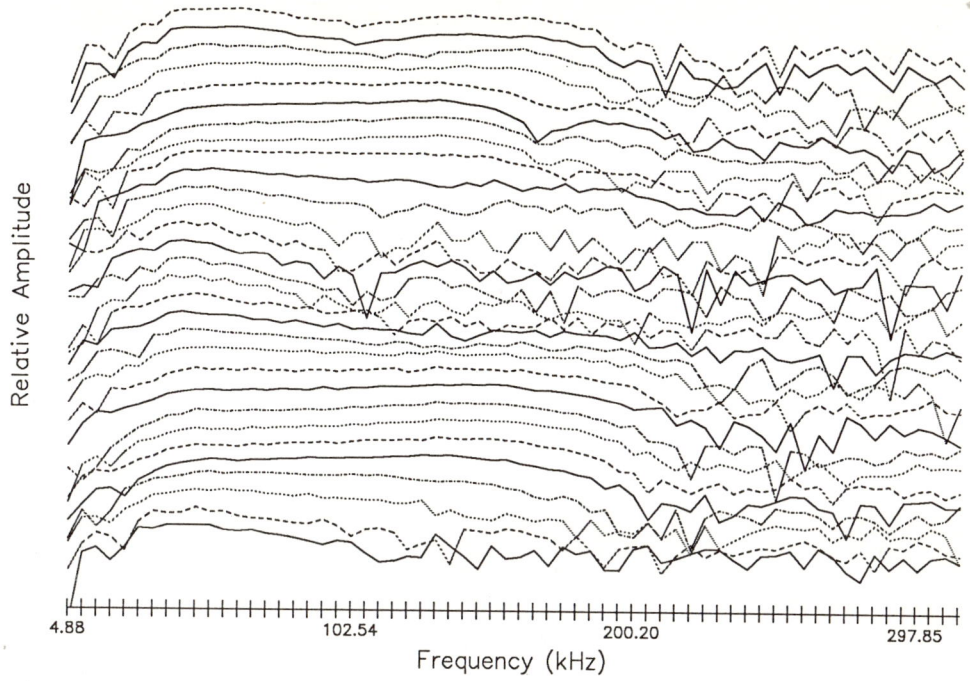

Fig. 3. The spectra from 30 successive clicks emitted to the cone sample. Click 1 is at the bottom of the figure, click 30 at the top. The frequency range is from 0 to 312.5 kHz. Amplitudes are in arbitrary relative units.

units corresponding to each of the 20 bins of spectral information. The level of activation of these units depends directly on the magnitude of the corresponding portion of the spectrum. The next layer contains 21 units. This layer normalized the inputs so that they were all contained in a vector of constant length (Hecht-Nielsen, 1987). The third layer consists of a so-called Kohonen layer. The units in this layer compete for activation. Each element in this layer has variable strength connections to every element in the preceding layer. The activation of a unit in this layer depends on the match between the levels of activation of the units in the preceding layer and the corresponding weights of the connections transmitting this activation to the units in the Kohonen layer. The unit that most closely matches the normalized input pattern wins the competition and produces an output of 1.0. The other units in this layer are then inhibited and produce an output of 0.0. The elements in the Kohonen layer "learn" to approximate the normalized input patterns. The winning element adjusts its weights to approximate more closely (and hence select more tightly) the input that caused it to win.

The Kohonen layer forms a Bayesian classification tesselation of the input patterns. Each element automatically adjusts its input connection weights to approximate the centroid or prototype of one category, and becomes active when

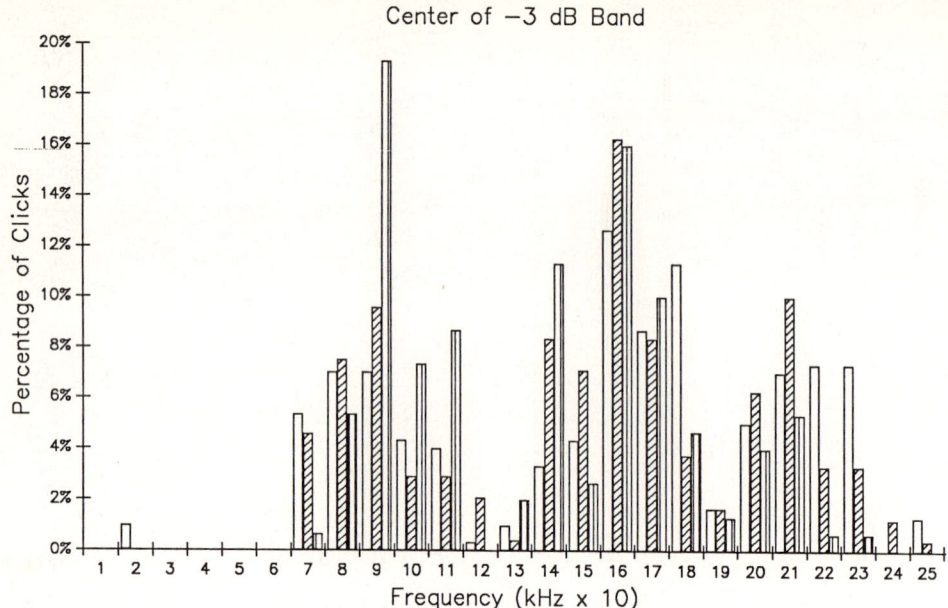

Fig. 4. Percentage of clicks with center frequencies in each range. The center frequency was defined as the center of a band 3 dB down from the peak amplitude. Notice that the distribution is trimodal.

the corresponding pattern or a similar pattern is presented to the input.

The output of the Kohonen layer leads to an output layer, containing one unit for each of the nominal target categories. The elements in this layer each connect to all of the elements in the Kohonen layer. The connection weights to these elements are adjusted during training to map the classifications described by the Kohonen layer onto the desired classes (i.e., tube, sphere, and cone). One of these units was activated when the cone was presented, a second when the sphere was presented and the third when the tube was presented.

This network "learned" to classify the spectral information from the echoes with considerable accuracy above 95% correct. This classification suggests two things. First, the spectral information present in the echoes was sufficient to identify the targets on which the dolphin was echolocating. Second, only a single echo was necessary to classify the target. Although the network could identify the target with only a single echo, the dolphin, concurrently performing the same task, emitted an average of about 35 clicks in identifying the same targets. Like the preceding experiment, however, this study also assumed that the echoes were derived from an essentially stationary distribution. A more detailed examination of the properties of the clicks emitted by the dolphin suggested that this assumption was wrong.

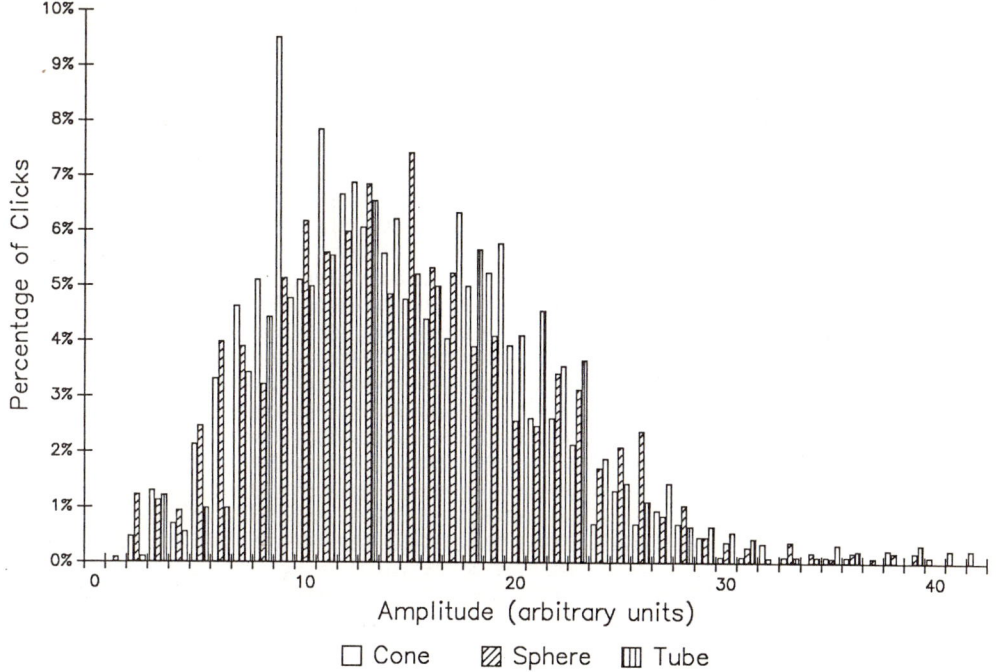

Fig. 5. Percentage of clicks with amplitudes in each range. The amplitude is given in arbitrary units.

EXAMINATION OF CLICK PROPERTIES

Fig. 3 shows the spectra of 30 successive clicks the dolphin emitted to a cone sample. Fig. 4 summarizes the spectrograms from all three samples. It shows the distributions of the center frequencies of the -3 dB energy bands of these clicks. An analysis of variance revealed that the center frequencies varied reliably as a function of the click's position in the train, but did not vary reliably across different targets. A separate analysis of the bandwidths of these same clicks showed that bandwidth was not related systematically to either position in train or to the identity of the target. Fig. 4 shows that the distribution of these clicks is multimodal. One set of clicks had center frequencies of about 90 kHz, another set had a mode of around 160 kHz, and a final set had a mode of about 210 kHz. The frequency distribution of the echoes these targets return probably is affected by the frequency of the particular click that happened to produce it.

Peak amplitudes and interclick intervals were collected from another set of two sessions. Peak amplitude, but not interclick interval varied reliably as a function of the click's position in the train, but did not depend on the identity of the target. The distribution of peak amplitudes is shown in Fig. 5. Unfortunately, these amplitudes were not obtained from the same sessions as the frequencies described above, so we do not know whether the unimodal distributions shown in Fig. 5 correspond to the trimodal distribution of click frequencies shown in Fig. 4. Nevertheless, it is clear that there is considerable variability in the amplitude of the clicks emitted by the dolphin in the delayed-matching task.

These results explain why the dolphin's actual performance was so much more variable than that of the simulated dolphin. The actual dolphin used a much more variable source of information than did the simulated dolphin. Clicks varied both in amplitude and in frequency and, therefore, the received echoes also varied in amplitude and frequency. The finding that click frequency varied reliably with the click's position in the train and the multimodal distribution of click frequencies suggests that the stationary distribution assumed by the sequential sampling model will need to be revised.

The causes of this variability remain unclear. If, for example, louder clicks return more easily discriminable echoes than quieter clicks, it remains obscure why the animal should emit these quieter clicks. One possibility is that the louder clicks are used to obtain information from the target, but the quieter clicks are not. For example, it might be difficult for the dolphin to start or stop emitting echolocation clicks, but relatively easy to vary their amplitude. The quiet clicks might play the role of place-holders, therefore, produced continuously while the dolphin processes information about previous echoes. This explanation does not receive overwhelming support from the data because it implies that click amplitudes should be bimodally distributed. One mode should reflect the information-gathering clicks, and the other should reflect the putative place-holding clicks. Instead, we observed a unimodal distribution of click amplitudes (Fig. 5).

The trimodal distribution of click frequencies, on the other hand, does raise the possibility that the dolphin intentionally varied the nature of its click in order to get different types of information from the object it is scanning. For example, low frequency clicks may be best suited to gathering information about the gross shape of the object and high frequency clicks may be better suited to getting information about finer structural details. The dolphin might employ both kinds of click to get adequate information about the objects to classify them. The present evidence also questions this conclusion, however. Although the dolphin varied the frequencies of his clicks in this experiment and other dolphins have been found to be able to control the frequencies of their clicks in response to specific experimenter-provided cues (Moore and Pawloski, 1987), there is no evidence that the dolphin in this experiment modified the frequency of his clicks in response to information returning from the object. There was no significant effect of target on the center frequency of successive clicks in a train. In fact, the only parameter of a click train that has been found to depend reliably on the stimulus being scanned is the number of clicks in the train (Roitblat et al., 1990).

In summary, these results indicate that the dolphin obtains information from each succeeding echo. Time is available between successive clicks to allow the animal to process the information returning in one echo before emitting the next (5-20 msec more then the two-way travel time of sound between the dolphin and the target). In principle, this time would allow the animal to process the information in one echo and control the production of the next click to obtain optimal information from the next. Dolphins do not appear to do this, however. They appear to attend to (i.e., emit clicks to) a target until sufficient information is obtained to identify the target, but they do not appear to modify the individual clicks in response to the presumed partial information.

REFERENCES

Dye, R. H., Jr. and Hafter, E. R., 1984, The effects of intensity on the detection of interaural differences of time in high-frequency trains of clicks, Journal of the Acoustical Society of America, 75: 1593-1598.

Gorman, R. P. and Sejnowski, T. J., 1988, Analysis of hidden units in a layered network trained to classify sonar targets. Neural Networks, 1: 75-89.

Green, D. M. and Swets, J. A., 1974, "Signal detection theory and psychophysics." Huntington, NY: Krieger.

Hecht-Nielsen, R., 1987, Counterpropagation networks. Applied Optics, 26: 4979-4984.

Hecht-Nielsen, R., 1988, Applications of counterpropagation networks. Neural Networks, 1: 131-139.

Lindsay, P. H., Taylor, M. M., and Forbes, S. M., 1968, Attention and multidimensional discrimination. Perception and Psychophysics, 4: 113-117.

McClelland, J. L., Rumelhart, D. E., eds.), 1986, "Parallel Distributed Processing." Cambridge, MA: MIT Press.

McClelland, J. L., Rumelhart, D. E., and Hinton, G. E., 1986, The appeal of parallel distributed processing. in: "Parallel Distributed Processing." ed. by D. E. Rumelhart and J. L. McClelland. Cambridge, MA: MIT Press. pp. 3-44.

Moore, P. W. B. and Pawloski, D. A., 1987) Voluntary control of peak frequency in echolocation emissions of dolphin (Tursiops truncatus). Paper presented at the 7th biennial conference on the biology of marine mammals. Miami.

Nachtigall, P. E., Patterson, S. A., and Bauer, G. B., 1985, Echolocation delayed matching-to-sample in a bottlenose dolphin. Paper presented at the Sixth Biennial Conference on the Biology of Marine Mammals, Vancouver, B.C., Canada. November.

Penner, R. H., 1989, Attention and detection in dolphin echolocation, in: "Animal Sonar: Processes and Performance," P. E. Nachtigall and P. W. B. Moore eds., New York, Plenum Press. 707-713.

Roitblat, H. L., 1980, Codes and coding processes in pigeon short-term memory. Animal Learning and Behavior, 8: 341-351.

Roitblat, H. L. Penner, R. H., and Nachtigall, P. E., 1990, Matching-to-sample by an echolocating dolphin. Journal of Experimental Psychology: Animal Behavior Processes, 16: 85-95.

Roitblat, H. L., Moore, P. W. B., Nachtigall, P. E., Penner, R. H., and Au, W. W. L., 1989, Dolphin echolocation: Identification of returning echoes using a counterpropagation network. Proceedings of the First

International Joint Conference on Neural Networks. Washington, DC: IEEE Press.

Roitblat, H. L., 1984, Representations in pigeon working memory, in: "Animal cognition." H. L. Roitblat, T. G. Bever and H. S. Terrace eds., Hillsdale, NJ: Erlbaum, 79-97.

Rumelhart, D. E. and McClelland, J. L., 1986, *Parallel Distributed Processing*. Cambridge, MA: MIT Press.

Smolensky, P., 1988, On the proper treatment of connectionism. *The Behavioral and Brain Sciences*, 11: 1-74.

Swets, J. A. and Green, D. M. , 1964) Sequential observations by human observers of signals in noise. In "Signal detection and recognition by human observers." J. A. Swets and D. M. Green eds., New York: Wiley.

STIMULUS EQUIVALENCE AND CROSS-MODAL PERCEPTION: A TESTABLE MODEL FOR
DEMONSTRATING SYMBOLIC REPRESENTATIONS IN BOTTLENOSE DOLPHINS

Ronald J. Schusterman

Long Marine Laboratory, University of California
100 Shaffer Road, Santa Cruz, California 95060, U.S.A.
and Department of Psychology, California State
University, Hayward, California 94542, U.S.A.

INTRODUCTION

Dolphins make a variety of decisions while swimming in schools. These decisions determine the way they navigate, avoid predation, forage, reproduce, care for young and otherwise engage in social interactions. These decisions are based on the ability of dolphins to process information from a variety of sensory avenues, including the active process of investigating objects via echolocation. Information from all sensory systems is most probably used in an integrated fashion. However, whether information from the various sensory modalities is also stored and retrieved, as well as used in an integrated way, i.e., whether dolphins are capable of intermodal stimulus equivalence or cross-modal perception, currently remains a hypothesis (Schusterman, 1988a).

Studies on the psychophysics of echolocation, hearing, vision, skin senses and taste suggest that dolphins have rather rich and detailed representations of the external world. (For detailed critical reviews on the psychophysics of dolphin sensory perception, see the following: Dawson, 1980; Fobes and Smock, 1981; Johnson, 1986; Madsen and Herman, 1980; Murchison, 1980; Nachtigall, 1980; 1986; Popper, 1980; Ridgway, 1986; Schusterman, 1980; Watkins and Wartzok, 1985). Research on rule learning and concept formation suggest that, like some primates, dolphins may be able to represent abstract relations, as well as perceptual relations in both the auditory and visual modalities (for reviews see Herman, 1980; 1986; Schusterman, 1988a; Seyfarth, 1986). Indeed, an important source of evidence for determining symbolic representation involves intermodal stimulus equivalence. This cognitive ability, which has been demonstrated bidirectionally in some anthropoid apes and monkeys using cues from visual and tactile modalities (Cowey and Weiskrantz, 1975; Davenport and Rogers, 1970; Davenport, Rogers and Cross, 1973), is thought by some to be essential for the emergence of language (Geschwind, 1965; Lancaster, 1968). It has also been suggested that cross-modal perception requires a "modality-free representation" of a stimulus pattern (Rumbaugh, et. al., 1982). Are dolphins capable of intermodal equivalence of sonar cues and reflected light cues emanating from common objects, or are they capable of intermodal equivalence of tactile cues and visual cues emanating from common objects? A dolphin's use of symbols within such intermodal tasks would be a good demonstration of the ability of these large-brained marine mammals to use modality-free or symbolic representations of stimulus patterns. In this paper, I present a simple model based on Sidman's notion of stimulus equivalence (1986) to test these ideas. However, before presenting the model as applied to cross-modal perception in dolphins, allow me to illustrate stimulus class equivalencies as applied to one of the most well-known and well-documented examples of semantic communication in animals--vervet monkey alarm calls (Seyfarth, 1986).

VERVET MONKEY ALARM CALLS AND STIMULUS EQUIVALENCES

In a conditional discrimination, the most commonly used procedure is called matching-to-sample (MTS) in which an animal's choice between two or more comparison stimuli is contingent on sample or conditional stimuli. For example, in the presence of A_1, B_1 is correct and reinforced, but not B_2 or B_3, etc., and in the presence of A_2, B_2 is correct and reinforced, but not B_1 or B_3, etc. In a concrete but totally hypothetical illustration, a naive vervet monkey may be shown two pictures simultaneously or played two different recordings of vervet monkey alarm calls in rapid succession; perhaps pictures of a leopard vs. a martial eagle in the first instance or a "loud bark" vs. "chuckle" in the second case. The monkey must use a third stimulus, the sample or instructional cue, that determines which picture or which alarm call should be responded to. In identity matching, the instructional cue and the appropriate comparison stimulus are physically the same, so the monkey would match a leopard comparison to a leopard sample and a martial eagle comparison to a martial eagle sample, etc. A different MTS procedure called arbitrary or "symbolic" matching, specifies a relation in which the sample and its matching comparison stimulus bear no physical resemblance to each other, and for that reason, the symbolic matching task has been of interest as a task that illustrates simple semantic relations (Catania, 1970). For example, a naive vervet monkey might learn to match animal pictures (comparisons) to vervet monkey call samples; with leopard a paired associate of the sample "loud bark" and martial eagle the paired associate of the sample "chuckle."

Note that several characteristics of the MTS procedure make it a suitable method for experimental studies of animal cognition including short-term memory, perceptual categorization, abstraction and various aspects of language, especially semantic comprehension (see Carter and Werner, 1978; Schusterman, 1988b; Schusterman and Gisiner, 1989; Sidman and Tailby, 1982 for reviews of various MTS paradigms in the study of animal cognition as it relates to semantic comprehension). Sidman and Tailby (1982) and others have pointed out that the term "MTS" sometimes refers to a procedure and sometimes it refers to the results of a procedure. These two different meanings of MTS have frequently been muddled in the interpreting of results which bear on fundamental issues in animal cognition [e.g. see the controversy between Herman (1988; 1989) and Schusterman and Gisiner (1988; 1989)]. For example, if a vervet monkey performs appropriately on a matching task and its opposite, a mismatching or oddity task, the behavior does not necessarily mean that the monkey has a "sameness" or "oddity" concept. The critical test of concept formation comes when the monkey must match novel stimuli solely on the basis of their identity relationship. As in identity MTS, symbolic MTS also tacitly assumes that each paired associate of sample and comparison stimulus is related not merely by an "if ... then ..." relationship, but by equivalence. Thus, in the vervet monkey illustration, it is easy to assume that each alarm call sample and each animal picture comparison stands in an equivalence relation to one another (e.g. the monkey makes both of these relationships: "if 'loud bark' then leopard", and "if leopard, then 'loud bark' "). However, as Sidman and Tailby (1982) have shown, like identity, the arbitrary relationship between so-called symbols and their referents remains in a unidirectional "if ... then ..." relation and can not be considered to form an equivalence class relationship unless there are explicit and independent tests. Simple behavioral variables may be mistakenly identified as evidence of complex cognitive processes, such as symbol manipulation, if the assumption of stimulus equivalence is in fact invalid (Mackay and Sidman, 1984).

If the training of a series of conditional discriminations with MTS paradigms (if A_1 then B_1; if A_2 then B_2; etc., or if "loud bark" then leopard; if "chuckle," then martial eagle, etc.), results in the emergence of untrained relationships between dissimilar stimulus patterns, then the equivalence of stimulus classes exists. Stimulus equivalence has three defining characteristics: reflexivity, symmetry and transitivity.

1. <u>Reflexivity</u>. Reflexivity emerges from generalized identity matching of the type: If A_1, then A_1; if A_2, then A_2; and if B_1, then B_1; if B_2, then B_2, etc. Thus, if it takes several trials before a naive vervet monkey can consistently match a "loud bark" call to itself, but then the monkey matches a "chuckle" to itself on the first trial and a "chittering" call to itself on the first trial, and if the monkey overcomes the difficulty in matching a picture of a leopard to itself, and immediately can match a picture of a martial eagle to itself and a picture of a python to itself, etc., then we can conclude that this vervet monkey who was taught a set of sample-comparison relations (vervet monkey alarm calls and pictures of animals) has demonstrated that these relations were reflexive by showing that it was capable of matching the two kinds of stimuli to themselves. Moreover, ideally, to develop additional critical tests of class equivalency, another set of sample-comparison relations are needed. These could consist of vervet monkey alarm calls and printed lexigrams. Reflexivity would also be demonstrated if the subject could match each lexigram to itself.

2. <u>Symmetry</u>. Symmetric relations are shown when two or more dissimilar stimuli are related bidirectionally or reciprocally (e.g., if A_1, then B_1; if B_1 then A_1). Figure 1 illustrates a basic equivalence paradigm. The vervet monkey who has learned to match comparison stimulus B_1 (e.g. leopard) to sample stimulus A_1 (e.g. "loud bark") or comparison stimulus C_2 [e.g. a lexigram consisting of a square shape with a dot in the middle (⊡)] to sample stimulus A_2 (e.g. "chuckle"), must then, without additional training, be able to match A_1 (comparison) to B_1 (sample) and A_2 (comparison) to C_2 (sample). Symmetry requires sample and comparison stimuli to be functionally interchangeable. Stated another way, within the context of semantics, symmetry occurs when "conditional cues have become more than conventional discriminative stimuli ... [i.e.], when signs and their referents are shown to be immediately interchangeable ..." (Schusterman and Gisiner, 1989). In this hypothetical experiment, symmetry could be tested indirectly with a vervet monkey by determining whether the test subject vocalizes appropriately to pictures of a leopard or to the printed lexigram ⊕ (see Fig. 1).

3. <u>Transitivity</u>. The emergence of transitive stimulus relations from conditional discriminations requires three stimulus types as illustrated in Fig. 1. Once "if A_1, then B_1" and "if A_1, then C_1" have been established, transitivity requires "if B_1, then C_1" to emerge without explicit training. Suppose, for example, our vervet monkey subject, having learned to select a picture of a martial eagle when it hears a "chuckle" alarm call, and having learned to select the lexigram ⊡ when it hears a "chuckle" alarm call, now without explicit training, chooses a picture of a martial eagle when presented with the lexigram ⊡, and chooses the same lexigram when shown a picture of a martial eagle. We may conclude that for the vervet monkey, the "chuckle" call, the martial eagle and the lexigram ⊡ form a single equivalent class despite no physical similarity. The monkey's emergent ability to perform new types of matching tasks, BC and CB, will have con-

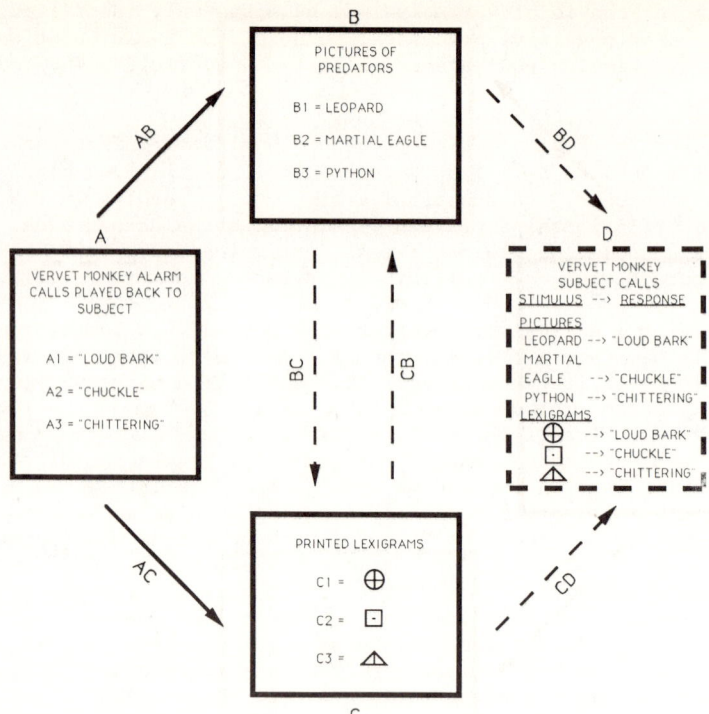

Fig. 1. An equivalence paradigm for teaching a "naive" vervet monkey subject semantic relations. Each of the three enclosed boxes A, B and C represent a set of three stimuli. Arrows AB, AC, BC and CB, each representing a set of conditional relations, point from sample to comparison stimuli. Solid arrows (AB and AC) represent relations that are explicitly taught to the monkey and broken arrows represent conditional relations that are expected to emerge subsequently. For a given sample stimulus, the appropriate comparison is designated by the same number. Broken box D represents calls by the monkey which name or label stimulus sets B and C. Broken arrows from these stimulus sets to vocal responses represent picture naming (BD) and printed lexigram naming (CD).

firmed the development of three novel, three-member classes of equivalent stimuli: $A_1B_1C_1$, $A_2B_2C_2$, and $A_3B_3C_3$, (see Fig. 1). Moreover, one could conclude that by passing the stimulus equivalence test, this monkey shows that the conditional relations between monkey calls and their referents as well as lexigrams and their referents involve semantic relations.

CROSS-MODAL PERCEPTION IN DOLPHINS AND STIMULUS EQUIVALENCIES

Figure 2 illustrates a model for testing stimulus equivalencies in bottlenose dolphins which includes a cross-modal perception task. As in the vervet monkey example, evidence of stimulus class equivalence requires three different types of stimuli. The stimuli in this paradigm consist of (A) acoustic signals, (B) a variety of shapes which can be inspected visually but are opaque to a dolphin's sonar signals, and (C) the same shapes which can be interrogated by a dolphin's sonar under water but are visually opaque. We are relatively safe in assuming that dolphins can do generalized MTS (see Herman and Gordon, 1974; Herman, Gory, Hovancik and Bradshaw, 1989) and thereby meet the reflexivity criterion.

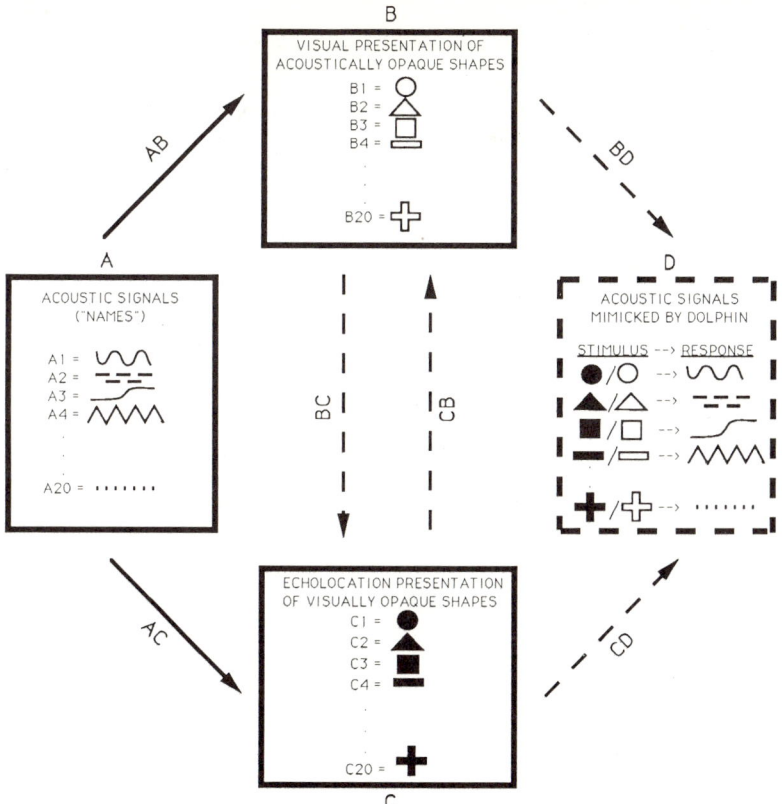

Fig. 2. An equivalence paradigm for teaching a dolphin subject bi-directional cross-modal transfer. Each of the three enclosed boxes A, B and C represent a set of twenty stimuli. Arrows AB, AC, BC and CB, a set of conditional relations, point from sample to comparison stimuli. Solid arrows (AB and AC) represent relations that are explicitly taught to the dolphin and broken arrows represent conditional relations ---- the bi-directional cross-modal perception ---- that are expected to emerge subsequently. For a given sample stimulus, the appropriate comparison is designated by the same number. Broken box D represents mimicked sounds by the dolphin which "name" or "label" stimulus sets B and C. Broken arrows from these stimulus sets to vocal mimicry of acoustic names or labels represent object shape naming in the visual mode (BD) and the echolocation mode (CD).

The dolphin first has to learn to select visually presented shapes (comparison stimuli) conditionally upon any of, for example, twenty acoustic signals (sample stimuli); AB in Fig. 2 represents 20 conditional relations (A_1B_1, A_2B_2 . . . $A_{20}B_{20}$). Next, the dolphin has to learn to select a stimulus shape interrogated by its sonar (comparison stimuli) conditionally upon the same twenty acoustic signals (sample stimuli); AC in Fig. 2 represents 20 new conditional relations (A_1C_1, A_2C_2 . . . $A_{20}C_{20}$). At the conclusion of AB and AC training, the dolphin should select any of 20 shapes inspected visually or by sonar conditionally upon a broadcasted signal.

It will then be feasible to determine whether AB and AC are equivalence relations by giving a combined test for symmetry and transitivity using cross-modal perception. Proof of equivalence and evidence for "modality-free" or "symbolic" representation requires a dolphin to select an appropriate shape inspected by echolocation (comparison stimulus) conditionally upon that same shape being investigated visually (sample stimulus) -- the BC relation, and to select an appropriate shape inspected visually (comparison stimulus) conditional upon that same shape being interrogated by the dolphins' sonar (sample stimulus) -- the CB relation. The above described cross-modal tasks have to be accomplished by the dolphin without explicit training; however, the dolphin will need some training in the mechanics of working in a cross-modal task of echolocation and vision using different stimulus objects. Furthermore, if we want to make inferences about dolphins having symbolic representations, a "control" animal should be given the cross-modal tasks (BC and CB) without having "names" or "tags" related to the shapes (AB and AC). Finally, based on previous research on vocal mimicry of computer-generated sounds and the vocal "labeling" of objects by a dolphin (Richards, 1986), it may be plausible for the dolphin to mimic the acoustic signals and "name" the shapes both visually (BD) and by echolocation (CD). The dolphins' emergent cognitive ability to do two new sets of matching tasks, BC and CB, will have confirmed the creation of 20 three-member classes of equivalent stimuli: $A_1B_1C_1$, $A_2B_2C_2$, ... $A_{20}B_{20}C_{20}$. Indeed, if the dolphin can mimic the names of the shapes presented to it for visual inspection (BD) or for echolocation (CD), then the original teaching of 40 conditional relations to the dolphin will have resulted in the creation of 40 novel conditional relations and 40 naming relations or a total of 80 novel performances.

SUMMARY AND CONCLUSION

Although the ability of dolphins, apes, monkeys and several other vertebrate taxa to respond to complex classes of stimuli are not in doubt, their ability to "refer" to objects, events and relations and, in general, to manipulate symbols is very controversial. The origin and nature of symbolic activity, which invariably involves the logical properties of reflexivity, symmetry and transitivity may be rooted in the way animals acquire rules to deal with social and nonsocial stimulus objects, events and relationships. Sidman (1986) has shown that in humans conditional discriminations can lead to a semantic correspondence between each sample and its matching comparison stimulus, i.e. a stimulus class equivalency within an MTS paradigm. I have attempted to show that cross-modal perception in dolphins and perhaps even semantic comprehension in vervet monkeys may be trained to reveal symbol manipulation in these mammals.

ACKNOWLEDGMENTS

The writing of this paper was fully supported by Office of Naval Research Contract N00014-85-K-0244. I thank Bob Gisiner, Brigit Grimm, Evelyn Hanggi, Maria Choy-Vasquez and Roberta Schusterman for help in preparing the manuscript.

REFERENCES

Carter, D. E. and Werner, T. J., 1978, Complex learning and information processing by pigeons: A critical analysis, J. Exp. Anal. Beh., 29:565.
Catania, A.C., 1980, Autoclitic processes and the structure of behavior, Behaviorism, 8:175.

Cowey, A. and Weiskrantz, L., 1975, Demonstration of cross-modal matching in rhesus monkeys, Macaca mulatta, Neuropsychologia, 13:117.
Davenport, R. K. and Rogers, C. M., 1970, Intermodal equivalence of stimuli in apes, Science, 168:279.
Davenport, R. K., Rogers, C. M. and Steele, R. I., 1973, Cross modal perception in apes, Neuropsychologia, 11:21.
Dawson, W. W., 1980, The cetacean eye, in: "Cetacean Behavior: Mechanisms and Functions," L. M. Herman, ed., Wiley, N.Y.
Fobes, J. L. and Smock, C. C., 1981, Sensory capacities of marine mammals, Psychol. Bull., 89:288.
Geschwind, N., 1965, Disconexion syndromes in animals and man, Brain, 88:237-294, 585.
Herman, L. M., 1980, Cognitive characteristics of dolphins, in: "Cetacean Behavior: Mechanisms and Functions," L. M. Herman, ed., Wiley, N.Y., 363.
Herman, L. M., 1986, Cognition and language competencies of bottlenosed dolphins, in: "Dolphin Cognition and Behavior: A Comparative Approach," R. J. Schusterman, J. Thomas, and F. G. Wood, eds., Erlbaum, Hillsdale, N.J., 221.
Herman, L. M., 1988, The language of animal language research: Reply to Schusterman and Gisiner, The Psychol. Rec., 38:349.
Herman, L. M., 1989, In which procrustean bed does the sea lion sleep tonight?, The Psychol. Rec., 39:19.
Herman, L. M. and Gordon, J. A., 1974, Auditory delayed matching in the bottlenosed dolphin, J. Exp. Anal. Beh., 21:19.
Herman, L. M., Gory, J. D., Hovancik, J. R. and Bradshaw, G. L., 1989, Generalization of visual matching by a bottlenosed dolphin (Tursiops truncatus): Evidence for invariance of cognitive performance with visual and auditory materials, J. Exp. Psychol., 15:124.
Johnson, C. S., 1986, Dolphin audition and echolocation capacities, in: "Dolphin Cognition and Behavior: A Comparative Approach," R. J. Schusterman, J. A. Thomas and F. G. Wood, eds., Erlbaum, Hillsdale, N.J.
Lancaster, J. B., 1968, Primate communication systems and the emergence of human language, in: "Primates: Studies in Adaptation and Variability," P. C. Jay, ed., Holt, Rinehart and Winston, N.Y.
Mackay, H. A. and Sidman, M., 1984, Teaching new behavior via equivalence relations, in: "Learning and Cognition in the Mentally Retarded," P. H. Brooks, R. Sperber and C. McCaulay, eds., Erlbaum, Hillsdale, N.J.
Madsen, C. J. and Herman, L. M., 1980, Social and ecological correlates of cetacean vision and visual appearance, in: "Cetacean Behavior: Mechanisms and Functions," L. M. Herman, ed., Wiley, N.Y.
Murchison, A. E., 1980, Detection range and range resolution of echolocating bottlenose porpoise (Tursiops truncatus), in: "Animal Sonar Systems," R. G. Busnel and J. F. Fish, eds., Plenum Press, N.Y.
Nachtigall, P. E., 1980, Odontocete echolocation performance on object size, shape and material, in: "Animal Sonar Systems," R. G. Busnel and J. F. Fish, eds., Plenum Press, N.Y.
Nachtigall, P. E., 1986, Vision, audition, and chemoreception in dolphins and other marine mammals, in: "Dolphin Cognition and Behavior: A Comparative Approach," R. J. Schusterman, J. A. Thomas and F. G. Wood, eds., Erlbaum, Hillsdale, N.J.
Popper, A. N., 1980, Sound emission and detection by delphinids, in: "Cetacean Behavior: Mechanisms and Functions," L. M. Herman, ed., Wiley, N.Y.
Richards, D. G., 1986, Dolphin vocal mimicry and vocal object labelling, in: "Dolphin Cognition and Behavior: A Comparative Approach," R. J. Schusterman, J. A. Thomas and F. G. Wood, eds., Erlbaum, Hillsdale, N.J.

Ridgway, S. H., 1986, Physiological observations on dolphin brains, in: "Dolphin Cognition and Behavior: A Comparative Approach," R. J. Schusterman, J. A. Thomas and F. G. Wood, eds., Erlbaum, Hillsdale, N.J.

Rumbaugh, D. M., Savage-Rumbaugh, E. S. and Scanlon, J. L., 1982, The relationship between language in apes and human beings, in: "Primate Behavior," J. L. Fobes and J. E. King, eds., Academic Press, N.Y.

Schusterman, R. J., 1980, Behavioral methodology in echolocation by marine mammals, in: "Animal Sonar Systems," R. G. Busnel and J. F. Fish, eds., Plenum, N.Y.

Schusterman, R. J., 1988a, Cognition and echolocation of dolphins, in: "Animal Sonar," P. E. Nachtigall and P. W. B. Moore, eds., Plenum, N.Y.

Schusterman, R. J., 1988b, Language and counting in animals: stimulus class and equivalence relations, Behav. Brain Sci., 11:596.

Schusterman, R. J. and Gisiner, R., 1988, Artificial language comprehension in dolphins and sea lions: The essential cognitive skills, The Psychol., Rec. 38:311.

Schusterman, R. J. and Gisiner, R. C., 1989, Please parse the sentence: Animal cognition in the procrustean bed of linguistics, The Psychol., Rec., 39:3.

Seyfarth, R., 1986, Vocal communication and its relation to language, in: "Primate Societies," B. B. Smuts, D. L. Cheney, R. M. Seyfarth, R. W. Wrangham, and T. T. Struhsaker, eds., The University of Chicago Press, Chicago.

Sidman, M., 1986, Functional analysis of emergent verbal class, in: "Analysis and Integration of Behavioral Units," T. Thompson and M. D. Zeiler, eds., Erlbaum, Hillsdale, N.J.

Sidman, M. and Tailby, W., 1982, Conditional discrimination vs. matching to sample: An expansion of the testing paradigm, J. Exp. Anal. Behav., 37:5.

Watkins, W. A. and Wartzok, D., 1985, Sensory biophysics of marine mammals, Mar. Mamm. Sci., 3:219.

THE ABILITY OF BOTTLENOSE DOLPHINS, TURSIOPS TRUNCATUS, TO REPORT ARBITRARY INFORMATION

Alexander V. Zanin

Karadag Department of Institute of Biology of Southern Seas
Sudak, Crimea 334876, USSR

Vladimir I. Markov, Irina E. Sidorova

Institute of Evolutionary Morphology and Ecology of Animals
33 Leninsky Prospect, Moscow 117071, USSR

INTRODUCTION

Investigations of the bottlenose dolphin sounds suggest that this species should have a highly developed hierarchically organized system of acoustic communication. This assumption is based on the high structural complexity of the dolphin signalisation, the peculiarities of its use, and the results of special mathematical and linguistical analysis (Markov, 1978; Markov et al.,1974; Markov and Ostrovskaya, 1978; Markov and Ostrovskaya in this volume). In our opinion, this system can be used for reporting complex arbitrary information. There are indirect corroborations of this standpoint such as some features of the dolphin behavior in the wild and captivity, the results of the experiments on the comprehension of the artificial language by the bottlenose dolphins (Herman, 1980; Herman et al., 1984), but the direct experimental evidence is absent.

At first, it seemed that Bastian's experiment (Bastian, 1967) gave enough convincing evidence of the bottlenose dolphin's ability to report arbitrary (i.e. biologically independent) information. However, further examination (Evans and Bastian, 1969; Herman and Tavolga, 1980) showed that the observed interactions between animals could be explained by the use of echolocation or the formed behavioral stereotypes. Due to these causes, the conclusions of Bastian are unconvincing although the method proposed by him undoubtedly has a good perspective.

We modified Bastian's method (the investigation of the dolphin's ability to co-ordinate actions of each other under conditions in which it was necessity to report arbitrary information through the acoustic channel) so that possible weak points were excluded. This modified scheme was used in our experiment to obtain evidence of the dolphin's ability to report arbitrary information.

MATERIALS AND METHOD

Two adult female bottlenose dolphins (Tursiops truncatus ponticus

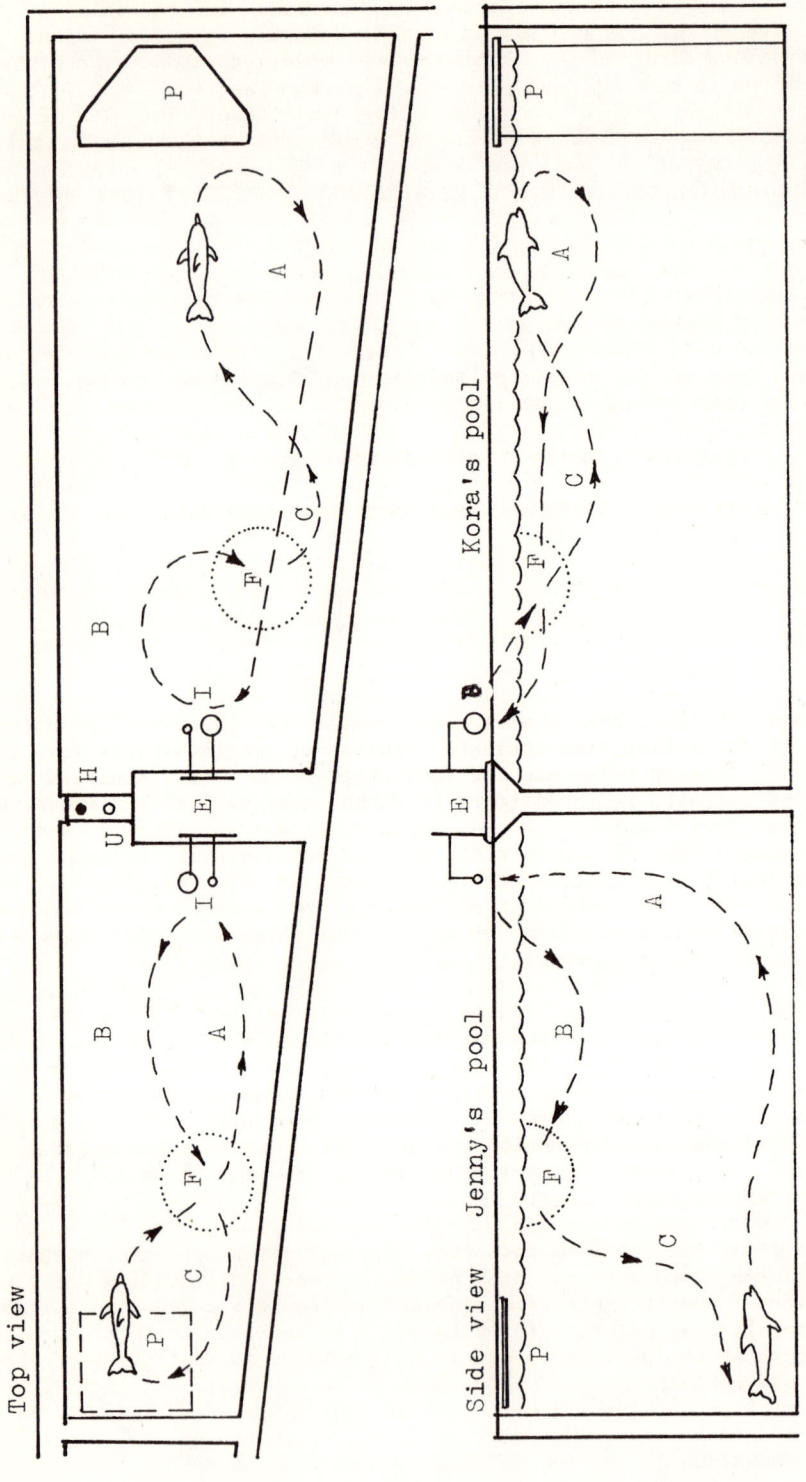

Fig. 1. The test situation: (P) platforms, (U) underwater speaker, (H) the hydrophone, (I) manipulandum, (E) experimenter, and (F) places of the feeding. Animal movement is shown by interrupting lines. (A) movement from the start position to manipulandum, (B) movement from manipulandum to the place of feeding and (C) return to start position.

Barabash, 1940) worked in the experiment. Both dolphins had been caught in the Black Sea on April 1982 in the waters near Anapa. For five weeks they lived together with other animals in the sea-net enclosure, then both females were transported to the Karadag oceanarium. The experiment was begun after the adaptation and taming of the dolphins when no forms of negative reactions directed on humans were observed and the dolphins actively aspired to contact and play with the experimenters. The females named Jenny and Kora were of the same age and size. In the Karadag oceanarium, they were kept in neighbouring pools separated by a concrete wall. The thickness of the wall was about 0.7 m. The depth of water in the pools was 4.5 m; the surface area of Jenny's pool was 34.1 m^2, the surface area of Kora's pool was 58.8 m^2. An opening 1.2m x 1.0m at the top corner of the wall connected both pools. The opening was covered by net with a mean size of 9cm x 9cm. In fact, both dolphins were visually isolated and the communication between them could only be realized throught acoustic channel provided by the opening in the wall. The dolphins did not approach the opening. The established avoidance of the opening zone had appeared due to the disposition of the underwater speakers and hydrophones in the centre of the opening aperture. The equipment frightened the animals and the dolphins developed the start position and movement paths which were convenient for the experimenters and simplified the work. The start positions of Jenny and Kora as well as their usual pattern of the movements in the experimental cycle of the sessions are shown in Fig.1.

The equipment used in the experiment (see Fig. 2) consisted of several signal generators, amplifiers, hydrophones, underwater speakers, apparatuses for the listening and recording dolphin signals. Later the factograph was included in the equipment. The operation was done from the central control desk. (Fig.3).

All events in the sessions were registered by the hand-written notes, each cycle in the session also could be registered automatically by the factograph. The session notes and the factograph bands were analysed and then they were compared to reconstruct a whole qualitative and temporal picture of each cycle in the sessions. We calculated the number of correct and incorrect reactions of the dolphins and classified the errors. The types of the errors were compared and tabulated. In the experiment, the continuous and selective recordings of sounds of the animals were done by the tape-recorder (the results of the analysis of the acoustic signals will be examined in a special paper).

Identical experimental installations were placed in the pools of the dolphins (Fig. 3). The installations were put on the top of the wall at the distance of 1.5 m from the inner corner of the opening. Each installation consisted of two boards with brackets. The manipulandum (two rubber balls with different diameters) were suspended at the ends of the brackets by thin kapron strings. The pairs of balls were identical for both installations. The distance between the balls in one pair was 0.4 m. The distance between the balls and the surface of the wall was about 1.0 m. The position of the manipulandum was always the same in both pools. The heigth of the balls above the water surface was 0.5 m. In any case, each dolphin could not see the position of the balls at the installation of the neighbouring pool due to the half a m heigth of the wall above the surface of water, the vizor at the top of the wall and a two-fold row of boards which screened the balls. The installations were put on a layer of the sound-absorbing rubber to reduce the noises, which could accompany the actions of the experimenter. Thus we took precautionary measures (such as: the clip of the balls above the water surface; the short distance between the balls; the necessity for the dolphin to come to a strictly determined point to act; and the reduction of possible outside noises) to exclude the

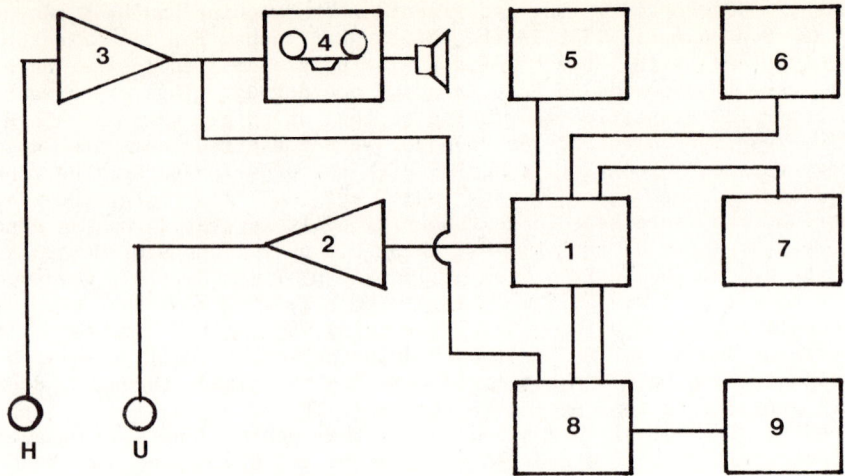

Fig. 2. Block diagram of experimental equipment: (H) hydrophone, (U) underwater speaker, (1) central control desk, (2) amplifier of sound generation system, (3) hydrophone amplifier, (4) tape recorder with the system of listening of the dolphin signals; (5), (6) the generators of the individual working signals, (7) the generator of the joint start signal, (8) the factograph (event recorder), and (9) the source of the supply.

mutual echolocation control of dolphins even if it would be possible to echolocate through the concrete wall.

A check-up showed that the attenuation of test pulses after their passage through the concrete wall was 35 - 39 dB. The calculation shows that the echolocation signal, which has the value 200 - 230 dyne per sq.cm, will be decreased to the value 0.025 - 0.063 dyne per sq.cm after the double passage through the concrete wall and the total reflection beyond it from a target. This value is higher than the hearing threshold of the bottlenose dolphin (Dubrovsky, 1976). It would seem that echolocation in these conditions is possible but, however, the calculation was made for the case of the normal incidence of sound waves, i.e. for the opposite disposition of the hydrophone and underwater speaker. In real conditions the angle of incidence varies due to movements of dolphins (see the trajectories of animal movements shown on Fig. 1). It is clear that part of the signal energy which penetrated through the wall, will be decreased correspondingly to the increase of that part of signal energy which will be reflected from this wall. The value of this part is determined by the position of the dolphin in relation to the wall and it varies due to movements of the dolphin. For example, the analysis of the movements of Jenny showed that the angle of incidence apparently always exceeds the critical angle. When the pulses pass through the concrete wall "an erosion" of direct and reverse signals distorts their spectral/temporal characteristics due to the forming of the lateral waves. The incomplete reflection of signals from the dolphin body decreases them even more and, due to all these causes, the conditions become unfit to localize the dolphin beyond the wall. The analysis of the trajectories and temporal characteristics of dolphin movements show that there was not any position in any cycle where the dolphins could directly locate each other. Our observations on the acoustic activity of the dolphins showed that they did not echolocate while approaching the manipulandum. Very few clicks were

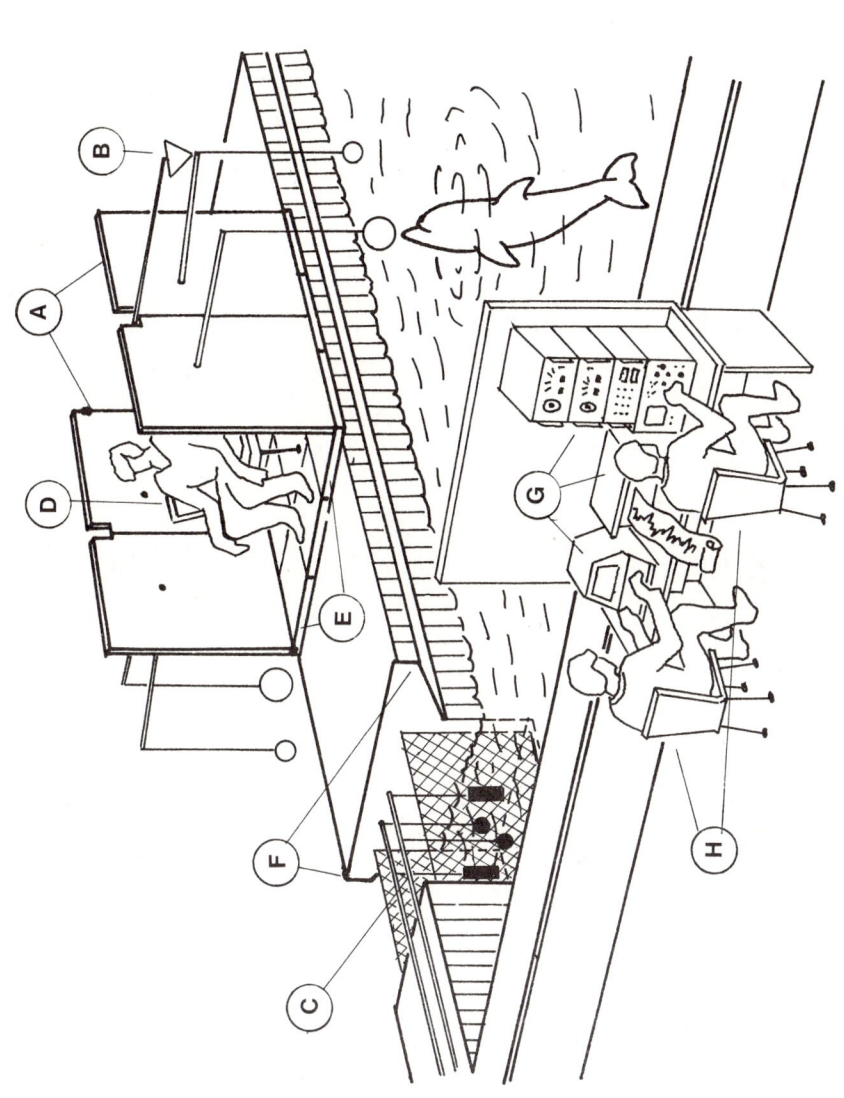

Fig.3. The experiment surroundings: (A) The boards with the brackets and the balls; (B) the prohibitive sign (a white plastic triangle); (C) the opening in the wall where the hydrophones and the underwater speakers were placed; (D) the position of the experimenter who controlled the prohibitive sign and gave the reward; (E) layer of the sound-absorbed rubber; (F) the visor of the wall; (G) the central control desk and the rest of the equipment; (H) the experimenters who controlled the equipment and registered the events of the session.

heard and only when the dolphins took their reinforcement at the end of the cycle.

We planned to divide the experimental procedure into three main stages. On the first stage, we planned to teach the dolphins to differentiate either between the acoustic and visual signals. We used three acoustic signals in the experiment:
1) the start signal (St) which was the same for both dolphins. This signal (pure tone, 3.8 kHz) indicated the beginning of the cycle;
2) the individual working signal for Jenny (Sj); a pure tone, 8.0 kHz;
3) the individual working signal for Kora (Sk); tone pulses, 3.8 kHz.

The individual working signal indicated the permission for the dolphin to come to the manipulandum, to choose one of two balls and to demonstrate the choice by pushing this ball. The visual differentiation consisted in the choosing of the ball to push. The prohibited ball was indicated by the visual sing (Vs) which was the white plastic triangle. Thus, at this stage of the experiment we worked with the following chains in the cycle:

$$St -- Sj+Vs -- Rj -- Fj$$
$$St -- Sk+Vs -- Rk -- Fk$$

where (Rj) and (Rk) are the reactions of Jenny and Kora; (Fj) and (Fk) are the food reinforcement of Jenny and Kora (as a reinforcement we used horse-mackerel, sprat and some other fish species). In the session such chains were repeated in random order. The sessions were carried - out two or three times a day. As errors we defined the following actions of a dolphin (in these cases the dolphin did not receive his reinforcement):
1. The refusal to act after his working signal ("total refusal").
2. The refusal to push the ball ("refusal to choose").
3. The push of the prohibited ball ("error").
4. The action after the start signal or the working signal of the other dolphin ("false-start").
5. The action before the start signal ("intersignal reaction").
6. The absence of the touch of the ball or the endeavour to push it ("unfinished reaction").

On the second stage, we were to connect the separate actions of the animals into the one chain:

$$St -- Sj(Sk)+Vs -- Rj(Rk) -- Sk(Sj)+Vs -- Rk(Rj) -- Fj+Fk$$

The succession of the displays of the working signals and the visual prohibitive sign for Jenny and Kora had a random order. It is important that the dolphins were to have a joint start and they were to receive their reinforcement simultaneously at the end of the cycle. The time delays of arbitrary durations were developed for different stages of a cycle. In the end of this stage of the experiment the dolphins were to receive the reinforcement, only if the actions of both animals were correct and they both did not receive the reinforcement in case of an error of one of them.

On the third stage, we planned to display the visual signal only to one dolphin, but both animals were to choose the manipulandum and the dolphin which had got the visual signal had to act in the second turn. Naturally, in this scheme, the transmission of the arbitrary information by the dolphins is the necessary condition for their correct reactions. We hoped to approach this stage by the special learning of the dolphins. It was supposed that on the third stage we would have enough means for the flexible operation by the animal behavior to study the character of the information reported by the dolphins. In the end of this stage we could carry-out special control sessions with completely isolated animals connected only by the electro-acoustic link.

The session was carried - out by three to four men. Usually one worked with the prohibitive sing, the others controled the equipment and registered the events of the session, working near the central control desk at a distance from the animals. The dolphins could not see the experimenters and had no contact with them during the session because one was hidden by two rows of boards and the others were outside the dolphins' field of vision. However, the possibility of the appearance of stereotypical behavioral elements, which could be used by the dolphins as cues, still remained due to the necessity of the preliminary learning procedure. We avoided this possibility owing to the initiative of our animals.

RESULTS AND DISCUSSION

Specialists who work with dolphins know that these animals begin to take part in the work actively when the contact between the experimenters and the dolphins has reached a certain level. In fact, the dolphins become partners of the experimenters and propose their own versions of actions to solve problems set by experimenters. The experiment is often accelerated due to these propositions which frequently satisfy the methodical scheme (the proposition is rejected by experimenters if it does not answer the purpose).

Before the beginning of the basic sessions of the experiment, we had good contact with Jenny and Kora and the dolphins proposed their versions of actions and solutions for experimental tasks. On the stage of the development of individual behavioral chains between other versions the dolphins had proposed themselves such a strategy which we hoped to develop as a result of the special learning far later: the strategy of co-operative indentical work. The dolphins suggested their strategy when we were working-out the chains St -- $Sj+Vs$ -- Rj -- Fj and St -- $Sk+Vs$ -- Rk -- -- Fk. The acoustic and visual signals were displayed for each animal separately in a random order, i.e. each dolphin fulfilled the acoustic and visual differentiations individually. The reward was given in the end of each chain (naturally, if the actions of the dolphins were correct).

From the start position Jenny did her passage to the installation on the depth about 3.5 m and came to the water surface just under the balls almost vertically. After she pushed the ball, she went to the place where she was to receive the reward and then returned to the start position. Note, that after some wrong choices Jenny went to the start position immediately without the delay of waiting for the reinforcement. It looks as if these cases were connected with a late decision about the choice and the impossibility to correct the error. It may be explained by the peculiarities of her trajectory as well as the features of her nature. When she could see the prohibitive sign only 1 - 2 seconds before the push she had no time (and a wish) to correct her actions.

Kora went to the installation practically on the water surface, she had enough time for the choice and worked in the experiment very stably and reliably. In other aspects her behavior was like Jenny's.

Jenny was more excitable and was more aspirated to the leadership than Kora; she had a very energetic personality with some sharpness of movements and some nervousness in her work. Kora was quiet and reasonable, her movements seemed slower and smoother, and the results of her work in the preliminary sessions were more stable.

To the start of the co-operative work, which was proposed by the dolphins, the acoustic differentiations had been generally formed. The dolphins knew their individual working signals and ignored the working signal of the neighbour. Kora and Jenny practically had no false-starts or refusals: in the morning session previous to the session when the dolphins proposed their variant strategy, Jenny had one false-start per 36 presentations of her signal. Kora had no false-starts or refusals. The results of the visual differentiation also were quiet satisfactory for both dolphins: Kora correctly determined the prohibited ball in 95% of cases, Jenny did it in 80% of the trials. We wanted to make the figures of Jenny's work better although we could begin the next stage of the experiment. Note, that the day before during one of the sessions the behavior of the dolphins was not usual. We tried to change the display of the prohibitive sign, but the number of the wrong reactions increased. The refusals appeared whenever small slope of the prohibitive triangle was observed although the dolphins had paid no attention to it earlier. It is possible that it was connected with the events of the next day. We shall use letter indexes for the sessions. The first session when the dolphins used the co-operative strategy we mark as session "A", the next sessions will be marked as session "B", session "C" and so forth.

Before the beginning of session "A" we paid attention to the unusually high level of acoustic activity and the diversity of the sounds of the animals. Just after the beginning of the experimental work in this session, it had turned-out that the behavior of the dolphins differed significantly from their behavior in the previous sessions. In the first cycle of session "A" the visual prohibitive and acoustic working signals were displayed to Kora. Kora came to the installation, emerged to the allowed ball and stood still under it like a float. In this position she waited some time and then returned to the start position. We interpreted this reaction as an unfinished reaction because the push of the ball was absent. The signal was repeated once more. The actions of Kora were the same. The third signal had been given to Jenny and she came to her installation and pushed the ball determined by the program of the experimenters. Immediately after the actions of Jenny, Kora came to her installation and made the same choice. In the sixth cycle the first signal had been given to Kora again and her behavior which had been observed in the first and second cycles was repeated. We did not understand Kora's refusals in these cases because she was hungry and her work was very reliable in the previous sessions. It seemed as if the preliminary agreement between Jenny and Kora about the joint work and the order of their actions was achieved. In fact, from the beginning of session "A", the earlier formed scheme of behavior had been destroyed completely. Independently from who was the addresser of the acoustic and visual signals, Jenny fulfilled the actions in the first turn and Kora was the second. In these conditions, both dolphins correctly fulfilled the visual differentiation task presented to one of them by the experimenters. Thus, the dolphins began to use the chain:
ST -- Sj(Sk)+Vs -- RJ -- RK. In the previous sessions the actions of the dolphin to the working signal of his neighbour were determined as false-starts. Therefore, during the first 12 cycles, the reward was not given to that animal who worked on the signal of the neighbour. However, the intensive signal interchange between the dolphins, the rapid increase of the qualitative diversity of these signals and the, on the whole, successful solution of the tasks by both animals had given us the possibility to suppose the transmission of some information from one animal to another. The decision was made to reinforce all cases of the correct joint work of both dolphins. Table 1 represents part of the data of the session "A".

In almost all cycles of this session Jenny was the first to choose a ball and Kora fulfilled her actions a few seconds later. When the visual signal was displayed to Jenny and she made her choice, correct or wrong, Kora acted immediately after the actions of Jenny and pushed the ball which coincided with the choice of Jenny. When Kora had got the visual signal and Jenny made a mistake, Kora could go away several times and come to the

Table 1. Work of Dolphins in Session "A"

	The number of the cycle	3	5	7	9	11	13	14	16	18	20	21	25	33	34
The visual signal is displayed to Jenny	The allowed ball	L	L	L	S	S	S	S	L	L	S	L	L	L	S
	Jenny's choice	L	L	L	S	-	L	L	L	L	S	L	L	L	S
	Kora's choice														
	The individual work														
	The refusals					▦									
	The co-operative work														

	The number of the cycle	1	2	4	6	8	10	12	15	17	19	22	23	24	26	27	28	29	30	31	32	35	36
The visual signal is displayed to Kora	The allowed ball	S	S	S	S	L	L	S	S	S	L	S	L	S	S	L	L	S	S	S	S	L	L
	Kora's choice	-	-	S	-	L	L	S	S	L	L	S	L	S	S	L	L	S	S	S	L	L	L
	Jenny's choice	-	-	-	-	-	-	S	S	L	L	L	L	S	L	L	L	S	S	L	S	L	L
	The individual work																						
	The refusals																						
	The co-operative work																						

(L) is the large ball; (S) is the small ball;
▨ the inversion; ■ own mistake; ▢ the mistake of the second dolphin

installation again; she could stay under the balls and turn from one to another; she could push the correct ball in the last moment or suddenly interrupt this movement, repeating it for several times. Usually, these "doubts" (which had been absent in her behavior before session "A") ended by a rapid decision and a push of the same ball as Jenny. If Jenny did the choice which coincided with the program set by the experimenters when the prohibitive sign was displayed to Kora, the actions of Kora followed immediately after the actions of Jenny; in opposite cases the described "doubts" of Kora were observed and her behavior became very hesitant. Apparently, Jenny imposed her decision on Kora and inverted her choice relative to the program presented by the experimenters. It is clear that in these conditions such imposition could be conducted only by the use of acoustic signals. In the cases of Kora's "doubts", we observed stereotypical signals of Kora. Such stereotypical signals usually are observed in stress situations (Sidorova et al., 1986). We interpreted such behavior of Kora as a display of a conflict between the tasks given by the experimenters and the actions dictated by Jenny.

Jenny's leadership was not only provided by herself, but Kora elicited it too. The latent time of her reactions (the time from the display of the signal to the push of the ball) was minimum in the previous sessions. In fact, it was the time which was necessary for the passage from the start position to the installation. In session "A" and the next session ("B" - "E") the latent time of Kora's reaction had singnificantly increased. Kora could increase the time by herself using the delays on her start position, near the installation, or even under one of the balls. The described "doubts" of Kora in the cases of Jenny's non-correct choice could increase the latent time still more. Moreover, several cases were observed when Kora had got the visual signal and after the error of Jenny she pushed the ball coincided with the program of the experimenters. However, just after the correct fulfilled action Kora came back to the installation, her signals became stereotypical, and after some period of "the doubts" and the intensive signal interchange with Jenny, Kora "corrected" her choice and pushed the other ball which coincided with the choice of Jenny.

Thus, the difference between session "A" and the preliminary sessions was the following:
- a significant increase of the intensity and the diversity of the acoustic signals used by the dolphins;
- an increase in number of reactions which we appreciated earlier as "a false-start" (86% for Kora and 82% for Jenny);
- the prevalence of joint actions of the animals (83% of all actions)and their work with the identical manipulandum (90% of the joint actions);
- the leadership of Jenny in the determination of the character of their joint actions and the demonstration of this leadership in the use of the signals;
- the appearance of new forms of individual behavior which were not observed earlier.

On the whole, these peculiarities permitted us to interpret the behavior of the dolphins in this session as a proposition of the co-operative strategy of the decision of the tasks by them, and we had begun to reward their joint correct work. However, the co-operative work of the dolphins which was well presented in the session "A", had begun to deteriorate in the sessions "B" and "C" and, in fact, had broken-completely in the sessions "D" and "E" (see Table 2). The disappearance of the joint work was the result of Jenny's behavior which determined Kora's actions to a considerable extent. During the stage of taming and preliminary learning Jenny preferred to work with the large ball and she came back to this preference in the described sessions. The real danger had appeared that Jenny will pass to a work with the only one (large) ball and, due to the

Table 2. Indices of Co-operative Work of Dolphins

Session	Addresser of visual program	Joint work N	Identical choice (%)			Different choice (%)
			Total	Program (correct)	Inversion	
A	Jenny	12	100.0	66.7	33.3	-
	Kora	18	83.3	66.7	16.7	16.7
B	Jenny	10	60.0	60.0	-	40.0
	Kora	31	77.4	74.2	3.2	22.6
C	Jenny	24	79.2	41.7	37.5	20.8
	Kora	22	59.1	50.0	9.1	40.9
D	Jenny	1	-	-	-	-
	Kora	28	60.7	60.7	-	39.3
E	Jenny	19	36.8	21.0	15.8	63.2
	Kora	22	45.5	36.4	9.1	54.5

leadership of Jenny, Kora will do it too, i.e. we shall obtain the identical one-sided work of both dolphins. This strategy can simplify the interactions between the animals and theoretically warrants them the reward in 50% of trials, but it leads the experiment in a deadlock. Therefore, we decided to hand-over the determination of the program (i.e. to display the prohibitive triangle) only to Kora. However, contrary to expectations it had been an abortive attempt because Jenny refused to take-part in the co-operative work under these conditions. Note that, as soon as we attempted to hand-over the determination of the ball to Kora, and Jenny refused to work, the acoustic signals disappeared rapidly.

Knowing the probability of the correct choice of the ball by the dolphins in the preliminary sessions, we calculated corresponding probabilities for the cases when the transmission of the information by the dolphins exists and when it is absent (see Table 3). In the case when the

Table 3. Theoretical Calculation of Probability of Coincidence of Choices

Conditions	Dolphin	Identical choice (%)			Different choice (%) (errors)
		Total	Program (correct)	Hidden errors	
Transmission of information is absent	Jenny	50.0	47.5	2.5	50.0
	Kora	50.0	39.8	10.2	50.0
Transmission of information exists	Jenny	79.6	75.5	4.1	20.4
	Kora	94.9	75.5	19.4	5.1
Data of session "A"	Jenny	83.3	66.7	16.7*	16.7
	Kora	100.0	66.7	33.3*	-

* - cases of inversions

dolphin, without the visual signal, chooses the ball by chance, the quantity of the coincidences with the correct and wrong choices of the other dolphin will be about 50%, i.e. the sum of the coincidences of correct and incorrect choices which are determined by the probability of the correct and wrong choices of the first dolphin. If the dolphins transmit the information to each other and choose the ball with the probability of the preliminary sessions, the probability of the coinciding choices will be determined by this probability and the coinciding correct choices will be determined by the product of the probabilities of individual correct choices of both animals (in our cases 75.5%). Thus, the transmission of the co-operative information is possible when the probability of the correct coinciding choices is more than the accidental probability (39.8% and 47.5%) and less or equal to 75.5%. The co-operative work, i.e. the agreement of the actions by the dolphins, is real when the probability of the coinciding choices is equal or is more than the probability of the individual correct choices. Note, that the total accordance between the calculated probability of the correct coinciding choices of both dolphins and the experimental data can be absent because errors in the transmission of the information may occur. It can be an impass of the information, the animals can misunderstand the report, a disinclination of dolphins to report or to use the information as well as to report the true information can be observed in some cycles of sessions. The experimental data show that the value of the probability of coinciding choices is more than the theoretical value, i.e. the existence of the co-ordination of the actions of the dolphins is undoubted. The probability of coinciding correct choices in the experiment is 88.3% of the theoretical maximum. The level of inversion is more than the theoretical level of hidden errors and, therefore, it allows us to conclude that Jenny forced her choices to Kora (the cases of inversion, i.e. the cases, when both dolphins fulfilled the same program determined by Jenny, and not by the experimenters).

Comparing the experimental data which were obtained in the sessions, with the results of the theoretical calculations, we conclude that the interchange by the information between the dolphins was observed in the session "A", and to a lesser extent in session "B"; the interchange was absent in the sessions "C", "D" and "E". The results of the observation on the acoustic activity of the dolphins and the data about their behavior in the sessions coincide: the acoustic activity, which was maximum in the session "A", became less in the session "B" and disappeared completely in the next session.

Thus, our results show that bottlenose dolphins can report arbitrary information and can co-ordinate the behavior of each other. It is difficult to define the concrete character of the information transmitted by the dolphins, but admittedly it would be the information either about the sizes of the balls ("large" - "small") or about the position of the allowed ball ("left" - "right").

ACKNOWLEDGMENTS

Our gratitude is due to Mr.P.Litvinov, Mrs.R.G.Zanina and Dr.V.A.Protasov for assistance provided.

REFERENCE

Bastian, J., 1967, The transmission of arbitrary environmental information between bottlenosed dolphins. in: "Animal sonar systems, biology and bionics", V.2 R.G.Busnel, ed.,Jouiene Josas, pp. 807-873.

Dubrovsky, N.A., 1976, Hearing. in: "Sensory foundations of cetacean orientation" V.M.Belkovich and N.A.Dubrovsky, ed.,Leningrad, Nauka, pp. 50-95.
Evans, W.E., Bastian, J., 1969, Marine mammals communication: social and ecological factors. in: "The biology of marine mammals" H.F.Andersen, ed., N.Y., Acad. Press, pp. 25-476.
Herman, L.M., 1980, Congnitive characteristics of dolphins. in: "Cetacean behavior, mechanisms and functions" L.M.Herman, ed., N.Y., Wiley Interscience Publ., pp. 363-421.
Herman, L.M., Tawolga, W.N., 1980, The communication systems of Cetaceans. in: "Cetacean behavior, mechanisms and functions" L.M.Herman,ed., N.Y., Wiley Interscience Publ., pp. 149-209.
Herman, L.M., Richards, D.G., Wolz, J.P., 1984,Comprehension of sentences by bottlenosed dolphins. Cognition, V.16, pp. 129-219.
Markov, V.I., 1978, The structural organization of the communicative system of bottlenosed dolphins. in: "Marine mammals", Moscow, pp. 216-217.
Markov, V.I., Ostrovskaya, V.M., 1978, About syntactical organization in signal successions produced by bottlenosed dolphins. in : "Marine mammals", Moscow, pp. 218-219.
Markov, V.I., Ostrovskaya, V.M. Organization of the communicative system in Tursiops truncatus. In this volume.
Markov, V.I., Tarchevskaya, V.A., Ostrovskaya, V.M., 1974, The organization of the acoustic signals in the bottlenosed dolphins of Black Sea. in: "Morphology, physiology and acoustics of marine mammals", Moscow, Nauka, pp. 173-189.
Sidorova, I.E., Markov, V.I., Ostrovskaya, V.M., 1986, Dynamics of the signalisation of the bottlenosed dolphins in the process of their adaptation to stress factors. in: "Marine mammals", Archangelsk, pp. 350-351.

CONCLUDING COMMENTS ON OTHER SENSORY ABILITIES

Herbert L. Roitblat

Department of Psychology, University of Hawaii, 2430
Campus Road, Honolulu, HI 96822

Although the most striking of cetacean sensory abilities
is their acoustic echolocation skill (Nachtigall and Moore,
1989), they are also endowed with sensitivities for a number
of other modalities. Among the senses considered at the
conference are included magnetic, tactile, electroreceptive,
pressure, and gravitational senses. Further, in addition to
the animals' sensitivity to these various modalities marine
mammals are also endowed with the cognitive capacities to make
use of this sensory information to represent their environment
and the events within it.

Of these senses, probably the most is known about the
animals' sensitivity to magnetic fields. The evidence for use
of a magnetic sense is largely circumstantial and based
primarily on field evidence. For example, whales and dolphin
strandings tend to occur at coastal points associated with
magnetic anomalies--areas in which the orientation of the
magnetic field is significantly different from surrounding
areas (Kirschvink, et al., 1986; Klinowska, 1988).

Considerably more evidence is needed to better establish
these phenomena and their dependence on magnetic properties of
the environment, rather than on other possibly correlated
features. This requires that we study the effects of magnetic
anomalies in situations where other geologic features are
uncorrelated with the magnetic features. For example, in the
Pacific Ocean there are more or less parallel bands of
reversing local magnetic fields (e.g., Vine, 1968). These
bands parallel the spreading ridges of the seafloor. As the
rock in these bands formed, it took on the dominant magnetic
orientation of the earth's magnetic field at the time. The
orientation of this field has reversed more or less regularly
in geologic history, thereby producing regular bands of
magnetic orientation in the seafloor. The orientation of the
magnetic field depends on the time at which the band was
formed, but the other geologic features of the area are
largely uncorrelated. Therefore, these bands provide an
opportunity to investigate the effects of magnetic fields
generally uncontaminated by other geologic features. If
animals respond to the patterns of the magnetic fields in this

area, for example, by following them, it would be good evidence that the magnetic fields per se are important factors in controlling the animals' behavior.

Substantially more field research is needed in this area to determine the extent to which cetaceans are responsive to natural and artificial magnetic fields. For example, the pattern of magnetic fields (as obtained from a towed magnetometer) may be related to the distribution of animals on the surface (as obtained from population surveys). Such a relationship would allow inferences regarding the effects of magnetic fields on animal distributions and on such questions as whether magnetic contours help to define population distributions. Detailed tracking studies would help to determine whether cetaceans follow contours. This research would also be aided by the establishment of more widely used registries for cross compiling stranding data and catch-locations of pelagic species.

Laboratory research could be directed fruitfully at investigations of the ability of cetaceans to discriminate the presence of magnetic fields under laboratory conditions. Use of isomagnetic contours in the field would imply that the animals are sensitive to field strength varitions, but this sensitivity has not been demonstrated adequately in the laboratory. The anatomy and mechanism of the animals' magentoreceptor could also be investigated in the laboratory. Psychophysical experiments are particularly valuable in identifying these mechanisms.

Magnetic fields also have other laboratory uses. For example, if cetaceans' magneto sense is not due to the inductive effects of passing physioelectric conductors through the magnetic field (to be established on the basis of behavioral and psychophysical research), then magnetic fields can be used to measure certain electrophysiological responses of the animal. This would provide a noninvasive means of investigating at least some neurophysiological functions in cetaceans. The use of such techniques, however, requires that the direct effects of the magnetic fields themselves be known so that they can be experimentally controlled in the most appropriate way.

Understanding of cetaceans' magnetosensitivity also has implications for their conservation. For example, if dolphins follow isomagnetic contours, then nets placed across these contours would be expected to capture more dolphins than nets placed parallel to these contours. Again, a great deal more research would be required to substantiate this prediction.

Electroreceptive sensitivity also may provide interesting opportunities for research. There is some suggestive evidence for electroreceptors in the dolphin's tongue and skin. The significance of these receptors remains obscure, but they could play a role in fish detection and, because moving currents of water can induce electrical fields, in navigation and orientation. These sensitivites could be investigated through conditioning experiments and through the use of psychophysical methods.

Other sensory systems that merit investigation include

the the tongue (e.g., Donaldson, 1977) and vibrissae (e.g., Bel'kovich, 1972; Ling, 1977). The tongue may provide important tactile information in addition to its more typical role as a chemoreceptor. Dolphins, for example, frequently manipulate objects in their mouths and respond strongly to oral tactile stimulation. The vibrissae also are potentially important sensory systems. All of the marine mammals have vibrassae at least at some stage in their life cycles. They are most prominent in pinnepedia, but are also present in humpback whales (<u>Megaptera</u> <u>novaengliae</u>), and vestigial remains of the their follicles can be found in bottlenose dolphins (<u>Tursiops</u> <u>truncatus</u>; Layne and Caldwell, 1969). The innervation of the vibrissal follicles (Kulikov, 1974) recommends them for study as a sensory system, even in dolphins, where they may serve as velocity receptors (Palmer and Weddell, 1964).

The sensory abilities of cetaceans and other animals have coevolved along with those processes that use the sensory information to control the animal's behavior. There are many interesting studies of dolphins' congnitive processes (e.g., Schusterman, et al., 1986; Roitblat, et al., in press), but many interesting questions remain. The interest in the cognitive capacities of cetaceans can be guided by a number of orienting questions including how animals use perception for representation and problem solving. What is the sensory basis for intelligence? Why do dolphins have such large brains? What role does attention play? How do animals integrate information from multiple sources?

These questions can be studied with some difficulty in the wild by examining the animals' behavioral strategies under varying environmental conditions. They are particularly suited to examination in the laboratory, however (see Roitblat, 1987 for examples of methods). A complete understanding of the sensory abilities of marine mammals requires that we have an overview of how that sensory information is used by the animal.

Many of these questions concerning the sensory and cognitive capacities of marine mammals are not unique to this group. The same questions can frequently be also be asked of other groups. Many of the issues are of interest acrosss a broad range of species and a broad range of ecological niches. Answering these questions will contribute substantially to our understanding of sensory and cognitive processes generally. Nevertheless, they have special applicability to studies of marine mammals. The environment in which marine mammals evolved is so different from our own that our intuitions about how sensory and cognitive processes must work may be of little value. As a result, the investigation of the comparative aspects of their behavior help to provide checks on our intuitions and interpretations that cannot help but provide new perspectives and insights into our own functioning. Such investigations are quite important for the information they can provide regarding the ways in which marine mammals interact with their environments and, thus, for suggesting possibly important avenues by which to gain information from the marine environment. The interest and the value of such investigations is only just beginning to be seen.

REFERENCES

Bel'kovich, V. M., 1972, Sight and other sense organs, in: "Kity i del'finy." (pp. 275-299). Moscow: Nauka.

Donaldson, B. J., 1977, The tongue of the bottlenosed dolphin (Tursiops truncatus). In R. J. Harrison (Ed.), "Functional anatomy of marine mammals." (pp. 175-197) New York: Academic Press.

Kirschvink, J. L., Dizon, A. E., and Westphal, J. A., 1986, Evidence from strandings for geomagnetic sensitivity in cetaceans. Journal of Experimental Biology, 120, 1-24.

Klinowska, M., 1988, Cetacean 'navigation' and the geomagnetic field. Journal of Navigation, 41, 52-71.

Kulikov, V. F., 1974, Topography of the trigeminal and facial nerves in the porpoise. Moscow: Nauka.

Layne, J. N. and Caldwell, D. K., 1969, Behaviour of the Amazon dolphin Inia geoffrensis in captivity. Zoologica, 49: 81-108.

Ling, J. K., 1977, Vibrissae of marine mammals. in: R. J. Harrison (ed.), "Functional anatomy of marine mammals." (pp. 387-415) New York: Academic Press.

Nachtigall, P. E. and Moore, P. W. B., 1989, "Animal sonar systems." New York: Plenum Press.

Palmer, E. and Weddell, G., 1964, The relationship between structure, innervation and function of the skin in the bottlenose dolphin (Tursiops truncatus). Proceedings of the Zoological Society of London, 143: 553-568.

Roitblat, H. L., 1987, "Introduction to comparative cognition." New York: Freeman.

Schusterman, R. J., Thomas, J. A., and Wood, F. G., 1986, "Dolphin cognition and behavior: A comparative approach." Hillsdale, NJ: Erlbaum.

Vine, F. J., 1968, Magnetic anomalies associated with mid-ocean ridges. in: R. A. Phinney (ed.), "The history of the earth's crust" (pp. 73-89). Princeton, NJ: Princeton University Press.

INDEX

Acoustic lens, 75
Adaptation,
  aquatic, 137-141, 363, 463, 520, 561
  auditory, 82-104, 363-384
Amazon river dolphin, see *Inia geoffrensis*
Ambient noise, 204, 317-319, 328, 338-339, 343, 395, 399
Anesthesia, 25
Arbitrary information, 685-698
Atlantic bottlenose dolphin, see *Tursiops truncatus*
Attention, 665-666
Audiogram, see separate species
Audition, see hearing
Auditory,
  adaptation, 82-104, 363-384
  bullae, 82
  cortical response, 385-387
  and echolocation, 81-104
  evoked potential, 237, 335, 385, 418
  meatus, 363-370
  nerve, 98
  subsystems, 233-254
  temporal delay-lines, 190
  thresholds, 238, 274, 400

Baiji, see *Lipotes vexillifer*
Balaena mysticetus, 67, 374, 552
Balaenoptera spp., 645
  acutorostrata, 35, 67, 348, 363, 371, 548, 571
  borealis, 67, 571
  edeni, 571
  musculus, 67, 548, 571

Balaenoptera (continued)
  physalus, 348, 548, 571, 644
Bangs, 295-301, 591
Basilar membrane, 91
Beaked whales, see *Mesoplodon* spp. and *Ziphius* spp.
Beampattern, 234, 255
Behavioral,
  mimicry, 457-458
  response to sound, 339-340
Beluga whale, see *Delphinapterus leucas*
Berardius bairdii, 15, 71
Bioluminescence, 551
Blowhole, 69, 71, 495
Blows, 572
Blue whale, see *Balaenoptera musculus*
Body postures, 528, 541-542, 553-554
Bottlenose dolphin see *Tursiops truncatus*
Bowhead whales, see *Balaena mysticetus*
Brain, 114
  activity, 166
  dolphin, 20, 31, 39
  mapping, 196
  whale, 39
Brainstem,
  auditory, 181-194, 385, 405
  geomagnetic sensing, 33
  medial nucleus trapezoid body, 181-194
  ventral cochlear nucleus, 181-194
Bubbles, 540, 576
Bursae, 9

California sea lion, see *Zalophus*

703

Caperea marginata, 571
Caudal nasal sacs, 14
Cephalorhynchus spp., 12, 427
  color patterns, 528, 550
  commersonii, 1-18
  sexual dimorphism, 548
  skull, 7-11
Cetacean, see dolphins or whales
Chemoreception, 435-570, 481-504, 537, 561, 563
Click sound, 295-304, 591
Climbing fiber, 34
Cochlea,
  anatomy, 93-95, 376
  basal diameter, 95
  computer reconstruction, 81-106
  frequency of sound, 385
  number turns, 363, 376
  scalae length, 95
Cognition,
  and ecology, 519-536
  information-processing, 678, 701
  in visual tasks, 455-462
Color patterns, 519-536, 546, 548
Commerson's dolphin, see *Cephalorhynchus commersonii*
Common dolphin, see *Delphinus delphis*
Communication,
  acoustic, 571-638, 599-622, 635
    dolphins, 599-622
    whales, 571-584
  non-acoustic, 537-544
  visual, 545-560
Computer,
  analysis, 39
  generated signals, 306, 396
  model, 255-268
  three-dimensional
    reconstruction, 1-18, 81-106
Conclusions,
  acoustic communication, 635-638
  anatomy, 195-202
  chemoreception, 561-570
  echolocation, 427-434
  hearing, 427-434
  other senses, 699-702
  physiology, 195-202
  tactile, 561-570
  vision, 561-570
Cortex,
  allocortex, 28
  auditory, 23
  cerebral, 18, 31-32, 107-136
  ratio to brain, 19
  sensory, 19-30
  visual, 20, 39-66, 110
  volume, 19
Critical interval, 210, 246, 263
Critical ratio, 321, 395, 400
Cross-modal perception, 667-684
CT scans, 3, 84

Dall's porpoise, see *Phocoenoides dalli*
Delay-lines, 181-194
Delphinapterus leucas,
  anatomy, 15
  cerebral cortex, 115
  echolocation range, 207
  frequency pulse, 298
  hearing, 321, 343, 363, 373, 395, 405
  melon, 15
  peak frequency, 305
  sexual dimorphism, 548
  source levels, 295, 305
  tongue, 482, 940
Delphinus delphis, 83, 148, 164, 363, 366, 482, 505, 521, 548, 644
Directivity Index, 206, 234, 264
Dolphin,
  attention and echolocation, 665-676
  blindfolded, 217-232
  chemoreception, 481-504
  detection of target, 203-216, 217-232, 305
  echolocation,
    amplitude, 297-298, 305-316
    beampattern, 234, 255
    and behavior, 255
    click interval, 234, 241, 263, 325, 673
    detection of nets, 269-294
    as energy detector, 210
    frequency, 191, 235, 255, 269, 283-294
    intensity of clicks, 295-304

Dolphin (continued)
  echolocation (continued)
    number of clicks, 275, 325, 656
    peak frequency, 95, 203, 206, 305, 327
    pulse control, 305-316, 317-320
    receptor, 255-268
    repetition rate, 263, 291, 299, 305, 327, 591
  foraging, 269-282, 301, 499, 526, 562
  habitat, 99, 150-155
  jaw geometry, 255-268
  prey disorientation, 295-301
  social behavior, 499, 519, 537
  sound production, 1, 67, 601
  tactile, 163-180, 537-544, 555
  target,
    detection, 203-216, 217-232, 305
      in noise, 191, 203-216
      in reverberation, 203-216, 258
    discrimination, 256
  taste, 447-454, 481-503
  touch, see tactile
  visual cortex, 20-23
Dusky dolphin, see *Lagenorhynchus obscurus*

Ear, see cochlea,
  inner, 81-105
  middle ossicles, 149
Echolocation,
  attention, 665-676
  amplitude, 283, 297-298, 305, 673
  bandwidth, 327
  center frequency, 672
  critical interval, 246, 263
  detection of nets, 269-294
  detection of targets, 203-216, 217-232, 305, 655-660
    ball bearing, 307
    cylinder, 670
    sphere, 204, 322, 670
  directionality, 295
  directivity index, 206, 234, 264

Echolocation (continued)
  discrimination, 255, 665-676
  duration, 287, 305
  frequency, 154, 191, 269, 283, 314, 385-394
  habitat, 82
  hearing, 363-384
  highlights, 235
  interclick interval, 234, 241, 263, 325, 673
  match-to-sample, 665-676
  mask signal, 207, 242
  noise, 191, 203, 317
  number of clicks, 656
  oscillogram, 318, 604
  peak frequency, 321, 327
  power spectrum, 298, 371, 589, 671
  pulse control, 305-316
  range detection, 280, 321-334, 321, 330
  repetition rate, 291-293, 299, 305, 327
  reverberation, 203, 258
  signal-to-noise, 203, 259, 657
  sound pressure level (SPL) 307-309, 311
  source levels, 283, 291, 295, 305, 321, 328-330
  spectrogram, 301, 602, 607
  waveform, 589
Electrocardiogram, 167, 485
Electromyogram, 167,173
Electrophysiology, 81, 164, 335, 405
Electroreceptivity, 700
Elliptic nucleus, 34
Embryo, 363
Energy Detector, 210
Eptesicus fuscus, 40, 48, 182
Erinaceus europaeus, 40, 48, 123, 128
Eschrichtius spp., 505, 527, 571, 577
Eubalaena spp., 70, 350, 548, 571
  E. australis, 571
  E. glacialis, 571
Evoked potential,
  auditory, 237,335,385 418,431
  sensory, 196,
  somatosensory, 167
Evolution,
  auditory system, 233

Evolution (continued)
  dolphin skull, 70, 137-162
  echolocation, 497
  eyes, 519
  nasal anatomy, 67-80
  olfaction, 495
  phylogeny, 69
  sensory neocortex, 107, 115-117
Eye,
  anatomy, 463-480, 505-518, 521
  chemical composition, 463
  histology, 464-465, 520
  iris, 470-472
  lens, 471, 521
  mucous, 463-464, 520
  refractive power, 463, 475, 520
  retina, 473, 520
  rods, 474, 520

False killer whale, see Pseudorca crassidens
Feresa attenuata, 83, 551, 565
Filter, 386
Fin whale, see *Balaenoptera physalus*
Fish, 481, 552
Fishing net, 280-283, 258, 347-362
Foraging behavior, 269-282, 301, 499, 526, 562
Forehead, 1-18
Fornix, 33
Fossil record, 69, 109, 137

Ganges river dolphin, see *Platanista spp.*
Ganglion cell,
  density, 94
  distribution, 505
Geomagnetic sense, 639-650, 651-663, 699
  anatomy, 31-38
  flux density, 653
  and live strandings, 639-650, 651-663
  and orientation, 639-650, 651-664
  topography, 639, 652
Gestural communication, 537
Globicephala spp., 39, 83, 115, 144, 523, 548, 644, 651
Golgi, 20, 39, 112
Grampus griseus, 75-76, 83, 144, 521-523, 548,

Grampus griseus (continued) 565, 644
Gray whale, see *Eschrichtius robustus*

Haircell, 31, 98
Harbor porpoise, see *Phocoena*
Hearing, 335-434
  critical interval, 246
  critical ratio, 321, 395
  dolphin, 233-254
  electrophysiological studies, 405-416
  experimental design, 395-404
  frequency discrimination, 385
  frequency resolution, 390,
  localization, 417-426, 429
  low frequency, 234,
  masked hearing, 242, 388-390, 395-404
  ontogeny, 200, 363-384
  receive directivity, 255-268
Hintere klappe, 7, 73
Hippocampus, 33
Histology,
  cochlea, 90
  dolphin eye, 463-480
  sensory cortex, 19-30
  tongue, 482, 490
Homo sapiens, 35, 94, 374, 463, 640
Humpback whale, see *Megaptera novaeangliae*
Hydrodynamics, 163, 172, 545
Hydrophone, 256, 296, 306, 317, 325, 357, 546, 578, 586, 687
Hyperoodon spp., 15, 72

Immunocytochemistry, 19-30, 39-66
Inia geoffrensis, 15, 67, 74, 145, 395, 405, 409-410, 420-422, 490, 505, 520, 546

Jaw-hearing, 258, 417-426, 417

Killer whale, see *Orcinus orca*
Kogia spp., 71-72, 145, 520, 644

Krill, 552

Lagenodelphis spp., 548
Lagenorhynchus spp., 528, 550
  L. acutus, 145, 644
  L. albirostris, 83, 283-294
  L. obliquidens, 168, 522, 529, 548
  L. obscurus, 526, 545 565, 577
Language, 599
  artificial, 530, 645
  body, 528, 541-542, 553
  gestural, 458-459
Lateral gyrus, 20-22
Learning,
  dolphins, 430
  neural networks, 671
Lipotes, 67, 482, 505, 546
Lissodelphis spp., 550

Magnetic induction sensitive neurons, 32
Magnetic sensitivity, *see* geomagnetic sense
Magnetite, 31, 640
Manatee, 405-416
Masked hearing, 395-404
Match-to-sample, 456-457, 529, 665-666, 678
Medial accessory olive, 34
Megaptera novaeangliae, 347-362, 527-528, 546, 571, 575-576, 700
Melon, 1, 72, 256, 317
Melon headed whale, *see Peponocephala spp.*
Memory, 665-667, 678
Mesoplodon spp., 71, 74, 547
Minimum angle of resolution, 522
Minke whale, *see Balaenoptera acutorostrata,*
Monodon monoceros, 83, 295-304, 482, 547
Mossy fiber, 33

Narwhal, *see Monodon monoceros*
Nasal,
  anatomy, 1-18, 67-80, 495
  diverticula, 77
  plug, 67
  system, 1-18
Nasofrontal sacs, 4-14, 69
Neocortex, 19, 39-45,

Neocortex (continued) 107-136, 197
  lamination, 108
Neophocoena phocoenoides, 505
Net detection, 464
Neural,
  artificial network, 670-672
  CCK reactive, 46-48, 59
  density, 22
  eye, 473, 505
  GABAergic, 23-24
  non-pyramidal, 22, 42, 112, 124
  NPY reactive, 42-46, 60
  perikarya, 42-48, 58
  pyramidal, 21, 43, 112, 117, 124
  substrate, 31-38
  synapse, 55
  taste, 482-483
  tongue, 498
  tooth, 260-261
  tracer substances, 25
  transmitters, 27, 39-66
  tyrosine hydroxylase, 52
Nissl, 20, 39, 183
Noise,
  ambient, 191, 203, 317-319, 335, 396
  rippled, 386
Northern right whale dolphin, *see Lissodelphis borealis*

Odontocetes, *see* dolphin, whale
Olfaction, 142, 481-503
Olfactory bulb, 482-483, 561
Ontogenesis, 363-385
Orcinus orca, 144, 395, 478, 524-525, 527-528, 538, 543, 545
Organ of Corti, 91, 381
Orientation, 499, 639
Oscillogram, 318, 604
Other Senses, 639-702

Panniculus carnosus muscle, 165
Pattern recognition, 670
Peponocephala spp., 551
Performance, 331, 457, 531, 674
Perianal gland, 482
Physeter catodon, 72, 83, 482, 547

Phocoena,
  audition, 83, 395, 410
  brainstem, 189
  cerebral cortex, 115
  color pattern, 548
  detection of nets, 269-294, 464
  echolocation, 269, 428
  eyes, 463-480, 521-522
  forehead anatomy, 1-18
  geomagnetic sense, 644, 653
  retina, 505
  sensory cortex, 19-30
  skull, 138
  tongue, 482, 490
  vibrissae, 566
  visual neocortex, 39
Phocoenoides dalli,
  color pattern, 538
  detection nets, 269-294
  echolocation, 269-294
  hearing, 269
  sexual dimorphism, 547
Photomicrographs, 42-66 111, 120, 126, 184
Physiology, 1-202
Pilot whale, see *Globicephala spp.*
Platanista spp., 15, 67, 74, 144, 520, 546
Pontoporia blainville, 15, 67, 72, 138, 482
Premaxillary sacs, 14, 69
Prey, 519, 545, 548
Pseudorca crassidens,
  echolocation, 321-334
  eyes, 478, 524
  hearing, 395-404
  range detection, 321-334
  vision, 565
Pulse, echolocation, see clicks
Pygmy killer whale, see *Feresa attenuata*

Quasi-olfaction, 481-503, 562

ROC plot, 214
Range detection, 278-280, 321-334
Reflexivity, 679
Retina, 466, 505-518
Reverberation, 258
Right whale, see *Eubalaena glacialis*
Risso's dolphin, see *Grampus griseus*
Sea lion, 436-440, 447-454

Sea lion (continued)
  California, see *Zalophus californianus*
  elliptic nucleus, 36
  sounds, 577
  tactile, 438
  taste, 447-454
  vision, 439
Seal,
  elliptic nucleus, 36
  retina, 505
Sexual dimorphism, 547
Skin,
  innervation, 163, 444
  sensitivity, 164-180, 170, 435, 555
  surface ridges, 167-168, 174
  vibrations, 165
Skull,
  anatomy, 496
  asymmetry, 71
  dolphin, 4-6, 70
  and hearing, 137
Sonar Signal,
  equation, 203-216
  shipboard, 283-294
  target identification, 256
  transducer, 284-286
Song, 572-575
Sotalia fluviatilis, 405, 409, 425-426
Source levels, 283, 291, 295, 305, 321, 328-330, 573
Spectrogram, 301, 573, 589, 602, 607
Sperm whale, see *Physeter*
Spermaceti organ, 69
Spinner dolphin, see *Stenella longirostris*
Spotted dolphin, see *Stenella spp.*, 144, 186, 520, 538, 550
  S. attenuata, 83, 363, 368, 528, 538, 540, 550, 565
  S. coeruleoalba, 39, 83, 112, 115, 126, 181, 482, 548, 644
  S. frontalis, 546, 645
  S. longirostris, 83, 259, 527-528, 540, 545
  S. plagiodon, 644
Steno spp., 148
Stereogram, 1-4, 85
Stimulus Equivalence, 667-684
Strandings, live, 31, 642

Stress, 623-634, 563, 617
Striped dolphin, see
    Stenella
    coeruleoalba
Symmetry, 674

Tactile,
  discrimination, 439
    in dolphins, 435-570,
        435-446, 541, 555,
        566
    in pinnipeds, 435-446
  of skin, 163-180
Target,
  ball bearing, 307
  cone, 670
  cylinder, 670
  detection, 203, 217
  discrimination, 256
  localization, 257
  phantom, 208, 244
  range, 321-344
  recognition, 217-232
  sphere, 204, 244, 321-322,
      670
  strength, 206, 278, 322
Taste,
  buds, 490, 494, 561
  discrimination, 447, 488
  dolphin, 447-454, 481-503
  pinniped, 447, 486, 490
Teeth, 258-264
Thalamus, 19-30, 32, 119
Tomograms, 3
Tongue, 447, 482, 490, 700
Transitivity, 679
Tursiops spp.,
  acoustic behavior, 577,
      585-598, 623-634
  activity pattern, 589
  audition, 81-104, 145,
      363, 385-394, 405,
      410, 422-424
  brainstem, 181
  cerebral cortex, 115
  click production, 305-316,
      656
  color pattern, 548
  communication, 599-622,
      685-698
  critical ratio, 321, 395
  cross-modal perception,
      667-684
  decision making, 665-666
  discrimination, 256
  echolocation, 235, 305-
      319, 321, 328
  evoked potential, 386-387
  eyes, 478, 505, 521
  ganglion cells, 50

Tursiops (continued)
  geomagnetic sense, 645
  jaw-hearing, 255-268, 417
  match-to-sample, 665-676
  movements, 585-587
  ontogeny, 363
  range detection, 331
  retina, 520
  sensory cortex, 19-30
  silence, 632-634
  skin, 164, 177, 180
  skull, 70, 496
  sounds, 587-688
  source levels, 295, 305,
      321, 331
  stimulus equivalence,
      667-684
  target detection, 203-216,
      217-232
  taste, 447-454, 481-503
  tongue, 482, 490
  tympanic membrane, 366
  visual cognition, 456
  visual neocortex, 40
  waveform, 298
  whistles, 623-634
Two-way travel time, 326
Tympanic bone, 152

Vestibular sacs, 4-6
Vibrissae, 435, 439-442,
    700
Vision, 464, 564
  acuity, 525, 564
  aerial, 526-527, 538, 564
  binocular, 477, 524, 538
  cognition, 455-462, 519-
      536
  retinae, 505-518
  sea lion, 438
  whale,
    of fishing gear 351-352
Visual,
  acuity, 525
  aerial displays, 539, 572
  color patterns, 527
  communication, 537, 539,
      545
  discrimination, 439, 529
  displays, 527, 537, 545-
      560

Waterfall display, 309
Waveform, 256, 274-276, 296,
    298, 318, 589
Wavelength, 99
Whales,
  behavior,
    dive, 340
    respirations, 339

Whales (continued)
  behavior (continued)
    response to sound, 339-340, 353-356
  clicks, 573
  communication, 571-584
  echolocation, 321-344
  frequency of sounds, 75, 572-573
  hearing, 199, 335-346, 353-354, 395-404
  predation, 321
  range detection, 321-334
  song, 572-575
  sounds, 571-584, 585-598
  vision, 527

Whistles, 305, 540, 588-591, 601, 623
White-sided dolphin, see *Lagenorhynchus obliquidens*
White whale, see *Delphinapterus leucas*

Zalophus californianus, 436-446, 447-454
Ziphius spp., 72, 551, 645